# Handbook of
# Silicon
# Semiconductor
# Metrology

# Handbook of Silicon Semiconductor Metrology

edited by
## Alain C. Diebold
*International SEMATECH*
*Austin, Texas*

CRC Press
Taylor & Francis Group
Boca Raton  London  New York

CRC Press is an imprint of the
Taylor & Francis Group, an **informa** business

CRC Press
Taylor & Francis Group
6000 Broken Sound Parkway NW, Suite 300
Boca Raton, FL 33487-2742

First issued in paperback 2019

© 2001 by Taylor & Francis Group, LLC
CRC Press is an imprint of Taylor & Francis Group, an Informa business

No claim to original U.S. Government works

ISBN-13: 978-0-8247-0506-0 (hbk)
ISBN-13: 978-0-367-39716-6 (pbk)

Visit the Taylor & Francis Web site at
http://www.taylorandfrancis.com

and the CRC Press Web site at
http://www.crcpress.com

*To Annalisa, Laura, and Gregory*

# Preface

The *Handbook of Silicon Semiconductor Metrology* is designed to serve as a reference to both the metrology and processing communities. The volume is intended for a broad audience, from research to development and manufacturing. Whenever possible, chapter authors have provided a description of the science of the measurement along with applications to the most recent materials and processes. Another appropriate title for this book would be *The Science and Technology of Silicon IC Metrology*. The more one knows about the science behind measurement technology, the better one is able to ensure that a given metrology method will provide useful and correct information. It can serve as an advanced reference and text for those studying the processes used in silicon integrated circuit (IC) manufacturing. The *Handbook of Semiconductor Manufacturing Technology*, edited by Yoshio Nishi and Robert Doering (Marcel Dekker, Inc., 2000) makes an excellent companion to this volume and vice versa.

Metrology itself is a wide-ranging field. An overview is provided in the introductory chapter. Trends in the logistics of measurement technology such as integrated metrology are also covered in the introduction. The introduction conveys the following philosophy: *Measurement data becomes a value for a physical parameter such as thickness only after a model is applied to the data and to the measurement instrument.*

Most of us think of metrology as measurement and its connection to process control. As the silicon wafer proceeds through the fabrication facility (FAB), metrology is used to ensure that the wafer has features that fall within a range of parameters that will result in ICs that have a specified set of electrical properties such as clock speed. When one looks at a generic process flow, the three most used measurements are overlay, critical dimension measurement, and defect detection. Equally critical measurements are done to ensure implant dose, the thickness of the gate dielectric, and thickness of the barrier layer for interconnect metal lines, among others. Whenever possible, the latest methods for measuring these properties have been included.

It is interesting to note that the industry has long emphasized precision over accuracy. This has changed, and an international viewpoint is one that emphasizes both aspects of a measurement. The increased need for accuracy can be traced to the shrinking of device size and the trend toward transfer of processes from process tool supplier to IC manufacturer. Integrated circuit manufacturers have traditionally found insightful ways to transfer processes from development to manufacture, including the use of golden wafers. The contributors to this book were asked to discuss measurement calibration and precision. Several chapters discuss calibration methods, including one chapter that is specific to the calibration of defect and particle sizes.

Measurement data becomes useful when it is turned into information. This topic is covered in chapters that describe data management for all measurements and the use of electrical test structures for statistical metrology. Another chapter on manufacturing sensitivity describes how the variation of the physical parameters, which metrology tools typically measure, is related to the variation of electrical properties of transistors.

The final part of this volume contains chapters describing reference measurements and the physics of optical measurements. In addition, the new area of ultraviolet (UV) ellipsometry is covered. This section should prove very useful to those seeking to calibrate their metrology tools, especially when reference materials are not available.

The chapter authors have done an excellent job of fulfilling the goals described above. In a separate acknowledgments section, I again thank them and many others for making this book possible. If a second edition is published, I hope to include several additional chapters that I had intended for this first edition. I believe that you will find that this volume covers the majority of topics that are referred to as metrology.

*Alain C. Diebold*

# Acknowledgments

With the impending publication of this book, I wish to express my thanks to a considerable number of people. My wife, Annalisa, and children, Laura and Gregory, have been very patient with my many late-night editing sessions. I am very grateful for their support. I hope that Laura and Gregory will now appreciate the effort involved in producing reference and textbooks as they progress through the rest of their education. Annalisa and I await their decisions on life's calling with joy and support. Perhaps there will still be metrology reference books then. I also thank my parents, Albert and Simone, who encouraged my scientific endeavors from a young age, and my brothers, Mark and Paul, who encouraged a nonscientist view of the world.

I wish to particularly acknowledge Russell Dekker for his encouragement and for suggesting that I edit a book on metrology. I also acknowledge Bob Doering of Texas Instruments, who suggested my name to Russell. All the authors who contributed chapters to this book have my sincere gratitude. Clearly, the book would not have been possible without their efforts and support. In addition, Barbara Mathieu at Marcel Dekker, Inc. has guided the publication process and production with great skill

My path to the field of metrology was circuitous. Clearly, everyone has a number of people whom they remember with great appreciation as significant influences in their career. Mentioning them all is not possible, but I will try. As an undergraduate, Wayne Cady stimulated my interest in chemical physics and theory. I still remember the research on gas collision theory. Steve Adelman, as my thesis advisor, continued this interest in theory with research in gas – solid surface scattering. Comparison with theory was inhibited by the need for experimental data from the scattering of gas atoms and molecules in well-defined energy states from perfect single-crystal surfaces. Some theoretical work was done in a single dimension for which no experimental data can exist. Nick Winograd provided the experimental postdoctoral fellowship in angle-resolved secondary ion mass spectrometry. Theory and experiment are a powerful combination for a metrologist. Allied–Signal was my first place of employment. There I was exposed to many characterization methods and to how chemical companies use measurement for process control. Ray Mariella introduced me to the world of III-V materials and epitaxial growth. He also introduced me to several optical characterization methods. Mark Anthony then hired me into SEMATECH's characterization group, where I learned the semiconductor industry's views on many of the same measurements. There have been many famous people here at SEMATECH and now International SEMATECH. Seeing folks like Bob Noyce was very inspirational. Turner Hasty once told a SEMATECH all-hands assembly that the FAB was where the action is and that we should learn from the great ones. There are many outstanding people here now. It is extremely rewarding to see Mark Mellier-Smith and Rinn Cleavelin show enthusiasm for metrology. International SEMATECH also has many assignees, such as George Casey and Ron Remke. George and Ron are a study in how to be both gentlemen and good managers. Ron and I co-manage a group covering metrology and yield management tools. Howard Huff has provided both technical and humorous discussions and unending enthusiasm. Howard's interest in the history of the industry provides inspiration to all of us. There are many council and working group members from International SEMATECH groups who continue to be a great joy to work with. I thank them all.

During the past 11-plus years at International SEMATECH, I have had the opportunity to meet technologists from many countries. In my view, Takeshi Hattori is an inspiration to all. He continues to set an example of how we all can keep the highest technical standards and encourage interaction with our international colleagues. Dick Brundle has also been an outstanding example of the highest level of technical accomplishment in fundamental research and in semiconductor manufacturing R&D. The U.S. National Laboratories have a considerable number of outstanding scientists and engineers such as Barney Doyle and Ray Mariella who apply their inventiveness to key problems. The American Vacuum Society (AVS) has provided an opportunity for professional growth. Some time ago, I remember hearing one of Gary Rubloff's postdoctoral students present at the AVS. He described the work that Gary's group was doing in the area of cluster tools and wafer cleaning, in other words, how R&D can effectively improve manufacturing. Gary Rubloff invited me to join the Manufacturing Science and Technology Group of the AVS, which has led to countless new contacts. Gary's technical excellence and wisdom have been a strong influence on me. More recently, Dave Seiler of the National Institute of Standards and Technology (NIST), decided to hold a series of conferences on characterization and metrology for ULSI Technology. These conference proceedings are a good reference for those who read this book.

*Alain C. Diebold*

# Contents

# Contents

# Contributors

**Richard A. Allen, M.S.**  Semiconductor Electronics Division, National Institute of Standards and Technology, Gaithersburg, Maryland

**Matthew Banet, Ph.D.**  Philips Analytical, Natick, Massachusetts

**Gabriel G. Barna, Ph.D.**  Process Development and Control, Texas Instruments, Inc., Dallas, Texas

**Laurie Bechtler, Ph.D.**  Boxer Cross Inc., Menlo Park, California

**P. Boher, Ph.D.**  R&D Department, SOPRA, Bois-Colombes, France

**Duane S. Boning, Ph.D.**  Department of Electrical Engineering and Computer Science, Massachusetts Institute of Technology, Cambridge, Massachusetts

**Peter Borden, Ph.D.**  Boxer Cross Inc., Menlo Park, California

**C. R. Brundle, Ph.D.**  Applied Materials, Inc., Santa Clara, California

**Brett Busch, Ph.D.**  Department of Physics and Astronomy, Rutgers University, Piscataway, New Jersey

**William W. Chism, Ph.D.**  International SEMATECH, Austin, Texas

**Taiheui Cho**  Microelectronics Research Center, University of Texas, Austin, Texas

**Michael W. Cresswell, Ph.D.**  Semiconductor Electronics Division, National Institute of Standards and Technology, Gaithersburg, Maryland

**C. Defranoux, M.S.**  Application Department, SOPRA, Bois-Colombes, France

**Richard D. Deslattes, Ph.D.**  Physics Laboratory, National Institute of Standards and Technology, Gaithersburg, Maryland

**Alain C. Diebold, Ph.D.**   International SEMATECH, Austin, Texas

**Thaddeus G. Dziura, Ph.D.***   Therma-Wave Inc., Fremont, California

**P. Evrard**   Application Department, SOPRA, Bois-Colombes, France

**Eric Garfunkel, Ph.D.**   Department of Chemistry, Rutgers University, Piscataway, New Jersey

**Michael Gostein, Ph.D.**   Philips Analytical, Natick, Massachusetts

**Joseph E. Griffith, Ph.D.**   Advanced Lithography Research, Bell Laboratories, Lucent Technologies, Murray Hill, New Jersey

**Torgny Gustafsson, Ph.D.**   Department of Physics and Astronomy, Rutgers University, Piscataway, New Jersey

**Clive Hayzelden, Ph.D.**   Film and Surface Technology Division, KLA-Tencor Corporation, San Jose, California

**Paul S. Ho**   Microelectronics Research Center, University of Texas, Austin, Texas

**Jimmy W. Hosch, Ph.D.**   APC Sensors and Applications Manager, Verity Instruments, Inc., Carrollton, Texas

**Chuan Hu**   Microelectronics Research Center, University of Texas, Austin, Texas

**Po-Fu Huang, Ph.D.**   Defect and Thin Film Characterization Laboratory, Applied Materials, Inc., Santa Clara, California

**Gerald E. Jellison, Jr., Ph.D.**   Solid State Division, Oak Ridge National Laboratory, Oak Ridge, Tennessee

**Michael A. Joffe, Ph.D.**   Philips Analytical, Natick, Massachusetts

**Walter H. Johnson**   Applications Department, KLA-Tencor Corporation, San Jose, California

**Ayman Kanan****   Therma-Wave Inc., Fremont, California

**Michael Kiene, Ph.D.**   Microelectronics Research Center, University of Texas, Austin, Texas

**Lawrence A. Larson, Ph.D.**   Front End Processes Division, International SEMATECH, Austin, Texas

---

* *Current affiliation*: KLA-Tencor, San Jose, California.
** *Current affiliation*: Lucent Technologies, Allentown, Pennsylvania.

**Karen Lingel, Ph.D.**   Boxer Cross Inc., Menlo Park, California

**Herschel M. Marchman, Ph.D.***   Department of Electrical Engineering, University of South Florida, Tampa, Florida

**Anna Mathai, Ph.D.**   Western Business Unit, KLA-Tencor Corporation, San Jose, California

**Richard J. Matyi, Ph.D.**   Department of Materials Science and Engineering, University of Wisconsin, Madison, Wisconsin

**Alex A. Maznev, Ph.D.**   Philips Analytical, Natick, Massachusetts

**Veena Misra, Ph.D.**   Department of Electrical and Computer Engineering, North Carolina State University, Raleigh, North Carolina

**Michael Morgen, Ph.D.**   Microelectronics Research Center, University of Texas, Austin, Texas

**Leonard Neiberg, Ph.D.**   Logic Technology Development, Intel Corporation, Portland, Oregon

**Keith A. Nelson, Ph.D.**   Department of Chemistry, Massachusetts Institute of Technology, Cambridge, Massachusetts

**Regina Nijmeijer, M.S.**   R&D Department, Boxer Cross Inc., Menlo Park, California

**J. P. Piel, Ph.D.**   R&D Department, SOPRA, Bois-Colombes, France

**Michael T. Postek, Ph.D.**   Precision Engineering Division, National Institute of Standards and Technology, Gaithersburg, Maryland

**Christopher J. Raymond, Ph.D.**   Scatterometry Department, Accent Optical Technologies, Albuquerque, New Mexico

**John A. Rogers, Ph.D.**   Condensed Matter Physics Research, Bell Laboratories, Lucent Technologies, Murray Hill, New Jersey

**Robin Sacco**   Philips Analytical, Natick, Massachusetts

**Wolf-Hartmut Schulte, Ph.D.**   Department of Physics and Astronomy, Rutgers University, Piscataway, New Jersey

**Alexander Starikov, Ph.D.**   Intel Corporation, Santa Clara, California

---

* *Current affiliation*: Advanced Metrology and Repair Department, IBM, Hopewell Junction, New York

**J. L. Stehle, M.S.**   SOPRA, Bois-Colombes, France

**Robert Stoner, Ph.D.**   Department of Engineering, Brown University, Providence, Rhode Island

**John C. Stover, Ph.D.**   ADE Corporation, Westwood, Massachusetts

**Kenneth W. Tobin, Jr., Ph.D.**   Image Science and Machine Vision Group, Oak Ridge National Laboratory, Oak Ridge, Tennessee

**Yuri S. Uritsky, Ph.D.**   Defect and Thin Film Characterization Laboratory, Applied Materials, Inc., Santa Clara, California

**Bradley Van Eck, Ph.D.**   Front End Processes, International SEMATECH, Austin, Texas

**András E. Vladár, Ph.D.**   Precision Engineering Division, National Institute of Standards and Technology, Gaithersburg, Maryland

**Eric M. Vogel, Ph.D.**   Semiconductor Electronics Division, National Institute of Standards and Technology, Gaithersburg, Maryland

**Peter M. Zeitzoff, Ph.D.**   International SEMATECH, Austin, Texas

**Jie-Hua Zhao**   Microelectronics Research Center, University of Texas, Austin, Texas

# Handbook of Silicon Semiconductor Metrology

# 1

# Silicon Semiconductor Metrology

**Alain C. Diebold**
*International SEMATECH, Austin, Texas*

## I. INTRODUCTION

Metrology is an exciting field that requires a background in both semiconductor fabrication (process) and measurement technology and physics. Although the official metrology roadmap of the semiconductor industry includes measurements done outside the cleanroom, this book attempts to cover metrology done inside the cleanroom of a semiconductor chip factory (1). It is divided into sections that focus on the process technology areas of lithography, transistor and capacitor (front-end processes), and on-chip interconnect. Other sections include metrology areas that are used by many process specialties, such as sensor-based process control, electrical test–based statistical metrology, data management, and how to obtain information from data. Overviews are provided of measurement technologies and calibration methods, such as visible and ultraviolet optical, ion beam, and x-ray reflectivity. Before presenting the overview in this introductory chapter, the definition of in situ, in-line, and off-line metrology are discussed.

In the *International Technology Roadmap for Semiconductors* (ITRS) (1), *metrology* is defined as measurements that are done in situ, in-line, and off-line. There are no universally accepted definitions of these terms, so it is useful to state how we define them here. *Off-line metrology* refers to measurements done outside of the cleanroom; the type of laboratories that perform these measurements include materials characterization and returned-part failure analysis. *In-line metrology* refers to measurements and process control done inside the cleanroom. *In situ metrology* refers to measurements and process control done using sensors placed inside the process chamber or as a part of a loadlock or wafer transfer chamber. Recently, the concept of *integrated metrology* was introduced. Integrated metrology is very similar in spirit to advanced process control, in which the sensor is either inside the process chamber or part of a cluster tool. In a cluster tool, several process chambers are connected directly to a central robot that moves wafers between process chambers. A clustered sensor or metrology station could be inside the central robot area, or it could be a separate chamber clustered to the central robot. There are many excellent books on materials characterization, therefore we refer you to those references instead of repeating the details here (2,3).

Predicting future metrology requirements is a difficult task. If one considers the industry-based ITRS projections of process technology, one can see that the crystal ball is clearer in the near term, where yearly requirements are projected. Longer-term needs are more difficult to predict, and three-year intervals are used. The design of transistor, capacitor, and interconnect structures is expected to change in the long term. Vertical, dual-gate transistor structures are one interesting example of how new designs can greatly alter metrology needs. In the traditional complementary metal oxide semiconductor (CMOS), the transistors are oriented so that the channel is horizontal. This results in many of the measured films also being horizontal. If vertical transistors are employed, then the channel and gate dielectrics are vertical and thus very difficult to measure. A critical dimension measurement for the transistor gate becomes a film thickness measurement instead of a top-down, linewidth measurement. Control of dopant regions may also be different. According to the processes used to make the first vertical transistors, source and drain extensions are produced by diffusion from doped glass instead of from implants. Thus, long-term predictions can easily be proved incorrect. Nevertheless, it is clear that there will always be a need for microscopy capable of "seeing" the features that are being manufactured. One hopefully correct prediction is that great advances will be required in this field. None of the chapters in this book are focused on predicting long-term metrology needs.

In this chapter, the topics of measurement precision and accuracy, evaluation of measurement capability, model-based measurement, and the relationship between process flow and metrology are all covered. The topic *"Ratio of Measurement Precision to Process Tolerance vs. Resolution and Accuracy"* is discussed in Sec. II. An overview of *model-based measurement* interpretation is provided in Sec. III. The fourth section of this chapter is about *enabling metrology, from research and development to mature process technology.* In the final section, *development trends and conclusions* are discussed, including the trend toward integrated metrology.

## II. RATIO OF MEASUREMENT PRECISION TO PROCESS TOLERANCE RATIO VS. RESOLUTION AND ACCURACY

Traditionally, process flows are kept under control by maintaining key process parameters, such as line width and film thickness, inside limits (or specifications) that are known to be critical for yielding integrated circuits that have required electrical properties (4). One of the statistical process control metrics is the process capability index $C_P$. The variation, $\sigma_{\text{process}}$, of a process parameter such as film thickness is related to the process quality metric, defined as follows:

$$C_P = \frac{\text{USL} - \text{LSL}}{6\sigma_{\text{process}}}$$

where the upper process specification and lower process specification limits are USL and LSL, respectively. Let us consider an example in which the allowed film thickness range is $\pm 5\%$ and the average variation of the process is at the midpoint of the range. Thus for a 2-nm-thick dielectric film, the process range is from 1.9 nm to 2.1 nm, with an average value of 2.0 nm. This requires that the film thickness measurement be capable of resolving the difference between 1.9 nm and 2.0 nm as well as that between 2.0 nm and 2.1 nm over a long period of time. For processes showing a Gaussian distribution of film thickness values, a well-designed and -controlled process will have $C_P$ values greater than or equal to 1, as shown in Figure 1. It is possible to have a process with the average value away from the center of the midpoint of the process range. The metrics that measures how

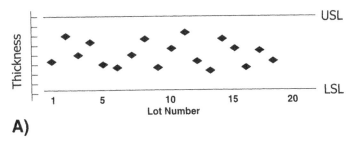

**A)**

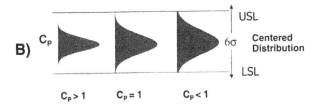

**B)**

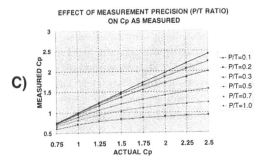

**C)**

**Figure 1**   Basic concepts of statistical process control (SPC). The metric for process variation is the process capability index $C_P$. Typically, one measures a variable such as thickness and plots the value for that lot. The thickness could be the value at the center of wafer, averaged over a wafer or lot of wafers. The goal for SPC is to keep the thickness between the upper and lower specified limits. Here we show the process variation for a process with the average value at the center of the specification limits in parts (A) and (B). The goodness of the measured value of $C_P$ is dependent on the variation of the measurement itself. The metric for metrology tools is the ratio of measurement precision to process tolerance, $P/T$. The change in measured $C_P$ with increased (worse) $P/T$ is also shown in part (C). Figure (C) is courtesy of Jerry Schlesinger.

well a process is centered are the process capability indexes $C_{PL}$, $C_{PU}$, and $C_{PK}$, defined as follows:

$$C_{PK} = \min\{C_{PL}, C_{PU}\}$$

$$C_{PL} = \frac{X_{\text{AVE}} - \text{LSL}}{3\sigma_{\text{process}}}$$

$$C_{PU} = \frac{X_{\text{AVE}} - \text{USL}}{3\sigma_{\text{process}}}$$

$X_{\text{AVE}}$ is the average value of the process. When a process is well centered, $C_{PL} = C_{PU}$ and $C_{PK} = C_P$. If measurements were always perfect, then determining process variation would be easy. Since measurements have their own variation and do not typically result in the true value, measurement equipment must itself be evaluated. The National Institute of Standards and Technology has a handbook on engineering statistics online at http://www.nist.gov/itl/div898/handbook/; Chapter 2, "Measurement Process Characterization," is of special interest.

The ratio of measurement precision to process tolerance is an accepted method of evaluating the ability of an automated metrology tool to provide data for statistical process control (4). The measurement precision or variation, $\sigma_{\text{measurement}}$, is a combination of the short-term repeatability of a measurement tool and the long-term reproducibility, $\sigma_{\text{measurement}} = (\sigma^2_{\text{repeatability}} + \sigma^2_{\text{reproducibility}})^{1/2}$. *Repeatability* is the variation of repeated measurements made under identical conditions (4). *Reproducibility* is the variation that results when measurements are made under different conditions, such as reloading the wafer on a different day (4). Precision is often determined by repeating a specific type of measurement on a reference sample over an extended period of time. One example of what is meant by a specific type of measurement would be determining the thickness of an oxide layer in the center of a wafer and repeating this using identical conditions (including not moving the wafer from the measurement position). Repeatability is the variation associated with repeated measurement at the same location on a wafer (4). Precision also includes measurement variation associated with reloading the wafer (4). Reproducibility is obtained by repeating the measurement procedure over a long period (for example, 15 days). As already mentioned, the measurement precision is the square root of the sum of the squares of the repeatability and reproducibility. Process tolerance is the range of allowed values of the process variable: upper process limit UL − lower process limit LL. The ratio of measurement precision, $\sigma$, to process tolerance $(P/T)$ is (1,2):

$$\frac{P}{T} = \frac{6\sigma_{\text{measurement}}}{(\text{UL} - \text{LL})}$$

Although $P/T$ should be less than 10%, a value of 30% is often allowed. The smaller the value of $P/T$, the closer the experimental value of $C_P$ or $C_{PK}$ is to the true value. As shown in Figure 1, it is difficult to know when a process is under control when a $P/T$ of more than 30% is used. It is important to note that $P/T = 3\sigma_{\text{measurement}}/\text{UL}$ when there is no lower process limit. Contamination limits represent the upper process limit.

Unfortunately, the measurement precision used to determine $P/T$ is often an approximation of the true precision. Determination of the true $P/T$ for the process range of interest requires careful implementation of the $P/T$ methodology. This means that many of the measurement conditions should be varied to determine measurement stability before setting final measurement conditions when determining precision (4). Varying the time between repeated measurement allows one to observe short-term issues with repeatability. In addition, reference materials should have identical feature size, shape, and composition to those of the processed wafer being measured. Often, there are no reference materials that meet this requirement, and $P/T$ and measurement accuracy are determined using the best available reference materials. One key example is the lack of an oxide thickness standard for $SiO_2$ and nitrided silicon oxides under 5 nm thick.

When the reference material has significantly different properties (e.g., thickness), then the precision may not be representative of the precision associated with the product wafer measurement, due to nonlinearity. Again, the example of critical dimension or film thickness measurements is useful. The precision associated with measurement of a 2-nm-gate oxide may be different than that associated with a 10-nm oxide film. If the true precision is large enough, it could mean that the metrology tool has insufficient resolution to distinguish changes over the process range of interest. One way to ensure that the metrology tool has adequate resolution to determine the true $P/T$ capability is to use a series of standardized, accurate reference materials over the measurement range specified by the upper and lower process limits. In Figure 2, we depict how the multiple reference materials approach might work. We note that some measurements may require only one suitable reference material for $P/T$ determination.

The measurement of the thickness of the transistor gate dielectric at the 100-nm technology generation is expected to be difficult in the manufacturing environment. If the gate is 2 nm thick and the process tolerance is 5% for $3\sigma$ (process variation), then

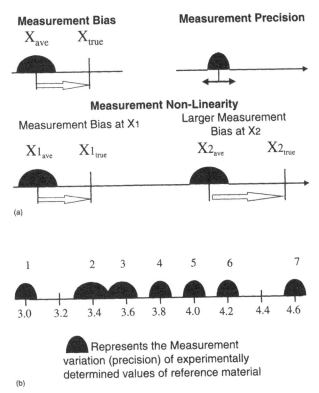

(a)

(b)

**Figure 2** (a) Reference materials and measurement linearity and resolution measurement non-linearity. Measurement nonlinearities can result in bias (difference between true and measured value) changes between the calibrated value and the range of interest. It is also possible that the bias can change inside the measurement range of interest. (b) Demonstration of resolution based on precision associated with measurement of a series of reference materials over the process specification range. For this example let us assume that the process tolerance (also called process specifications) is from 3.0 nm to 4.0 nm. The measurement precision at $3\sigma$ (variation) is shown for reference materials inside the process range. The experimental $P/T$ capability observed using reference materials 4, 5, and 6 indicates that a single measurement at 4.0 nm is different from one at 3.8 nm or 4.2 nm. Thus this fictitious metrology tool can resolve those values. According to the precision shown for reference materials 2 and 3, the tool is not able to resolve 3.4 nm from 3.6 nm at $3\sigma$.

$P/T = 10\% = 6\sigma/0.1$ nm, which gives a measurement variation $\sigma = 0.0017$ nm. The size of an atomic step on silicon is $\sim 0.15$ nm, and atomically flat terraces on specially prepared Si(001) are about 100–200 nm wide. The width of terraces after typical processing, such as sacrificial oxidation and gate oxide growth, is unknown. Thus, some have pointed to the issue of measuring film thickness to less than an atomic width. This is not an issue, because the measurement can be considered to be the determination of the average thickness of a perfectly flat layer, that is, this type of measurement precision requires analysis of a large area that averages over atomic steps at the interface and other atomic variations. Local variations in thickness of a 2-nm film will be a much larger percentage of total thickness than it would be for a 10-nm film. Therefore, the reproducibility of stage location affects the precision of a thickness measurement, especially for a small sampling area around 1 micron in diameter. Some metrologists have called for smaller area ($< 1$ m) measurements. The need for high precision makes this very difficult, and high precision with small-spot instruments (0.9 m) are achieved by averaging over a larger area. Even electrical measurements done using capacitors average over larger areas.

For contamination measurements and $P/T = 30\%$, $\sigma = UL/10$. Achieving a measurement variation of $UL/10$ requires that the detection limit be at or below $UL/10$ (5). Chemists have their own definitions of resolution, detection limit, and quantifiable detection limit, all of which must be considered (5). For convenience, these topics are illustrated by discussing off-line chemical analysis of trace contaminants in a liquid. Typically, the detection limits for trace amounts of contamination in a liquid vary due to changes in measurement sensitivity. Detection limits are ultimate values for optimum circumstances. When the liquid is analyzed by inductively coupled plasma mass spectrometry (ICP-MS), the detection of one element can be interfered with by the presence of another element. The quantifiable detection limit is the limit at which reliable, repeatable detection is possible (5). Resolution requirements are not well defined, and thus experience again provides some guidance.

## III.  MODEL-BASED MEASUREMENT

One of the 20th century's most famous scientific debates is over the effect of observation on the quantum property being measured. At the most fundamental level, interpretation of all observations is based on some type of a model. For example, we presume that our eyesight provides information that allows us to drive where we wish to go. With increased age, some people lose their ability to see clearly in dark conditions, and thus this model is tested when one drives at night. In this section, an overview of how metrology can be greatly improved by understanding and modeling the measurement process is provided. A very interesting and useful explanation of the relationship between modeling and measurement is already available in the literature (6).

Each measurement method is based on a model that relates the observed signal to a value of the variable being measured. One aspect of measurement is the physical model of the sample. As mentioned in the previous section, ellipsometric measurements assume that film can be approximated as a flat slab. A careful examination of this model in subsequent chapters on optical film thickness measurement will show that the model assumes average optical properties inside the slab. This a reasonable model for optical measurements of thin (nonabsorbing) dielectric films using light on the visible region, because the wavelength of this light is large compared to the film thickness. Another aspect of data interpretation is the algorithm used to extract values from the observed signal. Critical-dimension measurement by scanning electron microscopy can be done by using either an empirical or a more fundamental model. In the empirical model, the relationship between the secondary electron intensity and the sample shape are assumed to be given by assuming that some mathematical function (such as the maximum in intensity) determines the edge of the line, as shown in Figure 3. In the fundamental model approach, the relationship is determined by a Monte Carlo simulation of the secondary electron emission for a specific shape and material composition, such as a transistor gate with a slope, as shown in Figure 4. Chapter 14, by Postek and Vladar, and Chapter 15, by Marchman and Griffith, discuss this topic. The advantages of the latter method are evident when new materials such as porous, low-dielectric-constant insulators for interconnect metal lines are introduced. Thus, our discussion of modeling can be broken into two parts. First we describe the process determining the physical model of the sample that is required for physics of the measurement, and then we discuss modeling the metrology tool.

## A. Measurement Physics and the Physical Model of the Sample

The process of determining the appropriate physical model will be illustrated by building on the two examples described earlier. First we model thin silicon dioxide and silicon oxynitride dielectric films. Several questions must be answered before building the optical model that one uses for controlling the fabrication of these films by ellipsometry. For example, in addition to total film thickness, what else should be measured? Nitrogen concentration and profile (concentration vs. depth) are examples of quantities that control the electrical properties of oxynitride films. Can these quantities be measured? What is known about the detailed structure of these films from materials characterization methods such as high-resolution transmission electron microscopy? What is the relationship between physically measured properties and electrically measured properties? The answers to these questions come from a recent publication and a review article (7,8).

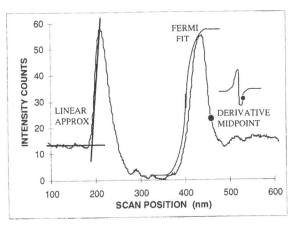

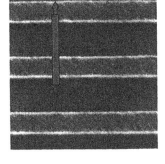

**CD-SEM**

**Signal ⇨ CD Value**

**Figure 3** Empirical CD measurement. As one can see from the change of secondary electron signal variation as the CD-SEM electron beam scans across the edge of a line, the edge position must be determined. Arbitrary, empirical relationships have been developed to determine edge position for line width measurement.

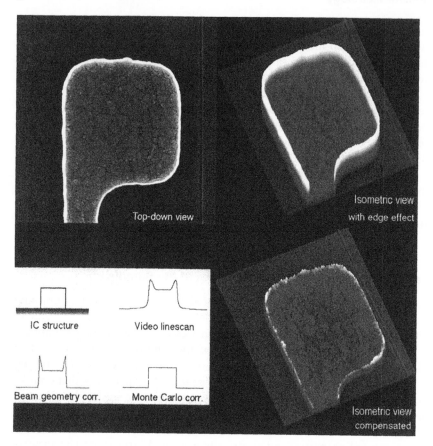

**Figure 4** Monte Carlo modeling of CD measurement. Monte Carlo modeling has been used to determine the secondary electron signal strength as the electron beam is scanned across a line or feature. The secondary electron intensity depends on electron beam energy, material composition, and line shape.

First let us examine the atomic structure of thin $SiO_2$ films grown on silicon. The system is known to include an interfacial layer that becomes a significant part of the total film thickness for sub-3-nm silicon dioxide or oxynitride films (8). The transition from bulk silicon to stoichiometric $SiO_2$ involves the atomic steps or terraces of the last silicon atoms of the substrate and substoichiometric oxides. If stress is present in the film, it may need to be included in the optical model. Since the dielectric constant is a function of the wavelength of light (or frequency of a capacitance–voltage measurement), it is referred to here as a dielectric function. Optical models consider the film to be made of flat layers (slabs) of material that typically have uniform dielectric functions—uniform in the sense that localized properties are naturally averaged by the fact that the wavelength of light is larger than the film thickness for modern transistor gate dielectrics under 3 nm. Thus an optical model that includes the interface is one that has a thin layer between the substrate and the bulk oxide. Several models have been used to include the interface. The atomic level roughness or step structure of the interface layer can be modeled as a mixture of the dielectric functions of silicon and silicon dioxide (7,8). This approach avoids determining the dielectric function of the substoichiometric oxide, and it leads to the question of what properties can be determined from very thin films through fitting the response of various models (7,8).

Ellipsometry in the visible wavelength range has limited capability to determine multiple film properties such as multilayer structure. For example, one cannot distinguish the difference between a silicon nitride layer on a silicon dioxide or vice versa from the quality of the fit of the measured data to an optical model that contains two "slab" layers, one of silicon dioxide and one of silicon nitride. This is discussed further in Chapter 25, by Jellison, in terms of the fundamental physics of ellipsometry of very thin layers (9). It has also been shown that simultaneously fitting several properties in an optical model results in a better fit but worse precision (7,8). Here we refer to an optical model that contains thickness for three layers ($SiO_2$/interface/silicon substrate) and variables such as the relative concentration of constituents for the interface layer (7,8). Thus one must carefully choose the properties that are fit during in-line metrology.

Correlation with electrical measurements requires use of the same dielectric functions (constant for each frequency) (8). Electrical measurements do not often consider the impact of the interfacial layer on the dielectric function of the total dielectric. Electrical measurements are further complicated by the processing that a capacitor or transistor structure receives after the dielectric is grown or deposited. Conventional wisdom indicates that the interfacial properties are strongly affected by the thermal treatment received during polysilicon or metal electrode deposition and other processing, such as annealing. In Chapter 2, Hayzelden discusses the optical models used to measure the thickness of gate dielectric materials.

## B. Modeling the Metrology Tool

The value obtained for the measurement of a specific property represents the result of a mathematical calculation or electronic calculation/evaluation of an operational model for the metrology tool. The details of a model are dependent on both the measurement method and the specific design of the system. Often, the advantage of one type of measurement system design can be found in the combination of hardware and model. The measurement of dimensions by a piezoelectric scanner–based scanned probe microscope is one example. Over macroscopic distances, the response of the piezoelectric actuator is not linear. Therefore the dimension ($x$ or $y$ distance) for each point in a scan is calculated from the applied voltage using a model that corrects for the nonlinearity. Typically this model is calibrated by scanning each new scanner using a reference material for distance. Thus the accuracy and repeatability of this type of scanned probe microscope depend on the hardware and the model. In this section, several examples will be discussed, and the remaining chapters will present this type of information whenever possible.

Critical-dimension measurement by scanning electron microscope (SEM) typically employs an empirical model that relates the observed signal to the linewidth, as shown in Figure 3. A more fundamental, Monte Carlo model that determines the true lineshape from the observed signal based on the interaction between the scanned electron beam and the sample has been developed. The value of this model is shown in Figure 4. The metrology community is hoping that the Monte Carlo model will replace the empirical model. The empirical model makes several assumptions that deserve mention. One assumption is that the SEM operates with exactly the same focus and resolution for each measurement. Another is that the secondary electron signal observed as one scans over the line does not change with sidewall angle or materials composition. Materials change with each level in the chip, from polysilicon at the transistor level to metal lines at the interconnect levels (for traditional metalization processes). Photoresist responds very differently to electron beam–based measurement. This is why CD-SEM measurements are calibrated for each process level. Another complication is the density of the features being measured. The measurement offset is different for isolated and dense lines. As one can see in Figure 4, the sidewall shape can be determined from the change in intensity as one scans across the edge of the line. Thus, one can see that a more sophisticated model might be able to reduce the use of calibration artifacts and increase the amount of information that one can obtain.

## IV. ENABLING METROLOGY, FROM RESEARCH AND DEVELOPMENT TO MATURE PROCESS TECHNOLOGY

The purpose of this section is to provide examples of a well-developed metrology program can help both process tool supplier and integrated circuit manufacturer. One might imagine that the amount of metrology activity, including materials characterization, is greatest during the research and development phase and decreases as one progresses toward a mature process flow. This idea is shown in Figure 5. It is useful to present the main concepts behind the implementation of metrology, including materials characterization. These concepts are as follows.

### 1.   Characterization and Metrology Activity Decreases with Process Maturity

When new materials, such as tantalum nitride barrier layers, for metal interconnects, the material properties, such as crystal structure, grain orientation, and elemental and chemical composition, are thoroughly characterized and process dependence determined. Process dependencies, such as growth rate vs. gas flow or etch rate vs. gas flow, are studied. Process integration issues, such as reactivity with previous or subsequent films or structures, are a part of this evaluation, and the impact of thermal processes on new materials is a critical piece of information. Particle generation mechanisms and contamination sources are characterized for each material and process tool design. Some new process tools do not fabricate new materials, but result from the new requirements associated with decreased feature size or circuit design. With the need to push new materials and process tools into production at an accelerated rate, patterned wafer testing can occur early in the evaluation. Testing of in-line metrology capability and

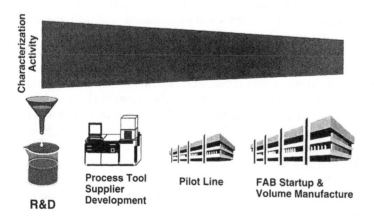

**Figure 5**   Characterization and metrology activity vs. process maturity. Measurement activity decreases as a new material or process moves from invention to pilot line to mature manufacturing.

testing by in-line metrology begin as soon as possible. The need for continued thorough characterization decreases as one gains knowledge about the properties of new materials and processes.

The next step is to evaluate the manufacturability of a new material or process tool using a process flow. This can be done with test structures or test chips. Test chips can be fabricated using part of the full process flow required for manufacture of a complete integrated circuit chip. This is known as a short-loop process. Occasionally, previous-generation memory chips serve as test chips. In-line metrology activity is typically greater during this phase of development. Some in-line metrology tools are aimed at short loop development and "pilot line" evaluation and are considered engineering tools. One example is SEM-based defect detection capability, which in 1999 required many hours to scan a patterned wafer. There is a drive to improve the throughput of these systems. During this period, particle detection capabilities are developed for use in volume manufacture.

When the process flow is transferred from pilot line to volume manufacture (or the pilot line is expanded into volume production), characterization and in-line metrology activity centers around comparing the physical and electrical properties to those found at the pilot line. Once a high yield is established, the amount of characterization and in-line metrology can decrease.

### 2. Metrology Should Help Reduce the Time for Learning Cycles at All Stages, from Research and Development to Manufacture

A learning cycle starts when a short-loop or full-flow process is begun and ends when analysis of electrical and physical data from this cycle is complete. One can also refer to the learning cycles that occur during research and development. One example would be fabricating a series of blanket (unpatterned) films using different process conditions and characterizing the changes in elemental composition or crystal structure. Typically one refers to the time required for a learning cycle as the *cycle time*. Metrology can add to the cycle time just through the time required for measurement. The point is to reduce this time whenever possible. This need has driven the introduction of materials characterization tools that are capable of measuring properties of whole wafer samples. Characterization and metrology can reduce the number of and time for learning cycles by providing information that either solves the problem or drives the conditions for the next learning cycle. For example, particle characterization cannot only identify the particle, but can also determine the source of the particle when an information system is used. Tobin and Neiberg discuss information systems in Chapter 23.

### 3. A Well-Characterized Process Tool Can Be Quickly Ramped to High-Yielding Production in a Pilot Line, and a Well-Characterized Process Flow Can Be Quickly Ramped to High Yield When a Pilot Line Process is Transferred to a Production Fab

The time required for ramping a new fab to high yield has decreased with subsequent technology generations. The time required for ramping pilot lines to adequate yield has also reduced, but not as dramatically. One way to ramp pilot lines to adequate yield more quickly is for process tool suppliers to provide thoroughly characterized process tools. This is quite a challenge considering that leading-edge integrated circuit manufacturers are purchasing certain key tools before the alpha version has been designed and tested. During 1999 and 2000, lithography tools operating at 157-nm wavelength are one example of this.

### 4. Typically, Physical Measurements Are Done to Ensure Electrical Properties, as Shown in Figure 6

During volume manufacture, one measures physical properties such as film thickness and flatness, critical dimensions, and dopant dose to ensure electrical properties. Key electrical properties for transistors include the distribution and values of leakage current, gate delay (or, equivalently effective, gate length), and threshold voltage. Interconnect electrical properties include contact resistance and transmission delay.

### 5. Metrology Strategies Usually Differ According to the Number and Type of Integrated Circuit Chips Produced in a Fab

The product lines and manufacturing strategies of a company help define metrology strategy. Some fabs manufacture mainly logic, memory, or application-specific integrated circuit (ASIC) chips, while others manufacture a wide variety of products. According to existing wisdom, a process tool that is used for a limited number of process recipes is thought to more easily remain inside the desired process range. One that uses a variety of process recipes often requires verification that the process is producing material within the process specification range (e.g., film thickness). Using this wisdom as guidance, more metrology is required in fabs that manufacture a variety of chip types. This wisdom may be more applicable to fabs using less mature process tools.

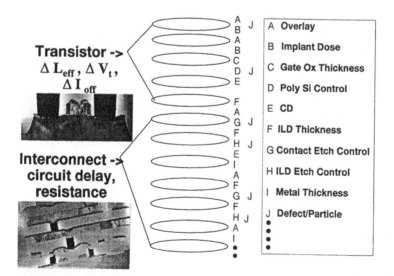

**Figure 6**  Physical measurements ensure electrical properties. The transistor gate dielectric thickness, dopant dose and junction, and physical gate length are measured to ensure the proper range of values for gate delay, leakage current, and threshhold voltage. The physical properties of interconnect structures are measured to ensure that the IC chip maintains clock speed and signal integrity.

## V.  DEVELOPMENT TRENDS AND CONCLUSIONS
## (INCLUDING INTEGRATED METROLOGY)

This introductory chapter concludes with a discussion of the trends that typify most metrology research and development. These topics can be used to guide one through the chapters that may seem to be unrelated to in-line metrology.

Reducing the time for development is always an important goal. This has driven the introduction of materials characterization tools that are equipped with sample stages that accept whole-wafer samples and the use of software that allows these tools to navigate to specific areas of interest and store data associated with these areas. The SEM defect review tool, the dual-column FIB-SEM (FIB is shorthand for "focused ion beam system"), and the whole-wafer-capable Auger-based defect review tool are examples of this trend. Recently, new measurement technology is being introduced with cleanroom and whole-wafer compatibility. Acoustic film thickness measurement is an example of this trend. Chapter 9, on impulsively stimulated thermal scattering (by Gostein et al.), describes one of the acoustic methods.

The use of sensor-based measurements for "integrated metrology" is another trend. From one point of view, this trend already existed and is known as advanced process control and advanced equipment control. Although this topic was briefly discussed in the introduction, it is useful to repeat it here. Chapter 22, by Barna, VanEck, and Hosch, presents a thorough discussion of sensor-based metrology. The idea is that the measurement takes places as a part of the process tool operation. When the information from these measurements is used to control the process tool after the process step is completed on a cassette of wafers, it is known as *run-to-run* control. When the information is used to control the process tool during its operation, it is known as *real-time* control. Real-time control includes detection of faults in process tool operation and control of process conditions during operations based on sensor inputs. This type of control will evolve from simple models of the process to sophisticated models that can learn from previous experience.

Precision requirements are becoming more difficult, if not impossible, for physical measurement methods to achieve. Solutions to this problem include: method improvement; invention of a new method; use of electrical methods only; and measurements using sophisticated models that learn from previous experience. The last solution could eventually overcome limitations such as inadequate precision. Electrical measurements provide statistically significant data from which information can easily be obtained. Chapter 24, by Boning, describes an electrical test methodology known as statistical metrology.

The increased sophistication of information systems that extract information from data is a critical part of metrology. Due to the greater number of devices on a chip and the greater number of chips/wafer per unit time, the amount of data increases. Storing this data is costly. In addition, the difficulty in gaining useful information from the data is becoming more difficult. The data itself is not useful, but the guidance obtained from the information extracted from the data is a critical enabler for manufacturing. Again, Chapter 23, by Tobin and Neiberg, describes information systems.

### A.  Integrated Metrology

Advanced equipment control (AEC) and advanced process control (APC) have been considered the ultimate goal of metrology by the advocates of these principles. In this volume, Chapter 22, by Barna, VanEck, and Hosch, devoted to sensor-based process control, describes these principles (11). A key premise of this approach is the goal of real-time process control. In other words, the feedback from the sensor is used to control the process tool during processing. The next-best method of AEC is run-to-run control, which feeds measurement results taken in a separate chamber or measurement tool immediately after processing directly back to the process tool. The fault detection mode is another aspect of AEC. In this case, the measurement data is used either to prevent processing or to shut the process tool down before running the process. One example would be detection of photoresist outgassing in the vacuum loadlock of an etch tool prior to insertion of the wafer into the main chamber. The wafers should be removed from the loadlock and the resist removed prior to processing. As discussed by Barna, VanEck, and Hosch in Chapter 22, this is one step beyond SPC. Figure 7 shows AEC/APC schematically. During 1999, industry observers have noticed that the driver for this approach has switched from the integrated circuit manufacturers to the suppliers of process equipment. With the switch in driver, this movement has recently emphasized the addition of measurement stations on clustered process tools. Frequently, the term *integrated metrology* refers to clustered metrology stations that are just beginning to provide feedback to control the process tool, and the first introductions appeared to be used for SPC or for fault detection. In this section, a brief introduction of integrated metrology is provided (see Figure 8).

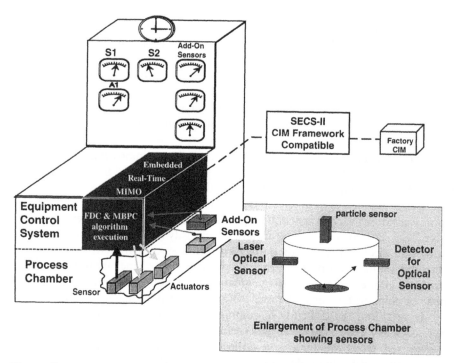

**Figure 7**  Advanced equipment control and advanced process control (AEC/APC). A schematic representation of a process tool is shown with an embedded "MIMO" equipment controller. The term MIMO refers to a multiple sensor input with multiple output. One example of this capability is control of wafer temperature along with the reactor gas flow. Sensors can be built into the chamber by the process tool supplier or added at the IC fab. The information from the process controller is transmitted to the factory computer-integrated manufacturing system to pass data into various data management systems.

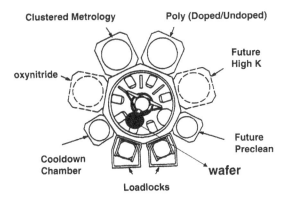

**Figure 8** Integrated metrology. A metrology cluster station is shown on a cluster tool for processing of the transistor gate stack. A wafer-transfer robot is located in the center of the system. This allows measurement of film thickness and uniformity after each deposition step without exposing the wafers to the ambient outside the cluster tool. In some cases, cluster tools may involve vacuum processes, and the wafer handler may also be under vacuum.

One of the stated goals of integrated metrology is to provide clustered measurement capability at low cost. Typically, the stand-alone metrology tool cannot be built as a low-cost cluster station without reducing measurement capability. The need for a high-precision measurement depends on the application. In some cases, one is only interested in knowing that a layer is present before proceeding to the next process step. This measurement can be done with a reflectometer instead of an ellipsometer, which greatly reduces the cost of the metrology station. In other cases, such as on wafer particle and defect detection, it seems likely that there is a need for a capability that matches stand-alone systems. Another approach is sensor-based in situ particle detectors located in the vacuum exhaust. There has been a considerable effort to correlate particle sensors to on-particle counts observed by defect detection done on a wafer surface.

## REFERENCES

1. Metrology Roadmap, 1999 International Technology Roadmap for Semiconductors. San Jose, CA: Semiconductor Industry Association, 1999.
2. H.W. Werner, M. Grasserbauer. Analysis of Microelectronic Materials and Devices. New York: Wiley, 1991.
3. C.R. Brundle, C.A. Evans, S. Wilson. Encyclopedia of Materials Characterization: Surfaces, Interfaces, and Thin Films. Boston: Butterworth-Heinemann, 1992.
4. S.A. Eastman. Evaluating Automated Wafer Measurement Instruments. SEMATECH Technology Transfer Document #94112638A-XRF. International SeMaTech 2706 Montopolis Drive, Austin, TX 78741.
5. L.A. Currie. Anal Chem 40:586–593, 1968.
6. H.R. Huff, R.K. Goodall, W.M. Bullis, J.A. Moreland, F.G. Kirscht, S.R. Wilson, NTRS Starting Materials Team. In: D.G. Seiler, A.C. Diebold, W.M. Bullis, T.J. Shaffner, R. McDonald, E.J. Walters, eds. Characterization and Metrology for ULSI Technology. Woodbury, NJ: AIP, 1998, pp 97–112.
7. A.C. Diebold, C. Richter, C. Hayzelden, D. Maher, et al. In press.
8. A.C. Diebold, D.A. Venables, Y. Chabal, D. Muller, M. Weldon, E. Garfunkel. Mat Sci Semicond Processing 2:103–147, 1999.

# 2

# Gate Dielectric Metrology

**Clive Hayzelden**
*KLA-Tencor Corporation, San Jose, California*

## I. INTRODUCTION AND OVERVIEW

With increasing clock speed, the scaling of CMOS transistor geometries presents ever more severe challenges to the maintenance of operational circuits. As the gate area is reduced, the thickness of the gate dielectric must also be reduced in order to maintain sufficient capacitive control of the MOS channel for adequate current flow. The thickness of the current generation of gate dielectrics is rapidly approaching the sub-20-Å level. Thus, gate dielectric metrology is a critical aspect of semiconductor fabrication. This chapter is concerned with gate dielectric metrology in the production environment, with an emphasis on optical metrology. The interested reader is directed to standard works that cover optical metrology and spectroscopic ellipsometry in depth (1). A SEMATECH-sponsored gate dielectric metrology research program at North Carolina State University, with strong industry participation, has provided much of the background to this chapter (2).

Section II discusses the silicon/silicon dioxide system that forms the basis for the current semiconductor industry. Section III provides a brief overview of the salient aspects of the 1999 Semiconductor Industry Association's *National Technology Roadmap for Semiconductors* (3). The emphasis here is on the continued scaling of gate dielectric thicknesses with the associated, and increasingly demanding, precision requirements for capable metrology. The expected introduction of high-dielectric-constant (high-$\varepsilon$, also commonly referred to as high-$\kappa$) gate dielectrics will then be discussed. Section IV describes the production application of optical metrology using spectroscopic ellipsometry. Section V describes a research investigation aimed at understanding the relative contributions of underlying variables to metrology variation, and focuses on modeling the complex interface of the $Si/SiO_2$ system. Attempts to model the formation of ad-layers to the $SiO_2$ surface from airborne molecular contamination are also described. The subject of airborne molecular contamination is of great concern to semiconductor fabrication (fab) metrology, and the results of thermal cleaning studies are presented in Section VI. Section VII discusses optical and (contact) electrical metrics that have been obtained as part of a research program to develop ultrathin-gate oxide standards. Section VIII describes the recent development of in-line noncontact production electrical metrology of nitrided-gate dielectrics and the information that can be obtained from a combination of optical and electrical metrology. In anticipation of the impending transition to high-dielectric-gate materials, the section concludes with an examination of some recent work in this important area.

## II. THE SILICON/SILICON DIOXIDE SYSTEM

### A. The Silicon/Silicon Dioxide Interface

Silicon dioxide ($SiO_2$) is composed of a covalently bonded random network with a resistivity in excess of $10^{16}\Omega$-cm and, in either the pure or nitrided form, forms the gate dielectric for the current generation of Si-based transistors. The silicon/silicon dioxide system has been studied most extensively (4). There is an abundance of physical (5,6), chemical (7,8), and optical (9–19) data that indicates that the interfacial region between the two phases differs from the bulk of either phase. Estimates of the extent of this interfacial region range from a few angstroms up to approximately 10 Å. For thick silicon dioxide films, this interfacial region represents a very small fraction of the entire film. However, for ultrathin-gate dielectric films, such an interfacial region could constitute 20–50% of the entire film. Thus, it is necessary to examine the extent to which this interlayer must be included in optical models to obtain the required measurement precision for ultrathin oxide films. Interfacial layers may also exist for other dielectrics. Moreover, stacked dielectrics consisting of individual layers of oxide and nitride present additional metrology challenges. In the ultrathin-film regime, the ellipsometric response of films with different optical constants is small. As a result, it can become difficult to obtain an independent measure of the thickness of individual layers. Advanced, high-dielectric-constant materials may include interfacial layers, either by design or through phase separation processes. Moreover, the optical constants of candidate materials may not be well known and are likely to be process dependent. Thus, metrology challenges exist for silicon dioxide, oxide/nitride stacks, and for high-dielectric-constant materials.

For spectroscopic ellipsometers (SEs), the optical model consists of values for the refractive index (which is generally a complex number) of the film, the substrate, and any possible interface layers, at each wavelength used in the measurements. For the $SiO_2$-on-silicon system, the starting point for any modeling is the index of refraction for $SiO_2$ and for the crystalline silicon substrate. Bulk $SiO_2$ optical constants are usually based on Malitson's work on fused silica (20). Bulk silicon optical constants are taken from various sources (21,22). For many years it has been recognized that the optical constants of thermally grown $SiO_2$ on silicon are actually slightly higher than Malitson's fused-silica values, apparently as a result of differences in density. As a practical matter, these density differences have been incorporated into optical models in several ways. One method is to introduce a two-phase model consisting of Malitson's $SiO_2$ and voids. Typically, the best fit between experimental and calculated data is obtained with a small, negative void fraction (16,17). Similar effects can be achieved by varying the offset parameter in Seillmeier representations of the index of refraction for $SiO_2$ (19) or by varying the oscillator number densities in harmonic oscillator models. Mixtures of amorphous silicon and $SiO_2$ have also been used (23). In addition to the apparent requirement that the index of refraction be slightly greater for thermal oxides, there exists abundant evidence that including an interface layer of intermediate refractive index between the silicon substrate and the oxide layer produces an improved goodness-of-fit. However, the precision with which the parameters in these interfacial models can be specified is rather poor. For example, Taft and Cordes (15) inferred the presence of an interface layer, approximately

0.6 nm thick, of intermediate index of refraction from single-wavelength ellipsometry measurements of progressively etched-back thermal oxides. However, they noted that their results were also consistent with interfacial layers of slightly different thickness and optical constants. Aspnes and Theeten (16,17) incorporated physical and chemical mixtures of silicon and silicon oxide to model this interface layer for thermal oxides characterized with spectroscopic ellipsometry. They concluded that the inclusion of almost any such mixture in the model improves the goodness of fit and that chemical mixtures produced somewhat better fits than physical mixtures. They report a best-fit model consisting of ~0.7 nm of chemically mixed silicon (0.8 atomic fraction) and silicon dioxide (0.2 atomic fraction) with error bars of 0.2 nm on the thickness and 0.1 on the atomic fractions. Nguyen et al. (18) used a combination of single-wavelength and spectroscopic ellipsometry to deduce the presence of a thin strained silicon layer beneath the oxide. Roughness at the interface between the silicon and the oxide film was modeled as a physical mixture of 88% silicon and 12% silicon dioxide. Herzinger et al. (19) recently re-examined the interface issue in detail with variable-angle spectroscopic ellipsometry measurements. They concluded that an interface layer in the subnanometer range with intermediate optical constants provided the best fit. However, they emphasize that correlation effects among the fit variables complicate an exact description of the nature of the interface layer. It is important to recognize that almost all of the ellipsometric data included in these studies comes from thermal oxides typically thicker than 10 nm and that the data analysis includes measurements from multiple samples with varying thickness. For ultrathin oxides, in the thickness range 2–3 nm, the interfacial layer might constitute nearly a third of the entire film. However, parameter correlation effects, such as those noted by Herzinger et al. (19), become even stronger with such thin films.

## B. Silicon Dioxide/Ambient Interface

A final complicating factor is the presence of contamination on the surface of the dielectrics to be measured. It is well established that organic contaminants are ubiquitous in the cleanroom environment. These contaminants can form an ad-layer on the surface of the dielectric film. Under these circumstances, the ellipsometric response includes the effects of the ad-layer. To eliminate these effects, it is necessary to develop a reproducible and easily implemented cleaning procedure, consistent with the precision requirements noted earlier. The analysis of ultrathin stacked dielectrics consisting of individual layers of oxide and nitride may also be complicated by parameter correlation effects. Advanced, high-dielectric-constant materials may include interfacial layers, either by design or through phase separation processes. Moreover, the optical constants of candidate materials may not be well known. Thus, it is necessary to investigate the optical properties of these materials and to determine the precision with which their thickness can be measured.

## III. GATE DIELECTRIC SCALING AND REQUIRED METROLOGY PRECISION

### A. Discussion of Metrology Metrics

*Static repeatability* ($\sigma_{SR}$) is the standard deviation of repeated measurements made on the same object under identical conditions over a very short period of time (typically seconds or minutes). Static repeatability represents measurement-to-measurement variation only. Static repeatability may be obtained by loading a wafer into the ellipsometer and making a series of consecutive measurements at one site. Focusing of the ellipsometer occurs prior to the first measurement only. Thirty measurements in rapid succession might typically be performed. Variation introduced by movement of the stage is thus eliminated.

*Dynamic repeatability* ($\sigma_{DR}$) is the standard deviation of measurements obtained from cycling the wafer into and out of the ellipsometer. Measurements are performed over a short period of time (typically tens of minutes) by loading and unloading the wafer 30 times. Dynamic repeatability emphasizes the variation introduced by stage accuracy and focus. When single measurements are made (such as center-site or a five-site pattern), dynamic repeatability inherently includes the variation associated with the static repeatability just discussed. To reduce the static repeatability contribution, one could use the average value of (e.g., 30) repeated measurements at each point. That is, to reduce (but not eliminate) the effect of static repeatability, one may perform static repeatability measurements at each site following each wafer load/unload operation. Dynamic repeatability measurements performed over several days lead to a measurement of stability, as described next.

*Stability* ($\sigma_S$) is the degree to which the measurement varies over time. Stability is typically evaluated by performing the dynamic repeatability measurements over a period of five days. Thus, stability represents the day-to-day variation in the measurements and may be identified with longer-term variation in the measurements.

*Reproducibility* ($\sigma_R$) is the total variation in the measurement system and includes the contributions from both static and dynamic repeatability. Reproducibility commonly includes variation over an extended time period in the form of a stability measurement:

$$\sigma_R = \sqrt{\sigma_{SR}^2 + \sigma_S^2} \tag{1}$$

Reproducibility is often determined from single measurements performed at five sites on a wafer with load/unload operations prior to each measurement. Such a measurement naturally includes both dynamic and static repeatability, as discussed earlier, and is often used in instrument qualification within a fab. For manufacturers of metrology equipment, however, there is considerable interest in understanding the individual contribution of static and dynamic repeatability to instrument precision. Equation (1) shows the sum of the root sum of squares. The root sum of squares difference between $\sigma_{SR}$ and $\sigma_S$ is a measure of the variability introduced by the wafer insertion process. When static and dynamic repeatability are determined, Eq. (1) expresses the reproducibility as Alain Diebold defines measurement precision, $\sigma_{measurement}$, in Chapter 1 of this volume.

*Precision* (*P*) of the metrology tool is widely expressed as $6\sigma_R$.

*Tolerance* (*T*) is the amount of variation acceptable in the manufacturing process—the upper control limit (UCL) minus the lower control limit (LCL). The tolerance for gate dielectric thickness is 8% ($6\sigma$) of the dielectric thickness (3).

*Precision-to-tolerance ratio* is simply the ($6\sigma$) precision divided by the tolerance.

*Accuracy* is commonly defined with respect to an artifact of known properties. Accuracy is a qualitative term measured by bias. In the field of ellipsometry, the most common artifact is a thermally grown oxide on Si, whose optical properties are NIST-traceable. The standard ellipsometric parameters of Psi ($\Psi$) and Del ($\Delta$) are reported, along with estimates of error in those values, often at the HeNe laser wavelength of 632.8 nm and, more recently, using SE. In practice, a reference material of known thickness is sought, and thus the thickness of the oxide is also calculated using standard (fixed) values of the optical properties of Si and $SiO_2$. Standard reference materials are inadequate for gate metrology purposes at the 2-nm level. For optical metrology, there exists a lack of consensus on the appropriate optical tables for simple oxides on Si, the model of the oxide/Si interface, and the best practice for dealing with the accretion of airborne molecular contamination on the surface of the artifact.

*Matching* between two ellipsometers is commonly determined by recording the difference in thickness measurements on the same artifact (usually an oxide) over a period of at least five days. It is common for a single measurement of thickness to be performed at five sites on the wafer used for matching. Measurements are repeated every eight hours, as with the dynamic repeatability test, and the mean and standard deviation are noted. Matching is then commonly defined as the difference between the means from one instrument and the means from the second instrument. It is important to note not only that the means must agree within specified limits, but also that the standard deviations from each tool must be comparable for each measurement site. In the production environment practice, matching is frequently defined with respect to a "golden tool." The tool thus designated is often the first such tool to be installed in a particular area of the fab, and subsequently added tools are expected to match it.

## B. Gate Dielectric Scaling

The current generation of multiprocessor unit (MPU) transistor gate dielectrics are rapidly approaching a thickness of 2 nm. For a 2-nm-thick gate oxide, the acceptable tolerance ($T$) i.e., process specification range (defined as the upper process limit minus the lower process limit), is typically $\pm 4\%$ of the film thickness. The tolerance for a 2-nm oxide is then 0.16 nm (i.e., 8% of the film thickness). Assuming that the acceptable process specification range follows a normal distribution, then the distribution may be described by a six-sigma standard deviation ($6\sigma$) about the mean (i.e., the target film thickness). Thus, for the 2-nm film described earlier, the $6\sigma$ standard deviation would be equal to the tolerance of 0.16 nm. While a $6\sigma$ tolerance of 0.16 nm is a remarkable achievement in gate fabrication, a 10-fold greater challenge is presented to the manufacturers of gate metrology equipment. To rephrase the old "watch makers' rule," it is necessary for the metrology tool to exhibit a 10-fold greater measurement precision than that sought for in the film under test. That is, we are left with the requirement that the metrology tool precision ($P$) be one-tenth of the tolerance, or

$$\frac{P(6\sigma)}{T} = 0.1 \qquad (2)$$

For a 2-nm gate, the $1\sigma$ precision of the metrology tool should be 0.0027 nm (i.e., [0.16 nm $\times$ 0.1]/6). Traditionally, such definitions of gate dielectric precision are typically presented as $3\sigma$ values (3), and the $3\sigma$ metrology precision would thus be 0.0081 nm. It should be noted that in some instances, $P/T$ ratios of up to 0.3 are found to be acceptable.

Device scaling for increased speed and lower power consumption will require that the thickness of gate dielectrics decrease to 1.9–1.5 nm (equivalent oxide thickness) at the 130-nm node, with an associated $3\sigma$ metrology precision of 0.006 nm (3). At the 100-nm node the dielectric thickness is expected to decrease further to the 1.5–1.0-nm range, with a $3\sigma$ precision of 0.004 nm. At the 70-nm node the gate thickness is expected to decrease to 1.2–0.8 nm, with a $3\sigma$ precision of 0.0032 nm. A 1.2-nm $SiO_2$ gate dielectric fabricated with a polysilicon electrode would be approximately 5 Si atoms thick. One of these Si atoms would be bonded to the silicon substrate, while a second Si atom would be bonded to the polysilicon electrode. Thus the "bulk" oxide would consist of only three covalently bonded $SiO_2$ "layers." Muller et al. (24), have demonstrated that three such layers (each approximately 0.4 nm thick) would be just sufficient to provide a working insulating layer. The required $3\sigma$ metrology precision (0.0032 nm) is then equal to approximately 1/125 the thickness of an $SiO_2$ layer. This is clearly a rather demanding requirement.

## C.  High-Dielectric-Constant Gates

It is expected that oxynitrides, and, possibly, stacked nitride/silicon dioxide layers, will be used at the 130- and 100-nm logic generations. Nitridation of the base oxide is used to reduce boron diffusion from the doped polysilicon gate, through the oxide, into the underlying gate channel. Furthermore, while maintaining a physical dielectric thickness that is consistent with acceptable levels of electron leakage, it is also possible to reduce the effective electrical thickness of the gate dielectric. High-dielectric-constant materials are expected to be used at and after the 70-nm logic node and possibly even at the 100-nm node. Wilk and Wallace have recently discussed the potential integration of $ZrO_2$ ($\varepsilon = 25$), $HfO_2$ ($\varepsilon = 40$) and the silicates $ZrSiO_4$ ($\varepsilon = 12.6$) and $HfSiO_4$ ($\varepsilon = 15$–25) as replacements for nitrided oxide gates (25). The increased physical thickness of the high-dielectric-constant layer, which reduces electron tunneling, can be calculated by multiplying the ratio of the dielectric constants ($\varepsilon$ high-$\kappa$/$\varepsilon$ ox) by the effective oxide thickness.

An example may be given for designing a gate for the 70-nm node having an effective oxide thickness of 1.2 nm using a dielectric constant of 15 (e.g., $HfSiO_4$). If used for the gate in the form of a single layer, then the physical thickness of such a dielectric would be 4.6 nm (i.e., 1.2 nm × 15/3.9). However, the precision listed in the NTRS roadmap is also based on equivalent oxide thickness. Therefore, for our hafnium silicate gate, the precision must be multiplied by the ratio of the dielectric constants (15/3.9) to obtain the required $3\sigma$ precision for the high-$\kappa$ gate 0.012 nm (i.e., 0.0032 nm × 15/3.9). Equally, we could determine the required metrology tool precision from the thickness and tolerance of the 4.6-nm high-$\kappa$ gate. Here the tolerance (8%) is 0.368 nm. The $1\sigma$ precision of the metrology tool should then be 0.0061 nm (i.e. [0.368 nm × 0.1]/6). We thus obtain a $3\sigma$ precision of 0.012 nm, as shown earlier. Clearly, metrology precision requirements would appear to be significantly relaxed for such physically thicker single-layer, high-$\kappa$ dielectrics. However, it remains to be seen from an integration perspective whether high-$\kappa$ films can be integrated successfully as a single layer. This concern stems from the potential high reactivity (compared to the stability of $SiO_2$ on Si) of high-$\kappa$ dielectrics with Si. Thus, a thin interfacial (oxidized) layer might be expected to form both at the dielectric/c-Si interface and at the dielectric/polysilicon electrode interface. Such interfacial layers would, in forming a multilayer "gate stack," have significant electrical and metrology ramifications. The total capacitance of the dielectric stack would then include that of the high-$\kappa$ layer, the interfacial layers (possibly engineered SiON), quantum-state effects at the channel interface, and that associated with depletion of charge in the polysilicon gate electrode.

## IV. IN-LINE PRODUCTION SPECTROSCOPIC ELLIPSOMETRY

### A. Fundamental Equations

In one production implementation of spectroscopic ellipsometry, analysis of the polarization state of the reflected light is accomplished by rotating a polarizer and using a fixed analyzer. Two spectra, $\tan \Psi_m(\lambda)$ and $\cos \Delta_m(\lambda)$, are the result of this measurement. The quantities $\tan \Psi$ and $\cos \Delta$ are the two standard ellipsometry parameters. They are derived from the intensities, $R_p$ and $R_s$, of the elliptically polarized reflected light and from the phase difference, $\Delta$, between the $R_p$ and $R_s$ components. They are expressed by the following equation:

$$\frac{R_p}{R_s} = \frac{|R_p|}{|R_s|} \cdot \exp(i\Delta) = \tan \Psi \cdot \exp(i\Delta) \tag{3}$$

where $R_p$ is the electric field reflection coefficient component with a polarization parallel to the plane of the incident and reflected beams, and $R_s$ is the component with a polarization perpendicular to that plane. As shown in Eq. (3), $\tan \Psi$ is the ratio of the $p$- and $s$-component intensities and $\cos \Delta$ is the real part of the complex quantity $\exp \cdot (i\Delta)$.

A good fit between measured and theoretical spectra requires knowing or calculating the values of all the layer thicknesses and of the refractive index, $n$, and the extinction coefficient, $k$, of each material at all the wavelengths in the measurement. Because this is not practically possible, mathematical approximation models—with a limited number of variables—are developed to describe the dispersion (variation of $n$ and $k$ with the wavelength) of the different materials that constitute the film stack. The material dielectric constant, $\varepsilon$, is related to the refractive index, $n$, and extinction coefficient, $k$, by

$$\varepsilon = (n - ik)^2 = n^2 - k^2 - 2ink \tag{4}$$

where $i$ is the imaginary unit. Note that the imaginary term $(ik)$ is negative in the definition used here.

In the optical system shown in Figure 1, the SE measures the reflected light intensity, $I(P, \lambda)$, as a function of the polarizer angle, $P$, and the wavelength, $\lambda$, for a fixed value of the analyzer angle, $A$. From the measured intensity, two spectra, $\alpha_m(\lambda)$ and $\beta_m(\lambda)$, are obtained. The quantities $\alpha$ and $\beta$ are related directly to $\tan \Psi$ and $\cos \Delta$:

$$\alpha = \frac{\tan^2 \Psi - \tan^2 A}{\tan^2 \Psi + \tan^2 A} \tag{5}$$

and

$$\beta = 2 \cos \Delta \frac{\tan \Psi \tan A}{\tan^2 \Psi + \tan^2 A} \tag{6}$$

To obtain the best fit between the theoretical and measured spectra, one needs to know or calculate the correct dispersia for all the materials in the film stack. The film stack to be measured is described in a film "recipe" where some of the film parameters are unknown and must be measured. The system uses this recipe to generate a set of ellipsometry equations. These equations, and the corresponding theoretical spectra, $\tan \Psi_t(\lambda)$ and $\cos \Delta_t(\lambda)$, represent the film stack well if the correct film parameter values are used. The calculations then consist in varying the unknown parameters in the equations until the best fit is obtained between the calculated and measured spectra.

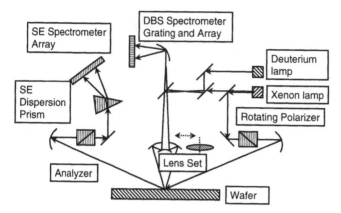

**Figure 1** Schematic layout of a production optical metrology tool showing the spectroscopic ellipsometer and spectroreflectometer subsystems.

## B. Dispersion Models

For a thorough discussion of optical models, the reader is referred to Chapter 25, by Jellison (26). A material dispersion is typically represented mathematically by an approximation model that has a limited number of parameters. In production ellipsometers, four major mathematical dispersion models are extensively used: the tabular model, the harmonic oscillator (HO) model, the Cauchy polynomial, and the Bruggeman effective medium approximation (BEMA). Variants on these and other specific models (such as the Tauc–Lorentz model) find some production application and considerable research use.

For some well-characterized common materials, the tabular model (or lookup table) provided by the system is typically used. One can also convert dispersia obtained from any of the other models into new tables. A dispersion table contains a list of values of $n$ and $k$ at evenly spaced wavelengths, and is used as the basis for (cubic spline) interpolation to obtain values at all wavelengths. The $n$ and $k$ values in the table are not varied during the calculations.

Perhaps the most commonly used model for initial characterization of new films is the HO model. This model is based on the oscillations of electrons bound to atoms in a material when they are under the effect of an electromagnetic field—the incident light. The number of peaks (resonances) in a material dispersion spectrum is an indication of the number of oscillators that represent this dispersion. If the material can be represented by $m$ oscillators, then the dielectric constant, $\varepsilon$, is given by

$$\varepsilon(E) = n_b + \frac{\sum_{s=1}^{m} H_s(E)}{1 - \sum_{s=1}^{m} v_s H_s(E)} \tag{7}$$

where, $n_b$ is the uniform background index (default value $= 1$) and $E$ is the electric field energy (in electron-volts), expressed as a function of the wavelength $\lambda$ (in nm) by

$$E = \frac{1,240}{\lambda} \tag{8}$$

$H_s$ is the contribution of the $s$th oscillator, as described later, and $v_s$ is the local field correction factor for the $s$th oscillator. The local field correction factor is equal to zero for metals, is close to zero for most semiconductors, and is equal to 0.333 for ideal dielectrics. $H_s$ is given by

$$H_s(E) = \frac{16\pi N_s \text{Ry}^2 r_0^3}{E_{ns}^2 - E^2 + iE_{gs}E} e^{-i\Phi_s} \tag{9}$$

where Ry is the Rydberg constant (Ry = 13.6058 eV), $r_0$ is the Bohr radius ($r_0 = 0.0529177$ nm), $N_s$ (or $N$osc) is the number density of the $s$th oscillator, in nm$^{-3}$, which represents the relative importance of this oscillator, $E_{ns}$ is its resonance energy, in eV, $E_{gs}$ is its damping energy, in eV, $\Phi_s$ (or Phi) is its (relative) phase (in radians). The HO model is widely applicable and especially to materials for which the dispersion curves exhibit multiple peaks. For a given material, one usually varies $N_s$, $E_{gs}$, and $\Phi_s$ only during calculation. Good starting parameter values are important in order to use this model successfully.

The HO model obeys the Kramers–Krönig relation (27,28), which is a physical limitation on the model based on the law of causality. For nonmagnetic systems, the two parts of the complex refractive index are related by:

$$n(e) = 1 + \frac{2}{\pi} P \int_0^\infty \frac{\xi k(\xi)}{\xi^2 - E^2} \, d\xi \tag{10}$$

$$k(e) = -\frac{2E}{\pi} P \int_0^\infty \frac{n(\xi) - 1}{\xi^2 - E^2} \, d\xi \tag{11}$$

The Kramers–Krönig relation imposes an important constraint on the complex refractive index. If either $n(E)$ or $k(E)$ is known over the entire range of $E$, then it is possible to calculate the other. Jellison (26) has discussed the applicability of the Lorentz and derivative models when the photon energy is close to, or above, the band gap of an amorphous material semiconductor. Since the optical transitions within an amorphous material lack **k**-vector conservation (27,28), the optical functions of amorphous materials are not expected to display the characteristically sharp peaks of their crystalline and microcrystalline counterparts. Near the band edge of an amorphous material, Tauc and coworkers (29) found that the imaginary part of the dielectric function is given by

$$\varepsilon_2(E) = A_T \Theta(E - E_g) \frac{E - E_g}{E^2} \tag{12}$$

where $\Theta(E - E_g)$ is the Heaviside function [$\Theta(E) = 1$ for $E \geq 0$ and $\Theta(E) = 0$ for $E < 0$] and $E_g$ is the bandgap of the amorphous material. Equation (12) has been used extensively to model the bandedge region of amorphous semiconductors, but it has not received great attention beyond this region. In particular, Eq. (12) gives no information concerning $\varepsilon_1$. To interpret optical measurements such as ellipsometry experiments, it is quite useful to have an expression for $\varepsilon(E)$ that corresponds to Eq. (12) near the bandedge but that also extends beyond the immediate vicinity of $E_g$. Furthermore, it is important that the expression be Kramers–Krönig consistent. One such parameterization that meets these criteria is the Tauc–Lorentz expression developed by Jellison and Modine (30). This model combines the Tauc expression [Eq. (12)] near the bandedge and the Lorentz expression for the imaginary part of the complex dielectric function. If only a single transition is considered, then

$$\varepsilon_2(E) = 2n(E)k(E) = \frac{A(E - E_g)^2}{(E^2 - E_o^2)^2 + \Gamma^2} \frac{\Theta(E - E_g)}{E} \tag{13}$$

The real part of the dielectric function is obtained by Kramers–Krönig integration:

$$\varepsilon_1(E) = \varepsilon_1(\infty) + \frac{2}{\pi} P \int_{R_g}^{\infty} \frac{\xi \varepsilon_2(\xi)}{\xi^2 - E^2} d\xi \tag{14}$$

which can be integrated exactly. There are five parameters that are used in this model: the bandgap $E_g$, the energy of the Lorentz peak $E_o$, the broadening parameter $\Gamma$, the value of the real part of the dielectric function $\varepsilon_1(\infty)$, and the magnitude $A$ (26). The Tauc–Lorentz parameterization describes only interband transitions in an amorphous semiconductor. Since additional effects (such as free carrier absorption or lattice absorption) that might contribute to absorption below the bandedge are not included in the model, $\varepsilon_2(E) = 0$ for $E < E_g$. Furthermore, it can be seen that $\varepsilon_2(E) \to 0$ as $E \to \infty$. This corresponds to the observation that $\gamma$-rays and x-rays are not absorbed very readily in any material (26).

Traditionally, the Cauchy polynomial model has been used for describing dielectrics and organic materials, mostly in the visible-wavelength range, because it provides a good approximation of monotonic dispersia. However, this model is not recommended for materials that have dispersion peaks in the UV range. The Cauchy polynomial model is written as:

$$n = N_0 + \frac{N_1}{\lambda^2} + \frac{N_2}{\lambda^4} \tag{15}$$

and

$$k = K_0 + \frac{K_1}{\lambda^2} + \frac{K_2}{\lambda^4} \tag{16}$$

where the $N_i$ and $K_i$ are constants, called Cauchy coefficients, and $\lambda$ is the wavelength in Å. During the spectral fitting calculation, these coefficients are treated as variables. The Cauchy model has no physical constraints: $n$ and $k$ can take nonphysically possible values, and there is no relationship between them. This limits the accuracy of the Cauchy model compared to the HO model.

The Bruggeman effective medium approximation (BEMA) model is applicable to materials that can be treated as alloylike mixtures or solid solutions of up to eight different components (in typical production analysis software). In the case of a BEMA composed of $m$ components, the model is described by

$$\sum_{i=1}^{m} f_i \frac{\varepsilon_i - \varepsilon}{\varepsilon_i + 2\varepsilon} = 0 \tag{17}$$

where $f_i$ and $\varepsilon_i$ are the volume fraction and the dielectric constant of component $i$, respectively. The volume fractions $f_i$ are within the range 0–1, and

$$\sum_{i=1}^{m} f_i = 1 \tag{18}$$

During the calculations, the volume fraction $f_i$ is treated as a variable, and the dispersia of the different components are first determined and saved as tables of values. If the dispersia of the $m$ components satisfy the Kramers–Krönig condition, then the resulting dispersion will also satisfy this condition.

During spectral fitting, some of the film parameters are specified as "known"; and their values will not be calculated. Other parameters are specified as "unknown"; and their values will be calculated. The "unknowns" are often given starting values in the film recipe. The calculation method (or regression) consists of varying the values of the "unknown" thicknesses and selected dispersion model parameters in the equations, until the best possible fit is obtained between the theoretical and measured spectra. A good fit and realistic $n(\lambda)$ and $k(\lambda)$ spectra indicate that the obtained parameter values are accurate. To assess the quality of the fit, several indicators of spectral fit quality are typically provided.

## C. Spectral Fit Quality

The fitting process compares the measured and calculated $\alpha(\lambda)$ and $\beta(\lambda)$ spectra. The quantity chi-squared (or $\chi^2$) is the residual difference between measured and calculated spectra at the end of the fitting process. The generic name of the indicators of spectral fit quality is *goodness of fit* or GOF. The range of the GOF is 0.0–1.0. A good fit is indicated by a GOF value close to 1 and a poor fit by a GOF value close to 0. The GOF calculated from $\chi^2$ (which is visually reported in one system as "CHI2") is given by:

$$CHI2 = 1 - \frac{2}{\pi}\tan^{-1}\left(\sqrt{\frac{\chi^2}{\eta\gamma^2}}\right) \tag{19}$$

where $\chi^2$ is the chi-squared value and $\gamma$ is a scaling constant. In Eq. (19), $\eta$ is given by

$$\eta = \sum_{i=1}^{N} 2 - \rho_i^2 \tag{20}$$

where $\rho_i$ is the correlation coefficient describing the interdependence of $\alpha$ and $\beta$ at the wavelength $\lambda_i$. This CHI2 value can be used to estimate the accuracy and robustness (i.e., the degree of reliability under normal process variations) of a particular measurement and to verify process consistency over time.

## V. MODELING THE SILICON/SILICON DIOXIDE INTERFACE AND ADSORBED LAYER

### A. Wafer Fabrication

Thermal oxides were grown on 150-mm ⟨100⟩ p-type silicon substrates with a resistivity of ~ 0.02 ohm-cm (2). Oxide growth was performed in a prototype vertical furnace in the temperature range of 675 to 820°C to produce oxides of nominal thickness 20, 30, 40, and 100 Å. These blanket structures were left in the as-grown condition for spectroscopic ellipsometry and Hg probe CV measurements. A second set of oxides was grown under the same conditions on patterned wafers to produce doped poly-Si gate capacitors. These samples experienced a thermal treatment of 900°C for 30 minutes during the poly-doping step. Thus, the thermal history of blanket and capacitor structures was different.

The specific sampling schemes used in this study included several short-term capability studies to obtain an initial evaluation of static repeatability, dynamic repeatability, and precision. These experiments were followed by a 14-day stability study. On each day, nominally 20-, 30-, 40-, and 100-Å oxide wafers were measured, with five repetitions on each of five cycles. Thus, for each oxide thickness, the repeatability was calculated from a total of 350 repeats, the reproducibility from a total of 70 cycles, and the stability over 14 days. To evaluate the overall precision and the components of variation that contributed to the precision, a gauge study was implemented based on Eastman's approach for evaluating automated measurement instruments (31). This experimental approach incorporates a series of short-term capability studies followed by a longer-term stability study. A nested sampling scheme is used, and statistical analysis methods (nested ANOVA) appropriate to this scheme allow evaluation of the linearity and stability of the repeatability, reproducibility, and precision of the measurements. The initial results from all of the studies were expressed in terms of the thickness derived from a simple bulk optical model consisting of bulk values of the index of refraction for the oxide film and for the silicon substrate. The fit variables in these determinations were the thickness of the oxide film and the angle of incidence. The data from the stability study were reanalyzed in terms of alternative optical models. The results of this analysis are presented in the next section.

## B.  SiO$_2$/Si Interface

The performance of alternatives to the bulk optical constants model were evaluated by reanalyzing all of the data from a 14-day stability study of 20-, 30-, 40-, and 100-Å oxides. The alternative models chosen for evaluation were based on studies reported in the literature for considerably thicker oxide films. The models fall into two basic classes:

Class I:    Single-Layer Models

    A.  Fixed index of refraction oxide
    B.  Variable index of refraction oxide
        1. Mixture of oxide and voids
        2. Mixture of oxide and amorphous silicon
        3. Variable harmonic oscillator number densities in the HO model

Class II:   Two-Layer Models

    A.  Interface layer consisting of a mixture of oxide and amorphous silicon
    B.  Contamination layer on surface of oxide modeled as amorphous carbon

The index of refraction of mixtures of materials in these models was obtained by means of the Bruggeman effective medium approximation (BEMA). The angle of incidence was always included as a fit variable, in addition to those associated with the models themselves.

The goodness-of-fit (GOF) parameter for the single-layer variable-index-of-refraction model using a BEMA mixture of oxide and voids is compared to the results of the simple bulk model in Figure 2. The oxide + void model resulted in an improved GOF at each oxide thickness; however, the trend toward a decrease in GOF with decreasing thickness remained. Although the GOF improved with this model, the results were physically unrealistic. The void fraction and oxide thicknesses derived from this model are shown in Figures 3 and 4.

As can be seen in Figure 3, the void fraction increases to about 80% for the 20-Å oxide. This resulted in an average extracted thickness of nearly 100 Å, with large errors, as can be seen in Figure 4. These physically unrealistic results are due to a very high correla-

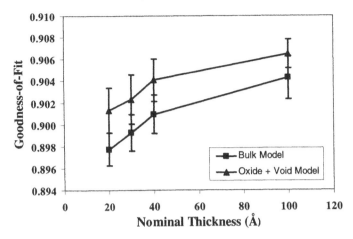

**Figure 2** Goodness-of-fit for bulk and variable-refractive-index (oxide + void) models.

tion between the fit variables. The oxide thickness and the void fraction typically have a correlation coefficient greater than 0.999 for the 20-Å oxide and 0.990 for the 100-Å oxide. Thus, the variables are not independent for ultrathin oxides, and the regression routines are unable to distinguish between the effects of the variables. Other variable-index single-layer models, including a BEMA mixture of oxide and amorphous silicon and a variable-oscillator-number-density model, yielded essentially the same unrealistic results for oxide thickness. Thus, variable index models in which no constraints are placed on the variables cannot be used for these ultrathin oxides due to severe parameter correlation effects. A two-layer model that included an interfacial layer between the oxide and the silicon sub-strate yielded considerably more promising results. The interface layer was modeled as a BEMA mixture of oxide and amorphous silicon. The fit variables for this model were the oxide thickness, the interlayer thickness, the oxide void fraction in the interlayer, and the angle of incidence. This interlayer model resulted in a dramatically improved goodness-of-fit compared to the simple bulk model, as shown in Figure 5. Moreover, in contrast to the

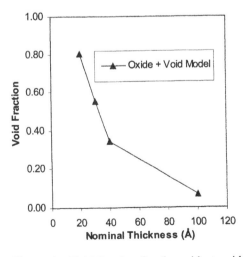

**Figure 3** Void fraction for the oxide + void model.

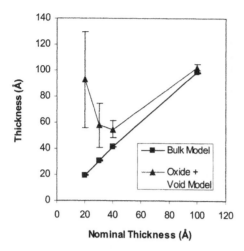

**Figure 4**  Thickness for the oxide + void model.

bulk model, the GOF was independent of oxide thickness. The interlayer thickness and the interlayer oxide fraction are shown in Figures 6 and 7. The interlayer thickness ranged from 1.5 to 2.0 Å; the interlayer oxide fraction was approximately −0.25. The negative oxide fraction suggests that, at least for these low-temperature oxides, the interlayer is best modeled with an absorbing optical model. The values of interlayer thickness shown here are considerably lower than most of those reported in the literature. This effect may be due to the low growth temperature of the oxides or to a lower interfacial roughness compared to the thicker, higher-temperature oxides studied in the literature. The oxide thickness and the total thickness (oxide thickness plus interlayer thickness) are shown in Figures 8 and 9. The interface model used here included three independent variables in addition to the angle of incidence. Somewhat improved results might be expected if the number of fit variables was reduced.

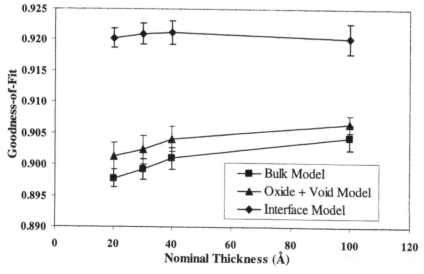

**Figure 5**  Goodness-of-fit for the bulk, oxide + void, and interface models.

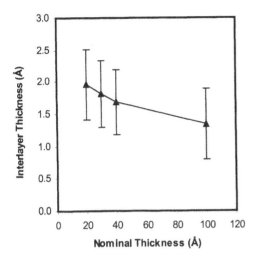

**Figure 6**   Interlayer thickness for the interface model.

## C.  Modeling of the Adsorbed Layer

The final model evaluated in this phase of the study was an attempt to account for a possible contamination layer on the surface of the oxide. The layer was modeled as amorphous carbon, so the fit variables were the thickness of the oxide and the thickness of the amorphous carbon layer above the oxide. The best-fit value of the carbon layer thickness was consistently zero. Thus, potential contamination layers cannot be modeled in this manner. In summary, a simple bulk optical model provides the best precision. Including an interfacial layer between the oxide and the silicon provided the best goodness-of-fit, at the expense, however, of poorer precision. Variable-refractive-index models yielded physically unrealistic results due to severe parameter correlation effects for ultra-thin oxide films. An attempt to develop an optical model for the surface contamination layer was unsuccessful.

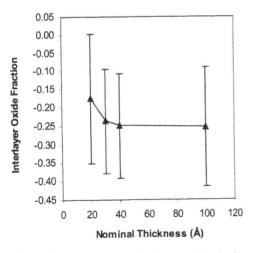

**Figure 7**   Interlayer oxide fraction for the interface model.

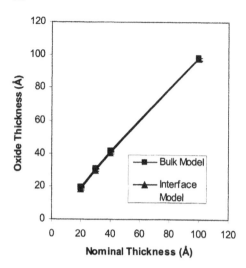

**Figure 8**   Oxide layer thickness for the bulk and interface models.

## VI.   EFFECT OF AIRBORNE MOLECULAR CONTAMINATION ON METROLOGY

### A.   Airborne Molecular Contamination Rates

The effect of accretion of airborne molecular contamination on thin oxides has been investigated in a production environment on oxides of approximately 27-Å thickness. Measurements of optical thickness were made using SE on two wafers, "immediately" after removal from the furnace. Following the initial measurements, each wafer was returned to a holding cassette for subsequent loading and unloading from the ellipsometer. The emphasis was on obtaining information about contaminants subject to airflow in the open environment rather than within the metrology tool itself. Oxide-thickness measurements were acquired every 10 minutes for the first 3 hours using a 49-site full-wafer map. Following the initial characterization of adsorbate growth rate, the wafers were cycled in

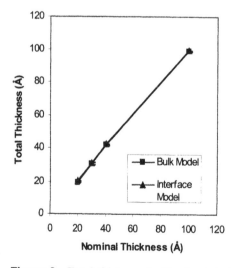

**Figure 9**   Total thickness (oxide plus interlayer) for the bulk and interface models.

and out of the SE tool for a period of 24 hours. The data acquired from the wafer stored in the horizontal position are shown in Figure 10. An initial growth rate of 0.026 Å/min was observed. After 3 hours, apparent oxide growth of 1.8 Å was observed (with a linear approximation of ~0.01 Å per minute). After 21 hours (~75,600 s), the regrowth of the ad-layer approached saturation. Similar results were obtained for the wafer that was stored vertically, with an observed growth rate of 0.009 Å per minute over the first 3 hours. Saturation also occurred after approximately 21 hours. It was concluded that storage orientation had little influence on the quantitative growth characteristics of the ad-layer.

## B. Removal of Airborne Molecular Contamination

The optimum thermal budget for the removal of the ad-layer was investigated using three as-grown oxides. The thickness of the oxides was measured "immediately" after removal from the furnace. The wafers were cycled through the ASET-F5 tool for a 3-hour period to monitor adsorption of contamination. The wafers, with oxides thicknesses in the range 40–42 Å, were heated using the IR hot stage of the Quantox. Three thermal budgets were used: 150°C/30 s, 200°C/30 s, and 200°C/60 s. The wafers were then cycled through the SE tool for a further 2 hours, followed by a second heating cycle. The 200°C/60 s thermal budget was considered optimal using this technique.

A third experiment was performed using the 200°C/60 s thermal budget to investigate the desorption characteristics of an ad-layer during multiple heat treatments. The thickness of the oxide, approximately 84 Å, was measured as grown and following each heat treatment. A full-wafer nine-site map of optical thickness was acquired using SE over the spectral range 240–800 nm. The time interval between measurements was ~ 130 seconds. The growth rates and projected intercepts (i.e., the thickness at time equal to zero for each heating/measurement sequence) were obtained from a linear model. The results are shown in Figure 11. The projected intercepts from the five heat treatments yielded an average thickness of 84.9 ± 0.045 Å. After each bake/measurement sequence, the adsorp-

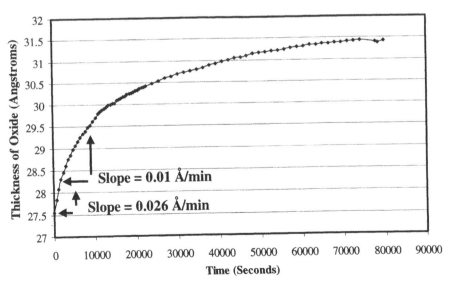

**Figure 10** Graph of oxide thickness plus an ad-layer vs. time.

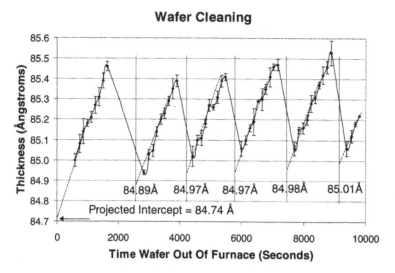

**Figure 11** Graph of oxide thickness plus an ad-layer vs. time before and after five thermal cleaning cycles.

tion rate was found to be quite consistent at $0.024 \pm 0.001$ Å per minute. It should be noted that a slight increase in thickness of 0.12 Å was observed between the first and last cleanings. This observation implies that a layer of "baked-on" contamination may be formed (possibly consisting of cracked hydrocarbon) and that an effective cleaning cycle may be developed using a considerably shorter heating time. In the production environment, a cleaning solution for airborne molecular contamination is required to have very little effect on wafer throughput. A successful solution is likely to be based, therefore, on a top-down optical heating methodology, which maximizes the capability of dual-finger robot wafer handling for cleaning, notch alignment, and wafer placement on the ellipsometer stage.

## VII. OPTICAL AND ELECTRICAL METROLOGY OF REFERENCE OXIDES

This section describes an approach to the establishment of standards for gate dielectrics that combines both optical and electrical metrology (2). The gate-oxide thickness ($T_{ox}$ in cm) can be calculated from the classical definition of the capacitance ($C_{ox}$) in accumulation by

$$C_{ox} = \frac{K \cdot \varepsilon_0 \cdot A}{t_{ox}} \tag{21}$$

where $\varepsilon$ (or $\kappa$ as shown in this equation) is the relative dielectric constant of oxide (3.9), $\varepsilon_0$ is the permittivity of free space ($8.854 \times 10^{-14}$ F/cm), and $A$ is the area (cm$^2$) of a parallel-plate capacitor. Conventional $CV$ and Hg-probe measurements were analyzed according to the algorithm developed by John Hauser (32), which may incorporate polydepletion and quantum mechanical corrections as appropriate. The interested reader is referred to Chapter 4, by Vogel and Misra, for a detailed discussion of electrical metrology (33).

**Table 1** Fabrication Conditions for Nominal Oxide Thicknesses

| Target thickness (nm) | Temperature (°C) | Time (min) |
| --- | --- | --- |
| 1.5 | 750 | 30 |
| 1.8 | 800 | 30 |
| 2.1 | 850 | 15 |
| 2.4 | 875 | 30 |

Three types of reference wafers were fabricated, with nominal thicknesses of 2 nm: blanket oxides, isolation-patterned oxides, and oxides subsequently processed for conventional MOSCAP electrical analysis. The essence of this program was that the metrics obtained from the blanket wafers be applicable to the sets of isolation-patterned and MOSCAP wafers. To investigate the effects of scaling, gate oxides of 15-, 18-, 21-, and 24-Å nominal thickness on n- and p-doped wafers were fabricated. The processing time and temperature for these oxides is shown in Table 1. A high furnace temperature, consistent with the growth of uniform ultrathin oxides, was used to reduce the effects of suboxides at the $SiO_2$/Si interface. The furnace ramp was performed in a $N_2$ atmosphere with 1% $O_2$ at atmospheric pressure; the oxidation was performed in $O_2$ at 985 milli-Torr.

The blanket reference wafers were characterized optically within 3 days of fabrication. Data sets were built from 121-point sampling schemes using polar coordinates to define each measurement site. All reference wafers with a nominal thickness of 20 Å returned a uniformity metric of <2%, where uniformity is defined as 1 standard deviation divided by the mean oxide thickness. The thickness metrics obtained from the scaled wafer set are shown in Table 2. The oxide-thickness metrics (target, mean, minimum, and maximum) are in angstroms and the GOF is a chi-squared value. The $CV$ data set for the blanket wafers was built from a 49-point sampling scheme using a polar coordinate system and a Hg-probe area of $7.3 \times 10^{-4}$ cm$^2$ (2). Analysis of the $CV$ curves using the NCSU CVC program yielded corrected values of electrical thickness, $T_{ox}$. The thickness metrics from the entire data set were $21.9 \pm 1.5$ Å. The corresponding metrics from SE measurement of optical thickness were $21.2 \pm 1.5$ Å, in excellent agreement.

For the isolation-patterned wafers, ellipsometric measurements of thickness were obtained first from central 1-cm-wide "stripes" of unpatterned oxide (19 sites in both $\pm x$ and $\pm y$ directions within 30 mm of the wafer center). Optical measurements of oxide thickness were recorded from 23 sites in each of the four quadrants ($Q_I$–$Q_{IV}$) of the patterned wafer, as shown in Table 3. Given a site-specific sampling methodology for the patterned reference wafers, it was possible to determine how well the optical measurements of oxide thickness correlated with electrical-thickness measurements extracted

**Table 2** Optical Characterization of Blanket Oxide Wafers

| Target | Mean | Minimum | Maximum | Std dev | GOF |
| --- | --- | --- | --- | --- | --- |
| 15 | 15.23 | 14.51 | 16.52 | 0.4139 | 0.901 |
| 18 | 18.65 | 17.91 | 20.39 | 0.5183 | 0.900 |
| 21 | 21.15 | 20.43 | 22.27 | 0.3798 | 0.903 |
| 24 | 24.62 | 23.86 | 26.05 | 0.6353 | 0.904 |

**Table 3** Optical Characterization of Patterned Oxide Wafers

| Sites | Target thickness | Mean thickness | Minimum thickness | Maximum thickness | Standard deviation | GOF |
|---|---|---|---|---|---|---|
| L-Strips: $\pm x$ & $\pm y$ | 21 | 23.34 | 22.95 | 24.05 | 0.27 | 0.904 |
| $Q_{II}$: 10E-5, $cm^2$ | 21 | 23.30 | 22.81 | 23.30 | 0.24 | 0.908 |
| $Q_I$: 6.4E-5 $cm^2$ | 21 | 24.10 | 23.58 | 24.65 | 0.29 | 0.894 |
| $Q_{III}$: 3.6E-5 $cm^2$ | 21 | 24.42 | 24.01 | 24.60 | 0.15 | 0.883 |
| $Q_{IV}$: 1.6E-5 $cm^2$ | 21 | 25.30 | 4.81 | 25.82 | 0.31 | 0.860 |

from the arrays. The results, shown in Figure 12, revealed an $R^2$ correlation of 0.997. From the data shown in Figure 12, use of Eq. (21) and the CVC analysis program yielded a corrected oxide thickness of 2.59 nm. The same analysis methodology was applied to $C$-$V$ data sets from n + /p polysilicon-gate capacitors on MOSCAP reference wafers, as shown in Figure 13. Figure 13 revealed an $R^2$ correlation of 0.999, and $T_{ox}$ corrected was determined to be 2.23 nm. As a result of the process flow for this wafer type, the effects of adsorbed contaminant layers and nonstoichiometric oxide layers should have a minimal impact on this oxide-thickness metric. The combination of optical and electrical metrology techniques has enabled the development of well-characterized and meaningful 2-nm gate oxide standards. A similar program is now under way to characterize advanced gate dielectrics.

## VIII. ADVANCED GATE DIELECTRIC METROLOGY

### A. Noncontact In-Line Electrical Gate Metrology

This section briefly describes the three noncontact technologies of the Quantox noncontact gate dielectric metrology tool. Corona ions, either positive or negative, are generated by a high-voltage source and are directed toward the dielectric surface. The corona ions emulate the function of a metal oxide semiconductor (MOS) electrical contact. The noncontact structure here is generally referred to as a corona oxide semiconductor (COS) system, as

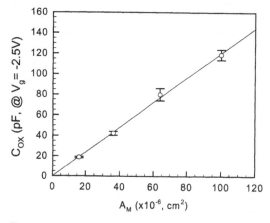

**Figure 12** Results of NCSU analysis methodology for $C$-$V$ data sets from four quadrants of a patterned wafer.

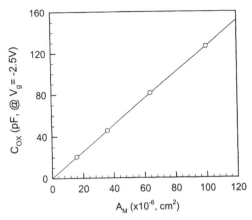

**Figure 13** Results of NCSU analysis methodology for *C-V* data sets from four quadrants of a MOSCAP wafer.

shown in Figure 14. A vibrating Kelvin probe provides capacitively coupled sensing of the wafer surface potential and performs as a nonintrusive voltmeter, with virtually infinite impedance. A schematic outline of the Kelvin probe is shown in Figure 15. A pulsed light source linked to the Kelvin probe enables the stimulus and detection of surface photovoltage (SPV), providing a measurement of band bending and a direct measurement of the flat-band voltage ($V_{fb}$). On a semiconductor, a surface photovoltage is generated by shining light on the surface. If the semiconductor has an internal electric field present at the surface, the electron-hole pairs generated by the incident light will separate in the presence of the field and will produce a counteracting photovoltage that is proportional to the band bending.

When measuring $T_{ox}$, a precisely known corona charge ($Q$) is applied to the oxide incrementally, and accumulated voltage ($V$) is measured. Taking the time derivative of the $Q = CV$ relationship, we obtain

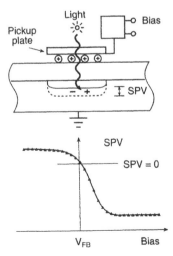

**Figure 14** Schematic outline of the Quantox corona oxide semiconductor (COS) system.

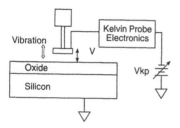

**Figure 15**   Schematic outline of the Quantox Kelvin probe.

$$I = \frac{dQ}{dt} = \frac{C_{kp}\, dV_{surf}}{dt} + \frac{V_{surf}\, dC_{kp}}{dt} \tag{22}$$

Since the surface voltage ($V_{surf}$) is constant, $dV/dt = 0$ and $I$ is a function of $V_{surf}$. By adjusting the bias potential until a current null is achieved, it is possible to determine $V_{surf}$ from the setting of the bias potential $V_{kp}$. In Figure 16, applied charge is plotted on the abscissa. Surface voltage is plotted on the left-hand ordinate, and surface photovoltage is plotted on the right-hand ordinate. When measuring $T_{ox}$, a precisely known corona charge ($Q$) is applied to the oxide incrementally, and the accumulated voltage ($V$) is measured. Since the parallel plate capacitor system is defined by a $Q = CV$ relationship, knowing $Q$ and $V$ defines the system completely, and the effective electrical $T_{ox}$ is readily derived from Eq. (23) with second-order corrections as described later.

$$T_{ox} = \varepsilon\varepsilon_o \frac{dV}{dQ} \tag{23}$$

where $\varepsilon$ is the dielectric constant and $\varepsilon_o$ is the permittivity constant $(8.854 \times 10^{-12}\,\mathrm{C^2/N \cdot m^2})$.

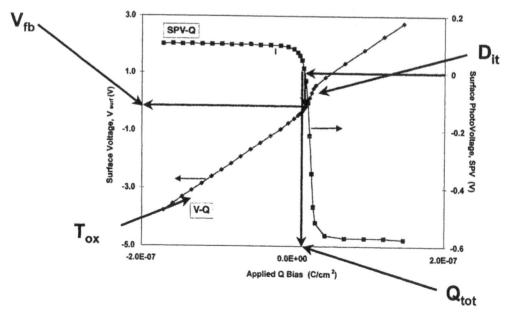

**Figure 16**   $Q$-$V$-SPV data resulting from a Quantox trace.

Because the Quantox deposits a calibrated charge per unit area, Eq. (23) results in a calculation that is independent of area. The net capacitance measured in a low frequency $C$-$V$ measurement trace will approach the true associated value of $T_{ox}$ at both strong inversion and accumulation. However, since silicon is nonmetallic, some field lines do penetrate into the substrate and so may introduce a systematic measurement error. While stronger biasing reduces the error, it is not possible to bias a sample to an unlimited level because of dielectric breakdown. Rather than impose extreme biasing to reduce this field penetration error, the system calculates a theoretical expected $Q$-$V$ trace and fits the measured $Q$-$V$ data to the theoretical trace using a nonlinear least squares fit, with $T_{ox}$ as one of the fitting parameters.

## B.  Nitrided-Gate Dielectrics

As device scaling for increased clock speed continues, a corresponding decrease in the equivalent oxide thickness (EOT) of gate dielectrics is necessary in order to maintain the required level of current flow for device circuit operation. Lucovsky has recently presented a thorough review of the materials and device aspects of nitrided $SiO_2$, with a strong emphasis on the application of remote plasma-assisted processing (34). Hattangady et al. have also recently discussed the introduction of this novel technique in the manufacturing environment (35). Ellipsometry has been widely used to determine the thickness and the index of refraction of plasma-enhanced chemical-vapor-deposited (PECVD) oxynitride antireflective layers. Variation of the deposition parameters has been correlated with changes in the optical properties of the films (36–43). The oxide-to-nitride ratio, hydrogen content, and amount of excess Si have all been shown to affect the index of refraction. Except for highly constrained experiments, therefore, care should be taken in drawing conclusions about film stoichiometry based upon the index of refraction. Tompkins et al. (44) have shown that the use of a simple effective medium model consisting of oxide and nitride overestimates the amount of oxide and underestimates the amount of nitride in silicon-rich oxynitrides designed to act as antireflective layers. Where Rutherford backscattering spectroscopy and Auger electron spectroscopy had indicated the films to be composed only of silicon nitride, the effect was observed to be particularly noticeable. In such instances, SE analysis using the effective medium approximation yielded (erroneous) results indicating that approximately one-quarter of the material was composed of silicon dioxide.

Recently, a combination of optical and electrical metrology using the KLA-Tencor ASET-F5 spectroscopic ellipsometer and the Quantox noncontact electrical metrology tool has been shown to be sensitive to variation in the remote plasma nitridation (RPN) process (45). To test the process sensitivity of the ASET-F5 and Quantox, a series of eight 200-mm p-type wafers was oxidized in a single furnace run. The starting oxide thickness was 26.5 Å. After oxidation, various combinations of the eight-wafer set were subjected to four different remote plasma nitridation (RPN1–4) and three different anneal (A1–3) processes. The process conditions used in this work are summarized in Table 4. Nine-site optical measurements were performed after each separate process step (oxidation, nitridation, and anneal). Spectroscopic ellipsometry measurements were made over the wavelength range of 240–800 nm. Standard optical tables for Si and $SiO_2$ were used throughout the analysis. Absolute reflectivity data was recorded over the wavelength range of 195–205 nm. Five-site electrical measurements were performed after the oxidation and anneal steps only.

**Table 4** Remote Plasma Nitridation Processing and Annealing Conditions

| Remote plasma nitridation | Post nitridation anneal |
|---|---|
| RPN1 = High power, medium time (std) | A1 = High temp., gas 1 (std) |
| RPN2 = Low power, maximum time | A2 = Low temp., gas 2 |
| RPN3 = Low power, medium time | A3 = High temp., gas 3 |
| RPN4 = High power, minimum time | |

The first split focused on the standard processing conditions of RPN1 and anneal A1. Figure 17 shows the optical-thickness measurements made on the eight-wafer set subjected to a single furnace firing and identical RPN nitridation and anneal. Figure 18 shows optical reflectivity measurements made on the same eight-wafer set. Again, clear correlation of process step to reflectivity is observed. Figure 19 shows the equivalent-oxide-thickness (EOT) measurements made by the Quantox on the same eight-wafer set. The starting oxide film thickness and post-RPN treatment/anneal film-thickness conditions are clearly distinguishable. A capability analysis was performed to determine whether these metrology instruments are capable of resolving differences in optical thickness, reflectivity, and EOT at various steps in the RPN process. Precision of the metrology tools (defined as $6\sigma$ variation) was determined from a 30-day gauge study. Capability was estimated by computing the difference in mean values of each data set, divided by the precision ($6\sigma$ variation) of the metrology tool. If the resulting value is greater than or equal to 1, the mean values are distinguishable by the metrology tools, at a confidence level of 99.9%. The significance calculation is

$$\left| \frac{Y1 - Y2}{6\sigma} \right| \geq 1 \tag{24}$$

where $Y1$ and $Y2$ are the mean values of the individual wafer means from each process set. Table 5 shows the significance of the differences for each of the process steps. The modulus of the significance value is not shown; instead the sign of the measurement differential is included to indicate the effect of processing on each measurement type. The data shows that the proposed optical and electrical metrology solution meets the statistical significance test.

To determine whether in-line measurements of the Quantox EOT correlated with end-of-line (EOL) polysilicon MOSCAP measurements of EOT, a range of RPN and annealing conditions was utilized in a second series of 26.5-Å furnace oxides. One set of four wafers was processed through each type of RPN treatment (RPN1–4) and subse-

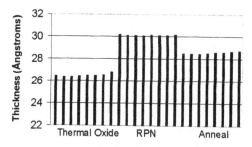

**Figure 17** Optical $T_{ox}$: initial oxide processed through nitridation and anneal.

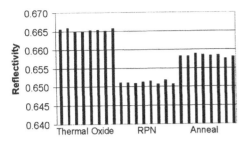

**Figure 18** Optical reflectivity: initial oxide processed through nitridation and anneal.

quently exposed to only one anneal treatment (A1). A second set of three wafers was processed using the standard RPN1 process only and then subjected to each of the three available annealing treatments, A1, A2, and A3. Finally, a replicate wafer was produced for the standard process conditions of RPN1 with A1 anneal. The wafers were subsequently processed to enable corresponding measurements to be made using conventional polysilicon MOSCAP structures. Six sites per wafer—two at the center and four at the edges—were probed to determine the EOL EOT.

Figure 20 shows the correlation observed between in-line Quantox determination of EOT for each RPN process (1–4) and the corresponding polysilicon MOSCAP EOL results. The standard annealing condition, A1, was used for all wafers. In Figure 20 the oxides processed using the RPN1-with-A1 conditions exhibited the smallest electrical thickness. The lower-power and shorter-duration RPN conditions yielded electrically thicker dielectrics. An offset was observed between the Quantox and MOSCAP measurement techniques. In order to make the MOSCAP EOL EOT measurements, the wafers had to be subjected to several additional processing steps to create the test structures, compared to the in-line Quantox measurements. In addition, polysilicon depletion was not taken into account for the p-doped polysilicon capacitors. Despite the offset, an $R^2$ correlation of 0.96 was observed between the Quantox and MOSCAP measurements of equivalent oxide thickness.

Figure 21 shows the correlation observed between in-line Quantox determination of EOT for each Anneal process (1–3) and the corresponding polysilicon MOSCAP EOL results. The standard RPN condition, RPN1, was used for all wafers. The replicate wafer processed using RPN1 and A1 is also shown on Figure 21. In this set, the replicate oxides processed using the RPN1 along with the high-temperature A1 anneal showed the thickest EOT. The lower-temperature A2 anneal and alternative annealing environment

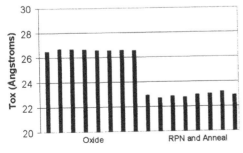

**Figure 19** Electrical $T_{ox}$: initial oxide processed through nitridation and anneal.

**Table 5**  Metrology Capability Results for an Oxide of Starting Thickness 26.5 Å

| Parameter | Significance |
| --- | --- |
| Optical thickness: | |
|     Initial oxide (Y1) – RPN (Y2) | −2.63 |
|     RPN (Y1) – anneal (Y2) | 1.13 |
|     Initial oxide (Y1) – anneal (Y2) | −1.50 |
| Reflectivity: | |
|     Initial oxide (Y1) – RPN (Y2) | 3.55 |
|     RPN (Y1) – anneal (Y2) | −1.80 |
|     Initial oxide (Y1) – anneal (Y2) | 1.75 |
| Electrical thickness: Initial oxide (Y1) – anneal (Y2) | 1.05 |

of anneal A3 lead to increasingly thinner EOT values. As observed with the previous data, an offset was observed between the Quantox EOT and the corresponding MOSCAP EOL electrical measurements of thickness. Likewise, a good correlation ($R^2 = 0.99$) was calculated between the Quantox and MOSCAP measurements of equivalent oxide thickness.

A combined optical and noncontact electrical metrology approach using the ASET-F5 and Quantox instruments was shown to provide statistically significant process metrology for remote plasma nitrided-gate dielectrics of 26.5-Å initial oxide thickness. Linear correlation values of Quantox EOT to polysilicon MOSCAP EOL EOT, despite the inherent changes to the gate dielectric during the MOSCAP fabrication process, indicate that an in-line metrology approach can be implemented with success.

## C. High-Dielectric-Constant Gate Materials

High-dielectric-constant materials are expected to be used at, and after, the 70-nm logic node and possibly even at the 100-nm node. In contrast to $SiO_2$, for which there exist standard tables of optical properties, relatively little is known about the properties of high-$k$ films on Si. Tantalum pentoxide ($Ta_2O_5$) exhibits a relative dielectric constant of approximately 25 and is widely used in DRAM applications. Although it is unlikely that $Ta_2O_5$ will be used for multiprocessor gate dielectric replacement, the $Ta_2O_5$ system

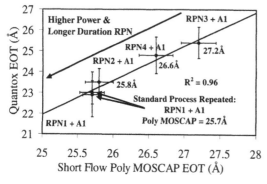

**Figure 20**  Correlation of Quantox EOT to MOSCAP EOL for a range of RPN conditions.

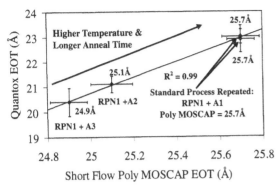

Figure 21 Correlation of Quantox EOT to MOSCAP EOL for a range of annealing conditions.

provides a prototypical material expected to exhibit many of the challenges associated not only with optical modeling, but indeed with fabrication. This section briefly discusses application of the HO and Tauc–Lorentz models to this system.

A $Ta_2O_5$ film of nominal thickness 5 nm was deposited on a Si wafer by chemical vapor deposition (CVD) for preliminary characterization (46). Prior to deposition of the $Ta_2O_5$ film, the Si wafer had undergone a rapid thermal anneal in an oxygen environment at 900°C for 30 seconds. Figure 22 shows the spectral data recorded from the center point of the wafer. A wavelength range of 240–800 nm was used, and the film was modeled as $Ta_2O_5$ on $SiO_2$ on Si. Standard tables of optical properties were used for the Si substrate and $SiO_2$ layer. Only the thickness of the oxide layer was allowed to vary. A simple two-

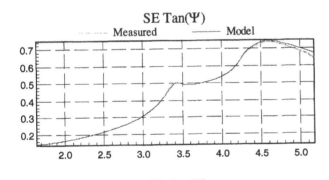

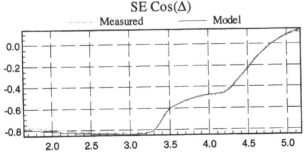

Figure 22 Tan$\Psi$ and cos $\Delta$ spectral data recorded from a $Ta_2O_5/SiO_2/Si$ film stack over the photon energy range 1.6–5.2 eV. The thickness of the $Ta_2O_5$ film was 4.4 nm. The $\chi^2$ goodness-of-fit was 0.893.

oscillator HO model (with oscillator energies of 7.8 and 4.47 eV) was used for this pre-liminary analysis of the $Ta_2O_5$ film. Only thickness and the oscillator densities were allowed to vary for the $Ta_2O_5$ film. Figure 23 shows the dispersion of the refractive index and extinction coefficient for the $Ta_2O_5$. The thickness of the $SiO_2$ was found to be 2 nm; the thickness of the $Ta_2O_5$ was found to be 4.4 nm.

Richter et al. have completed a more extensive study of CVD $Ta_2O_5$ films deposited on silicon (2,47). Spectroscopic ellipsometric analysis was performed over the spectral range 1.5–6.0 eV and analyzed using the Tauc–Lorentz model. In their analysis, the SE data was inverted to extract $\varepsilon_1$ and $\varepsilon_2$ at each photon energy. To perform the inversion, the starting thickness of the film was first determined from a Tauc–Lorentz fit to the SE data. Should the starting thickness be inaccurate, nonphysical results were reportedly obtained; for example, $\varepsilon_2$ might equal zero for photon energies less than the bandgap. The film thickness was then varied for each trial inversion until such nonphysical features were minimized. For example, CVD $Ta_2O_5$ films were deposited on a thin nitride layer on Si substrates. Of particular interest is the dispersia of the $Ta_2O_5$ films following annealing, since for this material an amorphous structure is desired to minimize leakage. Figure 24 shows $\varepsilon_1$ and $\varepsilon_2$ plotted against photon energy for films annealed for 5 minutes at 400 and 700°C. The film annealed at 400°C showed a smooth dielectric function and was believed to remain amorphous. In contrast, the film annealed at 700°C exhibited an additional peak when $\varepsilon_1$ and $\varepsilon_2$ were derived from inversion and was believed to be microcrystalline in structure. The extra peak in the dielectric functions was attributed to sharpening of the electronic band structure during transformation from the amorphous to the microcrystal-line state. The Tauc–Lorentz model developed by Jellison and Modine (30) is expected to be of particular value in the analysis of high-$\kappa$ materials.

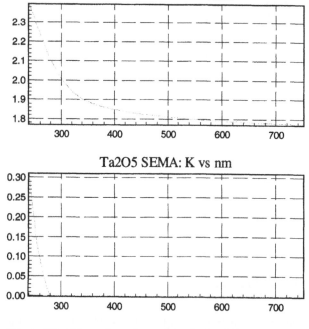

**Figure 23** Dispersion of $n$ and $k$ obtained from the $Ta_2O_5$ film shown in Figure 22. The wave-length range of 240–750 nm is shown for comparison with the photon energy range of Figure 22.

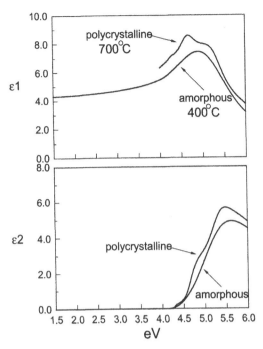

**Figure 24** Pseudo dielectric functions extracted by inversion for $Ta_2O_5$ samples following a low Temperature and a High Temperature Anneal.

## ACKNOWLEDGMENTS

The authors gratefully acknowledge the many contributions of their colleagues within KLA-Tencor Corporation. We would particularly like to thank Carlos Ygartua, Albert Bivas, John Fielden, Torsten Kaack, Phil Flanner, Duncan Mills, Patrick Stevens, David McCain, Bao Vu, Steve Weinzierl, Greg Horner, Tom Miller, Tom Casavant and Koichiro Kawamura. We express our gratitude to Alain Diebold, Dennis Maher, Rick Garfunkel, Robert Opila, George Brown, Dick Deslattes, Jim Erhstein, Curt Richter, Jay Jellison, Eric Vogel, Morgan Young, Rick Cosway, Dan Iversen, Stefan Zollner, Kwame Eason, Gene Irene, and Harland Tompkins for many informative discussions. The careful work of the NCSU gate dielectric metrology team, especially David Venables, Kwangok Koh, Shweta Shah, Chadwin Young, Mike Shrader, and Steven Spencer, is gratefully acknowledged.

## REFERENCES

1. See, for example: R.M.A. Azzam, N.M. Bashara. Ellipsometry and Polarized Light. Amsterdam: Elsevier, 1986; H.G. Tompkins, W.A. McGahan. Spectroscopic Ellipsometry and Reflectometry. New York: Wiley, 1999.
2. Gate Oxide Metrology Enhancement Project (FEPZ.001) managed by Alain Diebold (International SEMATECH).
3. The 1999 National Technology Roadmap for Semiconductors. San Jose, CA: Semiconductor Industry Association, 1999.

4.  A.C. Diebold, D. Venables, Y. Chabal, D. Muller, M. Weldon, E. Garfunkel. Characterization and production metrology of thin transistor gate oxide films. Materials Science in Semiconductor Processing. 2:103–147, 1999

5.  Y.P. Kim, S.K. Choi, H.K. Kim, D.W. Moon. Direct Observation of Si Lattice Strain and its distribution in the Si(001)-SiO$_2$ interface transition layer. Appl. Phys. Lett. 71:3504—3506, 1997.

6.  D.W. Moon, H.K. Kim, H.J. Lee, Y.J. Cho, H.M. Cho. Thickness of ultra-thin gate oxides at the limit. In: D.G. Seiler, A.C. Diebold, W.M. Bullis, T.J. Shaffner, R. McDonald, E.J. Walters, eds. Characterization and Metrology for ULSI Technology, 1998. New York: AIP, 1998, pp 197–200.

7.  T. Hattori. Chemical structure of the SiO$_2$/Si interface. CRC Crit. Rev. Solid State Mater. Sci. 20:339, 1995.

8.  H. Nohira, K. Takahashi, T. Hattori. Energy Loss of O1s photoelectrons in compositional and structureal transition layer at and near the SiO2/Si interface. Proceedings of UCPSS98. In press.

9.  S.D. Kosowsky, P.S. Pershan, K.S. Krisch, J. Bevk, M.L. Green, D. Brasen, L.C. Feldman, P.K. Roy. Evidence of annealing effects on a high-density Si/SiO$_2$ interfacial layer. Appl. Phys. Lett. 70:3119–3121, 1997.

10. N. Awaji, S. Ohkubo, T. Nakanishi, Y. Sugita. High-density layer at the Si/SiO$_2$ interface observed by x-ray reflectivity. Jpn. J. Appl. Phys. 35:L67–L70, 1996.

11. C.H. Bjorkman, T. Yamazaki, S. Miyazaki, M. Hirose. FTIR-ATR characterization of ultra-thin SiO$_2$ films on Si. Proc. Int. Conf. Advanced Microelectronic Devices and Processing, p 431, 1994.

12. J.T. Fitch, G. Lucovsky, E. Kobeda, E.A. Irene. Effects of thermal history on stress-related properties of very thin films of thermally grown silicon dioxide. J. Vac. Sci. Technol. B7:153, 1989.

13. T. Yamzaki, C.H. Bjorkman, S. Miyazaki, M. Hirose, Local structure of ultra-thin (3–25 nm) SiO$_2$ thermally grown on Si(100) and (111) surfaces. Proc. 22nd Int. Conf. on Physics of Semiconductors. Vol. 3 (Canada, 1994), pp 2653—2656.

14. M. Hirose, W. Mizubayashi, M. Fukuda, S. Miyazaki. Characterization of ultrathin gate oxides for advanced MOSFETs. In: D.G. Seiler, A.C. Diebold, W.M. Bullis, T.J. Shaffner, R. McDonald, E.J. Walters, eds. Characterization and Metrology for ULSI Technology. New York: AIP, 1998, pp 65–72.

15. E. Taft, L. Cordes. Optical evidence for a silicon–silicon oxide interlayer. J. Electrochem. Soc. 126:131–134, 1979.

16. D.E. Aspnes, J.B. Theeten. Optical properties of the interface between Si and its thermally grown oxide. Phys. Rev. Lett. 43:1046–1049, 1979.

17. D.E. Aspnes, J.B. Theeten. Spectroscopic analysis of the interface between Si and its thermally grown oxide. J. Electrochem. Soc. 127:1359–1365, 1980.

18. N.V. Nguyen, D. Chandler-Horowitz, P.M. Amirtharaj, J.G. Pelligrino. Spectroscopic ellipsometry determination of the properties of the thin underlying Si layer and the roughness at SiO$_2$/Si Interface. Appl. Phys. Lett. 64:2688–2690, 1994.

19. C.M. Herzinger, B. Johs, W.A. McGahan, J.A. Woollam, W. Paulson. Ellipsometric determination of optical constants for silicon and thermally grown silicon dioxide via a multi-sample, multi-wavelength investigation. J. Appl. Phys. 83:3323–3336, 1998.

20. I.H. Malitson. J. Opt. Soc. Am. 55:1205, 1965.

21. D.E. Aspnes, A.A. Studna. Dielectric Functions and Optical Parameters of Si, Ge, GaP, GaAs, GaSb, InP, InAs, and InSb from 1.5 to 6.0 eV. Phys. Rev. B. 27:985–1009, 1983.

22. G.E. Jellison Jr. Optical functions of silicon determined by two-channel polarization modulation ellipsometry. Opt. Materials. 1:41–47, 1992.

23. Y.Z. Hu, S.P. Tay, Y. Wasserman, C.Y. Zhao, K.J. Hevert, E.A. Irene, Characterization of ultrathin SiO$_2$ films grown by rapid thermal oxidation. J. Vac. Sci. Technol. A. 15:1394–1398, 1997.

24. D.A. Muller et al. Nature. 399:758–761, 1999.

25. G.D. Wilk, R.M. Wallace. Electrical properties of hafnium silicate gate dielectrics deposited directly on silicon. Appl. Phys. Lett. 74:2854–2856, 1999.

26. G.E. Jellison Jr. Chapter 25 of this volume and other references contained therein.

27. F. Wooten. Optical Properties of Solids. New York: Academic Press, 1972.

28. P. Yu, M. Cardona. Fundamentals of Semiconductors. New York: Springer-Verlag, 1996.

29. J. Tauc, R. Grigorovici, A. Vancu. Phys. Stat. Solidi. 15:627, 1966.

30. G.E. Jellison Jr, F.A. Modine. Parameterization of the optical functions of amorphous materials in the interband region. Appl. Phys. Lett. 69:371–373, 2137, 1996.

31. S.A. Eastman. Evaluating Automated Wafer Measurement Instruments. SEMATECH Technology Transfer Document #94112638A-XFR, 1995.

32. J.R. Hauser, K. Ahmed, Characterization of ultra-thin oxides using electrical C-V and I-V Measurements. In: D.G. Seiler, A.C. Diebold, W.M. Bullis, T.J. Shaffner, R. McDonald, E.J. Walters, eds. Characterization and Metrology for ULSI Technology. American Institute of Physics, pp 235–239, 1998.

33. E. Vogel, V. Misra. Chapter 4 in this volume and other references contained therein.

34. G. Lucovsky. Spatially-selective incorporation of bonded-nitrogen into ultra-thin gate dielectrics by low-temperature plasma-assisted processing. In: E. Garfunkel et al., eds. Fundamental Aspects of Ultrathin Dielectrics on Si-based Devices. Dordrecht: Kluwer Academic, pp 147–164, 1988.

35. S.V. Hattangady et al. Remote plasma nitrided oxides for ultrathin gate dielectric applications. SPIE 1998 Symp. Microelec. Manuf., Santa Clara, CA, Sept. 1998, p 1.

36. C. Hayzelden, C. Ygartua, A. Srivatsa, A. Bivas. Optimization of SiON anti-reflective layers for 248-nm and 193-nm lithography. Proc. 7th International Symposium on Semiconductor Manufacturing, Tokyo, 1998, pp 219–222.

37. J. Huran, F. Szulenyi, T. Lalinski, J. Liday, L. Fapso. Crystal Properties Preparation. 19 & 20:161, 1989; F. Szulenyi, J. Huran. Acta Phys. Slov. 41:368, 1991.

38. E. Dehan, P. Temple-Boyer, R. Henda, J. J. Pedroviejo, E. Scheid. Thin Solid Films. 266:14, 1995.

39. W.A.P. Claassen, H.A.J. Th. van der Pol, A.H. Goemans, A.E.T. Kuiper. J. Electrochem. Soc. 133:1458, 1986.

40. W.A.P. Claassen, W.G.J.N. Valkenburg, F.H.P.M. Habraken, Y. Tamminga. J. Electrochem. Soc. 130:2419, 1983.

41. W.R. Knolle. Thin Solid Films. 168:123, 1989.

42. S.-S. Han, B.-H. Jun, K. No, B.-S. Bae. J. Electrochem. Soc. 145:652, 1998.

43. L. Cai, A. Rohatgi, D. Yant, M. A. El-Sayed. J. Appl. Phys. 80:5384, 1996.

44. Harland G. Tompkins, Richard Gregory, Mike Kottke, David Collins. Determining the Amount of Excess Si in PECVD Oxynitrides, (and references contained therein). Submitted to JVST.

45. C. Hayzelden et al. Combined Optical and Electrical Metrology for Advanced Gates. SEMICON Japan, STS, December 3, 1999.

46. Samples courtesy of Alain Diebold, George Brown, and Mark Gilmer at International SEMATECH.

47. C.A. Richter, N.V. Nguyen, G.B. Alers. In: Huff, Richter, Green, Lucovsky, Hattori, eds. Proc. Mat. Res. Soc., 567. Warrendale, PA: Materials Research Society, 1999, p. 559.

# 3
# Metrology for Ion Implantation

**Lawrence A. Larson**
*International SEMATECH, Austin, Texas*

## I. INTRODUCTION

This chapter will describe the use of metrology tools for the measurement of ion implantation processes. The objective is to add to a generic description of the measurement tool and techniques the more detailed characteristics of these techniques as compared to the needs of standard ion implantation processes. The fundamental characteristics of each technique are described and detailed elsewhere in this book. The doped layers that they are used to measure have relatively specific characteristics. The characteristics of the metrology tool and, when applicable, the various techniques for using that tool can be described and evaluated as a function of the layers it is expected to measure.

The doped layers will be described as the parts of a standard CMOS process and the future trends of these layers through the trends described in the *International Technology Roadmap for Semiconductors* (ITRS) (1). As a method for organizing the techniques, they will be approached per the technology used for the measurement. These are:

1.  Resistivity measurement tools (four-point probe, eddy current, *C-V*)
2.  Damage measurement tools (thermawave, optical dosimetry)
3.  Profile measurement tools (SIMS, spreading resistance)
4.  New methods (Boxer Cross, acoustic emission)

## II. IMPLANT TECHNOLOGY TRENDS

The technology trends in ion implantation are summarized well by the sections of the *International Technology Roadmap for Semiconductors* (1) that describe the requirements and challenges for doping processes. The critical challenge for doping is the "Ultra-Shallow Junctions (USJ) with Standard Processing," which encompasses "Achievement of lateral and depth abruptness, Achievement of low series resistance, $< 10\%$ of channel sheet resistance, and Annealing technology to achieve $< 200\ \Omega/\text{sq}$ at $< 30$-nm junction depth." This summarizes a general trend for the doped layers of the devices to be generally shallower and, at the same time, more highly doped.

The leading-edge manufacturing companies are meeting this challenge. This is reflected in the observation that the requirements for shallow junctions are shrinking almost as fast per edition of the roadmap as they are driven by process technology changes. The challenge to doping technology echoes the metrology techniques used to measure the doped layers. The shallowness of the layers challenges virtually every measurement technology used to measure them, simply because the possible measurement interaction depth decreases. Similarly, many techniques have characteristic preferred dopant levels to measure. The trend to higher concentration layers changes the utility of the measurement techniques by this interaction.

Throughout this chapter a phase plot of layer junction depth versus dose will be used as a discussion tool to describe the characteristics of the technique in discussion. The first of these (Figure 1) is a description of the layers produced by the standard types of ion implanters and the trends in junction depth as described by the ITRS. Note that there are three major families of tool types represented: low energy, high energy, and medium current. These were chosen because the industry is in the midst of a change in emphasis from high current/medium current to high energy/low energy. At this time the high-current implanters have been pretty well absorbed by the description "low energy," for the newer machines are quite capable of high currents and of low energy. On the other hand, medium-current implanters cover a process space of low dose and medium depth (a few tenths of a micron) that are still difficult if not impossible for high-energy machines to process.

Perhaps more important than the operating regions of the machines is the trends in processes that are described by the arrows in Figure 1. Note that there is a compression of the dose range used. Although the machines are capable of about six decades of dose, only about three are commonly used. This is being constricted by the trends described in the ITRS. The "Deep Wells" arrow describes the deepest of the processes generally used. The marked trend for these processes is that the depth has decreased by the trend of the scaling factor, while the dose has just dropped a partial order of magnitude. This reflects the trend toward higher concentration layers that are markedly shallower. The midchannel implants represent implants such as the halo or anti-punchthrough implants. These have trended similar to the other layers of the transistor. The arrow marked "Drain Extension" represents the shallowest of the implants. This layer has had very aggressive scaling to shallower

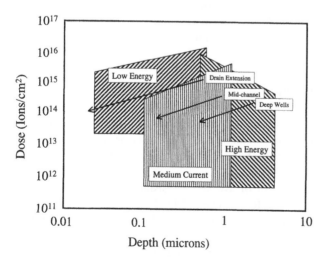

**Figure 1** Dose/depth characteristics of the implant tool families and of the layers produced by them.

depths, while the relative concentration has increased so much that they are no longer called LDD (for lightly doped drain) and are instead called either drain extension layers or sometimes HDD (highly doped drain)! While this is the layer where this trend is most markedly observed, the trending for shallow junctions is also reflected in the threshold adjust implant, which is also significantly shallower and has increased in dose by a similar amount.

The process control requirements for doping processes are becoming better known. Zeitzoff et al. (2) have calculated a number of process control requirements for the critical steps in the semiconductor manufacturing process. Within these, the implanted steps of most critical nature are those represented by the "Drain Extension" arrow in Figure 1. This is attributed to the threshold adjust step but also includes the interaction with the drain extension. The result of that calculation is that these critical implants can have a total deviation of roughly 5%, with the other steps allowed somewhat more. This figure should be considered the total tolerance for this process step. To achieve a precision/tolerance ratio of 10%, the implication is that a minimum precision of 0.5% is necessary for the successful application of a metrology technique. Zeitzoff covers this topic further in Chapter 6. In several places in this chapter, uniformity, performance, and resolution are important considerations. It is appropriate to note that uniformity is simply a measure of within-wafer statistical deviations. This is only one of the measures of the process distribution, all of which need to be included within the 5% allowance of the process distribution. The spatial nature of this measurement is an important consideration for the comparison of analytical tools. The two spatial dimensions of importance are the dimension of a few microns and the dimension of a centimeter. The dimension of a few microns is important because several designs rely on transistor matching; part of the precision of this matching is in the very precise matching of processes that comes through physical proximity. Similarly, the dimension of a centimeter is important because that is the spacing between individual chips. This again would show in final device performance as a strong variation chip to chip.

## III. RESISTIVITY TOOLS

The most common of the metrology tools used to measure implanted layers for process control are the resistivity measurement tools (four-point probe, eddy current, and $C$-$V$). These tools are described in Chapter 11, by Johnson. Of these, the four-point probe is used most for standard process control for silicon semiconductor production. Eddy current methods find utility where contact is not easily made to the implanted layer, such as with III–V material processing, and $C$-$V$ measurement refers to measurement of simple devices such as Van der Pau structures.

Four-point probe methods were commercialized in the late 1970s (3). They were successful due to inherent advantages of accuracy and ease of measurement. Key developments included the development of the dual-configuration mode, the development of wafer mapping, and easy and efficient user interfaces. Johnson describes the dual-configuration mode more fully in Chapter 11. This method uses measurement from alternating pairs of probes to deconvolve effects on the reading from nearby edges. This significantly increased the accuracy of the readings and allowed measurement to within a few millimeters of the edge of the wafer. Wafer mapping was developed in the early 1980s, partially as a result of the computerization of the measurement and its data analysis (4). This is a key development, for it enabled significant developments in the methods of troubleshooting semiconductor tools and processes (5). Being able to analyze the performance of the tool by its characteristic patterns on the wafer both eased the job of discovering problems in the process and allowed several new process issues to be discovered. These issues would most likely not be detected at all with single- or nine-point patterns that are analyzed by hand. A large-scale striping problem, where the implanter scan momentarily stops, is immediately apparent in a wafer map. It would be indicated in a small number of probes, but not so clearly. Consider a fine scan lockup pattern or perhaps a microuniformity issue. These would not be detected at all with a single- or nine-point probe pattern analyzed manually. As noted previously, the two spatial dimensions of primary concern are on the order of a few microns and on the order of a centimeter. Four-point probes are well positioned to monitor variations on the order of a centimeter, allowing chip-to-chip scale sensitivity. However, it is too broad a technique to be able to detect microuniformity type of issues that affect transistor-to-transistor matching.

Figure 2 illustrates the capability of four-point probes as described in relationship to the dose–depth characteristics of the processes to be measured. Only the activation level of the layer itself limits the highest dose that is measurable by the probe. The four-point probe is easily capable of measuring metallic layers, so the upper dose is left at the figure limit of $10^{17}/cm^2$. Similarly, the layer itself generally limits the depth of the layer measured; therefore the figure limit of 10 microns is used. The measurement is an average of the area of the penetration of the field lines of the measurement. A conductive layer will easily pass the current and thus only measure directly between the probes, while a resistive layer will have significant field and current penetration and will measure quite deep. In general, the measurement is an average over the entire depth within an isolated junction. The shallow depth limitation is that the probe must be correctly chosen to penetrate no more than the isolated junction. This limit is on the order of a few hundred angstroms and is deep enough that the shallow junctions under development for LDD and threshold-type processes have some difficulty due to this effect. The dose limitation as detailed on the figure is an issue that is somewhat harder to quantify. There are certain leakage currents that can affect the measurement when it is measuring minimum currents. These tend to be either instrumental leakage currents or, more likely, currents along surface conduction

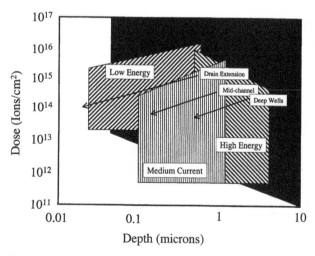

**Figure 2** Dose/depth characteristics of the four-point probe technique. The darker area in the upper right-hand quadrant represents the four-point probe measurement space.

paths on the wafer. Roughly, this effect is important as compared to the $10^{11}/cm^2$ to $10^{12}/cm^2$ dose measurement region for a standard layer that is 100 nm thick. It would then scale with the measurement current.

## IV. DAMAGE MEASUREMENT TOOLS

Another major family of techniques aimed uniquely at the measurement of "as-implanted" layers is the group that depends on the measurement of the damage produced by the implant as an indirect measure of the implant itself. There are two major variants of this technology. Optical dosimetry relies on the darkening of polymers that occurs as a result of exposure to the ion beam (6,7). Thermawave relies on the change of reflectivity of the silicon that has been implanted (8,9).

Optical dosimetry was "discovered" first in connection with an early method of aligning an implanter and testing beam shape involving doing a "beam burn," which was to expose a piece of paper (or plastic) to the beam and then to examine the burnt spot. This was never developed much past this simple test, and commercial versions of both techniques were developed and marketed in the early '80s (7,8). The two tools have markedly different advantages because they have quite different methods of creating and measuring their signals. Optical dosimetry is based on the darkening of polymers by the implanted ions. This method uses a special test wafer using the polymers that have their optical response quantified for dose and energy. The measurement tool is a specialized optical densitometer that reads and records the response of the wafer. This system has the characteristics that it is an in-fab measurement tool, rather than in-line, and that the test wafer may be read to a relatively high spatial resolution over the entire wafer. In contrast, the thermawave technique is based on the response of the implanted silicon to the ion implantation. It therefore is one of the few techniques that can measure product wafers in-line. The signal collection of the thermawave is somewhat slower than the densitometer though, so the resulting tests are generally of poorer spatial resolution.

Figure 3 illustrates the dose–depth characteristics of the damage measurement tools. As already noted, these tools evolved uniquely for the measurement of implants and would otherwise be useful only for measurement of other types of radiation damage. They share the benefit that there is normally two or three orders of magnitude higher damage than implanted ions, so the sensitivity of these techniques to lower doses is generally quite good (12). In the figure this is represented by the dose sensitivity extending down to the $10^{11}/cm^2$ range. In many cases it is actually much better than this, but the normal needs of CMOS processes do not extend this low. Note, however, that this is well beyond the capabilities of most of the other techniques described here. This sensitivity to the implanted dose extends to the upper part of the range also. The techniques are sensitive to dose, energy, and species, but as a very general trend the signal saturates in the $10^{14}/cm^2$ dose range. This is similarly well below the capability of most of the other techniques. The trend with depth is not at all clear, due to this dependence on species and energy. The part shown in Figure 3 illustrates the energy dependence of the signal. In general, the energy dependence part of the depth trend tends to produce more damage per ion and therefore saturates the signal at a lower dose. On the lower-depth side of the figure, the signal is lost through the layer's simply getting too small for interaction with the probing beam. This depth is on the order of 50–100 nm in depth. On the higher-depth side of the figure, the mechanism is similar. There the layer is sufficiently thick that it exceeds the capabilities of the probe beam. This point would be different for the two techniques. For optical densi-

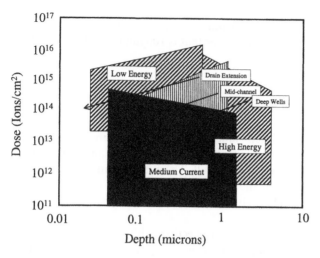

**Figure 3** Dose/depth characteristics of the damage measurement tools. The darker region in the lower center represents their measurement space.

tometry the thickness is determined by the layer thickness of the polymer layer, because the probe beam is transmitted through the wafer. In contrast, the thickness of the deepest thermawave measurement is determined by the depth characteristics of the probe laser in silicon. This is in the region shown in Figure 3, but it is also sensitive to dose, energy, and species of the implant.

The spatial characteristics of the techniques provide a similar interesting comparison. Thermawave has significantly better spatial resolution, on the order of microns, which enables performance at the spatial frequency of transistor matching, which is of strong benefit. Its signal acquisition time is long though, which makes it difficult to routinely check at this level. On the other hand, the optical densitometer has a resolution of slightly less than a millimeter. This is markedly larger than that needed for transistor matching and is somewhat smaller than the spacing between devices.

## V. PROFILE TOOLS

This section describes the use of SIMS and/or SRP to measure the implanted layer. SIMS, secondary ion mass spectroscopy, is a technique where the surface of the silicon is sputtered away by an ion beam. A portion of the surface is ionized by this beam and is detected as the signal. Spreading-resistance profiling, SRP, operates by exposing a tapered cross section of the layer. The layer is profiled with a two-point probe system, which measures resistivity through the current–voltage characteristics of individual points through that cross section.

These profile measurement techniques are more commonly used to measure details of transistor layers in the analytical lab but have been put to use in some facilities as a monitoring tool for fab processes. In most cases this is as an off-line measurement tool. But in at least one instance, SIMS has been put into place as an in-fab dose measurement tool (11). As a process monitoring tool, SIMS is a very powerful technique, because it allows both dose and depth measurement simultaneously. Its drawbacks center on the complex mechanisms involved with the SIMS measurement (13). Both the profile depth and the calculated concentration of the measured elements depend on the established knowledge of the system measured. Wittmaack and Corcoran (12) describe in some detail the controls that need to be put in place in order to operate this tool as a routine fab control tool. It is relatively expensive and time-consuming as compared to the other in-line techniques described here. One should also remain aware that the SIMS measurement gives relative atomic concentration, because it's a calculated signal, rather than electrical activation or chemical state. Similarly, SRP measures the profile of the electrical activation of the layer(s) and gives no information on the actual atomic concentration or chemical state of the dopants. Like SIMS, it is a powerful technique that allows measurement of both active dose and depth simultaneously. And similarly, it involves complex mechanisms that make precise measurement very difficult (14). In this case the state of the polished surface has a marked effect on the quality of the measurement; there are also several effects on the $C$-$V$ measurement, which makes quantification difficult. The resistivity is different depending on which dopant is present. Another well-known effect is that the $C$-$V$ measurement involves a volume related to the debye length when current is forced. When this volume is constrained by the depth of the layer being measured, then the calculated concentration is incorrect (15). This leads the SRP curve to cut off more abruptly than SIMS for shallow junctions. These techniques, SRP and SIMS, are the subject of technical conferences that focus on their mechanisms and on their use as analytical techniques (16–18). These references are highly recommended for in-depth knowledge of the attributes of the techniques.

The dose–depth characteristics of the SIMS measurement are described in Figure 4. This figure shows clearly why this technique, even though it may be more expensive and time-consuming, is used for in-fab measurement. It is more capable than many of the other techniques in several critical regions of this figure. It shows about four decades of dynamic range in the signal. The lower-dose limit of the curve is actually determined by the signal-to-noise characteristics of the tool. This would be the minimum signal present at the incident-beam current at which the depth can be sputtered in a reasonable time. On the lower-depth side of the figure this point is a delicate balance of setting beam current so that it doesn't sputter through the layer too fast to collect reasonable signal, versus a relatively high dose that would be needed to produce reasonable signal during that time.

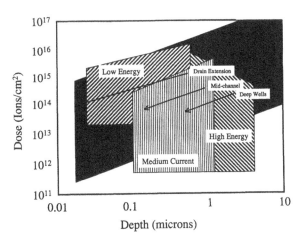

**Figure 4**   Dose/depth characteristics of SIMS analysis. The darker region in the center represents its measurement space.

This line is set, somewhat arbitrarily, between 10 and 20 nm in depth, indicating that there is some sort of limit in this balance. The upper part of the region is similarly determined. On the high-concentration/dose side of the region the main limitation is the choice of beam currents such that the profile is measured in a reasonable time. In general this region of the figure is well above any level of concern for doping applications. The limitations here tend to be in matrix effects that limit the ability of the system to realize accurate quantification.

## VI. NEW METHODS

There are new techniques continually under examination for applicability as a doping monitoring technique. As might be imagined, the dynamic range of the process, both in depth and in dose, makes this an extremely challenging application to approach.

An optical measurement technique has potential, for it could be a remote measurement technique. This might enable direct monitoring of the process within the implant tool. Good efforts have been made to use ellipsometry (19), both for shallow dose measurement and for amorphous layer measurement (20), but the shallowest layers and the lowest doping levels are difficult to measure.

The latest entry is the tool presented by Boxer Cross (21). This tool is an optically based technique where a probe laser is used to stimulate charges within the doped layer filling the electronic bands, while a measurement laser then measures reflectivity changes that are induced by this action. This technique then has quite a small area that is affected by the measurement, on the order of a few square microns. This enables measurement directly on product, which is a preference of the engineers. It also appears to have sensitivity to lower-concentration layers, which is the critical process need for measurement. However, the sensitivity characteristics of the technique are not yet well defined in terms of the dose–depth figures that have been used for the other techniques described here. This work is in process, but it will be some time before the technique is as well understood as the more established techniques. The goal would be to enable routine process monitoring on device wafers and to be able to do that at a speed where a majority of the wafers processed could be measured while not impacting process cycle times. Chapter 5, by Borden et al., describes this technique and much of our understanding of it to date.

## VII. SUMMARY

This chapter described the use and characteristics of several metrology tools for the measurement of ion implantation processes. The generic description of the measurement tool was given in reference to the more detailed characteristics of the techniques as compared to standard ion implantation processes. The doped layers that they are used to measure have relatively specific characteristics. The characteristics of the metrology tool and the various techniques of using that tool were evaluated as a function of the layers it is expected to measure. These are graphically shown as a plot of the dose–depth characteristics of the measurement technology and of the implanted processes. Similarly, the spatial resolution of the measurement can be compared to the process spatial characteristics. Important device spatial frequencies are on the order of microns for transistor matching characteristics and on the order of centimeters for die-to-die matching characteristics.

The doped layers were addressed as the parts of a standard CMOS process. The future trends for these layers are expected to follow those described in *International Technology Roadmap for Semiconductors*. The techniques were approached per the technology used for the measurement. These are resistivity measurement tools (four-point probe, eddy current, *C-V*), damage measurement tools (Thermawave, optical dosimetry), profile measurement tools (SIMS, spreading resistance), and new methods (Boxer Cross, acoustic emission).

# REFERENCES

1. Semiconductor Industry Association. International Technology Roadmap for Semiconductors. 1999 ed. Dec. 1999. [http://www.itrs.net/1999_SIA_Roadmap/Home.htm].
2. P.M. Zeitzoff, A.F. Tasch, W.E. Moore, S.A. Khan, D. Angelo. In: D.G. Seiler, A.C. Diebold, W.M. Bullis, T.J. Shaffer, R.M. McDonald, E.J. Walters, eds. Characterization and Metrology for ULSI Technology. New York: American Institute of Physics, 1998, pp 73–82.
3. D.S. Perloff, F.E. Wahl, J. Conragan. J. Electrochem. Soc. 124:58, 1977.
4. M.I. Current, M.J. Markert. In: Ion Implantation: Science and Technology. Orlando, FL: Academic Press, 1984, pp 487–536.
5. M.I. Current, L.A. Larson, W.A. Keenan, C.B. Yarling. In: J.F. Ziegler, ed. Handbook of Ion Implantation Technology. Amsterdam: North Holland, 1992, pp 647–674.
6. S. Wolf, R.N. Tauber, eds. Silicon Processing for the VLSI Era. Vol. 1—Process Technology. Sunset Beach, CA: Lattice Press, 1986, p 318.
7. J.R. Golin et al. Advanced methods of ion implant monitoring using optical dosimetry. Solid State Technol. (June 1985), 155.
8. W.L. Smith, A. Rosenwaig, D.L. Willenbourg. Appl. Phys. Lett. 47:584, 1985.
9. L. Zhou, N.W. Pierce, W.L. Smith. In: M.I. Current, C.B. Yarling, eds. Materials and Process Characterization of Ion Implantation. Austin, TX: Ion Beam Press, 1997.
10. B.J. Kirby, L.A. Larson, R.Y. Liang. Thermal-wave measurements of ion implanted silicon. Nucl. Instr. Meth. Phys. Res. B55(1–4): 550, 1987.
11. S.F. Corcoran, J. Hunter, A. Budrevich. Future Fab. Int. 5:337–339, 1998.
12. K. Wittmaack, S.F. Corcoran. J. Vac. Sci. Tech. B 16(1):272–279, 1998.
13. C.W. Magee, D. Jacobsen, H.J. Gossman. J. Vac. Sci. Tech. B 18(1):489–492, 2000.
14. R.J. Hillard, S.M. Ramey, C.W. Ye. J. Vac. Sci. Tech. B 18(1):389–392, 2000.
15. T. Clarysse et al. J. Vac. Sci. Tech. B 14(1):358–368, 1996.
16. USJ 1995 Conference Proceedings. J. Vac. Sci. Tech. B 14(1):191–464, 1996.
17. USJ 1997 Conference Proceedings. J. Vac. Sci. Tech. B 16(1):259–480, 1998.
18. USJ 1999 Conference Proceedings. J. Vac. Sci. Tech. B 18(1):338–605, 2000.
19. T.E. Tiwald, A.D. Miller, J.A. Woollam. In: D.G. Seiler, A.C. Diebold, W.M. Bullis, T.J. Shaffer, R.M. McDonald, E.J. Walters, eds. Characterization and Metrology for ULSI Technology. New York: American Institute of Physics, 1998, pp 221–225.
20. P. Boher, J.L. Stehle, J.P. Piel. In: D.G. Seiler et al., eds. Characterization and Metrology for ULSI Technology. New York: American Institute of Physics, 1998, pp 315–319.
21. P. Borden et al. J. Vac. Sci. Tech. B 18(1):602–604, 2000.

# 4

# MOS Device Characterization

**Eric M. Vogel**
*National Institute of Standards and Technology, Gaithersburg, Maryland*

**Veena Misra**
*North Carolina State University, Raleigh, North Carolina*

## I. INTRODUCTION

This chapter provides a survey and discussion of the electrical characterization techniques commonly used to determine the material, performance, and reliability properties of state-of-the-art metal-oxide-semiconductor (MOS) devices used in silicon semiconductor technology. The chapter is not intended as a comprehensive review of all electrical characterization techniques. Instead, the intent is to describe the electrical characterization techniques that are most common and directly relevant to fabricating and integrating high-performance, state-of-the-art MOS devices.

The remainder of the chapter has been broken into three sections. Section II describes MOS capacitance–voltage characterization ($C$-$V$), the most widely used device electrical characterization method for determining and monitoring such parameters as oxide thickness, substrate doping, and oxide charge. Section III describes the characterization of the parameters and properties related directly to the performance of state-of-the-art MOSFETs, including threshold voltage, drive current, and effective channel length. Section IV covers the characterization of the reliability of MOS devices under uniform voltage and hot-carrier stress. This chapter is written with the assumption that the reader has a basic knowledge of MOS device physics.

## II. MOS CAPACITANCE–VOLTAGE CHARACTERIZATION

Historically, capacitance–voltage measurements of the MOS capacitor have been used in both the monitoring and studying of the main parameters determining device operation.

**59**

In this section, the main experimental and theoretical considerations of MOS capacitance–voltage characterization are described. First, experimental considerations for measurements and test structures are reviewed. Second, a brief review of the relevant MOS theory is given. Finally, the main methods for extracting MOS device properties are reviewed.

## A.  Measurement and Test Structure Considerations

There have been a variety of techniques historically used to measure impedance of semiconductor device structures (1,2). The most popular method for measuring capacitance at frequencies from approximately 100 Hz to several MHz uses the inductance–capacitance–resistance ($LCR$) meter (2). An $LCR$ meter uses a small ac signal superimposed over a dc bias to measure the capacitance, conductance, or inductance. The accuracy of measuring capacitance depends strongly on instrument manufacturer but generally decreases with decreasing frequencies (2,3). The ac voltage used for measurements should generally be as small as possible while still allowing accurate measurements. Niccolian and Brews (1) suggest that the small signal range is the range over which capacitance and conductance are independent of the ac signal amplitude. This range may depend on the dc bias condition. Other researchers use an ac voltage signal less than or on the order of the thermal voltage ($\sim 25\,\text{mV}$) (4). Assuming that the device capacitance is properly extracted from the measured capacitance, both parallel and series equivalent circuit modes may be used, since the one can be derived from the other as (2)

$$C_{\text{series}} = C_{\text{parallel}}(1 + D^2) \tag{1}$$

$$D = \frac{1}{2\pi f R_{\text{parallel}} C_{\text{parallel}}} \tag{2}$$

There are several techniques used to measure the low-frequency capacitance. A lock-in amplifier can be used to measure the response of the device to a low-frequency ac input signal. However, these measurements are not used very often, because they require noise filters with long response times. The other methods used to measure low-frequency capacitance are those based on measuring either the displacement current or charge associated with a linear voltage ramp or voltage steps. The earliest techniques involved the measurement of the displacement gate current resulting from a pure voltage ramp. However, this technique has issues associated with maintaining equilibrium (1) and sensitivity to noise spikes. The most popular technique for measuring low-frequency capacitance is a quasi-static technique based on measuring the displacement charge associated with a small voltage step (1,5). This technique does not require calculating a derivative to determine capacitance and has greater noise immunity, because the charge measured is related to only one voltage step. The quasi-static techniques cannot be used for measurements of thin oxides in which the dc leakage current is much greater than the displacement current (3).

Most $LCR$ and quasi-static meters use coaxial connectors and have built-in correction functions to account for stray admittance and residual impedance. Cabling should be kept as short as possible, and meter corrections commonly available on $LCR$ meters should be used to improve measurement accuracy. The device to be measured should be in a lightproof, electrically shielded probe station with ambient control.

The capacitance–voltage ($C$-$V$) technique is arguably the most popular technique for determining many of the important parameters associated with a MOS capacitor or field effect transistor (FET). Differences between the two structures will be highlighted in the discussion of theory and parameter extraction that follows. The MOS capacitor or FET may be fabricated as part of a test chip with standard technology (e.g., polysilicon gate, LOCOS). A MOS capacitor may also be formed by depositing metal dots on the oxide via a shadow mask or by using a drop of mercury in a mercury probe configuration. Ideally, measurements should be performed on a device with an isolated active region to limit surface leakage and to ensure a known device area. Especially when measuring small-area devices, edge and stray effects such as field oxide capacitance must be included in the analysis. The effect of bulk and contact resistances should be considered in the measurements of thick-oxide MOS devices, and these effects can be magnified when measuring thin oxides. This resistance results in a measured capacitance that is different from the true device capacitance. Given the series resistance ($R_s$), the device capacitance ($C_c$) and conductance ($G_c$) can be obtained from the measured capacitance ($C_m$) and conductance ($G_m$) as (3,6,7)

$$C_c = \frac{C_m}{(1 - G_m R_s)^2 + \omega^2 C_m^2 R_s^2} \tag{3}$$

$$G_c = \frac{\omega^2 C_m C_c R_s - G_m}{G_m R_s - 1} \tag{4}$$

Several methods outlining the determination of and correction for series resistance have been described in the literature (1,3,7,8).

## B. MOS Theory

A picture of a simple MOS capacitor is shown in Figure 1(a). A simplified circuit for expressing the capacitance of this MOS capacitor with a metal gate electrode is shown in Fig. 1(b), where $C_{\text{ox}}$ is the oxide capacitance and $C_{\text{semi}}$ is the capacitance of the semiconductor. As shown in Fig. 1(c), $C_{\text{semi}}$ includes the inversion layer capacitance ($C_{\text{inv}}$), the interface state capacitance ($C_{it}$), the depletion layer capacitance ($C_d$), and the accumulation layer capacitance ($C_{\text{acc}}$). Tunneling, series resistance, and generation-recombination terms are neglected in this circuit. Figure 1(d) shows the equivalent circuit for the case of a polysilicon gate electrode. For this case, the polysilicon gate adds an additional capacitance ($C_{\text{gate}}$) in series with the $C_{\text{ox}}$ and $C_{\text{semi}}$. Determining the oxide capacitance accurately requires knowledge or assumptions for both the gate electrode capacitance and the semiconductor bulk capacitance.

For completeness, a comprehensive equivalent circuit for the measured capacitance of a MOS structure (1,7,9–11) is shown in Fig. 1(e). Tunnel and leakage conductance from the gate to conduction band ($G_{tc}$) and gate to valence band ($G_{tv}$) is in parallel with $C_{\text{ox}}$ and $C_{\text{gate}}$. Series resistance is included for both the top ($R_{\text{top}}$) and bottom ($R_{\text{bot}}$) contacts and the depletion layer ($R_{nb}$). Interface trap resistances are included for majority carriers ($R_{pt}$) and minority carriers ($R_{nt}$). Recombination resistances are included for majority carriers ($R_{pd}$) and minority carriers ($R_{nd}$).

The following will first describe the situation shown in Fig. 1(b), where the gate electrode capacitance, the tunneling conductance, and the series resistance are negligible and the semiconductor substrate energy bands are continuous in energy. This is generally the case for MOS devices with thicker oxides, where tunneling and leakage is small and electric fields are low. The impact of thin oxides, polysilicon depletion, tunneling/leakage current, and high electric fields will then be described. For the sake of discussion, a p-type substrate and n-type gate electrode are assumed. The capacitance and resistance values in

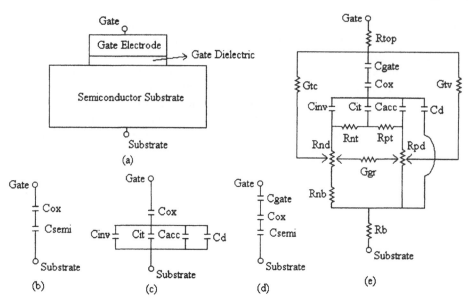

**Figure 1** Picture and equivalent circuits for a p-type MOS capacitor. (a) Picture of a simple MOS capacitor. (b) Simplified circuit for a MOS capacitor with metal gate and negligible interface states, tunneling, series resistance, and generation-recombination. (c) Simplified circuit for a MOS capacitor showing the capacitances associated with the semiconductor capacitance. (d) Simplified circuit for a MOS capacitor with a polysilicon gate electrode. (e) General circuit for a MOS capacitor in any bias condition, including tunneling, series resistance, interface states, and generation-recombination.

the following can be quantitatively related to the properties of an MOS capacitor as described in Refs. 1 and 11.

For negative gate voltages much less than the flat-band voltage, the semiconductor surface is heavily accumulated, so the accumulation capacitance ($C_{acc}$) is very high, thus dominating the other parallel capacitance. The total measured capacitance is therefore the gate dielectric capacitance ($C_{ox}$). For gate voltages between inversion and accumulation, the semiconductor is depleted and the total measured capacitance corresponds to the combination of $C_{ox}$ in series with the depletion capacitance ($C_d$), which in turn is in parallel with the interface state capacitance ($C_{it}$) and resistance ($R_{nt}$ and $R_{pt}$). A peak in the measured conductance is also present in depletion and is related to the interface state density (1,7,9,10,12,13). In inversion, if the signal frequency is low, the inversion layer capacitance ($C_{inv}$) dominates the other capacitance, so the total measured capacitance is again $C_{ox}$. If the ac frequency is high and the dc bias sweep rate is low, the inversion layer charge cannot follow the ac signal, and the semiconductor capacitance becomes the depletion capacitance associated with a pinned depletion layer width ($W_p$). Therefore, the high-frequency capacitance measured in inversion ($C_{min}$) is the combination of $C_{ox}$ in series with the depletion layer capacitance in inversion ($C_{inv}$). The difference between high and low frequency corresponds to whether the inversion layer charge can follow the applied test signal.

For a MOS capacitor, a low-frequency curve is usually generated using the quasi-static technique. With the device biased in strong inversion, a light is turned on to generate minority carriers. The voltage sweep is started (from inversion to accumulation) immediately after turning off the light. The quasi-static capacitance is then measured as a function of voltage. Another method to measure a low-frequency $C$-$V$ curve is using a MOSFET with the source-drain regions connected to the substrate. Since the source-drain regions have an abundance of minority carriers, a higher-frequency small signal may be used to measure the low-frequency capacitance. One difference between these two methods is that a voltage shift is present in the quasi-static method as compared to the MOSFET method (14). The reason for this is that the minority carriers in the capacitor are generated at the depletion layer edge and must travel to the surface, thereby causing a dc potential drop.

There are many additional effects associated with thin oxides and high electric fields that were not accounted for in the preceding discussion (15–25). Our analysis has assumed that the capacitance with the polysilicon gate was negligible so that the potential drop across the gate is small as compared to the other voltage drops. As the insulator capacitance increases and electric fields become higher, this may no longer be the case. This polysilicon depletion effect results in a reduction of the capacitance from the ideal case, as shown in Figure 2. The effect of polysilicon depletion becomes stronger with decreasing doping density.

Another effect associated with high electric fields is quantization of energy levels in the substrate. When high electric fields are present in the substrate, the potential varies quickly with position, resulting in a potential well. The energy bands are no longer continuous, but instead split into subbands. This quantum mechanical effect results in a shifting of the inversion layer electrons below the surface and a change in the surface potential. The shifting of the carriers from the interface further reduces the measured capacitance from the ideal case as shown in Figure 2. One way to include this effect is to model the potential in the substrate as linear, resulting in the triangular potential well approximation (15,19,26). Analytical expressions can then be found for the extra band bending and the average distance of the carriers from the interface (15,22). Both the polysilicon depletion effect and the quantum mechanical effects can also be included by self-consistently solving the Poisson and Schroedinger equations throughout the structure (18,22,23).

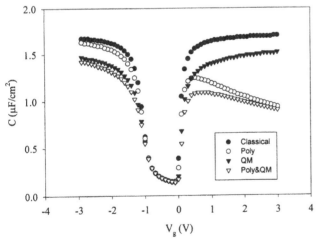

**Figure 2** Example of modeled capacitance–voltage curve illustrating the effects of energy level quantization (QM) in the substrate and polysilicon depletion (Poly) on a $C$-$V$ curve modeled assuming classical phenomena (Classical). In this simulation, the substrate doping density is $2 \times 10^{17}$ cm$^{-3}$, the oxide thickness is 2.0 nm, and the polysilicon doping density is $5 \times 10^{19}$ cm$^{-3}$.

The foregoing analysis has assumed that the tunneling and leakage current is negligible. As of yet, there have been no simple techniques developed to self-consistently calculate the effect of tunneling on potentials and carrier densities in the MOS system. However, if the current densities are small enough, the semiconductor–insulator interface will be in equilibrium with the semiconductor (10). Furthermore, the trap-assisted leakage and tunneling currents can be assumed to be in parallel with the interface state conductance (7,10). For very high tunnel currents, the interface will no longer be in equilibrium with the semiconductor (7,10).

## C. MOS Parameter Extraction

This section describes the methodologies used to extract properties such as oxide thickness, doping, flat-band voltage, oxide charge, work function, and interface state density from capacitance–voltage measurements of MOS structures (1,6,11). Unless otherwise stated, the following discussion assumes the ideal classical thick-oxide case, where effects such as polysilicon depletion and energy level quantization are not present and a uniform substrate doping profile is present. The most popular methods to date for including these effects, which are prominent for thin oxides and high electric fields, are computer models and simulations.

### 1. Oxide Thickness

Oxide thickness ($t_{ox}$) is determined from $C$-$V$ measurements by measuring the insulator capacitance and then relating this insulator capacitance to thickness. The relationship of insulator capacitance to thickness is simply

$$C_{ox} = \frac{\varepsilon_{ox}}{t_{ox}} A_g \tag{5}$$

where $\varepsilon_{ox}$ is the static dielectric constant of the insulator and $A_g$ is the device gate area. The dielectric constant for thick silicon dioxide is usually assumed to be $3.9\varepsilon_0$.

In reality, the determination of oxide thickness may not be this simple. A single homogeneous film with constant static dielectric constant is usually assumed (27). This may not be the case, for a variety of reasons, including a nonstoichiometric interfacial transition region, the presence of nitrogen, and the use of other dielectrics, paraelectrics, and ferroelectrics. Because of the possibility of an unknown dielectric constant, the thickness obtained from the gate dielectric capacitance ($C_{tot}$), assuming a dielectric constant of $3.9\varepsilon_0$ is often called *equivalent oxide thickness* (EOT). The gate dielectric capacitance in this definition is the true capacitance of the gate dielectric, including any corrections associated with quantum mechanical effects or polysilicon depletion. For a stacked gate dielectric made up of a thin oxide and a high-dielectric-constant (high-$\kappa$) dielectric, determining the thickness of the various layers directly from capacitance is impossible without knowledge of the dielectric constant for each layer. The EOT for a stacked dielectric comprising a layer of SiO$_2$ with thickness $T_{SiO_2}$, and a high-$\kappa$ layer with dielectric constant, $\kappa_{high-\kappa}$, and physical thickness, $T_{high-\kappa}$, is

$$\text{EOT} = \frac{3.9\varepsilon_0 A_g}{C_{tot}} = T_{SiO_2} + \frac{3.9}{\kappa_{high-\kappa}} T_{high-\kappa} \tag{6}$$

Figure 3 shows the EOT as a function of $T_{high-\kappa}$ for various values of $T_{SiO_2}$ and $\kappa_{high}$. As $\kappa_{high-\kappa}$ is increased or $T_{high-\kappa}$ is reduced, the EOT becomes increasingly dominated by the

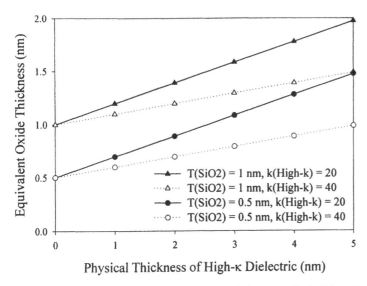

**Figure 3** Equivalent oxide thickness (EOT) for a stacked dielectric composed of a layer of SiO₂ and a layer of a high dielectric constant dielectric.

material having the lower dielectric constant (SiO₂ in this case). For a given EOT, there are any number of possible combinations for the physical thickness and dielectric constant of the layers. To determine the physical thickness or dielectric constant of any one layer requires additional metrology.

In the simplest classical approximation, the capacitance in strong accumulation ($C_{acc}$) or strong inversion at low frequency ($C_{inv}$) is assumed equal to the oxide capacitance. Because of errors associated with this approximation, the thickness obtained in this way is often called *capacitance equivalent thickness* (CET). This simple model does not account for a variety of effects, including the finite semiconductor capacitance associated with the semiconductor surface charge, the quantization of energy levels in the substrate, and the capacitance associated with the depletion of a polysilicon gate. Based on several physical arguments, some researchers suggest that quantization of energy levels in the substrate does not occur (28).

There are many analytical and extrapolation techniques that attempt to account for these effects (16,17,19,25,28–31). A comparison of some of these extrapolation algorithms is given in Ref. 28. The most often-cited analytical formulation for quantum effects is the model of van Dort (19). However, it is known that such formulations give results different from those of more rigorous Schroedinger-Poisson solvers (15–17). The method suggested by Maserjian et al. uses an extrapolation where the intercept is proportional to the gate dielectric capacitance (31). Maserjian's technique accounts for quantum effects; however, some error exists in accounting for polysilicon depletion (28). Furthermore, the extrapolation can give very different results depending on the range of bias used and the extrapolation assumed. The method developed by Riccò et al. uses higher-order derivatives of the $C$-$V$ curve at flat band to obtain insulator capacitance (29). Riccò et al. suggest that quantum effects, degeneracy, and effects of impurity ionization can be neglected by determining the oxide thickness at flat band. However, Walstra and Sah suggest that Fermi–Dirac degeneracy and impurity deionization may still be important at the flat-band voltage (28). Algorithms developed by McNutt and Sah (30) and Walstra and Sah (28) include effects such as Fermi–Dirac statistics, impurity deionization, and polysilicon depletion but neglect quantum mechanical effects.

## 2. Substrate Doping

There are two main analytical methods used to determine substrate doping information from $C$-$V$ curves. The first uses a high-frequency deep-depletion capacitance–voltage curve to obtain a depth profile of the substrate doping (6,32). The deep depletion can be obtained by using a high ramp rate for the voltage or by using pulsed gate voltages. Minority carrier generation, interface traps, and deep traps complicate the analysis. This method actually determines the majority carrier concentration and assumes that this is equal to the substrate doping concentration. This is a reasonable assumption for a slowly spatially varying dopant profile but becomes increasingly inaccurate with rapidly varying profiles. The Debye length ($L_D$) determines the deviation of the majority carrier profile from the dopant concentration and the spatial resolution of the profile. Assuming that the majority carrier concentration equals the doping concentration, and neglecting the effects of minority carrier generation and interface traps, the doping profile can be determined from the following:

$$N_A(W) \approx \frac{2}{q\varepsilon_{Si}A_g^2[d(1/C^2)/dV]} \tag{7}$$

$$W \approx \varepsilon_{Si}A_g\left(\frac{1}{C} - \frac{1}{C_{ox}}\right) \tag{8}$$

where $W$ is the depletion layer width at which the doping concentration is determined and $\varepsilon_S$ is the dielectric constant of the semiconductor.

Also extensively used for determining average substrate doping is the maximum–minimum capacitance technique (6,33). This simple technique is sufficient for uniformly doped substrates but gives only an average doping concentration for nonuniform doping concentrations. The average doping concentration is determined by transcendentally solving for $N_a$ in the following equation:

$$N_A \approx \frac{4(kT/q)\ln(N_A/n_i)}{q\varepsilon_{Si}A_g^2}\left(\frac{C_{min}}{1 - (C_{min}/C_{ox})}\right) \tag{9}$$

## 3. Flat-Band Voltage, Metal-Semiconductor Work Function, and Oxide Charge

The flat-band voltage ($V_{fb}$) is often determined from $C$-$V$ characterization of MOS devices because of its relationship to the metal-semiconductor work function difference ($\phi_{ms}$) and the charges present in the oxide (6,34). The classical definition for flat-band voltage, assuming a uniformly doped substrate, is

$$V_{fb} = \phi_{ms} - \frac{1}{C_{ox}}\sum_i \gamma_i Q_{ox,i}(\phi_s) \tag{10}$$

$$\gamma_i = \frac{\int_0^{t_{ox}} (x/t_{ox})\rho_{ox,i}(x)\,dx}{\int_0^{t_{ox}} \rho_{ox,i}(x)\,dx} \tag{11}$$

where $Q_{ox,i}$ is a type of charge in the oxide-per-unit area, $\gamma_i$ is the centroid associated with charge of type $i$ ($\gamma_i = 0$ when all of the charge is at the oxide–gate interface, and $\gamma_i = 1$ when the charge is at the oxide–semiconductor interface), $\phi_s$ is the surface potential and is introduced to account for charge such as interface-trapped charge whose occupancy depends on surface potential, and $\rho_{ox,i}(x)$ is the charge per unit volume. For the case of fixed oxide charge ($Q_f$) with centroid at the semiconductor–oxide interface and whose occupancy does not vary with surface potential, Eqs. (9) and (10) can be reduced to

$$V_{fb} = \phi_{ms} - \frac{Q_f}{C_{ox}} \tag{12}$$

A widely used method to determine the flat-band voltage uses the flat-band capacitance ($C_{fb}$) which can be approximated as

$$C_{fb} \approx \frac{\varepsilon_{si}}{L_D} \tag{13}$$

$$L_d \approx \sqrt{\frac{kT\varepsilon_{si}}{q^2 p_0}} \tag{14}$$

where $L_D$ is the extrinsic Debye length and $p_0$ is the equilibrium density of holes in the semiconductor. This definition of flat-band capacitance assumes that the doping profile is uniform, and neglects effects of interface traps (1,6).

Due to its relationship to oxide charge, the flat-band voltage is often used as a measure of oxide quality. Oxide charge is generally classifed into the four types: fixed charge, mobile ionic charge, oxide traps, and interface traps (35). Methods have been developed to determine the density of each of these charges and to determine the metal-semiconductor work function difference (1,6).

## 4.  Interface States

There are two primary methods commonly used to determine interface-state density from $C$-$V$ (36,37). The first method uses a comparison of a quasi-static $C$-$V$ curve that contains interface-trap capacitance with one free of interface-trap effects (37). The curve that is free of interface-trap effects can be determined theoretically but is usually determined from a high-frequency $C$-$V$ curve, where it is assumed that the interface states will not respond. The interface-state density ($D_{it}$) is determined as

$$D_{it} = \frac{1}{q}\left(\frac{C_{ox}C_{lf}}{C_{ox} - C_{lf}} - C_s\right) \tag{15}$$

where $C_{lf}$ is the measured low-frequency capacitance and $C_s$ is the semiconductor capacitance, which can be obtained either theoretically or from the high-frequency capacitance ($C_{hf}$) as

$$C_s = \frac{C_{ox}C_{hf}}{C_{ox} - C_{hf}} \tag{16}$$

This method assumes that $C_{hf}$ is measured at a high enough frequency that interface states do not respond, which is usually the case at 1 MHz for samples with reasonably low interface-state densities (1).

A second common method to determine interface state density is the Terman method (36). This method uses a high-frequency $C$-$V$ curve and the nonparallel shift or stretch-out associated with the changing interface-state occupancy with gate bias. A theoretical, ideal (no interface state) $C$-$V$ curve is used to find $\phi_s$ for a given $C_{hf}$. This $\phi_s$ is then mapped to the experimental curve for the same $C_{hf}$ to obtain a $\phi_s$ versus $V_g$ curve that contains the interface-state information. The interface-state density can then be determined by comparing the theoretical, ideal $\phi_s$ versus $V_g$ curve with the experimental $\phi_s$ versus $V_g$ by

$$D_{it} = \frac{C_{ox}}{q} \frac{d(\Delta V_g)}{d\phi_s} \tag{17}$$

where $\Delta V_g = V_g - V_g$ (ideal) is the voltage shift of the experimental curve from the ideal curve.

A variety of errors are present in both of these methods (1,6). If the theoretical $C_s$ is used, errors due to nonuniform doping and inaccurate calculations of band bending are an issue. Lateral nonuniformities in interface charge may distort $C$-$V$ curves and cause errors in extracted interface-state density. For larger substrate doping and thicker oxides, errors due to round-off of the measured capacitance become increasingly more significant. Finally, these techniques assume that the high-frequency $C$-$V$ curve is free of interface-trap response, which may not be the case for high interface-state densities. It should be mentioned here that there are a variety of less common and generally more complicated techniques that may be used to determine $D_{it}$. Some of these include conductance, the Gray–Brown Method, and deep-level transient spectroscopy (1,6).

## 5. Parameter Extraction Using Modeling and Simulation of *C-V*

Modeling and simulation of $C$-$V$ is becoming an increasingly popular method to extract parameters from experimental $C$-$V$ curves (15,23,38–42). Simulations can account for issues such as nonuniform doping profiles (38), polysilicon depletion (15,23,39,41,42), quantization of energy levels (15,23,39,41,42), and 2- or 3-dimensional effects that are present in state-of-the-art devices (41). The models and simulations range from fast analytical formulations (15) to rigorous self-consistent quantum simulations (23,39,42). At the time of this writing, we know of no 2-D rigorously self-consistent quantum simulators. However, some work is beginning to move in this direction (39,40).   .

## III. MOSFET CHARACTERIZATION

The MOS field effect transistor is the building block of the current ultralarge-scale integrated circuits (ICs). The size and complexity of MOS ICs have been continually increasing over the years to enable faster and denser circuits. Device measurements, characterization, and model parameter extraction are typically performed at the wafer level on test structures using probe stations to meet the requirements for large amounts of data. Wafer probe stations can be either manually operated or semi- or fully automated. In a manual case, the probes and the measurements are done manually for each site. In the semiautomated case, the probe station can be programmed with wafer stepping patterns and is controlled by a computer. Fully automated probe stations have the additional capability of automatic wafer loading.

There are many important parameters that are typically measured during transistor characterization; these will be discussed in the following section. These parameters can be classified as (a) technological parameters or (b) electrical parameters. The technological parameters are controlled by the design-and-fabrication process and include gate oxide thickness, channel doping, drawn channel length, and drawn channel width. The electrical parameters are determined by the electrical behavior of the device and include effective channel length and width, source/drain series resistance, threshold voltages, mobility, and short-channel effects. Both technological and electrical parameters can be obtained by measuring the MOSFET drain current data in linear and saturation regimes. In the linear regime, both weak and strong inversion are used. Channel doping, interface-trap density, and drain-induced barrier lowering (DIBL) are obtained from weak inversion data. Threshold voltage, low-field mobility, source/drain series resistance, channel doping, effective channel length and width are obtained from strong inversion data. In the saturation regime, other parameters can be extracted, such as the saturation drain voltage and saturation velocity.

Transistors with varying channel lengths and widths are generally available on a test chip. Typically the long and wide transistors are used as references to avoid short-channel effects, narrow-width effects, and series-resistance effects. However, large devices can suffer from various problems, such as gate leakage currents.

The following describes the measurement and extraction of technological and electrical parameters from transistor characteristics. It is assumed in the following discussion that the reader is familiar with the physics of MOSFET operation (43,44).

## A. Measurement Considerations

MOSFET drain current characteristics are typically measured using commercially available source measure units (SMU), such as model HP 4155 from Hewlett Packard. These units are both voltage sources for biasing the device terminals and ammeters for measuring the current flow. The standard specifications for a SMU include 100-μV to 100-V voltage sourcing and 10-fA to 100-mA current measurements. These units can generally be used in static mode or in sweep mode. The current readings can be taken using several integration times and averaging options. The connection leads can either be coaxial or triaxial in order to minimize the stray leakage currents between the output high and output low lead connections. The SMU can also provide a guard shield that surrounds the output high lead connection. Most of today's measurement equipment can be controlled by a computer via an IEEE bus connection.

## B. Linear Drain Current

The linear drain current is obtained with the gate voltage typically at $V_{DD}$ and the drain voltage at a low voltage (0.1 V) to ensure linear-region operation. The linear current can be used to extract parameters such as the effective channel length, the source/drain parasitic resistance, and the carrier mobility. It should be noted that scaled MOSFETs with ultrathin oxides with large gate leakages can corrupt the drain current measurement MOSFETs. This can be resolved by using smaller-length devices where the drain current is significantly larger than the gate current.

## C. Threshold Voltage

The linear threshold voltage is one of the most important parameters in a MOSFET. It signifies the turn-on point and separates the subthreshold region from the strong inversion region. Experimentally, the threshold voltage is determined by measuring the drain current for various values of gate voltage in the linear regime ( i.e., at low $V_{ds}$ values). There are three different ways in which $I_{ds}$–$V_{gs}$ data can be used to calculate $V_T$.

1. *Constant-Current Method*   In the first method, $V_{gs}$ at low drain voltage ($< 0.1$ V) and at a specified drain current $I_{ds} = I_T$ is taken to be the threshold voltage. Typically,

$$I_{ds} = I_T \frac{W}{L} = 10^{-7} \frac{W}{L} \tag{18}$$

is used. A typical measurement is shown in Figure 4a. Since only one drain voltage curve is required, this method is very fast and is often used for the purpose of process monitoring or to calculate $V_T$ from 2-D device simulators. This technique can be implemented easily through the use of an op-amp or by digital means (45).

2. *Linear Extrapolation Method*   In this method, threshold voltage is defined as the gate voltage obtained by extrapolating the linear portion of the $I_{ds}$–$V_{gs}$ curve, from maximum slope to zero drain current (46). Transconductance, $g_m$, is defined as the slope of the $I_{ds}$–$V_{gs}$ curve, $\partial I_{ds}/\partial V_{gs}$, and the point of maximum slope is where $g_m$ is maximum. This threshold voltage is often called the *extrapolated threshold voltage*. The drain current of an ideal MOSFET in the linear region is given by the following:

$$I_D = \frac{\mu C_{ox} W}{L} \left( V_{gs} - V_T - \frac{V_{ds}}{2} \right) V_{ds} \tag{19}$$

where $\mu$ is the minority carrier mobility. This equation will result in a straight-line plot of $I_{ds}$–$V_{gs}$ with an x-intercept of $V_{TE} = V_T + V_{ds}/2$. Threshold voltage, $V_T$, can then be calculated from

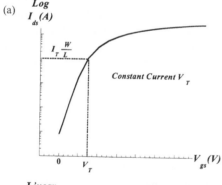

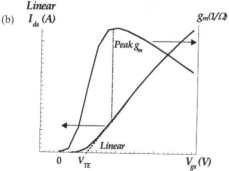

**Figure 4**   Typical $I_{ds}$–$V_{gs}$ curves used for the measurement of threshold voltage in the linear regime. (a) Constant-current method. (b) Peak $g_m$ method.

$$V_T = V_{TE} - \frac{V_{ds}}{2} \tag{20}$$

Equation (20) is valid for negligible series resistance, which is generally true at low drain currents (47). The foregoing linear extrapolation measurement technique is graphically shown in Fig. 4b. The reason for the extrapolation from the maximum slope is that $I_{ds}$ does not vary linearly with $V_{ds}$, because carrier mobility begins to fall as $V_{gs}$ increases. Therefore, extrapolation from the maximum-slope portion of the curve ensures that the degraded mobility is not included in $V_T$.

3. *Saturation Threshold Voltage*  Saturation threshold voltage can also be determined by extrapolating a $\sqrt{I_{ds}}$–$V_{gs}$ curve to $I_{ds} = 0$ at high drain voltage (45). The drain current of an ideal MOSFET at high drain voltage (i.e., in the saturation region) is given by

$$I_D = \frac{\mu C_{ox} W}{2L} (V_{gs} - V_T)^2 \tag{21}$$

It is clear that a plot of $\sqrt{I_{ds}}$–$V_{gs}$ gives a straight line that crosses the $V_{gs}$ axis at $V_T$.

The preceding three methods produce $V_T$ values that are not equivalent. Although the constant-current method is the simplest way to measure $V_T$ with reasonable accuracy, the $V_T$ obtained depends upon the chosen $I_{ds}$ value and on the subthreshold slope. Since the devices used in circuit design are normally operated in strong inversion, the two extrapolation techniques are more appropriate. For long-channel devices, the difference between linear and saturation $V_T$ is small; however, for short-channel devices, the difference can be quite large, owing to short-channel effects, as will be discussed. Also, the linear extrapolation technique is susceptible to source/drain series resistance and mobility degradation, both of which reduce transconductance, leading to a smaller slope of the linear extrapolated line which in turn leads to a lower extrapolated threshold voltage.

## D.  Subthreshold Swing

The drain current below threshold voltage, called the subthreshold current, varies exponentially with $V_{gs}$. The reciprocal of the slope of the $\log(I_{ds})$ vs. $V_{gs}$ characteristic is defined as the subthreshold swing, $S$, and is one of the most critical performance figures of MOSFETs in logic applications. It is highly desirable to have $S$ as small as possible, since this is the parameter that determines the amount of voltage swing necessary to switch a MOSFET from its "off" state to its "on" state. This is especially important for modern MOSFETs, with supply voltages approaching 1.0 V.

The drain current of a MOSFET operating below threshold voltage can be written as (48)

$$I_D = I_{D1} \exp\left(\frac{q(V_{gs} - V_T)}{nkT}\right)\left(1 - \exp\left(\frac{-qV_{ds}}{kT}\right)\right) \tag{22}$$

where $I_{D1}$ is a constant that depends on temperature, device dimensions, and substrate doping density and $n$ is given by

$$n = 1 + \frac{C_d + C_{it}}{C_{ox}} \tag{23}$$

where $C_d$, $C_{it}$, and $C_{ox}$ are the depletion, interface-trap, and oxide capacitance, respectively. In this expression, $n$ represents the charge placed on the gate that does not result in inversion charge. This charge originates from either the depletion charge or the interface charge. The parameter $n$ increases as doping density increases and as interface-trap density increases. To determine $S$, $\log(I_{ds})$ is plotted as a function of $V_{ds} > kT/q$, as shown in Figure 5. The slope of this plot is $q/[\ln(10)nkT]$. The inverse of this slope ($S$) is the amount of gate voltage necessary to change the drain current by one decade, given by

$$S = \frac{(\ln(10)nkT)}{q} \approx 60n\left(\frac{T}{300}\right)mV/decade \tag{24}$$

The lower limit of the subthreshold current is determined by the leakage current that flows from source to drain when the transistor is "off." This current, which is referred to as the *off-state leakage*, is determined by the source-to-channel potential barrier as well as the leakage currents of the source and drain junctions. For this reason, it is critical to be able to form low-leakage junctions.

## E.  Effective Mobility

The carrier effective mobility is an important parameter for device characterization and optimization. The Si–SiO$_2$ interface states and physical roughness have a strong effect on the mobility (49). Ideally, long-channel-length MOSFETs are used to minimize the parasitic series resistance and to reduce error in channel-length dimensions. However, MOSFETs with ultrathin gate oxides have larger gate leakage that affects the mobility measurements. Since the ratio of the drain to gate current increases with the square of the inverse of the channel length, short-channel-length devices can be used to measure mobility. However, accurate determination of the channel length and source/drain series resistance are issues that must be taken into account.

In order to determine carrier mobility, $I_{ds}$ is measured in the linear region; i.e., $V_{gs} > V_T$, and at low $V_{ds}$. In the linear region, the current at low $V_{ds}$ can be approximated as

$$I_D = \frac{\mu C_{ox} W}{L}\left(V_{gs} - V_T\right)V_{ds} \tag{25}$$

Differentiating this with respect to $V_{gs}$ and ignoring the $V_{gs}$ dependence of $\mu$ gives the transconductance, $g_m$, as

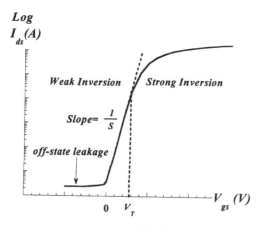

**Figure 5**   Typical $\log I_{ds}$–$V_{gs}$ curves for measuring subthreshold slope in the linear regime.

$$g_m = \frac{\mu C_{ox} W}{L} V_{ds} \tag{26}$$

The carrier mobility determined from the transconductance, called *field effect mobility*, is defined as

$$\mu_{FE} = \frac{g_m}{C_{ox} V_{ds}} \frac{L}{W} \tag{27}$$

As mentioned earlier, the field effect mobility is obtained using transconductance that does not include the dependence of $V_{gs}$ on $\mu$. A more appropriate value for mobility is obtained using the dc conductance method. The conductance, $g_{ds}$, is defined as $\Delta I_{ds}/\Delta V_{ds}$. The effective mobility, $\mu_{eff}$, is related to the conductance at constant $V_{gs}$ by

$$\mu_{eff} = \frac{g_{ds}}{C_{ox}(V_{gs} - V_T)} \frac{L}{W} \tag{28}$$

and is obtained by differentiating the linear-region current. At each $V_{gs}$, a $g_{ds}$ is calculated and is used to obtain $\mu_{eff}$. It should be noted that this method results in a significant drop of $\mu_{eff}$ near $V_{gs} = V_T$. Reasons for this include: (a) the expression $C_{ox}(V_{gs} - V_T)$ is only an approximation for the inversion charge and results in considerable error near $V_T$, and (b) $V_T$ is not accurately known. A direct measure of the inversion charge can provide more accuracy in extracting the effective mobility. This can be achieved by capacitance measurements where the gate-to-channel capacitance ($C_{gc}$) is measured and related to the inversion charge ($Q_n$) by

$$Q_n = \int_{-\infty}^{V_{gs}} C_{gc} \, dV_{gs} \tag{29}$$

This measurement is called the split *C-V* method and details can be found in the literature (50,51).

## F.   Short-Channel Effects

MOSFETs are continually downscaled for higher packing density, higher device speed, and lower power consumption. When physical dimensions of MOSFETs are reduced, the foregoing equations for drain current have to be modified to account for the so-called short-channel effects. One major short-channel effect deals with the reduction of the threshold voltage as the channel length is reduced. In long-channel devices, the influence of source and drain on the channel depletion layer is negligible. However, as channel lengths are reduced, overlapping source- and drain-depletion regions start having a large effect on the channel-depletion region. This causes the depletion region under the inversion layer to increase. The wider depletion region is accompanied by a larger surface potential, which makes the channel more attractive to electrons. Therefore, a smaller amount of charge on the gate is needed to reach the onset of strong inversion, and the threshold voltage decreases. This effect is worsened when there is a larger bias on the drain, since the depletion region becomes even wider. This phenomenon is called *drain-induced barrier lowering* (DIBL). The impact of DIBL on $V_T$ can be measured by sweeping $V_{gs}$ from below $V_T$ to above $V_T$ values at a fixed $V_{ds}$ (11,43,44). This is then repeated for $V_{ds} = V_{DD}$. The $I_{dso}$ value corresponding to the nominal threshold voltage is determined. The gate voltage interpolated from $I_{dso}$ at $V_{ds} = V_{DD}$ is obtained. The difference between the nominal $V_T$ and the $V_{gs}$ ($V_{DD}$) is defined as DIBL. It should be noted that this methodology is valid only as long as the subthreshold slope remains intact. A very large subthreshold swing implies that the device cannot be turned off. This phenomenon, called *punchthrough*, occurs roughly when the source- and drain-depletion regions meet. The effect of DIBL in $I_{ds}$ vs. $V_{gs}$ behavior is shown in Figure 6 for varying channel lengths. As shown, punchthrough occurs at very small channel lengths.

## G.  Series Resistance

The source and drain series resistance consists of the source/drain contact resistance, the sheet resistance of the source/drain, spreading resistance at the transition from the source diffusion to the channel, and any additional resistance associated with probes and wiring. The source/drain resistance and the effective channel length are frequently determined with one measurement technique, which is discussed in the next section. An accurate knowledge of both these parameters is required to avoid errors in the determination of other parameters, such as mobility.

## H.  Channel Length and Width

The channel length, $L$, and channel width, $W$, are the most basic parameters of a MOSFET and play a major role in the device characteristics (52). The channel length, $L$, differs from the drawn channel length, $L_m$, by a factor $\Delta L$ such that $L = L_m - \Delta L$. Similarly, the channel width, $W$, is smaller than the drawn channel width, $W_m$, by a factor $\Delta W$ such that $W = W_m - \Delta W$. There are two main methods of obtaining $\Delta L$ and $\Delta W$: (a) drain-current methods and (b) capacitance methods.

In methods involving the drain current, $I_{ds}$ is measured as a function of gate voltage, $V_{gs}$, at low drain voltages and zero substrate bias to ensure linear region operation. This technique is based on the fact that the channel resistance of a MOSFET in the linear regime is proportional to the channel length. The extraction of $\Delta L$ is based on the following drain-current equation, which is valid in the linear region:

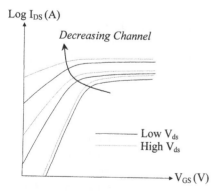

**Figure 6**   Effect of DIBL in $I_{ds}$ vs. $V_{gs}$ behavior varying channel lengths. Threshold voltage is measured at low and high $V_{ds}$ on various channel lengths. The change in threshold voltage provides a measure of DIBL.

$$I_{ds} = \frac{\mu_{\text{eff}} C_{\text{ox}} W}{L} \left[ (V'_{gs} - V_T - 0.5 V'_{ds}) V'_{ds} \right] \tag{30}$$

$$V_{gs} = V'_{gs} + I_{ds} R_S \tag{31}$$

$$V_{ds} = V'_{ds} + I_{ds} (R_S + R_D) \tag{32}$$

where $R_S$ and $R_D$ are the source/drain resistances and $V'_{gs}$ and $V'_{ds}$ are the voltage drops across the device. Using these equations, and given that in strong inversion $(V_{gs} - V_T)$ is large compared to $0.5 V_{ds}$, we get

$$I_{ds} = \frac{\mu_{\text{eff}} C_{\text{ox}} W}{L} (V_{gs} - V_T)(V_{ds} - I_{ds} R_{SD}) \tag{33}$$

$$R_{SD} = R_S + R_D \tag{34}$$

This can be rewritten in the following form:

$$I_{ds} = \frac{\mu_{\text{eff}} C_{\text{ox}} W (V_{gs} - V_T) V_{ds}}{(L_m - \Delta L) + W \mu_{\text{eff}} C_{\text{ox}} (V_{gs} - V_T) R_{SD}} \tag{35}$$

This equation is the basis for many techniques used to determine $R_{SD}$, $L$, $W$, and $\mu_{\text{eff}}$ and contains both the intrinsic channel resistance and the source/drain resistance terms. The total measured resistance of a MOSFET will include the channel resistance and the source/drain resistance and can be written as

$$R_m = \frac{V_{ds}}{I_{ds}} = R_{ch} + R_{SD} \tag{36}$$

Rearranging Eq. (36) we get

$$R_m = \frac{L}{\mu C_{\text{ox}} W (V_{gs} - V_T)} + R_{SD} \tag{37}$$

where

$$R_{ch} \approx \frac{L}{\mu C_{\text{ox}} W (V_{gs} - V_T)} \tag{38}$$

The following sections provide a brief description of some of these methods that use the foregoing analysis to extract $L$ and $R_{SD}$.

## 1. Channel Resistance Method

The most common method of extraction is called the channel resistance method, where $\Delta L$ is extracted by measuring the device channel resistance as a function of the gate voltage, $V_{gs}$, or gate drive, $V_{gs} - V_T$ (53,54). The channel resistance $R_{ch}$ of a MOSFET operating in the linear region as given by Eq. (38) can be further modified as

$$R_m = A(L_m - \Delta L) + R_{SD} \tag{39}$$

where

$$A = \frac{1}{\mu C_{\text{ox}} W (V_{gs} - V_T)} \tag{40}$$

and is the channel resistance per unit length.

To accurately determine $\Delta L$ it is important that (a) $R_{SD}$ be independent of the external bias and (b) $A$ be independent of channel length. Under these conditions, at a given $V_{gs}$, $R_m$ vs. $L_m$ should produce a straight line that has the following form:

$$R_m = A \cdot L_m + B \tag{41}$$

where the intercept $B$ is

$$B = R_{SD} - A \cdot \Delta L \tag{42}$$

Repeating this procedure for different gate voltages will ideally result in a family of straight-line curves that all intersect at one point, giving $R_{SD}$ on the $R_m$ axis and $\Delta L$ on the $L_m$ axis, as shown in Figure 7. However, in most real devices $R_m$ vs. $L_m$ lines do not intersect at a common point, in which case the foregoing procedure is typically modified. A second linear regression of the plot of $B$ vs. $A$ is obtained from different gate voltages. The slope and intercept of the $B$ vs. $A$ plot give $\Delta L$ and $R_{SD}$, respectively (55). To avoid narrow-width effects in determining channel length, wide transistors should be used. In addition, since $V_T$ is channel length dependent due to DIBL, a higher gate voltage is used to minimize the short-channel effects on $A$. A more accurate methodology is to adjust $V_{gs}$ such that the drivability term, $V_{gs} - V_T$, is equal for all transistors. It has been found that the $V_T$ determined by the constant-current method produces the most consistent results for $\Delta L$ and $R_{SD}$ extraction. This method is the one most commonly used for determining $\Delta L$ and has the ancillary benefit of providing the source/drain resistance. The method requires more than two devices with varying channel lengths. The extraction of $\Delta W$ can be performed similarly to the $\Delta L$ extraction.

### 2. Sucio–Johnston Method

Another representation of Eq. (41) has been adapted by Sucio and Johnston and has the following form (56):

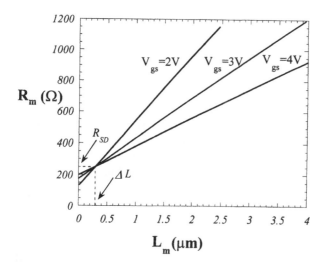

**Figure 7**   Measured output resistance, $R_m$, vs. mask length, $L_m$, as a function of gate voltage. Lines with different gate voltages intersect at a point, providing $\Delta L$ and $R_{SD}$.

$$E = R_m(V_{gs} - V_T) = \frac{(L_m - \Delta L)}{W \mu_{\text{eff}} C_{\text{ox}}} + (V_{gs} - V_T)R_{SD} \qquad (43)$$

where several forms of $\mu_{\text{eff}}$ can be used. Using the most common form of $\mu_{\text{eff}}$,

$$\mu_{\text{eff}} = \frac{\mu_0}{1 + \theta(V_{gs} - V_T)} \qquad (44)$$

where $\mu_0$ is the carrier mobility at zero transverse field, the expression for $E$ becomes

$$E = \frac{(L_m - \Delta L)[1 + \theta(V_{gs} - V_T)]}{W \mu_0 C_{\text{ox}}} + (V_{gs} - V_T)R_{SD} \qquad (45)$$

This equation can then be plotted vs. $(V_{gs} - V_T)$ and the slope, $m$, and the intercept, $E_i$, are used to obtain the relevant parameters. If this plot is created for varying channel lengths, then $m$ and $E_i$ will vary. A plot of the $m$ and $E_i$ vs. $L_m$ gives $R_{SD}$ and $\Delta L$ values. This method is shown in Figures 8 and 9.

### 3. De La Moneda Method

A method closely related to the Sucio–Johnston method is the De La Moneda method (57). The total resistance using the expression for $\mu_{\text{eff}}$ can be written as

$$R_m = \frac{L_m - \Delta L}{\mu_0 C_{\text{ox}} W(V_{gs} - V_T)} + \frac{\theta(L_m - L)}{\mu_0 C_{\text{ox}} W} + R_{SD} \qquad (46)$$

$R_m$ is plotted against $1/(V_{gs} - V_T)$. The slope, $m$, and the intercept, $R_{mi}$, are used to obtain the relevant parameters. If this plot is created for varying channel lengths, then $m$ and $R_{mi}$ will vary. A plot of $m$ and $R_{mi}$ vs. $L_m$ results in values for $\mu_0$, $\theta$, $\Delta L$, and $R_{SD}$. This method is displayed in Figures 10, 11, and 12.

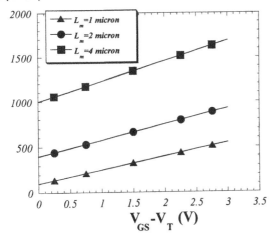

**Figure 8** Sucio–Johnston method of obtaining channel length. The quantity $E$ is plotted against $(V_{gs} - V_t)$ as a function of gate length.

$E_i$ (V-Ω), m (Ω)

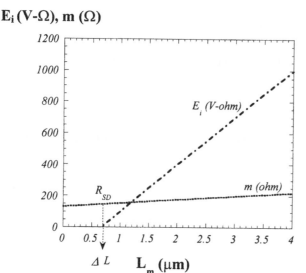

**Figure 9** Sucio–Johnston method of obtaining channel length. The y-intercept and the slope of each line in Figure 8 are plotted against mask length to obtain $\Delta L$ and $R_{SD}$.

### 4. Shift and Ratio Method

This method is also based on the channel resistance concept. In this technique, the mobility can be any function of the gate voltage and $R_{SD}$ (58). It uses a long device as a reference to calculate $L$ and $R_{SD}$ for a short device. Here, the total resistance is written as

$$R_m = R_{SD} + Lf(V_{gs} - V_T) \tag{47}$$

where $f$ is a general function of the gate overdrive. If this equation is differentiated with respect to $V_{gs}$, neglecting the weak dependence of $R_{SD}$ on $V_{gs}$, the following is obtained:

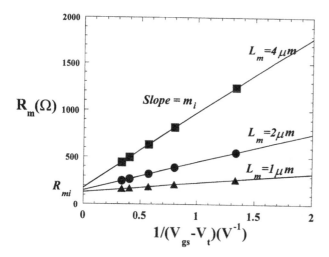

**Figure 10** De La Moneda method of obtaining channel length. The measured resistance, $R_m$, is plotted against $1/(V_{gs} - V_t)$.

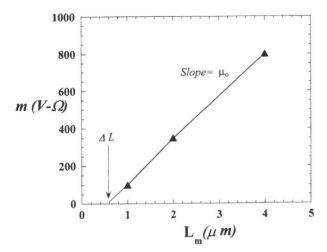

**Figure 11** De La Moneda method of obtaining channel length. The slope of each line in Figure 10 is measured and plotted against the mask length. The slope of this line provides the low-field mobility factor $\mu_0$, and x-intercept of this line results in $\Delta L$.

$$S = \frac{dR_m}{dV_{gs}} = L\frac{df(V_{gs} - V_T)}{dV_{gs}} \tag{48}$$

By using the derivative, $R_{SD}$ is eliminated from Eq. (47). The $S$ function is then plotted vs.. $V_{gs}$ for the large reference and the small-channel device. A typical $S$ vs. $V_{gs}$ curve is shown in Figure 13. The $S$ curve for the short-channel device is shifted horizontally by a varying amount $\delta$ to match with the $S$ curve for the long-channel device. The ratio $r$, given by

$$r = \frac{S(V_{gs})}{S(V_{gs} - \delta)} \tag{49}$$

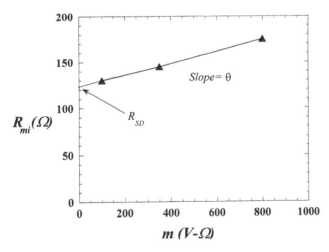

**Figure 12** De La Moneda method of obtaining channel length. The intercepts, $R_{mi}$, are plotted against the slopes, $m_i$. The slope of this curve provides the mobility degradation factor, $\theta$, and the y-intercept provides $R_{SD}$.

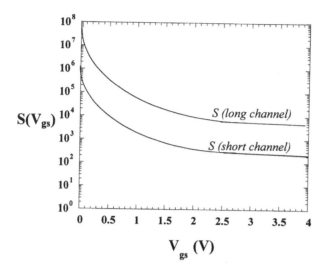

**Figure 13** Shift-and-ratio method of obtaining channel lengths. Typical $S$ vs. $V_{gs}$ curves obtained using the shift-and-ratio method. The $S$ curve for the short-channel device is shifted horizontally by a varying amount $\delta$ to match with the $S$ curve for the long-channel device.

is calculated for the two devices. The goal is to find the value of $\delta$ that results in a constant $r$ independent of the gate voltage. As shown in Figure 14, if $\delta$ is too small, then $r$ is a decreasing function of $V_{gs}$. On the other hand if $\delta$ is too large, then $r$ is an increasing function of $V_{gs}$. When $\delta$ is equal to the $V_T$ difference between the two devices, then $r$ is nearly constant. With constant overdrive, the mobility is identical and $r$ becomes

$$r = \frac{S(V_{gs})}{S(V_{gs} - \delta)} = \frac{L_o}{L} \qquad (50)$$

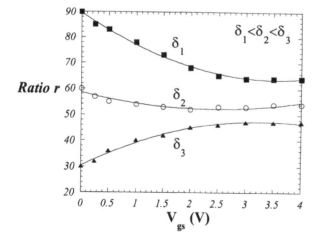

**Figure 14** Shift-and-ratio method of obtaining channel lengths. The effect of $\delta$ on the ratio of the $S$ (long channel) and $S$ (shifted short channel) curves is shown. When $\delta$ is equal to the $V_T$ difference between the two devices, then $r$ is nearly constant.

where $L_o$ and $L$ are the channel lengths of the large and small devices, respectively. Plotting the obtained $L$ vs. $L_m$ for several devices gives a line with intercept $\Delta L$ on the $L_m$ axis. One practical problem with the shift-and-ratio technique is the very small shift in current–voltage data required to accurately match the gate overdrive. In general, an empirical fit to the data is obtained to generate an equation that can be easily differentiated.

## J.  Breakdown of Junctions

Breakdown voltage of the junctions in the MOSFET is also an important device parameter, since a low breakdown voltage may indicate potential processing or device operation problems. Under certain conditions, such as a large applied drain voltage, device defects, or very small channel lengths, the electric field near the drain becomes very large. Grounding the gate electrode enhances this effect. If carriers near the drain attain sufficient energy, they can create electron-hole pairs that increase the drain current. As the drain voltage is increased further, the carriers become hotter and can cause an avalanche effect, at which point the drain current increases dramatically. The drain-source breakdown with gate shorted to source is called $BV_{DSS}$. For very short-channel devices, punch-through, rather than impact ionization, dominates the drain current. $BV_{DSS}$ is typically measured at a given drain current ($\approx 5\,\text{Na/\mu m}$).

## K.  Substrate Current

The substrate current in an n-channel MOSFET results from hole generation from impact ionization caused by electrons traveling from source to drain. This current is a good measure of the hot-electron activity. Assuming impact ionization occurs uniformly in the pinch-off region, the substrate current, $I_{bs}$, can be written

$$I_{bs} = I_{ds}\alpha\,\Delta L_{\text{pinchoff}} \tag{51}$$

where $\alpha$ is the ionization coefficient and $\Delta L_{\text{pinchoff}}$ is the length of the pinch-off region. With increasing $V_{gs}$, $I_{bs}$ increases, reaches a maximum value, and then decreases. The initial increase of $I_{bs}$ is due to the increase of $I_{ds}$ with $V_{gs}$. However as $V_{gs}$ is increased further, the lateral $(V_{ds} - V_{dsat})/L$ decreases, causing a reduction in $\alpha$. Thus, the peak substrate current occurs when the two competing factors cancel out and usually occurs at $V_{gs} \approx 0.5\,V_{ds}$.

## L.  Charge Pumping

Charge pumping has evolved into a reliable and sensitive method for measuring the interface-state density of small MOSFETs. In this technique, which was originally proposed in 1969 (59), the source and drain are tied together and slightly reverse-biased with respect to the substrate with a voltage $V_R$. A square wave is applied to the gate having sufficient amplitude so that the device can be alternately driven into inversion or accumulation. When an n-MOSFET is biased into inversion, the interface traps, which are continuously distributed throughout the bandgap, are filled with electrons. When the gate voltage changes from positive to negative potential, electrons in the inversion layer drift to both source and drain. In addition, electrons that were captured by interface traps near the conduction band are thermally emitted into the Si-conduction band and also drift to the source and drain. Those electrons that reside in interface traps deeper within the bandgap do not have sufficient time to be emitted and remain captured on interface traps. Once the hole barrier is reduced, holes flow to the surface, where some are captured by those interface traps that are still filled with electrons. Finally, most traps are filled with holes. When the gate returns to its positive value, the inverse process starts and electrons flow into the interface to be captured by traps. The time constant for electron emission from interface traps is given by

$$\tau_e = \frac{e^{E/kt}}{\sigma_n v_{\text{th}} N_C} \tag{52}$$

where $E$ is the interface trap energy measured from the bottom of the conduction band, with $E_C$ being the reference energy, $\sigma_n$ is the trap cross section, $v_{\text{th}}$ is the thermal velocity, and $N_C$ is the density of states in the conduction band of silicon. For a square wave of frequency $f$, the time available for electron emission is half the period, $\tau_e = 1/2f$. The energy interval over which electrons are emitted is obtained from Eq. (52):

$$E = kT \ln\left(\frac{\sigma_n v_{\text{th}} N_C}{2f}\right) \tag{53}$$

The hole-capture-time constant is

$$\tau_c = \frac{1}{\sigma_p v_{\text{th}} p_s} \tag{54}$$

where $p_s$ is the hole concentration at the surface and $\tau_c$ is very small for any appreciable hole concentration. Therefore, electron emission, not hole capture, is the rate-limiting process. During the reverse cycle, when the surface changes from accumulation to inversion, the opposite process occurs. Holes within an energy interval

$$E - E_v = kT \ln\left(\frac{\sigma_p v_{\text{th}} N_V}{2f}\right) \tag{55}$$

are emitted into the valence band, and the remainder recombine with electrons flowing in from the source and the drain. Those electrons on interface traps within the energy level $\Delta E$,

$$\Delta E = E_g - kT\left[\ln\left(\frac{\sigma_n v_{\text{th}} N_C}{2f}\right) + \ln\left(\frac{\sigma_p v_{\text{th}} N_V}{2f}\right)\right] \tag{56}$$

recombine with holes. Therefore, $Q_n/q$ electrons/cm$^2$ flow into the inversion layer from source/drain, but only $(Q_n/q - D_{it}\,\Delta E)$ electrons/cm$^2$ flow back. This difference $(D_{it}\,\Delta E$ electrons/cm$^2)$ recombine with holes. For each electron-hole pair recombination event, an electron and a hole must be supplied. Hence $D_{it}\,\Delta E$ holes/cm$^2$ also recombine. Hence, more holes flow into the semiconductor than leave, giving rise to the charge-pumping current $I_{cp}$. This current is dependent on the gate area $(A)$ and frequency $(f)$. The total charge pump current is given by (60)

$$I_{cp} = Af[qD_{it}\,\Delta E + \alpha C_{oc}(V_G - V_T)] \tag{57}$$

where $\alpha$ is the fraction of the inversion charge that does not drift back to the source and the drain. The basic charge-pumping technique gives an average value of $D_{it}$ over the energy interval $\Delta E$. A typical $I_{cp}$ vs. $V_{gb}$ curve is shown in Figure 15. Starting from the pioneering work of Brugler and Jespers, various methods have been employed to precisely characterize the energy distribution of interface states (60–62).

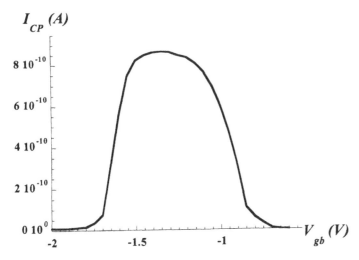

**Figure 15** Typical charge-pumping current versus gate pulse base level for an n-channel short-channel MOSFET.

The accuracy of charge pumping is limited by uncertainties in the effective area contributing to the charge-pumping current. In order to use charge pumping to characterize the interface of sub-2-nm-oxide MOSFETs, the effect of the large tunneling current must be considered. The use of a high-frequency trapezoidal gate pulse with the high pulse level just above threshold and the low pulse level slightly below flat band may provide reasonable measurements (63). The minimum frequency that can be used increases with either an increase in the low pulse level or a decrease in the oxide thickness.

## M.  MOSFET Capacitance

Both the junction capacitance and the overlap capacitance are important in modeling speed performance of a MOSFET. The junction capacitance is caused by the space charge in the junction depletion region and can be measured under reverse bias conditions using an $LCR$ meter. Both area- and field-intensive junction regions are typically measured. The measurement setup is shown in Figure 16. The gate overlap capacitance is typically measured by connecting the gate of a MOSFET to the high terminal of the $LCR$ meter. The source and drain are tied together and are connected to the low terminal of the $LCR$ meter (64). The substrate is connected to the ground terminal of the $LCR$ meter to eliminate the effects of drain-substrate and source-substrate capacitances on the measurement. This configuration is called the split-capacitance measurement method. The gate-to-source overlap region capacitance, $C_{gs}$, and the gate-to-drain overlap region capacitance, $C_{gd}$, are typically equivalent, and their sum is called the overlap capacitance, $C_{ov}$. The measured capacitance, $C_{gc}$, in accumulation is simply the overlap capacitance, since the channel is decoupled from the source and the drain. In inversion, the measured capacitance is the sum of the gate-channel capacitance plus the overlap capacitances. Since the overlap regions are very small for typical transistors, $C_{ov}$ is very small. This requires several transistors to be connected in parallel in order to obtain accurate measurements. The behavior of a typical measured capacitance, $C_{gc}$, is shown in Figure 17. The following equations can be obtained for $C_{inv}$ and $C_{acc}$:

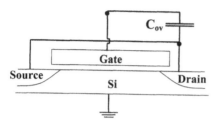

**Figure 16** Schematic diagram of making split-$CV$ measurements. The source and drain are shorted together, and the capacitance is measured between the gate and the source/drain.

$$C_{\text{inv}} = C_{\text{ox}} L_m W \qquad (58)$$

$$C_{\text{acc}} = C_{\text{ov}} = C_{\text{ox}} \Delta L W \qquad (59)$$

Rearranging Eqs. (58) and (59) results in

$$L = L_m \left( 1 - \frac{C_{\text{ov}}}{C_{\text{inv}}} \right) \qquad (60)$$

Equation (60) demonstrates another technique for obtaining the effective channel length (65). This measurement can be performed on a single device, or $C_{\text{inv}} - C_{\text{ov}}$ can be plotted against $L_m$. This plot produces a straight line with slope $= C_{\text{ox}} W$ and intercept $= \Delta L$. Advantages of this technique include the need to measure only a single device and the fact that the associated $\Delta L$ value is independent of the source and drain series resistances.

## IV. RELIABILITY CHARACTERIZATION

The gate dielectric is the most critical component affecting the reliable operation of the MOS device. The gate dielectric is susceptible to defect creation under either high fields or current injection. The following discussion of reliability characterization is based on MOS

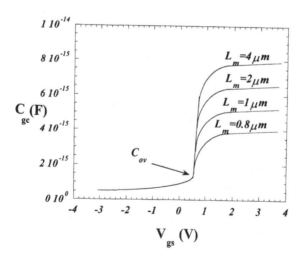

**Figure 17** Typical split-capacitance, $C_{gc}$ vs. $V_{gs}$, curves for a MOSFET for varying channel lengths.

devices having silicon dioxide as the gate dielectric. The discussion may or may not have application to devices with other gate dielectrics. The reliability of MOS devices has generally been classified into the two categories of constant-voltage reliability and hot-carrier reliability. Therefore, our discussion of reliability characterization will be separated into these two categories.

## A. Constant-Voltage Reliability

Historically, oxide breakdown and reliability have been characterized by applying a voltage to the gate of either a MOS capacitor or transistor and grounding all other terminals. The ramped voltage breakdown test uses a linearly ramped voltage applied to a MOS capacitor until the oxide breaks down and the current drastically increases (66,67). The voltage at breakdown ($V_{bd}$) or electric field at breakdown ($E_{bd}$) is sometimes referred to as the dielectric strength of an oxide. However, these values are a strong function of conditions such as ramp rate and do not relate directly to the breakdown physics. Ramped voltage tests are usually used to quickly flag major reliability problems and generally not used to determine device lifetime. However, there have been some methodologies developed to relate ramped voltage tests to device lifetime (68,69).

### 1. Constant-Voltage Reliability Tests and Statistics

The tests used to determine device lifetime are those of charge- or time-to-breakdown at either a constant voltage or constant current. These tests are sometimes referred to as time-dependent dielectric breakdown (TDDB). In these tests a large, constant voltage (current) is applied to a device until breakdown occurs, in the form of a current increase (voltage decrease) marking the time-to-breakdown ($t_{bd}$). The charge-to-breakdown per unit area ($Q_{bd}$) is determined by integrating the gate current density up to the time-to-breakdown:

$$Q_{bd} = \int_0^{t_{bd}} J_g(t) \, dt \qquad (61)$$

It has been suggested that for thicker oxides (> 7 nm) biased in the Fowler–Nordheim tunneling regime, constant current stress is more appropriate than constant-voltage stress (70). For oxides less than 5 nm, the constant-voltage stress should be used (70,71). It is known that for thick oxides (>∼ 7 nm), the energy of the electron is determined by the electrode field, approximately independent of thickness. The constant-current stress ensures that both the current density and the electric field (electron energy) are the same for different thicknesses and processes. Therefore, a fair comparison can be made of the reliability of thick oxides having different processing conditions used (70,71). For thin oxides, the applied gate voltage (not the electric field) determines the energy of the electron. Therefore, the evaluation of different process conditions is meaningful only when breakdown distributions are measured at a constant gate voltage (70,71).

Sometimes, various device parameters are measured intermittently as the device is being stressed. The characteristics that are measured include threshold voltage, fixed charge, interface-state density, transconductance, and stress-induced leakage current (SILC). Measured defects can be used to determine two other quantities, the defect generation rate ($P_g$) and the number of defects at breakdown ($N_{bd}$) (72,73). $P_g$ is extracted by taking the slope of the linear portion of a defect measurement, such as SILC or interface-state density, as a function of charge injected. $N_{bd}$ is the measured-defect density immediately prior to breakdown. $P_g$ describes the physics of defect generation, whereas $N_{bd}$ describes the statistics of breakdown. Issues such as saturation effects must be considered when performing these measurements.

The breakdown parameters ($Q_{bd}$, $t_{bd}$, $V_{bd}$, $N_{bd}$) are usually taken on a large number of samples. Oxides from a given process do not all have the same breakdown characteristic, but instead have a distribution of values. These breakdown parameters are usually described in terms of the cumulative distribution function $F(t)$ ($F(Q)$), which is the probability that a device will fail at a time before time $t$ (charge injected before charge $Q$). The two main cumulative distribution functions that have been used to describe oxide reliability are the Weibull and the log-normal (67). The Weibull distribution function is most commonly used and is described for $Q_{bd}$ as

$$F(Q_{bd}) = 1 - e^{-(Q_{bd}/\eta)^\beta} \tag{62}$$

where $\eta$ is the modal value of the distribution and $\beta$ is the shape factor (66,71,74). Plotting $\ln(-\ln(1-F))$ versus $\ln(Q_{bd})$, $\eta$ corresponds to the $Q_{bd}$ value where $\ln(-\ln(1-F))$ equals 0 and the $\beta$ value represents the slope. Using Weibull failure distributions and Poisson random statistics, the area dependence of breakdown is given as (71,75)

$$\ln(-\ln(1-F_2)) - \ln(-\ln(1-F_1)) = \ln\left(\frac{A_2}{A_2}\right) \tag{63}$$

$$\frac{\eta_1}{\eta_2} = \left(\frac{A_2}{A_1}\right)^{1/\beta} \tag{64}$$

Analyses of data using various statistical functions have indicated that the types of breakdown or oxide failures fall into three groups (67). The first group of oxide failures (A-mode) occurs instantly upon application of a small stress. These failures are generally due to gross defects such as pinholes in the oxide, which cause a short of the dielectric. The second group of oxide failures (B-mode) occurs under intermediate stresses, do not instantly short, but cause early failures in integrated circuits. These failures are believed to be due to weak spots or defects in the oxide. A-mode and B-mode are many times termed *extrinsic* failures. The final group of oxide failures (C-mode) is considered to be due to intrinsic properties of silicon oxide. These failures are believed to occur in defect-free oxides, and these oxides can withstand the highest stressing conditions. As shown in Figure 18, the B-mode failures typically show a lower Weibull slope than the C-mode failures. Burn-in is commonly used to screen out extrinsic failures (67). B-mode and C-mode failures are many times quantitatively modeled to determine if the circuit will have a long enough lifetime (usually considered to be ten years) under normal operating conditions (72,76,77).

## 2. Physical Models of Constant-Voltage Reliability

This section provides a short qualitative discussion of various physical models proposed to explain breakdown. The catastrophic breakdown of the gate dielectric is generally believed to be a two-stage process. First, defects are generated and the dielectric is degraded over time due to the passing of current or high electric fields. The second stage is a more rapid runaway process, where a final acceleration of damage leads to the formation of a conductive path through the insulator. The breakdown process itself will first be described, followed by a brief description of the models and phenomena associated with the generation of defects.

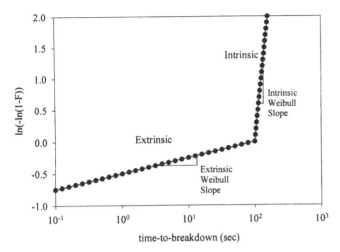

**Figure 18** Typical Weibull plot of modeled time-to-breakdown, illustrating the extrinsic and intrinsic portions of the distribution.

Independent of the assumed defect generation mechanism, most researchers agree that the final breakdown event is triggered by the buildup of a critical number of defects (75,78). The number of defects required to initiate the shorting of the dielectric ($N_{bd}$) has been shown to be a statistical process that becomes more random and area sensitive with decreasing dielectric thickness. Percolation theory has been used to successfully describe the number of defects needed for breakdown as a function of oxide thickness (72,75,78,79). Many researchers believe that the final breakdown event is related to either the power or the energy associated with the device under test and the measuring circuit (80). The final runaway and breakdown process has been shown to be highly localized (74). For thin oxides, it has been shown that the current following breakdown is not as large as for thicker oxides (80). At the time of the writing of this chapter, it has yet to be determined whether this soft breakdown event will result in device or circuit failure.

Models for uniform voltage reliability can generally be broken into *intrinsic* (related to dielectric) and *extrinsic* (related to other process-induced effects and contamination). There have been numerous studies attempting to determine the physical mechanisms of intrinsic degradation and breakdown, including trap creation models, hole-induced degradation models, and electric field–dependent models. The correct physical model describing intrinsic oxide degradation is still a matter of intense debate. The following will briefly describe the main models.

Some researchers suggest that anode hole injection dominates defect generation (76,77,81). The anode hole injection model suggests that an electron travels to the anode, where it promotes a valence-band electron via impact ionization. This hot hole that is created tunnels back into the oxide, creating damage. The model suggests that this injected hole fluence increases with time until a critical hole fluence is reached, marking the breakdown event. The classical anode hole injection model suggests that the logarithm of $t_{bd}$ should be inversely and negatively proportional to the oxide electric field. However, recent modifications to the model predict a linear dependence with negative slope of the logarithm of breakdown on field.

DiMaria and coworkers have suggested that defect generation can be described by several energy regimes (72,73,82,83). Electrons having energy greater than approximately

9 eV *in the oxide* result in bandgap ionization that dominates degradation. Electrons having energy greater than approximately 8 eV *in the anode* generate energetic holes, which then create damage. Electrons having energy greater than approximately 5 eV *in the anode* create damage by the process of trap creation. The degradation process for electrons having energy in the anode less than this 5 eV is simply related to the sub-threshold exponential tail of the trap-creation mechanism. The process of trap creation has been suggested to be due to hydrogen release or to the direct damage of the oxide lattice by the injected electrons. The logarithm of the charge-to-breakdown is predicted to show a linear proportionality with negative slope to the voltage in this regime.

The thermochemical E model describes defect generation as resulting from the interaction of electric field with intrinsic defects in the oxide (84–86). However, this model cannot describe a variety of experimental observations including the dependence of reliability on p- and n-type anodes (87). Combinations of the foregoing models have also been suggested.

The various models for intrinsic oxide degradation and breakdown have different physical origins. However, each of these models predicts, at least for oxide thicknesses less than approximately 5 nm and gate voltages less than approximately 5 V, that the logarithm of the breakdown time or charge is linearly proportional to the applied gate voltage or electric field. This is perhaps the most important observation concerning the characterization of the intrinsic lifetime of an oxide. As a reference, Figure 19 shows experimental data for $Q_{bd}$ for various oxide thicknesses (70,72,88–90) and Figure 20 shows experimental and modeled data for Weibull slope ($\beta$) as a function of oxide thickness (71,78,89,91,92). The $Q_{bd}$ data in Figure 19 was transformed to an area of $5 \times 10^{-4}\,\mathrm{cm^2}$ using a $\beta$ value appropriate for each thickness.

Extrinsic failures can result from a variety of defects (67). These defects include metal contaminants such as aluminum and iron, the presence of sodium, defects in the

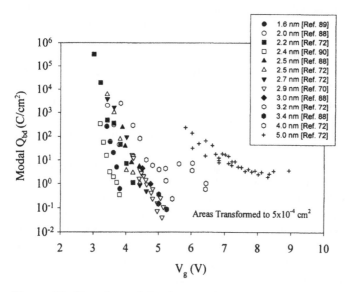

**Figure 19** Experimental $Q_{bd}$ data as a function of gate voltage for oxides thinner than 5.0 nm. The experimental data was transformed to an area of $5 \times 10^{-4}\,\mathrm{cm^2}$ using a $\beta$ value appropriate for each thickness. (Data from Refs. 70, 72, 88–90.)

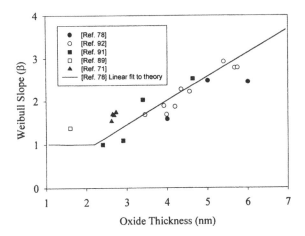

**Figure 20**  Experimental and calculated Weibull slope (β) as a function of oxide thickness. (Data from Refs. 71, 78, 89, 91, 92.)

crystalline substrate, and damage due to etching or implant. Plasma processing, such as that associated with reactive ion etching, can result in both increased extrinsic failures and reduction in intrinsic oxide lifetime. The mechanisms associated with plasma damage include charge buildup and tunneling caused by unbalanced ion currents and damage resulting from the presence of ultraviolet radiation. Failures from these processes can be characterized using both edge-intensive and area-intensive large-area capacitors and antenna structures.

## 3.  Constant-Voltage Stress Lifetime Characterization

The following will first describe intrinsic lifetime characterization, followed by methodologies for including extrinsic failures. The intrinsic lifetime characterization description assumes no extrinsic failures.

Because of the variety of physical models assumed to describe oxide degradation and breakdown, there have been a variety of techniques used to characterize intrinsic oxide lifetime for an integrated circuit (67,72,76,77). It is usually assumed that a chip will fail if one device on the chip fails, since it is unknown which device it will be. Therefore, the reliability of millions of devices in parallel is usually considered as one large device with some active chip area. A unit of failure used to describe reliability of an entire system of integrated circuits is the *failure unit* (FIT). A FIT is defined as 1 unit failure per billion hours of device operation. One must make the assumption that the test structure used in breakdown measurements will have the same lifetime as the device used in the circuit. As device dimensions are reduced, this assumption will become increasingly invalid. To measure millions of chips at operating conditions would be too costly and take too long. Therefore, reliability measurements are taken on a fixed number of smaller-area devices at accelerated stress conditions. These results are then extrapolated to large area, small cumulative failure percentage, and operating conditions. An example of a lifetime criterion definition is 100-ppm failure rate in 10 years at operating conditions for 0.1-cm$^2$ gate area. The following describes the methodology generally used.

First, breakdown measurements are taken on a statistically valid number of samples at a variety of accelerated voltage and temperature conditions. Often times an elevated temperature is used to allow for stressing at lower voltages. The modal or median, time- or charge-to-breakdown is then extrapolated to operating voltage and operating temperature. Extrapolating to operating voltage requires an assumption of the physical model or of the correct extrapolation law, as described in the previous section. Some researchers use the measured-defect generation rate as a function of voltage and assume that the critical defect density for breakdown is independent of voltage, to extend the experimental data to lower-voltage conditions. Extrapolating over temperature requires knowledge of the activation energy or temperature acceleration. This temperature acceleration may be non-Arrhenius and may depend on stress voltage. If the charge-to-breakdown is extrapolated, the time-to-breakdown must then be extracted using the experimental gate-current density. The statistics of both the charge-to-breakdown and the gate-current density must be considered to accurately extract chip lifetime. The operating condition modal breakdown obtained on the test structure area is extrapolated to the total active area of interest using Eq. (64). This value is then extrapolated to the cumulative failure rate of interest using the assumed statistical distribution. It is assumed for the final two extrapolations that the slope of the statistical distribution does not depend on the stress condition. This value of the operating condition breakdown time for the total active area and cumulative failure rate of interest is then compared to the required lifetime.

The preceding methodology assumes that all of the failures were intrinsic and could be explained using one cumulative distribution function. However, the lifetime of a circuit must be sufficient and characterized for both intrinsic and extrinsic failures. The most widely used method to include extrinsic failures is the oxide thinning model (69). This model assumes that the B-mode extrinsic failures can be included by assuming that this defective oxide of nominal thickness has reliability behavior of an effective thickness less than the nominal thickness. The effective oxide thickness may represent real oxide thinning, surface roughness, defects, or a reduced tunneling barrier height. The method, which is described in detail in Ref. 69, involves mapping each measured breakdown time/charge to an effective thickness associated with intrinsic breakdown.

Another technique used to handle extrinsic failures is burn-in or screening tests. The burn-in procedure is to operate devices at specified conditions so that extrinsic parts that would produce early fails actually fail during burn-in. This is preferred so that parts will not fail after going to the customer. Stress conditions are chosen that accelerate the failure of the extrinsic parts while minimizing the effect of reliability on the remaining parts (67). Methods have been developed for determining these optimal conditions and for including the effect of burn-in on the reliability of the remaining parts (69,93).

## B. Hot-Carrier Reliability

### 1. Phenomenon of Hot-Carrier Degradation

As the MOSFET device dimensions are reduced, the electric fields found in the device become increasingly high, resulting in reliability problems. Specifically, the larger electric fields found near the drain result in impact ionization and the creation of hot carriers. In an n-channel MOSFET, most of the generated electrons enter the drain and most of the holes enter the substrate, resulting in a measured substrate current (94–96). Some of these hot carriers, however, can be injected into the gate oxide, resulting in an increase of oxide charge and interface-state density, which can then cause threshold voltage instability and current drive degradation (94–96).

It has previously been observed on n-channel MOSFETs that hot-carrier injection, and the subsequent oxide degradation, has a strong dependency on biasing conditions (96). The change in charge-pumping current ($\Delta I_{cp,\max} \propto \Delta D_{it}$), change in threshold voltage ($\Delta V_t$), and change in peak transconductance ($\Delta g_m$) for n-channel MOSFETs with oxide gate dielectrics usually have a specific bias dependence. For low gate biases ($V_g \approx V_t$), hot hole injection is generally observed. This results in positive charge generation in the bulk of the oxide above the drain. Interface states are also generated, but this influence is masked by effective extending of the drain into the channel due to the positive charge. At high gate bias ($V_g \approx V_d$) it has been observed that electrons are injected into the oxide, resulting in a measured gate current. The dominant degradation mechanism for these gate biases was found to be electron trapping near the drain. For intermediate gate biases ($V_g \approx V_d/2$), both electrons and holes were observed to be injected. This condition approximately results in the maximum substrate current condition. It was observed that the main oxide degradation mechanism is interface-trap generation. These interface traps resulted in a positive shift of threshold voltage, suggesting that the charge associated with them was negative. The generation of these defects has been suggested to be due either to the breaking of a silicon–hydrogen bond by energetic electrons or to a two-step process in which an electron recombines with a previously trapped hole.

For p-channel MOSFETs with oxide gate dielectrics, it was observed that similar mechanisms are in operation (96). However, since holes are cooler than electrons at the same electric field, the gate bias dependencies of degradation were essentially shifted to higher electric fields. For low and intermediate gate biases ($V_g \approx V_t$ and $V_g \approx V_d/2$), hot electron injection is of primary importance. At these low gate biases, the electron gate current for p-channel MOSFETs is higher than the hole gate current for n-channel MOSFETs, due to the lower oxide barrier for electrons as compared to holes. Therefore, electron trapping in the bulk of the oxide is the dominant degradation mechanism in p-channel MOSFETs. Interface states are also generated, but their influence is masked by effective extending of the drain into the channel, due to the negative charge. For higher gate voltages ($V_g \approx V_d$), both electrons and holes are injected. Therefore, generation of interface traps occurs in p-channel MOSFETs.

The foregoing provided a general description of the phenomenon associated with hot-carrier stress as a background for describing stress tests and lifetime characterization. However, these phenomena depend strongly on processing conditions and device structure (67). For example, it has been observed that the interface endurance of oxynitrides is better than that of oxide gate dielectrics. Lightly doped drain (LDD) structures can be used to reduce the number of hot carriers and to tailor the position of injected carriers.

## 2. Hot-Carrier Reliability Tests and Lifetime Characterization

Characterizing MOSFET hot-carrier reliability involves applying stress bias conditions and intermittently monitoring device parameters (95,97). The parameters most often monitored include drive current, transconductance, threshold voltage, subthreshold swing, and interface-state density. The failure criterion for the device is defined in terms of the maximum acceptable change in one of these parameters. The device lifetime is defined as the amount of stress time to cause the defined change in a device parameter (67,98). The definition-of-failure criterion depends on the application and may be related to a defined acceptable degradation in circuit performance.

Stress conditions are usually chosen to represent the worse-case conditions. As described earlier, this usually involves stressing at the peak-substrate-current condition.

To accelerate failure time, the drain voltage is increased beyond the nominal design power supply voltage. The gate voltage at each drain voltage is chosen for the maximum-substrate-current condition. The lifetime is measured at several higher values of drain voltage and then extrapolated to operating conditions (95,97). This extrapolation is performed either as a function of the substrate current or as the inverse of the drain bias. This can be done because the logarithm of the substrate current is approximately linearly dependent on the inverse of the drain bias.

## ACKNOWLEDGMENTS

We would like to thank George Brown and Dave Berning for critical reviews of the manuscript.

## REFERENCES

1.  E.H. Nicollian, J.R. Brews. MOS (Metal Oxide Semiconductor) Physics and Technology. New York: Wiley-Interscience, 1982.
2.  Operation Manual for Model HP4284A Precision LCR Meter: Hewlett-Packard, 1988.
3.  W.K. Henson, K.Z. Ahmed, E M. Vogel, J.R. Hauser, J.J. Wortman, R. Datta, M. Xu, D. Venables. Estimating Oxide Thickness of Tunnel Oxides Down to 1.4 nm Using Conventional Capacitance–Voltage Measurements on MOS Capacitors. IEEE Elec. Dev. Lett. 20:179, 1999.
4.  G. Brown. Personal communication, 2000.
5.  Model 595 Quasistatic CV Meter Instruction Manual: Keithley, 1986.
6.  D.K. Schroder. Semiconductor Material and Device Characterization. 2nd ed. New York: Wiley, 1998.
7.  E.M. Vogel, W.K. Henson, C.A. Richter, J.S. Suehle. Limitations of conductance to the measurement of the interface state density of MOS capacitors with tunneling gate dielectrics. IEEE Trans. Elec. Dev. 47, 2000.
8.  K.J. Yang, C. Hu. MOS capacitance measurements for high-leakage thin dielectrics. IEEE Trans. Elec. Dev. 46:1500, 1999.
9.  T.P. Ma, R.C. Barker. Surface-state spectra from thick-oxide MOS tunnel junctions. Sol. St. Elecs. 17:913, 1974.
10. S. Kar, W.E. Dahlke. Interface states in MOS structures with 20–40-Å-thick $SiO_2$ films on nondegenerate Si. Sol. St. Elecs. 15:221–237, 1972.
11. S.M. Sze. Physics of Semiconductor Device. New York: Wiley, 1981.
12. E.H. Nicollian, A. Goetzberger, A.D. Lopez. Expedient method of obtaining interface state properties from MIS conductance measurements. Sol. St. Elecs. 12:937, 1969.
13. S.C. Witczak, J.S. Suehle, M. Gaitan. An experimental comparison of measurement techniques to extract Si-SiO2 interface trap density. Sol. St. Elecs. 35:345, 1992.
14. J.R. Hauser. Bias sweep rate effects on quasi-static capacitance of MOS capacitors. IEEE Trans. Elec. Dev. 44:1009, 1997.
15. J.R. Hauser, K.Z. Ahmed. Characterization of ultra-thin oxides using electrical $C$-$V$ and $I$-$V$ measurements. In: Characterization and Metrology for ULSI Technology. Gaithersburg, MD: American Institute of Physics, 1998.
16. S.A. Hareland, S. Krishnamurthy, S. Jallepalli, C.-F. Yeap, K. Hasnat, A.F. Tasch, C.M. Maziar. A computationally efficient model for inversion layer quantization effects in deep submicron N-channel MOSFETs. IEEE Trans. Elec. Dev. 43:90, 1996.
17. S.A. Hareland, S. Jallepalli, W.-K. Shih, H. Wang, G.L. Chindalore, A.F. Tasch, C.M. Maziar. A physically based model for quantization effects in hole inversion layers. IEEE Trans. Elec. Dev. 45:179, 1998.

18. T. Janik, B. Majkusiak. Analysis of the MOS transistor based on the self-consistent solution to the Schrodinger and Poisson equations and on the local mobility model. IEEE Trans. Elec. Dev. 45:1263, 1998.

19. M.J. v. Dort. A simple model for quantization effects in heavily doped silicon MOSFETs at inversion conditions. Sol. St. Elecs. 37:435, 1994.

20. R. Rios, N. Arora. Determination of ultra-thin gate oxide thicknesses for CMOS structures using quantum effects. 1994 IEDM Tech. Digest, p. 613.

21. F. Rana, S. Tiwari, D.A. Buchanan. Self-consistent modeling of accumulation layers and tunneling currents through very thin oxides. Appl. Phys. Lett. 69:1104–1106, 1996.

22. S.H. Lo, D.A. Buchanan, Y. Taur, W. Wang. Quantum-mechanical modeling of electron tunneling current from the inversion layer of ultra-thin-oxide nMOSFET's. IEEE Elec. Dev. Lett. 18:209, 1997.

23. W.-K. Shih, S. Jallepalli, G. Chindalore, S. Hareland, C.M. Maziar, A.F. Tasch. UTQUANT 2.0, Microelectronics Research Center, Department of Electrical and Computer Engineering, University of Texas at Austin, 1997.

24. K.S. Krisch, J.D. Bude, L. Manchanda. Gate capacitance attenuation in MOS devices with thin gate dielectrics. IEEE Elec. Dev. Lett. 17:521, 1996.

25. A. Schenk. Advanced Physical Models for Silicon Device Simulation. New York: Springer-Verlag, 1998.

26. T. Ando, A.B. Fowler, F. Stern. Electronic properties of two-dimensional systems. Reviews of Modern Physics. 54:437, 1982.

27. H. Reisinger, H. Oppolzer, W. Honlein. Thickness determination of thin $SiO_2$ on Silicon. Sol. St. Elecs. 35:797–803, 1992.

28. S.V. Walstra, C.-T. Sah. Thin oxide thickness extrapolation from capacitance–voltage measurements. IEEE Trans. Elec. Dev. 44:1136, 1997.

29. B. Riccò, P. Olivo, T.N. Nguyen, T.-S. Kuan, G. Ferriani. Oxide-thickness determination in thin-insulator MOS structures. IEEE Trans. Elec. Dev. 35:432, 1988.

30. M.J. McNutt, C.-T. Sah. Determination of the MOS oxide capacitance. J. Appl. Phys. 46:3909, 1975.

31. J. Maserjian, G. Petersson, C. Svensson. Saturation capacitance of thin oxide MOS structures and the effective surface density of states of silicon. Sol. St. Elecs. 17:335, 1974.

32. J. Hilibrand, R.D. Gold. Determination of the impurity distribution in junction diodes from capacitance–voltage measurements. RCA Rev. 21:245, 1960.

33. B.E. Deal, A.S. Grove, E.H. Snow, C.T. Sah. Observation of impurity redistribution during thermal oxidation of silicon using the MOS structure. J. Electrochem. Soc. 112:308, 1965.

34. R.F. Pierret. Field Effect Devices. 2nd ed. Reading, MA: Addison-Wesley, 1990.

35. B.E. Deal. Standardized terminology for oxide charges associated with thermally oxidized silicon. IEEE Trans. Elec. Dev. ED-27:606, 1980.

36. L.M. Terman. An investigation of surface states at a silicon/silicon oxide interface employing metal-oxide-silicon diodes. Sol. St. Elecs. 5:285, 1962.

37. C.N. Berglund. Surface states at steam-grown silicon–silicon dioxide interfaces. IEEE Trans. Elec. Dev. ED-13:701, 1966.

38. H.S. Bennet, M. Gaitan, P. Roitman, T.J. Russel, J.S. Suehle. Modeling of MOS capacitors to extract Si-SiO$_2$ interface trap densities in the presence of arbitrary doping profiles. IEEE Trans. Elec. Dev. ED-33:759, 1986.

39. R. Lake, G. Klimeck, R.C. Bowen, D. Jovanovic. Single and multiband modeling of quantum electron transport through layered semiconductor devices. J. Appl. Phys. 81:7845, 1997.

40. S.E. Laux, M.V. Fischetti. Issues in modeling small devices. 1999 IEDM Tech. Dig. p. 523.

41. User's Manual for TMA, MEDICI Version 2.3: Technology Modeling Associates, Inc., 1988.

42. S.-H. Lo, D.A. Buchanan, Y. Taur, L.-K. Han, E. Wu. Modeling and characterization of n+ and p+ polysilicon-gated ultra-thin oxide. In: Symposium on VLSI Technology, 1997.

43. Y. Tsividis. Operation and Modeling of the MOS transistor. 2nd ed. New York: WCB McGraw-Hill, 1999.

44. N.D. Arora. MOSFET Models for VLSI Circuit Simulation: Theory and Practice. New York: Springer-Verlag, 1993.

45. H.G. Lee, S.Y. Oh, G. Fuller. A simple and accurate method to measure the threshold voltage of an enhancement-mode MOSFET. IEEE. Trans. Elec. Dev. 29:346–348, 1982.

46. S.C. Sun, J.D. Plummer. Electron mobility in inversion and accumulation layers on thermally oxidized silicon surfaces. IEEE Trans. Elec. Dev. 27:1497–1508, 1980.

47. H.S. Wong, M.H. White, T.J. Kritsick, R.V. Booth. Modeling of transconductance degradation and extraction of threshold voltage in thin oxide MOSFETs. Sol. St. Elects. 30:953–968, 1987.

48. P.A. Muls, G.J. DeClerck, R.J. v. Overstraeten. Characterization of the MOSFET operating in weak inversion. Adv. Electron. Electron Phys. 47:197–266, 1978.

49. J.R. Schrieffer. Effective carrier mobility is surface space charge layers. Phys. Rev. 97:641–646, 1955.

50. M.S. Liang, J.Y. Choi, P.K. Ko, C.M. Hu. Inversion-layer capacitance and mobility of very thin gate-oxide MOSFETs. IEEE Trans. Elec. Dev. 33:409–413, 1986.

51. G.S. Gildenblat, C.L. Huang, N.D. Arora. Split $CV$ measurements of low-temperature MOSFET inversion layer mobility. Cryogenics. 29:1163–1166, 1989.

52. Y. Taur. MOSFET channel length: extraction and interpretation. IEEE Trans. Elec. Dev. 47:160, 2000.

53. J.G.J. Chern, P. Chang, R.F. Motta, N. Godinho. A new method to determine MOSFET channel length. IEEE Elec. Dec. Lett. 1, 1980.

54. K. Terada, H. Muta. A new method to determine effective MOSFET channel length. Jap. J. Appl. Phys. 18:953–959, 1979.

55. D.J. Mountain. Application of electrical effective channel length and external resistance measurement techniques to submicrometer CMOS process. IEEE Trans. Elec. Dev. 36:2499–2505, 1989.

56. P.I. Suciu, R.L. Johnston. Experimental derivation of the source and drain resistance of MOS transistors. IEEE Trans. Elec Dev. 27:1846–1848, 1980.

57. F.H.D.L. Moneda, H.N. Kotecha, M. Shatzkes. Measurement of MOSFET constants. IEEE. Elec. Dev. Lett. 3:10–12, 1982.

58. Y. Taur, D.S. Zicherman, D.R. Lombardi, P.R. Restle, C.H. Hsu, H.I. Hanafi, M. R. Wordeman, B. Davari, G.G. Shahidi. A new "shift and ratio" method for MOSFET channel-length extraction. IEEE Trans. Elec. Dev. 13:267–269, 1992.

59. J.S. Brugler, P.G.A. Jespers. Charge pumping in MOS devices. IEEE. Trans. Elec. Dev. 16:297–302, 1969.

60. G. Groeseneken, H.E. Maes. Analysis of the charge pumping technique and its application for the evaluation of MOSFET degradation. IEEE Trans. Elec. Dev. 26:1318–1335, 1989.

61. J.L. Autran, C. Chabrerie. Use of the charge pumping technique with a sinusoidal gate waveform. Sol. St. Elects. 39:1394–1395, 1996.

62. R.E. Paulson, M.H. White. Theory and application of charge pumping for the characterization of Si–SiO$_2$ interface and near interface oxide traps. IEEE. Trans. Elec. Dev. 41:1213–1216, 1994.

63. K. Ahmed. Electrical characterization of advanced CMOS devices. Ph.D. dissertation, North Carolina State University, Raleigh, NC, 1998.

64. N.D. Arora, D.A. Bell, L.A. Blair. An accurate method of determining MOSFET gate overlap capacitance. Sol. St. Elects. 35:1817–1822, 1992.

65. S.W. Lee. A capicitance-based method for experimental determination of metallurgical channel length and the drain and source series resistance of miniaturized MOSFETs. IEEE Trans. Elec. Dev. 41:1811–1818, 1994.

66. D.R. Wolters, J.J.V.D. Schoot. Dielectric breakdown in MOS devices. Philips J. Res. 40:115, 1985.

67. S. Wolf. Silicon Processing for the VLSI Era—The Submicron MOSFET. Vol. 3. Sunset Beach, CA: Lattice Press, 1995.

68. Y. Chen, J.S. Suehle, C.-C. Shen, J.B. Bernstein, C. Messick, P. Chaparala. A new technique for determining long-term TDDB acceleration parameters of thin gate oxides. IEEE Elec. Dev. Lett. 19:219, 1998.

69. J.C. Lee, I.-C. Chen, C. Hu. Modeling and characterization of gate oxide reliability. IEEE Trans. Elec. Dev. ED-35: 2268, 1988.

70. T. Nigam, R. Degraeve, G. Groeseneken, M.M. Heyns, H.E. Maes. Constant current charge-to-breakdown: still a valid tool to study the reliability of MOS structures? Proceedings for 36th Annual International Reliability Physics Symposium, Reno, NV, 1998, p. 62.

71. E.Y. Wu, W.W. Abadeer, L.-K. Han, S.H. Lo, G. Hueckel. Challenges for accurate reliability projections in the ultra-thin oxide regime. Proceedings for 37th Annual International Reliability Physics Symposium, San Jose, CA, 1999, p. 57.

72. J.H. Stathis, D.J. DiMaria. Reliability projection for ultra-thin oxides at low voltage. 1998 IEDM Tech. Dig., p. 167.

73. D.J. DiMaria, J.W. Stasiak. Trap creation in silicon dioxide produced by hot electrons. Appl. Phys. Lett. 65:1010, 1989.

74. D.R. Wolters, J.F. Verwey. Instabilities in Silicon Devices. Amsterdam: Elsevier, North Holland, 1986.

75. R. Degraeve, G. Groeseneken, R. Bellens, J.L. Ogier, M. Depas, P.J. Roussel, H.E. Maes. New insights in the relation between electron trap generation and the statistical properties of oxide breakdown. IEEE Trans. Elec. Dev. 45:904, 1998.

76. K.F. Schuegraf, C. Hu. Effects of temperature and defects on breakdown lifetime of thin $SiO_2$ at very low voltages. IEEE Trans. Elec. Dev. 41:1227, 1994.

77. K.F. Schuegraf, C. Hu. Hole injection $SiO_2$ breakdown model for very low voltage lifetime extrapolation. IEEE Trans. Elec. Dev. 41:761, 1994.

78. J.H. Stathis. Percolation models for gate oxide breakdown. J. Appl. Phys. 86:5757, 1999.

79. D.J. DiMaria, J.H. Stathis. Ultimate limit for defect generation in ultra-thin silicon dioxide. Appl. Phys. Lett. 71:3230, 1997.

80. M.A. Alam, B. Weir, J. Bude, P. Silverman, D. Monroe. Explanation of soft and hard breakdown and its consequences for area scaling. 1999 IEDM Tech. Dig., p. 449.

81. J.D. Bude, B.E. Weir, P.J. Silverman. Explanation of stress-induced damage in thin oxides. 1998 IEDM Tech. Dig., p. 179.

82. D.J. DiMaria, E. Cartier, D. Arnold. Impact ionization, trap creation, degradation, and breakdown in silicon dioxide films on silicon. J. Appl. Phys. 73:3367, 1993.

83. D.J. DiMaria, D. Arnold, E. Cartier. Degradation and breakdown of silicon dioxide films on silicon. Appl. Phys. Lett. 61:2329, 1992.

84. J. McPherson, V. Reddy, K. Banerjee, H. Le. Comparison of E and 1/E TDDB models for $SiO_2$ under long-term/low-field test condition. 1998 IEDM Tech. Dig., p. 171.

85. J.W. McPherson, H.C. Mogul. Underlying physics of the thermochemical E model in describing low-field time-dependent dielectric breakdown in $SiO_2$ thin films. J. Appl. Phys. 84:1513, 1998.

86. J.S. Suehle, P. Chaparala. Low electric field breakdown of thin $SiO_2$ films under static and dynamic stress. IEEE Trans. Elec. Dev. 44:801, 1997.

87. D.J. DiMaria. Explanation for the polarity dependence of breakdown in ultrathin silicon dioxide films. Appl. Phys. Lett. 68:3004, 1996.

88. E.M. Vogel, J.S. Suehle, M.D. Edelstein, B. Wang, Y. Chen, J.B. Bernstein. Reliability of ultra-thin silicon dioxide under combined substrate hot-electron and constant voltage tunneling stress. accepted IEEE Trans. Elec. Dev., 2000.

89. B. Weir, P. Silverman, M. Alam, F. Baumann, D. Monroe, A. Ghetti, J. Bude, G. Timp, A. Hamad, T. Oberdick, N.X. Zhao, Y. Ma, M. Brown, D. Hwang, T. Sorsch, J. Madic. Gate oxides in 50-nm devices: thickness uniformity improves projected reliability. 1999 IEDM Tech. Dig., p. 523.

90. K. Eriguchi, Y. Harada, M. Niwa. Effects of strained layer near $SiO_2$–Si interface on electrical characteristics of ultrathin gate oxides. J. Appl. Phys. 87:1990, 2000.

91. R. Degraeve, G. Groeseneken, R. Bellens, M. Depas, H.E. Maes. A consistent model for the thickness dependence of intrinsic breakdown in ultra-thin oxides. 1995 IEDM Tech. Dig., p. 863.

92. G.M. Paulzen. Qbd dependencies of ultrathin gate oxides on large area capacitors. Microelectron. Eng. 36:321, 1997.

93. D.L. Crook. Method of determining reliability screens for time-dependent dielectric breakdown. Proceedings for 17th Annual International Reliability Physics Symposium, San Francisco, CA, 1979.

94. C. Hu. Lucky-electron model of channel hot electron emission. 1979 IEDM Tech. Dig., p. 22.

95. C. Hu, S.C. Tam, F.-C. Hsu, P.-K. Ko, T.-Y. Chan, K.W. Terrill. Hot-electron-induced MOSFET degradation—model, monitor, and improvement. IEEE Trans. Elec. Dev. ED-32:375, 1985.

96. P. Heremans, R. Bellens, G. Groeseneken, H.E. Maes. Consistent model for the hot-carrier degradation in n-channel and p-channel MOSFETs. IEEE Trans. Elec. Dev. 35:2194, 1988.

97. E. Takeda, N. Suzuki. An empirical model for device degradation due to hot-carrier injection. IEEE Elec. Dev. Lett. EDL-4:111, 1983.

98. T.H. Ning. 1 μm MOSFET VLSI technology: part IV—hot electron design constraints. IEEE Trans. Elec. Dev. ED-26:346, 1979.

# 5

# Carrier Illumination Characterization of Ultra-Shallow Implants

**Peter Borden, Laurie Bechtler, Karen Lingel, and Regina Nijmeijer**
*Boxer Cross Inc., Menlo Park, California*

## I. INTRODUCTION

This chapter describes a new, rapid, noncontact optical technique for measuring activated doping depth of shallow implants. Called carrier illumination$^{TM}$ (CI) (1), it employs a 2-μm spot size, providing a measurement of fine-scale spatial uniformity in dimensions approaching those of individual devices. By measuring multiple sites, it can rapidly and nondestructively characterize uniformity over the wafer area from edge to edge. The CI method is described, and a number of applications are presented demonstrating its use in the characterization of junction depth uniformity and other process parameters, such as activated dose, sheet resistance, and anneal temperature uniformity.

One of the issues facing the fabrication of CMOS devices below the 180-nm node is the requirement for increasingly tight control of the depth and sheet resistance of shallow junctions, including source-drain (S/D) regions and S/D extensions. This has created a need for improved metrology to develop and control doping processes (2).

The S/D and S/D extension are shallow, highly doped regions. Typical junction depths are under 1000 Å, with doping concentrations on the order of $10^{20}/cm^3$, at depths reaching the 200–400-Å range over the next 5 years (2). Formation of these layers requires ultralow-energy (ULE) implantation at doses ranging from mid-14 to mid-15/cm$^2$ combined with fast-ramp anneals. The most common dopants are As (n-type) and B (p-type) (3).

The process issues in forming these layers lie both with the anneal and the implant. While it is relatively easy to implant a shallow layer, it is very hard to keep it at the surface after an anneal that provides sufficient activation to drop the sheet resistance to acceptable levels. The requirements are threefold:

1. *Junction depth control.* Excess diffusion of the dopant creates two problems. First, an excessively deep junction reduces the ability of the gate to control the source-drain current. This results in excess current flow in the off state. Second, lateral diffusion causes variation in the channel length. For example, 10-nm diffusion simultaneously from the source and the drain causes a 20% reduction in channel length in a 100-nm device.

2. *Sheet resistance.* Low sheet resistance must be maintained to minimize the voltage drop between the channel and the source or drain. This is an especially critical issue with the low operating voltage of deep submicron devices.

3. *Profile abruptness.* A graded profile places a higher-resistance region between the channel and the extension, increasing the voltage drop between the channel and the contacts.

The competing requirements for a shallow, abrupt profile and a low sheet resistance greatly reduce the size of the process window (allowable range of process parameters such as anneal time and temperature that result in fabrication of devices with an acceptable range of electrical parameters). For example, achieving low sheet resistance requires full activation of the implant, forcing use of a higher annealing temperature. This drives the junction deeper and yields a more graded profile. A few degrees variation in anneal temperature across the water can lead to unacceptable junction depth variation; similar problems may arise locally due to pattern effects.

This motivates the need for a sensitive, nondestructive profile depth measurement with high throughput to enable uniformity measurement and a small spot size to enable use in patterned structures. In-line measurement on patterned wafers then becomes feasible, providing the control necessary to maintain a process within a narrow window.

The strongest driver for this technology is the need for rapid turnaround in process development and debugging. The primary tools for profile measurement are analytical methods such as secondary ion mass spectroscopy (SIMS) (4) and spreading-resistance profiling (SRP) (5). These provide detailed profile information, but typical turnaround times are measured in days or weeks. As the process becomes more complex and requires more learning cycles to define an acceptable operating space, faster turnaround becomes essential. Secondary ion mass spectroscopy and SRP also require highly trained operators, especially for challenging shallow-junction measurements. This contributes to the slow turnaround and limits use for process control in volume manufacturing. Various other dopant profiling methods are reviewed in Ref. 6.

This chapter describes a profile depth measurement method called *carrier illumination* (CI). Nondestructive and requiring a few square microns of area, it provides a determination of the profile depth in a few seconds, offering a throughput sufficient for in-line process control and uniformity mapping.

Carrier illumination is sensitive primarily to the depth of the activated implant. However, in certain cases this depth relates to other parameters, such as sheet resistance and amorphous layer depth. Especially in process control applications, the correlation between profile depth and control parameters enables use of CI for in-line control of critical shallow-junction process parameters.

This chapter will often use the term *profile depth*. This is the depth at which the concentration reaches $10^{18}/cm^3$. It is often used in preference to the common term *junction depth* because the CI method can be used on $p^+/p$ or $n^+/n$ structures. In these cases, a conventional p/n or n/p metallurgical junction does not exist.

## II. DESCRIPTION OF THE CARRIER ILLUMINATION METHOD

The carrier illumination$^{TM}$ CI method derives its name from the use of photogenerated carriers to create contrast, so features of an active doping profile can be seen. This can be understood with an analogy from biology. Suppose a biologist wishes to see a transparent cell through a microscope. The clear cell has no contrast with respect to its aqueous medium, so the biologist adds a dye. The cell soaks up the dye, making all of its detail clearly visible.

Likewise, the active doping profile has no contrast because its index of refraction is nearly the same as that of the silicon substrate. The CI method uses a laser to generate a quasi-static distribution of excess carriers at the edge of the doped region (*excess carriers* being defined here as the difference between the dark distribution and the distribution under illumination). The index of refraction change due to the excess carriers is small, but their concentration is slowly modulated. This enables use of sensitive phase-locked methods to detect reflection from the excess carrier profile, providing contrast with a signal-to-noise ratio sufficient to measure characteristics of the doped region.

## A. Physics of the Measurement

The CI method is based on well-known pump-probe techniques. These have been used to study semiconductor properties since the 1970s (7). They also have been used to monitor damage in semiconductors and, through the use of thermal and carrier waves, as a process control tool for measuring implant dose in unannealed layers (8).

The CI method uses one laser to set up a quasi-static carrier distribution and a second laser to probe the distribution using interference methods. Interpreting this distribution enables extraction of a profile depth.

The carrier distributions obey the time-dependent diffusion equation. Under the assumption of sinusoidal excitation, this takes the following form in one dimension (9):

$$\frac{\partial^2 n}{\partial z^2} - n\left(\frac{1}{D\tau} + j\frac{\omega}{D}\right) + \frac{G}{D} = 0 \tag{1}$$

where $n$ is the excess carrier concentration, $D$ is the diffusion coefficient, $\tau$ is the carrier lifetime, $\omega$ is the excitation frequency, and $G$ is the rate per unit volume of excess carrier generation.

At high excitation frequencies, the solution to this equation is a wave. At low frequencies, the solution is a static carrier distribution. Carrier illumination operates in the latter regime, where $1/\tau \gg \omega$. In this regime, the excess carrier distribution is well defined in terms of the active doping profile, and may either analytically or empirically be related to the profile depth.

There are three components to a full model of the CI method. A semiconductor model describes how an index of refraction gradient arises within the doped layer as a result of illumination. An optical model describes how a signal is obtained by reflecting a probe beam from this index of refraction gradient. Finally, the algorithm describes how the reflection signal relates to a profile depth value.

## 1. Semiconductor Physics Model

The semiconductor model provides the excess carrier profile in the doped region arising from illumination. This profile gives rise to a gradient in the index of refraction. Optical probing of this gradient provides a signal related to the physical properties of the active profile.

The semiconductor model is found in two steps. First, the relationship between the index of refraction and the carrier concentration is found. Second, the carrier concentration as a function of depth is found. Combining these provides the index of refraction gradient used for the optical model. The first part arises from Maxwell's equations, which relate the index of refraction to the conductivity, and Drude theory, which relates the carrier concentration to the conductivity at optical frequency.

It can be shown that the propagation constant for a plane wave in a medium with a magnetic permeability $\mu = 1$, in the limit of poor conductivity ($\sigma/\omega\varepsilon_0\varepsilon \ll 1$), is

$$k \approx \sqrt{\varepsilon}\frac{\omega}{c} + i\frac{\sigma}{2c\varepsilon_0\sqrt{\varepsilon}} \tag{2}$$

where $\omega$ is the radial frequency of the probe laser light, $c$ is the speed of light, $\varepsilon$ is the dielectric constant of the substrate (usually silicon), and $\varepsilon_0$ is the dielectric constant of free space (10).

The first term in Eq. (2) is the propagation constant in a dielectric medium with an index of refraction equal to $\sqrt{\varepsilon}$. The second term is the perturbation due to the conductance of the medium, which is a function of the free carrier concentration. The conductance is commonly found using Drude theory (11), which models the carriers as a gas of free particles. The electron (hole) will experience scattering that will cause the drift velocity, $v_d$, of the carrier to go to zero in the absence of the driving field of light. This scattering is thus a damping term, and the drift velocity can be written as

$$m^*v_d + m^*gv_d = qEe^{i\omega xt} \tag{3}$$

where $m^*$ is the carrier effective mass, $g$ is a damping term, and $q$ is the electron charge (12).

Solving for the conductivity (using $J = \sigma E = qNv$) (13) gives

$$\sigma = \frac{q^2N}{m^*(g + i\omega)} \tag{4}$$

Combining Eqs. (2) and (3), the relationship between the real part of the index of refraction and the carrier concentration $N$ is

$$\Delta n = \frac{q^2N(ig + \omega)}{2\omega\varepsilon\varepsilon_0 m^*(g^2 + \omega^2)} \tag{5}$$

where $n \equiv k(c/\omega)$. The index of refraction varies linearly with the carrier concentration, and the excess carrier profile gives the induced index of refraction gradient.

The induced carrier profile is determined using finite element solutions to the carrier transport equations. These solutions are of limited accuracy, both because of poor knowledge of the carrier transport properties of shallow, heavily doped layers and because of the difficulty in determining an accurate active doping profile as a starting point for the calculation. However, a qualitative understanding of the behavior of the carriers can be gained by examining the carrier diffusion equation.

Figure 1 shows a laser beam illuminating the doped layer. The beam refracts to the normal, creating a column of light in the semiconductor. The wavelength is chosen so that the photon energy exceeds the bandgap energy. Photons are absorbed, creating excess electron-hole pairs. These carriers distribute themselves according to the diffusion equation. As described earlier, both wave and dc solutions are possible, depending on the laser modulation frequency. The analysis assumes operation in a regime where the dc solution dominates.

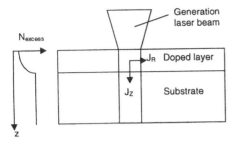

**Figure 1**   Excess carrier currents and concentration gradient upon illumination with a laser whose photon energy exceeds the semiconductor bandgap.

The excess carriers distribute within the semiconductor according to diffusion. Both radial and vertical diffusion currents flow, shown as $J_R$ and $J_z$, with the current density driven by the gradient of the carrier concentration,

$$J_z = qD\frac{\partial N}{\partial z}, \qquad J_R = -qD\frac{\partial N}{\partial r} \tag{6}$$

The full solution requires separate sets of equations for holes and electrons (9).

In the doped layer, the vertical component $J_z$ dominates, because the gradient occurs in a distance comparable to the profile depth ($< 0.1\,\mu$m). By comparison, the radial gradient occurs over a distance comparable to the beam dimensions, which is on the order of a few microns.

Because the beam radius is small compared to the diffusion length, recombination within the illuminated region is small, and most carriers generated in the implanted layer flow out through the vertical current. Assuming an approximately constant generation per unit volume, the vertical current rises linearly with depth and, from Eq. (6), the carrier concentration rises as the square of the depth.

In the substrate the situation is reversed. The absorption length determines the gradient in the vertical direction. The beam radius is small compared to the absorption length, so the radial current dominates. The substrate becomes flooded with carriers, and the concentration is almost flat. This leads to an excess carrier depth profile, as shown in Fig. 1, rising rapidly to the profile edge and then flattening out.

Modeling confirms the shape of the excess carrier distribution. Figure 2 shows the output of a PISCES model (14) showing the excess carrier concentration profile resulting from illumination of a thin p-doped layer with a 1-$\mu$m-radius, 10-mW, 830-nm laser beam. As described earlier, the excess carriers pile up steeply at the profile edge, and the concentration is relatively flat in the low-doped substrate.

The buildup of excess carriers to the profile edge creates a large index of refraction gradient, localized at the steeply graded portion of the profile. When a second laser beam illuminates this structure, reflection will occur at the depths where the gradient of the excess carrier profile is steepest, which is at the surface and the profile edge. Therefore, the reflection signal will contain information about the profile depth. The slow modulation of the generation laser beam (at a few hundred hertz to maintain the quasi-static distribution) allows the reflection of the second laser to be detected using phase-locked methods with very narrow noise bandwidth, providing the necessary contrast.

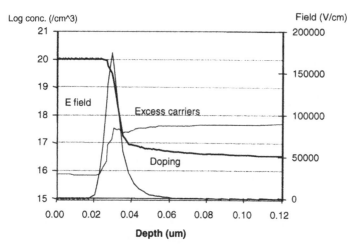

**Figure 2** Doping profile, electric field (V/cm), and excess carrier distribution (electrons + holes) in a thin p-doped layer under illumination. The beam radius is 1 μm, power is 10 mW, and wavelength is 830 nm.

The PISCES modeling also indicates that a surface recombination velocity on the order of $10^4$ cm/s does not significantly affect the excess carrier concentration. This suggests that the measurement may be used with unpassivated silicon surfaces.

Modeling also shows that the base region (layer below the activated implant) is generally in high-level injection, even at low laser power. The excess carrier concentration swamps the background concentration, and sensitivity to the effects of a doped base region are lost. This has been found to be true at mid- to high ($10^{17}$/cm$^3$) doping levels. The lack of sensitivity to base doping allows characterization of source/drain and extension regions formed in n- and p-wells.

## 2. Optical Model

With the excess carrier distribution known, the optical model is used to determine the signal as a function of the active doping profile and generation laser power. The model assumes there are three components to the reflection signal. The first comes from the index of refraction change at the air–semiconductor interface, and is essentially independent of the excess carrier concentration. The other two appear at the generation laser modulation frequency. The first comes from the steep gradient in excess carrier concentration at the surface of the semiconductor. The second comes from the index of refraction gradient caused by the excess carrier concentration profile. These two modulated components interfere with the unmodulated surface reflection to create an interference signal at the modulation frequency. This signal can be either positive or negative, depending on whether constructive or destructive interference occurs.

The optical model derivation assumes an excess carrier concentration consisting of a set of infinitesimally thin layers of constant carrier concentration $N_m$, each with thickness $\Delta$. The reflected electric field is

$$E_r = r_s E_0 + t^2 \sum_m r_m E_0 e^{j2n_m k\Delta} \tag{7}$$

where $E_0$ is the incident electric field, $r_s$ is the reflection coefficient of the surface, $t$ is the transmission coefficient of the surface, and $r_m$ is the reflection coefficient resulting from the step in the index of refraction between the $m$th and $(m+1)$th layer.

The signal power scales as the squared magnitude of the electric field given in Eq. (7). In the limit, as the layer thickness $\Delta$ approaches zero, this is

$$E_r^* E_r = r_s^2 E_0^2 \left\{ 1 + 2\beta \frac{t^2}{n_s r_s} \int_0^\infty \cos[2n(z)kz] \frac{dN}{dz}\, dz \right\} \tag{8}$$

where the index of refraction at a depth $z$ is $n(z) = n_x + \beta N(z)$, with $N(z)$ the excess carrier concentration at a depth $z$ and the coefficient $\beta$ found using Eq. (5). *The second term in the brackets of Eq. 8 gives rise to the measured signal.* This is the cosine transform of the gradient of the excess carrier concentration. As expected, the greatest signal comes from the depth at which the gradient is steepest, which is at the junction edge.

### 3. Conversion of Signals to a Profile Depth

The conversion of signals to a profile depth value employs an algorithm based on a fit to known values determined with SRP. The reference method SRP is used because it is the one most readily available to measure active doping.

In practice, signals are measured over a range of generation laser powers. This provides a *power curve*, which is a plot of the reflection signal as a function of generation laser power. Figure 3 shows a set of power curves obtained from 500-eV B[11] implants annealed over a range of temperatures. Both positive and negative signals are seen, representing constructive and destructive interference between the surface and profile reflections.

The power curve can be thought of as a bias sweep of the open-circuit photodiode created by the band bending of the high-doped implant into the low-doped substrate. As the generation power increases, the open-circuit voltage sweeps from zero to some maximum value. This voltage appears as a forward bias across the junction and changes the excess carrier distribution, giving rise to the shape of the power curves. These shapes are fit

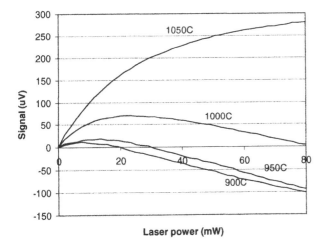

**Figure 3** Power curves showing the reflection signal as a function of generation laser power for a set of 500-eV B[11] implants RTA annealed at various temperatures.

to SRP measurements of profile depth in order to create an algorithm for conversion to a profile depth. This results in a formula relating the signal at various powers to a profile depth.

## 4. Accuracy, Precision, and Calibration

Any semiconductor metrology system must demonstrate precision significantly tighter than the process capability, a signal correlated to known standards, and good repeatability (gives the same answer on a run-to-run basis) and stability (gives the same answer over time).

Inherent measurement precision is high, because of the short wavelength of the probing laser beam in the silicon. For example, a 1.48-μm laser has a length of 4300 Å in silicon. The light reflecting off the profile travels a distance equals to twice the profile depth (from the surface to the steep edge of the profile and back), so a full $2\pi$ phase shift occurs for a profile depth of 2150 Å. Present systems have a noise-limited resolution of better than $1/3°$, corresponding to about 2 Å. Interatomic spacing does not limit resolution because the spot size is on the order of 1 μm$^2$, and the measurement is averaged over $> 10^6$ atoms in a typical application. Independent of absolute junction depth accuracy, this precision makes CI measurements very sensitive to small uniformity variation.

A difficulty in validating the measurement is the lack of traceable junction depth standards and weaknesses in the ability of readily available alternate methods (SIMS, SRP) to determine active profile depth. Figure 4 shows a correlation scatterplot between profile depths obtained with the algorithm and measured SRP depths at a concentration of $10^{18}/\text{cm}^3$. The data fits a straight line with a 45° slope. The error at 1 standard deviation is 100 Å. Most of this error is thought to arise from limitations in using SRP as a correlation standard. There are two sources of error. The most significant is that, because of the cost, SRP is done at only one site per wafer. This means the SRP depth is accurate only to the variation in profile depth in each sample. The CI data is wafer averages, greatly reducing the effect of across-wafer profile depth variation. Second, SRP may provide limited accuracy in determining the profile at concentrations well below the peak (15,16).

These effects are seen in Figure 5, showing correlation between profile depth measured on two lots of 500-eV, $1 \times 10^{15}/\text{cm}^2$ B$^{11}$ implants RTP annealed at various tem-

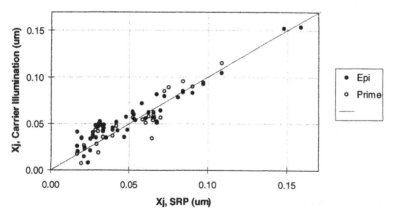

**Figure 4**   Correlation between profile depth measured at $10^{18}/\text{cm}^3$ using SRP and the result of the carrier illumination profile depth algorithm.

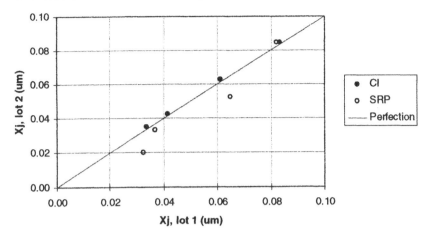

**Figure 5** Correlation between measured profile depth for two lots of wafers with identical processing.

peratures. Both lots had identical processing. The CI data is five site averages; the SRP data represents single sites per wafer. The SRP error is about 90 Å at 1 standard deviation.

It is important to emphasize that these results do not imply that the SRP measurement at any given site is inaccurate. They simply show that recognized limitations—a limited number of samples per wafer and profile errors below the peak—lead to errors in correlation between SRP and CI.

An alternate calibration method is to measure silicon layers deposited using molecular beam epitaxy (MBE). While diffusion during growth causes the depth of the doped region to be deeper than the thickness of the physically grown layer, the steps between layers for successive samples are well controlled an the profile is reasonably abrupt. This somewhat removes uncertainty about the shape of the profile, because the complex diffusion effects encountered in forming a shallow activated implant profile are avoided.

Figure 6 shows results of measurements on three samples grown with a physical layer thickness of 200, 300, and 400 Å. The doping within the layer was $2 \times 10^{20}/cm^3$ of boron. The doped-layer thickness at a concentration of $10^{18}/cm^3$ was measured using CI, SRP, and SIMS (Cameca 4F). The SRP data exactly matches the grown-layer thickness, showing that SRP provides an accurate profile depth measurement on a uniform sample. Due to carrier spilling, the SRP measurements tend to underestimate the true profile depth. The difference between the SRP and SIMS measured depths is consistent with earlier studies (17).

The CI results are obtained using two methods. The first (CI empirical) uses the algorithm based on an empirical correlation to SRP data. The second (CI model) uses a PISCES model to calculate the excess carrier distribution, assuming an abrupt doping profile, and the optical model to calculate the reflection signal as a function of generation laser power. The depth of the box profile is varied to fit the measured data. These fits are shown in Figure 7. Excellent correlation is seen, validating both the theoretical model and the empirical algorithm, and suggesting that the CI depths are probably close to the correct depths at $10^{18}/cm^3$.

Equally close agreement to the theory is not always obtained when modeling implanted layers. This is thought to be due to two factors. First, unannealed damage may cause the optical properties of some implants to differ from values used in the

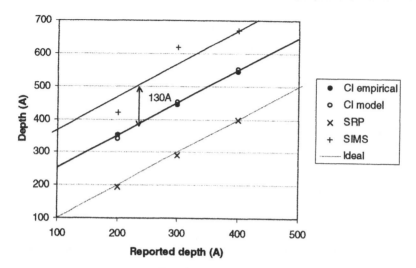

**Figure 6** Profile depths ($10^{18}/cm^3$) as measured by CI, SRP, and SIMS. Layers are grown by MBE and doped $2 \times 10^{20}$ with $B^{11}$, with as-grown thicknesses of 200, 300, and 400 Å. The substrate is p-epi, with an epitaxial doping resistivity of 10–20 Ω-cm.

model. Second, the theoretical model requires input of an accurate active doping profile edge, and SRP or SIMS may not always provide this. Conversely, the MBE material is probably of high quality with an abrupt profile, providing a more accurate basis for the model.

Both reproducibility and stability have been measured for CI systems operating in a factory environment. Junction depth reproducibility has been observed to be better than 1%, in tests where the same wafer is inspected 30 times with load/unload between measurements. Stability has been measured by determining the junction depth on a reference wafer over a four-month period. Drift over this time has been approximately 1 Å in a sample with a nominal profile depth of 500-Å. This level of reproducibility and stability is thought to obtain from the relative simplicity of the optical system and the fact that the system is installed in a vibration-free, climate-controlled cleanroom environment.

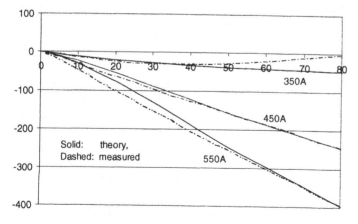

**Figure 7** Correlation between the theoretical model and measurement for three $B^{11}$ doped MBE silicon layers on p-silicon.

## III.  APPLICATIONS RELATED TO PROFILE DEPTH SENSITIVITY

The primary sensitivity of the CI method is to profile depth, with rapid, nondestructive, high-resolution profile depth mapping the primary application. Other parameters of the implant, such as sheet resistance, relate to the profile depth. In process control applications, the profile depth measurement can be used as the indicator of process stability. Variation in profile depth will then indicate the presence of variation in a related parameter. Measurement of profile depth and related parameters is discussed next.

### A.  Profile Depth and Sheet Resistance Uniformity Mapping

The most direct application is mapping of profile depth uniformity. Figure 8 shows a junction depth uniformity map of a 200-mm wafer with a 250-eV $1 \times 10^{15}/cm^2$ $B^{11}$ implant annealed 10 seconds at 1000°C. The map shows a radial pattern suggesting non-uniformity in the RTA heating pattern.

The profile depth measurement can also be used to control a process in which sheet resistance is a critical control parameter. This application is not the direct measurement of sheet resistance, but the measurement of a parameter that provides a high-sensitivity indicator of whether the sheet resistance is in control for a given process.

The sensitivity of the profile depth measurement to sheet resistance is understood with the aid of Figure 9. This graph shows the CI signal as a function of a sheet resistance

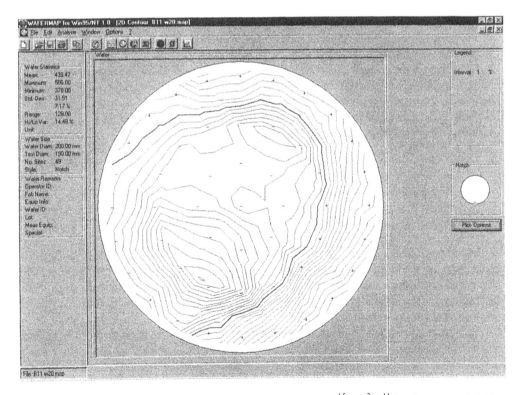

**Figure 8**  Junction depth uniformity map for a 250-eV, $1 \times 10^{15}/cm^2$ $B^{11}$ implant annealed 10 seconds at 1000°C. A radial pattern indicative of nonuniform RTA heating is clearly visible.

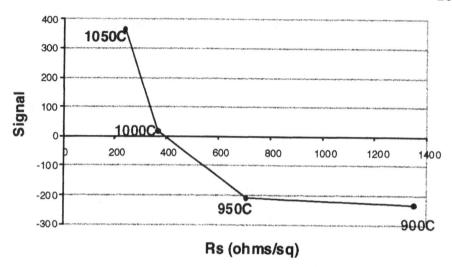

**Figure 9** Signal as a function of sheet resistance for four 500-eV $1 \times 10^{15}/cm^2$ $B^{11}$ implants annealed at the indicated temperatures.

for 500 eV, $1 \times 10^{15}/cm^2$ $B^{11}$ implants annealed at temperatures ranging from 900 to 1050°C. For lower anneal temperatures, the sheet resistance drops rapidly with increasing temperature, but the CI signal shows only a small change. At these temperatures, activation is increasing but the doping profile depth is not changing. Conversely, at higher temperatures, the sheet resistance drops slowly with increasing temperature, but the CI signal changes rapidly. At these temperatures, the activation is nearly complete, but the doping profile is diffusing rapidly.

These processes are used to form low-resistance source/drain extensions. Achieving sufficiently low sheet resistance requires an anneal at a temperature that causes movement of the profile. Consequently, these processes operate near the knee of the curve in Figure 9. This enables the CI measurement to provide a sensitive process control signal. Carrier illumination provides three benefits in this application:

1. The nondestructive, small-area measurement allows characterization of sheet resistance on product wafers without an edge exclusion.
2. The high resolution enables measurement of pattern-related effects.
3. The CI method does not require a junction to confine carrier flow, enabling characterization of boron implants into p-type layers.

An example of the relationship between profile depth and sheet resistance is shown in Figures 10a and b. Figure 10a shows a profile depth map of a wafer following an 800-eV $B^{11}$ implant into an n-well, annealed 10 seconds at 1000°C. Figure 10b shows a sheet resistance map from the same wafer. Figure 10c shows the correlation between profile depth and sheet resistance as a function of position on the wafer.

This data is consistent with the model of the sheet resistance decreasing as the depth of the profile increases. As Figure 10c shows, profile depth and sheet resistance correlate, with the sheet resistance being lowest for the deeper profile depths. These trends are readily visible in the wafer maps, Figures 10a and b.

Secondary ion mass spectroscopy and SRP were performed on the top, middle, and bottom of the same wafer. Figure 10d shows the resulting profiles. This confirms the depth trend observed in the CI map. The SRP profiles are considerably steeper than the SIMS profiles because of the presence of a $p^+/n$ junction. Consistent with the results shown in Figure 7, the SRP profile is considerably shallower than the SIMS profile.

## B. Active Dose in Extension Implants

Correlation has been observed between implant dose and signal for LDD implants after anneal. The aim has been to show that the CI measurement can be used as a dose monitor for process control.

Dopant redistribution during rapid thermal annealing is a complex process (18,19), well beyond the scope of this chapter. Nevertheless, assuming a diffusion-driven process to first order, then the concentration is a function of time and depth, given by

$$C(z, t) = \frac{\phi}{\sqrt{\pi Dt}} \exp\left[-\frac{z^2}{4Dt}\right] \tag{9}$$

where $\phi$ is the implant dose, $D$ is the diffusion constant, and $t$ is the anneal time (20). It is therefore reasonable to expect a relationship between the depth at a constant concentration and the dose.

Such a relationship is seen for $B^{11}$ LDD implants in the mid-$10^{14}/cm^2$ dose range. Other ranges have not been explored as of this writing. Figures 11a and b show the CI signal as a function of LDD dose over a split range of $4.5 \times 10^{14}/cm^2$ and $6.5 \times 10^{14}/cm^2$ for 800-eV implants annealed at 1000°C for 10 seconds. (Figure 11a) and after a second identical anneal to simulate a source/drain anneal (Figure 11b). In the former case, four wafers were run at each dose, two of which were then carried through the second anneal. The signal changes between the first and second anneal because the second anneal has deepened the profile, but the trend with dose remains.

## C. RTA Temperature Variation and Pattern Dependence

The sensitivity of profile depth to RTA temperature makes the CI method applicable for determining nonuniformity effects in RTA processes, due to effects such as temperature variation within the process tool and pattern effects. Pattern effects refer to local nonuniform annealing due to the lithographically patterned features on the wafer.

Figure 12 shows an example of the sensitivity of profile depth to temperature. This graphs the CI-measured profile depth as a function of RTA temperature for a matrix of wafers implants with $BF_2$ at an energy of 23 keV and dose of $1.6 \times 10^{15}/cm^2$ and annealed for 10 seconds each at 5°C steps. The profile depth is seen to increase with temperature, as expected. The resolution at the higher temperatures, where the profile movement is more pronounced, is better than 1°C.

The high spatial resolution combined with temperature sensitivity of the profile depth can be applied to determining pattern effects in RTA processes. One experiment was carried out in cooperation with Steag RTP Systems to investigate the effect of a backside pattern on front-side anneal uniformity. An aim of this study was to optimize twosided RTA heating configurations. Polysilicon and $SiO_2$ were deposited on the front side of a double-polished wafer. The poly was then patterned as shown in Figure 13. The back side was then implanted with 1-keV As and annealed under various conditions, as described in Ref. 20.

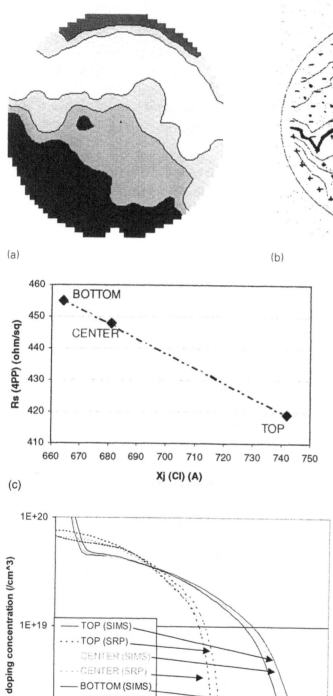

(a)

(b)

(c)

(d)

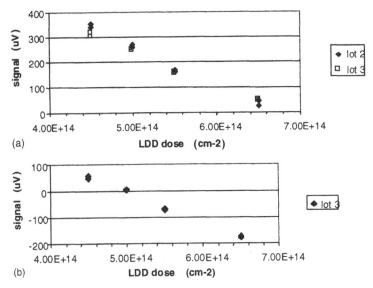

(a)

(b)

**Figure 11**   (a) CI signal as a function of LDD implant dose for two lots, each with two wafers at each dose split. Implant is 800-eV $B^{11}$, annealed at 1000°C for 10 seconds in $N_2$. (b) CI signal as function of LDD implant dose after lot 3 in Fig. 11a has undergone a second anneal at 1000°C for 10 seconds in $N_2$, simulating the source/drain implant anneal.

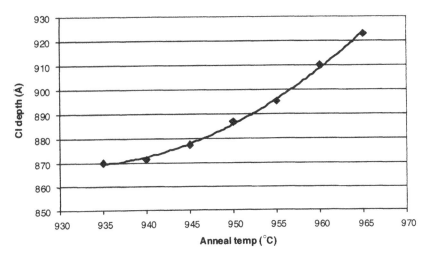

**Figure 12**   Measured profile depth as a function of anneal temperature for a matrix 23-keV, $1.6 \times 10^{15}$/cm$^2$ BF$_2$ implants annealed in 5°C increments.

**Figure 10**   (a) Profile depth map of 800-eV $B^{11}$ implant annealed 10 seconds at 1000°C, formed in an n-well of approximate concentration $3 \times 10^{17}$/cm$^3$. Depth varies from 664-Å (bottom) to 742-Å (top). Contours represent 20 Å. (b) Sheet resistance map of same wafer. $Rs$ varies from 455 $\Omega$/square (bottom) to 419 $\Omega$/square (top). (c) Correlation between CI-measured profile depth and 4PP-measured sheet resistance. (d) SIMS and SRP profiles taken from the top, center, and bottom of the wafer shown in Fig. 11, confirming that the doping profile is deepest at the top and shallowest at the bottom. The SRP profile drops rapidly due to the presence of a p/n junction.

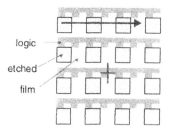

**Figure 13** Polysilicon pattern on the back side of pattern-effect wafers. Etched boxes have dimensions of 1 cm on a side.

The implanted (unpatterned) side was mapped using the CI method. Figures 14a, b, and c show the resultant signals with measurement points taken 0.5 mm apart (each map has 6000–10,000 profile depth measurements). The cross in Fig. 13 shows the position of the center of the scans. Line scans were also taken across three of the repeating patterns, as shown by the arrow in Fig. 13. Figure 14d shows the profile depth as a function of position for these scans.

The benefit of the high-resolution CI map is that it allows accurate determination of the effect of the pattern. Figures 14a, b, and c show a clear progression of improvement resulting from RTA system modifications. The patterns in this experiment are coarse, and the location of the pattern on the side opposite the implant tends to smear the pattern effect. However, the small spot size of the CI method enables probing of pattern effects on a micron scale. This is an area of ongoing investigation as of this writing.

## D. Amorphous Layer Depth

The CI method has been used to measure the depth of amorphous layers resulting from preamorphizing implants (PAI). It is thought that the amorphous layer creates a "dead layer." This pushes the interface at which the excess carrier concentration rises away from the surface by a depth equal to the thickness of the amorphous layer. The interference between the reflection from the excess carrier profile and the reflection from the front surface then becomes a measure of the amorphous layer thickness.

Figure 15 shows a plot of the correlation between the CI signal at a constant laser power and the thickness of the amorphous layer for a set of silicon implants. these included single, double, and triple implants of energies ranging from 8 to 30 keV, designed to provide a constant amorphous layer thickness. Table 1 gives the implant parameters. The layer thickness was calculated using deposited energy thresholds published by Prussin et al. (22), with the energy distribution calculated using TRIM (23). Similar correlation has been obtained with Ge PAI.

**Table 1** Implants Used for the PAI Study

| Energy (keV) | 8 | 20 | 8 + 20 | 30 | 8 + 30 | 20 + 30 | 8 + 20 + 30 |
|---|---|---|---|---|---|---|---|
| Total dose (/cm$^2$) | 1e15 | 1e15 | 2e15 | 1e15 | 2e15 | 2e15 | 3e15 |
| Depth (Å) | 140 | 333 | 349 | 497 | 508 | 523 | 525 |

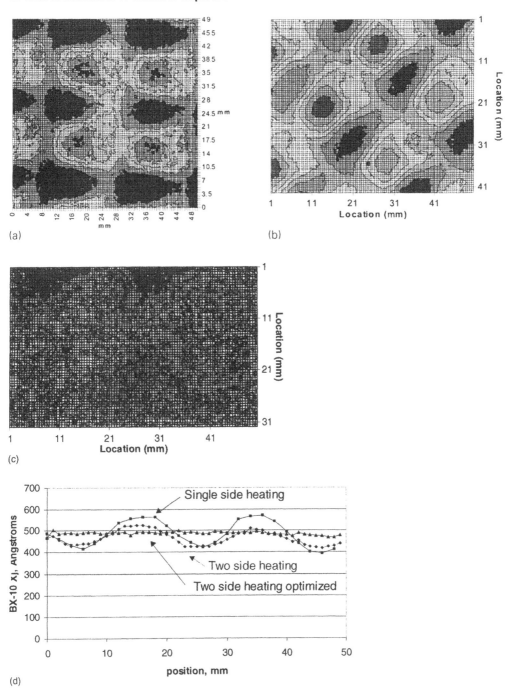

**Figure 14** (a) 5 × 5-cm area map in 0.5-mm steps (10,000 points) showing a CI signal map of the implanted but unpatterned side of the wafer after a single-sided anneal. (b) 5 × 5-cm area map in 0.5-mm steps (10,000 points) showing a CI signal map of the implanted but unpatterned side of the wafer after a double-sided anneal. (c) 3 × 5-cm area map in 0.5-mm steps (6,000 points) showing a CI signal map of the implanted but unpatterned side of the wafer and an optimized double-sided anneal. (d) Line scans across patterns of wafers shown in Figs. 14 a–c showing profile depth variation for various types of anneals.

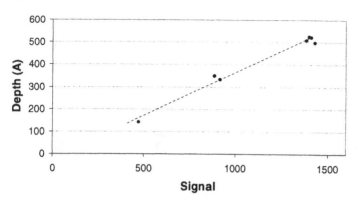

**Figure 15**  Amorphous layer depth as a function of signal for unannealed silicon PAI.

## IV.  USE OF THE MEASUREMENT

The carrier illumination measurement is intended for process development, control, and diagnosis. The capability of measuring nondestructively, quickly, and in small areas makes it particularly suitable as a tool for in-fab application. As such, the CI measurement has been implemented on a platform appropriate for use in the fab. Figure 16 shows a block diagram of how the system is used in-line (run by an operator) or for engineering (run by a process engineer). Typically, site maps and measurement recipes will have been preprogrammed. The primary difference between the two cases is the application of the resultant data. In the former case, the data is analyzed automatically in the tool to verify whether the signal or derived process parameter is within limits, and a limited set is then archived in the event further analysis is required. In the latter case, the data is archived immediately and analyzed off-line by an engineer.

## V.  CONCLUSION

Carrier Illumination is a nondestructive, fast, small-spot method for measuring the depth of activated implant profiles. It provides the capability to measure uniformity of active implants and enables measurement on patterned wafers. This makes it well suited as an in-fab tool for rapid process development and debugging. As the technology nodes move below 0.18 μm and the properties of shallow implants become more significant factors

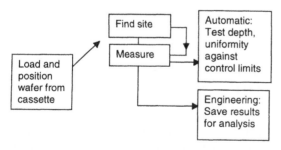

**Figure 16**  Method for in-line use.

determining CMOS device performance, this in-fab capability will become increasingly important.

## ACKNOWLEDGMENTS

The authors gratefully acknowledge the assistance of SEMATECH personnel who helped provide much of the data, including Larry Larson, Bill Covington, Billy Nguyen, Clarence Ferguson, Bob Murto, Mike Rendon, and Alain Diebold. We also thank Sing-Pin Tay and Jeff Gelpy of STEAG for the patterned wafer data, and the many customers who have worked with us to demonstrate the capabilities of the CI method.

## REFERENCES

1. US Patent 6,049,220. Others applied for.
2. Semiconductor Industry Association. International Technology Roadmap for Semiconductors, 1998 update.
3. C.R. Cleavelin, B.C. Covington, L.A. Larson. FEOL considerations for progression beyond the 100-nm-node ultra-shallow junction requirements. Proceedings of the Fifth International Workshop on Measurement, Characterization and Modeling of Ultra-Shallow Doping Profiles in Semiconductors, usj—99, Research Triangle Park, NC, March 28–31, 1999, pp 10–11.
4. G.R. Mount, S.P. Smith, C.J. Hitzman, V.K. Chiea, C.W. Magee. Ultra-shallow junction measurements: a review of SIMS approaches for annealed and processed wafers. In: D.G. Seiler, A.C. Diebold, W.M. Bullis, T.J. Shaffner, R. McDonalt, E.J. Walters, eds. Characterization and Metrology for ULSI Technology. AIP Conference Proceedings 449, American Institute of Physics Press, 1998, p 757.
5. M. Pawlik. J. Vac. Sci. Technol. B 10:388, 1992.
6. E. Ishida, S.B. Felch. J. Vac. Sci. Technol. B 14(1):397–403, Jan/Feb 1996.
7. M. Bass (ed.). Handbook of Optics. Vol. II. New York: McGraw-Hill, 1995, pp 36.67–36.72.
8. W.L. Smith, A. Rosencwaig, D.L. Willenborg. Appl. Phys. Lett. 47:584–586, Sept. 1985.
9. S.M. Sze. Physics of Semiconductor Devices. New York: Wiley, 1981, pp 50–51.
10. J.D. Jackson. Classical Electrodynamics. New York: Wiley, 1973, pp 222–224.
11. J.D. Jackson. Classical Electrodynamics. New York: Wiley, 1973, pp 225–226.
12. G. Burns. Solid State Physics. San Diego, CA: Academic Press, 1985, p 487.
13. G. Burns. Solid State Physics. San Diego, CA: Academic Press, 1985, p 196.
14. Atlas. Silvaco International, Santa Clara, CA.
15. L.A. Heimbrook, F.A. Baiocchi, T.C. Bitner, M. Geva, H.S. Luftman, S. Nakahara. A practical perspective of shallow junction analysis. Proceedings of the Third International Workshop on the Measurement and Characterization of Ultra-Shallow Doping Profiles in Semiconductors, Raleigh, NC, March 20–22, 1995, pp 3.1–3.15.
16. T. Clarysse, W. Vandervorst. Qualification of spreading resistance probe operations. Part I. Proceedings of the Fifth International Workshop on Measurement, Characterization and Modeling of Ultra-Shallow Doping Profiles in Semiconductors, usj—99, Research Triangle Park, NC, March 28–31, 1999, pp 79–95.
17. H. Jorke, J.J. Herzog. J. Appl. Phys. 60:1735–1739, Sept. 1986.
18. A. Agarwal, H.-J. Gossmann, D.J. Eaglesham, L. Pelaz, S.B. Herner, D.C. Jacobsen, T.E. Haynes, R. Simonton. Mater. Sci. Semicond. Processing 1:17–25, 1998.
19. D. Downey, S. Daryanani, M. Meloni, K. Brown, S. Felch, B. Lee, S. Marcus, J. Gelpy. Rapid Thermal Process Requirements for the Annealing of Ultra-Shallow Junctions. Varian Semiconductor Equipment Associates Report No. 298, 1997.
20. S.M. Sze (ed.). VLSI Technology. New York: McGraw-Hill, 1981, p 174.

21. L.H. Nguyen, W. Dietl, J. Niess, Z. Nenyei, S.P. Tay, G. Obermeier, D.F. Downey. Logic in RTP. Proceedings of the 7th International Conference on Advanced Thermal Processing of Semiconductors, RTP '99, Colorado Springs, CO, Sept. 8–10, 1999, p 26.
22. S. Prussin, D.I. Margolese, R.N. Tauber. J. Appl. Phys. 57(2):180, 15 January 1985.
23. J.F. Ziegler, J.P. Biersack. The Stopping and Range of Ions in Matter. Computer program version SRIM-2000.10, copyright 1998, 1999, by IBM Co.

# 6

# Modeling of Statistical Manufacturing Sensitivity and of Process Control and Metrology Requirements for a 0.18-μm NMOSFET

**Peter M. Zeitzoff**
*International SEMATECH, Austin, Texas*

## I. INTRODUCTION

Random statistical variations during the IC manufacturing process cause corresponding variations in device electrical characteristics. These latter variations can result in substantial reductions in yield and performance, particularly as IC technology is scaled into the deep submicron regime. In an effort to understand and deal with this problem, simulation-based modeling of the manufacturing sensitivity of a representative 0.18-μm NMOSFET was carried out, and the results were used to analyze the process control and metrology requirements.

In the sensitivity analysis, eight key device electrical characteristics (the threshold voltage, source/drain leakage current, drive current, etc.; these eight characteristics are also called the *responses*) were modeled as second-order polynomial functions of nine key structural and doping parameters (gate oxide thickness, gate length, source/drain extension junction depth, etc.; these nine parameters are also called the *input parameters*). The polynomial models were embedded into a special Monte Carlo code that was used to do statistical simulations. In these simulations, the mean values and the statistical variations of the nine structural and doping parameters were the inputs, and the resulting probability density functions (pdf's) of the eight key device electrical characteristics, as well as their mean values and statistical variations, were the outputs. Through numerous Monte Carlo statistical simulations with a variety of values for the input parameter variations, the level of process control of the structural and doping parameters required to meet specified targets for the device electrical characteristics was analyzed, including the tradeoffs. In addition, the metrology requirements to properly support the establishment and maintenance of the desired level of process control were analyzed.

## II. DESIGN OF AN OPTIMAL 0.18-$\mu$m NMOSFET

Figure 1 illustrates the overall structure and many of the important structural parameters for the 0.18-$\mu$m NMOSFET (1). The eight key electrical parameters listed and defined in Table 1 were chosen to characterize the optimal device's electrical performance. In the table, DIBL means "drain induced barrier lowering," which is a short-channel effect. The primary goal was to design a device that showed maximum drive current (at least 450 $\mu$A/ $\mu$m) while satisfying the targets in the table for peak off-state leakage, DIBL, peak substrate current (to ensure hot-carrier reliability), etc. The optimal device was meant to be broadly representative of industry trends, although this is a relatively low-power transistor due to the 50-pA/$\mu$m limit on the leakage current. Due to the short gate length of the device, it was necessary to include a boron "halo" implant as part of the device structure, in order to obtain an optimal combination of turnoff and drive current performance for the device. The effectiveness of the halo implant in suppressing short-channel effects as well as maintaining hot-carrier reliability has been previously reported (2,3). This implant, along with a boron $V_T$ adjust channel implant, was found to improve both the $V_T$ rolloff with decreasing channel length and the device reliability while maintaining acceptable $I_{dsat}$ vs. $I_{leak}$ characteristics of the device. Because of the 1.8-V power supply ($V_{dd} = 1.8$ V nominal) assumed for this technology, consideration was given to ensuring hot-carrier reliability. This was done through the use of a device in which shallow source-drain (S/D) extensions were doped with a peak concentration of $4 \times 10^{19}$ cm$^{-3}$ and were self-aligned to the edge of the oxide grown on the polysilicon gate (15 nm from the polysilicon edge). The deep S/D regions were self-aligned to the spacer oxide edge and had a junction depth of 150 nm, which was held constant throughout the analysis.

The process simulators, TSUPREM-3 (4) (one-dimensional) and TSUPREM-4 (5) (two-dimensional), were used to generate the doping profiles for the various regions of the device. Due to the uncertain accuracy of two-dimensional diffusion models for arsenic-implanted junctions with short thermal cycles, the one-dimensional vertical profile of both the shallow and deep S/D junctions was simulated using TSUPREM-3. For each junction, the two-dimensional profile was then generated by extending the vertical profile laterally using a complementary error function with a characteristic length corresponding to 65% of the vertical junction depth. Conversely, the two-dimensional halo implant profile was directly simulated using TSUPREM-4. The $V_T$ adjust implant vertical profile was simu-

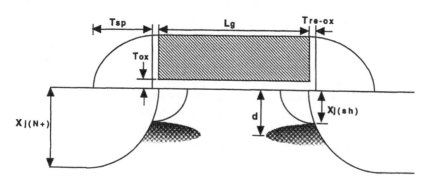

**Figure 1**   Schematic cross section of the 0.18-$\mu$m NMOSFET structure. The nominal values of the structure parameters and the maximum variations that were used in the sensitivity analysis are listed in Table 2. The polysilicon reoxidation thickness, $t_{re-ox}$, was fixed at 15 nm for all simulations.

**Table 1**   Key Device Electrical Characteristics and Target Values for the Optimal Device.

| Electrical characteristic | Target value |
|---|---|
| Threshold voltage (from extrapolated linear $I$-$V$, @ $V_d = 0.05$ V), $V_T$ (V) | $\leq 0.5$ |
| Drive current (@ $V_g = V_d = V_{dd}$), $I_{dsat}$ (μA/μm of device width) | $\geq 450$ |
| Peak off-state leakage current (@ $V_d = 2$ V, $V_g = 0$, $T = 300$K), $I_{leak}$ (pA/μm of device width) | $\leq 50$ |
| DIBL ($V_t$ @ [$V_d = 0.05$ V] $-$ $V_t$ @ [$V_d = V_{dd}$]), $\Delta V_T$ (mV) | $\leq 100$ |
| Peak substrate current (@ $V_d = V_{dd}$), $I_{sub}$ (nA/μm of device width) | $< 200$ |
| Subthreshold swing (@ $V_d = 0.05$ V), $S$ (mV/decade of $I_d$) | $\leq 90$ |
| Peak transconductance (@ $V_d = 2.0$ V); $g_m^s$ (mS/mm of device width) | $\geq 300$ |
| Peak transconductance (@ $V_d = 0.05$ V); $g_m^l$ (mS/mm of device width) | $\geq 30$ |

$V_{dd} = 1.8$ V (nominal).

lated using TSUPREM-3, and was then extended laterally without change over the entire device structure. A composite profile containing all the foregoing individual profiles was generated and imported to the device simulator, UT-MiniMOS (6) (where UT stands for University of Texas at Austin and UT-MiniMOS is a version of the MiniMOS device simulator with modifications from UT). UT-MiniMOS was chosen to simulate the device's electrical characteristics because it has both the UT hydrodynamic (HD) transport model based on nonparabolic energy bands and the UT models for substrate current (7), quantum mechanical effects (8–10), and mobility in the inversion layer (11–13). Also, UT-MiniMOS has adaptive gridding capability, and this capability was used to adapt the grid to the potential gradient and the carrier concentration gradients during the simulations.

The optimal device structure was determined by examining a large number of simulated devices with different halo peak depths and doses. For each value of the halo peak depth and dose, the boron $V_T$ adjust implant (also called the channel implant) dose was adjusted to satisfy the requirement that the maximum off-state leakage current is 50 pA/μm at room temperature (see Table 1). A number of simulations were performed to examine the ranges of variation. The result of these simulations was the selection of a boron halo implant dose of $1.5 \times 10^{13}$ cm$^{-2}$ with a peak doping profile depth of 80 nm and a boron channel implant dose of $5.65 \times 10^{12}$ cm$^{-2}$ in order to obtain maximum drive current while meeting all the other targets in Table 1.

In Table 2 the nine key structural and doping parameters for the optimal device are defined, and the optimal value for each is listed. For $L_g$, $T_{ox}$, $T_{sp}$, $X_{j(sh)}$, $N_{sh}$, and $R_s$, the optimal values were selected from technology and scaling considerations, and the values chosen are broadly representative of industry trends. For $N_{ch}$, $N_{halo}$, and $d$, the optimal values were determined from simulations aimed at defining an optimal device structure, as explained earlier.

## III.   DETERMINATION OF SECOND-ORDER MODEL EQUATIONS

A primary aim of this analysis was to obtain a set of complete, second-order empirical model equations relating variations in the structural and doping (input) parameters of the 0.18-μm NMOSFET to the resulting variations in the key device electrical characteristics listed in Table 1. (This technique is also known as *response surface methodology* (14).) In

**Table 2**  Nominal Value and Maximum Variation for Key Structural and Doping (Input) Parameters

| $i$ | Input parameter | Optimal, nominal value | Maximum variation |
|---|---|---|---|
| 1 | Gate length, $L_g$ (μm) | 0.18 | ±15% |
| 2 | Gate oxide thickness, $T_{ox}$ (nm) | 4.50 | ±10% |
| 3 | Spacer oxide width, $T_{sp}$ (nm) | 82.50 | ±15% |
| 4 | Shallow-junction doping profile depth, $X_{j(sh)}$ (nm), @ $N_D = 4.36 \times 10^{18}$ cm$^{-3}$ (where $N_D = N_A$ for the nominal device) | 50 | ±10% |
| 5 | Peak shallow-junction doping, $N_{sh}$ (cm$^{-3}$) | $4 \times 10^{19}$ | ±10% |
| 6 | Channel dose, $N_{ch}$ (cm$^{-2}$) | $5.65 \times 10^{12}$ | ±10% |
| 7 | Halo dose, $N_{halo}$ (cm$^{-2}$) | $1.5 \times 10^{13}$ | ±10% |
| 8 | Halo peak depth, $d$ (nm) | 80 | ±10% |
| 9 | Series resistance (external), $R_s$ (Ω-μm) | 400 | ±15% |

this analysis, the nominal or design center device was identical to the optimal NMOSFET from the previous section, and the variations were with respect to this nominal device. Hence, the "optimal" values of the input parameters in Table 2 are also the "nominal" values for this analysis. Also listed in Table 2 are the maximum variation limits for each input parameter. Since the model equations are accurate only for variations less than or equal to these maximum limits, these limits were intentionally chosen to be large to give a wide range of validity to the model equations. However, typical IC manufacturing lines have manufacturing statistical variations considerably less than the maximum variations listed in the table. In the next section, Monte Carlo simulations employing the model equations were used to explore the impact of smaller, more realistic variations.

A three-level Box–Behnken design (15) was performed in order to obtain the responses of the output parameters to the input parameters. Besides the centerpoint, where all factors were maintained at their nominal values, the other data points were obtained by taking three factors at a time and developing a $2^3$ factorial design for them, with all other factors maintained at their nominal values. The advantage of this design was that fewer simulations were required to obtain a quadratic equation as compared to other designs. A total of 97 simulations (96 variations plus the one nominal device simulation) was required for this analysis for the case of nine input factors. One drawback of this design, however, is that all of the runs must be performed prior to obtaining any equation, and it is not amenable to two-stage analyses. Hence, there is no indication of the level of factor influence until the entire experiment has been conducted.

In contrast to the nine input parameters that were varied (see Table 2), several device parameters, such as the deep S/D junction profile and its peak doping, were held constant throughout all of the simulations. In addition, a background substrate doping of $5 \times 10^{15}$ cm$^{-3}$ and an interface charge of $3 \times 10^{10}$ cm$^{-2}$ were uniformly applied. The eight key device electrical characteristics listed in Table 1 were the response variables. After the completion of the 97 simulations, two sets of model equations were generated for each response, one in terms of the actual values of the input parameters, and the other in terms of their normalized values. The normalized values were calculated using the following equation:

$$\Delta x_i = 2\frac{(\xi_i - \bar{\xi}_i)}{d_i}, \qquad i = 1, 2, \ldots, 9 \tag{1}$$

where $\bar{\xi}_i$ is the nominal value of the $i$th input parameter, $\xi_i$ is the actual value of the $i$th input parameter for any given run, $\Delta x_i$ is the normalized value of the $i$th input parameter, and $d_i$ is the magnitude of the difference between the two extreme values for $\xi_i$ used in the sensitivity analysis. Note that, from the definition in Eq. (1), it is clear that $-1 \le \Delta x_i \le 1$, and $\Delta x_i = -1$ corresponds to $\xi_i$ equal to its minimum possible value, $\Delta x_i = 0$ corresponds to $\xi_i$ equal to the nominal value, and $\Delta x_i = 1$ corresponds to $\xi_i$ equal to its maximum possible value. Furthermore, for any input parameter, $d_i$ is twice the maximum variation listed in Table 2 (for example, $i = 1$ for gate length, then $\Delta x_1 = \Delta L_g$ and $d_1 = 2 * 0.15 * 0.18\,\mu m = 0.054\,\mu m$). The equations that use the $\Delta x_i$'s for their variables will be called *normalized model equations*. In the remainder of this chapter, only the normalized equations will be considered.

After each of the 97 simulations was performed, the eight electrical responses were extracted for each device and then entered into a design matrix. Analysis of variance (ANOVA) methods were used to estimate the coefficients of the second-order model equations and to test for the significance of each term in the model (16). For the model as a whole, information such as the coefficient of determination, $R^2$, and coefficient of variation was generated using the data from the ANOVA. A diagnosis of the model was performed by using normal probability plots to evaluate the normality of the residuals. A transformation of the response was made if necessary. The ANOVA was performed again if a transformation was made, and comparisons of the normal probability plots and coefficients of determination were made to decide whether to use the transformed data or not. In addition, plots of Cook's distance vs. run order and Outlier-$t$ vs. run order were generated in order to check for the occurrence of any extraneous data (outliers). The corresponding model was then used to generate a set of reduced equations, i.e., equations including only those terms with at least a 95% level of significance (terms with a lower significance value were discarded from the ANOVA). As mentioned before, for input parameters outside of the range of values used in this analysis, the model equations are not guaranteed to be accurate, and they are expected to become less and less accurate as the values move further and further outside the range.

The final resulting normalized model equations for the eight key electrical responses are listed in the appendix at the end of this chapter. As mentioned earlier, all of the input parameters take on a value between $-1$ and $+1$, as determined by Eq. (1) . The advantages of the normalized equations are:

Because the input variables are dimensionless, the coefficients are independent of units.

The relative importance of any term is determined solely by the relative magnitude of the coefficient of that term. For example, in the model equation for the saturation drive current, $I_{dsat}$ is most sensitive to the normalized gate length ($\Delta L_g$), followed by the oxide thickness ($\Delta T_{ox}$), the shallow junction depth ($\Delta X_{sh}$), the spacer oxide width ($\Delta T_{sp}$), and the channel dose, $\Delta N_{ch}$. Also, this attribute of the normalized equations simplifies the generation of reduced equations by dropping less significant terms.

For all the normalized parameters, the mean value is zero and, as will be explained later, the maximum value of the standard deviation is $1/3$.

In some cases, the output responses were transformed into a form that yielded a more reliable model equation. Typically, if the residual analysis of the data suggests that, contrary to assumption, the standard deviation (or the variance) is a function of the mean, then there may be a convenient data transformation, $Y = f(y)$, that has a constant variance. If the variance is a function of the mean, the dependence exhibited by the variance typically has either a logarithmic or a power dependence, and an appropriate variance-stabilizing transformation can be performed on the data. Once the response is transformed, the model calculations and coefficients are all in terms of the transformed response, and the resulting empirical equation is also in terms of the transformation. In this study, both a logarithmic and an inverse square root dependence were used in transforming three of the output responses so that a better model equation fit was obtained (see the appendix).

## IV. MONTE CARLO–BASED DETERMINATION OF PROCESS CONTROL REQUIREMENTS

Monte Carlo simulations were utilized for several purposes.

1.  For a given set of input parameter statistical variations, to determine the probability density function (pdf) and the resulting mean value and statistical variation for each of the device electrical characteristics
2.  To determine the impact on the statistical variations of the device electrical parameters of changing the input parameter statistical variations by arbitrary amounts
3.  To analyze whether the device electrical targets (for example, the maximum leakage current specification) are reasonable
4.  To find an optimal set of reductions in the input parameter variations to meet device electrical targets
5.  To analyze process control and metrology requirements

The normalized model equations obtained in the previous section were utilized in Monte Carlo simulations to extract the pdf of each response for a specified set of statistical variations of the nine normalized structural and doping parameters. A special Monte Carlo code was written that treats each of these parameters as a random input variable in the normalized model equations. This is a good approximation, since any correlations among these parameters are weak and second order. Each of the random input variables comes from a normal distribution with a specified mean and standard deviation, $\sigma_{iN}$, where the "$N$" designates that this is for a normalized variable. (Note that, due to the definition of the normalized parameters, the mean value for all of them is zero. Furthermore, the maximum allowable value of all the standard deviations is set by the "maximum variations" in Table 2 and corresponds to $3\sigma_{iN} = 1$, or $\sigma_{iN} = 1/3$, for all of the input variables.) For each trial and for each of the nine normalized input parameters, the Monte Carlo code randomly selected a value by using a random number generator that returns a series of nonuniform deviates chosen randomly from a normal distribution that correctly predicts the specified standard deviation. For each trial, the set of randomly generated values for the nine input parameters was used as inputs to the second-order normalized model equations, which were evaluated to calculate the values of each of the eight electrical characteristics. A large number of such trials (typically 5000–10,000) was

**Table 3**  Monte Carlo Results with Maximum Input Parameter Statistical Variations ($3\sigma_{iN} = 1$ for all $i$)

| Response (key device electrical characteristics) | Critical parameter | Target value for critical parameter | Monte Carlo simulated value for critical parameter |
|---|---|---|---|
| $V_T$ (mV), extrapolated | $3\sigma$ variation | < 50–60 | **97.1** |
| $\Delta V_T$ (mV), DIBL | Maximum value | < 100 | **121.8** |
| $I_{dsat}$ (μA/μm) | Minimum value | > 450 | **430.2** |
| $I_{leak}$ (pA/μm) | Maximum value | < 50 | **95.5** |
| $I_{sub}$ (nA/μm) | Maximum value | < 200 | 55.6 |
| $S$ (mV/decade) | Maximum value | ≤ 90 | 86.8 |
| $g_m^s$ (mS/mm) | Minimum value | > 300 | 397.8 |
| $g_m^l$ (mS/mm) | Minimum value | > 30 | 48.0 |

*Note:* The bold font indicates critical parameter values which do not meet their respective target.

run to generate the pdf's of the characteristics. The pdf's were then analyzed to obtain the mean and standard deviation of each of the device electrical characteristics.

In Table 3, the definition of *critical parameter* and the target value for this parameter are listed for each of the responses. The target values are meant to be broadly representative of industry trends. The critical parameter is either the $3\sigma$ statistical variation or the maximum or minimum value of the response, where the maximum value is calculated as the mean value + [$3\sigma$ statistical variation], while the minimum value is calculated as the mean value − [$3\sigma$ statistical variation]. In this set of Monte Carlo simulations, all the input parameter variations, the $\sigma_{iN}$'s were set to the maximum value of $1/3$, corresponding to the "Maximum variations" in Table 2. Figure 2 shows a typical pdf, for the substrate current, $I_{sub}$. The mean value and the standard deviation, $\sigma$, are listed at the top. The crosses indicate the Monte Carlo–simulated pdf, while the solid curve is a fitted Normal probability distribution with the same mean and $\sigma$. The pdf is clearly not a Normal distribution, although the input parameter statistical distributions are Normal. The non-Normal, skewed pdf for $I_{sub}$ (and the other responses) is due to the nonlinear nature of the model equations (17), and the amount of skew and the departure from the Normal distribution vary considerably from response to response. The Monte Carlo simulation results are listed in Table 3, where the mean value and $\sigma$ from the Monte Carlo simulations were used to calculate the "simulated value" in the last column. The targets for the first four responses in the table ($V_T$, $\Delta V_T$ due to DIBL, $I_{dsat}$, and $I_{leak}$) were not met, but the targets for the last four parameters in the table ($I_{sub}$, $S$, $g_m^s$, and $g_m^l$) were met. To bracket the problem, and to determine whether the targets for the first four responses are realistic, all the input parameter variations were reduced in two stages, first to a set of more "realistic" values and second to a set of "aggressive" values. These sets are listed in Table 4; they reflect the judgment of several SEMATECH experts (18). In the table, both the statistical variations of the normalized input parameters, $3\sigma_{iN}$, and the corresponding statistical variations in percentage terms of the input parameters are listed. A Monte Carlo simulation was performed for each of these sets of variations, and the simulation results are listed in Table 5. The targets for the third and fourth responses, $I_{dsat}$ and $I_{leak}$, were satisfied with the "realistic" input variations, but the targets for $V_T$ and $\Delta V_T$ were satisfied only with the "aggressive" input variations. The conclusion is that the targets for all the responses can probably be met but that it will be especially difficult to meet them for $V_T$ and $\Delta V_T$ (DIBL).

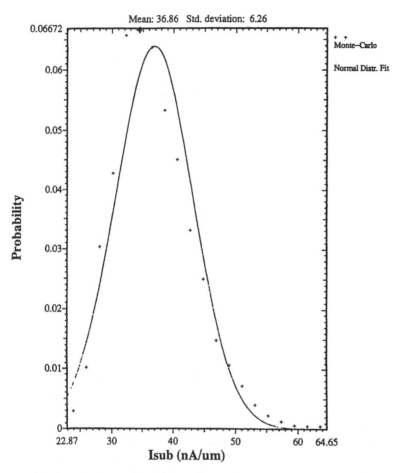

**Figure 2**  Probability density function (pdf) of substrate current ($I_{sub}$), from Monte Carlo simulation. The input parameter statistical variations are maximum: $3\sigma_{iN} = 1$ for all $i$.

**Table 4**  Sets of Normalized Input Parameter Statistical Variations, $3\sigma_{iN}$ [a]

| Input parameter | Maximum statistical variation | | Realistic statistical variation | | Aggressive statistical variation | |
|---|---|---|---|---|---|---|
| $\Delta L_g = \Delta x_1$ | 1 | ($\pm 15\%$) | 2/3 | ($\pm 10\%$) | 1/2 | ($\pm 7.5\%$) |
| $\Delta T_{ox} = \Delta x_2$ | 1 | ($\pm 10\%$) | 1/2 | ($\pm 5\%$) | 1/4 | ($\pm 2.5\%$) |
| $\Delta T_{sp} = \Delta x_3$ | 1 | ($\pm 15\%$) | 2/3 | ($\pm 10\%$) | 1/3 | ($\pm 5\%$) |
| $\Delta X_{sh} = \Delta x_4$ | 1 | ($\pm 10\%$) | 1/2 | ($\pm 5\%$) | 1/4 | ($\pm 2.5\%$) |
| $\Delta N_{sh} = \Delta x_5$ | 1 | ($\pm 10\%$) | 1 | ($\pm 10\%$) | 3/4 | ($\pm 7.5\%$) |
| $\Delta d = \Delta x_6$ | 1 | ($\pm 10\%$) | 1/2 | ($\pm 5\%$) | 1/4 | ($\pm 2.5\%$) |
| $\Delta N_{halo} = \Delta x_7$ | 1 | ($\pm 10\%$) | 3/4 | ($\pm 7.5\%$) | 1/2 | ($\pm 5\%$) |
| $\Delta N_{ch} = \Delta x_8$ | 1 | ($\pm 10\%$) | 1 | ($\pm 10\%$) | 3/4 | ($\pm 7.5\%$) |
| $\Delta R_s = \Delta x_9$ | 1 | ($\pm 15\%$) | 2/3 | ($\pm 10\%$) | 1/2 | ($\pm 7.5\%$) |

[a] For each $3\sigma_{iN}$, the corresponding statistical variation as a percentage of the mean value of the non-normalized input parameter is in parentheses.

**Table 5** Monte Carlo Results for Different Levels of Input Parameter Statistical Variations

| Responses (key device electrical parameters) | Critical parameter | Target value | Monte Carlo simulated value of critical parameter | | |
|---|---|---|---|---|---|
| | | | Maximum input parameter variation | Realistic input parameter variation | Aggressive input parameter variation |
| $V_T$ (mV), extrapolated | 3σ variation | < 50–60 | **±97.1** | **±66.1** | ±46.0 |
| $\Delta V_T$ (mV), DIBL | Maximum value | < 100 | **121.8** | **104.8** | 95.8 |
| $I_{dsat}$ (mA/μm) | Minimum value | > 450 | **430.2** | 457.8 | 482.6 |
| $I_{leak}$ (pA/μm) | Maximum value | < 50 | **95.5** | 30.3 | 15.3 |
| $I_{sub}$ (nA/μm) | Maximum value | < 200 | 55.6 | 49.4 | 46.2 |
| $S$ (mV/decade) | Maximum value | ≤ 90 | 86.8 | 85.7 | 85.1 |
| $g_m^s$ (mS/mm) | Minimum value | Maximize | 397.8 | 427.8 | 442.0 |
| $g_m^l$ (mS/mm) | Minimum value | Maximize | 48.0 | 52.1 | 54.1 |

*Note:* The bold font indicates critical parameter values that do not meet their respective targets.

An important point is that there is a difference between the nominal values for the eight key device electrical characteristics and the mean values for these characteristics from the Monte Carlo statistical simulations. This is undesirable, since the nominal is the optimal device center calculated in the design optimization section. Ideally, the difference between the mean and the nominal should be zero, and in any case it should be minimized. From an analysis of the model equations it can be shown that this difference is dependent on the size of the second-order terms. The form of the equations is

$$y_i = A_i + \sum_j B_{ij}\,\Delta x_j + \sum_{j,k;j\neq k} C_{ijk}\left(\Delta x_j\,\Delta x_k\right) + \sum_j D_{ij}\left(\Delta x_j\right)^2 \tag{2}$$

where $y_i$ is one of the eight key electrical device characteristics and $A_i$, $B_{ij}$, $C_{ijk}$, and $D_{ij}$ are coefficients. For the optimal value of $y_i$ from the design optimization (denoted by $y_{i,\text{opt}}$), all the $\Delta x_j$'s are zero, since $y_{i,\text{opt}}$ corresponds to all input parameters at their nominal values and hence all $\Delta x_j$'s set to zero. Then

$$y_{i,\text{opt}} = A_i \tag{3}$$

However, for the mean value of $y_i$, denoted by $\langle y_i \rangle$:

$$\langle y_i \rangle = A_i + \sum_j B_{ij}\langle \Delta x_j \rangle + \sum_{j,k;j\neq k} C_{ijk}\langle(\Delta x_j\,\Delta x_k)\rangle + \sum_j D_{ij}\langle(\Delta x_j)^2\rangle \tag{4}$$

However, for each $\Delta x_j$ the probability distribution is the Normal distribution centered about zero. Because this distribution is symmetric about its center, $\langle \Delta x_j \rangle = \langle(\Delta x_j\,\Delta x_k)\rangle = 0$, since $\Delta x_j$ and $(\Delta x_j\,\Delta x_k)$ are odd functions. On the other hand, $(\Delta x_j)^2$ is an even function; hence $\langle(\Delta x_j)^2\rangle \neq 0$(19). In fact, $\sigma_{jN}^2 = \langle(\Delta x_j)^2\rangle - (\langle \Delta x_j \rangle)^2$, and since $\langle \Delta x_j \rangle = 0$, $\langle(\Delta x_j)^2\rangle = \sigma_{jN}^2$. Hence,

$$\langle y_i \rangle - y_{i,\text{opt}} = \sum_j D_{ij}\sigma_{jN}^2 \tag{5}$$

Clearly, the nonlinear, second-order relationship between the responses and the input parameters causes a shift in the mean value of the responses from their optimal values. Using Eqs. (3) and (5):

$$\frac{\langle y_i \rangle - y_{i,\text{opt}}}{y_{i,\text{opt}}} = \frac{\sum_j D_{ij}\sigma_{jN}^2}{A_i} \tag{6}$$

The right-hand side of Eq. (6) can be used as a metric to evaluate the expected relative difference between the mean value and the optimal value for any of the responses. This metric, call it the *expected shift of the mean*, can be directly evaluated from the normalized model equation before a Monte Carlo simulation is run to determine $\langle y_i \rangle$. After such a simulation is run and $\langle y_i \rangle$ is determined, the expected shift of the mean can be compared to the "actual shift of the mean," $(\langle y_i \rangle - y_{i,\text{opt}})/y_{i,\text{opt}}$. These calculations were done for the case where all the normalized input parameter statistical variations are maximum, $3\sigma_{iN} = 1$. The results are listed in Table 6 for all the responses. For all the responses except the leakage current, the absolute value of the actual shift of the mean is small, at less than 5% in all cases and less than 1% in most, and the expected and actual shifts are quite close to each other. Even for leakage current, the actual shift of the mean is a tolerable 11%, but the expected and actual shifts are relatively far apart.

Next, Monte Carlo simulation was used to meet the targets for the output parameters with an optimal set of reductions in the input parameter statistical variations. Each input parameter statistical variation was reduced in steps, as listed in Table 7. (Note that,

**Table 6** From Monte Carlo Simulations, the Mean, Standard Deviation ($\sigma$), and Actual Shift of the Mean; From the Normalized Model Equations, the Expected Shift of the Mean

| $i$ | $y_i$ = response | $A_i = y_{i,\text{opt}}$ | Mean = $\langle y_i \rangle$ | $\sigma$ | Actual shift of mean $(\langle y_i \rangle - y_{i,\text{opt}})/y_{i,\text{opt}}$ | Expected shift of mean $(1/9)\Sigma_j(D_{ij}/A_i)$ |
|---|---|---|---|---|---|---|
| 1 | $V_t$ (mV) | 454.9 | 452.4 | 32.4 | −0.00550 | −0.00630 |
| 2 | $\Delta V_T$, DIBL (mV) | 72.3 | 74.1 | 15.9 | 0.02490 | 0.02751 |
| 3 | $\log(I_{\text{sat}})$ (uA/um) | 2.721 | 2.7243 | 0.03 | 0.00121 | 0.00135 |
| 4 | $\log(I_{\text{leak}})$ (pA/um) | 0.385 | 0.428 | 0.517 | 0.11169 | −0.08139 |
| 5 | $I_{\text{sub}}$ (nA/um) | 35.4 | 36.9 | 6.3 | 0.04237 | 0.04112 |
| 6 | $S^{-1/2}$ (mV/dec.)$^{-1/2}$ | 0.10916 | 0.10906 | | −0.00092 | −0.00080 |
| 7 | $g_m^s$ (mS/mm) | 470.6 | 467.4 | 23.2 | −0.00680 | −0.00633 |
| 8 | $g_m^s$ (mS/mm) | 58.63 | 59.2 | 3.7 | 0.00972 | 0.01006 |

for each input parameter, the maximum variation is the same as that used in the previous section [see Table 2] and earlier in this section [see Table 4], and the minimum is half or less than half of the maximum.) The straightforward approach is to run a series of Monte Carlo simulations covering the entire range of possible combinations for the input parameter variations. However, the number of simulations is 46,656 for each response (see Table 7), an unreasonably high number. In order to reduce the number of simulations to a more manageable total, the following procedure was used. For each response, the normalized model equation was examined to select those input parameters that are either missing from the equation or included only in terms with small coefficients. Since, as noted previously, these inputs are unimportant in influencing the response, the variation was held fixed at its maximum value for each of these selected parameters. As shown in Table 8, following this procedure, two parameters were selected for each response, and hence the number of Monte Carlo simulations was reduced to a more manageable 3888 or 5184. Since each Monte Carlo simulation took about 3 seconds to run on a Hewlett-Packard workstation, the total simulation time was about 3–4 hours for each response. (Table 8 does not include listings for $I_{\text{sub}}$, $S$, $g_{ml}$, or $g_m^s$, since those responses are within specification for the maximum values for the variation of all the input parameters, as shown in Table 3.) The outputs from the Monte Carlo simulations were imported to a spreadsheet program for analysis and display. By utilizing the spreadsheet capabilities, the input

**Table 7** Steps in Input Parameter Statistical Variation, $3\sigma_{iN}{}^a$

| Parameters | Maximum variation | Minimum variation | Step size | No. of steps | No. of combinations |
|---|---|---|---|---|---|
| $\Delta L_g$, $\Delta X_{\text{sh}}$, $\Delta N_{\text{sh}}$, $\Delta d$, $\Delta N_{\text{halo}}$, $\Delta N_{\text{ch}}$ | 1 (10%) | 1/2 (5%) | 1/4 (2.5%) | 3 | $3^6 = 729$ |
| $\Delta L_g$, $\Delta T_{\text{sp}}$, $\Delta R_s$ | 1 (15%) | 0.3 (4.5%) | 0.233 (3.5%) | 4 | $4^3 = 64$ |
| Total number of combinations | $64 \times 729 =$ | | | | 46,656 |

$^a$ For each $3\sigma_{iN}$, the corresponding statistical variation as a percentage of the mean value of the non-normalized input parameter is in parentheses.

**Table 8**  Number of Steps in Input Parameter Statistical Variation

| Response | $\Delta L_g$ | $\Delta T_{ox}$ | $\Delta T_{sp}$ | $\Delta X_{sh}$ | $\Delta N_{sh}$ | $\Delta d$ | $\Delta N_{halo}$ | $\Delta N_{ch}$ | $\Delta R_s$ | No. of combinations |
|---|---|---|---|---|---|---|---|---|---|---|
| $V_T$ | 4 | 3 | 4 | 3 | 3 | 1 | 3 | 3 | 1 | 3888 |
| $\Delta V_T$ | 4 | 3 | 4 | 3 | 3 | 3 | 3 | **1** | 1 | 3888 |
| $I_{dsat}$ | 4 | 3 | 4 | 3 | **1** | 1 | 3 | 3 | 4 | 5184 |
| $I_{leak}$ | 4 | 3 | 4 | 3 | 3 | 1 | 3 | 3 | 1 | 3888 |

*Note*: The bold font indicates that the number of steps has been reduced from the number in Table 7.

variations were then iteratively reduced from their maximum values to meet the targets for the responses.

As already discussed, it was most difficult to meet the targets for $V_T$ and for $\Delta V_T$ due to DIBL. Hence, these two were dealt with first. From the size of the coefficients in the normalized model equation for $V_T$, the terms containing $\Delta L_g$, $\Delta T_{ox}$, $\Delta X_{sh}$, and $\Delta N_{ch}$ are the most significant. Thus, the statistical variations of only these parameters were reduced to meet the $V_T$ target, while the variations of the other input parameters were held at their maximum values. Contour plots of constant $3\sigma$ variation in $V_T$ were determined using the spreadsheet program. The results are shown in Figure 3, where the statistical variations of $T_{ox}$ and $X_{sh}$ were fixed at their realistic values of 5% each (corresponding to $3\sigma_{iN} = 1/2$), and the statistical variations of $L_g$ and $N_{ch}$ were varied. Along Contour 1, the $3\sigma$ variation in $V_T$ is 50 mV, and the variations of both $L_g$ and $N_{ch}$ are less than 7.5%. Since these variations are quite aggressive (see Table 4), the 50-mV target will be difficult to meet. Along Contour 2, the $3\sigma$ variation in $V_T$ is 60 mV. This target is realistic because the variations of both $L_g$ and $N_{ch}$ on the contour are achievable, particularly in the vicinity of the point where the variations are about 9.5% for $L_g$ and 7.5% for $N_{ch}$ (see Table 4). Figure 4 also shows contours of constant $3\sigma$ variation in $V_T$; the only difference from Figure 3 is that the statistical variation of $X_{sh}$ is 7.5%, not 5% as in Figure 3. The 60-mV

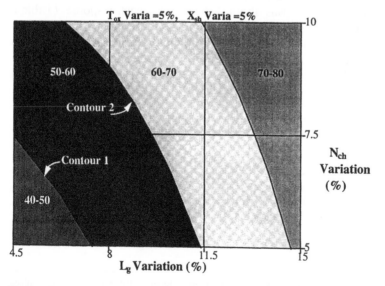

**Figure 3**  Contours of constant $3\sigma$ variation in $V_T$ (mV).

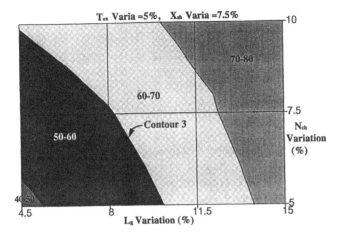

**Figure 4**  Contours of constant $3\sigma$ variation in $V_T$ (mV), with $X_{sh}$ variation of 7.5%.

contour here, labeled Contour 3, is shifted significantly to the left from the 60-mV contour in Figure 3 and hence is much more difficult to achieve. For the case where the variation of $T_{ox}$ is 7.5% while that of $X_{sh}$ is 5%, the 60-mV contour is shifted even further to the left than Contour 3. The contour plots can be utilized to understand quantitatively the impact of the statistical variations of the key input parameters and how they can be traded off to reach a specific target for $V_T$ variation. Looking particularly at Contour 2 in Figure 3, and utilizing "realistic" values of the variations as much as possible (see Table 4), an optimal choice for the variations is 5% for $T_{ox}$ and $X_{sh}$, 7.5% for $N_{ch}$, and 9.5% for $L_g$.

Next, the requirements to meet the target for $\Delta V_T$ due to DIBL were explored. From the size of the coefficients in the normalized model equation for $\Delta V_T$, the terms containing $\Delta L_g$, $\Delta X_{sh}$, $\Delta T_{sp}$, and $T_{ox}$ are the most significant. Thus, the variations of only these parameters were reduced to meet the $\Delta V_T$ target, while the variations of the other input parameters were held at their maximum values. Figure 5 shows contours of constant

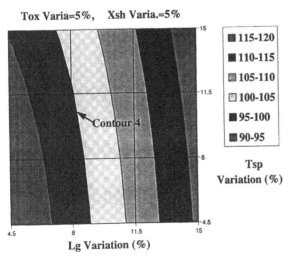

**Figure 5**  Contours of constant maximum $\Delta V_T$ (DIBL) (mV).

maximum value of $\Delta V_T$, where the variations of both $T_{\mathrm{ox}}$ and $X_{\mathrm{sh}}$ were held at 5%, as in Figure 3. Along Contour 4, the value is 100 mV, the target value. The variations of $L_g$ and $T_{\mathrm{sp}}$ on this contour are realizable, particularly in the vicinity of the point where the variations are about 9% for $L_g$ and 8% for $T_{\mathrm{sp}}$. Utilizing the same reasoning as discussed earlier for meeting the $V_T$ target, this point is about optimal.

Finally, the requirements to meet the targets for $I_{\mathrm{dsat}}$ and $I_{\mathrm{leak}}$ were explored. From the size of the coefficients in the normalized model equations, the terms containing $\Delta L_g$, $\Delta T_{\mathrm{ox}}$, and $\Delta X_{\mathrm{sh}}$ are the most significant for $I_{\mathrm{dsat}}$, while the terms containing $\Delta L_g$, $\Delta X_{\mathrm{sh}}$, $\Delta N_{\mathrm{ch}}$, and $\Delta T_{\mathrm{ox}}$ are most significant for $I_{\mathrm{leak}}$. Figure 6 shows the contours of constant minimum value of $I_{\mathrm{dsat}}$, with the variation of $T_{\mathrm{ox}}$ held at 5% as in Figures 2, 3, and 4. Along Contour 5, the minimum $I_{\mathrm{dsat}}$ is 450 $\mu$A/$\mu$m, the target value. Similarly, Figure 7 shows the contours of constant maximum value of $I_{\mathrm{leak}}$, with the variations of all input parameters except $X_{\mathrm{sh}}$ and $L_g$ held at their maximum variations. Along Contour 6, the maximum $I_{\mathrm{leak}}$ is 50 pA/$\mu$m, the target value. The input parameter variations along both Contours 5 and 6 are significantly larger than those required to meet the targets for $V_T$ and $\Delta V_T$ due to DIBL (see Contour 2 in Figure 3 and Contour 4 in Figure 5). Hence, if the $V_T$ and $\Delta V_T$ targets are met, then the targets for $I_{\mathrm{leak}}$ and $I_{\mathrm{dsat}}$ are also automatically met.

Tying all the foregoing results and discussion together, only five input parameters, $L_g$, $T_{\mathrm{ox}}$, $X_{\mathrm{sh}}$, $N_{\mathrm{ch}}$, and $T_{\mathrm{sp}}$, need to be tightly controlled (i.e., the statistical variation of each of them must be notably less than the "Maximum statistical variation" in Table 4) to meet the targets in Table 3 for the device electrical characteristics. The other four input parameters, $d$, $N_{\mathrm{halo}}$, $N_{\mathrm{sh}}$, and $R_s$, can be relatively loosely controlled (i.e., the statistical variation of each of them can be equal to or possibly larger than the maximum variation in Table 4), and the targets will still be met. An optimal set of choices that satisfies all the output response targets is: statistical variation of $L_g \leq 9\%$, statistical variation of $T_{\mathrm{ox}} \leq 5\%$, statistical variation of $X_{\mathrm{sh}} \leq 5\%$, statistical variation of $N_{\mathrm{ch}} \leq 7.5\%$, and statistical variation of $T_{\mathrm{sp}} \leq 8\%$. Note that this optimal set is challenging, since the statistical variation values range from "realistic" to "aggressive" according to the classification in Table 4. In particular, the 9% requirement on gate length control will be pushing the limits, since, according to the *1999 International Technology Roadmap for Semiconductors*

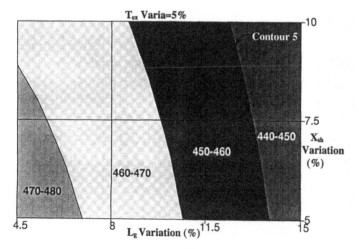

**Figure 6**   Contours of constant minimum $I_{\mathrm{dsat}}$ ($\mu$A/$\mu$m).

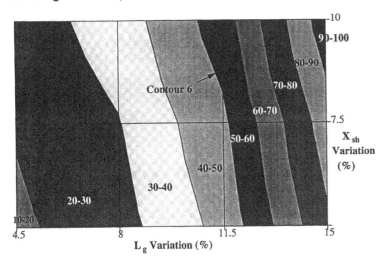

**Figure 7** Contours of constant maximum $I_{leak}$. Maximum statistical variation for all input parameters except $L_g$ and $X_{sh}$.

(20), the gate CD (critical dimension) control is 10%. This approach can also be used to determine tradeoffs. If, for example, the process variation of $T_{sp}$ can only be controlled to 12%, it is evident from Contour 4 of Figure 5 that the control of $L_g$ would have to be tightened so that its process variation is 8% or less.

For process control purposes, let UL be the upper specification limit for the process, let LL be the lower specification limit, and let MEAN be the mean value for the structural and doping parameters. A very important quantity is the "process capability," $C_p$. For the $i$th input parameter, $C_p = (\text{UL} - \text{LL})/6\sigma_i$, where $\sigma_i$ is the standard deviation of the $i$th (non-normalized) input parameter. The goal is to control the process variations and the resulting $\sigma_i$ so that $C_p \sim 1$. For $C_p$ much less than 1, a non-negligible percentage of the product is rejected (i.e., noticeable yield loss) because of input parameter values outside the process limits, as illustrated schematically in Figure 8. For $C_p \sim 1$, the statistical distribution of the input parameter values is largely contained just within the process limits, so the cost and difficulty of process control are minimized, but very little product is rejected because of input parameter values outside the process limits. Finally, for $C_p$ much larger than 1, the actual statistical distribution of the input parameter is much narrower than the process limits, and hence very little product is rejected, but the cost and difficulty of process control are greater than for the optimal case, where $C_p \sim 1$. In practice, because of nonidealities in IC manufacturing lines and the difficulty of setting very precise process limits, the target $C_p$ is typically somewhat larger than 1, with 1.3 being a reasonable rule of thumb (21,22). In practical utilization of the previous Monte Carlo results, especially the optimal set of variations, it makes sense to set UL = (MEAN + optimal $3\sigma_i$ variation) and LL = (MEAN − optimal $3\sigma_i$ variation) for all of the key structural and doping parameters. Using these formulas, the values of LL and UL are listed in Table 9 for all the input parameters.

Meeting the process control requirements for the statistical variation of the five key input parameters is dependent on the control at the process module level. For example, to meet the channel implant dose ($N_{ch}$) requirement, the channel implant dose and energy as well as the thickness of any screen oxide must be well controlled. As another example, to

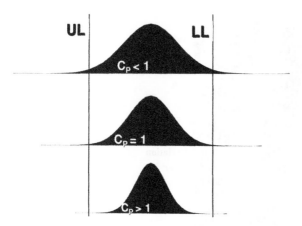

**Figure 8**   Impact of $C_p$ variation.

meet the $T_{ox}$ requirement, the gas flows, temperature, and time at temperature for a furnace process must be well controlled. Through empirical data or simulations, the level of control of the process modules necessary to meet the requirements on the input parameter statistical variations can be determined. Of course the tighter the requirements on the input parameter variations, the tighter the required level of control of the process modules.

## V.   METROLOGY REQUIREMENTS

The metrology requirements are driven by the process control requirements, i.e., the UL and LL in Table 9 for the input parameters. In-line metrology is used routinely in the IC fabrication line to monitor these parameters, to ensure that they stay between the LL and UL, or to raise an alarm if they drift out of specification. For in-line metrology on a well-established, well-characterized line, the most important characteristic is the measurement precision, $P$, where $P$ measures the repeatability of the measurement. For a measurement of a given parameter with a particular piece of measurement equipment (for example, a measurement of $T_{ox}$ using a particular ellipsometer), $P$ is determined by making repeated measurements of $T_{ox}$ on the same wafer at the same point. $P$ is defined to be $6\sigma_{METROL}$, where $\sigma_{METROL}$ is the standard deviation of the set of repeated measurements (23). A key parameter is the ratio of measurement precision, $P$, to the process tolerance, $T$, where $T = UL - LL$. Then $P/T = 6\sigma_{METROL}/(UL - LL)$. It is important that $P/T \ll 1$, or measurement errors will reduce the apparent $C_p$ (or, equivalently, the apparent process standard deviation will increase). This point is illustrated in Figure 9, where a simplified process control chart for one of the key doping and structural parameters, such as the gate oxide thickness or gate length, is shown. The position of each X indicates the value of an in-line measurement (or the mean value of multiple measurements on one or several wafers from a lot) if there were no metrology errors. The cross-hatched area around each X indicates the uncertainty introduced because of the random errors from the metrology, i.e., the influence of the nonzero $\sigma_{METROL}$. In the following, let $\sigma_{PROC}$ be the statistical standard deviation due to random process variations, as discussed in the previous sections. In Case 1 and Case 2 in the figure, $C_p \sim 1$, so $6\sigma_{PROC} \sim (UL - LL)$. In Case 1, $P/T \ll 1$,

**Table 9** Nominal Values and Optimal Statistical Variations for Input Parameters

| Input parameter | MEAN | Maximum $3\sigma_i$ variation (optimal case) [%] | Maximum $3\sigma_i$ variation (optimal case) [units] | UL (MEAN + $3\sigma_i$) | LL (MEAN − $3\sigma_i$) |
|---|---|---|---|---|---|
| Gate length, $L_g$ (μm) | 0.18 μm | ±9% | ±0.0162 μm | 0.196 μm | 0.164 μm |
| Gate oxide thickness, $T_{ox}$ (nm) | 4.50 nm | ±5% | ±0.225 nm | 4.73 nm | 4.28 nm |
| Spacer oxide width, $T_{sp}$ (nm) | 82.5 nm | ±8% | ±6.6 nm | 89.1 nm | 75.9 nm |
| Shallow-junction doping profile depth, $X_{j(sh)}$ (nm) | 50 nm | ±5% | ±2.5 nm | 52.5 nm | 47.5 nm |
| Peak shallow-junction doping, $N_{sh}$ (cm$^{-3}$) | $4 \times 10^{19}$ cm$^{-3}$ | ±10% | ±$4 \times 10^{18}$ cm$^{-3}$ | $4.4 \times 10^{19}$ cm$^{-3}$ | $3.6 \times 10^{19}$ cm$^{-3}$ |
| Channel dose, $N_{ch}$ (cm$^{-2}$) | $5.65 \times 10^{12}$ cm$^{-2}$ | ±7.5% | ±$4.24 \times 10^{11}$ cm$^{-2}$ | $6.07 \times 10^{12}$ cm$^{-2}$ | $5.23 \times 10^{12}$ cm$^{-2}$ |
| Halo dose, $N_{halo}$ (cm$^{-2}$) | $1.5 \times 10^{13}$ cm$^{-2}$ | ±10% | ±$1.5 \times 10^{12}$ cm$^{-2}$ | $1.65 \times 10^{13}$ cm$^{-2}$ | $1.35 \times 10^{13}$ cm$^{-2}$ |
| Halo peak depth, $d$ (nm) | 80 nm | ±10% | ±8 nm | 88 nm | 72 nm |
| Series resistance (external), $R_s$ (Ω-μm) | 400 Ω-μm | ±15% | ±60 Ω-μm | 460 Ω-μm | 340 Ω-μm |

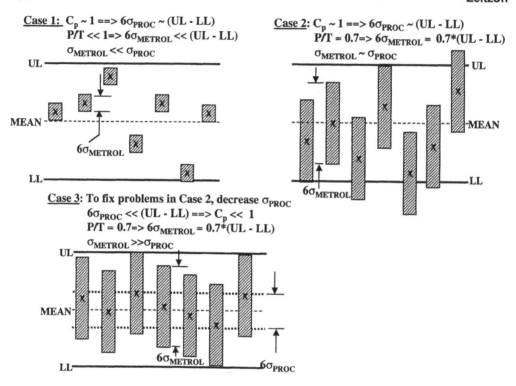

**Figure 9** Simplified process control charts showing the impact of different $P/T$ ratios.

so $6\sigma_{\text{METROL}} \ll (\text{UL} - \text{LL}) \sim 6\sigma_{\text{PROC}}$. Under those circumstances, as shown in the figure, for most of the measurements the error due to $\sigma_{\text{METROL}}$ does not impact whether a given measurement is within the process specification limits. For Case 2, however, where $P/T = 0.7$ and hence $6\sigma_{\text{METROL}} = 0.7(\text{UL} - \text{LL}) \sim 6\sigma_{\text{PROC}}$, the errors due to $\sigma_{\text{METROL}}$ are becoming comparable to $(\text{UL} - \text{LL})$. As shown in Case 2, the metrology error can cause the measured value to lie outside the process specification limits even though the actual parameter value lies inside these limits. If $\sigma_{\text{METROL}}$ cannot be reduced, then in order to ensure that the process parameter stays within the process specification limits, $\sigma_{\text{PROC}}$ must be reduced, as shown in Case 3. Depending on the amount of reduction required, this can be costly and difficult, since it requires more stringent process control.

These considerations can be quantitatively evaluated as follows (24). Since $\sigma_{\text{METROL}}$ and $\sigma_{\text{PROC}}$ are standard deviations due to independent randomly varying processes, the total standard deviation, $\sigma_{\text{TOTAL}} = (\sigma_{\text{METROL}}^2 + \sigma_{\text{PROC}}^2)^{1/2}$. Letting $C_{P,\text{PROC}} = (\text{UL} - \text{LL})/6\sigma_{\text{PROC}}$ and $C_{P,\text{TOTAL}} = (\text{UL} - \text{LL}/6\sigma_{\text{TOTAL}}$, $C_{P,\text{TOTAL}}$ is the apparent $C_p$ and $\sigma_{\text{TOTAL}}$ is the apparent standard deviation. From these equations and the definition of $P/T$, for given values of $P/T$ and $C_{P,\text{PROC}}$,

$$\frac{C_{P,\text{TOTAL}}}{C_{P,\text{PROC}}} = \left(1 + \left\{[C_{P,\text{PROC}}]\left[\frac{P}{T}\right]\right\}^2\right)^{-1/2} \tag{7}$$

Since $(\text{UL} - \text{LL})$ is fixed by specification, then $C_{P,\text{TOTAL}}/C_{P,\text{PROC}} = \sigma_{\text{PROC}}/\sigma_{\text{TOTAL}}$. A plot of $C_{P,\text{TOTAL}}/C_{P,\text{PROC}}$ versus $C_{P,\text{PROC}}$, with $P/T$ as a parameter varying from 0.1 to 1.0, is shown in Figure 10. This plot illustrates the impact of the measurement variation as

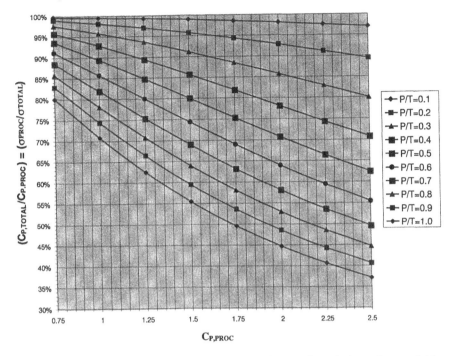

**Figure 10**   Impact of $P/T$ on $C_p$ and σ, where $C_{P,\text{PROC}}$ is an independent variable.

characterized by the parameter $P/T$. For a given value of $C_{P,\text{PROC}}$ (and hence of $\sigma_{\text{PROC}}$), $C_{P,\text{TOTAL}}$ decreases (and hence $\sigma_{\text{TOTAL}}$ increases) rapidly with $P/T$; and since the goal is to maximize $C_p$ and minimize σ, the increase of $P/T$ imposes a significant penalty.

An alternate way to evaluate the impact of $P/T$ variation is shown in Figure 11, where $C_{P,\text{TOTAL}}$ and $P/T$ are the independent variables and $C_{P,\text{TOTAL}}/C_{P,\text{PROC}}$ is plotted as in Figure 10, but versus the independent variable, $C_{P,\text{TOTAL}}$. The parameter $P/T$ varies from 0.1 to 0.8. As in Figure 10, since $(UL - LL)$ is fixed by specification, then $C_{P,\text{PROC}}/C_{P,\text{TOTAL}} = \sigma_{\text{PROC}}/\sigma_{\text{TOTAL}}$. Using the definitions of $P/T$, $C_{P,\text{TOTAL}}$, $C_{P,\text{PROC}}$, $\sigma_{\text{PROC}}$, and $\sigma_{\text{TOTAL}}$, the equation is

$$\frac{C_{P,\text{PROC}}}{C_{P,\text{TOTAL}}} = \frac{\sigma_{\text{PROC}}}{\sigma_{\text{TOTAL}}} = \left( 1 - \left\{ [C_{P,\text{TOTAL}}] \left[ \frac{P}{T} \right] \right\}^2 \right)^{1/2} \tag{8}$$

In general, for each key doping or structural parameter in an IC technology, there is a specified target for $C_p$, and since $(UL - LL)$ is also specified, there is a corresponding target for $\sigma_i$. Ideally, through process control techniques, $\sigma_{\text{PROC}}$ would be regulated to match the $\sigma_i$ target, and hence $C_{P,\text{PROC}}$ would equal the $C_p$ target. However, because of the random variations in the measurements, the total standard deviation is $\sigma_{\text{TOTAL}}$, not $\sigma_{\text{PROC}}$. As a result, $\sigma_{\text{TOTAL}}$ is matched to the $\sigma_i$ target and $C_{P,\text{TOTAL}}$ equals the $C_p$ target. The horizontal axis in Figure 11 then represents the $C_p$ target as well as $C_{P,\text{TOTAL}}$. The curves in the figure show, for any given $C_p$ target, the percentage reduction from the $\sigma_i$ target that is required in $\sigma_{\text{PROC}}$, as illustrated schematically in Case 3 of Figure 9. For example, for $C_p$ target = 1.3 and $P/T = 0.1$, $\sigma_{\text{PROC}}$ must be reduced only about 1% from the σ target. For $P/T = 0.2$, the reduction is about 3%; for $P/T = 0.3$, the reduction is about 7%; and the reduction is about 38% for $P/T = 0.6$. Clearly, for small $P/T$, up to

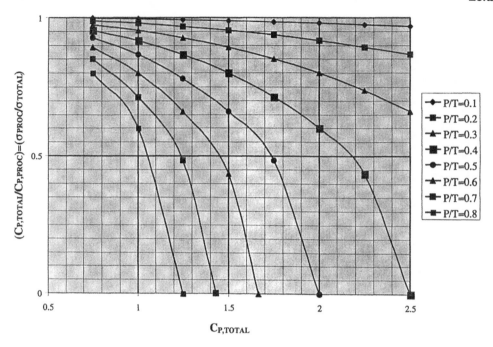

**Figure 11** Impact of $P/T$ on $C_p$ and $\sigma$, where $C_{P,\text{TOTAL}}$ is an independent variable.

perhaps 0.3 for this case, the reduction is relatively small and tolerable, but for large $P/T$, the required reduction is intolerably large. Note that the curves for $P/T \geq 0.4$ go to zero at some value of $C_{P,\text{TOTAL}} = C'_p$, where $C'_p < 2.5$, the maximum value plotted. This occurs for $(P/T)C'_p = 1$ [see Eq. (8)], which corresponds to $\sigma_{\text{METROL}} = \sigma_{\text{TOTAL}}$, and hence $\sigma_{\text{PROC}} = 0$, which is impossible to achieve. In this case, the entire budget for random variation is absorbed by the metrology variation, leaving none available for the random process variations.

Tying the foregoing considerations together with the process specification limits (UL and LL) in Table 9, the required metrology precision as reflected by the $\sigma_{\text{METROL}}$ values for $P/T = 0.1$ and for $P/T = 0.3$ are listed in Table 10. Note that in most cases, the required precision is quite high. For example, for $T_{\text{ox}}$, with $P/T = 0.1$, the $\sigma_{\text{METROL}}$ value of 0.0075 nm is less than 0.1 Å; even for $P/T = 0.3$, the 0.023-nm value is just over 0.2 Å. Similarly, for $P/T = 0.1$, the $\sigma_{\text{METROL}}$ value for $L_g$ is just over 5 Å and for $x_{j(\text{sh})}$, the $\sigma_{\text{METROL}}$ value is less than 1 Å.

The in-line metrology is limited by practical issues. Ideally, the in-line measurements are rapid, nondestructive, direct, and done on product wafers. However, in a number of cases, test wafers are used, and sometimes the measurement is destructive. Of the five key input parameters requiring "tight control" (i.e., $3\sigma_{\text{PROC}} < 10\%$, as discussed in the previous section and listed in Table 9), four parameters—the gate length ($L_g$), the gate oxide thickness ($T_{\text{ox}}$), the spacer oxide width ($T_{\text{sp}}$), and the channel dose ($N_{\text{ch}}$)—can be routinely and directly measured on product wafers. $L_g$ and $T_{\text{sp}}$ are measured optically or via SEM after etch, $T_{\text{ox}}$ is measured via ellipsometry, and $N_{\text{ch}}$ is measured via the thermal wave reflectance (25) technique. Alternatively, $N_{\text{ch}}$ is sometimes monitored on test wafers using secondary ion mass spectrosopy (SIMS) or four-point sheet resistance measurements, and $T_{\text{ox}}$ is sometimes measured on test wafers. The

**Table 10** $\sigma_{\text{METROL}}$ values for $P/T = 0.1$ and $P/T = 0.3$

| Input parameter | Mean | UL − LL | $\sigma_{\text{METROL}}$ $(P/T = 0.1)$ | $\sigma_{\text{METROL}}$ $(P/T = 0.3)$ |
|---|---|---|---|---|
| Gate length, $L_g$ (μm) | 0.18 μm | 0.0324 μm | 0.00054 μm | 0.0016 μm |
| Gate oxide thickness, $T_{\text{ox}}$ (nm) | 4.50 nm | 0.45 nm | 0.0075 nm | 0.023 nm |
| Spacer oxide width, $T_{\text{sp}}$ (nm) | 82.5 nm | 13.2 nm | 0.22 nm | 0.66 nm |
| Shallow-junction doping profile depth, $X_{j(\text{sh})}$ (nm) | 50 nm | 5 nm | 0.083 nm | 0.25 nm |
| Peak shallow-junction doping, $N_{\text{sh}}$ (cm$^{-3}$) | $4 \times 10^{19}$ cm$^{-3}$ | $8 \times 10^{18}$ cm$^{-3}$ | $1.33 \times 10^{17}$ cm$^{-3}$ | $4 \times 10^{17}$ cm$^{-3}$ |
| Channel dose, $N_{\text{ch}}$ (cm$^{-2}$) | $5.65 \times 10^{12}$ cm$^{-2}$ | $8.48 \times 10^{11}$ cm$^{-2}$ | $1.41 \times 10^{10}$ cm$^{-2}$ | $4.24 \times 10^{10}$ cm$^{-2}$ |
| Halo dose, $N_{\text{halo}}$ (cm$^{-2}$) | $1.5 \times 10^{13}$ cm$^{-2}$ | $3 \times 10^{12}$ cm$^{-2}$ | $5 \times 10^{10}$ cm$^{-2}$ | $1.5 \times 10^{11}$ cm$^{-2}$ |
| Halo peak depth, $d$ (nm) | 80 nm | 16 nm | 0.27 nm | 0.8 nm |
| Series resistance (external), $R_s$ (Ω-μm) | 400 Ω-μm | 120 Ω-μm | 2 Ω-μm | 6 Ω-μm |

other parameter requiring tight control, the $S/D$ extension ("shallow") junction depth ($X_{\text{sh}}$), cannot currently be routinely measured on product wafers. Consequently, $X_{\text{sh}}$ is typically monitored on test wafers via SIMS. However, in the future, the Boxer-Cross technique (26) shows promise of becoming practical for in-line measurements on product wafers. For the four input parameters requiring "looser control" (i.e., $3\sigma_{\text{PROC}} \geq 10\%$, as discussed in the previous section and listed in Table 9), none are routinely measured on product wafers. For $N_{\text{halo}}$ and the halo peak depth ($d$), given the loose control needed and the typically tight control on the halo implant dose and energy, monitoring test wafers via SIMS is generally adequate. The peak shallow-junction doping ($N_{\text{sh}}$) can be monitored using the same SIMS that is used to measure $X_{\text{sh}}$. Finally, at the end of the wafer fabrication process, routine electrical monitoring of test transistors on product wafers is used to measure $R_s$. Typically, for each lot in an IC manufacturing line, the in-line measurements are made on a reasonable sample size, and good statistics for the standard deviation ($\sigma_{\text{TOTAL}}$) and direct verification of meeting process control limits are obtained.

As mentioned earlier, for in-line metrology used for process control in well-established and well-characterized IC fabrication processes, the key requirement is precision (i.e., repeatability), since there is a well-established, optimal baseline, and the main requirement is to ensure that the process does not drift unacceptably far from the optimal. However, for establishment of new or strongly modified process modules or process flows, it is necessary to measure the values of key parameters, such as $T_{\text{ox}}$ and $L_g$, to a relatively high degree of absolute accuracy in order to understand and establish the optimal baseline. For example, in establishing the 0.18-μm NMOSFET process, it is important that $T_{\text{ox}}$ be accurately measured so that the real $T_{\text{ox}}$ of the process can be set close to 4.5 nm. (Note that this metrology with high absolute accuracy is often different from the in-line metrology. For example, transmission electron microscopy (TEM) measurements and/or electrical $C$-$V$ measurements with corrections for quantum effects and polysilicon depletion can be used to measure $T_{ox}$ on test wafers.) However, once an optimal process is well established, the in-line metrology can indicate that the nominal $T_{ox}$ is 4.8 nm as long as that corresponds to a real nominal $T_{ox}$ of 4.5 nm and as long as the in-line metrology has adequate precision. In other words, for in-line metrology the precision is critical, but adjustments can be made for consistent measurement bias.

## VI. CONCLUSIONS

A 0.18-μm NMOSFET device was designed and optimized to satisfy a specified set of electrical characteristics. This optimized device was the nominal design center for a simulated sensitivity analysis in which normalized second-order polynomial model equations were embedded within a special Monte Carlo code. Monte Carlo simulations with the code were used to correlate the random statistical variations in key electrical device characteristics to the random variations in the key structural and doping parameters. Using these simulations, process control tradeoffs among the different structural and doping parameters were explored, and the level of process control required to meet specified statistical targets for the device electrical characteristics was analyzed. It turns out that meeting these targets requires tight control of five key structural and doping parameters: the gate length, the gate oxide thickness, the shallow source/drain extension junction depth, the channel dose, and the spacer width. Making process control tradeoffs based on estimates of industry capability, an optimal set of $3\sigma$ statistical variations was chosen for the five parameters: 9%, 5%, 5%, 7.5%, and 8%, respectively. If the estimates of industry capability were different, the tradeoffs would be changed and hence the optimal set of variations would be changed.

The optimal set of parameter statistical variations drives the in-line metrology. The key requirement is that the metrology precision for any measured parameter be no more than 10–30% of the optimal statistical variation for that parameter. Also, the impact of not meeting the precision requirements was quantitatively analyzed, and it was found that the more the metrology precision departs from the requirements, the tighter the process control must be.

## ACKNOWLEDGMENTS

I am grateful to a number of people from International SEMATECH who were helpful in preparing this chapter. Jack Prins and Paul Tobias provided valuable inputs on the statistical modeling aspects of this work. Larry Larson provided important input on the process control techniques and capability of the industry, and Alain Diebold provided valuable input on metrology. My former colleague William E. Moore (now at AMD, Inc.) was very helpful in running the Monte Carlo simulations.

I am grateful to Professor Al F. Tasch of the University of Texas at Austin and to his former students, S. Arefin Khan, Darryl Angelo, and Scott Hareland. They were instrumental in developing the device optimization and the Monte Carlo code under a contract with International SEMATECH.

## APPENDIX:  NORMALIZED MODEL EQUATIONS

In the following equations, each of the eight output responses is given as a function of the normalized set of input factors. Each of these input factors, for example, $\Delta L_g$, will have a range of values between $-1$ and $+1$. Referring to Table 2, a $-1$ value for $\Delta L_g$ would correspond to a gate length, $L_g$, of $0.18\,\mu m - 15\% = 0.153\,\mu m$, while a $+1$ value for $\Delta L_g$ would indicate an $L_g = 0.18\,\mu m + 15\% = 0.207\,\mu m$. For the nominal case, each normalized input variable would have a value of 0.

$$
\begin{aligned}
V_T(\text{mV}) = {} & 454.9 + 56.1\,\Delta L_g + 52.4\,\Delta T_{\text{ox}} + 41.4\,\Delta N_{\text{ch}} - 33.6\,\Delta X_{\text{sh}} + 9.9\,\Delta N_{\text{sh}} \\
& + 9.7\,\Delta T_{\text{sp}} + 7.1\,\Delta N_{\text{halo}} - 1.6\,\Delta d - 19.9\,\Delta L_g^2 - 5.9\,\Delta T_{\text{ox}}^2 - 3.8\,\Delta T_{\text{sp}}^2 \\
& + 3.8\,\Delta d^2 + 21.9\,\Delta L_g \times \Delta X_{\text{sh}} - 6.9\,\Delta L_g \times \Delta T_{\text{sp}} - 6.2\,\Delta L_g \times \Delta N_{\text{sh}}
\end{aligned}
$$

$$
\begin{aligned}
\Delta V_T(\text{mV}) = {} & 72.3 - 39.4\,\Delta L_g + 17.8\,\Delta X_{\text{sh}} - 12.7\,\Delta T_{\text{sp}} + 8.4\,\Delta T_{\text{ox}} - 5.9\,\Delta N_{\text{halo}} \\
& - 4.7\,\Delta N_{\text{ch}} - 3.0\,\Delta N_{\text{sh}} - 1.5\,\Delta d + 12.4\,\Delta L_g^2 + 7.8\,\Delta T_{\text{sp}}^2 - 2.3\,\Delta N_{\text{sh}}^2 \\
& - 11.9\,\Delta L_g \times \Delta X_{\text{sh}} + 5.9\,\Delta L_g \times \Delta T_{\text{sp}} - 5.3\,\Delta L_g \times \Delta T_{\text{ox}} + 4.4\,\Delta T_{\text{sp}} \\
& \times \Delta d + 2.4\,\Delta L_g \times \Delta N_{\text{halo}} + 2.3\,\Delta T_{\text{sp}} \times \Delta N_{\text{halo}}
\end{aligned}
$$

*Note*: In the next two equations, a logarithmic transformation was used to achieve better normality and fit.

$$
\begin{aligned}
I_{\text{dsat}}(\log[\mu A/\mu m]) = {} & 2.721 - 0.060\,\Delta L_g - 0.052\,\Delta T_{\text{ox}} + 0.028\,\Delta X_{\text{sh}} - 0.019\,\Delta T_{\text{sp}} \\
& - 0.016\,\Delta N_{\text{ch}} - 0.008\,\Delta N_{\text{halo}} - 0.007\,\Delta R_s - 0.005\,\Delta N_{\text{sh}} \\
& + 0.015\,\Delta L_g^2 + 0.009\,\Delta T_{\text{ox}}^2 + 0.009\,\Delta T_{\text{sp}}^2 + 0.013\,\Delta X_{\text{sh}} \\
& \times \Delta N_{\text{ch}} - 0.011\,\Delta T_{\text{sp}} \times \Delta N_{\text{ch}} - 0.008\,\Delta L_g \times \Delta T_{\text{ox}} \\
& - 0.006\,\Delta L_g \times \Delta X_{\text{sh}}
\end{aligned}
$$

$$
\begin{aligned}
I_{\text{leak}}(\log[\text{pA}/\mu m]) = {} & 0.385 - 1.189\,\Delta L_g + 0.571\,\Delta X_{\text{sh}} - 0.508\,\Delta N_{\text{ch}} - 0.417\,\Delta T_{\text{ox}} \\
& - 0.241\,\Delta T_{\text{sp}} - 0.144\,\Delta N_{\text{sh}} - 0.127\,\Delta N_{\text{halo}} + 0.011\,\Delta d \\
& - 0.424\,\Delta L_g^2 + 0.104\,\Delta T_{\text{sp}}^2 - 0.080\,\Delta d^2 - 0.063\,\Delta N_{\text{sh}}^2 \\
& + 0.055\,\Delta T_{\text{ox}}^2 - 0.449\,\Delta L_g \times \Delta X_{\text{sh}} + 0.156\,\Delta L_g \times \Delta T_{\text{sp}} \\
& + 0.112\,\Delta L_g \times \Delta N_{\text{sh}} - 0.088\,\Delta L_g \times \Delta T_{\text{ox}}
\end{aligned}
$$

$$
\begin{aligned}
I_{\text{sub}}(\text{nA}/\mu m) = {} & 35.4 - 17.6\,\Delta L_g - 1.9\,\Delta T_{\text{sp}} - 1.7\,\Delta N_{\text{ch}} - 1.7\,\Delta R_s + 1.5\,\Delta N_{\text{sh}} \\
& - 1.5,\Delta T_{\text{ox}} + 1.2\,\Delta X_{\text{sh}} + 6.2\,\Delta L_g^2 + 2.0\,\Delta T_{\text{ox}}^2 + 1.7\,\Delta T_{\text{sp}}^2 \\
& + 1.7\,\Delta X_{\text{sh}}^2 + 1.5\,\Delta N_{\text{ch}}^2 + 3.2\,\Delta L_g \times \Delta N_{\text{ch}}
\end{aligned}
$$

*Note*: In the following equation, an inverse square root transformation was used to achieve better normality and fit.

$$S^{-1/2}(\text{mV/decade})^{-1/2} = 0.10916 - 0.00137\,\Delta T_{\text{ox}} - 0.00085\,\Delta L_g - 0.00050\,\Delta N_{\text{ch}}$$
$$- 0.00018\,\Delta N_{\text{halo}} - 0.00016\,\Delta T_{\text{sp}} + 0.00016\,\Delta X_{\text{sh}}$$
$$- 0.00030\,\Delta X_{\text{sh}}^2 - 0.00027\,\Delta L_g^2 - 0.00022\,\Delta N_{\text{halo}}^2$$
$$+ 0.00047\,\Delta L_g \times \Delta X_{\text{sh}} - 0.00028\,\Delta L_g \times \Delta T_{\text{sp}}$$

$$g_m^s(\text{mS/mm}) = 470.6 - 50.4\,\Delta T_{\text{ox}} - 41.7\,\Delta L_g + 14.4\,\Delta X_{\text{sh}} - 6.0\,\Delta N_{\text{halo}} + 2.9\,\Delta N_{\text{ch}}$$
$$- 14.3\,\Delta X_{\text{sh}}^2 - 12.5\,\Delta N_{\text{halo}}^2 - 20.2\,\Delta T_{\text{ox}} \times \Delta N_{\text{ch}}$$

$$g_m^l(\text{mS/mm}) = 58.63 - 9.44\,\Delta L_g - 3.93\,\Delta T_{\text{ox}} + 2.74\,\Delta X_{\text{sh}} - 1.69\,\Delta T_{\text{sp}} - 1.53\,\Delta R_s$$
$$- 0.85\,\Delta N_{\text{ch}} - 0.47\,\Delta N_{\text{sh}} + 0.43\,\Delta d - 0.34\,\Delta N_{\text{halo}} + 1.68\,\Delta L_g^2$$
$$+ 1.36\,\Delta T_{\text{ox}}^2 + 0.94\,\Delta N_{\text{halo}}^2 + 0.69\,\Delta N_{\text{ch}}^2 + 0.64\,\Delta T_{\text{sp}}^2 - 1.33\,\Delta L_g$$
$$\times \Delta X_{\text{sh}} + 0.78\,\Delta L_g \times \Delta T_{\text{sp}}$$

## REFERENCES

1. P.M. Zeitzoff et al. Modeling of manufacturing sensitivity and of statistically based process control requirements for a 0.18-μm NMOS device. Proc. of the 1998 International Conference on Characterization and Metrology for ULSI Technology, Gaithersburg, MD, pp. 73–81, 1998.
2. A. Hori, A. Hiroki, H. Nakaoka, M. Segawa, T. Hori. Quarter-micrometer SPI (self-aligned pocket implantation) MOSFET's and its application for low supply voltage operation. IEEE Trans. Electron. Devices, Vol. 42, No. 1, Jan. 1995
3. A. Chatterjee, J. Liu, S. Aur, P.K. Mozumder, M. Rodder, I.-C. Chen. Pass transistor designs using pocket implant to improve manufacturability for 256Mbit DRAM and beyond. IEDM, pp. 87–90, 1994.
4. Avanti Corp. User's Manual for TSUPREM-3.
5. Avanti Corp. User's Manual for TSUPREM-4.
6. UT-MiniMOS 5.2–3.0 Information Package. Microelectronics Research Center, The University of Texas at Austin, 1994.
7. V. Martin Agostinelli, T. James Bordelon, Xiaolin Wang, Khaled Hasnat, Choh-Fei Yeap, D.B. Lemersal, Al F. Tasch, Christine M. Maziar. Two-dimensional energy-dependent models for the simulation of substrate current in submicron MOSFETs. IEEE Trans. Electron. Devices, vol. 41, no. 10, Oct. 1994.
8. M.J. van Dort, P.H. Woerlee, A.J. Walker. A simple model for quantization effects in heavily doped silicon MOSFETs at Inversion Conditions. Solid-State Electronics, vol. 37, no. 3, pp. 411–414, 1994.
9. S.A. Hareland, S. Krishnamurthy, S. Jallepalli, C.-F. Yeap, K. Hasnat, A.F. Tasch, Jr., C.M. Maziar. A computationally efficient model for inversion layer quantization effects in deep submicron N-channel MOSFETs. IEDM, Washington, DC, pp. 933–936, December 1995.
10. S.A. Hareland, S. Krishnamurthy, S. Jallepalli, C.-F. Yeap, K. Hasnat, A.F. Tasch, Jr., C.M. Maziar. A computationally efficient model for inversion layer quantization effects in deep submicron N-channel MOSFETs. IEEE Trans. Electron. Devices, vol. 43, no. 1, pp. 90–96, Jan. 1996.

11. H. Shin, G.M. Yeric, A.F. Tasch, C.M. Maziar, "Physically based models for effective mobility and local-field mobility of electrons in MOS inversion layers. Solid-State Electronics, vol. 34, no. 6, pp. 545–552, 1991.

12. V.M. Agostinelli, H. Shin, A.F. Tasch. A comprehensive model for inversion layer hole mobility for simulation of submicrometer MOSFETs. IEEE Trans. Electron. Devices, vol. 38, no. 1, pp. 151–159, 1991.

13. S.A. Khan, K. Hasnat, A.F. Tasch, C.M. Maziar. Detailed evaluation of different inversion layer electron and hole mobility models. Proc. of the 11th Biennial University/Government/Industry Microelectronics Symposium, Austin, TX, p. 187, May 16–17, 1995.

14. G.E.P. Box, N.R. Draper. Empirical Model-Building and Response Surfaces. New York: Wiley, 1987.

15. G.E.P. Box, D.W. Behnken. Some new three-level designs for the study of quantitative variables. Technometrics, Vol. 2, No. 4, Nov. 1960.

16. G.E.P. Box, N.R. Draper. Empirical Model Building and Response Surfaces. New York: Wiley, 1987.

17. A. Papoulis. Probability, Random Variables, and Stochastic Processes. New York: McGraw-Hill, 1965, Chapter 4.

18. Private conversation with L. Larson of International SEMATECH.

19. B.W. Lindgren. Statistical Theory. New York: Macmillan, 1968, Chapter 4.

20. Semiconductor Industry Association. International Technology Roadmap for Semiconductors. Austin TX: International SEMATECH, 1999.

21. Private conversation with Jack Prins of International SEMATECH.

22. V.E. Kane. Process capability indices. J. Qual. Technol., Vol. 18, Jan 1986, pp. 41–52.

23. See Chapter 1, by Diebold, for a more detailed discussion of the definition of $P$ and $\sigma_{METROL}$.

24. Private communication from J. Schlesinger of Texas Instruments.

25. T. Hara, N. Ohno. Photo-acoustic displacement (PAD)/therma-wave methods. In: M.I. Current et al., ed. Materials and Process Characterization of Ion Implant. Austin, TX: Ion Beam Press, 1997, pp. 244–256.

26. Chapter 5, by Borden et al.

# 7
# Overview of Metrology for On-Chip Interconnect

**Alain C. Diebold**
*International SEMATECH, Austin, Texas*

## I. INTRODUCTION

The Interconnect section of this volume contains chapters that describe metrology used for characterization and measurement of interconnect materials and processes (see Chapters 8–13). Metrology for interconnects is a challenging area, and thus there are several methods that have been introduced since this volume was initiated. In an effort to include a description of in-line measurement capability whenever possible, this chapter provides an overview and briefly describes two methods not covered in separate chapters: in-line x-ray reflectivity and Metal Illumination[TM].

## II. OVERVIEW

Interconnect processing has evolved from the patterning of aluminum metal layers (including barrier metal) deposited on reflowed boron- and phosphorus-doped silicate glass. The first change was the introduction of chemical-mechanical polishing (CMP) for planarization of the insulator. Today, damascene trenches are patterned into low-dielectric-constant silicon dioxide glass, and then barrier layer and copper are deposited. The CMP technique is used to planarize Damascene structures. It is important to note that memory integrated circuits (ICs) have not yet required copper metal, because they use fewer layers of metal and the total length on the interconnect paths is shorter. Thus logic was the first to incorporate copper metal lines.

   It is useful to describe the ultimate goals of interconnect metrology. Monnig has championed a set of goals that reflects manufacturing experience with aluminum metalization as a basis (1). This author has found Monnig's insight extremely helpful, and it is the basis for the goals stated here. Patterned aluminum metal processing was a mature technology that required minimal metrology. As CMP and Damascene mature, they should also require fewer measurements for process control than at first introduction. The introduction of a new insulator material with nearly each generation of future ICs may require additional measurements as well as the extension of existing routine methods to that material. An example of a potential measurement requirement for low-κ dielectric materials is elastic constant determination after curing and other processing. An example of the extension of existing methods is the optical model development required for low κ and the addition of porosity (Chapter 8).

The ultimate goal for barrier layer/copper seed/electroplated copper is the measurement of barrier- and seed-layer thickness on the sidewalls and bottoms of damascene trenches and vias (1). This is shown in Figure 1. To date, the only way of verifying barrier/seed thickness on a sidewall or feature bottom is through destructive cross sectioning. Despite this goal, the need for this measurement when processing matures could initiate a move toward a process tool setup monitor such as is done for implant dose measurement.

The ultimate goal of CMP flatness control is to ensure defect-free planar surfaces that meet the depth-of-focus needs of the lithographic patterning capability (see Ref. 1 and Chapter 27 of this volume). The new, acoustic measurement methods discussed in this volume are capable of providing this type of control. Another approach is to use an optimal measurement for determining insulator thickness combined with profilometry for surface flatness measurement. Processes will be continually altered as each new low-κ insulator is introduced.

## III.  IN-LINE X-RAY REFLECTIVITY

Grazing incidence x-ray reflectivity (GI-XRR) is an off-line characterization method capable of measuring film thickness with great precision and accuracy (see Chapter 27). Although Deslattes and Mayti describe this method in great detail in Chapter 27, it is useful to give a brief description here. In GI-XRR, a collimated, monochromatic (single-wavelength) beam of x-rays is focused on the sample of interest, starting at angle that is glancing enough to result in total reflection (~<0.2°). The angle of incidence is then increased gradually in small increments (usually <0.01° steps) up to 2–3°, and the specular reflected intensity is plotted as a function of angle. In Figure 2, the x-ray reflectivity intensity vs. angle data demonstrates the oscillatory behavior of specular reflectivity data for a thin tantalum barrier film with seed copper. As indicated in Chapter 27, the

a                                        b

**Figure 1**  Measurement of sidewall thickness variation: A trench and via structure is shown with uniform barrier layer and seed copper films in (a) and with typical thickness variation in (b).

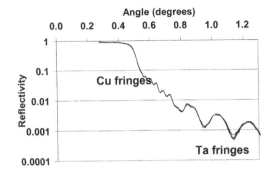

**Figure 2** Grazing incidence x-ray reflectivity characterization of TaN barrier- and seed-layer copper films: The intensity of the specularly reflected x-ray beam is plotted against the angle of incidence. The different oscillatory periods of the two films show the oscillatory behavior that results from the superposition of the reflections from each film.

key relationship that allows determination of film thickness from the angular difference in oscillation maximum is

$$\Delta\theta = \frac{\lambda}{2t}$$

or, more completely,

$$\theta_m^2 = \theta_c^2 + m^2\left(\frac{\lambda^2}{4d^2}\right) \qquad \text{and} \qquad \theta_{m+1}^2 - \theta_m^2 = (2m+1)\left(\frac{\lambda^2}{4d^2}\right)$$

where $t$ is the film thickness and $\lambda$ is the x-ray wavelength. $\theta_m$ and $\theta_{m+1}$ represent the angle of the maximum intensity of adjacent oscillations $m$ and $m+1$, respectively. Because film thickness can be calculated directly from the wavelength of the x-ray (whose value is known to many significant figures), GI-XRR provides an extremely accurate measurement of the thickness of barrier-layer and seed copper films over the illuminated area ($\sim$ cm$\times \sim$ cm). When several oscillations are observed, the thickness values from each angular difference can be averaged for greater precision and accuracy. Off-line GI-XRR apparatus can measure diffuse (nonspecular) scattering, which can be used to determine interfacial roughness. In addition, careful modeling of the decay behavior (decrease in intensity of the reflectivity oscillations) can be used to determine film density (Chapter 27). Information on film density and interfacial roughness using the experimentally observed critical angle as a measure of density $\theta_c \propto \lambda(\rho)^{1/2}$ and the slope of the decrease in reflectance signal with angle as a measure of interfacial roughness $\sigma$ (Å rms):

$$R = R_{\text{ideal}} \times \exp\left(-[4\pi\sin\varphi\sigma/1]^2\right)$$

X-ray reflectivity has been used to measure thickness, density, and roughness for film stacks with six or more layers, provided that adjacent layers exhibit at least a 10% density difference. Thickness of films from $<30$ Å to $\sim 2,000$ Å are routinely measured, while density surface roughness up to about 40 Å rms can also be determined. Because conventional XRR as described earlier requires many minutes to accumulate a reflectivity curve, it has been applied to off-line measurements of standards and R&D studies, but up to now it has not been considered a practical metrology technique.

Recently, Therma-Wave, Inc., introduced a metrology tool, the Meta-Probe-X, that utilizes a proprietary method for obtaining XRR data much more rapidly than conventional GI-XRR. The Meta-Probe-X method (Figure 3) uses a curved crystal monochromator and a position-sensitive x-ray detector to illuminate and collect x-rays from multiple angles simultaneously. With typical data collection times of 10 seconds per point, the Meta-Probe-X is a true XRR metrology tool.

## IV.  METAL ILLUMINATION

Metal Illumination[TM] (MI) is a newly introduced measurement technology that uses optical technology found in carrier illumination (see Ref. 4 and Chapter 5 of this volume). The resistance/length of fine metal lines or sheet resistance of large areas is determined from the observed heat transport properties of the metal. Patterned areas can be measured with a 2-μm spatial resolution without contacting metal, thus making this method ideal for process control during manufacture.

Metal Illumination[TM] measures resistance/length of lines and sheet resistance of pads. The first, red laser (830 nm) heats the metal lines, and the gradient in temperature across the illuminated area is due to thermal conductance. The maximum temperature is measured by monitoring the reflectance, R, with a second laser at 980 nm. The reflectivity of a metal changes $\sim 10^{-4}/K$. The red laser is pulsed and the probe laser signal's modulated intensity is used to determine the change in reflectivity $\Delta R/R$ (the Boxer-Cross signal) which is a measure of thermal conductivity. Thermal conductivity is related to electrical conductivity (resistance): $C_{TC} = \text{const} \times C_{EC}$. The Boxer-Cross signal is correlated to resistivity by a calibration with a measurement on test structures. This calibration is done once for the method. The optical path of Carrier Illumination is discussed by Borden et al. in this volume (12).

Borden and Madsen have shown the correlation of resistance determined by MI with traditional methods, such as electrical probing, 4-point probe, profilometry, and cross-sectional microscopy for copper structures (4). These first studies used < 200-nm, 50% fill (1/2-pitch lines) copper damascene structures to demonstrate correlation of MI with electrical probing of test structures (shown in Figure 4) and with profilometry (shown in Figure 5) (4). It is interesting to note that the resistance/length is a function of line cross section and is sensitive to both thickness variation (typically due to polishing) and line width variation (typically due to CD and fill). Because this method is designed to be an in-line monitor, application has been focused on variation induced by the integrated

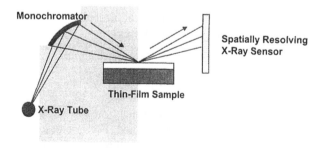

**Figure 3**  Optical path of the MetaProbe-X[TM] x-ray reflectivity apparatus: The multiangle detection method results in a substantial increase in data collection rate when compared to the traditional grazing incidence x-ray reflectivity systems. (From Ref. 2, provided by W. Johnson and used with permission.)

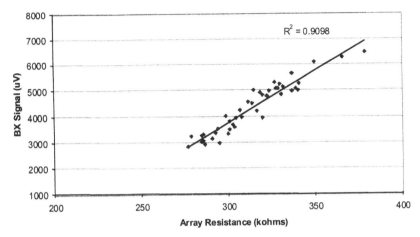

**Figure 4** Correlation of the resistivity of copper determined by Metal Illumination<sup>TM</sup> and four-print probe: The variation in the Metal Illumination<sup>TM</sup> determined resistivity of multiple arrays of 160-nm copper lines have a pitch of 320 nm is found to match that observed by electrical probing. (From Ref. 4, provided by Borden and used with permission.)

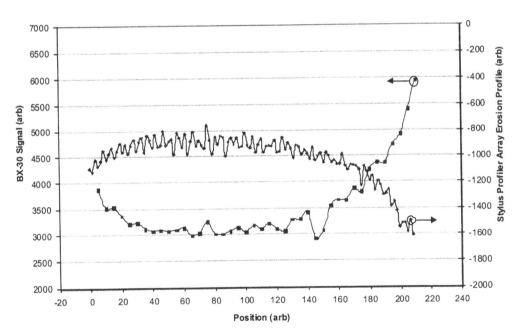

**Figure 5** Correlation of Metal Illumination<sup>TM</sup> determined resistivity with profilometry: The change in resistivity occurs with thickness variation for 160-nm copper lines across a single array. This demonstrates the local characterization capability of Carrier Illumination<sup>TM</sup>. (From Ref. 4, provided by Borden and used with permission.)

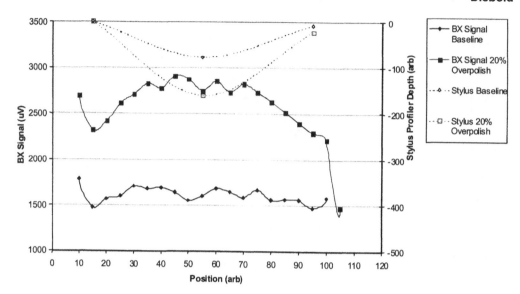

**Figure 6** Correlation of Metal Illumination$^{TM}$ determined resistivity with profilometry: The resistivity of copper pad structures is found to correlate with profilometry data. (From Ref. 4, provided by Borden and used with permission.)

Damascene process instead of high-precision thickness determination. The high resolution has also proven useful for CMP processing of pad structures to characterize global and local uniformity, as shown in Figure 6.

## V. CONCLUSIONS

Copper/Damascene processing and new low-$k$ insulators have driven the introduction of new measurement technology. The precision and accuracy of the new methods needs to determined in light of the manner in which each metrology tool is applied to process control. Thus, film thickness may be measured by one tool and resistivity by another. It will be interesting to revisit the status of interconnect metrology in five years to determine which measurement systems have become market leaders.

## REFERENCES

1. K.A. Monnig. The transition to Cu, damascene and low-k dielectrics for integrated circuit interconnects, impacts on the industry. In: D.G. Seiler, A.C. Diebold, M. Bullis, T.J. Schaffner, R. McDonald, eds. Characterization and Metrology for ULSI Technology. New York: AIP Press, 2000/2001.
2. W. Johnson. Semiconductor material applications of rapid x-ray reflectometry. In: D.G. Seiler, A.C. Diebold, M. Bullis, T.J. Shaffner, R. McDonald, eds. Characterization and Metrology for ULSI Technology. New York: AIP Press, 2000/2001.
3. W. Johnson, private communication.
4. P. Borden, J. Madsen. High resolution, non-contact characterization of fine pitch copper arrays for damascene process control. Japan LSI conference.

# 8

# Metrology for On-Chip Interconnect Dielectrics

**Alain C. Diebold and William W. Chism**
*International SEMATECH, Austin, Texas*

**Thaddeus G. Dziura[*] and Ayman Kanan[**]**
*Therma-Wave Inc., Fremont, California*

## I. INTRODUCTION

On-chip interconnect technology has undergone drastic changes since the 350-nm technology node. Materials and process technology has shifted from patterning aluminium metal and filling the open spaces with silicon dioxide insulator to patterning a low-dielectric-constant insulator and filling the connection line and via/contact openings with barrier metal and copper (Damascene processing). Planarization of each interconnect level is becoming increasingly important due to the shrinking depth of field (depth of focus) of lithographic patterning processes. Chemical mechanical polishing is now the process of necessity for Damascene processes. This chapter discusses many of the metrology methods used to determine the thickness of the insulator layer and briefly describes metrology used to ensure proper curing of the porous low-dielectric-constant materials. Determination of flatness is discussed in Chapter 13.

The insulator level used to separate metal connection lines is relatively thick when compared to other thin films in a typical integrated circuit (IC). If the insulator is one continuous material, then it is of the order of microns in thickness. In traditional metal patterning processes, reflectivity measurements were used for determination of silicon dioxide thickness. In older IC processes, boron- and phosphorus-doped silicon dioxide "glass") was heated and reflowed to achieve planarization of an interconnect level. Either Fourier transform infrared spectroscopy (FTIR) or x-ray fluorescence (XRF) was used to measure dopant concentration, depending on the process used to deposit the doped glass. Typically, XRF was used to control processes that resulted in a significant amount of water content in the unannealed glass (1). Because Damascene processes that employ CMP planarization are now typical, this chapter focuses on the measurement of dielectric films made from the lower-dielectric-constant materials used for these processes.

---

[*] *Current affiliation*: KLA-Tencor, San Jose, California.

[**] *Current affiliation*: Lucent Technologies, Allentown, Pennsylvania.

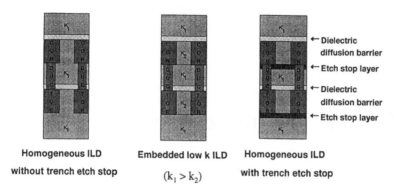

Homogeneous ILD
without trench etch stop

Embedded low k ILD
$(k_1 > k_2)$

Homogeneous ILD
with trench etch stop

**Figure 1** Three types of copper/low-$k$ stacked layers shown in the Semiconductor Industry Association's Interconnect Roadmap in the *International Technology Roadmap for Semiconductors*.

The drive for increased circuit speed has driven the use of copper and low-dielectric-constant ($k$) materials. In Figure 1, we show several generic configurations for the stack of materials layers that compose a single level of interconnect. Evaluating the effective dielectric constant of low-$k$ materials is complicated by the nature of high-frequency measurement technology itself (2). Furthermore, the effective dielectric properties require experimental determination of the dielectric constant in the plane and out of the plane of the film (3). Therefore, one must know the optical constants of each film in the stack of insulators used in each metal level. In order to further lower the dielectric constant, porosity is being introduced into materials. Here, pore size and the distribution of pore concentration can vary in a single layer of material. The optical properties of each layer must be known to determine film thickness. Moisture in the porous layers can cause significant reliability issues, and low-$k$ materials are annealed so that they become hydrophobic after curing. Fortunately, determination of water content in porous low-$k$ materials is relatively straightforward using FT-IR to probe vibrational modes of SiOH, located near 3500–3600 cm$^{-1}$. The foregoing background provides guidance for chapter organization.

This chapter is divided into two additional sections. First we cover thickness metrology for low-dielectric-constant materials, including Damascene-processed silicon dioxide. In this section, we also cover the measurement of multilayer interconnect stacks. In the last section, we describe metrology for porous low-dielectric-constant materials.

## II. THICKNESS DETERMINATION FOR LOW-DIELECTRIC-CONSTANT MATERIALS

Reflectivity and ellipsometry can both be used to determine the thickness of low-$k$ insulators and stacks of insulator films. One can determine the optical constants of thick films and then use these to determine film thickness for new samples. One important complication occurs for birefringent low-$k$ materials, which have different dielectric functions parallel and perpendicular to the film. The optical properties of these films depend on the direction of light propagation. This topic is also covered by Kiene et al. in Chapter 12. Once the optical constants of each insulator layer are known, most commercial optical metrology systems have optical modeling software that can build a multilayer optical model for a transparent film stack. It is strongly urged that Chapter 25, on the physics of optical measurements, by Jellison, be used as reference for this chapter. The fundamentals of dielectric functions, optical models, reflectance, and ellipsometry are all discussed in more detail in that chapter. This section is subdivided into: dielectric function and optical model determination, techniques for determination of optical constant and thickness, birefringent vs. isotropic materials, and multilayer stack measurement.

## A. Dielectric Function and Optical Model Determination

Although Jellison provides an in-depth discussion of optical methods and materials properties in his chapter (Chapter 25), information pertinent to this chapter is stated here for completeness. Reflectometry measures the ratio of the intensity of the reflected light to that of the incoming light. This can be done as a function of angle at a single wavelength or as a function of wavelength at a single angle. The reflectance can also be a function of the polarization state of the incoming light. The reflectance is the square of the reflection coefficient, and there is a coefficient for each polarization state: $p$ (for light polarized parallel to the plane of incidence) and $s$ (for light polarized perpendicular to the plane of incidence). The intensity of reflection is determined by the dielectric function of the materials being measured. Normal incidence reflectometry is most useful when applied to films thicker than 5–10 nm (see Chapter 25). Although the use of angular reflectance data improves the ability of reflectivity to measure these thin films, ellipsometry is typically the method of choice for thin gate dielectric films, which are discussed in Chapter 2, by Hayzelden. Ellipsometric data, $\psi$ and $\Delta$, are related to the ratio of $p$ reflection coefficient to the $s$ reflection coefficient, and ellipsometry essentially measures the effective dielectric function of a material at the frequency of the light used in the measurement.

It is well known that the dielectric function is a complex number $[\varepsilon(\lambda) = \varepsilon_1(\lambda) + i\varepsilon_2(\lambda)]$ and that a nonzero value for the imaginary part of the dielectric function indicates a nonzero light absorption. The complex index of refraction is related to the dielectric function as follows:

$$\varepsilon(\lambda) = \varepsilon_1(\lambda) + i\varepsilon_2(\lambda) = [n(\lambda) + ik(\lambda)]^2 \tag{1}$$

If one knows the complex index of refraction (dielectric function) vs. wavelength, then the thickness of a film can be calculated from a reflectance or ellipsometric measurement (see Chapter 24). The variation of the index of refraction with wavelength is known as the dispersion of the index. One approach is to determine exactly the optical constants vs. wavelength for the low-$k$ material and then to use these to determine thickness. This is not always possible, and in some instances the optical constants are expected to change with processing. Thus, it is very useful to consider using optical models that can be fit to the data for each measurement. There are several models that describe the dispersion properties, and the first property to determine in the selection of the appropriate model is light absorption. Silicon dioxide is transparent (i.e., $\varepsilon_2(\lambda) \cong 0$), and thus $k(\lambda) \cong 0$ in the visible wavelength range, while some of the new low-$k$ materials absorb some light in the visible region.

Commercial ellipsometers are equipped with software that can convert data into optical constants for the film of interest when the film thickness allows. This is often the case for 1-micron-thick films. The optical constants are determined by numerical inversion of the ellipsometric equation relating $\Psi$ and $\Delta$ to the complex reflection coefficients and the Fresnel reflection coefficients, which are a function of the film thickness and the complex index of refraction. These equations are described in Jellison's chapter (Chapter 25). In Figure 2, we show the ellipsometric data for a typical low-$k$ material and the resulting values for the complex refractive index.

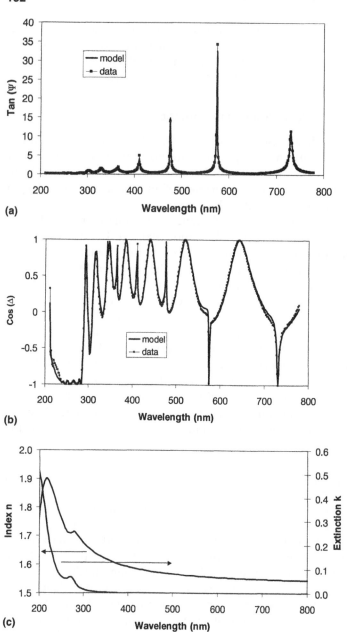

**Figure 2**  Opti-Probe OP5240 spectroscopic ellipsometer data (tan Ψ (a) and cos Δ (b)) and film index and extinction (c) as a function of wavelength, for $a \sim$ 1-μm BCB film on silicon.

Silicon dioxide is a relatively easy material to model optically. The dispersion of thick silicon dioxide films is often described by a Cauchy model (see Chapter 25) whose coefficients are fit to experimental data. The Cauchy dispersion relation is:

$$n = B_0 + \sum_j \frac{B_j}{\lambda^{2j}} \qquad (2)$$

The fit parameters can be stored in the software of the system or determined for each new film. The coefficients for the Cauchy model should remain constant for silicon dioxide.

Considerably more complicated optical models, such as several Lorentz oscillators, are also used for low-$k$ insulators. Porosity can be included in the models by using an effective medium approximation (EMA) in which a new dielectric function is formed by taking voids with the dielectric function of air together with the dielectric function of the base low-$k$ material to form an averaged dielectric function for the mixture (see Chapter 25). The void content can be varied until the EMA model fits the data, and thus the total amount of porosity determined. Such EMA models are often available in the software supplied with commercial ellipsometers and reflectometers. The Maxwell–Garnett EMA models the dielectric function of a material that is mainly one constituent by adding regions of the other constituent uniformly across the layer. The Bruggeman EMA is a formalism that models the dielectric function by combining the dielectric properties of two materials in nearly identical amounts. Voids are typically included using the Maxwell–Garnett EMA.

## B. Techniques for Determination of Optical Constant and Thickness

### 1. Reflectance Angular Spectrum (or Beam Profile Reflectometry)

One technique that is particularly powerful for the determination of thickness and optical constant of single-layer films and multilayer stacks is beam profile reflectometry (BPR) (4). The technique is essentially a measurement of the angular reflectance spectrum of a film on a substrate, but it is performed in such a way that results are obtained rapidly and in a very small spot, making BPR a practical tool for process control and film characterization. In the BPR method, polarized light from a single-wavelength visible laser diode is focused on a wafer sample with a high-power microscope objective. The beam reflected from the wafer is collected and recollimated by the same microscope objective, and then imaged on two linear CCD array detectors, which are oriented in two orthogonal directions. Because both the interface reflection coefficients and the round-trip phase of a film are dependent on the angle of propagation of the light beam, the film modifies the angular spectrum of the reflected light beam compared to that of the incident beam. The information about the thickness, index, and extinction of the film (and substrate) are contained in the details of the reflected angular spectrum. By examining Figure 3 one can see that the microscope objective performs a mapping of radial position along the incident beam to incoming angle of incidence, and then to pixel number on the CCD detector array. The reflected intensity vs. angle and polarization is referred to as the profile of the reflected beam. Films are measured, then, by detecting changes in the *profile* of the *reflected beam*. The measurement is performed in a diffraction-limited spot, and the reflected beam intensity profile can be read out rapidly by the CCD array, leading to a high-throughput measurement, with high spatial resolution. Although the light from the laser diode is emitted in a single polarization, by detecting the reflected profile in two orthogonal directions aligned parallel and perpendicular to the polarization direction the angular reflectance spectrum for both $p$-polarized and $s$-polarized light can be measured simultaneously. The maximum angle of incidence is determined by the numerical aperture of the microscope objective.

In practice the BPR profile measured on a sample is normalized by a BPR measurement made on a bare Si wafer. In this way the BPR profiles from all samples can be

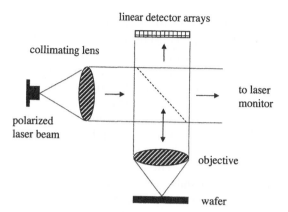

**Figure 3**  Optical system layout for beam profile reflectometry measurements.

referred to a common reflectivity level near $R = 1.0$. Film information is obtained by fitting the usual Fresnel equation models to the BPR profile data. Certain key features in BPR profiles can be used, however, to quickly extract information about the film stack. For low-index films on silicon substrates (the case considered here), the minimum reflectivity level in the BPR oscillation profile is determined by the film index, while the maximum reflectivity level is determined by the substrate index. For the ideal case of a single-layer dielectric film on silicon, the maximum theoretical normalized reflectivity is $R = 1.0$. Deviations from this condition indicate that the film is more complex in nature. For example, PECVD processes typically result in damage to the underlying silicon substrate, converting some of the near-surface region to polysilicon or amorphous silicon. Since the index of poly or $a$-Si at the BPR wavelength is higher than crystalline silicon, this results in a higher film-stack reflectance and a peak normalized BPR reflectivity $R > 1.0$. Considering a different situation, if the film index is not uniform throughout its depth, then the film may be modeled as a multilayer stack, in which the index of each sublayer is a constant value equal to the average index within that sublayer. The additional interfaces in the film stack cause additional film-stack reflectance, and the BPR profiles will be altered from the ideal case by an amount that depends on the magnitude of the index nonuniformity. The peak BPR reflectance will be greater than 1.0 (less than 1.0) if the upper sublayer index is higher (lower) than that in the lower portion of the film. Thus key features of BPR profiles for single-layer films can be used to rapidly diagnosis film deposition conditions. The BPR profiles for some typical interconnect dielectrics are shown in Figure 4.

When the complex index of refraction is not known, low-$k$ dielectric film thickness can be determined from the interference effects in the BPR angular spectrum. This is analogous to the interference pattern observed in wavelength-dependent reflectivity described next. This is shown in Figure 5. The real part of the index and the film thickness can be determined simultaneously.

## 2.   Wavelength-Dependent Reflectance ($R$)

Wavelength-dependent reflectance is typically done at normal incidence. This is in contrast to the BPR angular reflectance method described earlier. When one is determining the value of several unknowns, such as the thickness of multiple films, reflectance data at

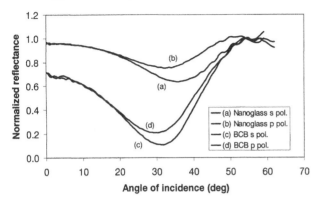

**Figure 4** BPR profiles for a 1026-nm-thick BCB film and an 832-nm-thick Nanoglass™ film (both on silicon substrates).

multiple wavelengths or angles provides the required additional information. Again, one is referred to Jellison's chapter (Chapter 25) for a thorough discussion of the fundamentals of reflectivity. Typically, commercial reflectometers work in the visible-wavelength range, and extension to the UV region of the spectrum has been motivated by the use of 193-nm and 157-nm light for lithography. One important feature of wavelength-dependent reflectivity data is the oscillatory behavior of the reflectance value vs. wavelength, as shown in Figure 6. This phenomenon allows easy determination of film thickness as discussed shortly.

When a single layer of sufficient thickness is present on a substrate such as silicon dioxide on silicon or low-$k$ films on metal, the light reflected from the top of the sample and the light reflected from the substrate–film interface can interfere. (Since metals absorb visible-wavelength light, a 1-micron-thick metal film can be considered an infinitely thick substrate.) Multiple intensity oscillations are possible over the visible-wavelength range, and the equations for the reflection coefficients found in Chapter 25 allow prediction of their occurrence. The equations for the reflectivity $\mathcal{R}$ of a single film on a substrate written using Jellison's (6) notation for the reflection coefficients $r$ are:

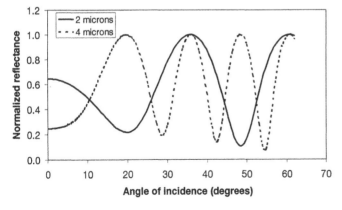

**Figure 5** $s$-Polarized BPR profiles for an oxide film of thickness $t = 2\,\mu m$ and $t = 4\,\mu m$ showing the interference-induced oscillations in reflectance vs. angle. The film thickness can be determined from the angular spacing of the reflectance minima or maxima.

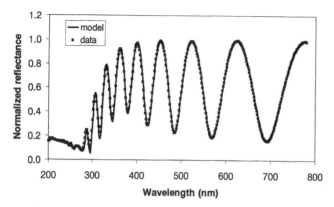

**Figure 6** Reflectance (normalized to bare silicon reflectance) versus wavelength for the BCB film of Fig. 2. The oscillatory nature of the reflectance value vs. wavelength can be used to determine thickness, as described in the text.

$$\mathcal{R} = |r_s|^2 \quad \text{or} \quad \mathcal{R} = |r_p|^2$$

$$r_p = \frac{r_{01,p} + r_{12,p}e^{-i\beta}}{1 + r_{01,p}r_{12,p}e^{-i\beta}} \quad \text{or} \quad r_s = \frac{r_{01,s} + r_{12,s}e^{-i\beta}}{1 + r_{01,s}r_{12,s}e^{-i\beta}} \tag{3}$$

$r_{01,p}$, $r_{12,p}$, $r_{01,s}$, and $r_{12,s}$ are the complex reflection coefficients for the first and second interface, and $\beta = 4\pi d_f \tilde{n}_f \cos(\phi_f)/\lambda$. $d_f$ is the film thickness and $\tilde{n}_f$ is the complex refractive index. The reflection coefficients, defined in Chapter 25, also contain a cosine dependency. At normal incidence the cosine terms are 1, and the reflection coefficients are functions of only the complex indices of refraction of the film and substrate (see Chapter 25). For nonabsorbing films, $\tilde{n}_f$ is real and the reflectivity will have maxima at $2\beta = 2\pi, 4\pi, 6\pi \ldots$ and minima at $\pi, 3\pi, 5\pi$, etc. (5). If the index of refraction $n$ is known for nonabsorbing dielectric films, the thickness can be determined from the wavelength of adjacent maxima:

$$nd = \frac{\lambda_i \lambda_{i+1}}{\lambda_{i+1} - \lambda_i} \tag{4}$$

This equation assumes that there is little or no change in refractive index with wavelength between $\lambda_i$ and $\lambda_{i+1}$. For films that do not show the wavelength-dependent oscillatory structure, the thickness of the film can be determined if one knows the complex index of refraction of both the film and substrate using the reflectance equations shown earlier. Absorbing films also show the oscillatory structure.

The thickness of an absorbing film and its optical constants can be determined by the procedure outlined in Tomkins and McGahan (5). The procedure involves first fixing the film thickness by locating a portion of the measured reflectance spectrum where the film is transparent (nonabsorbing) and then using this thickness and the optical constants of the substrate to determine the wavelength-dependent optical refractive index over the remainder of the wavelength range (6).

## 3. Ellipsometry

In this section, we contrast both beam-polarized reflectivity and wavelength-dependent reflectivity to ellipsometry. As already mentioned, Chapter 25 provides an outstanding overview of these topics. There are many similarities to reflectivity. Ellipsometric measurement of film thickness is based on the change in polarization of light after reflection from the dielectric on the wafer surface. Spectroscopic ellipsometry refers to the use of wavelength-dependent data. If the optical constants of the dielectric are known, the thickness can be determined. Typically, a model of the optical structure of the dielectric film or film stack on the silicon wafer is used.

Ellipsometric data is often reported as $\tan(\Psi)$ and $\cos(\Delta)$ values vs. wavelength. $\Psi$ and $\Delta$ are related to the ratio of the complex reflectivities as follows:

$$\rho = \frac{r_p}{r_s} = \tan \psi \exp(i\Delta) \tag{5}$$

The difference between ellipsometric measurements and reflectivity can be understood by comparing Eqs. (4) and (5). Ellipsometry has the advantage of having the information from both polarization states. The strengths of spectroscopic ellipsometry analysis include a characterization of multilayer samples that is not possible with single-wavelength ellipsometry or simple reflectivity. The addition of information from both polarization states makes BPR and SE both effective methods of multilayer sample analysis and metrology. This is especially useful for the new materials stacks used for low-$k$ insulator layers.

Plots of $\cos(\Delta)$ vs. wavelength show oscillatory behavior for thicker single-layer dielectric films that is reminiscent of reflectivity data, as seen in Fig. 2b. Optical models that describe the wavelength dependence of the dielectric properties of each layer are a critical part of spectroscopic ellipsometric determination of film thickness and other properties.

## C. Birefringent Versus Isotropic Materials

The preceding discussions of optical measurements assumed that the material to be characterized is *isotropic*; that is, the index and extinction do not vary appreciably with the direction of propagation of the light beam in the film. For certain interconnect dielectrics (typically organic materials), the deposition process can produce a film whose optical properties depend strongly on measurement orientation (*anisotropy*). For CVD films this is caused by the material's being evolved from the source as molecular chain fragments, which then deposit on the substrate with a preferred orientation. For spin-on films, thermomechanical stresses set up during the spin and cure cycles can cause an azimuthal symmetry in the layer. The practical consequence for interconnect performance is that if the in-plane dielectric constant ($\varepsilon_{\parallel}$ is greater than the out-of-the plane dielectric constant ($\varepsilon_{\perp}$), the line-to-line capacitance will be larger than expected, assuming an isotropic dielectric constant. The best dielectric candidates are those with an anisotropy that is either negligible or controllable. While the ideal measurement to make from the designer's point of view is a measurement of $\varepsilon_{\parallel}$ and $\varepsilon_{\perp}$ at operating frequency, the in-plane refractive index ($n_{\parallel}$) and out-of-plane refractive index ($n_{\perp}$) measured at optical frequencies provide at least a qualitative measure of the amount of anisotropy to be expected at device-switching frequencies.

Beam profile reflectometry is a particularly convenient technique for characterizing the anisotropy in materials, since measurements are made at a single wavelength and therefore are independent of material dispersion changes. In the case of a material exhibiting *uniaxial* optical symmetry, for which the index and extinction are independent of the light propagation direction for all polarization directions parallel to the substrate plane, the equations describing the film-stack reflectance simplify considerably. The effect of anisotropy on the BPR profiles is then a distortion of the $p$-polarized angular reflectance spectrum; the $s$-polarized reflectance is unaffected. The reflectance distortion can be expressed through an effective index of refraction $n_{\text{eff}}$ that is a function of propagation direction through

$$\frac{1}{n_{\text{eff}}^2(\theta)} = \frac{\cos^2\theta}{n_{\parallel}^2} + \frac{\sin^2\theta}{n_{\perp}^2} \tag{6}$$

where $\theta$ is the angle of propagation of the light in the film with respect to wafer normal. One can also define a parameter $\rho$ through

$$\rho = \frac{n_{\parallel}^2 - 1}{n_{\perp}^2 - 1} \tag{7}$$

which provides a way to quantify the amount of anisotropy.

An example of a BPR measurement on an anisotropic material (a 10-kÅ Parylene AF4$^{\text{TM}}$ film) on silicon is shown in Figures 7 and 8. The angular reflectance data and the model fits with and without film anisotropy are shown. Attempts to fit alternative multi-layer isotropic film models to the data were unsuccessful; only a model incorporating anisotropy effects was capable of simultaneously providing good fits to both the $s$-polarized and $p$-polarized angular reflectance spectra. The measured in-plane index was $n_{\parallel} = 1.5247$ and the anisotropy parameter $\rho = 1.27$. If this dielectric constant ratio was maintained down to device-operating frequencies, signal lines oriented horizontally would experience an $\sim 14\%$ increase in capacitance compared to vertically oriented lines of the same geometry. Any nonuniformity in $\rho$ across the chip would cause a signal skew that could severely limit the maximum operating speed. Clearly, characterization of film anisotropy and its variation across a wafer with optical techniques provides useful data for qualifying potential interconnect dielectric candidates.

## D.  Issues with Single-Layer and Multilayer Film Measurement

Measurement of the thickness and optical constants of single-layer dielectric films is straightforward if the film is well behaved. Rapid thickness and single-wavelength index and extinction information can be obtained with BPR; full-spectrum measurements of the

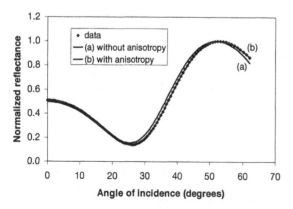

**Figure 7**  Fits to BPR $s$-polarization data for a 10.3-kÅ-thick Parylene AF4$^{\text{TM}}$ film on silicon, using models for which the material is assumed to be either isotropic (a) or anisotropic (b).

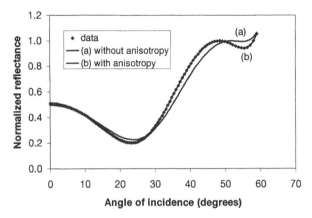

**Figure 8** Fits to BPR *p*-polarization data for a 10.3-kÅ-thick Parylene AF4™ film on silicon, using models for which the material is assumed to be either isotropic (a) or anisotropic (b).

dispersion can be obtained with spectroscopic tools (reflectance and ellipsometry). A variety of complications can arise, however, in the characterization of multilayer stacks and real-world single-layer films, requiring a more careful analysis of the data aided by knowledge of the film deposition process. In this section we illustrate in several case studies the detail that sometimes must be added to models in order to measure single dielectric layers accurately and the limitations of optical techniques to characterize multilayer stacks.

## 1. Single-Layer Film Case Studies

### *a. Nanoglass*

Nanoglass™ is a porous spin-on oxide film that has often been proposed as the ultimate interconnect dielectric. The dielectric constant is adjusted by changing the size and volume fraction of nanopores in the material. Thick Nanoglass films sometimes have thin cap and liner layers of low or zero porosity that are intentionally added as a barrier against contamination of surrounding layers during further wafer processing. Alternatively these layers may be an unintentional result of the film drying process. These barrier layers need to be included in the film-stack model to ensure accurate measurement of the thicker layer. The BPR data [combined with a single-wavelength ellipsometry measurement such as beam profile ellipsometry (BPE)] has some sensitivity for multithickness measurement of these stacks, but the data can also be well fit by a single-layer model in which the thickness and index are simultaneously optimized. Wavelength-dependent reflectance data provides a good visual clue that the film is multilayered in nature, since in the ultraviolet the wavelength is short enough to accumulate enough phase in the higher-index sublayers to perturb the reflectance spectrum (Figure 9). The film stack cannot be uniquely determined, however, since a single-layer model in which the material exhibits an absorption edge in the DUV can also provide an acceptable fit to the data. The spectroscopic ellipsometer measurements $\cos(\Delta)$ and $\sin(\Delta)$ provide the most information, since the primary effect of the index gradient in the film stack is on the phase of the reflected light. The presence of the cap and liner layers is revealed in a poor fit of a single-layer oxide model to the extrema of the $\cos(\Delta)$ and $\sin(\Delta)$ spectra (Figure 10). The optimum measurement

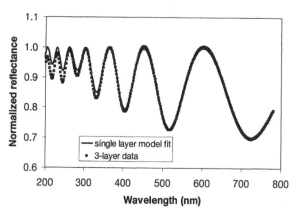

**Figure 9** Simulated data for an 8-kÅ-thick Nanoglass™ film (67.5% void fraction) with 50-Å-thick oxide liner and cap layers and a single-layer model fit.

recipe would probably be one that combined reflectance and ellipsometric data to constrain the dispersion of the sublayers of the film stack.

### b. Black Diamond

Black Diamond™ is an inorganic, silica-based CVD material also utilizing nanopores for dielectric constant reduction (11). The effects of various temperature cure cycles on the material have been studied, and in some cases these reveal a complex multilayer nature to the material. We have found that different processing conditions can result in the material's exhibiting either a high/low index variation from the top surface to the silicon substrate, or the reverse low/high index structure. In some cases the data reveals that the index gradient is wavelength dependent. Evidence for this unique case is shown in Figures 11–13. The BPR angular reflectance data exhibits a maximum reflectance $R > 1.0$, indicating that in the visible wavelength range ($\lambda = 673\,\text{nm}$) there is a high/low index variation throughout the depth of the film. From the spectral reflectance data (Fig. 12) we see that the maximum reflectance begins to decrease below unity, until in the DUV the reflectance drops significantly. This can be explained by a low/high index gradient structure, a bulk

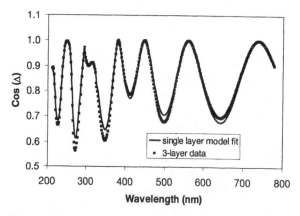

**Figure 10** Simulated ellipsometric data for an 8-kÅ-thick Nanoglass™ film (67.5% void fraction) with 50-Å-thick oxide liner and cap layers and a single-layer model fit.

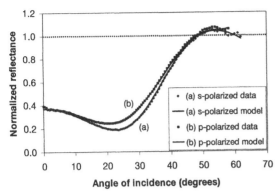

**Figure 11** BPR angular reflectance data and model fit for a Black Diamond™ film on a silicon substrate. The data is well fit by a model that assumes a high-to-low index gradient in the layer.

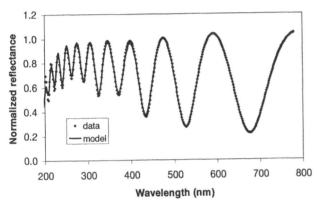

**Figure 12** Spectrometer data and model fit for a Black Diamond™ film on a silicon substrate. The data is well fit by a model that assumes a high-to-low index gradient in the layer.

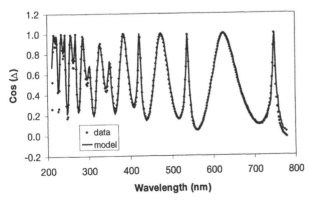

**Figure 13** OP5240 DUVSE cos(Δ) data and model fit for a Black Diamond™ film on a silicon substrate. The data is well fit by a model that assumes a high-to-low index gradient in the layer.

material absorption, or a combination of these effects. Model fits to the combined BPR, spectrometer, and DUVSE data (Fig. 13) using a critical-point functional form for the dispersion agreed with the index gradient model just described (Figure 14). Apparently whatever material changes occurring throughout the depth of the layer simultaneously lower the dielectric constant and increase the dispersion in the DUV for this particular process. The data was well fit by modeling the graded layer as a two-layer stack, with top sublayer thickness $t_1 \sim 1300\,\text{Å}$ and index $n_1 \sim 1.485$, and bottom sublayer thickness $t_2 \sim 7000\,\text{Å}$ and index $n_2 \sim 1.448$. It is not clear whether the small but measurable extinction in the material at the shortest wavelengths is due to bulk material absorption or scattering from the nanopores.

## 2. Information Content for Multilayer Measurement

From the preceding discussions it is clear that optical data is information-rich, and for some films and processes a detailed film-stack model must be used to accurately measure the thicknesses and optical constants of the layers (sublayers) in the film stack (film). Once the physical nature of the stack is determined, then the appropriate thicknesses and optical parameters to measure can be selected. The final step in the development of a metrology recipe is to choose the type of measurement to be used (BPR, BPE, reflectance, SE, or some combination of these). Each measurement technology will be sensitive to certain stack parameters and insensitive to others. A technique for quantifying the sensitivity of each measurement and of recipes that combine several measurements for determining selected film parameters is outlined in the next section.

### a. Multiparameter Measurement Sensitivity

The standard approach for determining the sensitivity of an optical measurement to a specific film-stack parameter is to plot a series of curves (BPR, reflectance, SE) as a function of the parameter varied over a certain range. Measurement technologies that exhibit some change (over an angle or wavelength range) are declared to be sensitive to that parameter. This approach is flawed for two reasons: (1) these curves are usually plotted without measurement noise, giving the reader no information on whether the sensitivity exceeds the noise or is limited by it; (2) few fab recipes these days measure

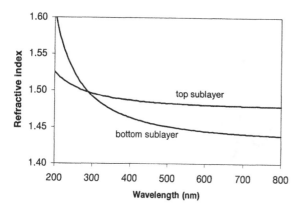

**Figure 14** Refractive index vs. wavelength for the two portions of a Black Diamond™ film on a silicon substrate, obtained by fitting a critical-point dispersion model to the BPR, spectrometer, and DUVSE data.

only one film-stack parameter, and it is more often the case that four or more thickness, index, or extinction variables are determined simultaneously. Single-parameter sensitivity plots provide no indication whether a given measurement is sensitive to a given film-stack parameter under these conditions. A metric that can be used to sort through these issues is one we call the *multiparameter sensitivity plot.*

If one considers how optical data is used to measure layer thickness, index, etc., one recalls that a film-stack reflectance model is compared to the data and the stack parameters are adjusted to provide the smallest residual (root-mean-square error). In the region near the solution point, the residual exhibits a quadratic dependence on the difference between the true stack parameter value and any trial value (Figure 15). For a measurement that is sensitive to the thickness of layer 2, for example, the residual will increase dramatically with deviation of $t_2$ from the correct value. For a measurement technology that is insensitive to $t_2$, the residual in the thickness range of interest will be determined primarily by measurement noise. A procedure for quantifying the sensitivity of a given measurement to determine a given parameter under multiparameter recipe conditions could then be: (1) Set up a metrology recipe to measure the film-stack parameters of interest, with a given measurement technology ($\equiv T_i$) or a combination of technologies ($\equiv C_i$). (2) Force a given stack parameter ($\equiv P_j$) to a value that is known to be incorrect (e.g., by adjusting the recipe fit range to only allow this value) and set ranges for the other recipe parameters so that a rather wide range of values for these parameters is allowed. (3) Record the residual from the recipe fit to the data under these conditions ($\equiv \rho_{ij}$), and when the recipe is allowed to calculate the correct values for all stack parameters ($\equiv \rho_{ij0}$) (4) then the "multiparameter information content" $C_{ij}$ can be defined for measuring $P_j$ with $T_i$ through (cf. Fig. 15)

$$C_{ij} = \frac{1}{\Delta P_j} \frac{\rho_{ij}}{\rho_{ij0}} \tag{8}$$

where $\Delta P_j$ is the amount of the forced change in $P_j$. (The units of $C_{ij}$ are then $nm^{-1}$ or $Å^{-1}$). $C_{ij}$ is a measure of the ability of a given measurement to determine the accurate value of a film-stack parameter under noise-influenced, multiparameter measurement conditions. In general it will be a function of the amount of measurement noise, any calibration errors, the quality of the fit to the data, and the value of the film-stack parameter (a given measurement may be more sensitive to a given parameter in certain ranges). For $M$

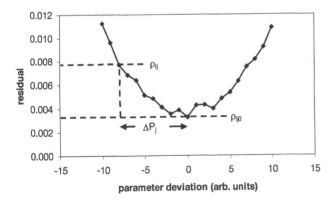

**Figure 15** Typical variation of the residual from a metrology recipe as the measurement parameter of interest is varied from its actual best-fit value.

considered measurements and $N$ desired measurement parameters, $C_{ij}$ is an $M \times N$ matrix of values.

### b. Dual Damascene Copper Measurement

As an illustration of information content analysis, we can apply the foregoing formalism to measurement of a dual Damascene copper stack. For fabrication of these structures, nitride-barrier and etch-stop layers are deposited alternately with oxide or dielectric layers in preparation for etching and filling a via contact to the underlying copper line. Often, for convenience or process flow considerations, the individual layers in the film stack (as well as any thin copper oxide or roughness layer between the copper and the nitride-barrier layer) must be measured after the full film stack has been deposited. The nitride layers may be silicon-rich, and their properties may significantly depend on wafer position in the deposition chamber, mandating a measurement of the optical properties of these layers as well. The large number of parameters to be measured in these stacks places a heavy burden on any optical measurement technique. We can determine the limitations to measuring these stacks by examining the content of the multiparameter information.

As an example we consider a rather typical Damascene stack consisting of a 500-Å-thick Si-rich oxynitride ARC layer, two 5-kÅ-thick oxide interlevel dielectric (ILD) layers, a 500-Å-thick Si-rich nitride etch-stop layer above and below ILD1, and a 50-Å-thick copper oxide interface layer between the nitride-barrier layer (NIT1) and the copper contact. The BPR, reflectance, and DUVSE data were simulated for this film stack using typical noise levels for each of the measurement technologies. A recipe was developed that measured the thicknesses only of the five dielectric layers and the copper oxide interface layer. The information content was calculated by forcing the recipe to a value that was in error by approximately 10% for each parameter. The corresponding plot for the DUVSE measurement technique is shown in Figure 16. It is seen that the measurement sensitivity is smallest for the two oxide ILD layers (ILD1 and ILD2). This is probably because the absorption edge in the DUV for the Si-rich oxynitride and Si-rich nitride layers provides a distinct spectral feature that allows the fitting algorithms to separate the contributions from the various layers. On the other hand, since the optical thickness of the nitride layers is significantly less than that of the oxide layers, the effect of the nitride layers on the optical phase will not change significantly even if the position of these layers in the stack is altered. This implies that the DUVSE data will have a reduced ability to

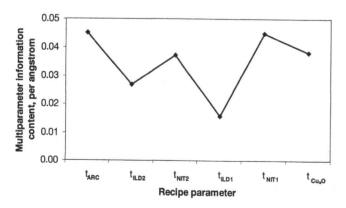

**Figure 16** Simulated multiparameter information content plot for a dual Damascene copper film stack six-thickness recipe.

determine the fraction that the top or bottom ILD layer contributes to the total ILD thickness. This feature manifests itself as a lower information content for those layers.

The ability of DUVSE to measure the composition of the oxynitride and nitride layers can be determined by adding the composition or refractive index as a recipe parameter and recalculating the model fit to the data. The information for measuring the top nitride layer index, e.g., can be determined by forcing $n_{NIT2}$ to a value that is incorrect and allowing the thicknesses of the six layers in the stack to take on values that give the best fit to the data. Since for complex dispersion models like critical point or Cauchy the results could be influenced by the robustness of the fitting algorithms and the number of dispersion parameters fitted, we have chosen to constrain the problem by modeling either the oxynitride or the nitride layers as an effective medium. The layer index was forced to an incorrect value by mixing the actual layer material with $\sim$ 10–20% voids. The result is shown in Figure 17. The information content for determining composition is high for the top oxynitride ARC layer and decreases monotonically as one moves deeper into the film stack. The DUVSE data contains the least amount of information on the composition of the copper–nitride barrier interface layer. This is probably because Si-rich oxynitride, Si-rich nitride, and $Cu_xO$ all possess a strong absorption in the DUV, and the topmost layers attenuate this portion of the spectrum before it can penetrate to the bottom of the stack, where it can be influenced by the NIT1 and copper interface layers. This dropoff in information content for the bottom layers in the stack was also observed for BPR and reflectance vs. wavelength data. This implies that the best method for measuring the full film stack is first to characterize the optical properties of the layers as the stack is built up one layer at a time.

Next-generation interconnect structures will utilize similar multilayers, with some or all of the oxide ILD layers replaced by low-$k$ dielectric. As an example of this we show in Figure 18 the multiparameter information content calculated from Opti-Probe data on an oxide (870-Å)/Si-rich nitride (475-Å)/SiLK$^{TM}$ (4180-Å)/oxide (8500-Å)/Si sample. The recipe measured all four layer thicknesses and kept the nitride and SiLK optical properties constant. It is evident that both BPR and spectrometer measurements have a greater ability to determine the top oxide and nitride layer thicknesses. The relative lack of accuracy in determining the SiLK$^{TM}$ and bottom oxide thicknesses may be due to the fact that their corresponding optical phase thicknesses (index × thickness) are comparable, forcing the fitting algorithms to use rather subtle features in the data to distinguish between the materials.

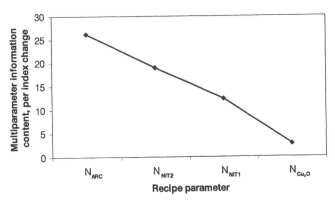

**Figure 17**   Simulated multiparameter information content plot for a dual Damascene copper film stack, for a recipe measuring six layer thicknesses and the index of a selected layer.

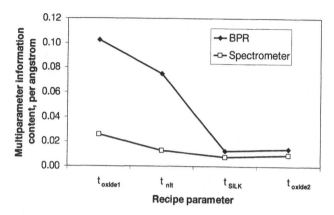

**Figure 18**  Information content in a BPR or spectrometer four-thickness recipe, taken from Opti-Probe data on an oxide/Si-rich nitride/SiLK$^{TM}$/oxide/Si sample.

It is evident then that examining the sensitivity of optical measurements under multi-parameter recipe conditions is a necessary step in the development of a process control methodology for any film stack. The results of the information content analysis may force a reconsideration of the places in the process flow at which wafers are measured, or may even be used to determine the allowed range on process control charts.

## REFERENCES

1.  K.O. Goyal, J.W. Westphal. Measurement capabilities of X-ray fluorescence for BPSG films. Advances in X-ray Analysis 40, CD-ROM Proceedings of the 45th Annual Conference on Applications of X-ray Analysis.
2.  M.D. Janezic, D.F. Williams. IEEE International Microwave Symposium Digest 3:1343–1345, 1997.
3.  A.L. Loke, J.T. Wetzel, J.J. Stankus, S.S. Wong. Low-Dielectric-Constant Materials III. In: C. Case, P. Kohl, T. Kikkawa, W.W. Lee, eds. Mat. Res. Soc. Symp. Proc. Vol. 476, 1997, pp 129–134.
4.  A. Rosencwaig, J. Opsal, D.L. Willenborg, S.M. Kelso, J.T. Fanton. Appl. Phys. Lett. 60:1301–1303, 1992. BPR is a patented technique available on the Opti-Probe® models.
5.  H.G. Tompkins, W.A. McGahan. Spectroscopic Ellipsometry and Reflectometry. New York: Wiley, 1999, pp 54–61.
6.  H.G. Tompkins, W.A. McGahan. Spectroscopic Ellipsometry and Reflectometry. New York: Wiley, pp 188–191.
7.  S.-K. Chiang, C.L. Lassen. Solid State Tech., October 1999, pp 42–46.

# 9

# Thin-Film Metrology Using Impulsive Stimulated Thermal Scattering (ISTS)

**Michael Gostein, Matthew Banet, Michael A. Joffe, Alex A. Maznev, and Robin Sacco**
*Philips Analytical, Natick, Massachusetts*

**John A. Rogers**
*Bell Laboratories, Lucent Technologies, Murray Hill, New Jersey*

**Keith A. Nelson**
*Massachusetts Institute of Technology, Cambridge, Massachusetts*

## I. INTRODUCTION

Impulsive stimulated thermal scattering (ISTS), also known as transient grating (TG) photoacoustics, is a noncontact, nondestructive optoacoustic technique for measuring the thickness and mechanical properties of thin films (1–9). In contrast to ellipsometry and reflectometry, which are used on transparent films, ISTS is ideally suited for measuring opaque films such as metals, because the laser light must be absorbed by the sample rather than transmitted through it. Since the typical spatial resolution of an ISTS instrument is a few tens of microns, the technique enables measurement of film thickness near the edges of sample films and on patterned wafers. This ability to measure film thickness nondestructively on patterned wafers makes ISTS ideal for process-monitoring applications.

In ISTS, the sample film is irradiated with a pulse of laser light from a pair of crossed excitation beams, which creates a transient optical grating that launches an acoustic wave in the film. By measuring the time-dependent diffraction of light from the sample surface and analyzing with a model of the acoustic wave physics, film thickness and/or other properties can be determined.

The ISTS technique is one member of a family of optoacoustic techniques that have been commercially developed for semiconductor metrology in the past several years (10). Another technique, described in Chapter 10, relies on femtosecond lasers to excite and detect acoustic waves reflecting from film interfaces, requiring picosecond time resolution (1,11,12). The fundamental distinguishing characteristics of ISTS are that two crossed excitation pulses are used to generate acoustic waves, and that time-dependent diffraction of probe light is used to monitor them. In addition, while ISTS may also be performed with picosecond time resolution to detect reflections from film interfaces (13), in its usual form discussed in this chapter it need be performed with only nanosecond time resolution.

In this lower-time-resolution form the technique detects waves propagating laterally in the plane of the sample film, over distances much larger than the film thickness, rather than reflections from film interfaces.

For nanosecond-time-resolved ISTS the experimental arrangement is simpler than for picosecond-time-resolved techniques. However, the tradeoff is that ISTS in its usual form cannot simultaneously measure as many layers in a multilayer structure.

Systems employing nanosecond-time-resolved ISTS for metal metrology are commercially available only from Philips Analytical at present. Noncommercial ISTS systems are also in use in research laboratories. Most of the discussion presented here is based upon the commercial systems available from Philips Analytical, and only nanosecond-time-resolved ISTS is discussed in this chapter.

This chapter is divided into two main sections. In Section II, the main principles of the ISTS method are discussed, including an overview of the experimental technique and the theory of measurement. In Section III, typical applications are described, including measuring film uniformity, measuring at sample edges, characterizing chemical-mechanical polishing (CMP) of metals, and measuring mechanical properties of materials.

## II. PRINCIPLES

### A. Experimental Technique

#### 1. Optical Apparatus

Figure 1 shows a schematic drawing of the optical setup of an ISTS measurement system. The compactness and simplicity of the apparatus are achieved by the use of miniature lasers and diffractive optics (4).

A miniature diode-pumped and frequency-doubled microchip YAG laser produces a subnanosecond excitation pulse with optical wavelength 532 nm. The beam is attenuated to a desired level by a filter and is focused onto a phase grating that produces many diffracted beams (4). The two ±1-order diffracted beams are then recombined at the sample surface by imaging optics. This yields a spatially periodic interference pattern of light and dark fringes. It results in the excitation of surface acoustic waves with acoustic wavelength equal to the fringe spacing, as will be discussed later. The excitation spot is elliptical, with a typical size being ∼ 300 microns along the long axis, perpendicular to the interference fringes, and ∼ 50 microns along the short axis, parallel to the fringes. The angle between the crossed excitation beams at the focal point on the sample is determined by the period of the phase mask grating, and this in turn determines the interference pattern fringe spacing and therefore the acoustic wavelength. Various phase mask patterns are used to set different acoustic wavelengths, typically from several to tens of microns.

Surface ripples caused by the acoustic waves and the thermal grating (discussed later) are monitored via diffraction of the probe laser beam. The probe beam is obtained from a diode laser, with optical wavelength typically ∼ 830 nm. It is focused to an elliptical spot in the center of the excitation area. The probe spot is smaller than the excitation area, with dimensions typically ∼ 50–100 microns along the long axis, perpendicular to the interference fringes, and ∼ 25 microns along the short axis, parallel to the fringes. The directly reflected probe beam is blocked by an aperture, while one of the diffracted probe beams is focused onto a fast photodetector whose signal is fed to a high-speed digital oscilloscope. A computer then records the output data for analysis.

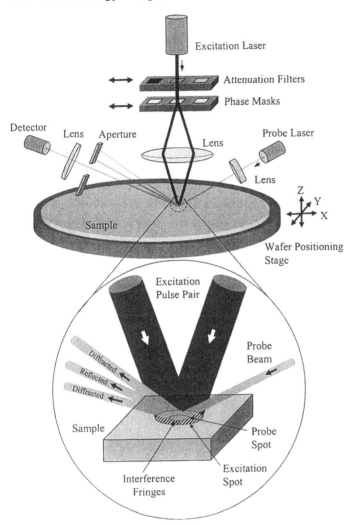

**Figure 1** Schematic diagram of experimental arrangement for ISTS. (Courtesy Philips Analytical.)

In some designs the optical scheme is enhanced by optical heterodyne detection (1,14). In heterodyne detection an additional laser beam, called a local oscillator or reference beam, propagates collinear to the diffracted signal beam, and the detector registers the result of coherent interference of the signal and reference beams. Heterodyne detection improves measurement reproducibility by yielding enhanced signal levels and suppressing the influence of parasitically scattered light.

## 2. Acoustic Wave Excitation

When the excitation laser pulse is fired, the sample area under each bright stripe of the interference fringe pattern absorbs light and heats slightly, a few tens of degrees Kelvin. The heating leads to sudden impulsive thermal expansion, and results in a stationary surface displacement pattern, called a *thermal grating*, whose period equals the period of the optical intensity pattern on the sample. This thermal grating decays slowly as heat flows away from the irradiated regions (15) and the film temperature equilibrates, typically within a hundred nanoseconds. However, the sudden initial impulsive expansion also launches a standing surface acoustic wave with acoustic wavelength $\lambda$ equal to the period of the optical intensity pattern, as shown at the top of Figure 2 (16,17). The standing acoustic wave is composed of two counterpropagating traveling waves. The film surface oscillates as these waves travel away from the excitation point in the plane of the film, rather like ripples on the surface of a pond, as illustrated in the figure. Note that the wave motion extends into the sample a distance on the order of the acoustic wavelength $\lambda$. It therefore may extend through the film layers all the way to the substrate.

The oscillation frequency of the ripple at any point on the surface is determined by the velocity and wavelength of the passing surface acoustic waves, according to the expression

$$F = \frac{v}{\lambda} \tag{1}$$

where $F$ is the oscillation frequency, $v$ is the wave velocity (the phase velocity), and $\lambda$ is the acoustic wavelength. The wave velocity is a function of the acoustic wavelength and the mechanical properties and structure of the film stack. Measurement of the oscillation

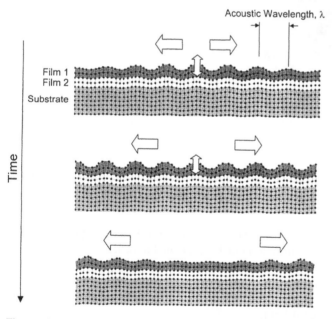

**Figure 2**   Schematic diagram showing propagation of surface acoustic waves induced in ISTS. The initial laser excitation causes the film to heat slightly and expand under each bright stripe of the excitation interference pattern, launching a standing acoustic wave with acoustic wavelength approximately equal to the interference fringe spacing. The standing wave is composed of two counterpropagating acoustic waves traveling in the plane of the film. The film surface oscillates as these waves travel away from the excitation point in the plane of the film, rather like ripples on the surface of a pond induced by a water wave. Note that the wave motion extends through the film layer all the way to the substrate.

frequency at a given acoustic wavelength therefore permits determining film thickness or other properties, as discussed later. For acoustic wavelengths ranging from several microns to tens of microns, the corresponding frequency of the ISTS waveform typically ranges from several gigahertz down to $\sim 100$ MHz.

As discussed later in Section II.B, at any given acoustic wavelength there are various possible oscillation modes characterized by different displacement patterns within the film and substrate. These modes each have a characteristic phase velocity and therefore result in different oscillation frequencies on the film surface. Figure 2 illustrates only the lowest-order oscillation mode, which is typically the strongest.

## 3. Wave Detection and ISTS Signal Waveform

Ripples on the film surface from the passing acoustic waves and the residual thermal grating cause the probe laser to diffract, as shown in Figure 1. As the acoustic waves propagate, the surface ripple is periodically modulated, causing the detected signal (i.e., the diffracted beam intensity) to oscillate. The left side of Figure 3 shows a typical ISTS waveform for a film stack of Cu on Ta on $SiO_2$ on a Si wafer. Before the excitation laser fires (at $\sim 6.5$ ns in the figure), there is no diffracted probe beam signal. When the excitation laser fires, the spatially periodic impulsive expansion of the film underneath the interference fringes causes strong probe beam diffraction. The diffracted intensity then oscillates as the acoustic waves propagate within and away from the excitation region, and it finally decays within approximately a hundred nanoseconds.

The digital oscilloscope acquires a complete signal waveform for each shot of the excitation laser and averages these data for typically 1 second per measurement site (e.g. several hundred laser shots) to yield a good signal-to-noise ratio. Data acquisition is therefore very rapid.

The signal waveform, as illustrated in Figure 3, can be approximately modeled (16) by the following equation:

$$S(t) \propto [A_T \exp(-t/\tau_t) + A_1 \cdot G_1(t) \cdot \cos(2\pi F_1 t) + A_2 \cdot G_2(t) \cdot \cos(2\pi F_2 t) + \ldots]^2 \quad (2)$$

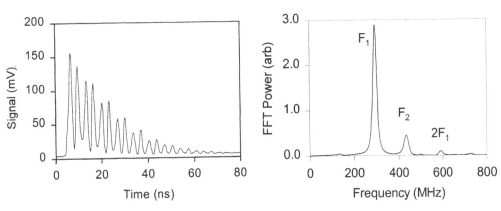

**Figure 3** (*Left*) Typical ISTS waveform from a $Cu/Ta/SiO_2/Si$ sample. (*Right*) The waveform's frequency spectrum. The frequency spectrum reveals signals from two thin-film acoustic modes (with frequencies $F_1$ and $F_2$), plus the second harmonic of one of these frequencies ($2F_1$). Very weak combinations of the frequencies are also apparent ($F_2 - F_1$ and $F_2 + F_1$). (Courtesy Philips Analytical.)

where

$$G_1(t) = \exp\!\left(-\Gamma_1' t - (\Gamma_1'' t)^2\right)$$

$$G_2(t) = \exp\!\left(-\Gamma_2' t - (\Gamma_2'' t)^2\right) \tag{3}$$

$$\vdots$$

The first term in Eq. (2) approximately describes the thermal grating contribution with amplitude $A_T$ and decay rate $\tau_T$. (For more detail, see Ref. 15.) The second term describes the contribution of the surface acoustic wave with amplitude $A_1$ and frequency $F_1$. The decay of this contribution represented by the function $G_1(t)$ is determined by two factors, $\Gamma_1'$ and $\Gamma_1''$. The former describes the acoustic damping in the material, while the latter accounts for the fact that eventually the acoustic wave travels outside the excitation and probing area due to finite laser spot sizes. The latter effect is usually dominant. Subsequent terms in Eq. (2) describe higher-order acoustic modes (see later) and have similar appearance. The number of acoustic oscillation periods observed in the signal waveform is usually on the order of the number of acoustic wavelengths within the excitation spot.

Note that the ISTS signal, as described in Eq. (2), is quadratic in terms of the individual oscillatory terms describing the surface ripple (16,17). The signal spectrum therefore contains not only the frequencies of the acoustic modes but also their harmonics and combinations. For example, the frequency spectrum in Fig. 3, which shows strong components at two acoustic mode frequencies, reveals a harmonic and weak combinations of the two modes. However, note that if a heterodyne detection scheme is used, then the time dependence of the signal may be dominated by the products of the heterodyne reference and the bracketed terms in Eq. (2) (1,14). This effectively linearizes the signal with respect to the surface ripple depth so that no harmonics or combination frequencies are present.

## B. Acoustic Physics in Thin Films

This section gives a brief overview of surface acoustic waves in thin films as relevant to the ISTS technique, described in detail in Refs. 16 and 17. For further details on acoustic mode propagation in thin films in general, see Refs. 18–21.

### 1. Acoustic Modes at Thin-Film Surfaces

ISTS excitation of a bulk sample surface, for example bare silicon, generates a surface acoustic wave, also called a Rayleigh wave (18–21). Its displacements decrease exponentially with depth and extend a distance comparable to the wavelength $\lambda$ into the sample. It has a characteristic velocity that depends on the sample mechanical properties (namely, the elastic moduli; e.g., for an isotropic material, Young's modulus and Poisson's ratio) and density.

ISTS excitation of a thin film similarly produces an acoustic wave that is localized within the film and substrate. The film acts as a planar acoustic waveguide, similar to a planar optical waveguide or an optical fiber, within which the acoustic energy is substantially localized.

The wave motion can be computed by applying the appropriate equations of motion and boundary conditions to the system (16,17). The displacements in each part of the system must obey (1) the following:

$$\frac{\partial^2 u_j}{\partial t^2} = \frac{c_{ijkl}}{\rho} \frac{\partial^2 u_k}{\partial x_i \partial x_l} \tag{4}$$

where $u$ is the displacement, $c$ is the stiffness tensor, and $\rho$ is the density. At free surfaces, the normal components of the stress tensor vanish, while at interfaces between tightly bound materials the displacements and normal components of the stresses are continuous. The solutions to the equations of motion are traveling waves with characteristic wavelength and phase velocity. For a thin film or thin-film stack there are multiple velocity solutions for each value of the acoustic wavelength. These individual solutions correspond to different modes of oscillation, each with a characteristic displacement pattern, as illustrated later. (Note that for film stacks the number of modes is not related to the number of layers in the stack.)

The acoustic wave velocities are calculated by solving for the zeroes of a characteristic determinant, which is given by the nontrivial solutions to the equations of motion and boundary conditions discussed earlier. For example, for the simple case of an isotropic film on an isotropic substrate, the velocities $v_m$ of the various modes are solutions to the following transcendental equation [written explicitly in Eq. (25) of Ref. 17]:

$$F\left(v_m, q, d^{(f)}, v_{tr}^{(f)}, v_{lg}^{(f)}, \rho^{(f)}, v_{tr}^{(s)}, v_{lg}^{(s)}, \rho^{(s)}\right) = 0 \tag{5}$$

Here, $F$ defines the boundary condition determinant, which is a function of the wavevector $q = 2\pi/\lambda$, the thickness of the film $d^{(f)}$, and $v_{tr}^{(f)}$, $v_{lg}^{(f)}$, $\rho^{(f)}$ and $v_{tr}^{(s)}$, $v_{lg}^{(s)}$, $\rho^{(s)}$ are the bulk transverse and longitudinal acoustic velocities and the densities of the film and the substrate, respectively. The bulk transverse and longitudinal sound velocities of each material are determined from the stiffness tensor $c$ used in Eq. (4), and they are equivalently expressed in terms of the Young's modulus $Y$ and Poisson's ratio $\sigma$, according to the following relations (22):

$$v_{lg} = \sqrt{\frac{Y \cdot (1-\sigma)}{\rho \cdot (1+\sigma) \cdot (1-2\sigma)}} \quad \text{and} \quad v_{tr} = \sqrt{\frac{Y}{2\rho \cdot (1+\sigma)}} \tag{6}$$

For the case of a multilayer film structure, Eq. (5) is generalized to read

$$F\left(v_m, q, d^{(f_1)}, v_{tr}^{(f_1)}, v_{lg}^{(f_1)}, \rho^{(f_1)}, d^{(f_2)}, v_{tr}^{(f_2)}, v_{lg}^{(f_2)}, \rho^{(f_2)}, \dots, v_{tr}^{(s)}), v_{lg}^{(s)}, \rho^{(s)}\right) = 0 \tag{7}$$

where $d^{(f)}$, $v_{tr}^{(f)}$, $v_{lg}^{(f)}$, and $\rho^{(f)}$ have been expanded into a set of parameters for all the $n$ film layers $f_1 \dots f_n$.

The surface acoustic modes are qualitatively different from the bulk acoustic modes. Their displacements include both shear and longitudinal distortions, and their velocities (unlike those of bulk acoustic waves) depend strongly on the acoustic wavelength, as will be discussed later.

Each mode velocity corresponds to a particular solution of Eq. (4) that yields the displacement patterns of the wave motion. Figure 4 shows the displacement patterns characteristic of the three lowest-order modes of a thin-film stack on a substrate, at a given acoustic wavelength. In ISTS measurements, the lowest-order mode usually dominates the signal since this mode's displacements cause substantial surface ripple that gives rise to strong ISTS signals. Film thickness measurements therefore use primarily this

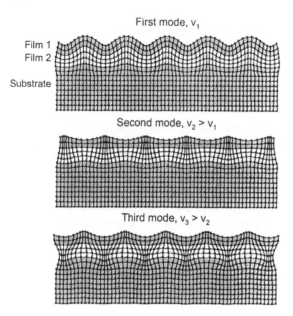

**First mode, $v_1$**

Film 1
Film 2

Substrate

**Second mode, $v_2 > v_1$**

**Third mode, $v_3 > v_2$**

**Figure 4** Sketches of the displacement patterns for the three lowest-order thin-film acoustic modes. Note that the wave motion penetrates into the surface a depth on the order of the acoustic wavelength (the lateral periodicity).

lowest-order mode. Higher-order modes may also be observed under some conditions, for example, as shown in Fig. 3.

Note that in many cases of interest in silicon semiconductor metrology the films and substrates are not isotropic, so the elastic properties and the corresponding acoustic velocities are functions of direction within the film. This applies for many metals that are deposited with preferred crystallographic orientations (26,27) as well as for Si substrates (23). In these cases, the Young's modulus and Poisson's ratio are not well defined, and Eqs. (5) and (7) do not strictly hold. They may nevertheless be used as approximations, with the mechanical properties treated as effective parameters to be determined empirically. More precise treatment requires consideration of the anisotropy of the materials (19, 23).

## 2. Acoustic Dispersion Curves

The velocities of the acoustic modes depend on the film-stack structure, on the mechanical properties of the films and the substrate, and on the acoustic wavelength. To examine the dependence, it is convenient to consider the case of a single film layer on a substrate. In the limit where the acoustic wavelength greatly exceeds the film thickness, the acoustic wave is contained primarily in the substrate. Its velocity (for the lowest-order mode) then approaches the Rayleigh velocity, $V_R$, of the surface acoustic wave for the substrate material, regardless of the film. In the opposite limit, where the acoustic wavelength is very small compared to the film thickness, the acoustic wave is entirely contained within the film. Its velocity then approaches the film material's Rayleigh velocity, regardless of the substrate. In between these two limits, the velocity of each acoustic mode varies continuously with the wavelength. Note that for most metals, which are softer than silicon, the Rayleigh velocity is considerably lower than for silicon. Therefore, for a typical metal film on a silicon substrate, the surface acoustic wave velocity decreases as the acoustic wavelength is decreased.

Figure 5 illustrates the variation of the acoustic velocity with the wavelength, referred to as the acoustic velocity *dispersion curve*, for a multilayer film stack, corresponding to a copper film on a tantalum barrier on oxide. Note that it is convenient to plot the dispersion curve in terms of both the wavelength $\lambda$ and the *wavevector*, $q = 2\pi/\lambda$. The dispersion curve shows the trends described earlier, namely, that the acoustic velocities decrease from the substrate limit at very long wavelength to the top-layer film limit at very short wavelength. Figure 6 illustrates how the acoustic wave velocity dispersion curves vary with film thickness, which is the foundation of the ISTS measurement technique. It shows the lowest-mode velocity dispersion curve for the film stack of Fig. 5, plotted for several different values of the thickness of the copper layer. The figure shows that, at a given wavelength, the acoustic velocity changes as the copper layer thickness changes. (In this example, the velocity progressively approaches the top-layer copper Rayleigh velocity as the copper thickness becomes large compared to the acoustic wavelength.) Experimentally, this would be observed as a decrease in the ISTS signal oscillation frequency as the copper thickness is increased, and it is the basis of the ISTS thickness measurement, discussed in Section II.C.

## C.  Single-Layer Film Thickness Measurement

### 1.  Principles

As discussed earlier, the acoustic velocities in a thin film stack, and therefore the experimentally observed ISTS frequency spectrum, are functions of the acoustic wavelength, the thicknesses of all film layers, and the mechanical properties of the film and substrate

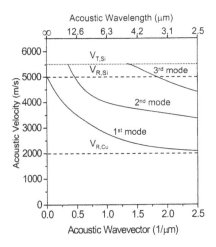

**Figure 5**  Simulated acoustic mode velocity dispersion curves for a $Cu/Ta/SiO_2$ film stack (10,000 Å/250 Å/4000 Å) on a silicon wafer. Dispersion curves are shown for several of the lowest-order acoustic waveguide modes. Wavevector $q = 2\pi/\lambda$. As the acoustic wavelength becomes larger compared to the total film thickness, the lowest-order acoustic waveguide mode velocity increases and approaches that of the silicon substrate Rayleigh velocity $V_{R,Si}$. As the acoustic wavelength becomes smaller compared to the top-layer Cu film thickness, the acoustic velocity approaches that of the copper Rayleigh velocity $V_{R,Cu}$. (Courtesy Bell Labs, Lucent Technologies.)

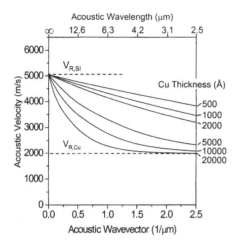

**Figure 6** Simulated variation of lowest-order acoustic mode velocity dispersion curve with Cu film thickness in a Cu/Ta/SiO$_2$ stack. See Fig. 5 for comparison. Ta and SiO$_2$ thicknesses are the same as in Fig. 5. (Courtesy Bell Labs, Lucent Technologies.)

materials. In particular, the lowest-order mode frequency varies continuously as the thickness of any one layer is varied. Therefore, if all properties of a film stack are known except the thickness of one layer, that thickness may be determined by analysis of the lowest-order mode frequency in the ISTS spectrum.

Figure 7 illustrates this with a particular example. The figure shows how the ISTS signal frequency (for the lowest-order oscillation mode) varies with the thickness of the Cu layer in a Cu/Ta/SiO$_2$ film stack on a Si substrate, for a given acoustic wavelength. The frequency is plotted for several combinations of underlying Ta and SiO$_2$ thicknesses. If the Ta and SiO$_2$ thicknesses are known, then measurement of the frequency yields directly the thickness of the Cu layer. Alternatively, if the Cu thickness is known, the frequency may be used to determine the thickness of Ta or SiO$_2$. Note in Figure 7 that the upper limit of the frequency, obtained when all the film-layer thicknesses are zero, is determined by the properties of the Si substrate. The lower limit of the frequency, obtained when the copper thickness is very large compared to the acoustic wavelength (i.e., large compared to the wave motion penetration depth), is determined by the properties of Cu alone, as discussed in Section II.2.

Figure 8 illustrates how the curve of frequency versus thickness varies for different metals. It shows the lowest-mode acoustic frequency for various metals on 1000 Å of SiO$_2$ on a Si wafer versus metal thickness. The high-thickness limiting value of each curve is determined by the mechanical properties of the respective metal, as discussed earlier. For small thickness, the curves are approximately linear, i.e. the frequency

$$F \approx F_0 - \frac{d}{C} \tag{8}$$

where $F_0$ is a constant determined by the underlying substrate and oxide layer, $d$ is the thickness of the metal film, and $C$ is a function of the metal film properties—primarily density—and determines the sensitivity of the frequency to film thickness (7,28). The figure illustrates that for small film thickness, the ISTS method is more sensitive to dense films than to light ones, because denser films yield a larger change in acoustic frequency per angstrom of film thickness.

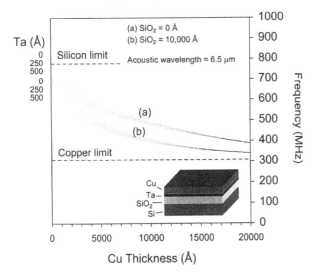

**Figure 7** Simulated dependence of acoustic wave frequency on layer thickness for a Cu/Ta/SiO$_2$/Si stack. As Cu is added to the film stack, the acoustic wave frequency gradually decreases and approaches a value determined by the Cu properties alone. As the Cu thickness approaches 0, the frequency is determined by the combination of the underlying Ta and SiO$_2$. When the Cu, Ta, and SiO$_2$ thicknesses are all 0, the frequency is determined by the Si substrate properties. (Courtesy Philips Analytical.)

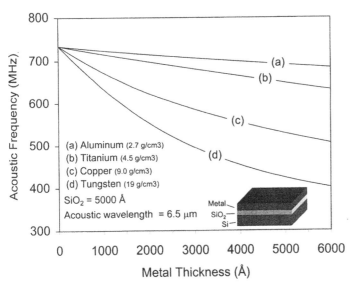

**Figure 8** Simulated dependence of acoustic frequency versus film thickness for several different metal film materials. When the metal thickness is 0, the frequency is determined by the Si and its SiO$_2$ overlayer. As metal is added, the frequency decreases and approaches (in the limit of high thickness) a value determined by the metal layer properties alone. For dense metals, the transition occurs more quickly, yielding a higher variation in acoustic frequency per angstrom of film thickness. (Courtesy Philips Analytical.)

To determine film thickness experimentally, the lowest-mode acoustic frequency is measured at a given acoustic wavelength, and the frequency is then analyzed with a mathematical model based on the principles discussed in Section II.B. Equation (1) is used to determine the acoustic velocity from the measured frequency and known wavelength. Equation (7) is then inverted to yield the thickness of a particular layer, using the wave velocity, the known mechanical properties of the layer, and the thicknesses and mechanical properties of the remaining layers. Alternatively, the film thickness may be determined by comparing the measured frequency to an empirically determined curve of frequency versus thickness.

### 2. Examples

Film thickness determined by ISTS correlates well with that determined by other methods. For example, Figure 9 shows the correlation between Cu film thickness measured by ISTS and by two different absolute reference techniques. The Cu films were deposited atop a 250 Å Ta barrier layer on 4000 Å of $SiO_2$ on Si wafers (24). For the thin films, grazing-incidence x-ray reflection (GIXR or XRR), a highly accurate technique based on interference of x-rays reflected from film interfaces, was used as the reference technique. (See, e.g., Chapter 27). For the thicker films, scanning electron microscopy (SEM) was used. The average agreement of $\sim 2\%$ is comparable to the accuracy of the two reference techniques used.

### 3. Precision

The thickness measurement precision depends on the measured frequency's repeatability and on the frequency sensitivity to thickness for the measured film, discussed earlier. With averaging times of $\sim 1$ second per measurement on samples with good optical quality, frequency repeatability of $\sim 0.05$ MHz is typically obtained in commercial instruments. Poor samples, such as those that yield low signal or have rough surfaces that scatter light

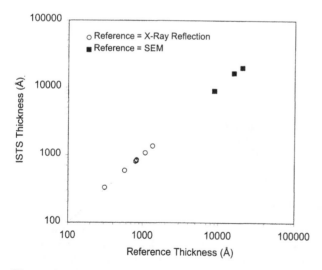

**Figure 9**  Correlation of Cu film thickness measured with ISTS to measurements with x-ray reflection and scanning electron microscopy. (Data from SEMATECH and Philips Analytical, after Ref. 24.)

into the signal detector, may exhibit poorer repeatability. The frequency sensitivity to thickness depends on the film material being measured (7,28). Sensitivity is usually greatest for dense materials, such as copper, since they cause the largest frequency change per angstrom of film thickness, as shown in Fig. 8. Sensitivity for a given film stack also depends on the acoustic wavelength selected for the measurement (28), since the wavelength determines the variation of velocity with thickness, as illustrated in Fig. 6. Other factors in the experimental setup, such as excitation laser intensity, can also be adjusted to optimize repeatability.

Table 1 shows single-layer measurement thickness precision for a commercial ISTS instrument. The table shows that the precision is typically better than ~1% for a range of film materials and that the corresponding precision over tolerance ($P/T$) ratios (defined in Chapter 1) are good. In the table the film thickness process tolerance is assumed to be ±10% for the sake of example. The precision of the commercial instruments continues to improve as the technology is further developed.

## 4. Calibration

Calculating film thickness from the observed acoustic frequency using a mathematical model requires knowledge of the film-stack structure and material properties, as discussed earlier. The density and mechanical properties (i.e., elastic moduli) of typical thin-film materials are usually well known and tabulated for their bulk single and poly-crystalline counterparts. These bulk property values can be used to approximately describe the thin films in the mathematical model for ISTS analysis, yielding good approximate thickness measurements. However, the mechanical properties of thin-film materials are often slightly different from those of their bulk counterparts, because the film microstructure depends on the deposition conditions. The density, grain structure, crystalline phase, and stoichiometry of materials can all vary with deposition conditions and subsequent processing, such as annealing (26, 27). See the references for just a few

**Table 1** Typical Precision of ISTS Single-Layer Measurement for a Variety of Film Stacks Using a Commercial Instrument

| Film-stack description | Measured film | Nominal thickness ($\text{Å}$) | Tolerance ± (%) | ± ($\text{Å}$) | $\sigma_{\text{Repeatability}}$ (approximate) ($\text{Å}$) | (%) | $\sigma_{\text{Measurement}}$ (approximate) ($\text{Å}$) | (%) | $P/T$ Ratio (%) |
|---|---|---|---|---|---|---|---|---|---|
| Si/SiO$_2$/ECD Cu | ECD Cu | 10000 | 10% | 1000 | 40 | 0.4% | 60 | 0.6% | 12% |
| Si/SiO$_2$/PVD Cu | PVD Cu | 1000 | 10% | 100 | 2 | 0.2% | 2 | 0.2% | 6% |
| Si/SiO$_2$/Ti/TiN/W | W | 4000 | 10% | 400 | 2 | 0.1% | 2 | 0.1% | 2% |
| Si/SiO$_2$/Ta | Ta | 250 | 10% | 25 | 3 | 1.2% | 3 | 1.2% | 36% |
| Si/SiO$_2$/Ti/Al/TiN | Al | 5000 | 10% | 500 | 40 | 0.8% | 60 | 1.2% | 24% |

$\sigma_{\text{Repeatability}}$ is the short-term standard deviation of repeated measurements at a single site. $\sigma_{\text{Measurement}}$ is the expected standard deviation of the measurement when all sources of variation, including both short-term repeatability and long-term reproducibility, are included. A process tolerance of ±10% of the mean film thickness is assumed for the sake of example. The table shows that the precision is typically better than 1% of the film thickness and that the precision over tolerance (P/T) ratios are good. Precision of commercial instruments continues to improve as the technology is further developed.
*Source*: Philips Analytical.

specific examples relevant to Cu (29–34), Al (35,36), Ta/TaN (37,38), and TiN (39) films. These variations change the effective material property values needed to analyze the materials (see Section II.B). Therefore, obtaining accurate thickness results requires calibration of the film material properties.

Effective mechanical properties for specific types of film materials used in an integrated circuit manufacturing process can be determined from acoustic wave measurements according to the methods described in Section II.E. These values can then be tabulated for thin films of practical interest as prepared under typical conditions.

For metrology applications, it is often convenient to further calibrate the thickness measurement by simply adjusting film property values as needed to obtain good agreement with desired reference data. For example, to measure the thickness of the copper layer in a Cu/Ta/SiO$_2$/Si stack, the frequency is measured at a particular acoustic wavelength for a few different samples with varying copper thickness. The film thickness of each sample is then determined by a desired reference technique (e.g., SEM, TEM, XRR, XRF, or four-point-probe). The property values needed to simulate the data are then adjusted to obtain best agreement between the measured and simulated frequencies for the film thicknesses determined by the reference technique. These property values are then used in subsequent thickness measurements of similarly prepared films.

Figure 10 illustrates the importance of calibration, by showing the effects of exaggerated calibration errors on the ISTS thickness calculation. The figure shows the ISTS measured thickness of the copper layer in a series of Cu/Ta/SiO$_2$/Si samples versus a reference thickness measurement made using XRR. (The data is from Fig. 9.) When the acoustic frequencies are analyzed with the correct model assumptions, the ISTS thickness values agree very well with the reference thickness. The solid lines on both sides of the figure show this correlation. The dashed lines show how the ISTS thickness values would change if the data were analyzed with (exaggerated) incorrect assumptions about

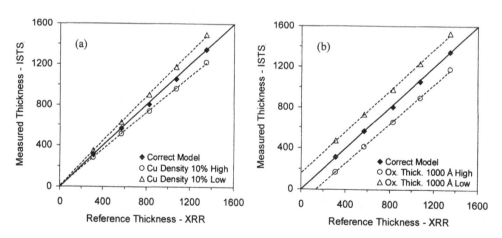

**Figure 10** Illustration of effect of incorrect model assumptions on ISTS calculated thickness. Data are from the Cu/Ta/SiO$_2$ film stacks on Si wafers of Fig. 9. (Ta thickness = 200 Å, SiO$_2$ = 4000 Å). With correct model assumptions, there is 1-to-1 agreement between ISTS and the reference method. An assumed 10% error in the Cu density (a), results in ~ 10% error in proportionality between ISTS and the reference measurement. In (b), an assumed 1000-Å error in the SiO$_2$ thickness (corresponding to 20%) leaves the measured-to-reference proportionality approximately unchanged, but produces an offset of ~ 150 Å in the measurement. The hypothetical errors were exaggerated for the sake of clarity in the figures. (Courtesy Philips Analytical.)

the film stack. Figure 10a shows that, if the Cu density is varied by $\pm 10\%$ from the correct value, the measured thickness values increase or decrease by $\sim \pm 10\%$. Figure 10b shows that, if the assumed oxide thickness is varied by $\pm 1000$ Å from its correct value of 4000 Å, the measured copper thickness is offset by $\sim \pm 150$ Å in this example. The two examples in Fig. 10 illustrate qualitatively the two main effects of incorrect calibration. Inaccurate property values for the measured layer generally result in an inaccurate proportionality between the measured and true thickness, as shown in Fig. 10a. Inaccurate property or thickness values for the other layers generally result in an offset in the curve of measured versus true thickness, as shown in Fig. 10b. The calibration procedure eliminates these two types of errors. The magnitudes of the effects depend on the specific film stack being measured and the acoustic wavelength used for the measurement.

## 5. Accuracy

Various systematic experimental errors may contribute to the measurement accuracy. For example, the spacing of the grating fringes imaged on the sample surface (sketched in Fig. 1) can usually be determined for a particular instrument only to within $\sim 0.3\%$. Furthermore, determination of the frequency of the transient ISTS waveform may show systematic absolute errors on the order of several megahertz, due to certain practical approximations in the Fourier analysis and the discrete nature of the data. Usually, these systematic errors can be mitigated by sweeping them into the calibration procedure discussed earlier. Once the recipe calibration is performed, any systematic experimental errors that remain usually result in film thickness measurement with accuracy better than a few percent. For example, the data in Fig. 9 shows an average accuracy of $\sim 2\%$.

## D. Bilayer Film Thickness Measurement

### 1. Principles

The discussion so far has focused on determining the thickness of a single layer in a film stack by measuring the stack's lowest-mode acoustic frequency at a particular acoustic wavelength. By analyzing additional information, the thickness of multiple film layers may be simultaneously determined.

For example, measuring the frequency of several of the acoustic modes and/or measuring the full spectrum of mode frequency versus acoustic wavelength provides enough information in principle to determine the thicknesses of several layers in the film stack simultaneously. Examining Fig. 4 reveals the physical basis for this. Different acoustic modes have different displacement patterns within the film stack, and their frequencies are therefore affected differently by the film layers at different depths. Therefore the relative changes in the mode frequencies for a given thickness change in a buried layer is different than for the corresponding change in a top layer. Similarly, waves propagating at different acoustic wavelengths probe the film to different depths, so short wavelength waves are more sensitive to the top film layers than are long-wavelength waves, as illustrated in Fig. 6. Therefore, combining frequency data from multiple modes and/or multiple acoustic wavelengths permits determining multiple unknowns about the film stack, such as multiple layer thicknesses. Note that this situation is analogous to that in spectroscopic ellipsometry (40), where measurement at multiple optical wavelengths provides the ability to determine more layer thicknesses than measurement at a single wavelength.

In addition to the frequency spectrum and wavelength dispersion of the acoustic modes, other types of information in the ISTS data can also be used. For example, the decay time of the thermal grating component in the signal waveform (see Section II.A.2) also depends on the combination of film-layer thicknesses, since different metals have different thermal properties. Similarly, the signal intensity or acoustic modulation depth yields information on the optical absorption and thermal expansion properties of the film stack, which also depend in detail on the film-layer structure.

Thus, analyzing the lowest-mode frequency at a particular acoustic wavelength, along with other mode frequencies, the frequency dispersion with wavelength, the signal decay time, and/or the signal amplitude, provides multiple pieces of information that can be used to determine two or more layer thicknesses (41). The particular combination of these parameters that yields greatest sensitivity for a given measurement application depends on the film stack.

These techniques have just recently been reduced to practice and are presently being refined. Currently available systems determine two layer thickness simultaneously for target applications. Because of the complexity of the physics, the bilayer measurement applications are presently better treated semi-empirically, rather than relying on a complete first-principles model of the optical, thermal, and acoustic behavior. The semi-empirical approach requires calibration for the specific types of film stacks to be measured, but it is adequate for applications where similar materials are measured routinely once calibration is performed.

## 2. Example: PVD Cu/Ta Stack

In Cu interconnect systems, a thin diffusion barrier layer, typically Ta or TaN, is applied before PVD Cu deposition and subsequent electroplating, to prevent Cu diffusion into the Si which would damage devices (26,48). The ISTS signal waveform can be analyzed to determine both the seed layer Cu and underlying barrier layer thickness, which permits detection of missing or misprocessed barrier films.

For thin seed layer Cu atop Ta, analysis of the lowest-mode frequency together with thermal decay time (see Eq. (2)) yields both Cu and Ta thicknesses, after comparison to calibration data from a suitable reference technique. Qualitatively, the frequency is used as a measure of the total metal mass in the stack, while the thermal decay time is used as a measure of the relative fractions of Cu and Ta. Since the thermal conductivity of Ta films is much lower than that of Cu, the Ta contributes more to the thermal decay time per angstrom of film thickness than does the Cu.

Figure 11 illustrates these correlations for a set of samples with systematic variations of Cu and Ta thicknesses. The ISTS frequency scales approximately linearly with the sum of $Cu + 2 \cdot Ta$, since the density of Ta is approximately twice that of Cu. The decay time scales approximately linearly with the fraction of Ta in the combined Cu/Ta stack.

Table 2 lists results of thickness measurements using this method on a commercial instrument and correlates these to reference measurements made with XRR. The table demonstrates average accuracy of $\sim 2\%$ for Cu and $\sim 3.5\%$ for Ta. Table 3 shows that the precision is typically 1–2% for both layers, and that the precision over tolerance ratio (P/T) is adequate for process-monitor application. Both accuracy and precision continue to improve as the technology is further developed.

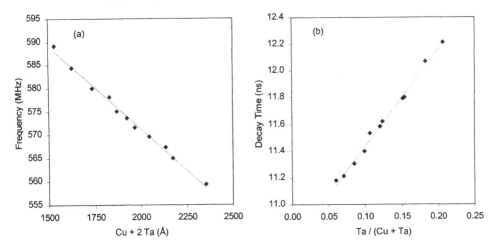

**Figure 11** Correlation of ISTS frequency and decay time with Cu and Ta film thickness for a systematic set of Cu/Ta/SiO$_2$ stacks. Ta thickness varies from 120–250 Å and Cu varies from 1200–1900 Å. The Ta and Cu are deposited atop 4000 Å oxide, and were measured independently with XRR. Frequency scales approximately with the total metal mass and decay time scales approximately with the fraction of Ta. The combination of frequency and decay time can be used to determine both thicknesses. (Courtesy Philips Analytical.)

The semi-empirical method can also be applied to other bilayer measurement problems, such as the measurement of copper and other diffusion barrier materials, or the measurement of TiN antireflective coatings on aluminum.

## E. Film Mechanical Property Measurements

The preceding sections focused primarily on the use of acoustic wave frequencies to determine the thicknesses of films in multilayer stacks. The dependence of the frequencies

**Table 2** Simultaneous Bi-layer Measurement of Cu Films and Buried Ta Diffusion Layers, Comparing ISTS Semi-empirical Method with XRR.

| | Cu | | | Ta | | |
|---|---|---|---|---|---|---|
| Wafer number | XRR (Å) | ISTS (Å) | Difference (Å) | XRR (Å) | ISTS (Å) | Difference (Å) |
| 1 | 1290 | 1284 | −6 | 121 | 124 | 3 |
| 2 | 1588 | 1627 | 39 | 121 | 114 | −7 |
| 3 | 1892 | 1868 | −24 | 120 | 131 | 11 |
| 4 | 1278 | 1284 | 6 | 176 | 184 | 8 |
| 5 | 1576 | 1618 | 42 | 174 | 173 | −1 |
| 6 | 1280 | 1301 | 21 | 228 | 232 | 4 |
| 7 | 1594 | 1608 | 14 | 226 | 224 | −2 |
| 8 | 1899 | 1820 | −79 | 227 | 244 | 17 |
| 9 | 1291 | 1326 | 35 | 289 | 282 | −7 |
| 10 | 1593 | 1560 | −33 | 289 | 296 | 7 |
| 11 | 1295 | 1301 | 6 | 337 | 333 | −4 |

The Cu and Ta were deposited on 4000 Å of SiO$_2$ atop Si wafers.

**Table 3** Precision of Bilayer Measurement of PVD Cu/Ta Film Stacks on Oxide, Using a Commercial Instrument.

| Film-stack description | Measured film | Nominal thickness (Å) | Tolerance ± (%) | Tolerance ± (Å) | $\sigma_{Repeatability}$ (approximate) (Å) | $\sigma_{Repeatability}$ (approximate) (%) | $\sigma_{Measurement}$ (approximate) (Å) | $\sigma_{Measurement}$ (approximate) (%) | P/T Ratio (%) |
|---|---|---|---|---|---|---|---|---|---|
| Si/SiO$_2$/Ta/Cu | Cu | 1000 | 10% | 100 | 8 | 0.8% | 10 | 1.0% | 30% |
| | Ta | 250 | 10% | 25 | 3 | 1.2% | 4 | 1.6% | 48 |

$\sigma_{Repeatability}$ is the short-term standard deviation of repeated measurements at a single site. $\sigma_{Measurement}$ is the expected standard deviation of the measurement when all sources of variation, including both short-term repeatability and long-term reproducibility, are included. A process tolerance of $\pm 10\%$ of the mean film thickness is assumed for the sake of example. The table shows that the precision is 1–2% of the film thickness and that the ratio of (P/T) precision over tolerance is adequate for process-monitor applications. (Precision continues to improve as the method is further developed.)
*Source*: Philips Analytical.

on acoustic wavelength, from which the velocity dispersion is calculated (see Section II.B.2), can also be used to determine other characteristics of films, such as their mechanical properties. Analysis of the acoustic dispersion can produce accurate values for the elastic moduli, e.g., Young's modulus and Poisson's ratio, of component films. Information on the elastic moduli of films is important in monitoring their reliability upon processing that may induce stress and subsequent delamination. The mechanical properties of a variety of polymer films have been assessed through ISTS in this manner (1,16,17,42,44–46). In addition, film elastic moduli information is needed to set up thickness measurement recipes using the techniques outlined earlier, as discussed in Section II.C.4.

To obtain the elastic moduli, e.g., Young's modulus and Poisson's ratio, of a film material, the ISTS frequency spectrum is measured over a range of acoustic wavelengths. This data is then used to plot the acoustic velocity dispersion curves for the sample, e.g., as illustrated by the example in Fig. 5. Simulated dispersion curves are calculated as described in Section II.B using known or assumed elastic moduli for each film layer. Elastic moduli values for the unknown layer are then optimized to fit the simulated theoretical dispersion curves to the measured data.

## III. APPLICATIONS

### A. In-Line Process Monitoring

Traditionally, process monitoring of metal deposition has been performed by using monitor wafers of blanket films or by selecting a representative fraction of product films to sacrifice for destructive SEM testing if necessary. The spatial resolution of optoacoustic techniques, however, permits metal-film thickness to be measured directly on patterned product wafers, quickly and nondestructively. This makes process monitoring much more economical. Test pads with suitable dimensions located within the streets or scribe lines in between dies on product wafers can be used for metrology. Because the ISTS method is rapid, with data acquisition times of typically 1 second per measurement site, it is ideally suited for use in high-throughput in-line process monitoring applications (47).

## B. Metal Film Uniformity

The simplest application of the ISTS method is the measurement of metal film uniformity achieved by deposition equipment. For example, Figure 12 shows an ISTS thickness contour map and diameter scan of a blanket PVD (physical vapor deposition) Cu film used as a seed layer for Cu electroplating. The contour map and diameter scan show that this film is thicker near the center than near the edge, a characteristic feature of some PVD systems. Similarly, Figure 13 shows an ISTS diameter scan across a blanket ECD (electrochemical deposition) Cu film. The scan shows that the film is thicker near the center as well as near the plating electrodes at the edges, both typical characteristics of some electroplating systems. The edge effects will be discussed further in Section III.C. For comparison, Fig. 13 also shows that a diameter scan of the same wafer made with a four-point-probe instrument agrees well with the ISTS measurement, though the four-point-probe cannot measure near the edge of the film. ISTS permits the film thickness uniformity to be measured directly on patterned wafers, as illustrated in Figure 14. The figure shows the thickness of ECD copper film on $\sim 100 \times 100$-micron test pads for each die on a patterned wafer. It shows that dies near the edge of the wafer have thinner metal than dies near the center. Interconnect resistance will therefore be higher for dies near the film edge. Metal-film uniformity information such as that shown in Figs. 12, 13, and 14 can be used to adjust deposition parameters and achieve process tolerances for film uniformity.

## C. Film Edges and Exclusion Zones

During deposition of certain metal films, a 1–2-mm region between the film's edge and the wafer's bevel is left exposed. This area, called an *edge-exclusion zone*, may be left intentionally or as an artifact of the deposition process. For tungsten films, for example, the edge-exclusion zone prevents film delamination that occurs when tungsten is deposited

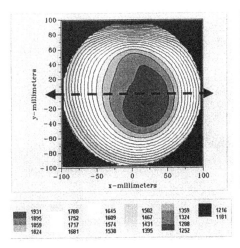
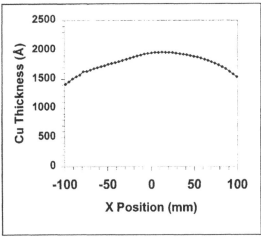

**Figure 12**   ISTS 49-point contour map (left) and diameter scan (right) of thickness of a blanket PVD Cu film. The film is deposited atop a 250-Å Ta barrier layer on $\sim$ 100-Å $SiO_2$ on a Si wafer. The profile shows the nonuniformity of the film due to the geometry of the PVD system. The contour map and the diameter scan were each completed with data acquisition times of a little over minute. (Courtesy Philips Analytical.)

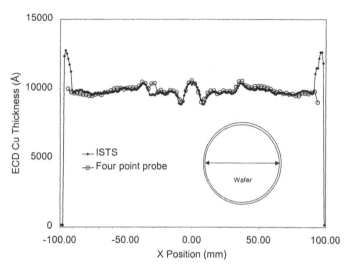

**Figure 13** Diameter scan of ECD Cu film thickness. The ECD Cu is deposited atop 250-Å Ta on 4000-Å $SiO_2$ on a Si wafer. The ECD Cu film thickness was measured with both ISTS and four-point-probe, which were previously calibrated against each other. The profiles show the nonuniformity of the Cu electroplating and also indicate good agreement between the two measurement techniques. With the ISTS measurement, large thickness nonuniformities near the plating electrodes at the wafer edges are evident. (Data from SEMATECH and Philips Analytical.)

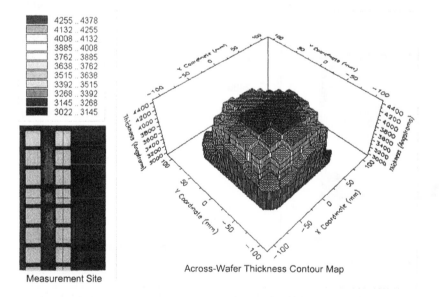

**Figure 14** Contour map of ECD copper film thickness on $\sim 100 \times 100$-micron test pads on a patterned wafer. The video image in the lower left corner shows the measurement site. The same site was measured on each die on the wafer. Thickness is in angstroms. Copper thickness is largest for dies near the wafer center. (Courtesy Philips Analytical.)

directly onto the wafer's bevel. In another case, during electrochemical deposition of copper films, electrodes used during the deposition process contact an underlying copper seed layer, thereby preventing metal deposition underneath the electrode and leaving an edge-exclusion zone. In both these cases, the thickness of the metal film often changes rapidly near the edge-exclusion zone. If the film requires chemical-mechanical polishing (CMP) to remove unwanted metal, relatively thick or thin metal regions near the edge can dramatically affect CMP removal rates and the film's resultant uniformity. This, in turn, can result in film-thickness values out of the process tolerances for devices fabricated near the edge.

Measurement of metal-film thickness in the edge-exclusion zone has traditionally been difficult to perform. For example, the four-point-probe does not permit adequate spatial resolution, while higher-resolution techniques such as stylus profilometry are cumbersome. Both require contacting the wafer surface. The ISTS method, however, permits a simple and noncontact measurement of metal-film thickness in the edge-exclusion zone, with spatial resolution typically tens of microns. Because data acquisition is rapid (typically 1 second per measurement site), detailed edge profiles can be quickly recorded and analyzed.

Examples of edge-profile line scan and high-density contour map measurements for copper films are shown in Figures 15 and 16. Figure 15, for example, shows edge-profile line scans from a relatively thick (approximately 1.5 microns) ECD copper film deposited on a thin (2500 Å) copper seed layer. The figure shows large nonuniformities in thickness near the film edge, with the thickness changing by thousands of angstroms over a region of film only a few hundred microns across. This dramatic nonuniformity is presumably due to preferential film deposition near the electrode used during the ECD process. Figure 16 illustrates this by showing a high-density thickness contour map measured from the same sample described earlier near 90° from the wafer notch. The contour map reveals several peaks in the ECD copper film that result from contact with the electrodes in the deposition process.

Thickness measurements near the edge-exclusion zone, similar to those shown in Figs. 15 and 16, can be analyzed and used for quality control during fabrication of semiconductor devices. For example, edge-profile measurements can be made periodically to determine a *usable film diameter*, defined as the diameter of the metal film within which the film's thickness is at least some target fraction of its maximum value. Similarly, measurements in the edge-exclusion region can identify undesirable nonuniformities or

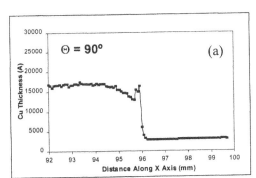
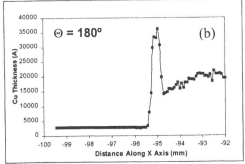

**Figure 15** Edge-profile line scans measured from an ECD copper film at (a) $\theta = 90°$ and (b) $\theta = 180°$ from the wafer's notch. (Courtesy Philips Analytical.)

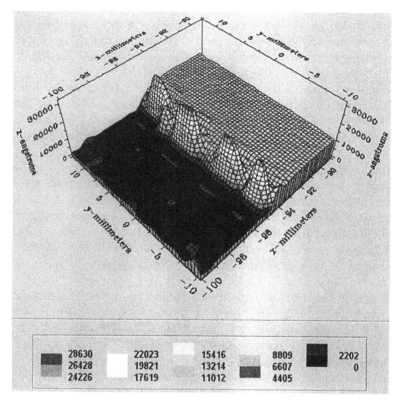

| 28630 | 22023 | 15416 | 8809 | 2202 |
| 26428 | 19821 | 13214 | 6607 | 0 |
| 24226 | 17619 | 11012 | 4405 | |

**Figure 16** High-density thickness contour map measured from the ECD copper film of Fig. 15 near $\theta = 90°$ from the wafer's notch. Thickness values are in angstroms. (Courtesy Philips Analytical.)

determine the degree of concentricity of the film with respect to the substrate wafer. Data such as this can be used to fine-tune process conditions for best performance.

## D. Chemical-Mechanical Polishing of Metals

### 1. Characterizing Chemical-Mechanical Polishing on Blanket Copper Films

Chemical-mechanical polishing is critical for Cu Damascene fabrication (48,43). Ideally, a CMP process step removes all excess metal (e.g., copper and tantalum or other barrier layers), stops when the endpoint is detected, and leaves all the insulator material (e.g., oxide) intact. In practice, most CMP systems remove all materials at varying rates. The challenge for a process developer is to find a combination of tool settings and consumables that matches the initial material topography with the target process window and yields an acceptable end result with reasonable throughput. The corresponding goals for metal CMP process control metrology are therefore to evaluate nonuniformity of incoming wafers, measure removal rates for target materials, and validate the product output for local and within-wafer nonuniformity (WIWNU). The examples of this section illustrate how characteristic features of the ISTS optoacoustic technique make it well suited for CMP process development and control (49–51).

The noncontact nature of the ISTS measurement combined with its short measurement time enables high-throughput full-wafer mapping of metal films. A reliable way to control nonuniformity of incoming (pre-CMP) wafers is to perform a detailed mapping of metal-film thickness. For electroplated copper films, nonuniformities near the film's edge (e.g., due to electrode positioning) and in the center (e.g., due to the flow of the electrolytic solution) can severely affect the CMP process. Figure 17 shows representative 225-point contour maps of pre- and post-CMP electroplated copper films, which were collected in about 6 minutes each (50). The maps reveal detailed information about film thickness variation across the wafer. Because the measurement is nondestructive, the measured wafers can be fed back into the manufacturing stream, thus eliminating the need to rely on indirect data from costly monitor wafers. For routine process monitoring, full or partial contour maps could be constructed with far fewer data points and correspondingly faster data acquisition times.

Another way of rapidly characterizing pre- and post-CMP wafers is diameter scanning, as illustrated in Figure 18. Line scans across the wafer visualize incoming wafer nonuniformity, as well as the resulting film topography. Furthermore, comparison of diameter scans after subsequent polishing rounds enables calculation of average and position-dependent removal rates. For example, the data shown in Fig. 18 illustrates undesirable conformal polishing, where the initial nonuniform thickness profile is reproduced, rather than corrected, by the CMP step.

As discussed in previous sections, electroplated copper films typically have significant non-uniformities in the near-edge regions, as shown in Figs. 13 and 15. These nonuniformities strongly affect subsequent polishing performance. However, detailed measurement of the edge profile with ISTS enables efficient CMP optimization.

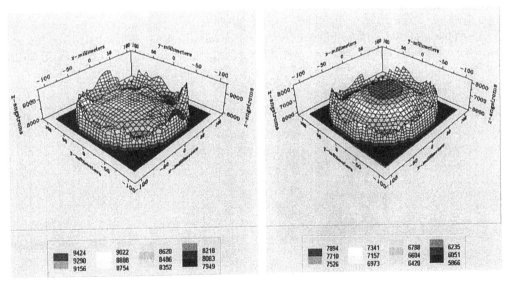

**Figure 17** 225-point contour maps of "as-plated" (left) and post-CMP (right) 200-mm ECD Cu wafers. Thickness values reported in angstroms. The detailed contour maps show incoming wafer nonuniformity and the results of the CMP process step. (Data from Philips Analytical, after Ref. 51.)

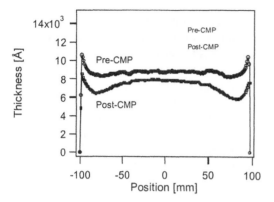

**Figure 18** Thickness diameter scans of the ECD Cu wafers shown in Fig. 17. The pre- and post-CMP scans illustrate conformal polishing, where the initial nonuniform thickness profile is reproduced, rather than corrected, by the CMP step. (Data from Philips Analytical, after Ref 51.)

## 2. Characterizing Chemical-Mechanical Polishing on Patterned Copper Structures

Polishing of patterned product wafers may differ significantly from polishing of blanket monitor wafers. Metal feature size and feature density both affect local CMP performance and the resulting local planarity (43). To adequately control the process, the metrology must be capable of direct measurement of small features. By virtue of its small spot size and nondestructive nature, ISTS is an effective thickness-measuring technique for in-line control of Damascene-patterned pads, lines, and vias.

Figure 19 shows an example of an ISTS thickness measurement made on a post-CMP patterned copper serpentine test area. In this case 100-micron-wide pads are cleanly resolved by the highly repeatable measurement.

ISTS can also be used to measure high-feature-density areas of micron and submicron Damascene-patterned structures. Figure 20 illustrates an example, where arrays of Damascene-patterned copper-filled trench structures have been calibrated for measurement with ISTS (52). The figure shows three examples, where the ratio of trench width

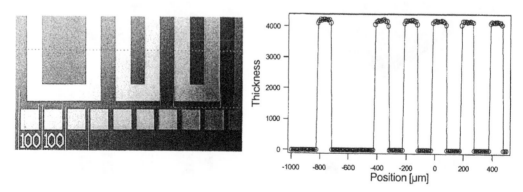

**Figure 19** Optoacoustic measurement scan across 100-μm-wide Cu Damascene trenches. The dotted line across the image (left) denotes the scan location. The scan indicates the ability of ISTS to measure film thickness directly on-product. (Data from Philips Analytical, after Ref. 51.)

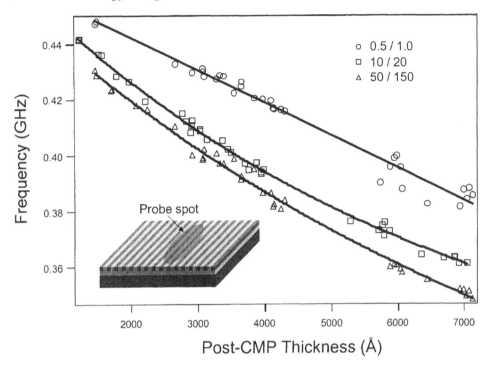

**Figure 20** Calibration curves of ISTS frequency versus thickness for Damascene-patterned copper-filled trench structures embedded in oxide. Numbers in the legend show the ratio of trench width to the array pattern repeat period, both in microns. Post-CMP copper thickness in the trenches was measured using high-resolution profilometry, with a known thickness of field oxide in a neighboring region serving as reference. The calibration curves permit measuring copper thickness in similar trench structures rapidly using ISTS. (Data from Philips Analytical, after Ref. 52.)

to repeat period varies from 0.5/1.0 microns to 50/100 microns. Array structures are irradiated with the excitation pulse, launching acoustic waves along the trenches. The acoustic waves propagating in the inhomogeneous, composite film composed of the alternating Cu and oxide bars are then measured the same way as described earlier. An average thickness for the irradiated region is calculated semiempirically by using effective values for the density and elastic moduli of the composite film. These are determined by a calibration procedure, illustrated in the figure, in which the ISTS frequency is measured for arrays of known thickness determined by a reference technique, and effective material properties are chosen that fit the measured data. The calibration can then be used to rapidly measure the copper thickness in similarly patterned structures using ISTS. The signal-to-noise ratio of the data is similar to that for blanket films, with precision approximately 1%.

Measurements such as these can be repeated on multiple die, either on Damascene array patterns as shown in Fig. 20 or on traditional solid test pads or bond pads. Such data can be used to generate a tile map of post-CMP thickness, similar to the map shown in Fig. 14. This allows a determination of within-wafer thickness variations in patterned films before and after a CMP process step. Data such as this can be used to control and adjust the CMP process, so variables such as pad pressure and slurry concentration (43) can be modified to improve within-wafer nonuniformity (WIWNU) and increase CMP yield.

### E.   Film Mechanical Properties

The preceding sections focused primarily on the use of ISTS to determine the thicknesses of films in multilayer stacks. ISTS can also be used to determine other characteristics of films, such as their elastic properties, by measurement and modeling of acoustic velocity dispersion curves, as explained in Section II.E.

An important emerging example of this application is the characterization of low-$k$ dielectric film materials being explored to replace silicon dioxide for high-speed interconnect systems (48). Although the low dielectric constants of these materials make their electrical performance superior to oxide, they are typically much weaker mechanically than oxide and have poor thermal conductivity and high coefficients of thermal expansion. These properties make the materials susceptible to failure during processing steps following deposition, particularly during chemical-mechanical polishing (43). To model and monitor the performance of these new film materials in a manufacturing process, information on their elastic properties is needed. ISTS has been explored as a technique for characterizing the elastic properties of these new low-$k$ films (53).

Figure 21 shows one example of ISTS data used to characterize a thin nanoporous silica low-$k$ dielectric material. The upper part of the figure shows typical data collected from an approximately 0.75-μm-thick layer of a film of nanoporous silica (porosity $\sim 65\%$) on silicon. (A thin overcoating of Al was deposited on top of the nanoporous silica to increase light absorption and improve signal quality.) The mismatch of the acoustic properties of the film and the substrate in this case yields a waveguide that supports many modes; several of these appear in the power spectrum in the inset at the top of the figure. Mapping out the wavelength dependence of the modes reveals the dispersion. Fitting this dispersion to an acoustic model that allows the elastic properties of the film to vary yields Young's modulus and the Poisson ratio of the film, as discussed in Section II.E. The lower part of Fig. 21 compares measurements and best-fit modeling results. The intrinsic longitudinal and transverse acoustic velocities were found, from this fitting, to be $610 \pm 50 \, \text{m/s}$ and $400 \pm 30 \, \text{m/s}$, respectively. From these values, Young's modulus and Poisson's ratio can be calculated by inverting Eqs. (6). Time-of-flight measurements performed with conventional contact-based acoustic transducers and bulk pieces of nanoporous silica with similar porosities reveal acoustic velocities in the same range (i.e., longitudinal velocities between 500 and 1000 m/s) (54–56). (However, precise quantitative comparsion is difficult, since thin films of nanoporous silica are typically formed with chemistries that are different than those used for production of bulk samples.)

## IV.   SUMMARY

Impulsive stimulated thermal scattering (ISTS) is a noncontact, nondestructive technique for measuring the thickness and/or properties of thin films, in particular, opaque films such as metals. The technique uses crossed excitation laser pulses to create a transient optical grating that launches acoustic waves in the sample surface, via absorption of laser light followed by mild but sudden, spatially periodic thermal expansion. The propagating acoustic waves are observed by recording the time-dependent diffraction intensity of a probe laser directed at the surface. The frequency spectrum of the detected signal is determined by the acoustic wavelength, which is imposed by the optical grating spacing, and by the velocities of the allowed oscillation modes in the film stack. These velocities are

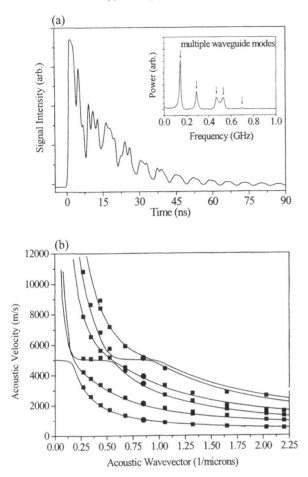

**Figure 21**  (a) Typical signal data collected from an approximately 0.75-μm-thick nanoporous silica layer on silicon. (A thin aluminum overcoat was used to increase signal quality.) The power spectrum of this data reveals multiple acoustic frequencies; each corresponds to a different acoustic waveguide mode of the structure. (b) Measured (symbols) and best-fit acoustic dispersion in this sample. The fit determines the elastic properties of the film. The data indicated by solid circles originated from the signal displayed in part (a). (Courtesy Bell Labs, Lucent Technologies.)

functions of the thickness and elastic properties of all layers in the film stack. In addition, the signal intensity, modulation depth, and decay time are all functions of the detailed stack structure and layer thicknesses. Analyzing the data with a model of the acoustic physics and/or using semiempirically determined calibration data permits determining film thickness and/or material property information.

The technique features high spatial resolution that permits measurement of thickness on test pads on patterned product wafers and rapid data acquisition of typically 1 second per measurement site. These features make it very attractive for routine process monitoring. Applications include measuring metal-film thickness uniformity, measuring thickness in and near the film edge-exclusion zone, characterizing metal chemical-mechanical polishing, measuring on arrays of finely patterned features, and characterizing mechanical properties of materials, such as emerging low-$k$ dielectrics.

## ACKNOWLEDGMENTS

The development of ISTS thin-film metrology stemmed from basic research at the Massachusetts Institute of Technology supported in part by National Science Foundation Grants nos. DMR-9002279, DMR-9317198, and DMR-9710140. We thank International SEMATECH, Novellus, and Applied Materials for providing samples and assistance in testing the first commercial systems, and the National Institute of Standards and Technology and Philips Analytical's Tempe applications lab for valuable reference data. We also thank Marco Koelink and Ray Hanselman, for proofreading the manuscript, and Alain Diebold and Chris Moore, for useful discussions.

## REFERENCES

1. J.A. Rogers, A.A. Maznev, M.J. Banet, K.A. Nelson. Annu. Rev. Mater. Sci. 30:117–57, 2000.
2. M.J. Banet, M. Fuchs, J.A. Rogers, J.H. Rienold Jr., J.M. Knecht, M. Rothschild, R. Logan, A.A. Maznev, K.A. Nelson. Appl. Phys. Lett. 73:169–171, 1998.
3. M.J. Banet, M. Fuchs, R. Belanger, J.B. Hanselman, J.A. Rogers, K.A. Nelson. Future Fab Int. 4:297–300, 1998.
4. J.A. Rogers, M. Fuchs, M.J. Banet, J.B. Hanselman, R. Logan, K.A. Nelson. Appl. Phys. Lett. 71:225-227, 1997.
5. R. Logan, A.A. Maznev, K.A. Nelson, J.A. Rogers, M. Fuchs, M. Banet. Microelectronic Film thickness determination using a laser-based ultrasonic technique. Proceedings of Materials Research Society Symposium, Vol. 440, 1997, pp 347–352.
6. J.A. Rogers, K.A. Nelson. Physica B:219 & 220:562–564, 1996.
7. R. Logan. Optical Metrology of Thin Films. Master's thesis, Massachusetts Institute of Technology, Cambridge, MA, 1997.
8. J.A. Rogers. Time-Resolved Photoacoustics and Photothermal Measurements on Surfaces, Thin Films, and Multilayer Assemblies. Ph.D. dissertation, Massachusetts Institute of Technology, Cambridge, MA, 1995.
9. J.A. Rogers. Real-Time Impulsive Stimulated Thermal Scattering of Thin Polymer Films. Master's thesis, Massachusetts Institute of Technology, Cambridge, MA, 1992.
10. R. DeJule, Semiconductor Int.: May 1998, pp 52–58.
11. H. Maris. Sci. Am. 278:86, 1998.
12. H.T. Grahn, H.J. Maris, J. Tauc. IEEE J. Quantum Electron. 25:2562, 1989.
13. T.F. Crimmins, A.A. Maznev, K.A. Nelson. Appl. Phys. Lett. 74:1344–1346, 1999.
14. A.A. Maznev, K.A. Nelson, J.A. Rogers. Optics Lett. 23:1319–1321, 1998.
15. O.W. Köding, H. Skurk, A.A. Maznev, E. Matthias. Appl. Phys. A 61:253–261, 1995.
16. J.A. Rogers, K.A. Nelson. J. Appl. Phys. 75:1534-1556, 1994.
17. A.R. Duggal, J.A. Rogers, K.A. Nelson. J. Appl. Phys. 72:2823–2839, 1992.
18. B.A. Auld. Acoustic Fields and Waves in Solids, Vols. I and II. 2nd ed. Malabar, FL: Krieger, 1990.
19. G.W. Farnell, E.L. Adler. In: W.P. Mason, R.N. Thurston, eds. Physical Acoustics, Principles and Methods, Vol. IX. New York: Academic Press, 1972, pp 35–127.
20. G.W. Farnell. In: W.P. Mason, R.N. Thurston, eds. Physical Acoustics, Principles and Methods, Vol. VI. New York: Academic Press, 1970, pp 109–166.
21. I.A. Viktorov. Rayleigh and Lamb Waves. New York: Plenum Press, 1967.
22. See, e.g., Eqs. (3.25) and (3.30) and p. 186 of Vol. I of Ref. 18.
23. A.A. Maznev, A. Akthakul, K.A. Nelson. J. Appl. Phys. 86:2818–2824, 1999.
24. M. Gostein, T.C. Bailey, I. Emesh, A.C. Diebold, A.A. Maznev, M. Banet, M. Joffe, R. Sacco. Thickness measurement for Cu and Ta thin films using optoacoustics. Proceedings of International Interconnect Technology Conference. Burlingame/San Francisco, CA 2000.

25. M. Banet, H. Yeung, J. Hanselman, H. Sola, M. Fuchs, R. Lam. All-optical, non-contact measurement of copper and tantalum films deposited by PVD, ECD, and CVD and in single Damascene structures. In: D.G. Seiler, A.C. Diebold, W.M. Bullis, T.J. Shaffner, R. McDonald, E.J. Walters, eds. Characterization and Metrology for ULSI Technology: 1998 International Conference. AIP Conference Proceedings 449. Woodbury, NY: American Institute of Physics, 1998, pp 419–423.

26. G.C. Schwartz, K.V. Srikrishnan, A. Bross, eds. Handbook of Semiconductor Interconnection Technology. New York: Marcel Dekker, 1998.

27. M. Ohring. The Materials Science of Thin Films. Boston: Academic Press, 1992.

28. A. Maznev, M. Gostein, M. Joffe, R. Sacco, M. Banet, K.A. Nelson. Precise determination of thin metal film thickness with laser-induced acoustic grating technique. Proceedings of Materials Research Society, Boston, 1999. In press.

29. H.M. Choi, S.K. Choi, O. Anderson, K. Bange. Thin Solid Films. 358:202–205, 2000.

30. E.M. Zielinski, R.P. Vinci, J.C. Bravman. J. Electronic Materials. 24:1485–1492, 1995.

31. S.H. Brongersma, E. Richard, I. Vervoort, H. Bender, W. Vandervorst, S. Lagrange, G. Beyer, K. Maex. J. Appl. Phys. 86:3642–3645, 1999.

32. J.M.E. Harper, C. Cabral Jr., P.C. Andricacos, L. Gignac, I.C. Noyan, K.P. Rodbell, C.K. Hu. J. Appl. Phys. 86:2516–2525, 1999.

33. A. Gangulee. J. Appl. Phys. 45:3749–4523, 1974.

34. J.M.E. Harper, K.P. Rodbell. J. Vac. Sci. Technol. B 15:763–779, 1997.

35. H. Mizubayashi, T. Yamaguchi, Y. Yoshihara. J. Alloys Compounds. 211/212:446–450, 1994.

36. H. Mizubayashi, Y. Yoshihara, S. Okuda. Phys. Stat. Sol. A 129:475–481, 1992.

37. R. Hoogeveen, M. Moske, H. Geisler, K. Samwer. Thin Solid Films. 275:203–206, 1996.

38. M. Stavrev, D. Fischer, C. Wenzel, K. Drescher, N. Mattern. Thin Solid Films. 307:79–88, 1997.

39. X. Jiang, M. Wang, K. Schmidt, E. Dunlop, J. Haupt, W. Gissler. J. Appl. Phys. 69:3053–3057, 1991.

40. Irving P. Herman. Optical Diagnostics for Thin Film Processing. Boston: Academic Press, 1996.

41. Patents pending, Philips Analytical.

42. J.K. Cocson, C.S. Hau, P.M. Lee, C.C. Poon, A.H. Zhong, J.A. Rogers, K.A. Nelson. Polymer 36:4069–4075, 1995.

43. J.M. Steigerwald, S.P. Murarka, R.J. Gutmann. Chemical Mechanical Planarization of Microelectronic Materials. New York: Wiley, 1997.

44. J.K. Cocson, C.S. Hau, P.M. Lee, C.C. Poon, A.H. Zhong, J.A. Rogers, K.A. Nelson. J. Mater. Sci. 30:5960–5966, 1995.

45. J.A. Rogers, L. Dhar, K.A. Nelson. Appl. Phys. Lett. 65:312–314, 1994.

46. L. Dhar, J.A. Rogers, K.A. Nelson, F. Trusell. J. Appl. Phys. 77:4431–4444, 1995.

47. M.A. Joffe, M. Gostein, A. Maznev, R. Sacco, R. Hanselman, M.J. Banet. Non-contact metal-film process control metrology. Proceedings of VLSI Multilevel Integration Conference, Santa Clara, CA, 1999, pp 468–473.

48. S. Wolf, R.N. Tauber. Silicon Processing for the VLSI Era, Volume 1—Process Technology, 2nd ed. Sunset Beach, CA: Lattice Press, 2000.

49. M. Banet, M. Joffe, M. Gostein, A. Maznev, R. Sacco. All-optical metrology for characterizing CMP of copper Damascene structures. Proceedings of CMP Multilevel Integration Conference, Santa Clara, CA, 2000, pp 537–540.

50. M.A. Joffe, H. Yeung, M. Fuchs, M. J. Banet. Novel thin-film metrology for CMP applications. Proceedings of CMP Multilevel Integration Conference, Santa Clara, CA, 1999, pp 73–76.

51. M. Banet, M. Joffe, M. Gostein, A. Maznev, R. Sacco, F. Queromes. Semiconductor Fabtech 11:77–82, 2000.

52. M. Joffe, R. Surana, D.Bennett, M. Gostein, S. Mishra, A.A. Maznev, R. Sacco, M. Banet, C. J. Moore, R. Lum, R. Bajaj. Non-contact optical measurement of post-CMP dishing and

erosion of high-feature-density damascene structures. Proceedings of the Advanced Equipment Control/Advanced Process Control (AEC/APC) Symposium XII, Lake Tahoe, NV, Sept 2000.

53. D. Nelsen, M. Gostein, A.A. Maznev. Proceedings of the Symposium on Polymers for Microelectronics, Wilmington, DE, May 2000. In preparation.
54. J. Gross, J. Fricke, L.W. Hrubesh. J. Acoust. Soc. Amer. 91:2004, 1992.
55. B.M. Gronauer, J. Fricke. Acoustica 59:177, 1986.
56. L. Forest, V. Gibiat, T. Woignier. J. Non-Cryst. Solids 225:287, 1998.

# 10

# Metal Interconnect Process Control Using Picosecond Ultrasonics

**Alain C. Diebold**
*International SEMATECH, Austin, Texas*

**Robert Stoner**
*Brown University, Providence, Rhode Island*

## I. INTRODUCTION

In this chapter, the measurement of metal-film thickness and density using picosecond ultrasonics is reviewed. The physics, technology, and materials science of picosecond ultrasonics are described in detail. Measurement precision is discussed in terms of variation with film thickness and film stack complexity. The versatility of picosecond ultrasonics makes it a very attractive method for in-line metrology. We present examples of the control of processes used to fabricate both aluminum- and copper-based on-chip interconnects. Because of its current technological relevance, picosecond ultrasonics capabilities for plated copper and copper seed layers on thin barrier layers are highlighted.

One of the most difficult measurement problems has been determining the film thickness and other properties of metal films used for on-chip interconnect. Recently, a new approach to metal-layer measurement was introduced: picosecond ultrasonics. Picosecond ultrasonics is based on measurement of film thickness based on timing the echo of an acoustic wave that travels into a layer and reflects from the interface with the layer below. The strength of picosecond ultrasonic laser sonar is its applicability to virtually all metalization steps, including those involving only single layers (for example, PVD Ti used for salicidation) and also multilayers (for example, interconnects and integrated liner-barriers such as TiN/Ti). Measurement time for each measurement site is short enough to make this method capable of measuring 30–60 wafers per hour at multiple points per wafer.

Metal-film stacks for aluminum-based interconnects comprise a thick (0.5–2-micron) layer of aluminum (often containing from 0.5 to 2% Cu to help suppress electromigration) clad by much thinner (i.e., 100–1000-angstrom) layers of PVD titanium nitride or titanium. In many instances, both cladding species may be present, so the complete stacks include as many as five separate layers. Although these stacks are usually deposited without breaking vacuum within a cluster tool, it has been common practice to monitor each deposition separately using films deposited individually on unpatterned monitor wafers.

More recently it has become possible using picosecond ultrasonics to measure all layers within the complete stack directly on product wafers. The picosecond ultrasonics approach brings with it numerous manufacturing advantages, including increased deposition tool utilization, and yield improvements resulting from a substantially increased frequency of process monitoring.

Currently copper interconnect layers are deposited in two steps designed to provide for filling of the high-aspect structures used in Damascene processes. The first step is to deposit a "barrier-seed" stack, usually comprising a thin layer (50–500 angstroms of Ta, TaN, or WN, for example) that is impervious to copper diffusion, followed by a PVD or CVD copper layer in the range 500–2000 angstroms. As with Al-based interconnect stacks, it is possible to measure both films in such bilayers using picosecond ultrasonics. The second step is to form a substantially thicker layer of copper on top of the seed layers by means of electroplating. As deposited, electroplated layers may range from 5,000 to 25,000 angstroms. As part of the damascene process, these layers are subsequently subjected to chemical-mechanical polishing, which in unpatterned areas of the wafer may reduce the copper to zero thickness, while leaving 5,000–10,000 angstroms of polished copper in regions where the underlying dielectric has been patterned to form trenches. Using picosecond ultrasonics it is possible to make measurements of the as-plated and polished films on product wafers over trenches and unetched interlevel dielectric layer (ILD). Accordingly, the technique has found widespread acceptance for copper deposition and polish rate control and for measurement of dishing and erosion of submicron structures.

In this chapter, we give a general description of the physics and technology of picosecond ultrasonics, including the optical system and underlying physics. This is followed by a discussion of the precision of picosecond ultrasonics and the effect of film thickness and film stacks on precision.

## II.  GENERAL DESCRIPTION OF PICOSECOND ULTRASONIC SONAR TECHNOLOGY

A detailed description of the physical processes involved in making measurements of this type is available in Refs. 1–3. The picosecond ultrasonics technique is extremely powerful, since it has high precision (a small numerical value), sampling speed (1–4 s/pt), and spatial resolution (less than a 10-μm-diameter spot size) and can be applied to single-layer and multilayer metal deposition processes. In a picosecond ultrasonics measurement for a sample consisting of a metal film deposited on a substrate (for example, Si), an acoustic wave first is generated by the thermal expansion caused by a subpicosecond laser light pulse absorbed at the surface of the metal. For a specific wavelength and pump pulse energy, the temperature rise caused by the laser depends on the thickness and type of metal, but it is typically from 5 to 10°. Thermal expansion within the heated surface region results in a depth-dependent isotropic thermal stress (2, 3) which gives rise to a sound pulse (or wave) that propagates normal to the sample surface into the bulk. This pulse is partially reflected at the interface between the film and substrate (1–3). For an acoustic wave, the reflection coefficient $R_A$ depends on the acoustic impedances $Z$ ($Z$ = density × sound velocity) of the film and substrate materials, and may be evaluated from the relation $R_A = (Z_{substrate} - Z_{film})/(Z_{substrate} + Z_{film})$ (1). When the reflected sound pulse (or "echo") returns to the free surface after a time $\tau$, it causes a small change in the sample's optical reflectivity ($\Delta R$). This change is monitored as a function of time by a second laser probe.

Figure 1 illustrates this process. Based on the sound velocities of the materials making up the sample (which for most materials are known from bulk measurements) and the echo time, the film thickness $d_{film}$ may be evaluated from the simple relation $d_{film} = v_s \tau$. Picosecond-ultrasonic thickness measurements rely on the echo of the sound pulse from the bottom of the layer being characterized. It is important to note that the timing of this echo is not influenced by the thickness or composition of substrate or film below. This is very useful for measurement of product wafers. The time-dependent optical reflectivity change $\Delta R(t)$ measured via picosecond ultrasonics does depend on the physical mechanisms underlying the sound generation and propagation and on sound-induced changes in optical properties of the metal. These may be simulated with very high fidelity, even for samples consisting of many layers with thicknesses ranging from less than 50 Å to several microns.

One way to analyze picosecond ultrasonics data is to use an iterative procedure in which the thicknesses of layers in a model for a sample are adjusted until a best fit to the measured $\Delta R(t)$ is obtained. The commercial picosecond ultrasonics system obtains typical thickness measurement precision of less than 0.1 nm (1 Å). This modeling methodology also allows other film properties, such as density, adhesion, and surface roughness to be obtained along with thickness, since these parameters have a predictable influence on the amplitudes and shapes of the optically detected echoes.

Figures 2 through 10 show the change in reflectance versus time for several interconnect films and stacked films. Figures 2 and 3 show the reflectivity changes with time

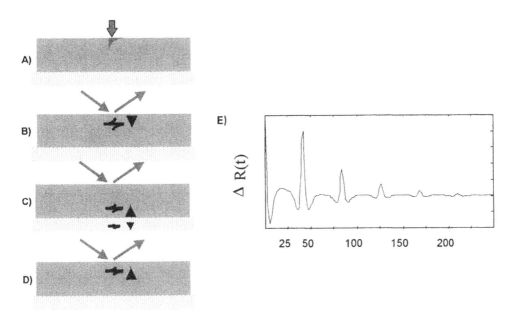

**Figure 1** Picosecond ultrasonics technology: (A) Laser pulse launches acoustic wave into the film in a direction perpendicular to the surface. (B) Acoustic wave travels into top film as a second "probe" laser beam begins to monitor the change in surface reflectivity. (C) Acoustic pulse partially transmits and partially reflects to the surface. (D) As the echo (reflected acoustic pulse) hits the surface, the change in reflectivity is monitored by the probe laser. (E) Reflectivity vs. time shows the effect of several echos from repeated reflection of the acoustic wave from the interface below the top film.

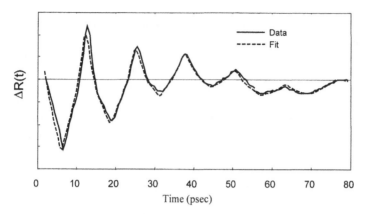

**Figure 2**  Picosecond ultrasonics analysis of 60.6-nm PVD Ti/ILD.

associated with single-layer films of Ti and TiN (respectively) on thick $SiO_2$-based inter-level dielectric films. The time-dependent reflectivity of a TiN/Ti/ILD film stack is shown in Figure 4. The composition of the substrate layer changes the shape of the time-dependent reflectivity, as shown by comparing Figure 5 for TiN on Al with Fig. 3 for TiN on $SiO_2$-based ILD. The TiN/Ti/ILD stack structure also shows this change. The presence of a thick Al film results in a reflectivity change later in time, as shown in Figure 6.

Barrier-layer thickness can be determined below copper films having thickness in the range of a few hundred nanometers or less. Sound reflections from a buried tantalum (or other barrier layers) film give additional echoes from which the tantalum thickness can be determined along with the copper-film thickness. Figure 7 shows a picosecond ultrasonics measurement obtained for a sample consisting of a copper film (PVD seed layer) deposited on top of a thin Ta layer (less than 20 nm thick) with a substrate consisting of a thick tetraethoxysilane (TEOS) layer (about 600 nm).

The thicknesses of thin barrier layers below thick (> 1 μm) electroplated Cu films are difficult to determine as a result of echo broadening associated with the typically large roughness. Data obtained for a thick ECD copper film on a thin tantalum layer is shown in Figure 8. The sharp feature observed at a time of about 800 ps is the echo caused by sound that has reflected from the bottom of the 2.1-μm-thick copper layer and returned to the surface of the sample. To determine the film thickness from this data, only the echo component of the signal is used; the background is discarded. From the product of the one-way travel time for sound through the film (406 ps) and the sound velocity (51.7 Å/ps),

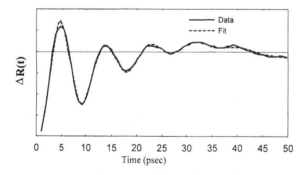

**Figure 3**  Picosecond ultrasonics analysis of 60.6-nm PVD TiN/ILD.

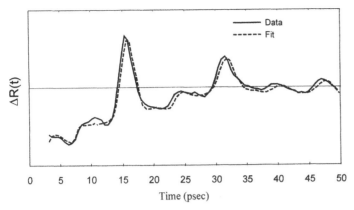

**Figure 4**   Picosecond ultrasonics analysis of 39.3-nm PVD TiN/23-nm PVD Ti/ILD.

the film thickness is found to be 2.10 μm. The sharp feature observed in a range of times less than about 50 ps is associated with relaxation of the electrons in the copper film, which initially gain energy from the ultrashort light pulse. This energy is transferred to thermal phonons in the film, which gives rise to an increase in its temperature by $\sim 1°$. The subsequent flow of heat out of the film and into the underlying TEOS occurs on a time scale of hundreds of picoseconds.

Picosecond ultrasonics can be used to measure very thin barrier-layer films at the surface of thicker films. The limit to this thickness is about 2.5 nm for Ta, TaN, or Ti. Films in this thickness range vibrate at very high frequencies. The vibrational frequency of a 2.5-nm-thick film is $v_s/(2 \times \text{thickness})$ for a metal film on silicon dioxide (and most other insulators) or silicon. The vibrational frequency will be one-half this value for films having lower impedances than the substrate, as discussed later in the details section. Although the commercial system can measure reflectivity data at 0.006-ps intervals, there is a limit to the highest frequency that can be measured, imposed by the duration of the laser pulse. This limit is of the order $1/2\pi \times$ pulse width, which for a typical pulse width gives a limit of about 1.25 THz.

An example of this is shown in Figure 9. In this example, a 25-Å Ti film on $SiO_2$ or Si would have a frequency of $(60.7 \text{ Å/ps})/2(25 \text{ Å}) = 1.214$ THz, which is very close to the minimum measurable thickness. Notice that this thickness determination is completely

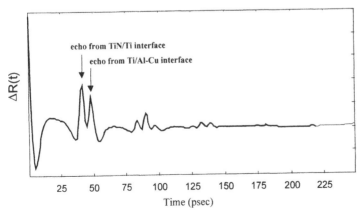

**Figure 5**   Simulated picosecond ultrasonics analysis of 200-nm TiN/20-nm Ti/Al.

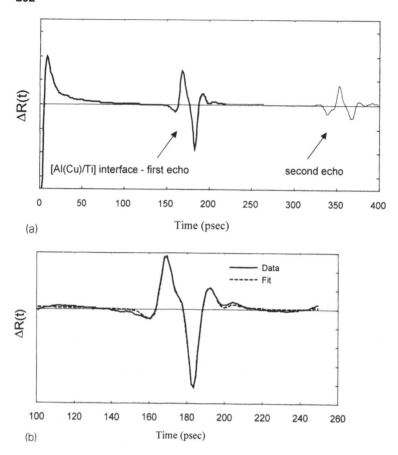

**Figure 6**  Picosecond ultrasonics analysis of 547.3-nm Al(Cu)/44.6-nm PVD Ti/ILD: (a) First and second echo; (b) expanded view of first echo.

independent of the densities of the film and substrate. This is an important distinction between picosecond ultrasonics and other acoustical techniques for which the sound velocity and density both affect the detected frequency.

The ability to measure film density has considerable practical significance, especially for $WSi_x$, TiN, WN, TaN, and other materials whose composition and structure depend on many aspects of the deposition process, such as pressure, target composition, temperature, and gas mixture. Picosecond ultrasonics has been used to distinguish between TiN films differing in density (or composition) by only a few percent. Picosecond ultrasonics also has been used to detect silicide phase transitions for Ti and Co reacted with silicon at different temperatures based on changes in sound velocity, density, and optical properties (4).

## III.  THE PICOSECOND ULTRASONICS' OPTICAL SYSTEM

The commercially available MetaPULSE™ system, which employs many of the basic principles just discussed uses a solid-state, compact, ultrafast laser as a source for both the generating (or "pump") and detecting (or "probe") beams, dividing the single beam into two by means of a simple beam splitter. The sensitivity of the technique can be

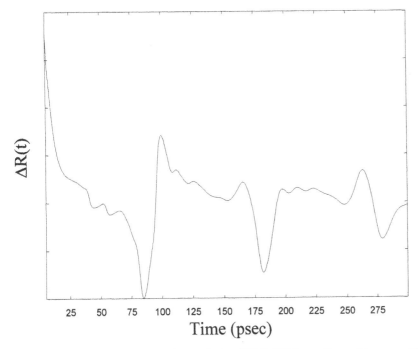

**Figure 7**   Picosecond ultrasonics analysis of PVD 209.3-nm Cu seed layer on 29.2-nm Ta, showing Ta reflections.

optimized to give improved throughput and sensitivity for specific materials: a high-throughput copper system was released in late 1998. The MetaPULSE$^{TM}$ system is currently available in a highly automated, standalone configuration equipped with an optical microscope, pattern recognition software, and precision sample stage so that patterned wafers can be monitored. Because of its long working distance and high throughput, picosecond ultrasonic technology also can be applied to in situ measurements.

Picosecond ultrasonic measurements of copper required modifications to the measurement system first introduced commercially. The first commercial system used a pump laser at 800 nm. Copper is highly reflective at this IR frequency. Therefore, there is only

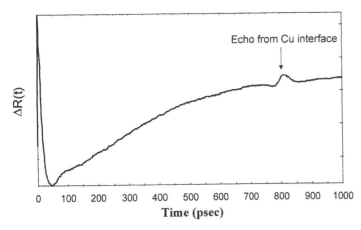

**Figure 8**   Picosecond ultrasonics measurement of 2.1-μm Cu/ ~ 20-nm Ta/~ 600-nm TEOS.

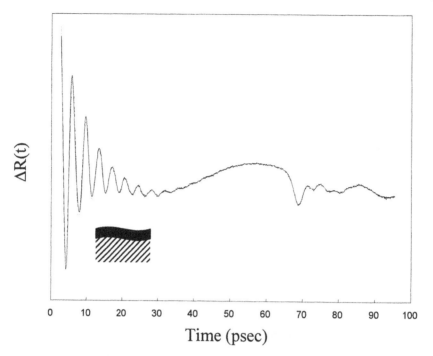

**Figure 9**   Picosecond ultrasonics analysis of 7.7-nm Ta film on oxide. The analysis determined thickness, density, and also the 400.2-nm thickness of the underlying oxide. The Ta density was 11.3 g/c$^3$, which is rather low for Ta and probably means the Ta is porous.

weak absorption of light to create sound waves. A frequency doubler was added to produce 400-nm light pulses. This is shown in Figure 10. *The new system can operate at both 400 nm and 800 nm, making it more versatile than the original system.*

## IV.  DETAILS OF PICOSECOND ULTRASONICS TECHNOLOGY

In this section, the physical principles of picosecond ultrasonics are examined in more detail to further understand their origin. Following Maris and coworkers (2), elastic theory is used to describe how the ultrasonic pulse is generated and then evolves in

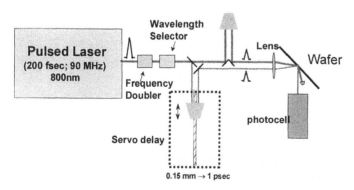

**Figure 10**   MetaPULSE-Cu optical system.

time. The excitation beam covers an area ($A$) on the surface that has linear dimensions much larger ($\sim 10 \, \mu m$) than the thickness ($d$) of the films being measured (10 nm to $\sim 2 \, \mu m$). This results in a uniform pulse over the area characterized by the probe beam. The probe beam's light pulse has an energy $Q$ that is deposited in the sample with a depth dependence $W(z)$ determined by the reflectivity ($R$) of the sample surface and its absorption length $\xi$ (2):

$$W(z) = (1 - R)\left(\frac{Q}{A\xi}\right)e^{-z/\xi}$$

The temperature of the sample increases by 5–10 degrees according to the specific heat per unit volume $C$ (2):

$$\Delta T(z) = \frac{W(z)}{C}$$

One can estimate the increase in temperature by assuming that all energy in the laser pulse is absorbed uniformly in depth. *This method estimates the increase in temperature of 5°, when the pump pulse is absorbed.*

$$\Delta T = \frac{Q}{A\xi C} \sim 5K, \qquad Q = 0.1 \, \text{nJ}, \qquad \xi \sim 10 \, \text{nm},$$

$$A = \pi r^2 = \pi(20 \, \mu m)^2 = 1.3 \times 10^{-5} \, \text{cm}^2, \qquad C \sim 2 \, \text{J/cm}^3\text{K}$$

The cycle time between pump pulses is long enough to allow all of the sound generated by one pulse to dissipate before the arrival of the next.

Before discussing the stress that this induces in the sample, it is useful to recall the definitions of stress and strain. *Strain* is the change in local position $u_{zz}$ from that expected at equilibrium, and *stress* is the force that causes the strain. Thus strain is a dimensionless quantity, $\eta_{33} = \partial u_{zz}/\partial z$. The sample is assumed to have isotropic elastic properties, and the isotropic thermal stress for a sample having bulk modulus $B$ and linear expansion coefficient $\beta$ (2) is

$$-3B\beta \, \Delta T(z)$$

The thermal stress is combined with the stress that restores the sample to equilibrium to give the total relationship between the stress $\sigma_{zz}$ and strain $\eta_{zz}$. The time- and depth-dependent equation for the strain is shown in eq. (7) in Ref. 2. *The stress (force) is only in the direction z perpendicular to the surface, and thus the pulse travels toward the surface and toward the bottom of the top layer.* This relationship can be simplified to estimate the strain when the pump pulse is absorbed:

$$\eta_{zz}(z \text{ close to surface}, t = 0) \sim \Delta T \beta (1 + \nu)/(1 - \nu) \sim 10^{-5}$$

The thermal expansion coefficient $\beta$ is around $10^{-6} \, \text{K}^{-1}$ and Poisson's ratio $\nu \sim 0.25$. The optical properties of the sample are a function of the depth-dependent change in strain, as indicated in Eqs. (19) and (20) in Ref. 2. The probe beam detects a change in reflectivity each time the sound pulse returns to the surface.

For a thick film, the width of an ideal sound pulse is approximately twice the optical absorption length (i.e., $2\xi$). Since the pump light pulse has a finite duration $\tau_0$, during the interval in which the pump pulse is present the sound pulse travels a distance equal to the length of the pump pulse times the longitudinal velocity of sound in the solid, $v\tau_0$. This distance should be small compared to the pulse length, i.e., $\xi \gg v\tau_0$ (2). For metals with a short absorption length of $\sim 10$ nm and a sound velocity of $4 \times 10^5$, cm/s, the *pulse length must be much less than 2.5 ps*. Picosecond ultrasonics uses a pump light pulse length of 0.1 ps.

In high-conductivity metals such as copper and aluminum, the width of the sound pulse generated by the pump may be significantly greater than in low-conductivity metals (such as titanium) as a result of rapid transport of energy away from the surface by electrons. This effect may be modeled, and under some conditions it is possible for such materials to make a determination of the electronic mean free path from a suitable analysis of echo data (e.g., see Ref. 5). As a result of this redistribution of energy by electrons, the energy density at the metal surface may be lower by as much as a factor of 2 than it would be otherwise. Generally, the reduction in the amplitude of the sound wave that is launched from the surface is insignificant compared with the scale of variations that result between different metals as a result of other material properties, such as optical reflectivity, specific heat, and thermal expansion. Consequently, sharp, sizable echoes are easily generated and detected in these high-conductivity materials, such as those illustrated in Fig. 8 (seed copper on Ta).

*Determination of sample density is also possible*. For a single thick film deposited on a substrate, the echo size is determined by the amount of the pulse that is reflected at the film–substrate interface. The ratio of the size of any two successive echoes can be used to calculate the film density if the substrate density and sound velocity are known. This makes use of the relation given earlier, which gives the ratio of any two successive echoes as $R_A = (Z_{substrate} - Z_{film})/(Z_{substrate} + Z_{film})$. The procedure is illustrated in Figure 11.

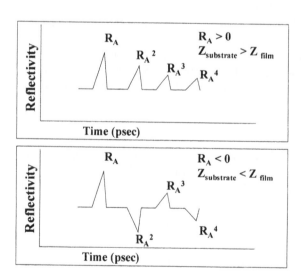

**Figure 11** Origin of general shape of data: The signal intensity ratio of an echo to its subsequent echo is a function of the acoustic impedance ($Z =$ density $\times$ sound velocity) of the film and substrate: $R = (Z_{substrate} - Z_{film}/(Z_{substrate} + Z_{film})$. When the substrate impedance is less than the film impedance, $R$ is negative. $R$ is also negative (and equal to unity) for sound incident on the free surface of the film from the bulk. Therefore in any round trip through the film (i.e., the sound pulse experiences reflections at both surfaces), the overall sign of the echo is positive, and so successive echoes have the same sign. On the other hand, if the substrate impedance is greater than that of the film, then $R$ is negative for the film–substrate interface, and so a sign reversal is observed for successive echoes.

The sign of the acoustic reflection coefficient $R_A$ can be used to understand the shape of the optical reflectivity curve $\Delta R(t)$. Note that when the substrate impedance is less than the film impedance, $R_A$ is negative. $R_A$ is also negative (and equal to unity) for sound incident on the free surface of the film from the bulk. Therefore in any roundtrip through the film (i.e., the sound pulse experiences reflections at both surfaces), the overall sign of the echo is positive, and so successive echoes have the same sign. On the other hand, if the substrate impedance is greater than that of the film, then $R_A$ is negative for the film–substrate interface, and so a sign reversal is observed for successive echoes.

Surface roughness and grain boundary scattering may affect the shape of the sound pulse as it propagates within the film. Both cause broadening, although the rate and nature of the broadening is different in each case. Suitable models can be constructed for these two effects, which, when applied to picosecond ultrasonics data, may be used to deduce the root mean square (rms) roughness and grain size distribution. Grain boundary scattering is operative only for metals exhibiting substantial elastic anisotropy. *Copper is the only example among metals commonly used in semiconductor manufacturing in which elastic anisotropy gives a significant effect for picosecond ultrasonics measurements.*

For a thin film having thicknesses of the order of the penetration depth of the pump probe, the time between successive echoes becomes small compared to the width of each echo, and so the train of decaying echoes takes the form of a damped sinewave. As a result of the sign reversals on reflection from the top and bottom surfaces of the film described earlier in relation to thick films, the period of the oscillations is $2d/v$ if $Z_{\text{substrate}} < Z_{\text{film}}$ or $4d/v$ if $Z_{\text{substrate}} < Z_{\text{film}}$. Reference 2 describes this in greater detail in eq. 47 and on p. 4135 as well. Figure 10 also shows this phenomenon.

Most metals have approximately isotropic sound velocities; however, the speed of sound in copper varies by $\sim 20\%$ along different crystallographic directions. Although copper films can have a preferred texture (a large percentage of the grains are oriented to nearly the same crystallographic direction), at microscopically different positions on the order of the grain size, sound will travel through textured and untextured copper films at unequal rates. *Thus, the returning sound echo (pulse) in copper is spread out in time as compared to the initial pulse, and the echo is further spread in time with each subsequent reflection through the film.* Seed layers of copper show this effect after each echo, and the thicker electroplated layers are strongly affected after a single echo. The length of an echo in thick copper layers can be $\sim 100$ ps, as shown in Fig. 11. *Films with rough surfaces or interfaces also show a spread in echo length, and this is used to estimate film roughness based on an appropriate distribution in roughness dimensions.*

## V. RESOLUTION AND PRECISION OF ACOUSTIC FILM THICKNESS MEASUREMENTS

In Chapter 1 of this volume, *precision* is defined as the square root of the sum of the squares of repeatability and reproducibility. *Repeatability* is the short-term variation observed when the measurement is done several times without moving the wafer. *Reproducibility* is the measurement variation due to loading and reloading the wafer and instrumental instabilities over the course of many days. The precision of picosecond ultrasonics measurements is related to the precision with which the echo times within a sample may be determined. For a hypothetical measurement in which a single echo is analyzed, the precision is equal to the product of the sound velocity in the film (typically about 50 Å/ps) and the uncertainty in determining the centroid (time) of the echo

Δτ. The two most important factors that influence Δτ are the system noise (from mechanical, electrical, and optical fluctuations) and the linearity and precision of the mechanism used to establish the time base (i.e., the delay between the generating and measuring laser pulses). The latter is a mechanical delay line consisting of a retroreflector mounted on a linear stage, so the time base linearity and precision are determined by the linear stage characteristics. In practice, a Δτ of about 0.006 ps can be achieved using readily available components over a range of echo times extending from 0 to over 2 ns. This gives a precision for thickness of the order 0.3–0.6 Å for measurements based on a single echo, and this is typical of observed static measurement repeatability for films thicker than a few thousand angstroms, giving a precision of the order 0.1% or better. For thinner films, it is generally possible to observe several echoes within the finite delay range of the measuring system, and this gives a corresponding improvement in precision by a factor that is proportional to the number of echoes observed. *Thus, the precision in dimensions of length remains approximately constant for thicknesses ranging from a few hundred angstroms to several microns.* For very thin films, the signal-to-noise ratio tends to degrade due to the increasing transparency of the films, and the echo period becomes comparable to the probe pulse width, as described earlier, so measurements become impractical for films thinner than about 25 Å (depending on the material). Therefore for films up to a few hundred angstroms, the measurement precision may vary between 0.1 and 1%, depending on the specific combination of optical properties and film thickness. An example of the reproducibility of single-layer measurements for physically deposited "seed" layer copper and for plated copper are shown in Figures 12 and 13.

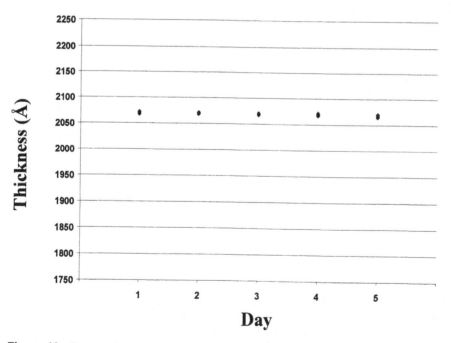

**Figure 12** Repeated thickness measurements of a PVD copper film of nominal thickness close to 2000 Å on five successive days. The wafer was removed from the system after each measurement. The mean center thickness was 2068.7 Å with a standard deviation of 1.7 Å.

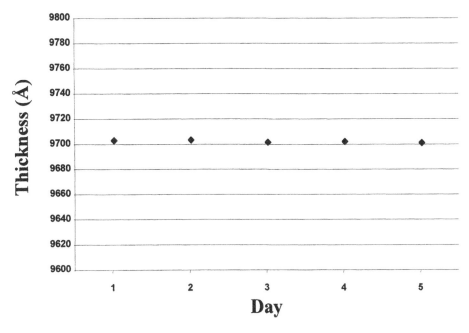

**Figure 13.** Repeated thickness measurements of a plated copper film of nominal thickness close to 1 micron on five successive days. The wafer was removed from the system after each measurement. The mean center thickness was observed to be 9704.4 Å with a standard deviation of 5.5 Å.

For multilayer films, the measurement precision is layer-specific and also dependent on the proximity of each layer to the free surface as well as the ideality of the interfaces between layers. For samples with perfect interfaces between all layers, the measurement uncertainty (in dimensions of length) for any single layer increases relative to the equivalent single-film case by an amount proportional to the number of films between it and the free surface. This increase in uncertainty is usually negligible since, in general, more than one echo is observed within a buried layer less than about 1000 Å thick. Examples of the reproducibility of multilayer film measurement is shown in Figures 14 and 15.

## VI. APPLICATIONS

Picosecond acoustics has been applied to process control of film thickness for both aluminum and copper, chemical-mechanical polishing (CMP), and cobalt silicide. In this section, CMP and cobalt silicide are discussed.

Chemical-mechanical polishing defects can be classified as either local or global. Local defects include scratches, local nonplanar polishing of the copper (dishing), and across-the-chip polishing variation due to line density. Global defects include overpolishing at the edges of the wafer. Dishing can be observed in large copper lines as shown in Figure 16. An example of global CMP nonuniformity is shown in Figure 17.

Stoner (6) has studied the effect of titanium nitride and titanium capping layers on the smoothness of cobalt silicide formed using two rapid thermal-anneal steps. A 120-Å cobalt film was capped with either 150-Å PVD TiN or 100-Å PVD Ti and subjected to annealing. The cap layers help prevent formation of an oxide layer between the cobalt and

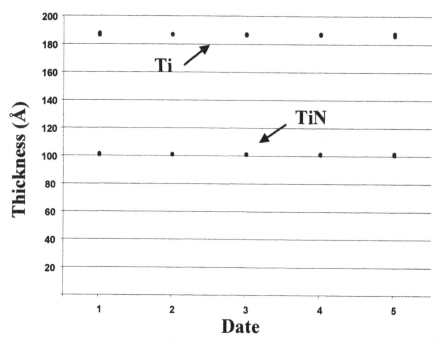

**Figure 14** Repeated thickness measurements of a nucleation (or adhesion) stack for CVD tungsten. The stack consists of a 100-Å CVD TiN layer deposited on top of a nominally 150-Å PVD Ti layer. The measurements were made simultaneously for both films at the wafer center on five successive days, and the sample was unloaded after each measurement. For the Ti layer the mean thickness was 186.9 Å with a standard deviation of 0.34 Å. For the CVD TiN the mean was 100.9Å with a standard deviation of 0.24 Å.

silicon. The samples from each group were then annealed over a range of temperatures from 460 to 700°C for 60 seconds. The capping layers were then chemically removed from all samples using standard chemical etches SC1 and SOM. SC1 is mixture of water, hydrogen peroxide, and ammonia, and SOM is a sulfuric acid/ozone mixture. All samples were then exposed to a final anneal at 800°C for 30 seconds. All anneals were performed in a nitrogen ambient.

The cobalt silicide thickness obtained for the TiN- and Ti-capped films is a measure of how well the capping layer protected the cobalt film so that it could react with the silicon. The results are shown in Figure 18 as a function of the first anneal temperature. Since the cobalt silicide thickness for the TiN-capped samples is ~ 400 Å, for most of the first anneal temperatures it clearly prevented oxygen diffusion. The Ti-capped samples did not protect the cobalt, as shown by the decrease in thickness with decreasing temperature. Only the 660°C sample reached a thickness comparable to the TiN-capped series. The reason the thickness is smaller for certain Ti-capped samples annealed at lower temperatures is that the chemical etch removed the unreacted cobalt.

Film roughness can also be evaluated, and the cobalt silicide study is an interesting example (6). The decay of the signal intensity is given by $R = (Z_{substrate} - R_{film})/(Z_{substrate} + Z_{film})$, where $R$ is the ratio of one signal peak to the next for a smooth film. Comparing this rate of decay of signal intensity with the observed decay is a measure of film roughness. At the time this chapter was written, it was not possible to distinguish between the top and bottom of film roughness. Using an exponential decay model, the

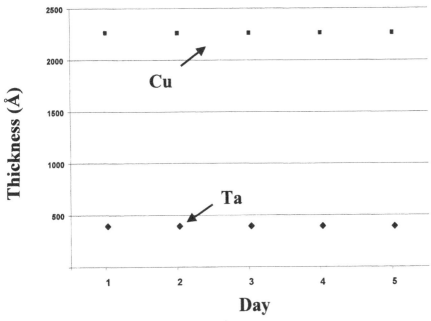

**Figure 15**  Repeated thickness measurements of a barrier-seed stack comprised of a PVD copper film of nominal thickness close to 2000 Å deposited on 400 Å of PVD Ta. The wafer was removed from the system after each measurement. For the copper layer the mean center thickness was observed to be 2265.1 Å with a standard deviation of 1.8 Å. For the Ta, the mean center thickness was 405.5Å with a standard deviation of 1.3 Å.

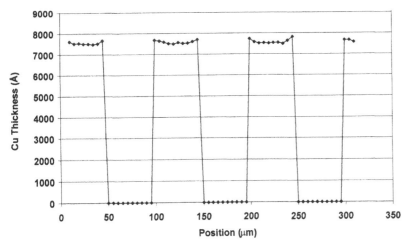

**Figure 16**  Line scan of copper thickness measured for an array of 50-micron-wide Damascene lines separated by 50 microns. The dishing of the copper near their centers is an outcome of the difference in the polishing rates of the copper and the surrounding material.

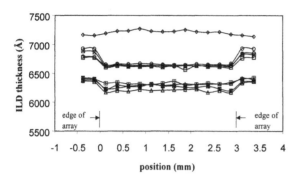

**Figure 17** Picosecond ultrasonic measurement of the mean ILD and copper thickness (assumed to be the same) within an array of Damascene copper lines of width 0.5 microns separated by 2.5 microns. The array is close to 3 mm wide. Results of line scans across nominally identical arrays located at nine different sites on the wafer are shown. The steps at 0 and 3 mm show that the region containing the lines is significantly eroded in all but one case due to the higher polishing rate for copper.

decay rate can be expressed in terms of the roundtrip time for an echo to travel from the surface to the interface and back (2 × thickness/film sound velocity $v_{film}$):

$$R = \exp\left(-\frac{2d_{film}\Gamma}{v_{film}}\right) \quad \text{or} \quad \Gamma = -\left(\frac{v_{film}}{2d_{film}}\right)\ln R$$

The density of the final $CoSi_2$ (i.e., the density following the second anneal at $800°C$) is assumed to be approximately the same for all samples. It was also assumed that the films are homogeneous (which is reasonable in this case, since the final anneal temperature is well above the threshold for conversion of any residual monosilicide into $CoSi_2$), so any differences between the damping rates observed for different samples may be attributed to

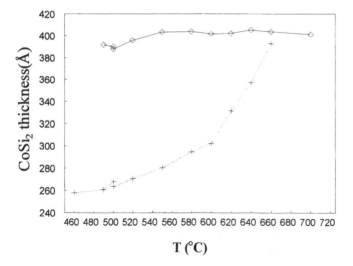

**Figure 18** Picosecond ultrasonic evaluation of TiN and Ti capping layers for cobalt silicidation. The average $CoSi_2$ is shown for films capped with TiN (solid line) and Ti (dashed line) cap layers versus first anneal temperature. (Courtesy Rob Stoner, reproduced with permission. From Ref. 6.)

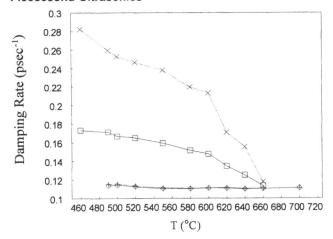

**Figure 19** Determination of film roughness using picosecond ultrasonics: Comparison of mean measured (dashed lines) and predicted ideal oscillatory damping rates for reflectivity signal. The measured damping rate is the dashed line. The predicted damping rate is shown by the solid lines, with the $CoSi_2$ films formed under TiN being the lower two curves and those under Ti the upper two curves versus first anneal temperature. Note that the two curves for the TiN-capped samples are nearly perfectly overlapped. (From Ref. 6.)

variation in the roughness. The results are shown in Figure 19. Further work is expected to provide a quantitative measure of roughness.

Picosecond acoustics can also be used to characterize bond pad etch processes. This final example is shown in Figure 20. The data proves that the etched bond pads were "overetched" into the aluminum in certain areas of the wafer. This resulted in a variation in electrical performance between bond pads. This is reflected in the layer thickness information shown in Fig. 20. It is interesting that the sublayers do not affect the measurements. Thus picosecond acoustics can be used as a yield characterization method.

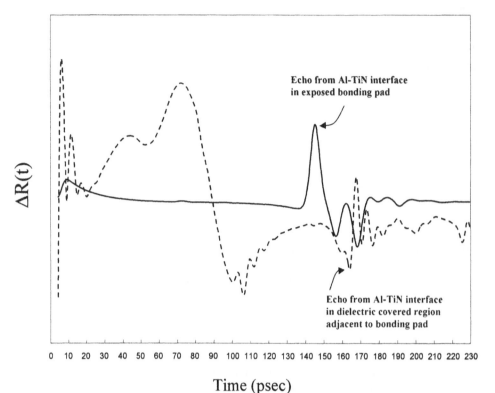

Time (psec)

**Figure 20** Bond Pad Characterization: Echoes measured using PULSE in a bonding pad (solid curve) and adjacent to a bonding pad elsewhere on the same wafer (dashed curve). The metal stack in the pad area is 464.5-nm Al/53.8-nm TiN/35.2-nm Ti/ILD layer 5. The stack in the unetched area is 289.9-nm SiN/299.9-nm SiO2/27.1-nm TiN/517.4-nm Al/61.9-nm TiN/32.7-nm Ti/ILD layer 5. All thicknesses were determined by iterative modeling. All thicknesses were measured simultaneously, including the dielectric layers on top of the TiN layer for the measurement outside the pad area. This shows how the process used to etch the opening in the dielectric layers to make the pad removes not only the TiN layer but also several hundred angstroms of Al.

## REFERENCES

1.  H.J. Maris, R.J. Stoner. Non-contact metal process metrology using picosecond ultrasonics. Future FAB Int. 3:339–343, 1997.
2.  C. Thomsen, H.T. Grahn, H.J. Maris, J. Tuac. Surface generation and detection of phonons by picosecond light pulses. Phys. Rev., B34:4129–4138, 1986
3.  H.J. Maris, H.N. Lin, C.J. Morath, R.J. Stoner, G. Tas. Picosecond optics studies of vibrational and mechanical properties of nanostructures. American Society of Mechanical Engineers, Applied Mechanics Division, Vol. 140. Acousto-Optics and Acoustic Microscopy:134–148, 1992.
4.  H.N. Lin, R.J. Stoner, H.J. Maris, J.M.E. Harper, C. Cabral, J.M. Halbout, G.W. Rubloff. Appl. Phys. Lett. 61:2700–2702, 1992.
5.  G. Tas et al. Electron diffusion in metals studied by picosecond ultrasonics. Phys. Rev. B49:15046, 1994.
6.  R. Stoner, G. Tas, C. Morath, H. Maris, L.-J. Chen, H.-F. Chuang, C.-T. Huang, Y.-L. Hwang. Picosecond ultrasonic study of the electrical and mechanical properties of CoSi$_2$ formed under Ti and TiN cap layers. Submitted to IITC 2000, San Jose, CA, May 2000.

# 11

# Sheet Resistance Measurements of Interconnect Films

Walter H. Johnson
*KLA-Tencor Corporation, San Jose, California*

## I. INTRODUCTION

Sheet resistance is a quick and simple measurement that provides not only process control information abut also information that directly impacts the device yield. Often the power of sheet resistance measurements is in the ability to make many measurements over the surface of a water and graphically map the results. These maps paint a picture that can often identify or fingerprint the source of a problem. Although identifying these finger-prints is not discussed in this chapter, methods for obtaining quality measurements will be so that you can trust the picture that the data paints. Much of the data presented in this chapter is unpublished data collected in the author's lab over the last 15 years. Its inclusion is not meant as scientific proof but as a guideline to be used by the reader to optimize the information obtained from sheet resistance measurements on interconnect films used in the semiconductor process.

## II. DERIVATION OF SHEET RESISTANCE

What follows is an overview of how sheet resistance is derived and the major parameters that influence the resulting value. First the bulk properties (conductivity and resistivity) will be discussed for conductor layers. Next the conversion to sheet resistance will be described. Finally sheet resistance will be used to calculate the resistance of a line.

### A. Conductivity

Free valence electrons are the carriers, which allow conduction in metals. Conductivity, $\sigma$ is the ability of the metal to conduct electron flow (1). Conductivity ($\sigma$) is given by the expression

$$\sigma = \left[ ne^2 \mathrm{L}/2\,m\mu \right]$$

where $e =$ charge of an electron, $n =$ number of electrons, $L =$ mean free path, $\mu =$ thermal velocity, and $m =$ mass of an electron.

## B. Resistivity

Resistivity is a measure of a material's ability to oppose electrical conduction that is induced by an electrical field that has been placed across its boundaries. Resistivity, $\rho$, is defined as

$$\rho = 1/\sigma = \sigma^{-1}$$

Materials will often be divided into three classifications:

Insulator: $\rho > 10^3$ ohm-cm
Conductor: $\rho < 10^{-3}$ ohm-cm
Semiconductor: $10^{-3} < \rho < 10^3$ ohm-cm

Table 1 (2) lists some typical resistivity values for films used in semiconductor manufacturing. Many of the pure metals are listed in standard reference textbooks, but most of the bimetal numbers are measured values for typical processed films. This chapter will address conductor films only.

## C. Resistance

Once the resistivity of a sample is known, its resistance may be determined from its physical dimensions. Resistance, $R$, is defined as

$$R = \rho L/A = \rho L/t * W$$

where $\rho$ = resistivity (ohm-cm), $L$ = length (cm) of the sample, $W$ = width (cm) of the sample, $A$ = cross-sectional area = thickness (cm) × width (cm) = $t$ (cm) × $W$ cm.

**Table 1** Typical Resistivity Values of Several Metal Films

| Metal | Electrical resistivity ($10^{-6}$ ohm-cm) | Metal | Electrical resistivity ($10^{-6}$ ohm-cm) |
|---|---|---|---|
| Aluminium (sputtered) | 2.7* | Niobium | 13* |
| Antimony | 41.8* | Osmium | 9* |
| Cadmium | 704* | Platinum | 10.5* |
| Chromium | 13* | Selenium | 141.4* |
| Cobalt silicide | 18–20** | Silicon (doped) | 1e3-1e9** |
| Copper | 1.673* | Silver | 1.59* |
| Gold | 2.35* | Sodium | 4.2* |
| Iron | 9.7* | Tantalum silicide | 50–55** |
| Lead | 20.6* | Titanium nitride (CVD) | $\sim$ 80** |
| Magnesium | 4.45* | Titanium nitride (reactive PVD) | $\sim$ 250** |
| Manganese | 185* | Tungsten silicide | 13–16** |
| Mercury | 98.4* | Tungsten silicide | $\sim$ 70** |
| Molybdenum silicide | $\sim$ 100** | Vanadium | 25* |
| Nickel silicide | $\sim$ 50** | Zinc | 5.92* |

*Source*: *Ref. 2; **Ref. 21.

## D.  Sheet Resistance

Thin layers are described in terms of sheet resistance ($R_s$), rather than by resistivity or resistance. This is more descriptive since it involves the thickness, $t$, of a layer. Sheet resistance (3) is calculated from the resistance of the sample by the formula

$$R_s = \rho \text{ (ohm-cm)}/t \text{ (cm)} = \text{ohms (more commonly referred to as ohm/square,}$$
$$\text{because this is the resistance of a square of any size)}$$

where $\rho$ = resistivity of the thin layer (ohm-cm), $t$ = thickness of the layer (cm), and $W = L$. The value of the concept of sheet resistance lies in the ability to calculate the resistance of large areas of lines when only the size of the sample and its sheet resistance are known.

## E.  Line Resistance

The line resistance can easily be calculated by using the following equation:

$$R = R_s * L/W$$

where $R_s$ = sheet resistance, $L$ = length of the line, $W$ = width of the line.

As an example, if the sheet resistance of a multilayer film stack is 0.015 ohms, then the resistance of a 0.1-micron-wide and 1000-micron-long line would simply be

$$0.015 * 1000/0.1 = 1.5 \text{ ohms}$$

The line could be thought of as a series of squares, with the line resistance being the sheet resistance (ohms/square) times the number of squares. Of course this does not take into account the contact and via resistances, which is beyond the scope of this chapter. One should also keep in mind that the adhesion and barrier layers will affect the line resistance. With the traditional aluminum process, the total resistance is the parallel resistance of all the layers:

$$R_{total} = 1/(1/R_{aluminum} + 1/R_{barrier})$$

where $R_{total}$ is the sheet resistance of the combined layers, $R_{aluminum}$ is the sheet resistance of the aluminum film, and $R_{barrier}$ is the sheet resistance of the barrier layer. In via contacts the barrier resistance will add directly as a series resistance. In a dual damascene-type process there will be two sides to consider as parallel resistances to the main copper conductor line. These barrier layers may also affect the assumed line thickness or width and should be accounted for.

## III.  MEASUREMENT METHODS

There are three common methods of measuring sheet resistance: the four-point probe, Van der Pauw structures, and the eddy current method (4). Van der Pauw structures require photolithography steps, are commonly used for device performance rather than process monitoring, and will not be discussed here. Four-point probe technology has been used for around 40 years and is the most common method of obtaining sheet resistance values. The traditional approach to making four-point probe measurements has been to use four independent pins with individual springs for each one or a single spring for all four in order to provide a constant downward force. How close one could measure to the edge of a sample was limited to approximately 6 times the probe spacing, until Dr. Perloff introduced a geometric correction technique (5) that reduced this restriction to around 3 times the probe spacing. The introduction of a probe orientation perpendicular to the edge and the appropriate equations by the author in 1998 (6) further reduced this edge exclusion to around 1 probe spacing.

With the introduction of the M-Gage® by Tencor Instruments™ (7) in 1980, the eddy current method gained popularity. Although the introduction, by Prometrix™, of dual configuration on the four-point probe and the increase in applications knowledge provided by the mapping capability diminished the popularity of the M-Gage®, there are still many in use today. In 1994 Prometrix™ introduced a small-spot eddy current system (8). Due to its limited range and the broad capabilities of the four-point probe, it had limited acceptance and ceased production in 1997.

## A. Four-Point Probe

The four-point probe is the most common tool used to measure sheet resistance. Many processes, including ion implantation, metal deposition, diffusion, and epitaxial silicon growth, use sheet resistance to help control the process and to help predict device yields. The normal range for sheet resistance in semiconductor processing is from less than 0.01 ohms/square for copper films to about 1 million ohms/square for low-dose implants into silicon. This ability to measure over more than eight orders of magnitude makes the four-point probe a valuable tool that has survived despite its destructive nature. For the four-point probe to work requires some isolating junction or blocking layer to the dc current used for this technique. For metal films this is usually an oxide layer, for other processes it is a p/n junction.

### 1. Four-Point Probe Arrangement

*a. Pin Arrangement*

The probe assemblies for the four-point probe historically have the probe tips that induce the current and measure the resulting voltage drop in the sheet resistance measurement in one or two arrangements: either linear or square arrays. The linear probe-tip array that is perpendicular to the edge of the water became the common orientation around 1985.

Sheet resistance is measured by passing a current between two of the probes and measuring the resultant voltage drop across the other two. In the linear array it is customary for the outer probes to carry the current and the inner two to measure the voltage (9). Geometric effects that interfere with the sheet resistance measurement, such as when the probe is very close to the edge of the water (a result called *current crowding*), can be corrected for by a technique called *dual configuration* (10). This technique requires that two sheet resistance measurements be made with different geometrical configurations.

The first measurement, $R_A$, is obtained in the traditional manner, as shown in Figure 1a. The outer probes carry the current and the inner probes sense the voltage. The second measurement, $R_B$, is obtained by having the first and third probes carry the current and sensing the voltage on probes 2 and 4 as shown in Figure 1b. The use of both arrangements allows for the correction of discrepancies in the probe spacing and any current-crowding effect (10).

*b. Probe Orientation*

    1. *Parallel to a radius*: As mentioned earlier, the traditional probe orientation for a linear array is parallel to a radius (10). This arrangement was standardized

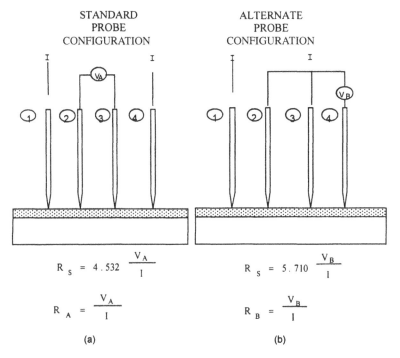

**Figure 1**   Two separate methods of configuring a four-point probe measurement can be used to correct for edge effects. (From Ref. 8.)

when the dual-configuration method was established using this configuration. In this orientation the ratio of $R_A$ to $R_B$ has the greatest sensitivity to current crowding.

2. *Perpendicular to a radius*: Although Swartzendruber (11) published edge correction tables for the perpendicular orientation back in 1964 for the National Bureau of Standards (NBS), it wasn't until 1998, when KLA-Tencor introduced the RS100, that this orientation became standard on a system. This required new algorithms to correct for edge effects while in the perpendicular-to-a-radius mode. Figure 2 shows the error in the sheet resistance measurement as the probe approaches an abrupt boundary in both the parallel and perpendicular modes.

## 2.   Choosing a Four-Point Probe

Because most probes can achieve ohmic contact to metal films, the major consideration is usually damage to the film. Even this is not a problem if the film is to be measured only once. A probe with a tip radius of 0.0016 in. and a 100-gram spring loading is typical for metal-film measurements and will cause sufficient damage (described later in the chapter) that it becomes a significant factor in repeatability studies with as few as 15 repeated measurements. By using blunter tips, such as a 0.008-in. tip-radius probe, standard deviations of less than 0.5% one sigma have been achieved on films as thin as 150 angstroms (Figure 3).

The gradual drift up in resistance may be a combination of film property changes and damage caused by repeated measurements. The same increase in resistance was noted

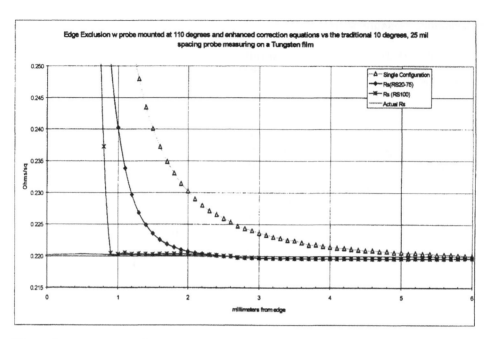

**Figure 2** Current crowding as the probe approaches the edge of a film will cause the apparent resistance to rise. Probe orientation and edge correction algorithms can be used to compensate for this error. (From Ref. 25.)

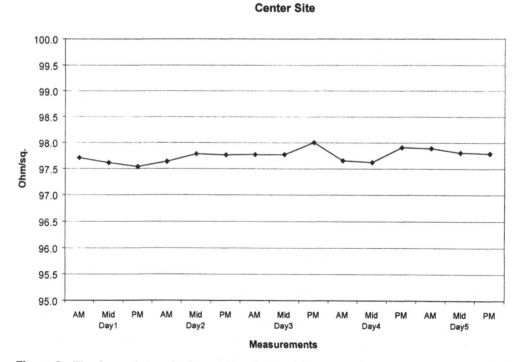

**Figure 3** The four-point probe is capable of repeated measurements on films as thin as 150 angstroms of Tantalum, as shown in this graph. (From Ref. 25.)

on a second four-point probe, supporting the conclusion that the film is actually changing resistance.

## B.  Eddy Currents

Eddy currents owe their existence to mutual induction, which was first described in 1820 by Hans Christian Orsted. Many instruments have been made for testing the resistivity of materials, but it wasn't until the design of the oscillating ratio frequency tank circuit by G. L. Miller in 1976 (12) that eddy currents became commonly used in the semiconductor industry. As a startup company, Tencor Instruments further commercialized this approach, including the addition of an acoustic thickness measurement, and introduced the M-Gage® and Sono-Gage® in 1980.

Eddy current systems determine the sheet resistance of a film by creating a time-varying magnetic field from a coil. The coil radiates energy, and, when placed close to a conductive layer, eddy currents are generated by absorbed energy in the conductive film. The eddy currents in turn create a magnetic field (13) that then interacts with the primary coil, creating a change in electrical parameters of the coil (Figure 4). The more conductive the film, the more the energy gets trapped in the film to create eddy currents. The field drops off with distance or depth, but it will still penetrate through the thin films typical of semiconductor processing and into the underlying films and substrate. The undesirable field developed in the substrate can have significant effects on the measurement and is the biggest limiting factor for eddy current use. Electrical eddy currents, just like their wet counterparts in the streams, require sufficient room to create a circular path. Interruptions in this circular path will disrupt the eddy currents, preventing their formation in that layer. Transformers, for example, have laminated cores so that eddy currents do not have a continuous path in which to flow. This reduces the energy lost and the heat generated by the transformer. Therefore little to no signal is produced from patterned films where the conductive features are significantly small compared to the coil diameter.

## 1.  Tank Circuit

Miller introduced the resonant tank circuit approach in 1976 by designing a system where the coil was split into two sections and the sample placed between the two halves. This

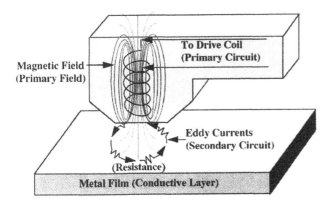

**Figure 4**   Eddy currents are induced in a nearby film by placing a coil with a time-varying electrical current near the sample. (From Ref. 25.)

approach simplified the analysis of a magnetic field gradient interacting with a resistivity gradient through the depth of the sample, as would be the case with a single coil.

## 2. Front-Side Mutual Induction Probe

Although the single-sided technique has a long history, its use in the semiconductor field did not take place until the introduction of the OmniMap NC110® in 1994 by Prometrix Corp (8). This technique is heavily used in the aerospace and nuclear power industries for metal thickness and defect detection. But in those fields the metal is typically hundreds of microns to several millimeters thick. For semiconductor applications the small signals obtained from the thin films required improvements to these existing techniques.

### a. Coil Arrangement

The principle of the mutual inductance method with a single-sided coil is described as follows: If an alternating current is flowing in the drive coil a magnetic field is developed (Fig. 4) and is called the *primary field*.

The impedance of the drive coil, $Z$ is expressed by the following equation:

$$Z^2 = R^2 + X_L^2 \tag{14}$$

where $R$ is the ac resistance of the coil and $X_L$ is the inductive reactance.

The inductive reactance is a function of the frequency used and the coil inductance and is calculated from:

$$X_L = 2\pi f L_0$$

where $f$ = frequency in hertz and $L_0$ = coil inductance in henrys.

### b. Skin Depth

The eddy currents produced in the film (secondary circuit) are dependent on the film resistance (or sheet resistance) and magnetic permeability. Because the films that are normally used in the manufacturing of semiconductor devices have a relative permeability of 1, we will ignore that parameter for the purposes of this application. The magnetic field created by these eddy currents, and their effect on the primary circuit, will depend on the sheet resistance of the film. The density of eddy currents will vary with depth, as indicated in Figure 5.

The depth at which the eddy current density has decreased to 37% of its value at the surface is called the *standard skin depth* or *standard depth of penetration* (15). The standard skin depth for nonmagnetic films can be approximated by the following equation:

$$S = 50,000(\rho/f)^{1/2} \tag{16}$$

where $S$ is the standard skin depth in micrometers, $\rho$ is the resistivity in $\Omega$-cm, and $f$ is the frequency in hertz.

The standard skin depth curve for aluminum is charted in Figure 6 as a function of frequency. Typical commercially available eddy current systems have operated at a frequency of 10 MHz or lower. For these frequencies the standard skin depth will be much larger then the thickness of the metal films typical in semiconductor processing. In the case of a metal thin film on a conductive substrate, such as aluminum film on a highly doped silicon substrate, substantial eddy currents will also form in the silicon substrate. However, if the metal-film resistivity is significantly lower than the substrate resistivity, the effect of substrate can often be ignored.

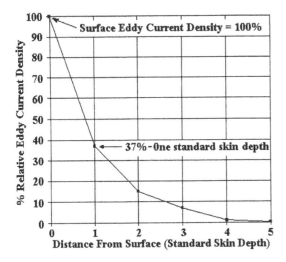

**Figure 5** Eddy current strength will vary with depth below the film surface. (From Ref. 15.)

### c. Liftoff Effect

Typically the changes in coil values are measured rather then the absolute coil parameters. This greatly simplifies the procedure. To obtain a reference point, the coil is placed sufficiently far away from the film so as to have no significant effect, and then the coil resistance and reactance are measured. This is the reference value, commonly called the *open coil value.* As the coil is brought close to the film, the resistance and reactance will gradually change. To prevent the probe head from crashing into the substrate, the measurement is started at or near the surface and the probe pulled away or lifted off the surface; this is commonly called a *liftoff.* Plotting these resistance and reactance values creates a curve (liftoff curve), illustrated in Figure 7.

### d. Liftoff Family of Curves

Each resistance will create a unique liftoff curve. Several liftoff curves can be collected on standard samples to create a calibration matrix (Figure 8).

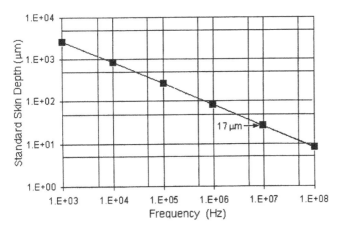

**Figure 6** Skin depth decreases with increasing frequency and for aluminum is approximately 17 microns for a frequency of 10 MHz. (From Ref. 15.)

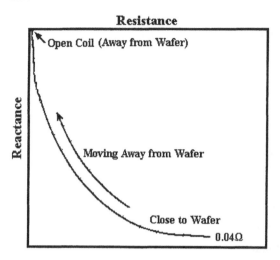

**Figure 7** The reactance and resistance of an eddy current sensing coil will change as the coil is moved away from a sample. (From Ref. 15.)

If the reactance and resistance for the standard samples, collected at a constant height, are connected, they form a continuous curve called an *isodistance curve*. A series of these isodistance curves can be overlaid on the liftoff curves, as shown in Figure 8. The sheet resistance of an unknown test wafer can then be interpolated from the sheet resistance values of the two adjacent calibrated liftoff curves at any given height. Figure 9 shows two contour maps of sheet resistance data collected by both four-point probe and eddy current methods.

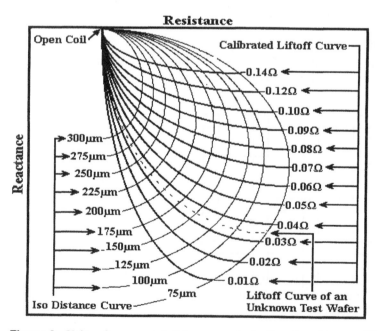

**Figure 8** Values from several distances are typically referred to as a *liftoff curve*. (Stylized representation from Ref. 25.)

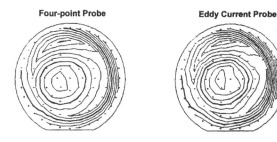

**Four-point Probe**                    **Eddy Current Probe**

Titanium 200Å                    Titanium 200Å
Mean: 62.904 OHMS/SQ            Mean: 62.560 OHMS/SQ
Std. Dev. 2.548%               Std. Dev. 2.940%

**Figure 9** Under conditions where the wafer resistance can be ignored, the eddy current probe will produce results similar to those of a four-point probe. (From Ref. 25.)

In some cases it is possible to measure directly on product wafer with the eddy current approach. The following examples will indicate some of the similarities and some of the differences between measuring sheet resistance of a metal film on a monitor wafer and on a product wafer.

Figure 10 shows measurements, for a tungsten layer, taken across the diameter of both a monitor and a product wafer, where the monitor wafer was measured by the eddy current probe as well as a four-point probe. In this case the DRAM product wafer has a substrate resistance that is sufficiently high and, along with the absence of underlying topography, yields product wafer results that match the monitor wafer results. The 3% lower sheet resistance on the product wafer was determined to be a result of an increased tungsten thickness on the product wafer.

The substantially lower substrate resistance of a wafer used in a typical logic process can cause a marked influence on the estimated film sheet resistance, as shown in Figure 11. Here, there is about a 3% difference in a 1-micron aluminum film, which was attributed to the addition of the substrate resistance, which adds as a parallel resistance:

$$R_{\text{total}} = 1/\left(1/R_{\text{film}} + 1/R_{\text{substrate}}\right)$$

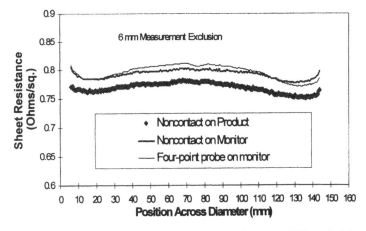

**Figure 10** In the case of a product wafer that uses high-resistivity substrates, the eddy current method gives equivalent results to the four-point probe. (From Ref. 25.)

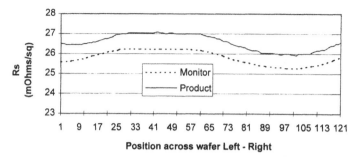

Figure 11   When the resistivity of the substrate is significant relative to the film to be measured, it will influence the eddy current measurement. (From Ref. 18.)

where $R_{total}$ is the sheet resistance of the combined layers, $R_{film}$ is the sheet resistance of the film, and $R_{substrate}$ is the sheet resistance of the substrate.

By measuring the substrate before the film deposition and performing a parallel subtraction of the substrate from the total resistance measured after film deposition, the use of eddy current may be practical. This approach is illustrated in Figure 12, where the initial 5% offset on the 1-micron aluminum film was compensated for through parallel subtraction (17).

When underlying topography is introduced into the equation, one might get striking differences between traditional monitor wafers and product wafers. Figure 13 shows the increased resistance due to the area increase and subsequent decrease in thickness caused by the underlying topography. Notice the change in resistance across the die due to the topography differences between the die and the scribe lines.

Even with the correction for substrate resistance, the pattern across the die can result in a difficult interpretation of the process. Figure 14 shows a wafer in which the pattern across the die varies from one side of the water to the other.

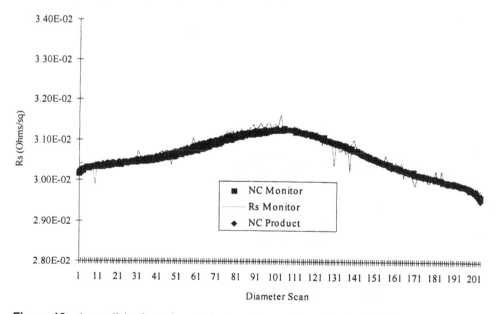

Figure 12   A parallel subtraction method can be used to minimize the influence of a conductive substrate. (From Ref. 17.)

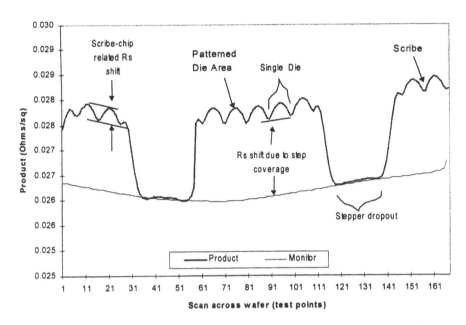

**Figure 13** Underlying device topography can influence the film resistance in a product wafer. (From Ref. 17.)

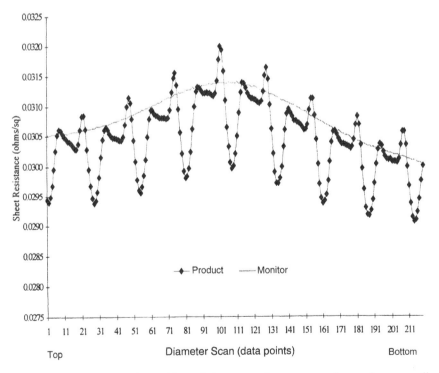

**Figure 14** Product wafers with varied topography across a die produce complicated resistance patterns. (From Ref. 17.)

In order to determine the effect of underlying wiring on the eddy current technique, a four-point probe and eddy current probe were used to measure a third-level metal layer on a product wafer. The close match of the two techniques shown in Figure 15 seem to support the absence of any unique effect of the underlying wiring on the eddy current method.

In cases where there is a substantial amount of residual metal left in the underlying layers, such as is sometimes the case at the edge of the wafer where there are no printed die, both metal layers may be detected by an eddy current probe. Notice in Figure 16 how the four-point probe measurements on the product wafer match the monitor readings at the wafer edge but the eddy current readings drop dramatically. Although the readings are off-scale in the graph, the resultant indicated sheet resistance was equal to that of the parallel addition of the first and second metal layers.

Although the wafer shown in Figures 17 and 18 was scrapped prior to the measurement due to a visual inspection, it does show the potential of the eddy current method in detecting large voids in the film. In this case the voids were due to missing poly gates from a prior etch step.

Noncontact measurements lead to other possibilities, such as mapping of the complete reactor (18), as shown in Figure 19. Here the wafer maps of a full product batch were placed in their respective positions in the reactor.

## IV.   FILM THICKNESS EFFECTS ON RESISTIVITY

Gillham et al. (19) reported resistivity as a function of deposited thickness as far back as 1955. The estimation of copper resistivity can be further complicated due to the use of

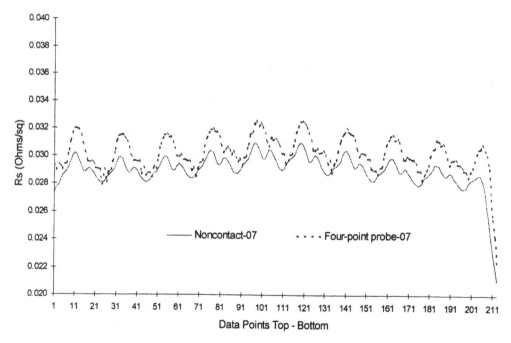

**Figure 15**   The metal film resistance change across the product die can be measured with a four-point probe as well as an eddy current probe. (From Ref. 18.)

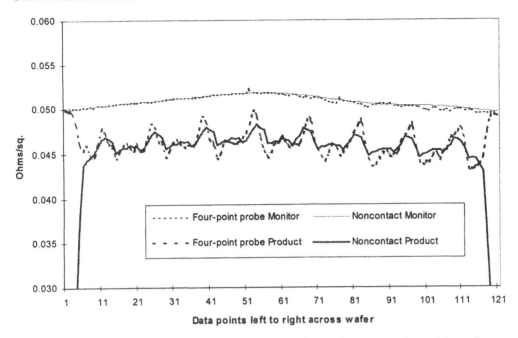

**Figure 16** When measuring higher-level metals, large pieces of unpatterned metal from the previous layer can influence the eddy current readings. (From Ref. 25.)

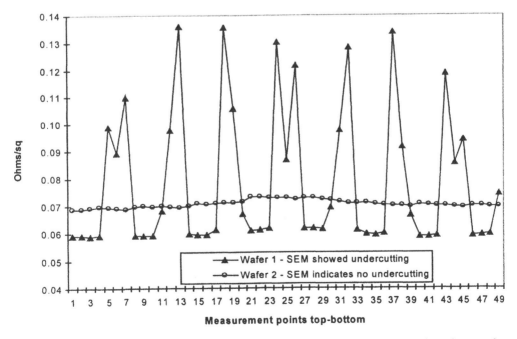

**Figure 17** These two product wafers are plannarized, but one shows large rises in resistance that were traced to missing poly gates. (From Ref. 25.)

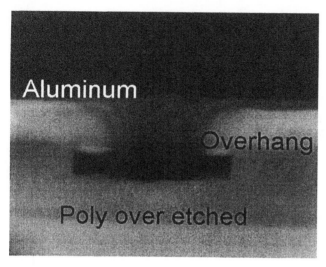

**Figure 18**  Cross section of the missing poly gate from Fig. 17. (From Ref. 25.)

chemical-mechanical polishing (CMP). As the following data shows (Figure 20), the resistivity is dependent on the deposited thickness, not the actual thickness when measured. The properties of polished and plated copper films of matching thickness values are compared. Wafers were prepared with a Ta barrier and 100-nm seed layer, then plated to various thickness values ranging from 200 nm to 1500 nm. Five of the 1500-nm films were then annealed to eliminate the effects from resistivity shifts due to self-annealing (20). Sheet resistance and temperature coefficient of resistance (TCR) measurements were made on all samples with a four-point probe. Steps were etched in these films, and step height and surface roughness measurements made.

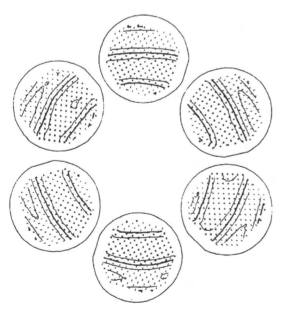

**Figure 19**  Arrangement of wafer contour maps indicating the reactor uniformity. (From Ref. 18.)

**Polished and Electro-Plated Films**

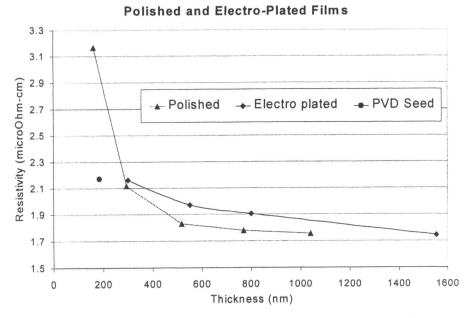

**Figure 20** Resistivity of a copper film is dependent on not only the final thickness but also the initial thickness. (From Ref. 17.)

Focused ion beam (FIB) images (Figure 21) were taken of the deposited and polished films. These images clearly supported the increase in grain size with increasing thickness. The thickness values ($t$) and sheet resistance values ($R_S$) were used to calculate the resistivity values ($\rho$) using the following equation:

$$\rho = R_S * t$$

The influence of the Ta barrier layer was removed by using a parallel subtraction equation and the average sheet resistance value of 57 ohms/sq. for the Ta layer. The data indicates that the resistivity rises as the thickness drops, which is well supported in the literature (3,19,20).

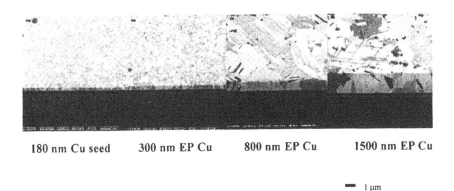

180 nm Cu seed    300 nm EP Cu    800 nm EP Cu    1500 nm EP Cu

1 μm

**Figure 21** These focused ion beam images show the increase in grain size as the thickness increases. (From Ref. 21.)

It is clear that the polished films have a distinct curve, which indicates lower resistivity than the equivalent as-deposited thickness. This is believed to be due to the grain size remaining at the same in-plane dimension as the originally deposited film.

## V.  TEMPERATURE COEFFICIENT OF RESISTANCE

Sheet resistance also varies with temperature. Temperature correction values may be somewhat hard to find or calculate for thin or compound metal layers and may need to be measured. Figure 22 is a typical plot of sheet resistance as a function of temperature, which was used to determine the TCR values for the various copper films.

The effects of thinning copper on the grain size and bulk resistivity cause the temperature coefficient of resistance (TCR) of the film to change. Figure 23 shows how the TCR rises as the thickness increases, up to about 0.8 µm, for both an electroplated and a polished set of films (21). The polished films have a different curve, presumably due to the larger grain size, from the electroplated-only films.

The TCR correlates well with the resistivity. This may provide a novel method of estimating the film thickness. To illustrate this point, the measured TCR was used to estimate the Cu film resistivity, and then this resistivity value was used along with the sheet resistance to calculate the thickness. Figure 24 shows the correlation between the measured Cu thickness (by profilometer) values and the Cu thickness values derived from the four-point probe measurements of the Ta/Cu film stack.

## VI.  MEASUREMENT DAMAGE

It should always be kept in mind that the four-point probe is a destructive technique and will have some effect on subsequent measurements. By measuring sheet resistance values along a diameter that cuts through a repeatedly tested area, the effect can be measured. Figure 25 shows the typical effect of 50 repeated probe qualification tests on both a thick aluminum film and a thin titanium film, made in the same areas. These areas corresponded to 95 and 105 mm on the chart.

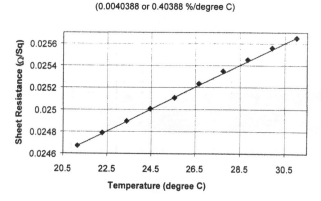

**Figure 22**  Most materials have a positive temperature coefficient of resistance (TCR), where the resistance rises with a rise in temperature. This copper film shows a 0.4%/°C TCR. (From Ref. 25.)

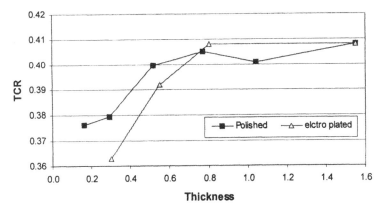

**Figure 23** Much like the resistivity value, the TCR is dependent on the thickness. Because the grain size is the major contributor to the change in TCR, we can see a difference in TCR between two films of the same thickness but different grain sizes. (From Ref. 25.)

The areas in the center and just to the right and left of center show a rise in resistance relative to the neighboring values. The softer and thinner the film, the greater effect probing will have on subsequent sheet resistance measurements. Probe-head parameters will also affect the level of damage. It is the damage between probe pins rather then the damage under the pin that causes the increase in sheet resistance. This damage should be kept in mind when monitoring the long-term repeatability of a sample.

## VII. SELF-ANNEALING

Jiang (20) has described the self-annealing of copper films. This work shows the increase in grain size as a function of time and how the relaxation rate is dependent on the original thickness, as seen in Figure 26. Note in the figure that the final resistivity varies with the deposited thickness, as indicated in Section IV.

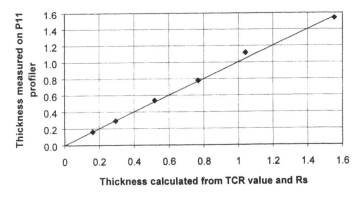

**Figure 24** A four-point probe can be used to measure both the sheet resistance and TCR values. In a novel use of the four-point probe, one can use the TCR value to estimate the resistivity and a thickness can be calculated. (From Ref. 21.)

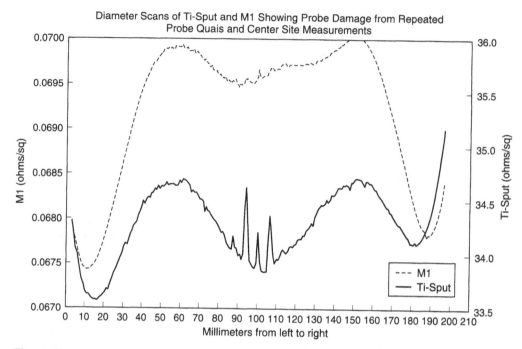

**Figure 25** The four-point probe will damage the film with repeated testing, causing a rise in resistance, as shown in the areas in the center and to the right and left of center. As might be expected, the thinner films, such as the titanium film shown here, are more easily damaged. (From Ref. 25.)

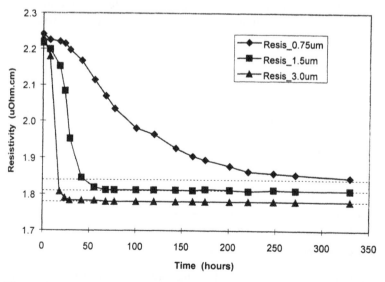

**Figure 26** Grain size (and therefore resistivity) may change following deposition. Here, three copper films show resistivity changes at room temperature. (From Ref. 20.)

## VIII. FILM STABILITY ISSUES

Minienvironments such as the standard mechanical interface (SMIF) have been used to decrease the running cost of cleanrooms in some semiconductor manufacturing companies (22). Wafers can be isolated from particulate contamination, but they are not always isolated from organic, ambient oxygen and moisture. Unless they are hermetically sealed, they will be affected by the ambient.

Most of the metal films used in semiconductor manufacturing can change resistance over time when exposed to ambient conditions, which cannot be attributed to grain size changes. This drift with time might occur for several reasons, ranging from oxidation to interface reactions. With thin films and in particular highly reactive thin films, these drifts can be quite significant. We can see from Figure 27 that there was a 3.5% drift over 5 days for a 40-angstrom TiN film.

Koichiro et al. (22) described how the drift in cobalt silicide sheet resistance could be reduced by storing the wafer in a nitrogen-purged pod to stabilize the film. He stored wafers in the nitrogen-filled pod for 1 week. Some of these stored wafers were exposed to ambient for several hours, and some wafers exposed to cleanroom air for 1 week were sputter-etched to clean the surface. Cobalt and titanium nitride layers were deposited on the wafers. All the wafers were annealed for 30 s at 550°C in nitrogen for CoSi formation. After any unreacted cobalt and titanium nitride were selectively removed, the wafers were annealed for 30 s at 700°C in nitrogen for $CoSi_2$ formation.

The average sheet resistances of the films are plotted as a function of the time for air exposure of the wafers and closed-pod storage prior to the Co deposition (Figure 28). The increase in sheet resistance of samples exposed to air is observed for those with a 4-h exposure or more. The sheet resistance then increases to more than 40 ohms/sq after a 1-week exposure. In contrast, the sheet resistance of the samples stored

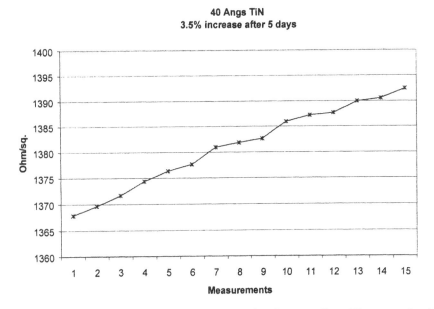

**Figure 27** Resistivity can also increase with time for some films. The reason for this increase is not clear but may involve adsorption of water vapor. (From Ref. 25.)

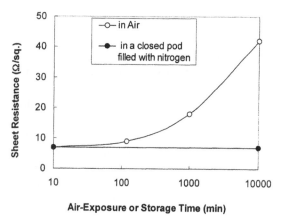

**Figure 28**  Air exposure prior to deposition may also affect the resistance of the deposited film. Here the increase in a cobalt silicide film is shown. (From Ref. 22.)

in the closed box for 1 week still remains at 5.9 ohms/sq. The authors attributed this to oxide growth.

Figure 29 shows the cumulative probability of the sheet resistance of the source/drain in the device structure on wafers stored in two different atmospheres for 1 week prior to Co deposition. The increase in the sheet resistance is prevented for samples subjected to sputter etching after exposure to air (Figure 29a) as well as those stored in a closed pod filled with nitrogen (Figure 29b). It is assumed that native oxides are reduced by physical etching so that cobalt can react with silicon. This effect was also noted on device structures, resulting in increased source/drain sheet resistance and junction leakage for the air-exposed samples. They speculated that damage induced by sputtering is concentrated in the peripheral region of the junction and leakage current flows through localized points in

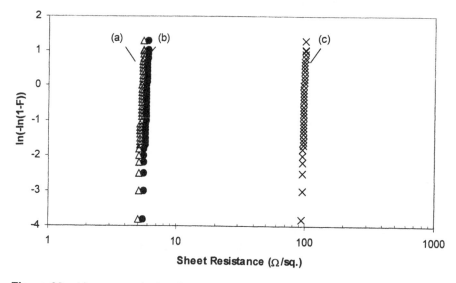

**Figure 29**  Air exposure is also shown to increase the source-drain resistance on a device structure. (From Ref. 22.)

the peripheral region. Note that storing wafers in a closed box filled with nitrogen is more efficient in terms of the formation of high-quality and uniform cobalt silicides than removing native oxides on the silicon surfaces with sputter etching after exposure to air.

## IX.  SUBSTRATE EFFECTS

### A.  Short to Substrate

Due to the low resistivity of most materials used in semiconductor processing, it is a common practice to deposit directly on the silicon surface of a monitor wafer. This can lead to problems for thin films where the substrate sheet resistance is not greater then 100 times the film sheet resistance. The solution to these problems is to deposit an insulating layer, usually silicon dioxide, between the substrate and the metal layer. This works for the four-point probe method but does not affect the substrate influence for eddy current systems.

## X.  ULTRATHIN-FILM MEASUREMENTS

As feature sizes shrink, the interconnect delay times are becoming the dominant factor in chip speed, putting emphasis on reducing interconnect resistances. One aspect of lowering resistance is the need to reduce the thickness of the adhesion and barrier layers. Measuring the sheet resistance of these layers can present a challenge to the four-point probe, which is destructive by nature. Methods for optimizing measurements on these very thin metal films down to 40 angstroms include using larger tip radii, as discussed earlier. Due to the high sheet resistance of these thin films, eddy current methods are not used because of the influence of the silicon substrate resistance. In some cases it may be convenient to use a nonconductive substrate, such as quartz, to circumvent this limitation.

## XI.  EFFECTS OF UNDERLYING TOPOGRAPHY ON FILM RESISTANCE: A UNIQUE APPLICATION OF RS

A small experiment was conducted to look at film resistance changes with topography changes (23). Silicon wafers were coated with a specified thickness of oxide, then patterned. The test pattern consisted of repeating line/space pairs. Figure 30 shows the layout of the test pattern and the test structure on the wafer. The oxide films were etched and aluminum films were deposited on the matrix of oxide line/space test structures. Table 2 shows the three-factor sample matrix. The variables used for this experiment are the metal thickness, $t_m$, the oxide step height, $t_{ox}$, and the pitch of the line/space pairs.

In each group a blank control wafer was processed with the patterned test wafers, to produce the metal film with identical microstructure and electrical properties. The sheet resistance of this blank control wafer was used as a reference for a step resistance calculation. The *step resistance* is the ratio of the resistance of a metal film over a step relative to the resistance of the film over a planar surface.

Diameter scans of the sheet resistance on the step coverage test wafers were collected and step resistance values were calculated. As an example, Figure 31 shows the step resistances of Group 1 (Wafers A, B, C, and D in Table 2). Only the center values of

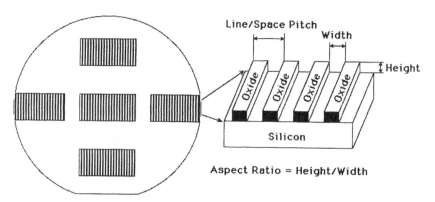

**Figure 30**  This test pattern was used to investigate step coverage and its effect on metal-film resistance. (From Ref. 23.)

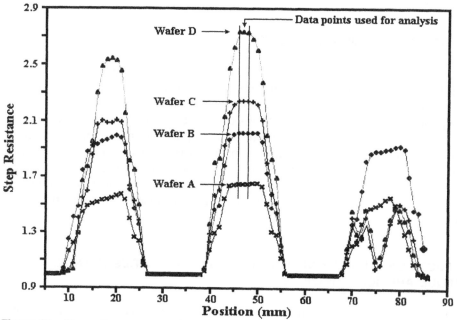

**Figure 31**  These diameter scans show the increase in resistance caused by poor step coverage of several step/space combinations. (From Ref. 23.)

**Table 2**  Matrix of Linewidths and Spacing Used to Evaluate Step Resistance as a Process Monitor

|  | Group 1 | | | | Group 2 | | | |
|---|---|---|---|---|---|---|---|---|
| Wafer ID | A | B | C | D | E | F | G | H |
| $t_m$ | 1.35 | 1.35 | 1.35 | 1.35 | 0.88 | 0.88 | 0.88 | 0.88 |
| $t_{ox}$ | 0.79 | 1.16 | 0.79 | 1.16 | 0.81 | 1.17 | 0.79 | 1.17 |
| Pitch | 3.0 | 3.0 | 2.0 | 2.0 | 3.0 | 3.0 | 2.0 | 2.0 |
| Aspect ratio | 0.54 | 0.86 | 0.71 | 1.10 | 0.55 | 0.87 | 0.72 | 1.09 |

All measurements were in micrometers and obtained from SEM cross section of the test wafer.
*Source*: Ref. 33.

the structure were used for comparisons in this study. This is because the side test patterns were affected by unexpected photolithography problems.

In Figure 32, the step resistance acquired from the center of the wafer was plotted against the relative aspect ratio. The *relative aspect ratio* is defined as aspect ratio divided by pitch ratio, where the pitch ratio is the test structure pitch divided by the reference pitch. The reference pitch used here is 2 μm. The relative aspect ratio is introduced to compensate the line/space pair density difference caused by different line/space pair pitch. Figure 32 shows that the step resistance is a function of metal thickness, aspect ratio, and line/space pair pitch. Since the step resistance measurement results are an overall measurement of the metal conformity on the test structure, it is very difficult to describe the exact relationship between the step resistance and critical elements of the metal step coverage, such as side wall, bottom fill, and cusping. The test wafers with better metal sidewall coverage and bottom fill have lower step resistance (Fig. 32, points A and B). In contrast, the wafers with the highest aspect ratio have the highest step resistance within their groups (Fig. 32, points D and H). The longer deposition time (thicker metal film Fig. 32D) provided very limited improvement in side wall thickness and bottom fill thickness, due to the closing of the gap between two metal cuspings, which virtually prevent metal deposition inside the trench and resulted in higher step resistance.

Figure 33 illustrates the relative location of a used sputter target erosion profile and the step resistance of a test wafer processed prior to the target change. The large variation in the step resistance curve (approximately 13 mm from the center) coincides with the location on the target with the maximum erosion. This example demonstrates that the across the water metal step coverage can easily be measured with this technique.

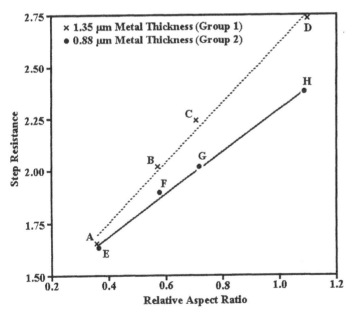

**Figure 32** Relative resistance, or step resistance as it is called here, can be a useful aid in monitoring step coverage. (From Ref. 23.)

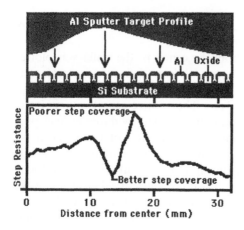

**Figure 33** Here the step resistance is used to identify an area of poor step coverage due to an eroded track in a sputtering target. (From Ref. 23.)

As an example of the application of the step resistance value, a retrograde structure was used. Poor metal step coverage over the retrograde polysilicon sidewall (Figure 34a) resulted in higher step resistance (step resistance = 1.746). Better metal step coverage over the normal polysilicon sidewall (Figure 34b) resulted in lower step resistance (step resistance = 1.573).

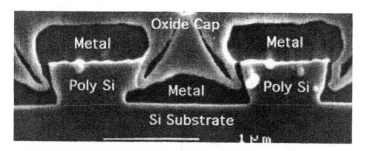

A.

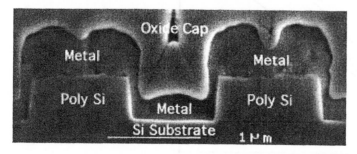

B

**Figure 34** Poor metal step coverage (A) over the retrograde polysilicon resulted in higher step resistance (step resistance = 1.746). Better metal step coverage over the normal polysilicon sidewall (B) resulted in lower step resistance (step resistance = 1.573). (From Ref. 23.)

## XII.  DETERMINING QUALITY READINGS FROM A FOUR-POINT PROBE

In order to obtain the best possible ratio of measurement precision to process tolerance ($P/T$) it is important to monitor the system noise or variation. Because the four-point probe technique is destructive by nature and the probe head wears with use, its repeatability can vary substantially depending on the application. One of the main sources of noise in the sheet resistance measurement by four-point probe is the probe contact to the sample (24). The accumulation of metal on the tip surface and subsequent oxidation is the most likely cause of contact deterioration for metal measurements. To get accurate and repeatable results, the probe-head pins must make good electrical contact each time they touch the surface of the film. A special probe qualification procedure (described next) was devised to check the short-term repeatability of a probe head.

For example, if a probe used to measure a process with a uniformity spec of 0.75% has a short-term repeatability (noise) of 0.5%, and the true wafer uniformity is 0.65%, then the net result is

$$\sigma = (\sigma_{water}^2 + \sigma_{probe}{}^2)^{1/2} = (0.65^2 + 0.5^2)^{1/2} = 0.82\%$$

where $\sigma$ = resulting standard deviation, $\sigma_{water}$ = real standard deviation of the water uniformity, and $\sigma_{probe}$ = probe qualification standard deviation. The standard deviation of 0.82% is larger than the uniformity specification and therefore would be out of specification.

From this example we may conclude that the tighter the process monitor specification, the smaller should be the allowable noise induced by the probe head. As the probe repeatability degrades, the quality of the maps degrades first. The second parameter to degrade is the standard deviation; the last is the average sheet resistance. In most cases (all that the author has encountered), the sheet resistance value will drop as the probe degrades.

In much the same manner, the short-term repeatability will affect the measurement precision when added to the reproducibility:

$$\sigma = (\sigma_{repeatability}^2 + \sigma_{reproducibility}{}^2)^{1/2}$$

where $\sigma$ is the measurement precision, $\sigma_{repeatability}$ is the variation of repeated measurements made under identical conditions, and $\sigma_{reproducibility}$ is the variation that results when measurements are made under different conditions.

To be sure of the continued quality of the measurements and maps, check the probe repeatability regularly. If a fab runs three shifts and the FPP system is in constant use, check probe repeatability at the beginning of each shift.

It is also essential to use the right probe head for each application (24). Metals, with the exception of ultrathin films, are not generally sensitive to probe type. Probe repeatability is best checked on the process of interest. Sometimes monitor wafers specially chosen for this purpose are used. Depending on the process monitored, it is essential to keep a few monitor wafers aside. These backup monitor wafers can be used when the normal process monitor measurements indicate an out-of-limits value and it is not known if the wafer or probe is the source of the error. A common probe qualification procedure consists of five measurements, each $1/4°$ apart. This grouping of five measurements is repeated at each of the four sites. A good rule of thumb is to try to maintain the standard deviation (STDV) of the $R_S$ measurements within each site at 0.2%. This procedure was deigned to take into account the damage caused by the pin contacts. Of course, the ability to make good contact to the surface is highly application sensitive and may not be obtainable for difficult layers or for cases in which the probe head is not the proper type for the layers being measured. This approach also assumes that the true sheet resistance variation over the one-degree arc is negligible compared to the probe contact contribution. Repeated measurements in the same area can cause the sheet resistance and repeatability to increase due to the accumulated damage to the film from prior measurements.

## XIII.   MAINTAINING QUALITY READINGS

When a qualification test fails and the probe head is suspected as the cause of the problem, a conditioning process should be followed. The intent of this process is to clean and resurface the tips of the probe head. Several materials have been used for the purposes of conditioning probe tips, including diamond paste, sapphire substrates, alumina substrates, and even the backside of the silicon wafer. The alumina substrate has gained the most widespread use.

### A.   Checking Repeatability to Determine the Probe Count

If the standard deviation is greater than 0.2% but less than 0.5%: Probe about 25 times on a ceramic substrate as described shortly, then run the probe qualification procedure twice. The second set of STDV values should be within specification. If the standard deviation is larger than 0.5% or does not meet the desired specification after a conditioning attempt, increase the conditioning count.

### B.   Probe-Conditioning Routine

The manufacturer of the system should provide a routine to condition the probe pins in order to refinish the tips and clean off any loose debris. Although it might provide more consistent results to remove the probe head from the system and possibly the pins from the probe, this is not a very practical procedure, in that it could take several hours or a couple of days to disassemble and condition (polish) the tips and reassemble the probe.

## XIV.   CONCLUSIONS

There are two main methods for measuring sheet resistance of interconnect film, the four-point probe and eddy current. Each has its own advantage. The four-point probe is a direct measurement and covers a very large range of sheet resistances. The source of difficulty with this method stems from the requirement for a good ohmic contact to the film. Although this method is destructive, many films can be measured multiple times before the damage affects the sheet resistance value to substantial level. The eddy current method is nondestructive and allows for repeated measurements but generally has a more limited measurement range and cannot easily be separated from any substrate signal. Once quality measurements have been ensured, sheet resistance values can provide valuable insight to the thickness and resistance of an interconnect film.

## REFERENCES

1. D.D. Pollock, ed. Physical Properties of Materials for Engineers. 2nd ed. Boca Raton, FL: CRC Press, 1993, pp 128–130.
2. R.C. Eeast, M.J. Astle, eds. CRC handbook of Chemistry and Physics. 63rd ed. Boca Raton, FL: CRC Press, 1982–1983, p E-81.
3. L.I. Maissel, R. Glang, eds. Handbook of Thin Film Technology. New York: McGraw-Hill, 1983, p 13-7.
4. D.K. Schroder. Semiconductor Materials and Device Characterization. New York: Wiley, 1990, pp 1–40.
5. D.S. Perloff. J. Electrochem. Soc. 123:1745–1750, 1976.
6. Patent pending.
7. K. Urbanek, G.J. Kren, W.R. Wheeler. Non-Contacting Resistivity Instrument with Structurally Related and Distance Measuring Transducers. U.S. Patent 4,302,721, Nov. 1981.
8. C.L. Mallory, W.H. Johnson, K.L. Lehman. Eddy Current Test Method and Apparatus for Measuring Conductance by Determining Intersection of Liftoff and Selected Curves. U.S. Patent 5,552,704, Sept. 1996.
9. ASTM. Standard method for measuring resistivity of silicon slices with a collinear four-probe array. Annual Book of ASTM Standards. Vol. 10.05. ASTM F84.
10. D.S. Perloff, J.N. Gan, F.E. Wahl. Dose accuracy and doping uniformity of ion implantation equipment. Solid State Technol. 24(2):Feb. 1981.
11. L.J. Swartzendruber. Correction Factor Tables for Four-Point Probe Resistivity Measurements on Thin, Circular Semiconductor Samples. Technical Note 199. April 1964.
12. G.L. Miller, D.A.H. Robinson, J.D. Wiley. Method for the Noncontacting Measurement of the Electrical Conductivity of a Lamella. U.S. patent 4,000,458, Dec. 1976.
13. J. Vine. Impedance of a Coil Placed Near a Conductive Sheet. J. Electron. Control 16:569–577, 1964.
14. H.E. Burke. Handbook of Magnetic Phenomena. New York: Van Nostrand Reinhold, 1986, p 183.
15. ASM. ASM Handbook, Nondestructive Evaluation and Quality Control. Vol. 17, 1992, p 169.
16. J.M. Davis, M. King. Mathematical Formulas and References for Nondestructive Testing. Itasca, IL: The Art Room Corp., 1994, p 5.
17. W.H. Johnson, B. Brennan. Thin Film Solids 270:467–471, 1995.
18. W.H. Johnson, C. Hong, V. Becnel. Application of electrical step resistance measurement technique for ULSI/VLSI process characterization. Proceedings of the Int. Conf. on Characterization and Metrology for ULSI Technology, 1998, pp 321–325.
19. E.J. Gilliam, J.S. Preston, B.E. Williams. Phil Mag. 46:1051, 1955.
20. Q.T. Jiang, R. Mikkola, B. Carpenter. Conference Proceedings ULSI XIV Mat. Res. Soc., 1999.
21. W.H. Johnson, C. Hayzelden. SEMI CMP Technology for ULSI Interconnection, SEMI, 1999, sec. I.
22. S. Koichiro, K. Hitoshi, H. Takeshi. Electrochem. Solid-State Lett. 2(6):300–302, 1999.
23. W.H. Johnson, C. Hong, V. Becnel. Step coverage measurements using a non-contact sheet resistance probe. VMIC Conf., June 1997, pp 198–200.
24. W.A. Keenan, W.H. Johnson, A.K. Smith. Advances in sheet resistance measurements for ion implant monitoring. Solid State Technol. 28(6): June 1985.
25. W.H. Johnson (unpublished data).

# 12

# Characterization of Low-Dielectric Constant Materials

**Michael Kiene, Michael Morgen, Jie-Hua Zhao, Chuan Hu, Taiheui Cho, and Paul S. Ho**
*Microelectronics Research Center, University of Texas, Austin, Texas*

## I. INTRODUCTION

It is widely accepted that the continuing decrease in feature sizes and consequent increase in performance of advanced microelectronics products will be limited by stray capacitances and circuit delays in the metal interconnect structure of these devices. The resistive delay, as well as the power dissipation, can be reduced by replacing the standard Al(Cu) interconnect metal with lower-resistivity Cu wiring. Likewise, interline capacitance and crosstalk can be minimized by the use of low-dielectric-constant (low-$k$) materials in place of the standard $SiO_2$ ($k = 3.9$–$4.3$) interlayer dielectric (ILD). The industry is well on its way to implementing the change to copper metallization (1,2), and first implementations of low-$k$ ILDs have been announced (3,4). The *1999 International Technology Roadmap* (5) has identified two generations of low-dielectric-constant materials to be integrated in future high-density interconnects. The first generation will have dielectric constants of 2.7–3.5, (2001, 2002), while for the second generation the dielectric constant has to be reduced to 2.7–2.2 (2003, 2004).

The interconnect structure for Cu metallization is fabricated using the damascene process, where Cu is deposited into wiring channels patterned into the ILD layer and then planarized using chemical-mechanical polishing (CMP). The damascene structure introduces a new set of materials and processes distinctly different from the standard AlCu interconnect technology, making the implementation of low-$k$ dielectrics a challenging task. At this time, low-$k$ integration is proceeding with a number of candidate materials. In spite of the intense effort, no clearly superior dielectric with $k < 3$ has emerged. The lack of a clear choice of low-$k$ dielectric can be largely attributed to the many challenges associated with the successful integration of these materials into future on-chip interconnects. In addition to low dielectric constant, candidate intra- and interlevel dielectrics must satisfy a large number of diverse requirements in order to be successfully integrated. The requirements include high thermal and mechanical stability, good adhesion to the other interconnect materials, resistance to processing chemicals, low moisture absorption, and low cost (6). In recent years there have been widespread efforts to develop low-$k$ materials that can simultaneously satisfy all of these requirements. A particularly difficult challenge for material development has been to obtain the combination of low dielectric constant and good thermal and mechanical stability. Generally, the types of chemical structures that impart structural stability are those having strong individual bonds and a high density of such bonds. However, the strongest bonds often are the most polarizable, and increasing the bond density gives a corresponding increase in polarization.

Given the variety of low-$k$ candidates, an effective selection scheme for materials prior to process integration tests becomes necessary. The scaling of layer thickness with device dimensions and the often-inferior mechanical properties of low-$k$ materials present challenges for the characterization metrology itself. Thinner layers require increased sensitivity for many of the techniques used, and surface- and interface-related effects, such as substrate confinement, layer homogeneity, and surface contamination, become more and more important. Because many materials are substantially softer than $SiO_2$ or more prone to cracking, material handling often becomes extremely difficult, in many cases, free-standing films, required for many standard characterization techniques, cannot be prepared at all.

Since it is difficult to reduce the dielectric constant below 2.5 with fully dense materials, it may be necessary to introduce micro- or mesoporosity to achieve very low-$k$ values ($\leq 2.0$). *Mesoporous* typically encompasses materials having voids 2–50 nm in diameter, while *microporous* is used for materials containing voids equal to or less than 2 nm in diameter. The introduction of voids decreases the dielectric constant by lowering the density of the material. In principle, one can vary the percentage of the porosity and therefore the material density and dielectric constant. Because the introduction of voids will compromise other material properties, such as mechanical strength, the total porosity and the pore size distribution must be carefully characterized in addition to the chemistry and structure of the matrix.

In this chapter, the most important material properties and some requirements for process integration of low-$k$ dielectrics are first discussed. This is followed by a description of material characterization techniques, including several recently developed for porous materials. Finally, the material characteristics of selected candidate low-$k$ dielectrics will be discussed to illustrate their structure–property relations.

## A.  Critical Properties for Low-$k$ Integration

Interlevel dielectric materials need to meet stringent material property requirements for successful integration into the conventional or damascene interconnect structures. These requirements are based on electrical properties, thermal stability, thermomechanical and thermal stress properties, and chemical stability. The desired electrical properties are low dielectric constant, low dielectric loss and leakage current, and high breakdown voltage, but $RC$ delay and crosstalk are determined primarily by the dielectric constant, which is $\sim 4$ for a typical CVD $SiO_2$ film. While many polymeric materials satisfy these electrical criteria, the dimensional stability, thermal and chemical stability, mechanical strength, and thermal conductivity of polymers are inferior to those of $SiO_2$.

The fabrication of multilevel structures requires as many as 10–15 temperature excursions, which may reach 400–425°C. Therefore, thermal cycling during processing requires future ILD materials to have high thermal stability and properties that are relatively insensitive to thermal history. Outgassing of volatiles due to decomposition or to trapped solvent or reaction products can cause via poisoning and delamination and blistering in the ILD. Changes in the degree of crystallinity or crystalline phases lead to properties dependent on thermal history. Thermal cycling also causes stresses in the interconnect structure due to the coefficient of thermal expansion (CTE) mismatch between the ILD material and the metal or substrate. Such stresses may cause delamination if adhesion is poor. Adhesion may be improved with an adhesion promoter to enhance wetting and chemical bonding at the interface, but this is undesirable for a manufacturing process since it adds processing steps. Also, if the adhesion promoter thermally degrades, it may lead to adhesion failures or create a leakage path.

Adhesion strength is determined by chemical bonding at the metal/ILD interface and the mechanical interaction between the metal and the ILD. Thus, future ILDs should ideally have good mechanical properties, such as a large Young's modulus ($E$), tensile strength, and elongation at break, although it is not yet clear what constitutes sufficient mechanical strength for successful integration into a manufacturable process. The elongation at break should be as large as possible to sustain the deformation and impart crack resistance. A high modulus retention at elevated temperatures, $E(T)$, is required for the ILD to maintain its structural integrity and dimensional stability during subsequent processing steps. Related to $E(T)$ is the glass transition temperature, $T_g$. Exceeding $T_g$ causes a large decrease in the modulus and yield stress in amorphous, noncrosslinked polymers, so a $T_g$ greater than or equal to the highest processing temperature is desired. For example, residual compressive stresses in capping layers can cause buckling and delamination of the capping films due to the compliance of an ILD above its $T_g$ (7,8). Buckling has also been observed in capping layers deposited below the ILD's $T_g$ if the capping film is highly compressive (9).

Other processing concerns include chemical resistance to the solvents and etchants commonly used during chip fabrication, chemical interaction with the metal lines that causes corrosion, and moisture uptake. Moisture is a primary concern since a small amount can have a large impact on the dielectric constant. The ILDs should also be free of trace metal contaminants, have long shelflives, and, preferably, not require refrigeration. Metal contamination, which can compromise the device and provide a leakage path between lines, is often a problem for polymers synthesized using metal catalysts. Other processing issues include the ability to pattern and etch the film, etch selectivity to resist, and good thickness uniformity.

The long-term reliability of chips fabricated using low-$k$ materials must also be evaluated. Electromigration and stress voiding are primary failure mechanisms in integrated circuits (10–12), and these are reliability concerns when replacing $SiO_2$ with an alternative ILD that has thermal and mechanical properties inferior to those of $SiO_2$. Discussions of these topics are beyond the scope of this chapter, but we do note that both processes have Arrhenius-like temperature dependencies and are, therefore, accelerated by Joule heating. Thus, Joule heating effects necessitate an ILD with a sufficiently large thermal conductivity to remove heat from the metal lines (13,14), although alternative packaging designs and integration architectures, such as an embedded insertion scheme, have been proposed to partially mitigate this problem (15).

The thermal conductivity of most organic polymeric materials is typically about four to five times smaller than that of PE-TEOS ($K_{th} = 1.1 \, \mathrm{W \, m^{-1} \, K^{-1}}$) (16–20), making Joule heating effects in a polymeric ILD a major reliability concern. Also, since the mechanical strength of polymers is substantially less than that of $SiO_2$, the polymer passivation will not suppress hillock formation in the metal lines as effectively as a rigid, encapsulating $SiO_2$ layer. While this has led to shorter electromigration lifetimes in Al lines in some studies (21–23), others have shown larger electromigration lifetimes due to morphology changes in the metal lines during the polymer cure (24,25). Clearly, more work is required to fully understand the impact of low-$k$ material integration on interconnect reliability.

## B.  Screening Analysis of Low-*k* Candidates

Because a final evaluation of a low-dielectric-constant material for process integration involves extremely time- and cost-intensive fabrication of one and multilevel Cu damascene test structures, an effective screening of materials properties is important to select materials for further evaluation. The following is a list of properties and techniques for materials screening to identify promising materials. All measurements can be performed for 1 µm-thick blanket films, requiring a minimum of sample processing. Although some of the techniques employ relatively advanced equipment, they allow evaluation of a large number of new materials in a relatively short period.

> Dielectric constant (1 MHz): MIS capacitance
> Thickness uniformity and anisotropy: prism coupling
> Thermal stability: TGA
> Moisture uptake: quartz crystal microbalance or SANS
> Adhesion to $SiO_2$, $Si_3N_4$, Ta, TaN: m-ELT
> Young's modulus: dual-substrate bending beam
> CTE: dual-substrate bending beam
> Chemical signature and chemical resistance: FTIR
> Density, porosity, and pore structure: SXR, SANS, PALS

## II.  CHARACTERIZATION TECHNIQUES

Since ILD materials will need to be used as thin films ($\leq 1$ µm thick), and since thin film properties can differ appreciably from bulk properties or even from thick-film properties for a given material, it is important to have characterization techniques applicable to thin-films. In fact, many materials, such as organosilicate glasses (OSGs) and porous silica, cannot even be prepared in thicknesses much greater than about 1 µm without extensive cracking. Because it is difficult to remove such thin films from a substrate for free-standing film measurements, it is usually necessary to perform thin-film characterization using on-wafer techniques. As a result, there has been a great deal of work done to develop material-testing methods capable of measuring the properties of very thin films on-wafer. Some of those will be described later. Table 1 lists a number of methods used for characterization of thin dielectric films that will not be discussed in further detail. The reader is referred to textbooks or review articles cited in the table.

## A.  Adhesion

Adhesion of ILDs to the surrounding materials is one of the most critical issues for process integration. Although a wide variety of adhesion tests exist, the correlation of test data to the actual integration requirements remains difficult. The measured adhesion strength consists of contributions from several mechanisms, such as intrinsic adhesion, plastic deformation, and interface roughness, and depends strongly on sample and load geometry as well as on the test environment (26–28). Because debond driving force and locus of failure are difficult to determine in a complex interconnect structure, no clear choice for an adhesion test has evolved. Even comparisons among adhesion strengths from different methods or between materials with different plasticity, e.g., polymers and OSGs are difficult. Among the commonly used methods are the dual cantilever beam technique, the four-point bend technique, stud pull-and-peel tests (29).

**Table 1** Characterization Methods for Thin-Film Dielectrics

| Property | Methodology (refs.) |
|---|---|
| Refractive indices (in-plane and out-of-plane) | Prism coupling; optical measurement $\lambda = 632.8$ nm (95); ellipsometry (53,73). |
| Dielectric constant (out-of-plane) | Metal–insulator–semiconductor (MIS) or metal–insulator–metal (MIM) parallel-plate capacitance measurement; typ. 1 MHz |
| Dielectric constant (in-plane) | Calculated based on the measured in-plane refractive index; measured using line structures and capacitance simulation |
| Dielectric breakdown voltage | MIS or MIM capacitors |
| Leakage current | MIS or MIM capacitors, or serpentine line structure |
| Thermal stability | Isothermal thermogravimetric analysis (TGA) at 350 and 420°C for 8 h (96,97); change in thickness after 8 h anneal at 350 and 425°C. |
| Chemical resistance | Treatment with chemicals, including: boiling water, NMP, BOE, 2% NaOH, and photoresist stripper; measured are chemical signature (FTIR), changes in thickness, and adhesion to the Si wafer |
| Adhesion tests | Scotch tape pull, ASTM standard testing protocol |
| Lateral thermal expansion coefficient | Thermal-mechanical analysis (TMA); freestanding films with a thickness of 5–10 μm (96,97) |
| Vertical thermal expansion coefficient | Home-built differential capacitance system; measurements made on a Cu/dielectric/Si structure with a 0.2°C/min ( ). heating rate; film thickness: 5–20 μm |
| Glass transition temperature | TMA and/or thermal stress measurement (96,97) |
| Thermal stress | Wafer curvature or bending beam measurement (37,38) |
| Tensile strength (MPa) | Stress–strain characterization performed using a custom-built micro–tensile tester, 5-μm-thick freestanding films (98) |
| Young's modulus (GPa) | Stress–strain measurement using microtensile tester, 5-μm-thick freestanding films (98); nanoindentation, on wafter (99) |
| Young's modulus as a function of temperature | Dynamic mechanical analysis (DMTA), > 5-μm-thick, freestanding films (96,97) |
| Chemical signature | FTIR (100,101), NMR ( ), XPS ( ) |
| Phase transformations | Different scanning calorimetry (DSC) (96,97) |
| Wide-angle x-ray diffraction | Crystallinity as a function of temperature ( ) |

The m-ELT developed by Ed Shaffer (30,31) of Dow Chemical Company was adopted by our laboratory to quantify the adhesion strength of dielectric films to various substrates. The test involves applying a thick (50–200 μm) layer of epoxy to the top of the film and inducing failure by lowering the temperature of the sample until the stored strain energy in the epoxy layer exceeds the adhesion energy of one of the interfaces. From the temperature at which debonding is observed by optical inspection, the fracture toughness can be calculated, if the stress temperature profile of the epoxy layer is known. After failure, the sample is examined to determine the interface at which debonding occurred. This method allows one to measure the adhesion strength of a material to various substrates or capping layers using blanket films with a minimum of sample preparation.

## B.  Moisture Uptake

On-substrate moisture absorption experiments for thin-film ILD materials are difficult, because direct gravimetric methods cannot be applied due to the extremely small mass of the dielectric films as compared to the substrate mass.

One method for measuring moisture uptake is to perform a SANS analysis as described in the next paragraph on a sample immersed in deuterated water. The difference in SANS intensity as compared to a dry sample can be converted into a volume fraction of the absorbed water (32).

In our lab the amount of moisture absorbed by low-$k$ dielectric candidate materials is measured using a quartz crystal microbalance (QCM). In the QCM experiment, a quartz crystal is coated with a thin film of the ILD material and the resonance frequency of the crystal is measured while exposing the specimen to water vapor. Changes in resonance frequency can be correlated with the mass of the absorbed moisture using the Sauerbrey (33) equation:

$$\Delta m = \frac{\sqrt{\rho_Q \mu_Q}}{2} \left( \frac{\Delta f}{f_0^2} \right) \tag{1}$$

where $\rho_Q$ is the density of quartz and $\mu_Q$ is the shear modulus of quartz.

Because this technique relies on the measurement of the resonance frequency (here $\approx 6$ Mhz) of a quartz oscillator, the required resolution of $< 0.1$ weight % is easily achieved. However, one of the main disadvantages of this technique is that the material has to be coated onto quartz oscillators. Typical oscillators are $\frac{1}{2}$-in. quartz disks, with gold electrodes.

To obtain the moisture uptake relative to the mass of the dielectric film, the resonance frequency of a blank quartz crystal, $f_0$, is measured, and then the crystal is coated with an ILD film. By measuring the resonance frequency for the dry film, the mass of the coating is determined. Then the chamber is back-filled with water vapor from a constant-temperature vapor source, and the frequency of the crystal with the ILD material is monitored as a function of time until saturation is observed. By subsequent evacuation of the chamber and back-filling with water vapor, the total moisture uptake and its time dependency can be measured. Assuming a Fickian diffusion mechanism, the initial moisture uptake can be quantified as

$$M_t / M_\infty = 2 \left( \frac{D}{\pi} \right)^{1/2} \left( \frac{t^{1/2}}{l} \right) \tag{2}$$

where $M_t$ is the mass change at time $t$, $M_\infty$ is the mass change at saturation, $l$ is the film thickness, and $D$ is the diffusion coefficient (cm$^2$/s). Linear regression of the initial moisture uptake data yields a line with a slope $2(D/\pi)^{1/2}$, from which the diffusion coefficient can be calculated.

## C.  Dual-Substrate Bending Beam

Traditionally, the modulus of polymer thin films has been determined by stress–strain curves obtained from microtensile testing of freestanding films. This type of experiment normally requires films at least 3–5 microns thick. In order to make measurements on materials that cannot be prepared in freestanding form, because they are either too thin or too brittle, wafer curvature methods have been developed to obtain the modulus from thermal stress measurements. The substrate curvature as a function of temperature is normally measured optically by monitoring the reflection of a laser off the wafer. The stress is calculated from the curvature using Stoney's equation (34) :

$$\sigma = \frac{M_s t_s^2}{6 R t_f} \tag{3}$$

where $M_s$ is the biaxial modulus of the substrate ($M \cong E/1 - \nu$, where $\nu$ is the Poisson ratio), $t_s$ is the substrate thickness, $t_f$ is the film thickness, and $R$ is the radius of curvature. A discussion of the limits of the Stoney equation have been described in the literature (35,36). Generally, the relationship of Eq. (3) is applicable to thin films on thicker, rigid substrates.

In our laboratory, we use a bending beam variation of the curvature method (37,38), as shown in Figure 1. The curvature of a narrow beam of a rigid substrate (3 mm $\times$ 48 mm) due to the thermal stress of a thin ($\sim$ 1-$\mu$m) film coated on top is optically measured. We use thin substrates (125–250 $\mu$m) to amplify the curvature change that occurs with thermal cycling, and we monitor changes in curvature using position-sensitive detectors. A long light path (1–4 m) offers improved sensitivity compared to commercial instruments, and allows measurement of the relatively small stresses (5–50 MPa) present in low-$k$ films.

The slope of the stress vs. temperature curve due to the CTE mismatch between a film and a rigid underlying substrate, such as Si, is

$$k = \frac{d\sigma}{dT} = M_f(\alpha_s - \alpha_s) \tag{4}$$

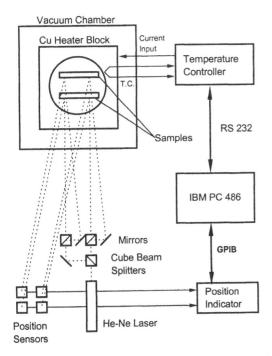

**Figure 1** Bending beam system for measuring thermal stress of thin films. Two beams can be measured simultaneously.

where $\alpha_s$ and $\alpha_f$ denote the substrate and film CTE, respectively, and $M_f$ is the film's biaxial modulus. By measuring the slope of the stress curve for film samples deposited on two substrates having different CTEs, one obtains two equations that can be solved simultaneously to determine both $M_f$ and $\alpha_f$ (39):

$$a_f = \frac{k_2\alpha_{s1} - k\alpha_{s2}}{k_2 - k_1} \tag{5}$$

$$M_f = \frac{k_2 - k_1}{\alpha_{s2} - \alpha_{s1}} \tag{6}$$

where $k_{1,2}$ are the stress vs. temperature slopes on the two different substrates and $\alpha_{1,2}$ are the associated substrate CTEs. In our lab, Ge (CTE 5.8 ppm/°C) has been used as a substrate in conjunction with Si (CTE 2.6 ppm/°C) (40).

It should be noted that high precision and accuracy in the slope measurement are needed for a reliable measurement, particularly if the difference in slopes or in substrate CTE is small. Figure 2 shows the intervals where the calculated CTE and biaxial modulus have errors of 10% or less, using our current setup and data for different materials that have been measured in our lab. It should be noted that, the method is still applicable beyond the intervals shown in Fig. 2, but extra care has to be taken to enhance the precision of the slope measurement. This can be achieved by taking data over an extended temperature range or by enhancing the sensitivity of the instrument by enlarging the light path. However, the repeatability and precision of the slopes have to be checked carefully. In data analysis, the temperature dependence of the various parameters in Eqs. (4)–(6) should be accounted for, or, if possible, the stress measurement should be taken over only a limited temperature range to minimize the variation of these parameters. In any case the CTE and modulus determined are implicitly average (temperature-independent) values. The technique has been used successfully in our laboratory to measure the properties of

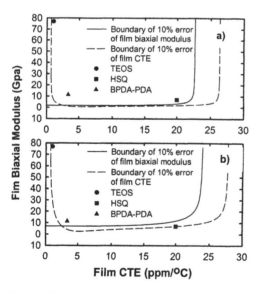

**Figure 2** Applicable regions of dual-substrate bending beam method for 1-μm-thick films. (a) 125-μm-thick Si and Ge, (b) 250-μm-thick Si and 300-μm-thick GaAs. Plot assumes 0.5% error in film, substrate thickness, and substrate modulus and a CTE, curvature error of $2.1 \times 10^{-4}$ (1/m)

polymers, OSGs, and several porous silica–based films. Results for HSQ (38) are shown in Figure 3.

## D.  $3\omega$ Method for Thermal Conductivity

The thermal conductivity ($K_{th}$) of the ILD is important, because if not enough heat is transported away from the interconnect, Joule heating can cause temperature-induced reliability problems. Because most low-$k$ materials are expected to have a significantly lower $K_{th}$ than TEOS-deposited oxide ($K_{th} = 1.2$–$1.4$ W/m°C) (41,42), measurement of $K_{th}$ is prudent to estimate the severity of potential problems associated with heat dissipation.

In our laboratory we have developed a variation of the $3\omega$ technique (41,43–45), for measuring $K_{th}$. This technique determines the thermal conductivity by measuring the thermoelectrical response of a thin-film conductor patterned on the dielectric film. The experimental setup of the $3\omega$ method used is schematically shown in Figure 4. The sample was formed using a dielectric film deposited on a Si substrate, and a metallic heater element was patterned on top of the dielectric film by a lithographic technique. The heater element was processed as a 300-nm-thick Al line structure with a 5-nm Cr adhesion promoter underneath. To minimize the edge effect at the end of heater, the length of the heater, typically 3 mm, was chosen to be about 100 times larger than its width (30 μm). The heater, which served also as a thermometer to sense the increase in temperature, was incorporated as one arm of a Wheatstone bridge circuit. Three other resistors completed the circuit, which were chosen to have a low temperature coefficient of resistivity in order to minimize their contributions to the $3\omega$ signal. When a current at fre-

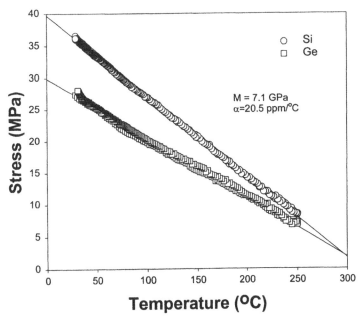

**Figure 3**  Thermal stress data from dual-substrate bending beam measurement of 0.5-mm-thick HSQ films on Ge and Si substrates. The biaxial modulus and TEC were determined to be 7.1 GPa and 20.5 ppm/°C, respectively.

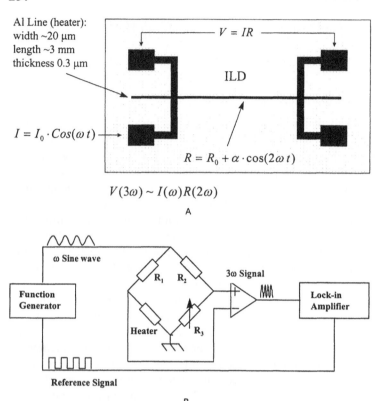

Al Line (heater):
width ~20 μm
length ~3 mm
thickness 0.3 μm

$V = IR$

ILD

$I = I_0 \cdot Cos(\omega t)$

$R = R_0 + \alpha \cdot \cos(2\omega t)$

$V(3\omega) \sim I(\omega)R(2\omega)$

A

ω Sine wave

Function
Generator

$R_1$  $R_2$

3ω Signal

Heater  $R_3$

Lock-in
Amplifier

Reference Signal

B

**Figure 4** (A) Top view of 3ω sample showing the metal line structure. (B) Experimental configuration for 3ω measurement. The sample (heater) is incorporated into a Wheatstone bridge to
eliminate noise, while the signal is isolated using a lock-in amplifier.

quency ω is applied to the heater element, the Joule heating causes a small increase in
resistance modulating at a frequency of 2ω:

$$R = R_0 + \alpha \, \Delta T \cos(2\omega t + \phi) \tag{7}$$

where α is the temperature coefficient of the electrical resistivity, $\Delta T$ is the change in
temperature, and $R_0$ is the resistance at a reference temperature. By measuring the voltage
drop $I(\omega)R(2\omega)$ across the heater element, there is a small 3ω component directly corresponding to the temperature increase:

$$V = I \cdot R = I_0 R_0 \cos(\omega t) + I_0 \alpha \, \Delta T \cos(\omega t) \cos(2\omega t + \phi) \tag{8}$$

$$= I_0 R_0 \cos(\omega t) + \tfrac{1}{2} I_0 \alpha \, \Delta T \cos(\omega t + \phi) + \tfrac{1}{2} I_0 \alpha \, \Delta T \cos(3\omega t + \phi) \tag{9}$$

By measuring the 3ω amplitude $\tfrac{1}{2}\alpha\Delta T \, I_0$, we can determine the temperature change of the
heater as a function of the frequency ω. The temperature coefficient of electrical resistivity
is measured separately for each sample to calibrate the temperature measurement. To
extract the 3ω component, an operational amplifier is coupled with lock-in amplification
to achieve the precision required.

The experiment is performed as a function of input frequency. The sensitivity of the
technique results from measuring only the incremental resistance rise rather than the much
larger total resistance of the line. Also, the use of a small line makes the line resistance a

sensitive function of temperature, so small temperature changes ($< 0.1°C$) can be readily measured. As a result, the thermal conductivity of very thin films ($\sim$ 50 nm) having only a small temperature gradient can be measured. The measured thermal resistance $R_{th}$,

$$R_{th} = \frac{\Delta T}{P} \tag{10}$$

where $P$ is applied electrical power, has two contributions: that from the Si substrate and that from the ILD film. The film contribution can be isolated by measuring $R_{th}$ of a line deposited on bare Si (with its thin native oxide layer for insulation) and subtracting it from the measurement of the ILD-on-Si sample.

Normally, the film thickness and modulation frequency ($3\omega$) are chosen so that the thermal diffusion length inside the film ($L_{th}$),

$$L_{th} = \sqrt{\frac{\Lambda}{\omega}} \tag{11}$$

where $\Lambda$ is thermal diffusivity, is much larger than the film thickness. For films of about 1-μm thickness this is usually satisfied if the frequency is chosen below 1kHz. In this case the thermal conductivity of the film is easily calculated from that part of the thermal resistance contributed by the film ($R_{th}^{f}$):

$$K_{th} = \frac{t}{R_{th}^{f} l w} \tag{12}$$

where $t$ is the thickness of the film, $l$ is the length of the heater, and $w$ is the width of the heater. Under these conditions $R_{th}^{f}$ is frequency independent. The calculation leading to Eq. (12) assumes a 1D vertical thermal flow, which should be a good approximation when the film thickness is small compared to both the linewidth and the thermal diffusion length in the film, $L_{th}$, $w \gg t$. Also, the line-end effects are ignored, which is reasonable when $l \gg w$. These assumptions have been verified by finite element analysis, showing that our geometry corrections for 2D heat flow are negligible for thermal conductivities greater than 0.06W/mK (41).

We have recently made $K_{th}$ measurements on a number of ILD films (Table 2). The result for TEOS is in good agreement with measurements of $SiO_2$ made by Refs. 42 and 47, and the BPDA-PDA value agrees well with a photothermal measurement made in our laboratory (48). The $K_{th}$ values for xerogel samples suggest that the conductivity decreases quickly as a function of porosity. The thermal resistance data for 48% porous xerogel is shown in Figure 5. Both the in-phase and out-of-phase components are shown. The in-

**Table 2**  Thermal Conductivities of Low-$k$ Thin Films ($\sim$ 1 micron)

| Material | $k_{th}$ (W/m°C) |
| --- | --- |
| Xerogel[a] 48% porous | 0.250 |
| 63% | 0.157 |
| 70% | 0.108 |
| 77% | 0.065 |
| BPDA-PDA | 0.209 |
| TEOS | 1.26 |

[a] Porosity determined by RBS.

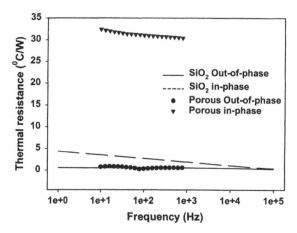

**Figure 5** Experimental 3ω data for in-planar and out-of-phase thermal resistance of a ~ 1-mm-thick 48% porous xerogel film. The calculated thermal resistance for the substrate and a SiO₂ which is in agreement with experimental measurements made in our laboratory. Note that the thermal conductivity is calculated from the in-phase component of the response.

phase component measures the conductivity, while the out-of-phase component is due to dissipation. When $L_{th} \gg t$, there is negligible dissipation in the film, and the out-of-phase component of the 3ω signal of the film goes to zero. (There *is* dissipation in the substrate, but the substrate contribution is subtracted from the data). Also shown in the figure for comparison are results for SiO₂.

## E.  New Methods for Porous Materials

Several on-wafer techniques that were recently applied to characterize porous thin films are summarized in Table 3 and are illustrated using the results from ~ 0.9-µm thick mesoporous silica films (a xerogel) from AlliedSignal, Nanoglass™ K2.2-A10B (49). Three of the techniques listed in Table 3 are standard methods: MIS dot capacitors for dielectric constant measurements, as described later, Rutherford backscattering (RBS), and variable-angle spectroscopic ellipsometry (VASE). Both RBS and VASE have previously been used to measure the total porosity of thin films on Si substrates (50–52). Rutherford backscattering directly measures the areal density of the film, so the film thickness must be accurately known to determine the average film density, ρ. The chemical composition (stoichiometry) is also necessary for RBS to determine the area density, but RBS can provide the composition of heavier elements such as silicon, oxygen, and carbon. RBS cannot accurately quantify light elements such as hydrogen, and hydrogen content is often ignored when calculating the areal density of oxide-based materials. The porosity is calculated from the average film density relative to the assumed density of its nonporous analog (the density of the connecting material or the pore wall density, $\rho_w$). For oxide-based materials such as xerogels, the total porosity is commonly calculated using the average film density relative to the density of the connecting material ($\rho_w$), which is assumed to be the density of thermal oxide, using the relationship: $1 - \rho/\rho_w$. VASE can also measure the total porosity by modeling the porous material using an effective medium approximation such as the Lorentz–Lorentz, Maxwell–Garnet, or Bruggman model (53). However, extreme care is necessary for data analysis because the porosity values can

**Table 3** Characterization Data Reported for a Mesoporous Silica Film from AlliedSignal, Nanoglass™ K2.2-A10B

| Material property | Method | Nanoglass™ K2.2-A10B | TEOS |
|---|---|---|---|
| Dielectric constant (1 MHz) | MIS | 2.29 | ~ 4.2 |
| Film density | SXR | 0.548 g/cm$^3$ | 2.0–2.25 g/cm$^3$ |
| Matrix (connecting material) density | SANS | 1.16 g/cm$^3$ | 2.0–2.25 g/cm$^3$ |
| Total porosity | SXR | 75.6% | |
| | RBS | 76–79% | 0 |
| | VASE | 71% | |
| Mesoporosity | SXR/SANS | 52.9% | |
| Average pore size | SANS | 6.48 nm | No pores |
| | PALS | 7.5 nm | |
| Moisture uptake | SANS | 2.99 wt% | ~ 3–4 wt% |
| Modulus, CTE, $T_g$ | Bending beam | film too soft, $E < 1$ GPa (estimated) | $E \sim 77$ GPa CTE ~ 1 ppm/°C |
| CTE | SXR | 62.2 ppm/°C | |
| Pore structure | SANS | 22.4% of pore connected | No pores |
| | PALS | 100% of pores connected | |

Literature values are reported for TEOS oxide as a comparison.

significantly vary depending on the model used and the operator. As shown in Table 3, the RBS and VASE measurements on Nanoglass films are in reasonable agreement, but the total porosity measured by VASE is often somewhat lower than both the RBS and SXR results for porous silica films in general.

## 1. Specular X-Ray Reflectivity

Specular x-ray reflectivity (SXR) (54,55) is a very precise and direct measurement of thin-film density, and it has only recently been applied to characterize porous low-$k$ film up to 1 μm thick (32,49,56). Like RBS and VASE, SXR alone can measure the porosity by comparing the average film density to an assumed density of the nonporous material. And SXR has an advantage over RBS in that the film thickness is not necessary to determine the average film density. Also, no modeling is needed to determine the porosity as is the case with VASE.

In SXR, an incident beam of x-rays is directed toward the sample surface at an angle θ to the surface normal. When x-rays cross an interface between two media, they are refracted according to Snell's law: $n_1 \cos \theta_1 = n_2 \cos \theta_2$. The refractive index is $n = 1 - \delta + i\beta$, where δ is proportional to the electron density of the sample and β is related to absorption of x-rays. For most materials, the refractive index is less than 1 for x-rays with wavelengths of a few tenths of a nanometer. Hence, there is a critical angle, named the grazing-incidence angle, at which a well-collimated beam of x-rays is no longer totally reflected off the free surface but starts penetrating into the sample. Below the critical angle, the x-rays undergo total external reflection, and the critical angle is given by

$$\theta_c = \lambda \sqrt{\frac{\rho_c \gamma_c}{\pi}} = \sqrt{2\delta} \tag{13}$$

where $\lambda$ is the wavelength, $\rho_c$ is the electron density, and $\gamma_c$ is the classical electron radius. Thus, the average electron density of a film is determined directly from the critical angle, and the average mass density of the film can be calculated if the sample stoichiometry (or chemical composition) is known. We typically use RBS (for quantifying heavier elements) and forward recoil elastic scattering (FRES, for quantifying hydrogen) to measure the chemical composition of the film (57). The SXR results are typically presented as the logarithm of the ratio of the reflected beam intensity to the incident beam intensity versus $Q_z$, as shown in Figure 6, which displays SXR results from a 0.9-μm-thick Nanoglass film on a silicon substrate. $Q_z$ is the magnitude of the x-ray momentum transfer in the film thickness direction and is defined as $(4\pi/\lambda)\sin\theta$, where $\lambda$ is x-ray wavelength and $\theta$ is the grazing-incident angle. The data in Fig. 6 shows two critical angles. The first critical angle is that of the Nanoglass film, and an average film density of 0.548 g/cm$^3$ is determined from this critical angle and the sample stoichiometry. The average film density is simply $\rho_w(1 - P)$, where P is the porosity. Assuming that nonporous Nanoglass has the same density as thermal oxide (assume $\rho_w = \rho_{oxide}$), then the total porosity of Nanoglass is calculated to be 75.6%, which is in good agreement with the RBS measurement. The second critical angle corresponds to the electron density of the silicon substrate.

The oscillations between the first and second critical angles are due to optical coupling, where the Nanoglass film acts as an x-ray waveguide. The oscillations at angles above the silicon critical angle are due to the interference of x-rays reflected from the Nanoglass film surface and the film/substrate interface. The film thickness is accurately determined from the spacing of these interference fringes. The vertical CTE is measured by measuring the film thickness at several temperatures (58). The thickness of the

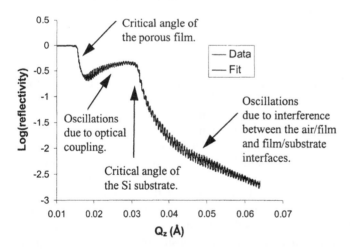

**Figure 6**  Specular x-ray reflectivity curve showing the log of the reflectivity versus the momentum transfer normal to the sample surface, $Q_z$, for 0.9 μm-thick film of Nanoglass coated on a silicon wafer substrate. Also shown in the plot is the best fit to the data by modeling the film as a series of layers using a one-dimensional Schrodinger equation that gives an electron density depth profile for the film shown in Fig. 7. The steep drop near $Q_z = 0.0155$ Å is the critical edge of the film and corresponds to the average electron density of the film. The oscillations in the curve provide a sensitive measure of the film thickness.

Nanoglass film was measured at several temperatures up to 225°C, from which an out-of-plane CTE of 62.2 ppm/°C was determined. The CTE measurement is corrected for the Poisson effect and substrate confinement using an assumed Poisson's ratio of 0.34. It is important to note that the interference fringes are attenuated or not present if the film is too rough over the 2–5 cm² spot size. Thus, the thickness and CTE cannot be measured for films that have severe striations after spin coating. However, the density and porosity can still be determined.

The SXR technique can also measure the density and porosity as a function of depth through the film thickness by modeling the electron density as a series of layers through the film using a one-dimensional Schrodinger equation (59). Figure 7 displays the depth profile results for the Nanoglass film. The results are consistent with a film having a uniform density, except at the first few hundred angstroms at the top and bottom interfaces. The modeling results in Fig. 7 also ensure that the average film density determined by the Nanoglass critical angle is representative of the density through the entire film thickness.

Both SXR and RBS determine the total porosity from the average film density relative to the assumed density of the nonporous analog. However, the density of the connecting material between pores (pore wall density) is not necessarily the same as the density of the nonporous analog. Thus, for a mesoporous solid, SXR and RBS alone measure the total porosity in the film, which includes any addition free volume (or micropores) in the connecting material that may not be present in the nonporous analog. For example, the density of the connecting material in Nanoglass may be much lower than that of thermal oxide.

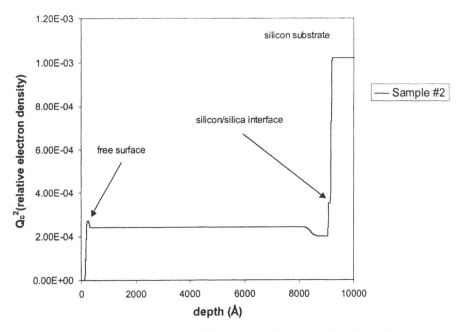

**Figure 7** Electron density space profile corresponding to the best fit to the data. At the free surface, a surface "skin" about 5 nm thick with an electron density slightly higher than the bulk film is observed. A similar layer is observed at the film/silicon interface. The rest of the film has a uniform density.

## 2.  Small-Angle Neutron Scattering

Small-angle neutron scattering (SANS) is a technique complimentary to SXR, in that it can directly measure the density of the connecting material (32,60). The porosity calculated from the average film density (measured by SXR) relative to the pore wall density (measured by SANS) is the mesoporosity due only to the mesopores in the film.

The details of SANS experiments have been previously described (32,60). Briefly, a neutron beam is incident along the surface normal of the thin film sample on a silicon substrate. The resultant scattering vector, $q$, ranges from $0.01\text{Å}^{-1}$ to $0.18\ \text{Å}^{-1}$, where $q = (4\pi/\lambda)\sin(\theta/2)$, $\lambda$ is the wavelength, and $\theta$ is the scattered angle from the incident beam path. For thin-film samples ($\sim 1\ \mu\text{m}$ thick), the scattering signal is enhanced by stacking several pieces of the sample together. The silicon wafer that supports the porous thin-film sample is transparent to the neutrons, and the scattering angles of interest are due only to the porous film. The scattering contrast arises from the difference in the neutron-scattering length of the connecting material and the pores themselves (the neutron-scattering length is assumed to be zero).

To analyze the SANS data, the porous film is modeled using a density correlation function developed by Debye and coworkers (61) for random, two-phase structures: $\gamma(r) = \exp(-r/\xi)$, where $\xi$ is the correlation length. Using this model, the average pore dimension (average pore size) or chord length is $\xi/(1 - P)$, where $P$ is the porosity. The SANS intensity based upon the Debye model takes the form of

$$\frac{1}{I(q)^{1/2}} = \frac{1}{(c\xi^3)^{1/2}} + \frac{\xi^2 q^2}{(c\xi^3)^{1/2}} \tag{14}$$

The quantities $c$ and $\xi$ can be determined from the slope and zero-$q$ intercept of the SANS data plotted as $I)q_{-1/2}$ versus $q^2$, as shown in Figure 8. The SANS data in Fig. 8 follows the relationship in Eq. (14) except in the low-$q$ region, which indicates that the Debye model is a valid description of the Nanoglass film. The deviation in the low-$q$ region is commonly observed in silica gels, crosslinked polymers, and porous materials such as shale and is referred to as a *strong forward scattering*. The quantity $c$ in Eq. (14) is related to the porosity $P$ and the mass density of the connecting material (pore wall density), $\rho_w$, as $P(1 - P)\rho_w^2$. And SXR measures the average film density, $\rho_w(1 - P)$. Thus, from the slope and zero-$q$ intercept of the $I(q)^{-1/2}$ versus $q^2$ plot and the average film density from SXR, the mass density of the connecting material (pore wall density) and the average chord length (a measure of the average pore size) can be calculated. For the Nanoglass films, SANS measured the pore wall density, $\rho_w$, as $1.16\ \text{g/cm}^3$, the mesoporosity as 52.9%, and the average pore size (chord length) as 6.48 nm. The mesoporosity is much less than the total porosity. The total porosity is calculated assuming that the pore wall density was the same as that of thermal oxide. However, the pore wall density measured by SANS is much less than the density of thermal oxide ($2.25\ \text{g/cm}^3$), indicating that the pore wall material has a lot of free volume or microporosity relative to thermal oxide. The pore size determined using SANS (6.48 nm) is in excellent agreement with a BET gas absorption measurement using powdered samples (60). Gas absorption is the conventional method to measure pore size and pore size distribution, but it is difficult to apply to 1-$\mu$m-thick films coated on industry standard 8-in. silicon wafers (63).

The SANS technique can also measure moisture uptake and the fraction of mesopores that are connected by immersing the sample at 25°C in deuterated water and deuterated toluene (d-toluene), respectively (32). As the mesopores are filled with a solvent, the scattering contrast changes, which, in turn, changes the scattering intensity. If all of the

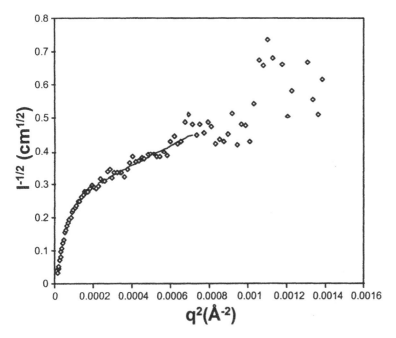

**Figure 8** Debye plot derived from a small-angle neutron scattering experiment on a stack of Nanoglass thin films, each coated on silicon substrates. The slope of the fitted line provides the correlation length of the porous structure from which the chord length (average pore size) is calculated. The intercept of the fitted line at $q^2 = 0$ gives a measure of the porosity and the mass density of the connecting material between the pores (pore wall material).

mesopores are filled with d-toluene (or d-water), the SANS intensity is much larger than, and is easily related to, that of the dry sample. However, the scattering intensity of Nanoglass films immersed in d-toluene is much less than expected if all of the mesopores are filled. A two-layer model was developed to calculate the fraction of pores filled, and the analysis showed that 22.4% of the mesopores filled with d-toluene. This suggests that only 22.4% of the mesopores are connected to the Nanoglass surface, but the size of the d-toluene molecule may severely limit its ability to enter all the pores. In fact, a positronium annihilation lifetime spectroscopy (PALS) study demonstrated that all of the pores in Nanoglass are connected to the film surface. Similarly, SANS measures moisture uptake by measuring the fraction of voids that are filled with deuterated water and then calculating the weight percentage of water uptake. Nanoglass showed only 3.00% uptake of moisture.

## 3. Positronium Annihilation Lifetime Spectroscopy

Positronium annihilation lifetime spectroscopy (PALS) (64–66) using radioactive beta-decay positrons, is commonly used to study void volume in polymers. However, the high-energy beta-decay positrons are too penetrating ($\sim 0.3$ mm) for thin-film studies. Low-energy-beam PALS varies the beam transport energy to implant positrons to depths ranging from about 50 to 2000 Å (64). Low-energy-beam PALS was used to determine the average pore size and pore structure and to measure metal diffusion into Nanoglass (67).

Implanted positrons will react a number of ways in a dielectric film, as illustrated in Figure 9. One fate of the positron is to form a hydrogen-like atom with an electron, called

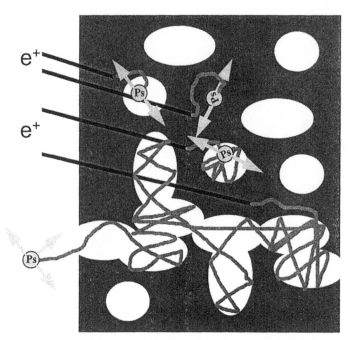

**Figure 9** Illustration of positronium annihilation lifetime spectroscopy (PALS). The incident positrons thermalize in the sample and positronium (Ps) forms, primarily in the voids. The vacuum Ps lifetime ($\sim$ 142 ns) is shortened in the voids due to collisions with the void walls that increase the probability of annihilation into gamma rays. The Ps lifetime is measured by monitoring the gamma rays, and the lifetime is related to the average pore size. If the material has pores that are connected to the surface, Ps can diffuse out of the sample and into the vacuum. The detection of Ps in the vacuum chamber provides a way to determine whether the pores in the film are connected or isolated.

positronium (Ps). Positronium has a vacuum lifetime of $\sim$ 142 ns before it annihilates into gamma rays that can be detected. In porous films, the formation of Ps occurs preferentially in the voids of the film, and the Ps annihilation lifetime is shortened from the vacuum lifetime due to collisions with the pore walls. This effect provides a means of determining the average pore size from the Ps lifetime in the porous film. If all of the pores are connected, the Ps has a single lifetime as it samples all of the pores in the film, and a single average pore size is measured. If the pores are isolated, the Ps has many lifetime components, each corresponding to a different pore size. Thus, in closed-pore materials, PALS has the potential to provide pore size distribution information. The potential of PALS to measure pore size distribution in materials with isolated pores is very exciting, especially since gas absorption measurements of pore size distribution are limited if the gas cannot access the pores from the surface of the film.

The Ps lifetime in a Nanoglass film capped with a 100-nm TEOS oxide layer was 98 ns, which corresponds to an average pore size of 7.7 nm. The average pore size measured by PALS is actually a mean free path, and it is somewhat different from the chord length measured by SANS. Nevertheless, the PALS result agrees reasonably well with the SANS and BET gas absorption results (49).

When the Nanoglass is not capped, all of the Ps diffuses out of the film and into the vacuum chamber, giving a Ps lifetime nearly equal to the 142-ns lifetime in a vacuum. This observation demonstrates that all of the pores in Nanoglass are connected to the surface of the film. When Nanoglass films are capped using sputtered Al, keeping the film at $\sim 25°C$ during the metal deposition, the sample exhibits the same 98-ns lifetime observed in the TEOS oxide capped samples. However, another 3 ns lifetime component appears with the 98 ns component when the sample is annealed. The 3-ns component indicates that Al is diffusing into the material and coating and closing some of the pores. The onset of Al diffusion into Nanoglass is at 450°C, as determined from the appearance of the 3-ns-Ps-lifetime component (Figure 10). At higher temperatures, the 3-ns component begins to dominate the 98-ns-lifetime component as more Al diffuses into the film. Interestingly, if the Al cap is thermally evaporated (sample temperature $\ll 450°C$), both lifetimes are present even without annealing the sample, which indicates that some pores were coated and closed by mobile Al during deposition.

## III.  PROPERTIES OF SELECTED LOW K MATERIALS

### A.  Dielectric Constant

In order to reduce the $k$-alue relative to that of $SiO_2$, it is necessary either to incorporate atoms and bonds that have a lower polarizability or to lower the density of atoms and bonds in the material, or both. With regard to the first effect, there are several components to the polarizability that must be minimized in reducing the dielectric constant. The polarization components usually considered are the electronic, atomic, and orientational responses of the material. The last two components constitute the nuclear response and are important at lower frequencies ($10^{13}\,s^{-1}$), while the electronic response dominates at higher frequencies. At typical device operating frequencies, currently $10^9\,s^{-1}$, all three components contribute to the dielectric constant and should be minimized for optimum performance.

Some typical electronic polarizabilities and the associated bond enthalpies are shown in Table 4. The data indicates that single C–C and C–F bonds are among those having the lowest electronic polarizability, making fluorinated and nonfluorinated aliphatic hydrocarbons potential candidates for low-$k$ applications. Incorporation of fluorine atoms is particularly effective in lowering the polarizability (68) due to their high electronegativity,

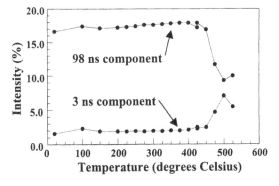

**Figure 10**  Positronium lifetime intensity as a function of 1-hour anneal temperature for positrons implanted into a Nanoglass film capped with sputtered aluminum. The 98-ns component is characteristic of the 7.7-nm pores that were measured in TEOS-capped Nanoglass films. The 3-ns component is characteristic of pores coated and closed off by diffused aluminum.

**Table 4**  Electronic Polarizability and Bond Enthalpies

| Bond | Polarizability[a] ($\mathring{A}^3$) | Average bond energy[b] (kcal/mole) |
|------|--------------------------|------------------------------|
| C—C | 0.531 | 83 |
| C—F | 0.555 | 116 |
| C—O | 0.584 | 84 |
| C—H | 0.652 | 99 |
| O—H | 0.706 | 102 |
| C=O | 1.020 | 176 |
| C=C | 1.643 | 146 |
| C≡C | 2.036 | 200 |
| C≡N | 2.239 | 213 |

*Sources:* [a] Ref. 102; [b] Ref. 103.

which leads to tight binding of electrons. Conversely, the electronic polarizability is high for materials having less tightly bound electrons. For example, materials containing a large number of carbon double and triple bonds can be expected to have a large polarization due to the increased mobility of the $\pi$ electrons. Conjugated carbon double bonds in aromatic structures are a common source of extensive electron delocalization leading to high electronic polarizability. Note, however, that there is a tradeoff in achieving low dielectric constant and high bond strength, for the low-polarizability single bonds are among the weakest, while the double- and triple-bonding configurations have much higher bond enthalpies.

The nuclear dielectric response will result from polarization due to both permanent and transition dipoles in the material. The response is often dominated by polar substituents, such as hydroxyl and carbonyl groups, that can increase the orientational component of the polarizability. An indication of the relative importance of the electronic, relative to the nuclear, response for a material can be found by examining the differences between the $k$-values measured at high vs. low frequencies. The high-frequency value (633 nm, or $4.74 \times 10^{14}$ Hz), reflecting the electronic component, can be obtained through the optical index of refraction, according to Ref. 69, $k = n^2$. This relationship assumes there is no absorption at the optical frequency used in the measurement. The low-frequency $k$-value, representing both the electronic and nuclear components, can be determined from capacitance measurements of metal–insulator–semiconductor (MIS) or metal–insulator–metal (MIM) structures at 1 MHz.

Table 5 shows the high-frequency electronic response, obtained from the optical index at 633 nm, and the total low-frequency $k$-value at 1 MHz for a number of proposed low-$k$ dielectrics. The difference between the two measurements represents the nuclear components to the dielectric constant. The data indicates that for many low-$k$ materials under consideration, the nuclear components are small relative to the electronic part of the response. In contrast, $SiO_2$ and other oxide-based materials have a large nuclear component, largely due to the strong atomic polarization. The $k$-value of $SiO_2$ can be reduced to 3.3–3.7 by incorporating fluorine into the material. Yang and Lucovsky have shown that the decrease is largely due to weaker atomic (infrared) activity (70).

Adsorbed moisture is particularly troublesome in raising the orientational component of the dielectric constant in thin films. Since water has a large permanent dipole moment, a small quantity of moisture can substantially impact the dielectric constant.

**Table 5** Dielectric Properties of Low-$k$ Materials

| Material | $n_{in}$ | $n_{out}$ | $k_{out}$ | $n_{out}^2$ | $\Delta = k - n_{out}^2$ |
|---|---|---|---|---|---|
| PTFE | 1.350 | 1.340 | 1.92 | 1.796 | 0.12 |
| BPDA-PDA | 1.839 | 1.617 | 3.12 | 2.615 | 0.50 |
| PMDA-TFMOB-6FDA-PDA | 1.670 | 1.518 | 2.65 | 2.304 | 0.35 |
| PAE #1 | 1.676 | 1.669 | 3.00 | 2.787 | 0.11 |
| FLARE #2 | 1.671 | 1.672 | 2.80 | 2.796 | 0.00 |
| BCB™ | 1.554 | 1.554 | 2.65 | 2.415 | 0.24 |
| SiLK™ | 1.630 | 1.624 | 2.65 | 2.637 | 0.01 |
| OXZ (fluor. poly.) | 1.566 | 1.551 | 2.406 | 2.493 | 0.08 |
| MSQ/HSQ hybrid | 1.374 | 1.373 | 2.52 | 1.886 | 0.63 |
| OSG (13% C) | 1.433 | 1.435 | 2.059 | 2.69 | 0.63 |
| OSG (30% C) | 1.438 | 1.482 | 2.195 | 2.60 | 0.40 |
| SiO$_2$ | 1.47 | 1.47 | 4.0 | 2.16 | 1.8 |

As a result, when designing low-$k$ materials it is desirable to avoid the use of highly polar substituents that attract and bind water. However, many dielectric films absorb water to some extent. As a result, the dielectric constant and the loss factor depend strongly on moisture exposure. When comparing dielectric constants, measured in the megahertz range, it is therefore important to specify sample treatment and humidity, to account for the moisture uptake. Oxide-based materials especially, such as OSGs, tend to absorb moisture. This can be observed in the form of a weak silanol absorption band around 3200–3700 cm⁻1 by FTIR, although FTIR usually lacks the sensitivity for trace amounts of water, which already have strong effects on the dielectric constant. Porous oxide-based materials are usually surface treated to obtain hydrophobic surfaces. Silylation has been shown to provide hydrophobic surfaces by replacing terminating–OH groups with non-polar groups such as–Si(CH$_3$)$_3$ (71,72). However, many of the materials tested so far still showed increasing dielectric constants when exposed in the lab ambient. This is illustrated in Figure 11, where dielectric constants for the same OSG material are shown, measured after different sample storage conditions. The data for one porous material is shown as well. Note that the relative increase in dielectric constant is even larger for the porous material.

The dielectric constant is determined not only by the type of atoms and bonds but also by the atom and bond densities. The dielectric constant of any material can be reduced by decreasing the density. The density can be lowered by using lighter atoms and/or by incorporating more free space around the atoms. For example, the lower dielectric constant of organic polymers relative to SiO$_2$ is partly due to the lighter C and H atoms vs. Si and O and to the low packing density of most polymer chains relative to the crosslinked silica network. Likewise, the incorporation of light, space-occupying groups such as H or CH$_3$ into the silica network can significantly lower the material density, and therefore the dielectric constant, of materials such as spin-on glass (SOG) relative to dense oxide.

The introduction of nanometer-sized pores into the material is a natural extension of this strategy to increase the free space and decrease the material density. The effect of porosity on dielectric constant can be predicted using a simple model, such as the Bruggemann effective medium approximation (73):

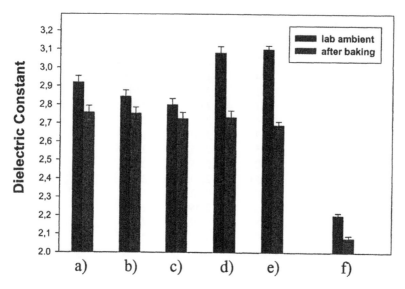

**Figure 11** Dielectric constant of two OSGs and one porous material before and after baking for 60 min in nitrogen ambient: (a) as received; (b) 2nd measurement, 39 days later; (c) 3rd measurement, 10 more days; (d) and (e) second sample from same wafer, measured the same time as (c); (f) MSQ-based porous material.

$$f_1 \frac{k_1 + k_e}{k_1 + 2k_e} + f_2 \frac{k_2 - k_e}{k_2 + 2k_e} = 0 \qquad (15)$$

where $f_{1,2}$ represents the fraction of the two components, $k_{1,2}$ represents the dielectric constant of the components, and $k_e$ is the effective dielectric constant of the material. The model assumes two components to the film: the solid wall material and voids. Figure 12 shows the dielectric constant as a function of porosity predicted by the model for $SiO_2$ ($k = 4.0$) and for a lower-$k$ ($= 2.8$) wall material. The plots show that the $k$-value decreases slightly faster than linearly. Although the model is simple, the predicted results appear to be in reasonable agreement with recent experimental measurements on methyl silsesquioxane (74) and oxide porous films. Differences between the theoretical prediction and experimental results are likely related to surface chemistry, such as the presence of terminating OH groups and adsorbed water, and to the pore geometries.

One point demonstrated by Fig. 12 is that to obtain a given $k$-value, significantly less porosity would have to be incorporated into the lower-$k$ material than into the $SiO_2$. For example, to get to $k = 2.0$, about 55% porosity would be needed in an oxide material, whereas only $\sim 35\%$ would be needed in the lower-$k$ material. Since a high percentage of porosity in the film can be expected to give rise to a number of reliability concerns, there is a definite advantage using a lower-$k$ starting material to minimize the amount of porosity needed.

## B. Thermomechanical Properties

In addition to low dielectric constant, candidate ILDs must have sufficient thermal and mechanical stability to withstand the elevated processing temperatures and high stresses that can occur in the interconnect structure. Stability with regard to decomposition can be

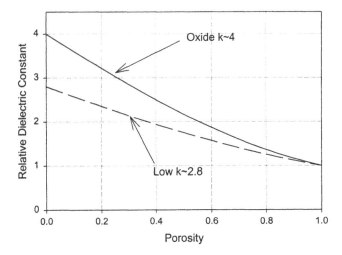

**Figure 12** Bruggemann's effective medium approximation showing dielectric constant vs. porosity for oxide and a low-*k* material.

increased by using strong individual chemical bonds and by incorporating rings, cross-linking, and networked structures so that multiple bonds would have to be broken in order for molecular fragments to be released. Mechanical strength is another important requirement, because if the film cannot withstand stresses occurring during processing, structural integrity of the interconnect can be compromised. The value of the elastic, or Young's, modulus ($E$) is often used as an indication of mechanical stability for low-*k* candidate materials. The Young's modulus of most organic and inorganic low-*k* candidate materials is at least an order of magnitude lower than that of standard $SiO_2$ films prepared from tetraethoxysilane (TEOS) ($E \sim 59$ GPa) (38). As a result, the mechanical reliability of these alternative dielectrics is an important integration concern.

In addition to the modulus, the film's thermal expansion coefficient (CTE) is also of importance, since most of the stresses that occur in the interconnect are thermally induced, resulting from CTE mismatches between various materials in the interconnect structure. For example, the CTE of $SiO_2$ is 0.5 ppm/°C, that of Cu is 16.5 ppm/°C, and that of Al is 23.1 ppm/°C (75). The CTE of many organic dielectrics is over 50 ppm/°C, which can lead to high tensile stresses in the film following high-temperature processes. It would be desirable to minimize the thermal mismatches, especially for dielectrics with a low modulus, by using a low-CTE dielectric. Table 6 summarizes data on thermal stability, mechanical strength, and thermal expansion for a variety of materials.

## C. Thermal Conductivity

In addition to low dielectric constant, a high thermal conductivity is desirable to minimize joule heating, which poses a reliability concern for high-density interconnect structures. Table 7 summarizes thermal conductivities measured, in our lab, by the 3ω and photo-thermal technique. The thermal conductivities of all candidate materials tested so far are significantly lower than for TEOS. Many polymers have thermal conductivities of about 0.2 W/mK, and heat transport seems to be best in the dense and crosslinked polymers. Thermal conductivities for the OSGs and porous materials scale to a first approximation with their density. However, the thermal conductivity decreases much faster than the

**Table 6** Selected Dielectric and Thermomechanical Properties of Low-k Thin Films

| Material | $k$ | Young's modulus (GPa) | Lateral TEC 25–225°C (ppm/°C) | $T_g$ (°C) | TGA % weight loss (425°C, 8 h) |
|---|---|---|---|---|---|
| PTFE | 1.92 | 0.5 | 135 | 250 | 0.6 |
| BPDA-PDA | 3.12 | 8.3 | 3.8 | 360 | 0.4 |
| Crosslinked PAE | 2.8–3.0 | 2.7 | 52 | 350 | 2.5 |
| Fluorinated PAE | 2.64 | 1.9 | 52 | > 400 | x |
| BCB | 2.65 | 2.2 | 62 | — | 30 |
| SiLK | 2.65 | 2.3 | 54 | — | 2.1 |
| Parylene-N | 2.58 | 2.9 | 55–100 + | 425 (melt) | 30 |
| Parylene-F | 2.18 | 4.9 | 33 | — | 0.8 |
| HSQ | 2.8–3.0 | 7.1* | 20.5 | — | x |

*: Biaxial modulus
x: Not measured
—: None observed.

density at high porosity, with the thermal conductivity being lower by a factor of 20 as compared to TEOS for 77% porosity.

## D. Structure Property Relations

### 1. Polymers

The low value of dielectric constant of fluorinated organic materials such as PTFE is due to the use of molecular bonds with the lowest polarizability (from the top of Table 4). PTFE, which consists of singly bonded carbon chains saturated with fluorine atoms, has one of the lowest $k$-values ($\sim 1.9$) of any nonporous material. One drawback of PTFE is that the flexible and uncrosslinked chain structure limits the thermomechanical stability of

**Table 7** Thermal Conductivities for Several Low-Dielectric-Constant Materials

| Material | | Dielectric constant | Thermal conductivity (W/mK) |
|---|---|---|---|
| TEOS | | 4.0 | 1.26 |
| Polymers: | PMDA-ODA | 3.2 | 0.22 |
| | BPDA-PDA | 3.12 | 0.21 |
| | PTFE-based | 2.15 | 0.19 |
| | BCB | 2.65 | 0.24 |
| OSG 13% | | 2.69 | 0.34 |
| 28% | | 2.64 | 0.20 |
| 30% | | 2.60 | 0.21 |
| Porous MSQ | matrix I | 2.82 | 0.30 |
| | | 2.36 | 0.22 |
| | | 2.20 | 0.17 |
| | matrix II | 2.23 | 0.22 |
| | matrix III | 2.39 | 0.26 |
| | matrix IV | 2.41 | 0.19 |

the material. For example, one PTFE material evaluated in our laboratory was found to have a low yield stress (12 MPa), low elastic modulus (0.5 GPa), low softening temperature ($\sim 250°C$), and high TEC ($> 100\,ppm/°C$). A second issue for PTFE, which is a concern for all fluorine-containing materials, is the potential release of fluorine atoms that can cause corrosion of metals or other reliability problems in the interconnect structure.

Better thermomechanical properties compared to PTFE can be obtained by incorporating a stiffer polymer chain or by crosslinking the chains. The classic example of the former type of material are the aromatic polyimides, which have a rigid backbone due to the many aryl and imide rings along the chain. This can raise the modulus to 8–10 GPa and the $T_g$ to 350–400°C (76). However, the rigid chain structure causes the PI chains to align preferentially parallel to the substrate, especially when deposited as thin films, which results in anisotropic material properties (77–82). The thermomechanical properties are likewise anisotropic. For instance, the TEC of thin films of rigid PIs is often $\leq 10\,ppm/°C$ in the plane of the film, but can be more than 10 times larger in the out-of-plane direction (78).

The spin-on poly-aryl-ether (PAE) materials result from attempts to balance the dielectric and thermomechanical properties of the material. The aryl rings in these materials provide better thermomechanical properties than PTFE, but the flexible aryl linkages allow bending of the chains that results in a more isotropic material than is obtained for PIs. PAEs typically have a $k$-value of 2.8–2.9, while typical values for the modulus and TEC are 2.0 GPa and 50–60 ppm/°C, respectively. Resistance to thermal decomposition can be quite good for PAEs; weight losses of only $\sim 2\%$ over 8 hours at 425°C has been observed in our isothermal TGA experiments. One drawback of uncrosslinked PAEs is that they have a relatively low $T_g$ of $\sim 275°C$. In order to raise the $T_g$, reactive functional groups can be incorporated along the backbone, which allows for crosslinking of the film during cure. Experiments conducted in our laboratory have shown that crosslinking of PAEs can increase the $T_g$ to over 450°C. The effects of crosslinking on $T_g$ are shown in the stress curves of Figure 13. Unfortunately, crosslinking can also lead to a corresponding increase in the $k$-value if the crosslinking is not optimized. Therefore, careful control of the crosslinking chemistry is necessary to improve the material stability without sacrificing the dielectric properties.

Because it is difficult to obtain high $T_g$ values in uncrosslinked polymers, efforts have been made to use polymers that crosslink into a three-dimensional network upon curing. Two spin-on examples of these organic thermosets are the bis(benzocyclobutenes) (BCB) (83,84) and SiLK$^{TM}$ thermosets (85) produced by Dow Chemical, both of which are highly aromatic. The thermomechanical properties of these films ($E = 2$–3 GPa, TEC $\sim 60$) are comparable to the PAEs, but they have a somewhat improved $k$-value of 2.65. In the case of BCB, the low $k$-value results partly from the incorporation of relatively nonpolarizable aliphatic rings in addition to the aromatic rings. The large number of rings and crosslinking make these materials quite rigid, so they do not show a $T_g$ below 450°C (85), although the relatively weak Si–C bond in BCB results in thermal decomposition of this material above 350°C. The thermal stability of SiLK$^{TM}$ is significantly better, because the material shows very low thermal weight loss ($\sim 2\%$ over 8 hours at 425°C).

## 2. Organosilicate Glasses

Since mechanical and thermal stability is difficult to obtain using purely organic materials, there has been much interest in inorganic or inorganic/organic hybrid materials. The strategy behind the silica-based materials is to use the much stronger and more rigid

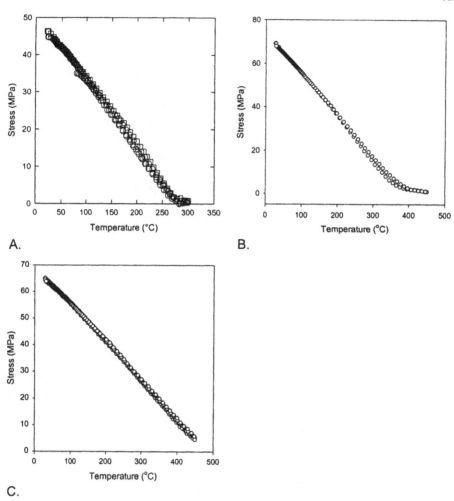

**Figure 13**  Thermal stress measurement of (A) uncrosslinked, (B) partially crosslinked, and (C) fully crosslinked PAE film, showing an increase in the $T_g$ from 275°C to over 450°C. A corresponding rise in $k$ from ~ 2.9 to 3.1–3.2 was observed for these films.

$SiO_2$ network as a framework and to lower the $k$-value by lowering the density through the incorporation of organic chemical substituents or voids into the film. The organosilicate glasses (OSGs) are one such class of hybrids. Thin films of OSG can be deposited by a CVD process using methylsilane and an oxidizing agent or by a spin-on process using silsesquioxanes, then often called spin-on glasses (SOGs). In the silsesquioxanes each Si atom is bonded to one terminating group, such as hydrogen, methyl, or phenyl, resulting in a nominal stoichiometry $SiO_{1.5}R_{0.5}$. Both the crosslinking density and the material density are reduced due to these terminating groups. The organic content in the films can be varied by the CVD process conditions or by the curing conditions after spin-on. The dielectric constant of these materials is in the range of 2.5–3.0. The reduction in $k$ from that of $SiO_2$ is thought to be due mainly to the reduction in density. For example, the density of hydrogen silsesquioxane (HSQ) ($k = 2.8$–3.0) can be about one-third less (1.4 g/$c^3$) (86,87) than that of TEOS (2.2–2.4 g/$c^3$).

The mechanical properties of the OSG materials are generally superior to those of most organic polymers. For instance, our laboratory has determined a biaxial modulus of 7.1 GPa and TEC of 20.5 ppm/(C for HSQ. The latter value is in good agreement with x-ray reflectivity measurements (58). Black Diamond, a CVD film from Applied Materials (version April '99) analyzed in our lab showed FTIR spectra similar to those of an HSQ/MSQ hybrid. The material had a carbon content of 13%. It should be noted that the carbon contents for the OSGs reported here were measured by x-ray photoelectron spectroscopy (XPS) and therefore do not account for the hydrogen present in the films. Typical for these materials are stress vs. temperature curves that bend toward smaller slopes at elevated temperatures, as shown in Figure 14. At temperatures below 100°C the CTE for this version of Black Diamond was 23 ppm/°C and the biaxial modulus was 7.7 GPa. Above 100°C the CTE was 22 ppm/°C and the biaxial modulus dropped to 6.1 GPa. Because the oxide network provides the rigidity of the structure, the mechanical properties decay with increasing organic content. Bending beam experiments on OSGs, with systematically varied carbon contents but otherwise similar process conditions, are shown in Figure 15. The data shows a clear trend toward larger negative products of $E'\Delta\alpha$ with increasing carbon content. The negative sign of $E'\Delta\alpha$ is due to CTEs larger than silicon; the increasing absolute numbers show an increasing tendency to build up thermal stresses. Assuming that the biaxial modulus of the material does not increase with carbon content, the data shows an increase of CTE with carbon content. Actually this assumption is very conservative, for it seems reasonable that the modulus of the material decreases rather than increases when the Si–O bonds, providing the rigidity of the structure, are replaced by more and more organic groups. This indicates a tradeoff between lower dielectric constant, which can be achieved by increasing the number of organic groups, and the thermomechanical properties of these materials.

Many types of SOGs experience a large amount of shrinkage during curing that leads to high tensile stresses that can crack the films. This often limits the maximum

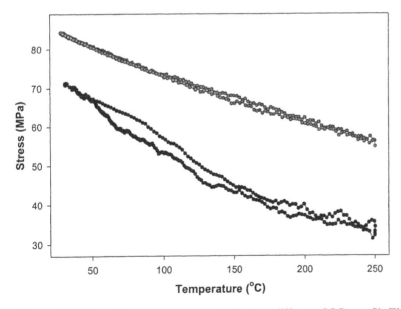

**Figure 14** Thermal stress measurements for two different OSGs on Si. The lower curve is more noisy because it was obtained from a film on a 700 μm-thick substrate, after extending the light path to 6 m.

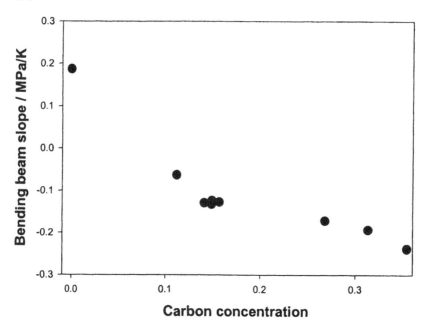

**Figure 15** Slopes ($E\Delta\alpha$) of stresses vs. temperature for OSGs as function of carbon content (excluding H). TEOS is shown at 0% for comparison.

thickness of these films to only a few thousand angstroms. The silsesquioxanes undergo much less shrinkage, are less prone to cracking, and therefore can be prepared in thicknesses of ~ 1 micron (89).

Resistance to thermal decomposition for silsesquioxanes is generally better than that of most organic films, due to the increased stability of the silica network. For example, AlliedSignal has reported no outgassing in TDMS experiments conducted on an MSQ film at 450°C (89). Conversely, HSQ is susceptible to oxidation at temperatures over 350°C, which produces a denser, higher-$k$ film having a reduced number of Si–H bonds (86,90). Oxidation can also form terminating OH groups that bind water. Oxidation appears to be less of a concern for OSGs with a higher content of carbon, due to the better stability of the Si–C bond relative to Si–H. Still, OSG films are prone to oxidation, leading to a small increase in weight in TGA experiments when traces of oxygen are present. As for HSQ, oxidation usually leads to a decrease in the number of Si–H bonds as seen in FTIR and to a larger dielectric constant as the material becomes more oxide-like. As a result OSG films are often cured in an inert environment and capped with oxide.

### 3. Microporous Materials

Since very few dense materials have a $k$-value less than 2.5, there has been much interest in fabricating porous materials in order to reach ultralow-$k$ values ($\leq 2.0$). Most of these materials are not as well developed as the fully dense dielectrics. Nevertheless, a wide variety of porous materials have been produced to date, and some initial integration studies have been done (49).

There are a number of reliability concerns that are particularly prominent for porous ILDs. One of these is mechanical strength, since films with voids will undoubtedly be weaker than similar, fully dense materials. Another set of potential problems involves the size and distribution of the pores in these materials. It is critical to the reliability that these parameters be carefully controlled. For instance, if pores are too large, feature sizes and shapes will not be well defined during integration. Also, if the pores percolate together to form channels, the material may have a low breakdown voltage and mechanical strength.

As already mentioned, an important reliability concern for porous materials is their mechanical strength. Studies of highly porous bulk aerogel materials show that the Young's modulus scales with at least the second or third power of the density (92,93). This suggests that the mechanical strength of ILD films will deteriorate very quickly as voids are introduced. Evaluation of the mechanical properties as a function of porosity is clearly desirable for potential ILD materials. Unfortunately, characterization of these properties for thin porous films is difficult. Indentation studies are one option (94). However, interpretation of the results is not straightforward.

At our laboratory at the University of Texas, we have investigated the mechanical properties of porous films by thermal stress measurements using the two-substrate curvature method mentioned earlier. Initial experiments confirm the concerns about mechanical integrity, for a number of the films tested appear to be mechanically weak, exhibit extensive cracking, and do not appear capable of supporting very high stress levels. Some of the highly porous gel-type films can be especially susceptible to these problems. More recent versions of MSQ-based materials show more promising mechanical properties; one example is given in Figure 16. The data shows a decrease of slopes in the stress vs. temperature curves, indicating that higher stress levels are not supported by the material. However, it should be noted that the curve reproduces in subsequent temperature cycles, showing that this effect cannot be attributed to stress relaxation by cracking or collapse of the cell

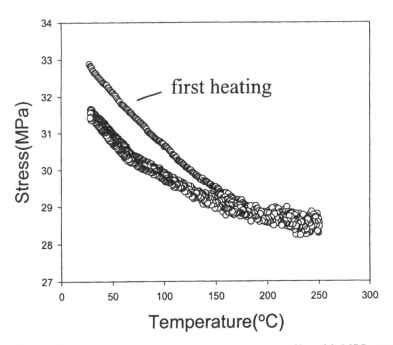

**Figure 16** Thermal stress measurement for a porous film with MSQ matrix material. Shown are three heating and cooling cycles.

structure. At this point it is not clear whether high porosity per se is limiting mechanical strength, or whether the structure in these films is just not optimized. For example, a structure having a large number of free pendent molecular "arms" in the film is probably an inefficient use of material, whereas use of a more highly crosslinked network can be expected to maximize material strength for a given porosity.

In Figure 17, the thermal conductivity data for various porous materials, given in Table 7, is plotted against the dielectric constant. For a given matrix material the thermal conductivities follow a linear relationship with the dielectric constant. Using Bruggemann's EMA (see earlier) the dielectric constant scales approximately linearly for small porosities. This indicates that the thermal conductivity is related mainly to the density of the oxide skeleton for small ($< 30\%$) porosities. Because lower density is the main mechanism to lower the dielectric constant of these materials, there is a fundamental tradeoff between low dielectric constants and thermal conductivity. The sharp drop in thermal conductivity for higher porosities as observed for the xerogels indicates mechanisms that further reduce the thermal conductivity. It should also be noted that thermal conductivities differ significantly for samples with similar dielectric constants, depending on the matrix material. Together with the data on thermal conductivity, this points out that a carefully tailored matrix material as well as a controlled pore structure are important for optimizing the properties of porous low-dielectric-constant materials. Clearly, a great deal of work will be required to correlate the structures with the thermal and mechanical properties.

## IV.  SUMMARY

Low-$k$ dielectrics have to meet stringent requirements in material properties in order to be successfully integrated. A particularly difficult challenge for material development is to obtain a combination of low dielectric constant with good thermal and mechanical properties. The basic difficulty is that the strong chemical bonds that impart structural stability are the most polarizable and thus yield a high dielectric constant. There are a variety of

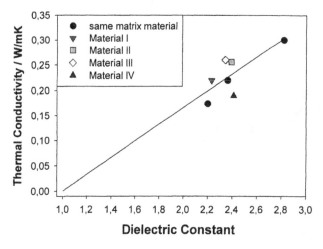

**Figure 17**  Thermal conductivity vs. dielectric constant for different porous materials with MSQ matrix.

characterization techniques available for thin-film characterization and several of them that allow for on-wafer characterization have been discussed in detail in this chapter. Several promising candidate materials with $k$-value of 2.7–3.0 have been developed, and intense efforts are under way to develop process integration.

The next-phase development of porous materials with $k < 2.5$ seems more challenging, since the incorporation of porosity degrades further the thermal and mechanical properties. The scaling of layer thickness with device dimensions presents challenges for the characterization metrology itself. Several methods have been developed to characterize the structure and properties of thin porous films, but much work remains to be done to understand how the porosity affects the macroscopic properties of the films. First characterization results for several types of porous and low-density silica films indicate that there is not a simple relationship between density and dielectric constant, especially at high porosities. The porosity, pore structure, surface area, and surface chemistry on the pore walls all affect the dielectric constant and also the mechanical strength and uptake of moisture or other solvents. These observations suggest that the amount of porosity should be minimized, and with pores preferably in closed form. Thus there is a definite advantage of using starting materials with a lower $k$-value to develop porous dielectrics. Clearly, further development of metrology for the characterization of porous films is necessary to understand the properties of these materials and their impact on process integration.

## ACKNOWLEDGMENTS

The authors acknowledge SEMATECH and SRC for supporting this work.

## REFERENCES

1. Edelstein, D., Heidenreich, J., Goldblatt, R., Cote, W., Uzoh, C., Lustig, N., Roper, P., McDevitt, T., Motsiff, W., Simon, A., Dukovic, J., Wachnik, R., Rathore, H., Schulz, R., Su, L., Luce, S., Slattery, J. In IEEE International Electron Device Meeting, 1997, p. 773, Washington, DC.
2. Venkatesan, S., Gelatos, A.V., Misra, V., Smith, B., Islam, R., Cope, J., Wilson, B., Tuttle, D., Cardwell, R., Anderson, S., Angyal, M., Bajaj, R., Capasso, C., Crabtree, P., Das, S., Farkas, J., Filipiak, S., Fiordalice, B., Freeman, M., Gilbert, P. V., Herrick, M., Jain, A., Kawasaki, H., King, C., Klein, J., Lii, T., Reid, K., Saaranen, T., Simpson, C., Sparks, T., Tsui, P., Venkatraman, R., Watts, D., Weitzman, E.J., Woodruff, R., Yang, I., Bhat, N., Hamilton, G., Yu, Y.. In IEEE International Electron Device Meeting, 1997, p. 769, Washington, DC.
3. IBM Press release, IBM.com/Press, April-3-2000.
4. Lammers, D., eet.com, April-7-2000.
5. Semiconductor Industy Association. International Technology Roadmap for Semiconductors: 1999 edition. Austin, TX: SEMATECH, 1999.
6. Lee, W.W., Ho, P.S. MRS Bull. 22:19, 1997.
7. Ray, G.W. Mat. Res. Soc. Symp. Proc. 511:199, 1998.
8. Fox, R., Pellerin, J.P. Unpublished observations.
9. Hummel, J. P. Advanced multilevel metallization materials properties issues for copper integration. In: Schuckert, C.S. ed. DuPont Symposium on Polyimides in Microelectronics, Wilmington, DE. Vol. 6, 1995, p. 54. 18.
10. Ho, P.S., Kwok, T. Rep. Prog. Phys., 52:301, 1989.
11. Hu, C.-K., Rodbell, K.P., Sullivan, T.D., Lee, K.Y., Bouldin, D.P. IBM J. Res. Develop. 39:465, 1995.

12. Wilson, S.R., Tracy, C.J. eds. Handbook of Multilevel Metallization for Integrated Circuits. Park Ridge, NJ: Noyes Publications, 1993.

13. Shih, W.-Y., Chang, M.-C., Havemann, R.H., Levine. J. Symp. VLSI Tech. Dig. p. 83, 1997.

14. Shih, W., Levine, J., Chang, M. 1996. In: Havemann, R., Schmitz, J., Komiyama, H., Tsubouchi, K., eds. Advanced Metallization and Interconnect Systems for ULSI Applications in 1996. Pittsburgh, PA: Materials Research Society, 1996, p. 479.

15. Jeng, S., Taylor, K., Lee, C., Ting, L., Chang, M., Seha, T., Eissa, M., Nguyen, H., Jin, C., Lane, A., McAnally, P., Havemann, R., Luttmer, J. In: Ellwanger, R., Wang, S., eds. Advanced Metallization and Interconnect Systems for ULSI Appications in 1995. Pittsburgh, PA: Materials Research Society, 1996, p. 15.

16. Leu, J., Lu, T.-M., Ho, P. S. Low Dielectric Constant Materials for Deep-Submicron Interconnects Applications. Materials Research Society: Fall 1996 Meeting Tutorial, Boston, 1996.

17. Krevelen, D. Properties of Polymers. 3rd ed. New York: Elsevier, 1990.

18. Jin, C., Ting, L., Taylor, K., Seha, T., Luttmer, J. Proc. 2nd Int. Dielectric for VLSI/ULSI Multilayer Interconncect Conf. (DUMIC) Tampa, FL, 1996, p. 21.

19. Griffin, A., Brotzen, F., Loos, P. J. Appl. Phys. 76:4007, 1994.

20. Hurd, A.. MRS Bulletin 21:11, 1996.

21. Jeng, S., Chang, M., Kroger, T., McAnally, P., Havemann, R. In: VLSI-Technology Systems and Applications Proc., 1994.

22. Graas, C., Ting, L. Mat. Res. Soc. Sym. Proc. 338:429, 1994.

23. Jeng, S.-P., Taylor, K., Chang, M.-C., Ting, L., Lee, C., McAnally, P., Seha, T., Numata, K., Tanaka, T., Havemann, R. H. Mat. Res. Soc. Symp. Proc. 381:197, 1995.

24. Wang, P.-H., Ho, P.S., Pellerin, J., Fox. R. J. Appl. Phys. 84:6007, 1998.

25. Wang, P.-H., Ho, P.S. Mat. Res. Soc. Symp. Proc. 511:353, 1998.

26. Cao H.C., Evans A.G. Mech. Mater. 7:295, 1989.

27. Liechti K.M., Hanson, E. C. Int. J. Fracture Mech. 36:199, 1988.

28. Rice J.R. J. Appl. Mechanics. 55:98, 1988.

29. Kinloch A.J. Adhesion and Adhesives. London: Chapman Hall, 1987.

30. Shaffer E.O., Townsend P., Im J.-H. MRS Conference Proc. ULSI XII, San Francisco, CA, 1997, p. 429.

31. Shaffer E.O., McGarry F.J., Hoang L. Polymer Engineering and Science, 36:2381; 1996.,

32. Wu W.-L., Wallace W.E., Lin E.K., Lin G.W., Glinka C.J., Ryan E.T., Ho H.-M. J. Appl. Phys. 87:1193, 2000

33. Sauerbrey, G.Z. Fuer Physik. 155:206, 1959

34. Stoney, G.G. Proceedings of the Royal Society London, Ser. A. 82:172, 1909.

35. Evans, A.G., Hutchinson, J.W. Acta Metallugica Materialia. 43:2507, 1995.

36. Freund, L.B., Floro, J.A., Chason, E. Appl. Phys. Lett.. 74:1987, 1999

37. Zhao, J.-H., Malik, I., Ryan, T., Ogawa, E.T., Ho, P.S., Shih, W.-Y., Mckerrow, A.J., Taylor, K.J. Appl. Phys. Lett. 74:944, 1999.

38. Zhao, J.-H., Ryan, E.T., Ho, P.S., Mckerrow, A.J., Shih, W.-Y. J. Appl. Phys. Lett. 85:6421,

39. Retajczyk, T.F., Sinha, A.K. Appl. Phys. Lett. 36:161, 1980.

40. Reeber, R.R., Wang, K. Materials Chemistry Physics. 46:259, 1996.

41. Hu C., Morgen M., Ho P.S., Jain A., Gill W., Plawsky J., Wayner JR P. Mat. Res. Soc. Symp. Proc. 565:87, 1999.

42. Goodson, K.E., Flik, M.I., Su, I.T., Antoniadis, D.A. IEEE Electonic Device Letters, 14:490, 1993.

43. Cahill D.G. Rev. Sci. Instrum., 61(2):802, 1990.

44. Cahill D.G., Allen T.H. Appl. Phys. Lett. 65(3):309, 1994.

45. Birge N.O., Nagel S.R. Phys. Rev. Lett. 54:2674, 1985.

46. Birge N.O. Phys. Rev. B34:1631, 1985.

47. Kading, O.W., Skurk, H., Goodson, K.E. Appl. Phys. Lett. 65:1629, 1994.

48. Hu, C., Ogawa E.T., Ho, P.S. J. Appl. Phys. 86:6028, 1999.

49. Ryan E.T. Ho H.-M., Wu W.-L., Ho P.S., Gidley D.W., Drage J. Proceedings of the IEEE 1999 International Interconnect Technology Conference, San Francisco, CA, 1999, p. 187.
50. Smith, D.M., Anderson, J., Cho, C.C., Gnade, B.E. Mat. Res. Soc. Symp. Proc. 371:261, 1997.
51. Jin, C., List, S., Zielinski, E. In: C. Chiang, P. Ho, T.-M. Lu, J.T. Wetzel, eds. Materials Research Society Symposium Vol. 511. San Francisco: Materials Research Society, 1998, p. 213.
52. Bakhru, H., Kumar, A., Kaplan, T., Delarosa, M., Fortin, J., Yang, G.-R., Lu, T.-M., Kim, S., Steinbruchel, C., Tang, X., Moore, J.A., Wang, B., McDonald, J., Nitta, S., Pisupatti, V., Jain, A., Wayner, P., Plawsky, J., Gill, W., Jin, C. Mat. Res. Soc. Symp. Proc. 511:125, 1998.
53. Tompkins H.G. A User's Guide to Ellipsometry. San Diego, CA: Academic Press, 1993.
54. Chason, E., Mayer T.M. Crit. Rev. Solid State Mat. Sci. 22:1, 1997.
55. Dietrich, S., Haase, A. Phys. Rep. 260:1, 1995.
56. Wallace, W.E., Wu, W.L. Appl. Phys. Lett. 67:1203, 1995.
57. Tesmer, J.R., Nastasi, M. Handbook of Modern Ion Beam Materials Analysis. Pittsburgh, PA: Materials Research Society, 1995.
58. Wu, W.-L., Liou, H.-C. Thin Solid Films. 312:73, 1998.
59. Parratt, L.G. Phys. Rev. 95:359, 1954.
60. Higgins, J.S., Benoit, H.C. Polymers and Neutron Scattering. Oxford, UK: Clarendon Press, 1994.
61. Debye, P., Anderson, H.R., Brumberger, H. J. Appl. Phys. 28:679, 1957.
62. Ramos ,T., Rhoderick, K., Roth, R., Brungardt, L., Wallace, S., Drage, J., Dunne, J., Endisch, D., Katsanes, R., Viernes, N., Smith D.M. Mat. Res. Soc. Symp. Proc. 511:105, 1998
63. Gregg, S.J., Sing, K.S.W. Adsorption, Surface Area and Porosity. 2nd ed. New York: Academic Press, 1982.
64. DeMaggio, G.B., Frieze, W.E., Gidley, D.W., Zhu, M., Hristov, H.A., Yee, A.F. Phys. Rev. Lett. 78:1524, 1997.
65. Hristov, H.A., Bolan, B., Yee, A.F., Xie, L., Gidley, D.W. Macromolecules 29:8507, 1996.
66. Xie, L., DeMaggio, G.B., Frieze, W.E., DeVries, J., Gidley, D.W. Phys. Rev. Lett. 74:4947, 1995.
67. Gidley, D.W., Frieze, W.E., Dull, T.L., Yee, A.F., Ho, H.-M., Ryan, E.T. Phys. Rev. B60:R5157, 1999.
68. Hougham, G., Tesoro, G., Viehbeck, A., Chapple-Sokol, J.D. Macromolecules. 27:5964, 1994.
69. Fowles, G.R. Introduction to Modern Optics. New York: Holt, Rinehart and Winston, 1968.
70. Yang, H., Lucovsky, G.. Mat. Res. Soc. Symp. Proc. 511:371, 1998.
71. Prakash, S.S., Brinker, C.J., Hurd, A.J. J. Non-Cryst. Solids. 190:264, 1995.
72. Smith, D.M., Anderson, J., Cho, C.-C., Gnade, B.E. Mat. Res. Soc. Symp. Proc. 371:261, 1995.
73. Azzam, R.M.A., Bashara, N.M. Ellipsometry and Polarized Light. Amsterdam: Elsevier, 1977.
74. Nguyen, C.V., Byers, R.B., Hawker, C.J., Hendrick, J.L., Jaffe, R.L., Miller, R.D., Remenar, J.F., Rhee, H.-W., Toney, M.F., Trollsas, M., Yoon, D.Y. Polymer Preprints. 40:398, 1999.
75. Lide, D.R. Handbook of Chemistry and Physics. Boca Raton, FL: CRC Press, 1993.
76. Auman, B.C. In: T.-M. Lu, S.P. Murarka, T.-S. Kuan, C.H. Ting, eds. Materials Research Society Symposium. Vol. 381. San Francisco: Materials Research Society, 1995, p. 19.
77. Molis, S. E. In: C. Feger, M.M. Khojasteh, J.E. McGrath. eds. Polyimides: Materials, Chemistry and Characterization. Amsterdam: Elsevier, 1989.
78. Ree, M., Chen, K.J., Kirby, D.P. J. Appl. Phys. 72:2014, 1992.
79. Chen, S.T., Wagner, H.H. J. Electronic Materials. 22:797, 1993.
80. Lin, L., Bastrup, S.A. J. Appl. Polymer Sci. 54:553, 1994.
81. Boese, D., Lee, H., Yoon, D.Y., Rabolt, J.F. J. Polymer Sci.: Polymer Phys. 30:1321, 1992.
82. Hardaker, S.S., Moghazy, S., Cha, C.Y., Samuels, R.J. J. Polymer Sci.: Polymer Phys. 31:1951, 1993.
83. Kirchhoff, R.A., Carriere, C.J., Bruza, K.J., Rondan, N.G., Sammler, R.L. J. Macromolec. Sci.-Chem. A28:1079, 1991.
84. Burdeaux, D., Townsend, P., Carr, J., Garrou, P. J. Electron. Materials. 19:1357, 1990.

85. Townsend, P.H., Martin, S.J., Godschaix, J., Romer, D.R., D.W. Smith, J., Castillo, D., DeVries, R., Buske, G., Rondan, N., Froelicher, S., Marshall, J., Shaffer, E.O., Im, J.-H. Mat. Res. Soc. Symp. Proc. 476:9, 1997.

86. Liu, Y., Bremmer, J., Gruszynski, K., Dall, F. VLSI Multilevel Interconnection Conference, Santa Clara, CA, 1997, p. 655.

87. Bremmer, J.N., Liu, Y, Gruszynski, K.G., Dall, F. C. Mat. Res. Soc. Symp. Proc. 476:37, 1997.

88. Hacker, N. MRS Bulletin. 22:33, 1997.

89. Hacker, N.P., Davis, G., Figge, L., Krajewski, T., Lefferts, S., Nedbal, J., Spear, R. In: C. Case, P. Kohl, T. Kikkawa, W.W. Lee, eds. Materials Research Society Symposium. Vol. 476. San Francisco: Materials Research Society, 1997, p. 25.

90. Kim, S.M., Yoon, D.Y., Nguyen, C.V., Han, J., Jaffe, R.L. In: C. Chiang, P.S. Ho, T.-M. Luy, J.T. Wetzel, eds. Materials Research Society Symposium. Vol. 511. San Francisco: Materials Research Society, 1998, p. 39.

91. Jin, C., Luttmer, J.D., Smith, D.M., Ramos, T.A. MRS Bulletin. 22:39, 1997.

92. Gross, J., Reichenauer, G., Fricke, J. J. Phys. D Appl. Phys. 21:1447, 1988.

93. Gross, J., Fricke, J. J. Non-Crys. Solids. 145:217, 1992.

94. Nix, W.D. Metallurgical Transactions A 20A:2217, 1989.

95. Ulrich, R., Torge,R. Appl. Opt. 12: 2901.

96. Robert Speyer. Thermal Analysis of Materials. New York: Marcel Dekker, 1994.

97. Hatakeyama, T., Liu, Z. Handbook of Thermal Analysis. New York: Wiley, 1998.

98. Kang, Y.-S. Microstructure and strengthening mechanisms in Al thin films on polyimide film. PhD Dissertation, Univ. of Texas, Austin, 1996.

99. Oliver, W.C., Pharr, G.M. J. Mater. Res. 7:1564, 1992.

100. Socrates, G. Infrared Characteristic Group Frequencies, Tables and Charts. 2nd ed. New York: Wiley, 1994.

101. Garton, A.R. Infrared Spectroscopy of Polymer Blends, Composites and Surfaces. Vol. 1. New York: Oxford Univ. Press, p. 279.

102. Miller, K.J., Hollinger, H.B., Grebowicz, J., Wunderlich, B. Macromolecules 23:3855, 1990.

103. Pine, S.H. Organic Chemistry. 5th ed. New York: McGraw-Hill, 1987

# 13

# High-Resolution Profilometry for CMP and Etch Metrology

**Anna Mathai and Clive Hayzelden**
*KLA-Tencor Corporation, San Jose, California*

## I. INTRODUCTION TO CHEMICAL-MECHANICAL POLISHING AND ETCH SURFACE METROLOGY REQUIREMENTS

The desire to increase the clock speed of integrated circuits (ICs) has led toward smaller, denser geometries and higher levels of metallization. Such scaling creates a formidable set of fabrication challenges. Smaller device geometries are currently being generated by extending optical lithography with high-numerical-aperture lenses and shorter-wavelength light sources. The use of shorter-wavelength light sources comes at a high cost—lower depth of focus. Hence, it is then necessary to minimize surface topography to achieve optimal lithography, and global planarization has become an essential process step. Global planarization meets the requirement for shallower depth of focus for submicron lithography, improves step coverage, and enables higher device yield.

Chemical-mechanical polishing (CMP) processes are used widely in the semiconductor industry to provide global planarity (1). However, within-wafer, wafer-to-wafer, and lot-to-lot process variations in CMP can lead to many failure modes and the need for extensive process monitoring. For example, most polishing processes lead to recessed metal features because of the differential polishing at the dielectric/liner/metal interface. In addition, CMP performance is highly dependent on the consumable set used: The polishing pad, slurry, carrier film, and conditioning of the pads will dramatically affect the erosion, dishing, and recess of metal features. As the feature size of ultralarge-scale-integration technology decreases, stringent requirements are placed on the spatial resolution of the CMP metrology tools.

Copper is being widely introduced at the 130-nm technology node as an interconnect metal (2). Copper is an attractive substitute to aluminum/tungsten due to its lower resistivity and improved resistance to electromigration. The lower resistivity of copper also allows for greater scalability, leading to higher device densities. The implementation of copper has required the use of novel processing techniques such as damascene processing. In this approach, oxide trenches are etched and then filled with copper, followed by chemical-mechanical planarization to remove the excess copper. The damascene process may be repeated for as many metallization levels as necessary. However, multilevel metallization (MLM) requires highly planarized surfaces. The greater the attenuation of topography at the lowest dielectric and metal levels, the lower the stack-up effects will be at the subsequent levels. The success of MLM systems using subquarter-micron geometries and copper interconnects, therefore, depends on highly planarized, well-characterized surfaces. Shrinking design rules and MLM have made CMP one of the most important process steps in manufacturing of microelectronic devices.

The drive toward smaller device geometries has also placed much tighter control limits on key plasma etch processes, making etch metrology increasingly important but also more difficult. Plasma etching is used in a variety of processes, including the manufacture of contacts and vias, shallow trench isolation, DRAM, and dual damascene. Etch metrology revolves around two key parameters: critical dimension (CD) and depth of the etched features. Etch depth monitoring is important to avoid poor electrical performance (underetch) or underlying structural damage (overetch).

The subsequent sections further detail CMP and etch surface metrology requirements and solutions that meet these requirements.

## II. CHEMICAL-MECHANICAL POLISHING SURFACE METROLOGY

Chemical-mechanical polishing process sectors have commonly used profilometry to measure the post-CMP planarization of the wafer. Stylus profilometers are robust instruments that are simple and easy to use and have good vertical resolution (3). With a lateral resolution of a few micrometers, they are eminently suitable for measuring global surface structure on the macroscopic scale. However, as the feature size of ULSI technology has decreased, it has become increasingly difficult to use a profiler to resolve individual surface features. Atomic force microscopes (AFMs) overcome the lateral resolution limitation of the stylus profilometer and have $\sim$ 1-nm resolution in both the $x$ and $y$ directions (4). However, AFMs typically have a scan length of less than 100 μm, with a limited $z$ range of $\sim$ 5 μm and hence cannot be used for macroscopic wafer measurements. To overcome these limitations, a new class of instrument, the high-resolution profiler (HRP), has been invented to provide surface metrology measurements on length scales from submicron to the entire wafer (5–7). Over the past year, the HRP has become the tool of choice for developing and monitoring advanced CMP processes.

## A. High-Resolution Profilometry

### 1. Dual-Stage Technology

The HRP has micro- and macroimaging capabilities: the ability to position the stylus with nanometer resolution and a lateral scan range of up to 300 mm. This is accomplished with a dual-stage system as shown in Figure 1. The sample stage, used for macro-imaging, is a mechanical stage with a repeatability of 1 μm (1σ). The drive mechanism of the stage is based on a mechanical screw with a pitch of 1 mm. The carriage slides on a glass reference flat that constrains the motion of the sample stage to a horizontal plane.

The sensor stage, used for micro-imaging, has a range of 90 μm and 1-nm resolution. It consists of piezoelectrics mounted in a mechanical (flexure) frame. The flexure frame is designed to remove the undesirable bending motions of the piezoelectrics and to only allow the desired x–y rectilinear expansion. The sensor stage also contains two capacitance sensors, both with resolution of 1 nm, which determine the x and y positions of the stage.

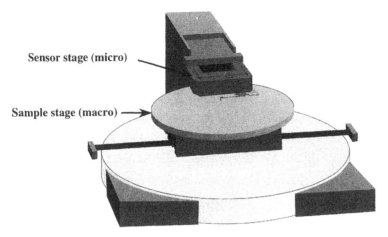

**Figure 1** HRP dual-stage technology enables micro- and macroimaging. The sample stage provides long-scan capability of up to 300 mm. The sensor stage has a range of 90 µm, with 1-nm lateral resolution.

These sensors are calibrated against a commercially available waffle pattern from VLSI Standards (8) and use a feedback loop to eliminate hysteresis and creep.

## 2. Sensor and Stylus Design

The HRP sensor consists of an electromagnet, a flexure hinge, a linear variable-differential capacitor (LVDC) sensor, and a sensor vane, as shown in Figure 2. The stylus moves up and down as it lightly scans over the contours of the surface in contact with it (contact mode). The flexure hinge constrains the stylus arm to pivot about the hinge in a seesaw motion and changes the relative position of the sensor vane between the capacitor plates. By measuring the change in the capacitance, one can measure any changes in the height of the surface being scanned. This sensor design enables tracking vertical changes in the sample surface topography with better than 0.1% linearity over the entire range of 130 µm.

The HRP also a provides a low, constant force independent of the stylus. When a new stylus is placed in the HRP, the electromagnet swings the pivot through its entire range. The corresponding force vs. distance calibration curve for the pivot is recorded. A

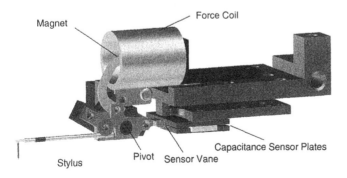

**Figure 2** HRP ultralite sensor provides a low, constant force between the stylus and the surface being imaged.

digital signal processor–based control system is then used to cancel out the spring force of the pivot so that the force is constant at all times. Providing a low, constant force between the stylus and sample is the key to nondestructive imaging. It has been demonstrated that soft metals such as Al and Cu can be imaged at the typical HRP scanning force of 0.05 mg $(5 \times 10^{-7}\,\mathrm{N})$ with no impact on yield (9).

The styli used in the HRP in contact mode are made from bulk diamond, which makes them robust and long lasting. They are available with variety of radii of curvature as appropriate for applications ranging from measuring the recess of submicron tungsten vias to the height of solder bumps.

### 3. Step Height Calibration

An extensive array of step height calibration standards is available commercially from VLSI Standards (8). These step height standards range in heights from 8 nm to 50 μm, and it is recommended that the user calibrate the profiler to a standard that represents the typical application.

### B. Applications of Chemical-Mechanical Polishing

### 1. Global Planarity

One key metric for CMP performance is global planarity, which is a measure of the lithography depth-of-focus budget that has been consumed by the polishing process. Wafer surface topography or flatness is a measure of CMP performance; however, any wafer-chucking effects or thickness variations also contribute to this measurement. Global planarity can be characterized in terms of full-die or across-wafer scans using 2D line scans or 3D images (10). Figure 3 shows an HRP scan of a full die, approximately 6 mm × 5 mm, postoxide CMP.

**Figure 3**   HRP image of a full die, showing postoxide CMP die-level planarity.

## 2. Tungsten Via Recess

Contacts are holes in the interlevel dielectric (ILD) that, when filled with metal, make contact between the transistors and the first metallization layer. Vias are similar to contacts except that they connect two metallization levels. The tungsten/aluminum metallization steps begin with etching contacts or vias in the ILD. Next a CVD tungsten film is deposited and the excess tungsten removed using CMP, leaving the contacts and vias filled with tungsten. Then PVD Al is deposited and etched to complete that wiring layer. The process is then repeated, starting with depositing the oxide and a CMP step to provide a planar starting point for subsequent steps.

It is necessary to overpolish at the tungsten CMP step, leaving the tungsten vias slightly recessed with respect to the ILD. This ensures that there is no residual tungsten on the surface to cause an electrical short. However, if the recess is too large, this will lead to increased circuit resistance. In addition, via recess depends on via size and polishing parameters as well as on how many vias are surrounding it. Roughly speaking, since the tungsten polishes faster than the ILD, its pattern becomes recessed with respect to the ILD. The tungsten area continues to recede until it is no longer in good contact with the pad. With the tungsten recessed, the pressure on the ILD is higher relative to unpatterned areas of the wafer. This increased pressure increases the polishing rate, causing a more rapid polishing of the patterned area, known as *erosion*. Hence, for isolated vias, the via recess is large, since the ILD polishes slower than the tungsten. In areas with high via density, the ILD polishes more rapidly, due to erosion, and hence the recess is smaller. Understanding pattern density effects is essential to developing the models needed for control and optimization of metal polishing.

Figure 4 shows an HRP image of recessed tungsten vias. In addition to via recess, there is significant oxide erosion over a long scan of an array of tungsten vias, as shown in Figure 5. In addition, high-resolution scans of tungsten vias in different regions of the array measure via recess. The HRP uses micro- and macroimaging to measure both tungsten via recess and oxide erosion, post-CMP.

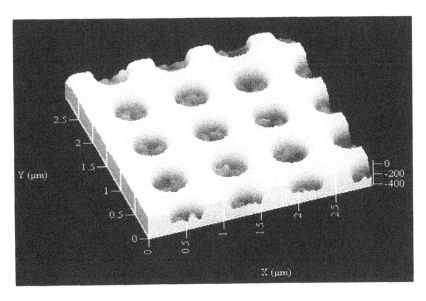

**Figure 4** Tungsten CMP plug recess.

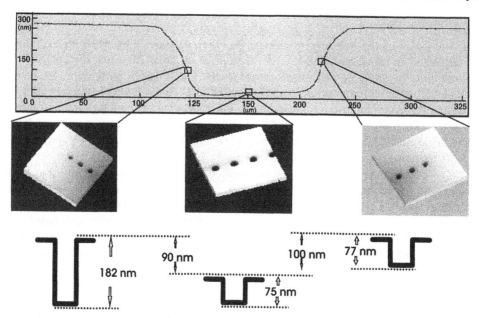

**Figure 5** HRP provides a combination of micro- and macroimaging for characterization and monitoring of recess and oxide erosion.

## 3.  Oxide Erosion and Metal Dishing

Currently the aluminum/tungsten metallization process has been used very successfully down to 0.25-μm technology. For future generations, copper has been identified as an attractive substitute due to its lower resistivity and improved electromigration resistance. In addition, the dual-damascene architecture reduces the processing steps by about 20% by combining the implementation of the interconnects and wiring simultaneously.

In the dual-damascene process (see Figure 6), the trench and via features are etched into the ILD, which is then lined with a barrier metal such as Ti or TiN. A copper seed layer is deposited, and then the lines are electroplated with Cu. Then the excess Cu is polished, leaving a planar surface to repeat these steps to provide subsequent metallization levels. As in the case of oxide surrounding tungsten vias, the oxide surrounding Cu lines and pads will be polished at a different rate from the field oxide and can be characterized using the HRP. As illustrated in Figure 7, critical control parameters for metal CMP are metal dishing and recess and oxide erosion.

In order to ensure that the CMP planarity variation results in an electrical performance compatible with design criteria, one way to monitor the process and the consumable set is to incorporate a test structure in the chip design. Figure 8 shows the schematic of such a test structure and HRP scans from three wafers polished under different conditions. It is clear that polishing condition C gives the best results, with the least dishing and erosion.

Figure 9 shows a high-resolution scan of array of 0.5-μm-wide lines with 1-μm spacing. As expected, the Cu lines are recessed with respect to the oxide.

## 4.  Surface Texture

Another key metric of the CMP process is microscale surface texture or roughness. The HRP is used to measure the surface roughness of films such as AlCu and Cu, as shown in

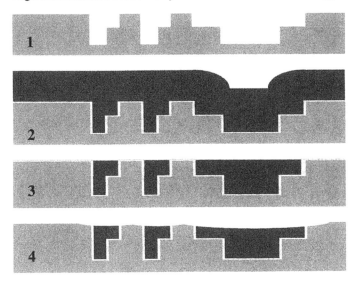

**Figure 6**   Dual-damascene process: (1) etching, (2) barrier layer deposition and copper fill, (3) copper CMP, and (4) possible overpolish.

Figure 10. It can also be used to measure pits, microscratches, and residual slurry particles. Figure 11 shows good measurement correlation between AFM and HRP data on surface roughness.

## 5.   Accuracy and Precision

All systems are calibrated to certified step height standards from VLSI Standards. There are three measurement ranges available, and it is recommended that the user choose the appropriate measurement range and also calibrate the system to a VLSI standard close to the typical measurement. The short-term repeatability of the systems is guaranteed to be the larger number of 7.5 Å or 0.1%, while the long-term repeatability is guaranteed to be the larger number of 20 Å or 0.25%.

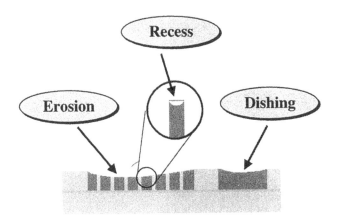

**Figure 7**   Erosion, recess, and dishing in Cu CMP.

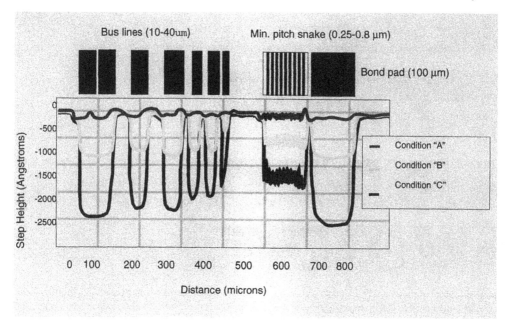

**Figure 8**   Test structure for monitoring planarity performance of a Cu CMP process.

## 6.   Limitations on Stylus Life

In a tip-life study measuring the recess of a single 0.25-μm tungsten plug repeatedly, the Durasharp stylus was still sharp after over 250 mm of travel (see Figure 12). While stylus life is application dependent, the HRP styli being made of diamond are robust and long lasting.

In contact-mode scanning, the stylus–sample interaction generates lateral or shear forces. The need to limit, as well as the ability to withstand, this shear force necessitates that one use a robust stylus with a relatively large included angle. In turn, the stylus geometry limits the aspect ratios of the features that can be resolved by the stylus. To overcome this limitation, a new mode of profiler operation called *dipping mode* has been developed. This is described further in Section III.

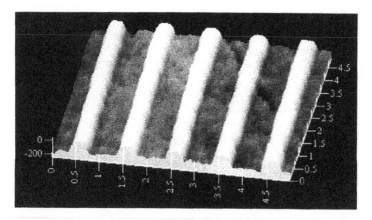

**Figure 9**   Copper CMP recess in dense lines.

(a) (b)

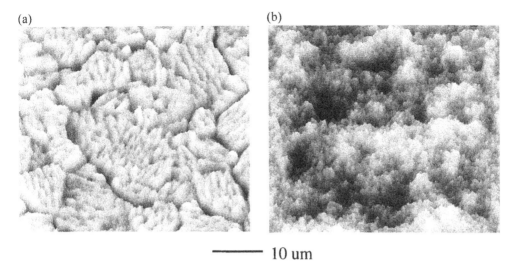

——— 10 um

**Figure 10** HRP images of: (a) AlCu with Ra = 19.7 nm and (b) Cu with Ra = 6.3 nm. (Courtesy of Texas Instruments.)

## III. ETCH SURFACE METROLOGY

The drive toward smaller-design rules imposes new challenges on etch processes. Of these, the largest barrier to consistent device performance is maintaining critical dimension uniformity (lateral and vertical). Currently there are three ways to monitor etch depth. Scanning electron microscopy (SEM) is the most reliable and hence the most commonly used technique. While appropriate for critical dimension (CD) measurements, this technique is not an efficient way to monitor etch depths in production. It is costly and undesirable, since it requires the cross-sectioning of product wafers. It is also time consuming, and the longer time to results means more wafers will have gone through process prior to depth-monitoring feedback, which increases the risk to those wafers. The second solution is to conduct electrical tests on the wafers after the interconnect level is complete.

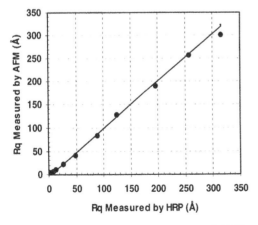

**Figure 11** Measurement correlation of RMS roughness (Rq) between AFM and HRP. (Courtesy of Texas Instruments.)

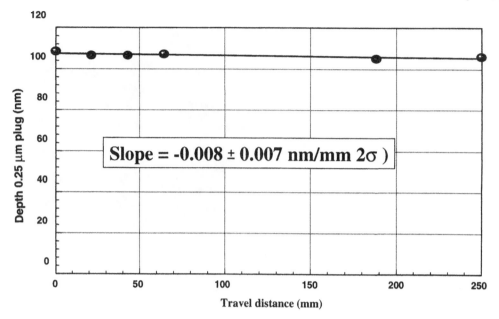

**Figure 12**  Tip-life study during measurement of 0.25-µm tungsten plug recess. The Durasharp stylus reveals no apparent change after 250 mm of travel.

This is also time consuming, leaving untested wafers in process for longer periods of time. Third, atomic force microscopes (AFMs) are used occasionally to monitor etch depth; however, concerns about tip quality, reliability, and ease of use have limited their effectiveness as a solution in a production environment. Clearly, there is a need for a nondestructive, reliable, and easy-to-use technique to measure etch depths with quick time to results.

## A.  High Aspect Ratio Measurements—Dipping Mode

To enable high aspect ratio measurements, a new mode of profiler operation called *dipping mode* has been developed (see Figure 13). In dipping mode, the sample is scanned in a series of discrete steps as follows: The stylus is lowered into contact with the sample surface, a data point is obtained, and the stylus is then lifted from the sample surface to a user-specified height and moved laterally prior to beginning the next cycle. Since the stylus is not in contact with the sample surface during its lateral motion, the shear forces it is subjected to are minimized. This enables the use of fine styli with included angles of less than 10° and radii of curvature of 10–20 nm and hence the ability to measure the depth of high-aspect-ratio features.

## B.  Etch Applications

### 1.  Contacts

The electrical connections between the transistor level to the first and subsequent metallization levels are made using contacts and vias, respectively. Contact and via depth are

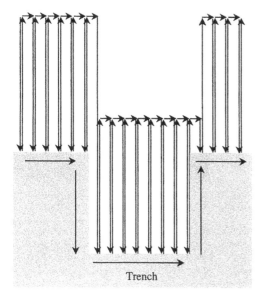

**Figure 13** Schematic profiling of a trench in dipping mode.

important to monitor to avoid poor electrical performance resulting from underetching or damage to the underlying structure resulting from overetching.

Figure 14a shows a dipping mode image of a 0.35-μm contact near the center of the wafer. The image was built up in a raster fashion, with a dipping mode profile scan along the $x$-direction and stepping over the $y$-spacing between subsequent profiles. Figure 14b shows a cross-sectional profile through the centers of the contacts in Figure 14a, with the depth measured as 1.105-μm. Figure 14c shows a SEM cross section of a contact in approximately the same location—the measured depth of 1.028-μm agrees well with the HRP data. The correlation results for a series of local interconnects (0.27-μm diameter) and vias (0.6-μm diameter) are shown in Figure 15. There is excellent agreement between the HRP and SEM depth measurements.

Figure 16 shows the HRP image of a 0.18-μm CD via that is 0.96 μm deep, showing system capability to measure aspect ratios as high as 5:1 at 0.18-μm width.

## 2. Shallow Trench Isolation Depth

Isolation technology is rapidly evolving away from local field isolation (LOCOS) to shallow trench isolation (STI) (11). This is the result of the inevitable requirements to further shrink design rules while maintaining sufficient oxide thickness for isolation. The STI technique process offers the benefits of smaller isolation size and a planar starting surface, both of which are conducive to achieving higher device densities per unit area.

In the STI process, one of the first steps is to etch a trench into the silicon, with a pad oxide and nitride stack defining the active area (see Figure 17). The trench is then filled with a deposited oxide, followed by an oxide CMP step to remove the oxide over the active area. This is one of the most technically demanding etch applications, since, unlike other etch steps, the STI trench does not have an etch-stop layer. The etch depth of the trench is

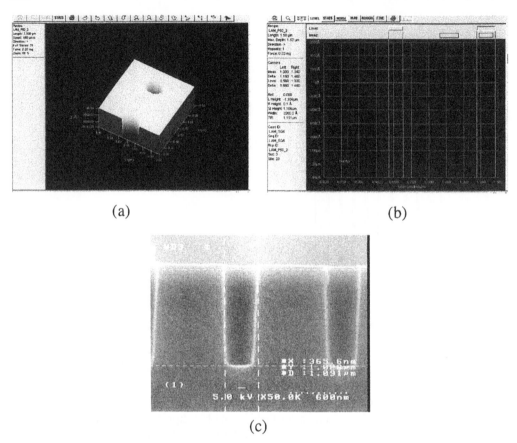

(a)                                                          (b)

(c)

**Figure 14** 0.35-μm geometry contacts, postetch: (a) HRP image, (b) cross-sectional profile through centers of contacts, and (c) SEM cross section, showing good agreement.

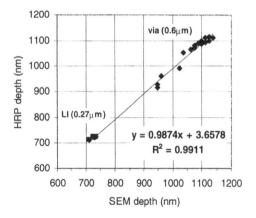

**Figure 15** Correlation between the HRP-240$^{ETCH}$ and the SEM for etch depth measurements of local interconnects and vias.

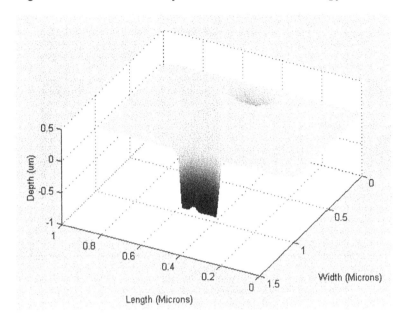

**Figure 16**  Postetch HRP image of a via, 0.96-μm deep and 180 nm wide.

a key process parameter—trenches that are etched too shallow will lead to poor isolation and parasitic transistor effects, while if too deep they can induce stress in the wafer.

Figure 18 shows an HRP image of a 0.25-μm SRAM pattern, after shallow trench isolation etch processing. An analysis algorithm automatically levels the data based on the user's choice of highest, lowest, or most populous plane. For postetch metrology, a histogram analysis typically reveals a bimodal distribution, as shown in Figure 18b. The final

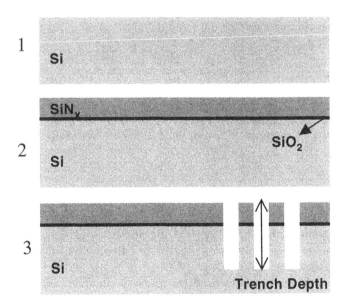

**Figure 17**  Shallow trench isolation: (1) substrate, (2) pad oxide with nitride deposition, and (3) trench etching. The trench depth is indicated.

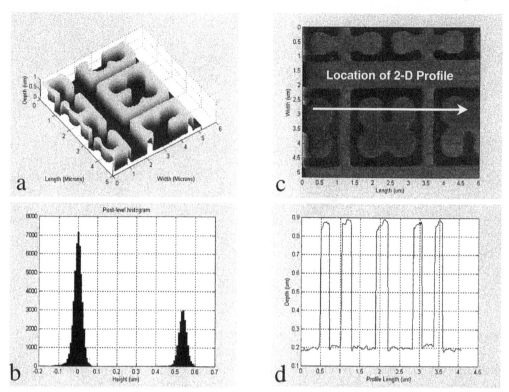

**Figure 18**  0.25-μm SRAM pattern, after shallow trench isolation: (a) HRP image, (b) histogram of all data, (c) top-down view of HRP image showing location of 2D profile, and (d) 2D profile.

depth is calculated by taking the difference between the means of the two modes in the data. Figure 18c shows a top-down view of the HRP data and the location of a 2D profile through the data, shown in Fig. 18d.

### 3. Damascene Trench Depth

In the dual-damascene process (see Fig. 6), the first step is to etch the trench and via features into the ILD. Critical control parameters for this etch process are the trench and via depth. Figure 19 shows results of a production-type measurement, with long-term repeatability of better than 1.5% ($3\sigma$).

## IV. SUMMARY

The manufacturing of microelectronic devices based on 0.18-μm and below design rules and multilevel metallization has made CMP a necessary process step. In addition, there is a transition to copper as the interconnect metal. Because CMP is a timed and feature-density-dependent process, it relies heavily on advanced metrology techniques, such as high-resolution profiling, for its success. The increased packing densities of today's integrated devices also places further demands on the etching process steps and necessitates nondestructive high-aspect-ratio depth metrology.

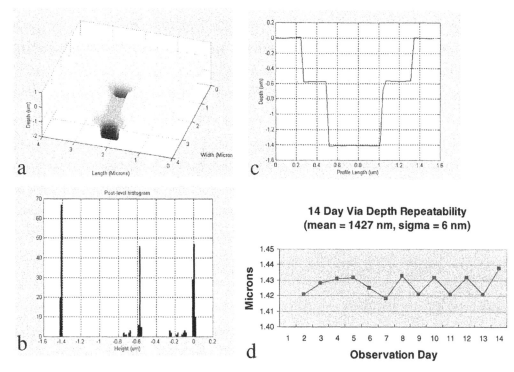

**Figure 19** HRP measurement of dual-damascene etch depth: (a) HRP image, (b) histogram of all data, (c) 2D profile through trench, and (d) 14-day via depth repeatability data showing long-term reproducibility of better than 1.5% (3σ).

## ACKNOWLEDGMENTS

The authors would like to thank their colleagues Kelly Barry, Marcus Afshar, Michael Young, Amin Samsavar, and John Schmidt for their input and for many fruitful discussions.

## REFERENCES

1. J.M. Steigerwald, S.P. Murarka, R.J. Gutmam. Chemical Mechnical Planarization of Microelectronic Materials. New York: Wiley, 1997.
2. National Technology Roadmap for Semiconductors, 1999 edition. Semiconductor Industry Association, San Jose, California.
3. L. P. Howard. Long range constant force profiling for measurement of engineering surfaces. Review of Scientific Instruments. 63(10):4289–4295, 1992.
4. D. Rugar, P. Hansma. Atomic force microscopy. Physics Today: pp 23–30, 1990.
5. J. Schneir, R.T. Jobe, V.W. Tsai. High-resolution profilometry for CMP process control. Solid State Technology. 6:203–206, 1997.
6. J. Farkos, M. Freeman. New requirements for planarity and defect metrology in soft metal CMP. Proceedings of the Fourth Workshop on Industrial Applications of Scanned Probe Microscopy, Gaithersburg, MD, 1997.

7.  J. Schneir, R. Jobe, V.W. Tsai, A. Samsavar, D. Hetherington, M. Moinpour, Y.C. Park, M. Maxim, J. Chu, W. Sze. High-resolution profilometry for CMP process control. Advanced Metallization and Interconnect Systems for ULSI Applications XII, MRS Conf. Proc., 1997, pp. 555–559.
8.  VLSI Standards, Inc. www.vlsistd.com.
9.  A. Mathai, J. Schneir. Force control for non-destructive HRP imaging of dual damascene post-CMP. KLA-Tencor Applications Note, 1998.
10. A. Mathai, Y. Gotkis, D. Schey, J. Schneir. New method for characterizing post-CMP surface topography. CMP-MIC '98.
11. P.V. Cleemput, H.W. Fry, B. van Schravendijk, W. van den Hoek. HDPCVD films enabling shallow trench isolation. Semiconductor International: pp 179–186, 1997.

# 14

# Critical-Dimension Metrology and the Scanning Electron Microscope

**Michael T. Postek and András E. Vladár**
*National Institute of Standards and Technology, Gaithersburg, Maryland**

## I. INTRODUCTION

Metrology is a principal enabler for the development and manufacture of current and future generations of semiconductor devices. With the potential of 130-nm, 100-nm, and even smaller linewidths and high-aspect-ratio structures, the scanning electron microscope (SEM) remains an important tool, one extensively used in many phases of semiconductor manufacturing throughout the world. The SEM provides higher-resolution analysis and inspection than is possible by current techniques using the optical microscope and higher throughputs than scanned probe techniques. Furthermore, the SEM offers a wide variety of analytical modes, each contributing unique information regarding the physical, chemical, and electrical properties of a particular specimen, device, or circuit (3). Due to recent developments, scientists and engineers are finding and putting into practice new, very accurate and fast SEM-based measuring methods in research and production of microelectronic devices.

## II. FUNDAMENTAL SEM ARCHITECTURE

Scanning electron microscopes, whether general-use laboratory or specialized instruments for inspection and dimensional measurement of integrated circuit structures, all function essentially the same. The SEM is named in this way because in it, a finely focused beam of electrons is moved or scanned from point to point over the specimen surface in a precise, generally square or rectangular pattern called a *raster* pattern (Figure 1). The primary beam electrons originate from an electron source and are accelerated toward the specimen by a voltage usually between 200 V (0.2 kV) and 30,000 V (30 kV). For semiconductor production applications, an automated critical-dimension SEM (CD-SEM) tool typically operates between 400 V and 1000 V. The electron beam travels down the column, where it undergoes an electron optical demagnification by one or more condenser lenses. This demagnification reduces the diameter of the electron beam from as large as several micrometers to nan-

---

* See Refs. 1 and 2.

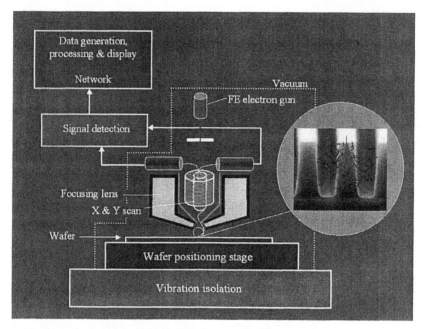

**Figure 1** Schematic drawing of the SEM column.

ometer dimensions. Depending on the application, magnification range, resolution required, and specimen nature, the operator optimizes the image by choosing the proper accelerating voltage and amount of condenser lens demagnification. Generally, depending upon operating conditions and type of instrument, the electron beam diameter is few nanometers when it impinges on the sample.

Scanning coils in the microscope precisely deflect the electron beam in a raster pattern controlled by a digital or analog X and Y scan generator. This deflection is synchronized with deflection of the visual and, where provided, record cathode ray tubes (CRTs) so there is a point-by-point visual representation of the signal being generated by the specimen as it is scanned. The smaller the area scanned in the raster pattern relative to the fixed size of the display CRT, the higher the magnification. Alternatively, the smaller the area on the sample represented by one pixel, the higher the effective magnification. The proper calibration of either the raster pattern (i.e., magnification calibration) or pixel size is essential (4). The SEM is capable of extremely high magnification and resolution, depending upon the instrument design, (5,6). In comparison, many optical microscopes, depending upon their illumination source, have their best resolution limited by the effects of diffraction to about 0.25–0.5 μm.

Figure 1 describes essentially the design of a "top-down" type of instrument. In this type, which is optimized for throughput, there is no tilting of the sample possible. Other instrument designs, incorporating tilt, are also available (especially in the laboratory-type and defect-review designs), thus enabling the observation of sidewall structure, cross-sectional information, and optimization for x-ray collection. Some of the newer instruments also provide the ability to obtain stereoscopic information.

One of the major characteristics of the SEM, in contrast to the optical microscope, is its great depth of field, which is 100–500 times greater than that of an optical microscope. This characteristic allows the SEM to produce completely in-focus micrographs of relatively rough surfaces even at high magnifications. Even with this large depth of field, some high-aspect-ratio semiconductor structures are larger than the depth of field afforded by this instrument. The SEM is also parfocal; it can be critically focused and the astigmatism corrected at a magnification higher than that needed for work, so the magnification can be reduced without a loss of instrument sharpness.

The proper operation of a scanning electron microscope requires maintaining the electron microscope column under high vacuum, since electrons cannot travel for any appreciable distance in air. The requirement for vacuum does tend to limit the type of sample that can be viewed in the SEM. The mean free path (MFP), or the average distance, an electron travels before it encounters an air or gas molecule must be longer than the SEM column. Therefore, depending upon the type of instrument design, ion pumps, diffusion pumps, or turbomolecular pumps are utilized to achieve the level of vacuum needed for the particular instrument design. An alternative design, utilizing lower vacuum in the specimen environment, has been used in laboratory applications but has not been applied to semiconductor production (See Sec. VIII, Environmental SEM). This type of instrument may have greater application in the near future.

## III.  SEM SIGNALS

The interaction of an energetic electron beam with a solid results in a variety of potential "signals" being generated from a finite interaction region of the sample (3). A signal, as defined here, is something that can be collected, used, or displayed on the SEM. The most commonly used of the SEM signals are the secondary and backscattered electron signals. The intensity distribution of these two signal types is shown in Figure 2. The electron beam can enter into the sample and form an interaction region from which the signal can originate. The size of the interaction region is related directly to the accelerating voltage of the primary electron beam, the sample composition, and the sample geometry. Those signals that are produced within the interaction region and leave the sample surface can be potentially used for imaging, if the instrument is properly equipped to collect, display, and utilize them.

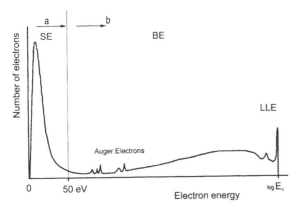

**Figure 2**  Intensity distribution of some of the typical SEM signal types. The arrows denote the energy ranges of (a) secondary electron signal and (b) backscattered electron signal.

## A.  Electron Range

The primary electron beam can enter into the sample for some distance, even at low accelerating voltage. Thus, it is important to understand and define this interaction volume. The maximum range of electrons can be approximated several ways (7). Unfortunately, due to a lack of fundamental understanding of the basic physics underlying the interaction of an electron beam in a sample (especially at low accelerating voltages), there are no equations that accurately predict electron travel in a given sample (8). Over the past several years, work has improved this situation a great deal. It is important to have an understanding of the electron beam penetration into a sample. One straightforward expression derived by Kanaya and Okayama (9) has been reported to be one of the more accurate presently available for approximating the range in low-atomic-weight elements and at low accelerating voltages. The calculated trajectories of electrons in a photoresist layer are shown for 5 keV and 800 eV (Figure 3). The electron ranges shown approximate the boundaries of the electron trajectories as if the material was a continuous layer. In the case of a multifaceted structure, such as the one illustrated with three lines, the electrons follow far more complex paths in the higher-accelerating-voltage case. They penetrate (and leave and potentially penetrate again) the surfaces in every direction, making the signal formation process more difficult to accurately model.

## B.  Secondary Electrons

The most commonly collected signal in the CD-SEM is the secondary electron (SE); most micrographs readily associated with the SEM are mainly (but not exclusively) composed

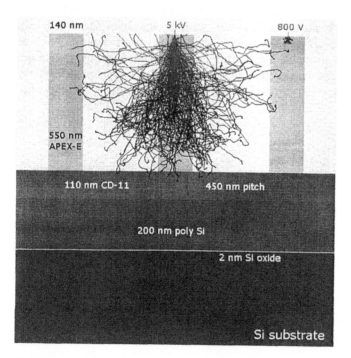

**Figure 3**  Monte Carlo electron trajectory plots for high (5 kV) (left) and low (800 V) (right) accelerating voltages.

of SEs (Figure 4). Some of the SEs, as discussed already, are generated by the primary electron beam within about the first few nanometers of the specimen surface, their escape depth varying with the accelerating voltage and the atomic number of the specimen. Typically, this depth ranges from 2–10 nm for metals to 5–50 nm for nonconductors. Secondary electrons also result from backscattered electrons as they leave the specimen surface or collide with inside surfaces of the specimen chamber (10). The number of secondary electrons emitted from a particular sample relates to the secondary electron coefficient of the materials comprising the sample and other factors, such as surface contamination (11–15).

The secondary electrons are arbitrarily defined as those electrons generated at the sample that have between 1 and 50 eV of energy. The secondary electrons are the most commonly detected for low-accelerating-voltage inspection due to their relative ease of collection and since their signal is much stronger than any of the other types of electrons available for collection. However, due to the low energy of the secondary electrons, they cannot escape from very deep in the sample, and thus their information content is generally surface specific. Consequently, the information carried by secondary electrons potentially contains the high-resolution sample information of interest for metrology.

Unfortunately, the reality of the situation is not that straightforward. Secondary electrons do not originate from the location of the primary electron beam only. The nature of the imaging mechanism has a direct effect upon the secondary electron image observed. This is an extremely important point, because this means that the "secondary" electron signal, usually collected in the SEM, is a composite of a number of signal mechanisms. It has been calculated that the number of remotely generated electrons (i.e., energy less than 50 eV) is much larger than of those generated from the primary electron beam interaction by a factor greater than 3 (16). It should be clearly understood that because of electron scatter in a sample, secondary electrons can and do originate from points other than the point of impact of the primary electron beam (17–19).

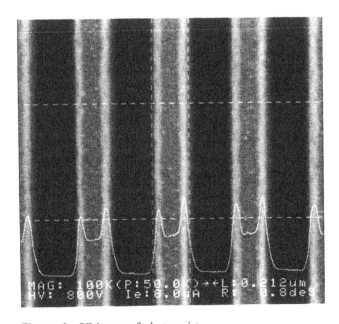

**Figure 4**  SE image of photoresist.

1.  SE Signal Components

There are four potential locations from which the electrons composing the SE image can be generated (10). The SE signal is composed not only of those secondary electrons generated from initial interaction of the electron beam as it enters the sample (SE-1), but also from secondary electrons generated by the escape of elastically and inelastically scattered backscattered electrons (BSEs) when they leave the sample surface (SE-2). The BSEs can have multiple interactions with other structures on the sample or other internal instrument components and generate more secondary electrons (SE-3). Stray secondary electrons (SE-4) coming from the electron optical column itself may also enter the detector (Figure 5). Backscattered electrons can also be collected as a component of the secondary electron image if their trajectory falls within the solid angle of collection of the electron detector. This is one reason why the modeling of the secondary electron signal is so difficult.

Peters (19) measured the components of the secondary electron signal from gold crystals. He found that, depending upon the sample viewed, for the total SE image the contribution of the SE-2 electrons is approximately 30% and the contribution to the image of the SE-3 electrons is approximately 60%, as compared to approximately 10% of the image contributed by the SE-1–derived signal. The standard secondary electron detector does not discriminate among these variously generated electrons, and thus the collected and measured secondary electron signal is composed of a combination of all of these signal-forming mechanisms. For metrology, the difficulties in interpreting this composite signal can lead to interpretation errors. These errors can be highly variable, and they have a strong dependence upon sample composition, sample geometry, and to a lesser or greater extent (depending on instrument design) other physical factors, such as an instrument's internal geometry that induces anomalies in the detector collection field (i.e., stage motion). Furthermore, since this signal is highly variable and often instrument specific, it is extremely difficult to model.

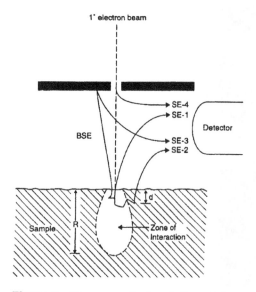

**Figure 5**   Diagrammatic description of the possible four derivations of secondary electrons in a typical laboratory SEM.

## 2. Collection of Secondary Electrons

In most SEMs, secondary electrons are generally collected by the use of a scintillator type of detector of the design of Everhart and Thornley (20), often referred to as an E/T detector, or a modification of that design. Other detector types are also possible, including the microchannel-plate electron detector (21,22). Due to the low energy of the secondary electron signal, the electron paths are easily influenced by any local electric or magnetic fields; therefore, this detector is equipped with a positively biased collector to attract the secondary electrons. The collection efficiency of an SE detector relates directly to its position, its potential, and the field distribution at the sample. Detectors that have a location at some off-axis angle, as in many laboratory instruments designed to accept detectors for x-ray microanalysis, show asymmetry of detection with respect to the orientation of feature edges; i.e., the right and left edges are different for a narrow vertical structure (e.g., resist line). In these cases, it is not possible to achieve symmetrical video profiles. For metrology, the symmetry of the video profile is very important. Asymmetry of the image is a diagnostic that indicates some sort of misalignment, whether it is specimen, detector, or column. Furthermore, it is not easily determined if the video asymmetry demonstrated is derived from the position of the detector, from other problems introduced by the instrument's electronics, by column misalignments, by specimen/electron beam interactions, by specimen asymmetries, or by a random summing of all possible problems.

## C. Backscattered Electrons

Backscattered electrons (BSEs) are those electrons that have undergone either elastic or inelastic collisions with the sample atoms and are emitted with an energy that is larger than 50 eV. A significant fraction of the BSE signal is composed of electrons close to the incident beam energy. This means that a 30-keV primary beam electron can produce a larger amount of backscattered electrons of 24–30 keV. A 1-keV primary electron beam can produce close to 1-keV BSEs that can be collected and imaged or interact further with the sample and specimen chamber. The measured backscattered electron yield varies with the sample, detector geometry, and chemical composition of the specimen, but it is relatively independent of the accelerating voltage above about 5 kV. Because of their high energy, backscattered electrons are directional in their trajectories and are not easily influenced by applied electrostatic fields. Line-of-sight BSEs striking the E/T detector contribute to all SE micrographs.

Backscattered electrons have a high energy relative to secondary electrons, and thus they are not affected as greatly by surface charging. Thus, optimization of collection using sample tilt and collector bias can often enable observation of uncoated, otherwise-charging samples.

## 1. Collection of Backscattered Electrons

Backscattered electrons emerge from the sample surface in every direction, but the number of electrons in any particular region of the hemisphere is not equal. Because of their higher energies, BSEs have straight-line trajectories; consequently they must be detected by placing a detector in a location that intercepts their path. This may be accomplished by the use of a solid-state diode detector (23), a microchannel-plate electron detector (24–27), or a scintillator detector positioned for this purpose (28). The size and position of the detector affect the image and thus affect any measurements made from it. Therefore, the particular characteristics of the detector and its location must be taken into account when analyzing the observed backscattered electron signal for any application.

Backscattered electrons can also be collected through the use of energy-filtering detectors or low-loss detectors (29,30). Energy filtration has the advantage of detecting those electrons that have undergone fewer sample interactions (low-loss) and thus have entered and interacted with the sample to a lesser degree (i.e., over a smaller volume of the specimen) and thus carry higher-resolution information. This type of detector has been used successfully at low accelerating voltages, although it does suffer from signal-to-noise ratio limitations (30). The energy-filtering detector holds promise in assisting our understanding of the generation of the signal measured in the CD-SEM. Many of the input parameters are well known for this detector, therefore, electron-beam interaction modeling becomes more manageable. This is especially helpful in the development of accurate standards for CD metrology. Other types of electron detectors are available in the SEM, and these detectors have been reviewed by Postek (31,32). The reader is directed to these publications for further information.

## IV. CD-SEM METROLOGY

In 1987, when a review of SEM metrology was done (10), the predominant electron sources in use were the thermionic-emission type of cathodes, especially tungsten and lanthanum hexaboride ($LaB_6$). The SEM columns were also much less sophisticated at that time. CD-SEM metrology was in its infancy, and these instruments were essentially only modified laboratory instruments. In a later review (31,32), many major changes and improvements in the design of SEMs were introduced, especially the predominance of field emission cathodes and new, improved lens technology. The reader is referred to those publications for details regarding those improvements. Table 1 outlines some of the general characteristics and requirements for CD-SEMs.

### A. Low-Accelerating-Voltage Inspection and Metrology

Low-accelerating-voltage operation for production and fabrication is presently of great interest to the semiconductor industry (33–36). At low accelerating voltages (200 V to 2.5

**Table 1**  Typical CD Scanning Electron Microscope Metrology Instrument Specifications

| | |
|---|---|
| Minimum predicted feature size | $< 0.1\,\mu m$ |
| Image resolution (@1 kV) | $< 4$ nm |
| Accelerating voltage range | General purpose: 0.5–30 kV |
| | In-line: 0.5–2.5 kV |
| Magnification | $100\times$–$500,000\times$ |
| Wafer size capabilities | 300 mm |
| Cleanliness | $\ll 1$ particles added/pass |
| Mean time between failure | $\gg 1000$ h |
| Availability | $> 95\%$ |
| $3\sigma$ repeatability (lines and spaces) | Static $< 2$ nm |
| | Dynamic $< 10$ nm |
| Wafer throughput | $> 50$ h |
| Pattern recognition—probability of detection | $> 99\%$ |

kV), it is possible to inspect in-process wafers in a nondestructive manner; with the advent of submicrometer geometries, it is imperative that on-line inspection take place for many processing steps. Modern, clean vacuum technology using turbomolecular and ion pumps enables nearly contamination-free inspection to screen wafers for proper development, registration, etching, resist removal, and the absence of visual contaminants before the next processing step.

Low-accelerating-voltage operation is not only restricted to wafer fabrication, photomask inspection can also take place in the SEM. Defects in the photomask, either random or repeating, are sources of yield loss in device fabrication. Defects may occur in the glass, photoresist, or chrome and appear as pinholes, bridges, glass fractures, protrusions, solvent spots, intrusions, or even missing geometrical features. Many of the techniques developed for the semiconductor industry are being applied elsewhere, such as in the polymer industry and for biological applications.

## 1. Nondestructive SEM Inspection and Metrology

All CD metrology is currently done under "nondestructive" SEM conditions. Nondestructive inspection in an SEM implies that the specimen is not altered before insertion into the SEM and that the inspection in the SEM itself does not modify further use or functions of the sample.

The techniques used in "nondestructive" operation of the SEM, in this context, have been applied in practice for only about the past decade and a half. Historically, scanning electron microscopy was done at relatively high (typically 20–30 kV) accelerating voltages in order to obtain both the best signal-to-noise ratio and image resolution. At high accelerating voltages, nonconducting samples require a coating of gold or a similar material to provide conduction to ground for the electrons and to improve the SE signal generation from the sample. Further, early instruments were designed to accept only relatively small samples, so a large sample such as a large wafer, typical of the semiconductor industry, needed to be broken prior to inspection. This was not cost effective, since for accurate process monitoring it was necessary to sacrifice several rather expensive wafers during each processing run. As wafers became larger this became even more undesirable. Therefore, in recent years this procedure, for production CD inspection and metrology, has been abandoned. Modern on-line inspection during the production process of semiconductor devices is designed to be nondestructive, which requires that the specimen be viewed in the scanning electron microscope without a coating and totally intact. This required a substantial overhaul of the fundamental SEM design, thus driving the incorporation of field emission sources for improved low accelerating performance, large chamber capability, improved lens designs, clean pumping systems, and digital frame storage. Therefore, the semiconductor industry is and has been the primary driver for many of the modern improvements to the SEM.

On-line inspection in an SEM potentially has another concern: that high-energy electrons interacting with the sample can also damage sensitive devices (37–39). However, more modern testing and review of these concerns needs to be done. Low-accelerating-voltage inspection is also thought to eliminate, or at least to minimize, such device damage. In order to accomplish this in the SEM, the sample is typically viewed at low accelerating voltages. Low-accelerating voltage operation is generally defined as work below 2.5 kV—generally within a range of about 0.4–1.2 kV. Further advantages derived from operating the SEM at low accelerating voltages are that the electrons impinging on the surface of the sample have less energy and therefore penetrate into the sample a shorter distance. The electrons also have a higher cross section (probability) for the production of secondary electrons near the surface, where they can more readily escape and be collected. Thus, in this context, nondestructive evaluation requires that the sample not be broken and that it be viewed in an instrument at an accelerating voltage below the point where electron beam damage will become a problem. Hence, an understanding of the sample's electrical characteristics at that stage of manufacture is useful prior to examination.

For low-accelerating-voltage operation, it is imperative to keep the primary electron beam accelerating voltage at the minimum practical values. This may necessitate variable accelerating voltages in fully compensated steps of as little as 10 V within the practical operating range for photoresist samples of 200 V to 2.5 kV. In this way, an ideal beam energy usually occurs where the secondary electron emission volume results in a near-zero secondary electron buildup on the surface of the sample (see the next subsection, Total Electron Emission). This ideal beam energy can vary from sample to sample (and from location to location), depending on the incident current, the nature of the substrate, or the type and thickness of the photoresist. Variation of only 100 V in accelerating voltage or of a couple of degrees of specimen tilt may change a uselessly charging image to one relaying useful sample information. The nature of conductivity in several photoresists were recently studied by Hwu and Joy (40), thus underscoring the capricious nature of this type of sample.

## 2.   Total Electron Emission

The behavior of the total electrons emitted ($\delta$) from a sample per unit beam electron is shown in Figure 6. This graph is extremely significant to nondestructive low-accelerating-voltage operation. The principles demonstrated here are crucial to the successful imaging of insulating specimens such as shown in Figure 7. The points where the total emission curve crosses unity (i.e., $E_1$ and $E_2$) are the points where there is, in principle, no net electrical charging of the sample (i.e., number of emitted electrons equals number of incident electrons). During irradiation with the electron beam, an insulating sample such as photoresist or silicon dioxide can collect beam electrons and develop a negative charge. This may result in a reduction in the primary electron beam energy incident on the sample. If the primary electron beam energy of the electrons impinging on the sample is

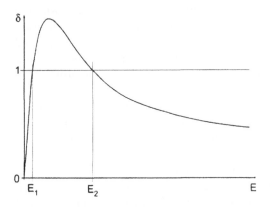

**Figure 6**   Typical total electron emission curve for nondestructive SEM metrology and inspection. $E_1$ and $E_2$ denote the points where no charging is expected to occur on the sample.

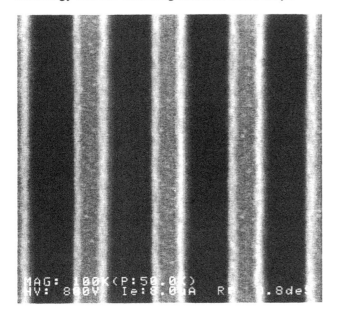

**Figure 7** Low-accelerating-voltage image of uncoated photoresist.

2.4 keV and the particular sample has an $E_2$ point at 2 keV, then the sample will develop a negative potential to about $-0.4$ kV to reduce the effective incident energy to 2 keV and bring the yield to unity. This charging can have detrimental effects on the electron beam and degrade the observed image. If the primary electron beam energy is chosen between $E_1$ and $E_2$, there will be more electrons emitted than are incident in the primary beam, and the sample will charge positively. Positive charging is not as detrimental as negative charging, since positive charging is thought to be limited to only a few volts. However, positive charging does present a barrier to the continued emission of the low-energy secondary electrons. This reduction in the escape of the secondary electrons limits the surface potential but reduces signal, since these electrons are now lost to the detector. The closer the operating point is to the unity yield points $E_1$ and $E_2$, the less the charging effects. Each material component of the specimen being observed has its own total emitted electron/beam energy curve, and so it is possible that, in order to eliminate sample charging, a compromise must be made to adjust the voltage for all materials. For most materials, an accelerating voltage in the range of about 0.2–1 kV is sufficient to reduce charging and minimize device damage. Specimen tilt also has an effect on the total electron emission, and it has been reported that increasing the tilt shifts the $E_2$ point to higher accelerating voltages (41,42). This is a very complex signal formation mechanism, because the number of detected electrons depends not only on the landing energy of the primary electrons, but also on the number and trajectories of the emitted electrons, which is strongly influenced by the local electromagnetic fields.

## B. Linewidth Measurement

Linewidths and other critical dimensions of device structures must be accurately controlled to ensure that integrated circuit performance matches design specifications. However, traditional light-optical methods for the linewidth measurement of VLSI and

ULSI geometries are not able to attain the accuracy or precision necessary for inspection of these devices. Since present wafer fabrication methods employ resist exposure radiation of very short wavelength, such as x-rays and electron beams, it follows that testing and measuring the fabricated structures would involve similar short-wavelength optics and high resolution.

Two measurements critical to the semiconductor industry are *linewidth* and *pitch*, or *displacement* (Figure 8). *Linewidth* is the size of an individual structure along a particular axis, and *pitch*, or *displacement*, is the measurement of the separation between the same position on two or more nearly identical structures (5,31,41–44).

Unlike the optical microscope, the range in SEM magnification can continuously span more than four orders of magnitude. All SEM linewidth measurement systems rely on the accuracy of this magnification, computed from numerous internal instrument operational factors, including working distance and acceleration voltage. Although acceptable for most applications, the magnification accuracy of many typical SEMs may be inadequate for critical measurement work, since the long-term magnification may vary with time (4). For critical, reproducible linewidth measurement, all sources of magnification instability must be minimized to achieve long-term precision and measurement accuracy. Other sources of error in image formation and linewidth measurement have been outlined by Jensen and Swyt (43,44) and Postek (6,31,42) and must be corrected before the SEM can make accurate, reproducible measurements.

## C.  Particle Metrology

Particle metrology and characterization is now a growing field. Particles are a significant problem for semiconductor manufacturing (34). Particle metrology can be considered a special case of CD metrology, in that the same issues relating to the measurement of the width of a line apply to the measurement of the size of particles. Particles are produced by the processing steps and equipment as well as by the inspection process itself. The SEM has numerous moving parts. Each can generate particles through wear mechanisms. As the wafer is transferred into and out of the system, particles can be generated from contact with the wafer transfer machinery. The evacuation (pumping) process causes some degree of turbulence, which can mobilize particles, possibly depositing them on the wafer surface.

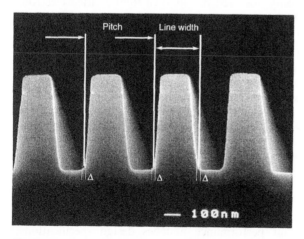

**Figure 8**  Pitch and linewidth comparison.

Particles can also be formed by temperature and pressure changes during the sample exchange process, leading to water vapor condensation, droplet formation, and liquid-phase chemical reactions. Clearly, the size of the specimen, as well as the size of the particles, must also be considered in such a specification in order to make it meaningful to a specific process. Reduction of particle generation is also important to the performance of the instrument, since a charged particle landing on a sensitive portion of the instrument can rapidly compromise the resolution of the SEM.

## D. Overlay Metrology

The resolution provided by visible light and ultraviolet light optics is currently adequate for overlay metrology. However, just as in CD metrology, as the overlay metrology structures shrink, the SEM will be used to a greater extent in this form of metrology (45). The SEM has also been used in the control of bipolar integrated circuit technology where the emitter-to-base overlay was measured (46). The SEM has been explored by Rosenfield (47) and Rosenfield and Starikov (48) as a means to obtain the information necessary for the next-generation semiconductor devices.

## E. Automated CD-SEM Characteristics

The major characteristics of the automated CD-SEM outlined in Table 1 are only a needed guideline. The Advanced Metrology Advisory Group (AMAG), comprised of representatives from the International SeMaTech consortium member companies; the National Institute of Standards and Technology (NIST); and International SeMaTech personnel, joined to develop a unified specification for an advanced scanning electron microscope critical-dimension measurement instrument (CD-SEM) (49). This was deemed necessary because it was felt that no single CD measurement instrument (other than the SEM) or technology would provide process engineers, in the near future, with the tools they require for providing lithographic and etching CD measurement/control for sub-180-nanometer manufacturing technology. The consensus among AMAG metrologists was that CD-SEMs needed improvement in many areas of performance. The specification addressed each of these critical areas, with recommendations for improvement and a testing criterion for each. This specification was designed to be a "living document," in that as the instruments improved the specifications must also be improved. The critical areas that were targeted for improvement in that document were as follows.

### 1. Instrument Reproducibility

Confidence in an instrument's ability to repeat a given measurement over a defined period is key to semiconductor production. The terms *reproducibility* and *repeatability* are defined in a general way in ISO documentation (50). The new SEMI document E89-0999 expanded on these definitions and includes the term *precision* (51). The various components of reproducibility are useful in the interpretation of and comparison of semiconductor fabrication process tolerance.

### 2. CD-SEM Accuracy

The semiconductor industry does not yet have traceable linewidth standards relevant to the kinds of features encountered in VLSI fabrication. A great deal of focused work is progressing in that area and is discussed in a later section of this chapter. Achieving accuracy requires that important attributes necessary in a measurement system must be evaluated (52). Currently, some of the measurable entities include beam steering accuracy, linearity and sensitivity testing of working algorithms, analysis of instrument sharpness, and the apparent beam width.

### 3. Charging and Contamination

Contamination and charging are two of the most important problems remaining in SEM-based IC metrology. While charging starts to show up as soon as the electron beam hits the wafer and contamination tends to build up somewhat more slowly, they act together to change the number, trajectory, and energy of the electrons arriving into the detector and to make it difficult to make precise measurements. It is very difficult to measure contamination and charging independently.

### 4. System Performance Matching

System matching refers to measurement output across several machines. The matching of instrumentation typically applies to all machines that have the same hardware by virtue of the fact that they have the same model number. Matching between manufacturers and different models is deemed desirable, but due to instrument design variations this is seen as unattainable at this time. Matching error is a component of reproducibility, and within the ISO terminology it is the measurement uncertainty arising from changing measurement tools. The matching specification targeted is < 1.5-nm difference in the mean measurement between two tools for the 180-nm-generation CD-SEMs, or 2.1-nm range of the means across more than two tools.

### 5. Pattern Recognition/Stage Navigation Accuracy

Pattern recognition capture rate is characterized as a function of pattern size and shape characteristics, layer contrast, and charging, and must average > 97% on production layers. Errors need to be typed and logged so that they are available for an analysis of pattern recognition failures. Stage accuracy and repeatability for both local and long-range moves must be measured for each of 5-micrometer, 100-micrometer, and "full-range" across-wafer stage movements. And CD-SEMs must be able to measure features that are 100 micrometers from the nearest pattern recognition target.

### 6. Throughput

Throughput is an important driver in semiconductor metrology. The current throughput specification is found in Table 1. The throughput CD-SEM specification is designed to test the high-speed sorting of production wafers by a CD-SEM. Throughput must be evaluated under the same condition as the testing of precision, contamination and charging, linearity and matching, which is using the same algorithm and SEM configuration, and the same wafers.

### 7. Instrumentation Outputs

Critical-dimension control at 180 nanometers and below demands sophisticated engineering and SEM diagnostics. There are a number of outputs that metrologists require from an advanced tool, in addition to the output CD measurement number itself. These include raw line-scan output, total electron dose, signal-smoothing parameters, detector efficiency, signal-to-noise ratio, pattern recognition error log, and others outlined in the text of the AMAG document (49).

Once this AMAG specification was circulated to manufacturers, rapid improvements in many of the targeted areas occurred, resulting in drastically improved CD-SEM instrumentation.

## V. INSTRUMENT AND INTERACTION MODELING

It is well understood that the incident electron beam enters into and interacts directly with the sample as it is scanned (Fig. 3). This results in a variety of potential signals being generated from an interaction region whose size is related to the accelerating voltage of the electron beam and the sample composition (3). The details of this interaction are discussed in Sec. III. For historical and practical reasons, the two major signals commonly used in SEM imaging and metrology are divided into two major groups: backscattered and secondary electrons. Transmitted electrons have also been utilized for specific metrology purposes (53,54). However, it must be understood that the distinction between secondary and backscattered electrons becomes extremely arbitrary, especially at low beam energies. Other commonly used signals include the collection and analysis of the x-rays, Auger electrons, transmitted electrons, cathodoluminescence (light), and absorbed electrons, these will not be discussed here but can be found elsewhere (55–57).

### A. Modeling of the SEM Signal

The appearance of a scanning electron micrograph is such that its interpretation seems simple. However, it is clear that the interaction of electrons with a solid is an extremely complex subject. Each electron may scatter several thousand times before escaping or losing its energy, and a billion or more electrons per second may hit the sample. Statistical techniques are appropriate means for attempting to mathematically model the interactions. The most adaptable tool now is the so-called Monte Carlo simulation technique. In this technique, the interactions are modeled and the trajectories of individual electrons are tracked through the sample and substrate (Fig. 3). Because many different scattering events may occur and because there is no a priori reason to choose one over another, algorithms involving chance-governed random numbers are used to select the sequence of interactions followed by any electron (hence the name *Monte Carlo*). By repeating this process for a sufficiently large number of incident electrons (usually 1000 or more), the effect of the interactions is averaged, thus giving a useful idea of the way in which electrons will behave in the solid. The Monte Carlo modeling techniques were initially applied to x-ray analysis in order to understand the generation of this signal. Today the Monte Carlo technique is being applied to the modeling of the entire signal-generating mechanisms of the SEM.

The Monte Carlo modeling technique provides many benefits to the understanding of the SEM image. Using this technique, each electron is individually followed; everything about it (its position, energy, direction of travel) is known at all times. Therefore, it is possible to take into account the sample geometry, the position and size of detectors, and other relevant experimental parameters. The computer required for these Monte Carlo simulations is modest, so even current high-performance desktop personal computers can produce useful data in a reasonable time. In its simplest form (58–64), the Monte Carlo simulation computes the backscattered electron signal. Since this requires the program to follow and account for only those electrons that have energy higher than 50 eV, it is relatively fast. More time-consuming simulations that calculate secondary electron signal generation as well as the backscattered electron signal take somewhat longer, depending upon the input parameters and the detail of the required data.

By further dividing the electrons based on their energy and direction of travel as they leave the sample, the effect of the detection geometry and detector efficiency on the signal profile can be studied. While the information regarding the backscattered electrons is a valuable first step under most practical conditions, it is the secondary electron signal that is most often used for imaging and metrology in the SEM, and recent work to improve this model is being done (60). Simulating the secondary electron image is a more difficult problem, because many more electrons must be computed and followed. While this is possible in the simplest cases, it is a more difficult and time-consuming approach when complex sample geometry is involved. The importance of being able to model signal profiles for a given sample geometry is that it provides a quantitative way of examining the effect of various experimental variables (such as beam energy, probe diameter, choice of signal used) on the profile produced. This also gives a way of assessing how to deal with these profiles and determine a criterion of line-edge detection for given edge geometry and, thus, how better to calculate the linewidth. However, efficient, more accurate Monte Carlo techniques are now available, but are still under development and as such have some limitations.

## 1.  Accurate SEM Modeling

Accurate SEM metrology requires the development of an appropriate, well-tested computer model. Early Monte Carlo models for metrology were based on the pioneering work of Drs. David Joy, Dale Newbury, and Robert Myklebust and others (65). More recently, a Monte Carlo computer code specifically designed for CD metrology has been developed at NIST (60,61) and is undergoing continual development and improvements. With this program, the location of an edge in a patterned silicon target has been determined to an error below 6 nm from comparisons between computed and measured backscattered and secondary electron signals in a scanning electron microscope. The MONSEL (Monte Carlo for Secondary Electron) series of Monte Carlo computer codes are based on first-principles physics (61). The code simulates the backscattered, secondary, and transmitted electron signals from complex targets in the SEM. Measurements have been made on a specially made target composed of a 1-m step in a silicon substrate in a high-resolution SEM (66). By overlaying the measured data with the simulation (which predicts the expected signal for a given target geometry), it is possible to determine the position of a measured feature in the target to a low level of uncertainty (67). This work proved that it is possible to obtain agreement between theoretical models and controlled experiments.

An example of a comparison between measured and simulated data is shown in Figure 9 for the secondary electron signal for a 1-keV electron beam energy near the edge of the step in the silicon target (66). The modeled edge starts at the zero position in the figure and extends to 17 nm because of the measured approximately 1-degree wall slope. The solid line shows the simulation result for the signal around the edge, which agrees well in shape with the experimental data, given by the dashed lines. Without the simulation, one would not be able to determine the edge location accurately and be forced to use a "rule of thumb." All one could conclude is that the edge occurred somewhere within the region of increased signal. With the simulation, one can determine the edge location within < 3 nm, which is a reduction in edge uncertainty by at least a factor of 4

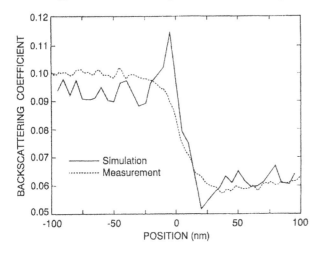

**Figure 9** Monte Carlo modeled and experimental data comparison.

due to the modeling. The results of Fig. 8 were produced by the Monte Carlo code named MONSEL-II, which is a variation for two-dimensional targets. An extension of this code, named MONSEL-III, has been written to compute three-dimensional targets. All of these codes are available from NIST (68).

### 2. Comparison Metrologia/MONSEL Codes

The NIST MONSEL modeling code is primarily a research tool, and it was designed for researchers. Metrologia (SPECTEL Co.) is a commercially available modeling program employing a Monte Carlo computer code (69). Collaboration between NIST and developers of Metrologia enabled the testing and comparison of the two programs in order to determine the level of agreement. A number of modeling experiments were run with both codes. The operations and results of the two codes are now mutually compatible and in general agreement.

### 3. Simulating the Effects of Charging

The buildup of a positive or negative charge on an insulating specimen remains a problem for SEM metrology. Charging can affect the electron beam and thus the measurements (41). Accurate metrology in the SEM requires that an accurate charging model be developed. Ko et al. (70–72) have quantitatively investigated the effects of charging utilizing Monte Carlo modeling. Specimen and instrumentation variations make charging difficult to reproduce, and thus, to study in the quantitative manner necessary for accurate metrology. The deflection of the electron beam by charge buildup was also studied by Davidson and Sullivan (73), and a preliminary charging model was developed and published. Studies of the induced local electrical fields were also done by Grella et al. (74).

### B. Inverse (Simulation) Modeling

Monte Carlo modeling of a defined structure is a very valuable metrology tool. Taking the data derived by the model and forming an image analog (inverse modeling) is an even more powerful tool, since all the measurement parameters are fully known. This image can then be used to test metrology algorithms and compare measurement instruments. Figure 10 is a Monte Carlo–modeled linescan of a 1-μm silicon line with nearly vertical edges. The pixel spacing is 1.15 nm. This was modeled using the NIST MONSEL code on the NIST Cray computer. Once the data was obtained, various beam diameters were convoluted to add beam diameter effects. The next step converts the linescan into a SEM image analog, as shown in Figure 11. Once noise and alphanumerics have been added, the image appears similar to any SEM image. Compare the simulated Fig. 11 with the actual image in Fig. 7. The emulated image is a standard image file and thus can be imported into any commercial measurement programs or input to other metrology instruments. This form of modeling has been used to test various metrology algorithms, including the apparent beam width algorithm (see the next subsection).

## C.  SE vs. BSE Metrology

Closer scrutiny and understanding regarding the differences between the secondary and backscattered electron measurements must be done. The majority of the current CD-SEMs measure the secondary electron image because of the larger signal afforded by that mode of operation. In the past some tools used BSE image–based measurements, but the method has lost favor mainly due to the poorer signal-to-noise ratio, which led to slower measurements and reduced throughput. Measurement of the backscattered electron signal offer some distinct advantages, such as the much easier modeling of the signal because the BSE trajectories are so well understood. It was found (75) that there is an apparent difference in the width of the resist lines between the secondary and backscattered electron image measurements in the low-accelerating-voltage SEM. This difference must be accounted for in the modeling. It has been observed and documented that there is difference in the measurements between the SE and the BSE images in both laboratory and production line instruments (76,77). These differences are identical to the results published earlier in a similar experiment during the microchannel-plate electron detector development (24,25). Figure 12a is a secondary electron image and Figure 12b is a backscattered electron image. In this example, there is a measurement difference of 17 nm between the two modes of electron detection. The SE image taken with highly tilted (but not cross-sectioned) samples revealed that the walls in the imaged structures are sloped. This could account for some of the difference between the imaging modes; however, scanning probe microscope linescans demonstrated a more vertical wall profile in the measurement region.

The discrepancy between the secondary electron image and the backscattered electron image is not explainable by the currently available electron beam interaction models.

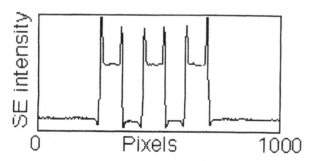

**Figure 10**  Monte Carlo–modeled linescan of the secondary electrons of photoresist.

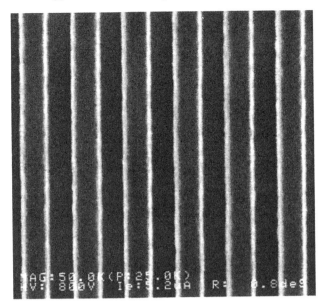

**Figure 11** Completed simulated SEM image based upon the same data shown in Fig. 10. Compare the simulated image to the actual image shown in Fig. 7.

Well-characterized conductive samples are needed to exclude the charging effects produced by the sample. There are three components to this experiment: instrument, sample, and operator. The operator component is eliminated by automation, and the sample issues can be excluded by proper sample selection, thus leaving the instrument effects to be studied.

## D. Modeling and Metrology

The semiconductor industry requires fully automatic size and shape measurements of very small, three-dimensional features. These features are currently 180 nm and smaller

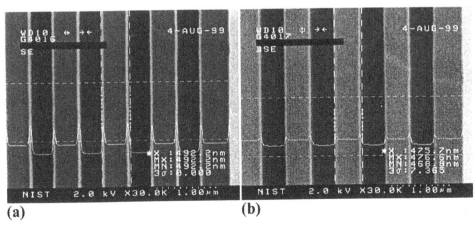

(a)    (b)

**Figure 12** Secondary electron and backscattered electron image comparison of uncoated photoresist. (a) Secondary electron image demonstrating a width measurement of 492.2 nm. (b) Backscattered electron image showing a width measurement of 475.7 nm.

in size, and these measurements must be done in seconds, with an accuracy and precision approaching atomic levels. One main problem is that the measurements are done with primitive edge criteria (regression algorithm, threshold crossing) and certain, not necessarily justified presumptions and beliefs (41). Table 2 shows the results of the application of several algorithms to the measurement of a simulated line image as just described. A simulated image is extremely valuable in this measurement because all the input parameters to the simulated image are known; hence, the pitch, linewidth and space width are known. A similar discrepancy among width measurements was demonstrated in the SEM Interlaboratory Study (41). To accurately determine where the measurement of width should be made on the intensity profile, an accurate model is required.

The images and linescans taken with the CD-SEM contain much more information than is generally being used (78). For example, these images and the individual linescans differ for "same-width" photoresist lines taken at various focus and dose settings. Beyond the fact that this is a source of measurement error, further information can be extracted, like whether or not the given line was exposed with the intended dose and focus. This information is essential in current and future UV lithography. Modeling the possible cases can help to draw correct conclusions and makes it possible to use more accurate, customized measuring algorithms.

## E.  Shape Control in the Lithography Process

The lithography process is expected to create the integrated circuit features (conductive and nonconductive lines and areas, contact holes and interconnects, etc.) in certain shapes with designed sizes and tolerance. In many steps, millions of transistors and connections among them are being made at the same time on a single chip on wafers that contain many chips.

The overall goal is to find a reliable link, some kind of transfer function between the designed and developed resist features and the final etched, e.g., polysilicon (polySi), structures. Figure 13 shows an isolated resist line after development of the resist (upper image), after bottom antireflecting coating (BARC) removal (middle), and after polySi etch (bottom image). After development, the resist line has a certain height, width, and wall angle, and the top and bottom corners are rounded somewhat.

Through BARC removal, the shape and especially the height of the line substantially change. After this step the resist features still have to have enough material to withstand the polySi etch process. If these process steps go through acceptably, polySi features are fabricated with the desired sizes and tolerances (see polySi line at the bottom).

**Table 2**  Comparison of Measurement Algorithms

| Algorithm | Space width (nm) | Linewidth (nm) |
|---|---|---|
| Peak | 109.52 | 91.15 |
| Threshold | 91.65 | 110.6 |
| Regression | 75.63 | 125.9 |
| Sigmoid | 92.95 | 110.52 |

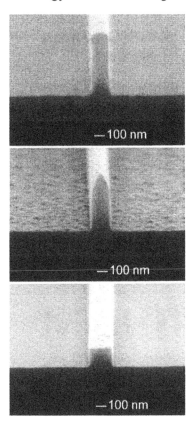

**Figure 13** Lithography process illustrated with resist line after development (upper image), after BARC etch (middle), and after final polySi line etch (lower image).

## 1. Focus Exposure Matrix

The lithography tool (stepper or scanner) works with a certain focus setting and the amount of light applied to the photosensitive resist material. The focus is essential to produce the right light intensity distribution, and the amount of light applied is usually controlled by the number of laser pulses. These parameters are kept under extremely tight control by the lithography tools. Similar conditions apply to other technologies using sources of illumination such as electrons and x-rays.

A focus-exposure matrix (FEM) is used to simulate focus and dose variations occurring in the lithography process. The FEM is a mechanism to obtain the best possible settings of the focus and exposure and to find the range of these parameters that provide the "process window." Within this window the lithography process is thought to produce features with the specified size and shape. Generally, during the process development the engineers follow the formation and shape change of isolated and dense lines and contact holes. The relationships among the designed and developed resist features are traditionally established through the FEM. Figure 14 shows two sets of images of isolated and dense resist lines. The nominal linewidth in this case is 250 nm. Those lines that lead to acceptable polySi linewidths after final etch are shown with marked values. In certain cases this assessment is not necessarily enough; in the cases of other-than-line features, for example, memory cell structure, things may work out quite differently.

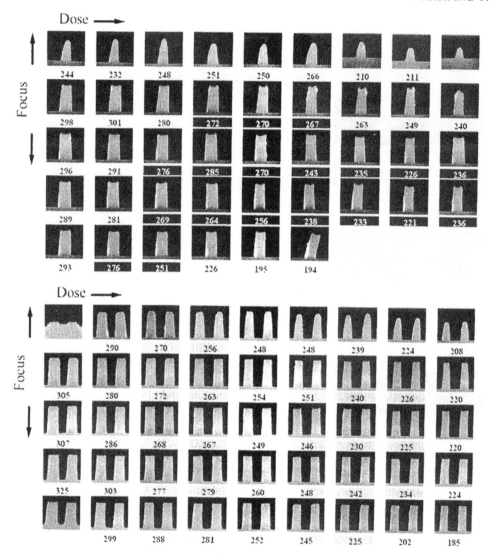

**Figure 14** Focus-exposure matrices of isolated and dense resist lines. (Courtesy of Dr. Alain Diebold of International SEMATECH.)

In the fabrication facility the speed of the processes is very important, and dimensional measurements of these parameters with linewidths close to 100 nanometers must be carried out in seconds with atomic level of uncertainty. Based on good-quality resist measurements, the engineer can decide to let the process go further or to rework the resist formation steps (illumination, develop, and bake). Lack of adequate characterization or flawed dimensional measurements lead to large losses; therefore good metrology is becoming an enabling, essential technology. It is essential, but it is not an easy task.

## 2. Dimensional Measurements During Process Development

During the development or modification of the lithography process, many times a large number of measurements must be taken in order to investigate the consequences of the intentional or unintentional changes of the process. Since the various measurement methods deliver results with limitations, it is essential that the engineer choose the right technique. For example, to survey intrawafer or wafer-to-wafer variations, electrical linewidth measurements may be much better than any other method, even if these must be done on special, so-called Van der Pauw test patterns. These measurements are typically limited to the conductive layers or steps of the IC fabrication process, e.g., polySi structures. In other cases, SEMs or scanning probe measurements will yield the necessary information. In most cases, a combination of methods is needed.

## 3. Shape Analysis

Measurement of linewidths of isolated and dense lines is not necessarily enough; in a number of cases the linewidth values are within specifications, but other areas are not developed well enough to result in properly sized features after final etch. Figure 15 shows an example of three sites of the same features from three different locations on the same wafer. The features with shapes like those on the left did not yield at all; in the center, the yield was acceptable; while the areas with resist structures no different from the right yielded excellently. Similar effects can be seen at the tip of the lines that can "pull back"; i.e., the resist lines are shorter, with less well-defined tips, instead of longer, with properly developed endings and with close-to-vertical walls. On the other hand, resist lines that are out of specification occasionally may result in acceptable lines after final etch.

In the future, due to the real three-dimensionality of the structures, shape measurements have to take place beyond or instead of mere linewidth, i.e., one-dimensional size measurements. These measurements have to account for all possible changes in the shape and size of various structures and patterns. This requirement will lead to more frequent use of image-based measurements. The throughput of CD-SEMs may get worse, because instead of a few dozen lines, several hundred lines have to be collected in a much longer process of acquisition of images with good signal-to-noise ratios. Clever imaging methods, for example, selected area imaging, can alleviate these problems.

The height, width, wall angle, and top and bottom corner and tip rounding of resist features must be correctly measured; otherwise defective circuits will be manufactured. All of these contribute to the image formation in cross-sectional and top-down SEM imaging. Any change results in somewhat different images, and this fact gives the possibility for determination of these parameters. In the future, especially in the case of 300-mm wafers, wafer cleaving will be less than desired because of the cost of the wafers is too high. Imaging methods that can provide information about the size and real, three-dimensional

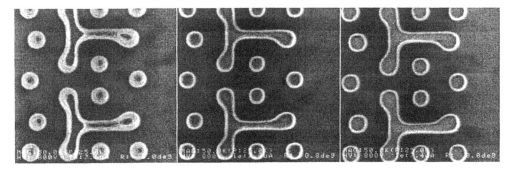

**Figure 15**  Three sites of the same features from three different locations on the same wafer.

shape of the structures will be more favorable. Fortunately, the top-down images do show the difference of various shapes. There is information that is currently not being used in the linewidth calculations. Figure 15 images clearly show differences even if the bottom contours are quite similar. Methods that are sensitive to the shape changes can deliver better and much more useful results.

Recently introduced methods aim to provide better sensitivity to shapes. These methods may rely on comparisons or correlation to previously saved optimal images or linescans, or use image or template libraries of known good structures, or take advantage of information provided by more than one angle view of the same structure. Modeling of linescans and images is more promising than just better linewidth measurements; with it it seems feasible to perform real three-dimensional measurements with much better accuracy. Eventually the quality of this model-based metrology method can be so good that it will allow for reconstruction of the real 3-D structure with height information included.

The accuracy and reliability of these advanced dimensional metrology methods will deliver adequate information to calculate the optimal focus and exposure settings for the lithography process.

## F.  Model-Based Metrology

Model-based metrology is a new concept, one that will ultimately combine a number of currently developing areas into a single approach. These five areas are as follows.

### 1.  Modeling/Image Simulation

The various new Monte Carlo methods, especially those developed at NIST (MONSEL series), are providing better and more accurate data than ever. The modeled results are very closely matching the results of real measurements. Simulated images from modeled data can now begin to approach actual data in appearance and can be used to compare to real-world samples during the measurement process.

### 2.  Adaptive Monte Carlo Modeling

Adaptive Monte Carlo modeling utilizes a database of measured video waveforms from production samples and a library of modeled waveforms (79). Any waveform coming from a new structure is compared to a number of waveforms or linescans in the library. The result can be extremely accurate, not just the width of the line but the top-corner rounding, the wall angle, and even the height of the resist line is correctly reported. Figure 16 shows a cross section of a photoresist line and the computed lineshape superimposed. The shape of the line was calculated from the top-down CD-SEM image of the same line. The adaptive Monte Carlo model is evolving rapidly and getting ready for practical applications in industrial CD metrology.

### 3.  Specimen Charging

International SeMaTech and the Advanced Metrology Advisory Group (AMAG) have identified charging as the biggest problem in accurate SEM-based metrology and the successful modeling of the geometry of real integrated circuits. Further work incorporating an accurate charging model into the basic specimen interaction and signal generation modeling must be done. Alternatively, work to increase the electrical conductivity of resist has also been done (80).

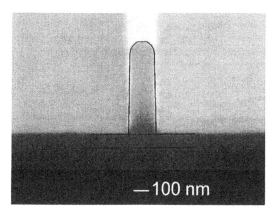

**Figure 16** Cross section of a photoresist line, with the structure calculated from the top-down view and modeled data.

### 4. Signal Path Modeling

The modeling of the signal path, including the electronics and signal processing of the SEMs, remains an essential area of research and is critical to the successful development of an integrated and accurate model.

### 5. Algorithm Development and Analysis

Algorithms currently used in metrology instruments have no physical basis for analysis of the data. Once an integrated model has been developed, an algorithm for the accurate measurement of a structure will follow.

## G. Critical-Dimension Measurement Intercomparison

From the foregoing discussions it becomes clear that inferring a line's width from its image requires assumptions about how the instrument interacts with the sample to produce the image, and how, quantitatively, the resulting apparent edge positions differ from the true ones. For the initial steps to understand this "probe-sample" interaction quantitatively and to verify models of instrument behavior, Villarrubia et al. (81) employed highly idealized samples, fabricated in single-crystal silicon (82). The lines were electrically isolated from the underlying wafer by a 200-nm-thick oxide to permit electrical critical dimension (ECD) measurements. Table 3 summarizes the results of that work.* For three measurement techniques, SEM—atomic force microscopy (AFM), and ECD—an uncertainty budget was developed according to NIST and ISO guidelines (83,84); it listed the major components contributing to the measurement uncertainty. Listing these components in an honest manner provides a tool for the determination of opportunities for improving the measurement accuracy. In the SEM, the scale (i.e., magnification) component accounted for about one-half of the overall uncertainty. This is an area where improvements can be made. This work generated confidence in the ability to determine

---

* The numbers in Table 3 reflect the total uncertainty of the measurements. Certain components of it (e.g., precision) can be negligible, but others contribute to the large uncertainty. The reader is encouraged to review Ref. 81 in order to understand how these numbers were obtained.

**Table 3**  Linewidth Results on Single-Crystal Silicon Sample

| Technique | Width (nm) | 3σ Uncertainty (nm) |
|-----------|------------|---------------------|
| SEM       | 447        | 7                   |
| AFM       | 449        | 16                  |
| ECD       | 438        | 53                  |

*Source*: Ref. 81.

edges and thus to provide a meaningful linewidth measurement. This raises a question: To what extent will corner rounding, deviations from ideally vertical sidewalls, or surface and edge roughness—all imperfections likely to be encountered in actual production samples—affect linewidth measurement accuracy? It is therefore important to continue to test our understanding for samples that approximate as closely as possible the product samples of greatest industrial interest.

## VI.  INSTRUMENT CALIBRATION

The accuracy of measurements and the precision of measurements are two separate and distinct concepts (85–86). Process engineers want accurate dimensional measurements, but accuracy is an elusive concept that everyone would like to deal with by simply calibrating their measurement system by using a standard developed and certified at the National Institute of Standards and Technology (NIST). Unfortunately, it is not always easy either for NIST to calibrate submicrometer standards or for the engineer to use standards in calibrating instruments. Accurate feature-size measurements require accurate determination of the position of both the left and right edges of the feature being measured. The determination of edge location presents difficulties for all current measurement techniques, for the reasons discussed in earlier sections. Since linewidth measurement is a left-edge–to–right-edge measurement (or converse), an error in absolute edge position in the microscopic image of an amount $\Delta L$ will give rise to an additive error in linewidth of $2\Delta L$. Without an ability to know the location of the edges with good certainty, practically useful measurement accuracy cannot be claimed. For accurate SEM metrology to take place, suitable models such as discussed earlier must be developed, verified, and used.

Recently, the need has been identified for three different standards for SEM metrology. The first standard is for the accurate certification of the magnification of a nondestructive SEM metrology instrument; the second standard is for the determination of the instrument sharpness; and the third is an accurate linewidth measurement standard.

### A.  Magnification Certification

Currently, the only certified magnification standard available for the accurate calibration of the magnification of an SEM is NIST SRM 484. SRM 484 is composed of thin gold lines separated by layers of nickel, providing a series of pitch structures ranging from nominally 1 to 50 μm (87). Newer versions have a 0.5 μm nominal minimum pitch. This standard is still very useful for many SEM applications. During 1991–1992 an interlaboratory study was held using a prototype of the new low-accelerating-voltage SEM magnification standard (41). This standard was initially fabricated (88–89) and released as a reference material (RM 8090) (90). This RM was rapidly depleted from stock, and a second batch of the artifacts, identified as NIST SRM 2090, is currently being fabricated.

## B.  Definition and Calibration of Magnification

In typical scanning electron microscopy, the definition of magnification is essentially the ratio of the area scanned on the specimen by the electron beam to that displayed on the photographic CRT. The size of the photographic CRT is fixed, therefore, by changing the size of the area scanned on the sample; the magnification is either increased or decreased. Today, where SEM metrology instruments are concerned, the goal is not necessarily to calibrate the magnification as previously defined and discussed, but to calibrate the size of the pixel in both the $X$ and the $Y$ directions of the digital measurement system. Since the digital storage system is common to the imaging, the "magnification," therefore, is also calibrated. It should be noted that because of the aspect ratio of the SEM display screen, the number of pixels in $X$ may differ from the number in $Y$, but the size of the pixel must be equal in both $X$ and $Y$. This is an important concept, because in order for a sample to be measured correctly in both $X$ and $Y$ the pixel must be square. The concept of pixel calibration and magnification is essentially identical, and pitch measurements can be used to adjust either. Adjustment of the calibration of the magnification should not be done using a width measurement until an accurate model (as discussed earlier) is available. This is because width measurements are especially sensitive to electron beam/specimen interaction effects. This factor cannot be ignored or calibrated away.

Fortunately, this factor can be minimized by the use of a pitch-type magnification calibration sample, such as SRM 484 or the new standard SRM 2090 when it is issued (Figure 17). These standards are both based on the measurement of "pitch." *Pitch* is the distance from the edge of one portion of the sample to a similar edge some distance away from that first edge. In a pitch standard, that distance is certified, and it is to that certified value that the magnification calibration of the SEM is set. Under these conditions, the beam scans a calibrated field width. That field width is then divided by the number of pixels making up the measurement system, thus defining the measurement unit or the pixel width. If we consider two lines separated by some distance, the measurement of the distance from the leading edge of the first line to the leading edge of the second line defines the pitch. Many systematic errors included in the measurement of the pitch are equal on both of the leading edges; these errors, including the effect of the specimen beam interaction, cancel. This form of measurement is therefore self-compensating. The major criterion for this to be a successful measurement is that the two edges measured be similar in all ways. SEM pixel/magnification calibration can be easily calibrated to a pitch.

The measurement of a width of a line, as discussed earlier, is complicated, in that many of the errors (vibration, electron beam interaction effects, etc.) are now additive. Therefore, errors from both edges are included in the measurement. The SEM magnification should not be calibrated to a width measurement, since these errors vary from specimen to specimen due to the differing electron beam/sample interaction effects. Effectively, with this type of measurement we do not know the accurate location of an edge in the video image, and more importantly it changes with instrument conditions. Postek et al. (41), in an interlaboratory study, demonstrated that the width measurement of a 0.2-μm nominal linewidth ranged substantially among the participants. Calibration based on a width measurement requires the development of electron beam modeling, as described previously. This is the ultimate goal of the program at NIST and recently has been shown to be successful for special samples, such as x-ray (53–54) and SCALPEL (scatter-

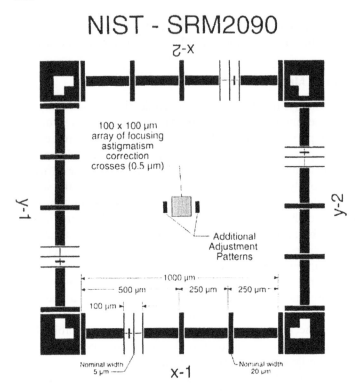

**Figure 17**   SRM 2090.

ing with angular limitation in projection electron beam lithography) (91–92) masks measured in the SEM and in the linewidth correlation study (81).

## B.  Linewidth Standard

During the past several years, three significant areas directly related to the issuance of a linewidth standard relevant to semiconductor production have significantly improved. The first area is modeling. Collaborative work between NIST and International SeMaTech has led to substantial improvements in the modeling of the electron beam–solid-state interaction. International SeMaTech support in the modeling area has been crucial to the progress that has been made. The International SeMaTech cosponsoring with NIST of several Electron Beam/Instrument Interaction Workshops at the SCANNING International meetings over the past several years has provided a forum that for the first time drew model builders from all over the world. This has resulted in significant and more rapid progress in the area of electron beam interaction modeling. The NIST MONSEL computer codes have been significantly improved and experimental verification of the modeling has produced excellent results on certain well-defined structures.

Second, confidence in the model has been fostered by comparison to commercial code through a NIST/Spectel Company model comparison, also fostered by International SeMaTech. This forward-looking project facilitated the third component that was a linewidth correlation project partially funded by International SeMaTech (81). For the first time, three metrology methods were carefully applied to a given, well-characterized structure, and more importantly an uncertainty of the measurement process was thoroughly assessed. This remains an ongoing project, and it ties directly to the development of a linewidth SEM standard.

A linewidth test pattern was recently developed and will be placed upon the next AMAG test wafer set (Figure 18) (93). The wafer set will be composed of semiconductor-process-specific materials, and it has been designed to be measured with several metrology techniques, including the NIST length-scale interferometer (94). The AMAG test wafer will serve as the prototype for a traceable integrated-circuit-production-specific linewidth standard.

## VII. CD-SEM INSTRUMENT PERFORMANCE

In industrial applications, such as semiconductor production, users of automated SEM metrology instruments would like to have these instruments function without human intervention for long periods of time and to have some simple criterion (or indication) of when they need servicing or other attention. No self-testing is currently incorporated into these instruments to verify that the instrument is performing at a satisfactory performance level. Therefore, there is a growing realization of the need for the development of a procedure for periodic performance testing. A number of potential parameters can be monitored, and some of them have been reviewed by Joy (95) and others (49). Two measures of instrument performance have been suggested for incorporation into future metrology tools. These are a measure of sharpness and a measure of the apparent beam width.

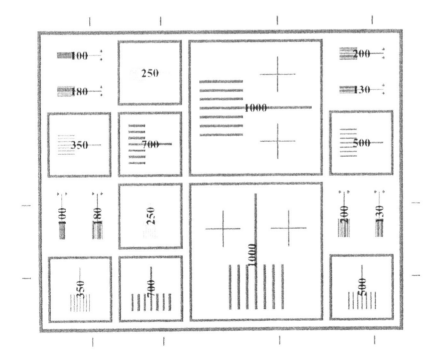

**Figure 18** Drawing of linewidth sample prototype.

## A. The Sharpness Concept

SEMs are being utilized extensively in semiconductor production, and these instruments are approaching full automation. Once a human operator is no longer monitoring the instrument's performance and multiple instruments are used interchangeably, an objective diagnostic procedure must be implemented to ensure data and measurement fidelity. The correct setting of the sharpness and the knowledge of its value are very important for these production-line instruments. A degradation of the sharpness of the image of a suitable test object can serve as one of perhaps several indicators of the need for maintenance. Postek and Vladar (96) first published a procedure based on this sharpness principle, and it has subsequently been refined into a user-friendly stand-alone analysis system (97,98). This concept was based on the objective characterization of the two-dimensional Fourier transform of the SEM image of a test object for this purpose and the development of appropriate analytical algorithms for characterizing sharpness. The major idea put forth in those papers was that an instrument can be objectively tested in an automated manner, and the solution provided was one approach to the problem. A third paper outlining, in the public domain, a statistical measure, known as multivariate kurtosis, that was also proposed to measure the sharpness of SEM images (99–100).

Figure 19 shows a summary of 25 linewidth measurements and their standard deviation for photoresist lines viewed on an automated CD-SEM tool. The measurement of the lines with this instrument resulted in an average of 247.7 nm for the width of the lines. The

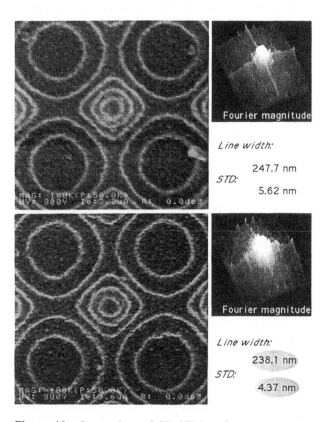

**Figure 19** Comparison of CD-SEM performance and linewidth.

top left micrograph is an image of the photoresist test sample. Under these conditions the instrument was functioning more poorly than evidenced by the lower left figure. The performance of the instrument was improved after changing the final lens aperture and correctly adjusting the electron optical column. The same photoresist lines were measured again, and their width was 238.1 nm on average, with a smaller standard deviation. The greater than 9-nm difference in the measurement, resulting from the performance difference of the same instrument, is important to the production engineers. With smaller and smaller linewidths on the horizon, the correct setting of the SEM becomes indispensable.

It is known that the low-frequency changes in the video signal contain information about the large features and the high-frequency ones carry information of finer details. When a SEM image has fine details at a given magnification—namely, there are more high-frequency changes in it—we say it is sharper. A procedure based on the Fourier transform technique on SEM images can analyze this change. Other procedures based on the Fourier transform technique can also be found (101–103). Since a SEM image is composed of a two-dimensional array of data, the two-dimensional Fourier transform generates a two-dimensional frequency distribution. Based on the computed frequency spectra of selected SEM images, it can be observed that when a SEM image is visibly sharper than a second image, the high-spatial-frequency components of the first image are larger than those of the second. The important point is not the particular technique of how the SEM image is analyzed for sharpness, but the fact that it can and should be analyzed.

## 1. Reference Material (RM) 8091

Reference Material (RM) 8091 is one type of specimen in a class of samples appropriate for testing the sharpness of the scanning electron microscope (104). The original targeted material was a specimen identified as "grass," in the semiconductor manufacturing vocabulary. Grass proved to be too difficult to have manufactured, and so an alternative approach was taken. The tungsten-coated silicon sample was the first reference material able to be manufactured that is fully compatible and manufacturable with state-of-the-art integrated circuit technology. RM 8091 is composed of a microcrystalline tungsten layer deposited on a silicon wafer. This RM can be used at either high accelerating voltage or low accelerating voltage (Figure 20). In the current version, the standard is available as a diced 10-mm × 10-mm square sample (capable of being mounted on a specimen stub) for insertion into either a laboratory or wafer inspection scanning electron microscope. Also available are 200-mm or 150-mm special (drop-in) wafers with recessed areas for mounted chip-size samples. These can easily be loaded as any other wafer. An alternative candidate for the sharpness standard is an evaporated tin-sphere sample. This sample has the distinct advantage of greater "$Z$" depth than the tungsten sample; thus it is less prone to contamination.

## 2. Performance Monitoring

The sharpness technique can be used to check and optimize two basic parameters of the primary electron beam; the focus and the astigmatism. Furthermore, this method makes it possible to regularly check the performance of the SEM in a quantitative, objective form. The short time required to use this method makes it possible that it can be performed regularly before new measurements take place. To be able to get objective, quantitative data about the resolution performance of the SEM is important, especially where high-resolution imaging or accurate linewidth metrology is important. The Fourier method, image analysis in the frequency domain, summarizes all the transitions of the video signal constituting the whole image, not just one or several lines in given

**Figure 20**   Tungsten sample.

directions. This improves the sensitivity (signal-to-noise ratio) and gives the focus and astigmatism information at once. The best solution would be if this and other image processing and analysis functions were incorporated as a built-in capability for all research and industrial SEMs.

## B.   Increase in Apparent Beam Width (ABW) in Production CD-SEMs

Along with the NIST sharpness approach, the measurement of apparent beam width (ABW) is one other potential diagnostic procedure for the periodic determination of the performance of on-line CD-SEM measurement tools. The ABW procedure is a quantitative measure of the sum of all the factors contributing to the apparent electron beam size as it scans across the edge of the sample (Figure 21). Like sharpness, the measurement of ABW provides a single comparable number. Archie et al. (105) have reviewed this concept and using experiments and Monte Carlo modeling demonstrated the value of this procedure.

Measurement instruments in the production environment have demonstrated that the apparent beam width is greater than expected for the performance level of a given instrument. Under a given set of conditions, an instrument, which is potentially capable of demonstrating 4-nm "resolution," often will demonstrate 20-nm or greater beam width on production samples. It would be expected that, due to electron beam interaction effects, the guaranteed "resolution" and apparent beam width would not be identical, but they should be much closer than the 5× (or greater) difference being observed. This phenomenon does not appear to be isolated to any particular instrument type or manufacturer. This presents a serious problem for metrology at 0.25 μm and below. The ABW can be altered in a number of ways. Figures 22a and b are micrographs both taken with the same

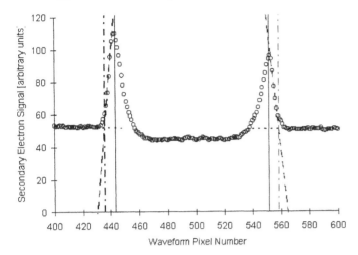

**Figure 21**  ABW analysis of the left edge and the right edge. (From Ref. 105.)

instrument operational conditions. The SEM was optimized for this mode of operation, and the "resolution" capability was the best possible for the given conditions. Both of the images were identically adjusted for contrast and brightness within the computer, and averaged linescans were taken through the two figures to demonstrate the apparent beam width problem. The dotted lines of Figure 23 are from an integrated linescan through the micrograph of Fig. 22a. The solid line of Fig. 23 is a linescan through the micrograph of Fig. 22b. The contrast of Fig. 22b appears saturated because of the identical image processing done on both micrographs. But note that even under the same instrumental conditions, the apparent beam width changes, and the measured width is larger for the stored image. The measurement of apparent beam width in advanced metrology tools has been recommended in the Advanced SEM specification developed by the Advanced Metrology Advisory Group (49).

The increase in apparent beam diameter is a function of a number of factors, including: beam diameter, wall angle, sample charging, sample heating, vibration, and the image-capturing process. The examples shown in Figs. 22 and 23 appear to be the experimental demonstration of a statement made in 1984: "It can be further postulated that severe negative charge storage by the sample can, if the conditions are met, result in an electrostatic effect on the primary beam as it is sampling the effected area. This effect would become more severe the smaller and closer together the structures of interest become" (41). The deflection of the electron beam was also studied by Davidson and Sullivan (73), and a preliminary charging model has been developed.

Based on the current knowledge of the ABW situation, one possibility is that sample charging is apparently affecting the electron beam as it scans the sample. The electron beam is potentially being deflected as the beam scans the dynamically charging and discharging sample. The sample then appears to be "moving" as the image is stored in the system. The image-capturing process is averaging what appears to be a moving sample, and thus the edges become enlarged. Another possibility is an environmental effect—vibration. Vibration would have a similar detrimental effect on the image by increasing the measurement. This can be tested only with a fully conductive sample.

One problem associated with this issue is that when a pitch is used as a sanity check on instrument calibration, that measurement will be correct due to the self-compensation

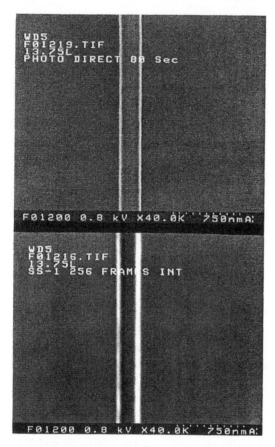

**Figure 22** Uncoated photoresist micrographs demonstrating the difference in ABW relative to scan speed. (a) Slow single-frame-scan image acquisition. (b) Integrated fast frame-scan image.

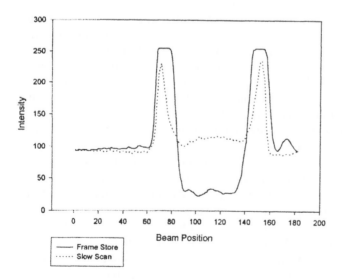

**Figure 23** Fast single linescan (solid) compared to slow single-linescan (dotted) image acquisition for the images of Fig. 22.

characteristics of the measurement. However, subsequent linewidth measurements may be detrimentally affected by this effect, because of the additive nature of the errors in this type of measurement. Thus, erroneous conclusions may be potentially drawn from these data.

## C. Contamination Monitoring

The deposition of contamination on the surface of any specimen in the SEM is a pervasive problem. The low surface roughness of SRM 2091 makes this standard useful in the determination of specimen contamination deposition. Since this standard is susceptible to the effects of contamination, care must be taken always to operate the instrument on a clean area and not to dwell too long on any particular area. For this reason SRM 2091 is also a good sample of the measurement of contamination.

## VIII. ENVIRONMENTAL SEM

New developments in SEM design have not been restricted only to electron sources and lens designs. Concepts in the management of the vacuum have also evolved. Not all applications of the SEM require high vacuum in the specimen chamber, and many samples are damaged or distorted during the specimen preparation processes. Recent innovations in "environmental" SEM have changed the rules somewhat. Many specimens can be viewed "wet" in special instruments. This provides an advantage for sample preparation, as well as reducing specimen charging. Specimen charging can be dissipated at poor vacuums. Environmental scanning electron microscopes have been introduced in several areas of general SEM applications in order to look at samples generally prone to charging. For many years, scanning electron microscopy has routinely been done at relatively high vacuum in the specimen chamber, and now a possibility exists for electron microscopy at low vacuum. Environmental SEM is relatively new to the overall SEM field, and a great deal of work is being done to understand the mechanisms of operation. The reader is directed to the work of Danilatos (106, 107) for further information. Environmental SEM has the potential of solving the charging problem associated with the measurement of semiconductor structures. Some technical complications do exist in the application of this technology. Currently, no application of low-vacuum scanning electron microscopy to CD metrology has occurred in the production environment; however, this methodology holds some promise for the future.

## IX. TELEPRESENCE MICROSCOPY

Telepresence microscopy is an application of the currently available telecommunications technology to long-distance scientific endeavors (108). *Long distance* is a relative concept. This can refer to collaboration across the country or from one distributed location within a single company to another. Telepresence is currently being applied to electron microscopy in several locations where unique analytical facilities (such as those at NIST) can be used via Internet connection. Potentially this can provide tremendous savings to a company where asset sharing can now be rapidly and effectively used or where remote unique facilities can be reached without the requirement of expensive and time-consuming travel. This also has tremendous potential for the wafer fabrication facility, since the engineer can monitor the process remotely without having to enter the cleanroom. NIST, Texas

Instruments, and Hewlett-Packard worked together to develop a microscopy collaboratory test bed to demonstrate the value of telepresence microscopy in the industrial environment. This test bed shows the value of this technology for technology transfer to organizations having distributed manufacturing facilities, such as Texas Instruments, and between organizations, such as NIST and Texas Instruments or Hewlett-Packard (109).

## X.  CONCLUSION

The first commercial scanning electron microscope was developed in the late 1960s. This instrument has become a major research tool for many applications, providing a wealth of information not available by any other means. The SEM was introduced into the semiconductor production environment as a CD measurement instrument in the mid-to-late 1980s, and this instrument has undergone a significant evolution in recent years. The localized information that SEMs can provide quickly will be needed for the forseeable future. New methods often substantially improve the performance of these metrology tools. Digital image processing, fully computerized advanced measuring techniques can overcome obstacles that seem unmanageable at a glance. New, shape-sensitive, 3-dimensional, model-based measuring methods are being pursued and will soon be implemented. This instrument now holds a significant role in modern manufacturing. The evolution is not ended with the improvements provided by newer technologies, such as modeling, and the potential afforded by improved electron sources; this tool will continue to be the primary CD measurement instrument for years to come.

## ACKNOWLEDGMENTS

The authors would like to thank International SEMATECH and the NIST Office of Microelectronics Programs for their support of this program.

## REFERENCES

1. Contribution of the National Institute of Standards and Technology. This work was supported in part by the National Semiconductor Metrology Program at the National Institute of Standards and Technology; not subject to copyright.
2. Certain commercial equipment is identified in this report to adequately describe the experimental procedure. Such identification implies neither recommendation nor endorsement by the National Institute of Standards and Technology, nor does it imply that the equipment identified is necessarily the best available for the purpose.
3. M.T. Postek. In: J. Orloff, ed. Handbook of Charged Particle Optics. New York: CRC Press, pp. 363–399, 1997.
4. M.T. Postek, A.E. Vladar, S.N. Jones, W.J. Keery. NIST J. Res. 98(4):447–467, 1993.
5. M.T. Postek. In: K. Monahan, ed. SPIE Critical Review 52:46–90, 1994.
6. M.T. Postek. NIST J. Res. 99(5):641–671, 1994.
7. S.G. Utterback. Review of Progress in Quantitative Nondestructive Evaluation:1141–1151, 1988.
8. D.C. Joy. Inst. Phys. Conf. Ser. No. 90: Ch. 7. EMAG 175–180, 1987.
9. K. Kanaya, S. Okayama. J. Phys. D. 5:43–58, 1972.

10. M.T. Postek, D.C. Joy. NBS J. Res. 92(3):205–228, 1987.

11. L. Reimer. SEM Inc. 1979/II:111–124, 1979.

12. L. Reimer. Scanning 1:3–16, 1977.

13. L. Reimer. SEM Inc.299–310, 1984.

14. L. Reimer. Electron Beam Interactions with Solids.299–310, 1984.

15. L. Reimer. Scanning Electron Microscopy. Physics of Image Formation and Microanalysis. New York: Springer-Verlag, 1985.

16. H. Seiler. Z. Angew. Phys. 22(3):249–263, 1967.

17. H. Drescher, L. Reimer, H. Seidel. Z. Angew. Phys. 29:331–336, 1970.

18. K.-R. Peters. SEM/1982/IV, SEM Inc., pp 1359–1372, 1982.

19. K.-R. Peters. Scanning Electron Microscopy/1985/IV SEM Inc., pp 1519–1544, 1985.

20. T.E. Everhart, RFM Thornley. J. Sci. Instr. 37:246–248, 1960.

21. M.T. Postek, W.J. Keery, N.V. Frederick. Rev. Sci. Inst. 61(6):1648–1657, 1990.

22. M.T. Postek, W.J. Keery, N.V. Frederick. Scanning 12:I-27–28, 1990.

23. S. Kimoto, H. Hashimoto. In: T.D. McKinley, K.F.J. Heinrich, D.B. Wittry, eds. The Electron Microscope. New York: Wiley, pp. 480–489, 1966.

24. M.T. Postek, W.J. Keery, N.V. Frederick. Scanning 12:I-27–28, 1990.

25. M.T. Postek, W.J. Keery, N.V. Frederick. EMSA Proceedings, pp 378–379, 1990.

26. P.E. Russell. Electron Optical Systems, SEM, Inc., pp 197–200, 1990.

27. P.E. Russell, J.F. Mancuso. J. Microsc. 140:323–330, 1985.

28. V.N.E. Robinson. J. Phys. E: Sci. Instrum. 7:650–652, 1974.

29. O.C. Wells. Appl. Phys. Lett. 19(7):232–235, 1979.

30. O.C. Wells. Appl. Phys. Lett. 49(13):764–766, 1986.

31. M.T. Postek. SPIE Critical Review 52:46–90, 1994.

32. M.T. Postek. NIST J. Res. 99(5):641–671, 1994.

33. T. Ahmed, S.-R. Chen, H.M. Naguib, T.A. Brunner, S.M. Stuber. Proceedings SPIE 775:80–88, 1987.

34. M.H. Bennett. SPIE Critical Review 52:189–229, 1993.

35. M.H. Bennett, G.E. Fuller. Microbeam Analysis: 649–652, 1986.

36. F. Robb. Proceedings SPIE 775:89–97, 1987.

37. P.K. Bhattacharya, S.K. Jones, A. Reisman. Proceedings SPIE 1087:9–16, 1989.

38. W.J. Keery, K.O. Leedy, K.F. Galloway. SEM/1976/IV IITRI Research Institute, pp 507–514, 1976.

39. A. Reisman, C. Merz, J. Maldonado, W. Molzen. J. Electrochem. Soc. 131:1404–1409, 1984.

40. J.J. Hwu, D.C. Joy. Scanning 21:264–272, 1999.

41. M.T. Postek. SEM/1984/III SEM Inc., pp 1065–1074, 1985.

42. M.T. Postek. Review of Progress in NDE 6(b):1327–1338, 1987.

43. S. Jensen. Microbeam Analysis. San Francisco: San Francisco Press, 1980, pp 77–84.

44. S. Jensen, D. Swyt. Scanning Electron Microscopy I:393–406, 1980.

45. A. Starikov. Metrology of Image Placement. (Chapter 17 of this volume.)

46. H.J. Levinson, M.E. Preil, P.J. Lord. Proc. SPIE 3051:362–373, 1997.

47. M.G. Rosenfield. Proc. SPIE 1673: 157–165, 1992.

48. M.G. Rosenfield, A. Starikov. Microelectronics Engineering 17: 439–444, 1992.

49. J. Allgair, C. Archie, G. Banke, H. Bogardus, J. Griffith, H. Marchman, M.T. Postek, L. Saraf, J. Schlessenger, B. Singh, N. Sullivan, L. Trimble, A. Vladar, A. Yanof. Proc. SPIE 3332:138–150, 1998.

50. International Organization for Standardization. International Vocabulary of Basic and General Terms in Metrology—ISO, 1993, 60P. Geneva, Switzerland, 1993.

51. SEMI. Document E89–099—Guide for Measurement System Capability Analysis, 1999.

52. W. Banke, C. Archie. Proc. SPIE 3677:291–308, 1999.

53. M.T. Postek, J.R. Lowney, A.E. Vladar, W.J. Keery, E. Marx, R.D. Larrabee. NIST J. Res. 98(4):415–445, 1993.

54. M.T. Postek, J.R. Lowney, A.E. Vladar, W.J. Keery, E. Marx, R. Larrabee. Proc. SPIE 1924:435–449, 1993.
55. M.T. Postek, K. Howard, A. Johnson, K. McMichael. Scanning Electron Microscopy—A Students' Handbook. Burlington, VT: Ladd Research Industries, 1980.
56. O.C. Wells. Scanning Electron Microscopy. New York: McGraw Hill, 1974.
57. J.I. Goldstein, D.E. Newbury, P. Echlin, D.C. Joy, C. Fiori, E. Lifshin. Scanning Electron Microscopy and X-Ray Microanalysis. New York: Plenum Press, 1981.
58. G.G. Hembree, S.W. Jensen, J.F. Marchiando. Microbeam Analysis. San Francisco: San Francisco Press, pp 123–126, 1981.
59. D.F. Kyser. In: J.J. Hren, J.I. Goldstein, D.C. Joy, eds. An Introduction to Analytical Electron Microscopy. New York: Plenum Press, 1979, pp 199–221.
60. J.R. Lowney. Scanning Microscopy 10(3):667–678, 1996.
61. J.R. Lowney, M.T. Postek, A.E. Vladar. SPIE Proc. 2196:85–96, 1994.
62. E. Di Fabrizio, L. Grella, M. Gentill, M. Baciocchi, L. Mastrogiacomo, R. Maggiora. J. Vac. Sci. Technol. B. 13(2):321–326, 1995.
63. E. Di Fabrizio, I. Luciani, L. Grella, M. Gentilli, M. Baciocchi, M. Gentili, L. Mastrogiacomo, R. Maggiora. J. Vac. Sci. Technol. B. 10(6):2443–2447, 1995.
64. E. Di Fabrizio, L. Grella, M. Gentill, M. Baciocchi, L. Mastrogiacomo, R. Maggiora. J. Vac. Sci. Technol. B. 13(2):321–326, 1995.
65. D.C. Joy. Monte Carlo Modeling for Electron Microscopy and Microanalysis. New York: Oxford University Press, 1995.
66. M.T. Postek, A.E. Vladar, G.W. Banke, T.W. Reilly. MAS Proceedings (Edgar Etz, ed.), 1995, pp 339–340.
67. J.R. Lowney, M.T. Postek, A.E. Vladar. MAS Proceedings (Edgar Etz, ed.), 1995, pp 343–344.
68. Contact Dr. Jeremiah R. Lowney at the National Institute of Standards and Technology.
69. M.P. Davidson. SPIE Proc. 2439:334–344, 1998.
70. Y.-U. Ko, M.-S. Chung. Proc. SPIE 3677:650–660, 1999.
71. Y.-U. Ko, S.W. Kim, M.-S. Chung. Scaning 20:447–455, 1998.
72. Y.-U. Ko, M.-S. Chung. Scanning 20:549–555, 1998.
73. M.P. Davidson, N. Sullivan. Proc. SPIE 3050:226–242, 1997.
74. L. Grella, E. DiFabrizio, M. Gentili, M. Basiocchi, L. Mastrogiacomo, R. Maggiora. J. Vac. Sci. Technol. B 12(6):3555–3560, 1994.
75. M.T. Postek. Rev. Sci. Instrum. 61(12):3750–3754, 1990.
76. M.T. Postek, W.J. Keery, R.D. Larrabee. Scanning 10:10–18, 1988.
77. N. Sullivan, personal communication.
78. J. McIntosh, B. Kane, J. Bindell, C. Vartuli. Proc. SPIE 3332: 51–60, 1999.
79. M.P. Davidson, A.E. Vladar. Proc. SPIE 3677:640–649, 1999.
80. P. Romand, J. Panabiere, S. Andre, A. Weill. Proc. SPIE 2439:346–352, 1997.
81. J.S. Villarrubia, R. Dixson, S. Jones, J.R. Lowney, M.T. Postek, R.A. Allen, M.W. Cresswell. Proc. SPIE 3677:587–598, 1999.
82. R.A. Allen, N. Ghoshtagore, M.W. Cresswell, L.W. Linholm, J.J. Sniegowski. Proc. SPIE 3332:124–131, 1997.
83. B. Taylor, Kuyatt. NIST Technical Note 1297, 1994.
84. International Organization for Standardization. Guide to the expression of Uncertainty in Measurement (corrected and reprinted 1995), 1997. This document is also available as a U.S. National Standard NCSL Z540-2-1997.
85. R.D. Larrabee, M.T. Postek. SPIE Critical Review 52:2–25, 1993.
86. R.D. Larrabee, M.T. Postek. Solid-State Elec. 36(5):673–684, 1993.
87. J. Fu, T.V. Vorburger, D.B. Ballard. Proc. SPIE 2725:608–614, 1998.
88. B.L. Newell, M.T. Postek, J.P. van der Ziel. J. Vac. Sci. Technol. B13(6):2671–2675, 1995.
89. B.L. Newell, M.T. Postek, J.P. van der Ziel. Proc. SPIE 2460:143–149, 1995.
90. M.T. Postek, R. Gettings. Office of Standard Reference Materials Program NIST, 1995.

91. R.C. Farrow, M.T. Postek, W.J. Keery, S.N. Jones, J.R. Lowney, M. Blakey, L. Fetter, L.C. Hopkins, H.A. Huggins, J.A. Liddle, A.E. Novembre, M. Peabody. J. Vac. Sci. Technol. B 15(6):2167–2172, 1997.

92. J.A. Liddle, M.I. Blakey, T. Saunders, R.C. Farrow, L.A. Fetter, C.S. Knurek, A.E. Novembre, M.L. Peabody, D.L. Windt, M.T. Postek. J. Vac. Sci. Technol. B 15(6):2197–2203, 1997.

93. M.T. Postek, A.E. Vladar, J. Villarrubia. Proc. SPIE (in press), 2000.

94. J.S. Beers, W.B. Penze. J. Res. Nat. Inst. Stand. Technol. 104:225–252, 1999.

95. D.C. Joy. Proc. SPIE 3332:102–109, 1997.

96. M.T. Postek, A.E. Vladar. Proc. SPIE 2725:504–514, 1996.

97. M.T. Postek, A.E. Vladar. Scanning 20:1–9, 1998.

98. A.E. Vladar, M.T. Postek, M.P. Davidson. Scanning 20:24–34, 1998.

99. N.-F. Zhang, M.T. Postek, R.D. Larrabee. Proc. SPIE 3050:375–387, 1997.

100. N.-F. Zhang, M.T. Postek, R.D. Larrabee. Kurtosis. Scanning, 21:256–262, 1999.

101. K.H. Ong, J.C.H. Phang, J.T.L. Thong. Scanning 19:553–563, 1998.

102. K.H. Ong, J.C.H. Phang, J.T.L. Thong. Scanning 20:357–368, 1998.

103. H. Martin, P. Perret, C. Desplat, P. Reisse. Proc. SPIE 2439:310–318, 1998.

104. M.T. Postek, A.E. Vladar. Proc. SPIE (in press), 2000.

105. C. Archie, J. Lowney, M.T. Postek. Proc. SPIE 3677:669–686, 1999.

106. G.D. Danilatos. Adv. Electronics Electron Physics 71:109–250, 1998.

107. G.D. Danilatos. Microsc. Res. Tech. 25:354–361, 1993.

108. M.T. Postek, M.H. Bennett, N.J. Zaluzec. Proc. SPIE 3677:599–610, 1999.

109. NIST Telepresence videotape is available through the Office of Public and Business Affairs, Gaithersburg, MD 20899.

# 15

# Scanned Probe Microscope Dimensional Metrology

**Herschel M. Marchman\***
*University of South Florida, Tampa, Florida*

**Joseph E. Griffith**
*Bell Laboratories, Lucent Technologies, Murray Hill, New Jersey*

## I. INTRODUCTION

Since their introduction almost two decades ago, scanning probe microscopes (SPMs) have already deeply impacted broad areas of basic science and have become an important new analytical tool for advanced technologies, such as those used in the semiconductor industry. In the latter case, the metrology and characterization of integrated circuit (IC) features have been greatly facilitated over the last several years by the family of methods associated with proximal probes. As IC design rules continue to scale downwards, the technologies associated with SPM will have to keep pace if their utility to this industry is to continue. The primary goal of this chapter is to discuss the application of SPM technology to dimensional metrology. In the past, critical-dimension (CD) metrology has mostly involved the measurement of linewidth using either light or electron optical microscopes. However, increased aspect ratios (height/width) of today's IC features have created the need for measurement information in all three dimensions.

The definition of linewidth is not clear for structures typically encountered in IC technology, such as the one shown in Figure 1. Features may have edges that are ragged and side walls that can be asymmetric or even re-entrant. A more appropriate term than *linewidth* is perhaps *line profile*, which describes the surface height ($Z$) along the $X$ direction at a particular $Y$ location along the line. Alternatively, the profile could be given by the width $\Delta X$ at a particular height $Z$ of the line (the line being a three-dimensional entity in this case). The choice of height at which to perform a width measurement will probably depend on the context of the fabrication process in which the feature is to be measured. For instance, the linewidth of interest may be measured at the bottom (foot) of a resist structure that is in a highly selective etch process. However, the width measurement point should probably be shifted further up the profile for less selective etch processes. Edge roughness is also a major factor in the uncertainty of dimensional measurements, because the measured value will differ according to where along the line (i.e., $Y$ coordinate) one samples.

---

\**Current affiliation*: IBM, Hopewell Junction, New York.

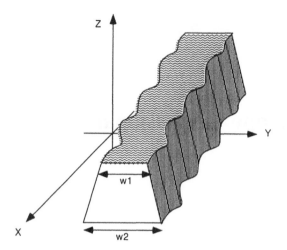

**Figure 1**  Three-dimensional representation of a line profile.

Scanning probe microscopes (SPMs) offer an attractive alternative to optical and electron microscopes for complete three-dimensional imaging of structures. And SPMs are capable of achieving atomic-level resolution for a wide range of materials, even in ambient conditions. Unlike optical and electron microscopes, SPMs do not use optics or waves to obtain images. In an SPM, a needlelike probe is brought very close (< 3 nm) to the sample surface and typically traversed back and forth in a raster fashion. The probe rides up and down at a constant height above the sample, so a topographic image of the surface is obtained. High resolution in all three dimensions is achieved simultaneously. Accurate profiles of the surface can be extracted at any position in a completely nondestructive manner, and a complete rendering of the surface can be created. The main characteristic that distinguishes SPMs from optical and electron-based microscopes is that a solid body, the probe tip, is intimately involved in the measurement process.

## A.  Scanning Tunneling Microscopy

Scanning tunneling microscopes (STMs) were the first types of SPM to be widely used (1–3). In this technique, a metallic probe is brought to within the electron tunneling distance of a conductive sample surface (see Figure 2), such that a small tunneling current flows when a voltage bias is applied. The tunneling current (typically < 1 nA) flows to or from a single atomic cluster at the apex of the probe. Tunneling current density decays exponentially with distance between the two electrodes (tip and sample). Essentially, this can be thought of as a form of "near-field" SEM. There is no need for vacuum or electron lenses because the source and sample are brought to within the near-field distance from each other and scanned while tunneling occurs. Tip-to-sample separation is adjusted by moving either the sample or tip with a piezoelectric transducer that is controlled by a feedback system that maintains constant current and, hence, distance. In this way, the tip is made to track the height of the surface. An image is obtained by rastering the probe across the field of view and plotting the change in tip height position as it tracks the surface. It is this exponential dependence of the current on tip-to-sample separation that gives the STM its extremely high vertical resolution (subangstrom). Lateral resolution on the atomic scale is achieved if the current emanates from a single atomic cluster (the ideal situation) near the apex of a tip. Unfortunately, the types of samples that can usually be imaged with the STM are limited to only those having highly conductive surfaces.

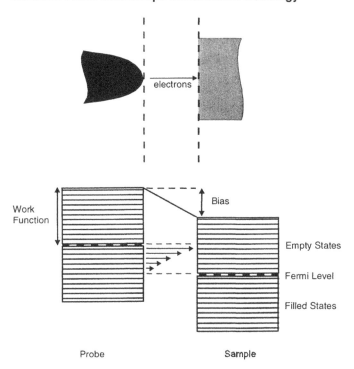

**Figure 2** Energy band diagram for metal–vacuum–metal tunneling in STM. The arrows pointing to the right between the probe and the sample represent tunneling electrons.

## B. Atomic Force Microscopy

Most samples encountered in integrated circuit (IC) metrology contain poorly conducting or insulating surface regions. Photoresists and oxides are two common examples. The surface proximity sensor for probe microscopes must therefore be able to operate over all regions uniformly. For this reason, force sensors are the most commonly used type of detection for proximity of the probe tip to the surface. The atomic force microscope (1) (AFM) was developed shortly after the STM, as a result. Such AFMs measure the topography of a surface by bringing a sharp probe very close (within angstroms) to the sample surface to detect small forces due to the atoms on the surface instead of passing a current. Figure 3 shows the force plotted against distance from the surface. A strong repulsive force is encountered by the tip at distances very near to the surface atoms, due to the exclusion principle. Contact-mode scanning is said to occur when these repulsive forces are used for regulating the tip-to-sample distance. If the tip is moved away from the surface, a small attractive force is encountered several nanometers back. Feedback control with these longer range forces is employed with the noncontact, attractive, mode of imaging. It should be reiterated that the main advantage of AFM is that insulating samples can be imaged just as well as conductive ones.

A simplified diagram of an AFM is shown in Figure 4. Forces between the tip and sample surface are sensed as the probe tip is brought towards the sample surface, as stated earlier. The probe-to-sample distance is regulated by maintaining a constant force, via a feedback servo-system, once the initial approach has been completed (i.e.,

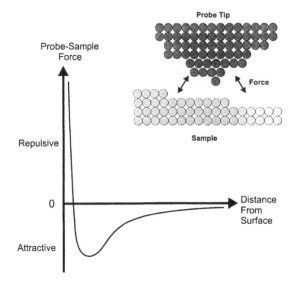

**Figure 3**  Diagram of the tip–sample junction and a force curve for the AFM. Details of the force curve may depend on whether the tip is approaching or withdrawing from the surface.

the surface has been found). The vertical displacement of the cantilever acquired during scanning is then converted into topographic information in all three dimensions. The AFM is analogous to a stylus profilometer, but scanned laterally in $x$ and $y$, with a constant amount of tip-to-sample force maintained in order to obtain images. Although resolution is not at the same level as STM, atomic-scale imaging can be performed by AFM.

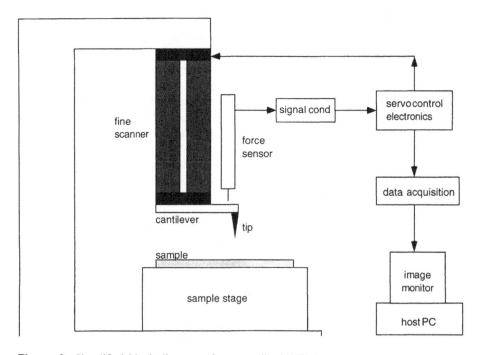

**Figure 4**  Simplified block diagram of a generalized AFM system.

## II. ATOMIC FORCE MICROSCOPY IMAGING MODES

It is clear that the characteristics of different force modes have significant implications on their application in dimensional metrology. In the repulsive-force mode, damage to the tip is accelerated by constant contact with the surface during scanning. The tip-to-sample distance in an attractive-mode AFM is usually an order of magnitude larger than that of the repulsive mode. Increased tip-to-sample separation helps to minimize unwanted tip-to-sample contact. Lateral forces from approaching sidewalls can also be sensed in the attractive mode, which makes this type of AFM ideal for imaging high-aspect features that are commonly found in semiconductor processing. However, contact mode may track the actual surface more acurately under some environmental conditions, such as during static charge or fluid layer buildup on the surface. There is an amazing variety of sensors used by AFMs for detecting the presence of surface forces in either mode. The large number of surface-proximity detectors invented in recent years attests to the importance, and difficulty, of surface detection in scanned probe microscopy (3–5). We will briefly discuss five different force sensors commonly used for measurements on wafers: (1) one-dimensional resonant microcantilever, (2) two-dimensional resonant microcantilever, (3) piezoresistive microcantilever, (4) electrostatic force balance beam, and (5) and lateral/shear-force resonant fiber.

### A. One-Dimensional Resonant Microcantilever

In most atomic force microscope designs the probe is mounted on a microcantilever, which serves as a force sensor (6,7). Forces from the surface are measured by monitoring the flexing of a cantilever using an optical lever. This optical lever is implemented by reflecting a laser beam from the end of the cantilever onto a position-sensitive detector (8). To improve both the sensitivity and stiffness, most designs exploit resonance enhancement. Because of their small size, a typical resonant frequency for the cantilever is around 300 kHz. Interactions between the probe and sample introduce variation in the amplitude, frequency, and phase of the cantilever oscillation. Any of these signals can be fed back into the microscope and used to maintain the vertical position of the scan head. One drawback of this design is that the cantilever is typically tilted about $10°$ with respect to the sample (see Figure 5). Oscillation thus causes high-frequency lateral motion of the probe tip as well. The actual force of interaction is not directly available from the system, but Spatz et al. (9) have obtained an estimate through modeling of typically a few tenths of a micro-newton. In standard-mode operation, a simple raster scan is used and servo control is performed using the one-dimensional resonant attractive-mode technique to maintain a constant tip-to-sample spacing of approximately 2–3 nanometers. This type of scanning algorithm makes uniformly spaced steps across the sample surface, and the vertical tip position is recorded at each step.

### B. Two-Dimensional Resonant Microcantilever

Martin and Wickramasinghe have developed a sophisticated two-dimensional force sensor whose design is also based on a microcantilever (10–12). This system is supplied with a flared probe tip, shown in Figure 6, called a boot tip, which allows improved access to vertical sidewalls and undercut regions (11). The force sensor operates in a noncontact fashion that exploits resonance enhancement of the microcantilever in two directions. A microcantilever with a spring constant of $\approx 10\,N/m$ and $Q \approx 300$ is vibrated both verti-

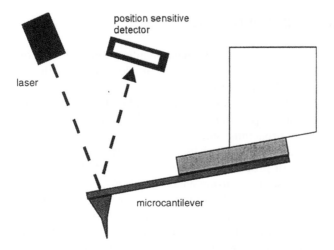

**Figure 5**  Schematic diagram of an AFM microcantilever with its optical-level force monitor.

cally and laterally at different frequencies with an amplitude of $\approx 1$ nm. Cantilever vibration is detected using an optical interferometer. The force sensitivity is approximately $3 \times 10^{-12}$ N, which allows it to detect the presence of the much weaker attractive forces (not necessarily van der Waals only). Separation of the vertical and horizontal force components is accomplished through independent detection of the each vibration frequency. Servo control of both the horizontal and vertical tip-to-sample distances now occurs via digitally controlled feedback loops with piezo actuators for each direction. When the tip encounters a sidewall, their separation is controlled by the horizontal feedback servo system. Conversely, tip height ($z$ direction) is adjusted by the vertical force regulation loop.

The scan algorithm uses force component information to determine the surface normal at each point and subsequently deduce the local scan direction (11). This allows the scan direction to be continually modified as a function of the actual topography in order for the motion to stay parallel to the surface at each point. In this way, data is not

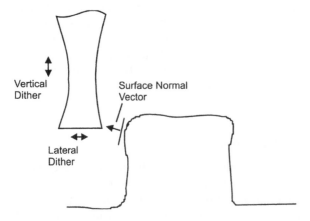

**Figure 6**  Dithering the microcantilever in two dimensions allows this microscope to determine the local surface normal vector.

acquired at regular intervals along $x$ anymore, but at controlled intervals along the surface contour itself. This is essential for scanning undercut regions, because in addition to having a flared tip, it is necessary to have a two-dimensional (2-D) scanning algorithm that can provide servoed tip motion in both the lateral and vertical scan axis directions. One could imagine what would happen if a flared tip were to try to move straight up while under a feature overhang. The data set collected is stored as a three-dimensional mesh that maintains a relatively constant areal density of data points, irrespective of the sample slope. It is especially well suited for measuring the angle of a sidewall or its roughness. Another unique capability of this system is the ability to automatically adjust scan speed to surface slope so that it slows when scanning a nearly vertical topography and hastens when scanning a nearly flat surface. Data point density (in the $x$ direction) also increases at sudden changes in surface height, such as feature edges (of most interest), because the sampling rate is fixed.

## C. Piezoresistive Microcantilever

One strategy for improving AFM throughput is to design a microscope in which a large array of probes operates simultaneously. Implementing such an array presents two technical problems. First, optical lever systems are awkward to assemble in large arrays. Second, the height of the probe tips must be independently controllable because most samples are not perfectly flat. Tortonese et al. developed a novel piezoresistive-force sensor suitable for use in large arrays (13). The microcantilever is supported by two parallel arms, each of which contains a piezoresistive film, B-doped $\langle 100 \rangle$ Si. The force on the probe flexes the arms, producing a small change in resistance, which serves as the feedback signal. Minne et al. developed the independent actuator for this sensor (14). They built into each cantilever a piezoelectric ZnO actuator, which is capable of raising or lowering the probe by several micrometers. Thus, each probe tip in the array can independently follow the sample surface at scan speeds up to 3 mm/s (15).

## D. Balance-Beam Force Sensor

Because of their tiny size, the probe tips of microcantilevers are typically manufactured along with the cantilever, which limits the methods and materials used. It is possible to achieve high sensitivity in a larger sensor. An example is a centimeter-long beam held balanced on a weak pivot by electrostatic force (Figure 7). The two sides of the balance

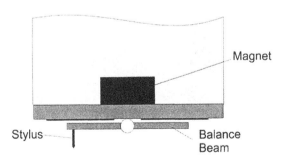

**Figure 7** Balance-beam force sensor. The magnet holds the beam on the base plate. A force-balance circuit balances the beam with electrostatic forces.

beam form capacitors with the base. This balance-beam force sensor (16), also known as the interfacial force microscope (17), is an inherently unstable mechanical system stabilized by a servo loop. By using force balance rather than a weak spring, this method of force sensing uncouples the sensitivity from the stiffness. The pivot, developed by Miller and Griffith (18), is a pair of steel ball bearings, one on each side of the beam, held to the substrate with a small magnet. The ball bearings are rugged, so a beam can be used almost indefinitely. The magnetic constraint suppresses all degrees of freedom except the rocking motion, so the sensor is quiet. This system has been used to measure surface roughness with RMS amplitude less than 0.1 nm.

The capacitors serve two purposes: They both detect and control the position of the beam. The detector is a 500-kHz bridge circuit, which compares the relative values of the two sides. The bridge circuit controls the dc voltage, $V \pm \Delta V$, across the two capacitors to keep the beam balanced. The capacitors set the size of the balance beam, since the electronics are designed for a capacitance of a few picofarads. The size of the beam is $5 \, mm \times 10 \, mm$. The force $F$ impressed on the tip can be calculated directly from the dimensions of the sensor and the difference $\Delta V$ in the voltage on the two sides. For a typical sensor, $\Delta V = 1 \, mV$ implies $F = 2 \times 10^{-8} \, N$.

The balance-beam force microscope has several advantages. The most important of these arises from its large size. It can accept almost any probe tip. The sensor consequently does not constrain either the probe material or its mode of fabrication. The tip can be several millimeters long, so the sensor does not have to be extremely close to the surface. The sensor can be mounted on the scanner piezos, so the fixed sample can be large. The sensor performs well in vacuum. It can operate simultaneously as a tunneling microscope. Magnetic constraint allows easy tip exchange, and the pivot is exceptionally rugged. The sensor also allows the tip to be accurately aligned perpendicular to the sample, which is important when measuring vertical sidewalls.

### E.  Lateral Force Resonant Fiber

In lateral force microscopy, a tapered fiber or wire whose length axis is normal to the surface (Figure 8) is vibrated at mechanical resonance laterally as it is brought close to the surface (19,20). Surface forces are indicated by changes in either the phase ($\theta$) or amplitude ($A$) of the fiber resonance. In many cases a combination of the two, known as the quadrature component, or $A \cos(\theta)$, is actually used for regulation. This technique offers the

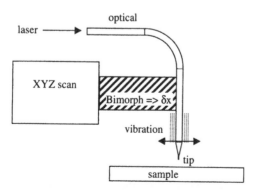

**Figure 8**  Head assembly for the lateral/shear-force fiber SPM.

stability of an amplitude signal with the enhanced speed and sensitivity of a phase component. Actually, force sensing in all directions is achieved. Distance to the sample surface is sensed vertically through shear forces and horizontally to approaching side walls through noncontact damping (similar to resonant mode). Lateral forces between the tip and approaching side walls are attractive, so there is less likelihood of accidental contact (i.e., damage). Noncontact lateral-force sensing also enables quicker scanning. Shear-force sensing is advantageous for vertical height regulation in that it is not as susceptible to effects caused by fluid layers or static charges as the attractive mode.

Lateral-force probes are typically made from optical fibers that are composed mainly of glass. The actual imaging probe is etched from the optical fiber's endface. In addition to yielding high-aspect cylindrically shaped probes for metrology, optical fibers possess several intrinsic properties that can be exploited for simplifying SPM instrumentation. The fiber body's natural resonance provides a means of force sensing and the ability to emit light provides a signal for tracking the probe's motion. An optical microscope with high NA (> 0.9) objectives are used to position the tip as the sample is translated by long-range (> 200-mm) motion stages. The entire probe scan assembly is mounted to the optical viewing microscope. The lateral-force SPM essentially looks and operates like an ordinary optical microscope with a scanned probe in the same field of view as the feature to be measured. Simultaneous viewing of the tip (emitting light) and feature is achieved because the optical fiber axis is oriented normally to the sample surface. Throughput and probe lifetime are greatly enhanced with this configuration, since the need for SPM imaging to locate and center the feature of interest is eliminated. There is relatively immunity to Abbe' offset error because the light signal for position sensing is emitted at the proximal point. The apparatus can also be used to collect optical intensity within the near field once a confinement layer is added to the probe walls, thus allowing optical measurements to be made with nanometer resolution (21).

## III.  SOURCES OF ERROR

As with other microscopes, undesired artifacts are also present in probe microscope images, which adversely affect dimensional measurement performance. Two elements of probe microscopes exibit strongly nonlinear behavior that can seriously affect measurement accuracy and precision well before atomic resolution has been reached. The first is the piezoelectric actuator that is used for scanning the probe. The second, and probably more serious, problem arises from interaction between the probe and the sample.

### A.  Scan Linearity

Piezoceramic actuators are used to generate the probe motion because of their stiffness and ability to move in arbitrarily small steps. Being ferroelectrics, they suffer from hysteresis and creep, so their motion is not linear with applied voltage (22). Therefore, any attempt to plot the surface height data verses piezo scan signal results in a curved or warped image and does not reflect the true lateral position of the tip. A variety of techniques have been employed to compensate for the nonlinear behavior of piezos. In many instruments, the driving voltage is altered to follow a low-order polynomial in an attempt to linearize the motion. This technique is good to only several percent and doesn't really address the problem of creep. Attempting to address nonlinearities with a predetermined driving algorithm will not be adaquate for dimensional metrology because of the complicated and nonreproducible behavior of piezoelectric materials. Another approach is to independently monitor the motion of piezoactuators with a reliable sensor. Several types of systems using this approach have been reported. One monitored the motion of a flexure stage with an interferometer (23). Another measured the position of a piezotube actuator with capacitance-based sensors (24). A third group employed an optical slit to monitor the piezotube scanner (25). Electrical strain gauges have also been used for position monitoring as well.

These techniques monitor the position of the piezoactuator and not the actual point of the probe that is in closest proximity to the surface (known as the proximal point). The error associated with sensing position in a different plane from that of the proximal point, referred to as Abbe' offset, is illustrated in Figure 9. Abbe' offset error is given by $D \sin(\alpha)$, which increases with the tilt of the scan head as the probe moves. Unless special precautions are taken, the position sensors will not respond to this tilting. There is a tendency in the case of SFM systems to dismiss this error as negligible because of the submicrometer dimensions being measured. As an example of Abbe' offset error, the angular tolerance for a 1-cm displacement between the scan-head motion and sample planes can be computed. If we demand less than 1-nm Abbe' offset, the allowable angular error is $10^{-7}$ radians, or 1/50 arc-second. In systems built around piezoceramic tubes, the head tilting induces an $\approx 1\%$ nonlinearity into the monitor response (26–28). In tube scanners, the magnitude of the tilting is not a very stable function of the lateral motion of the tube, which makes calibration of the tube unreliable. To minimize this error the sensors are designed to be as close as possible to the probe. One can, alternatively, add extra sensors to measure the tilt.

Because of these problems with tube scanners, a design being developed at the National Institute for Standards and Technology employs a flexure stage to produce more reliable motion (29,30). A design goal of this tool is to reduce the Abbe' offset as much as possible. Ideally, the position of the probe apex should be directly measured; a new tool developed by Marchman et al. achieves this ideal with an optical-position sensor and light-emitting optical fiber probe (31,32). Errors intrinsic to the position monitors are not constant and vary with sample topography as well as lateral scan range—for example, errors in capacitance measurements when the plates become far apart at the lateral scan extremes. In the optical slit method, "$1/f$" noise of the detector at low frequencies (scanning) imposes limitations on resolution as well as complications involved with the slit (e.g., tilting of the slit as the tube scans, or aperture edge irregularities).

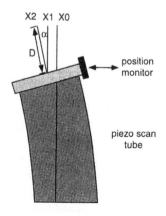

**Figure 9**  Abbe' offset error for a tube-based scanner.

A problem common to one-dimensional scanning techniques, even optical and SEM, arises from the imaging algorithm itself. As seen in Figure 10, data points in a single linescan are equally spaced in increments of $\Delta X$ in the lateral direction. The size of each $\Delta X$ increment is constant, regardless of the topography, so there is insufficient pixel density when abrupt changes in surface height (e.g., feature sidewall) are encountered. Scanning re-entrant profiles with even a flaired probe is also forbidden, because the probe always moves forward or upwards and is not able to reverse its direction without a 2-D algorithm.

## B. Probe Shape

Regardless of the specific force used to sense proximity to the surface, all SPMs share a common working element: the tip. The probe tip is where the rubber meets the road in probe microscopy. While there have been, and continue to be, important advances in other aspects, such as detection schemes and position monitoring, improvements in the tip offer the greatest potential for increasing SPM metrology performance. As described earlier, current-generation probe tips are made of etched or milled silicon/silicon nitride (these make up by far the bulk of commercial tips) or etched quartz fibers, or else are built up from electron-beam-deposited carbon. While all these tips have enabled significant metrological and analytical advances, they suffer serious deficiencies, the foremost being wear, fragility, and uncertain and inconsistent structure. In particular, the most serious problem facing SPM dimensional metrology is the effect of probe shape on accuracy and precision. Very sharp probe tips are necessary to scan areas having abrupt surface changes (33). Most commercially available probes are conical or pyramidal in shape, with rounded apexes. When features having high aspect ratios are scanned (see Figure 11), such as those encountered in IC fabrication, they appear to have sloped walls or curtains. The apparent surface is generated by the conical probe riding along the upper edge of the feature. Even if we had known the exact shape of this probe, there is no way to recover the true shape of the sidewalls. The fraction of the surface that is unrecoverable depends on the topography of the surface and sharpness or aspect ratio of the tip. There exists no universal probe shape appropriate for all surfaces. In most cases, a cylindrical or flaired tip will be the preferred shape for scanning high-aspect features (34). The cross-sectional diagram in Figure 12 demonstrates the effect of even a cylindrical probe shape (i.e., the ideal case) on pitch and linewidth measurements. Pitch measurement is unaffected by the probe width, but the linewidth and trench-width values are offset by the width of the probe. Simple probe shapes provide greater ease of image correction, and the unrecoverable regions are vastly reduced. It is essential to make the probe as long and slender as possible in order to increase the maximum aspect ratio of feature that can be scanned.

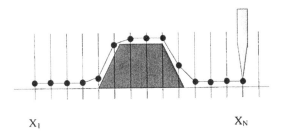

$X_1$ $X_N$

**Figure 10** One-dimensional raster scan algorithm.

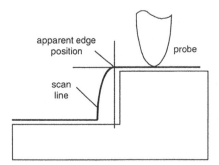

**Figure 11**   Apparent surface generated by probe shape mixing.

Conical tips are still useful for measurements on surfaces with relatively gentle topography. A cone having a small radius of curvature at the apex of the structure, shown in Figure 13, can perform surface roughness measurements that cover regions of wavelength–amplitude space unavailable to other tools (35,36). Though a conical probe tip may not be able to access all parts of a feature, its interaction with an edge has some advantages over cylindrical probes (33,37,38). At the upper corners of a feature with rectangular cross section, size measurement becomes essentially equivalent to pitch measurement. The uncertainty in the position of the upper edge becomes comparable to the uncertainty in the radius of curvature of the probe apex, which can be very small. If the sidewalls are known to be nearly vertical, then the positions of the upper edges give a good estimate for the size of the feature. To produce sharper conical tips one can employ a focused ion beam (FIB) technique, in which a Ga beam is rastered in an annular pattern across the apex of an etched metal shank (39). This method routinely produces tips having a radius of curvature at the apex of 5 nm and widening to no more than 0.5 μm at a distance of 4 μm from the apex. Occasionally the FIB sputtering generates a tip with nearly cylindrical shape. AFM linescans of periodic photoresist lines acquired using pyramidal and FIB-sharpened tips are displayed in Figures 14(a) and (b), respectively. The inability of either conical probe to measure the sidewalls of high-aspect features is clearly reflected by the trapezoidal image profiles. It is interesting to see that even though the FIB-sharpened tip is sharper, it still exhibits strong probe shape mixing.

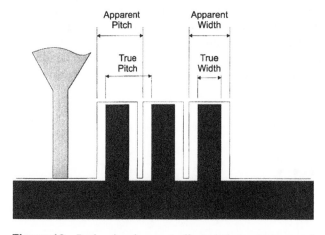

**Figure 12**   Probe size does not affect pitch measurement, but it does affect width measurement.

**Figure 13** A conical SPM tip etched on the underside of a silicon microcantilever.

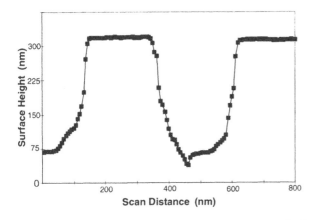

**Figure 14(a)** Profile of a square grating structure obtained with a pyramidal tip.

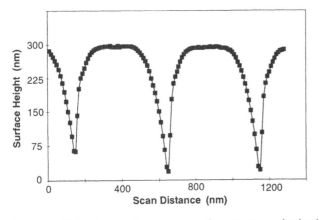

**Figure 14(b)** Profile of a square grating structure obtained with an FIB-sharpened tip.

A cylindrical probe can be formed by chemical etching of an optical fiber with buffered hydrofluoric acid (40,41). This procedure produces a nearly perfect cylinder that can have a diameter less than 0.1 μm. Since the elastic modulus of glass is typically 70 GPa, the probes are stable against flexing as long as the slender section is not too long. During scanning, these probes have proven to be very durable. An optical microscope image of a fiber probe before final etch is shown in Figure 15. Upon final etch, the same shape is maintained and the diameter is reduced to as low as 100 nm. The fiber's cylindrical shape is preserved throughout the isotropic HF etch. This technique also allows flared quartz probes to be made. A scan taken with the quartz cylindrical probe is superimposed on the corresponding SEM cross section shown in Figure 16.

A cylindrical probe that widens, or flares, at the apex provides the most complete rendition of sidewall shape. An example of such a probe tip is shown in Figure 17; it was developed by T. Bayer, J. Greschner, and H. Weiss at IBM Sindelfingen (7). This probe is fabricated on a silicon microcantilever. The tolerances in the fabrication of this probe are very tight. The two-dimensional scan algorithm described earlier must also be employed in order to use the flared tips on features with undercut regions. With all cylindrical or flared probes, the accuracy of a width measurement depends on knowledge of the probe width. The linescan from a two-dimensional resonant AFM and flared tip of the same periodic photoresist lines shown earlier in Figure 14 is shown in Figure 18. We can now see from the profile that the photoresist structures were actually re-entrant, or undercut, near the feature tops.

## C. Probe Stiffness

In semiconductor manufacturing, one routinely encounters features, such as vias, less than 0.25 micrometers wide and over 1 micrometer deep. Features with such extreme aspect ratios pose a challeging problem for probe microscopy. The probe must be narrow enough to fit into the feature without being so slender that it becomes mechanically unstable. To achieve an accurate measurement, the apex of the probe must remain fixed relative to the

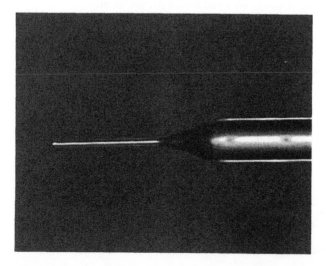

**Figure 15**   Optical microscope image of an etched optical fiber probe.

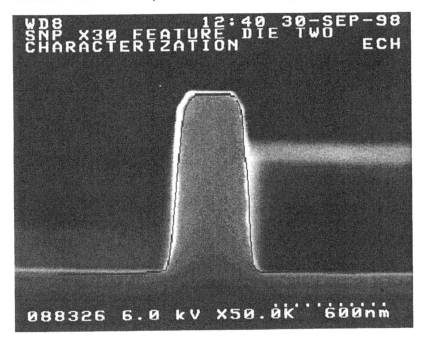

**Figure 16** Comparison of a cross-sectional SEM measurement and a scan line taken with the surface/interface SNP. The line superimposed on the image is the SNP scan.

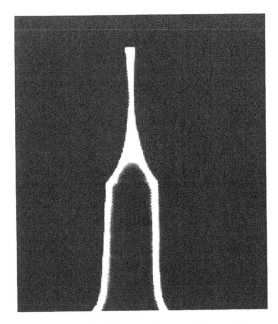

**Figure 17** Flared probe tip etched from silicon cantilever.

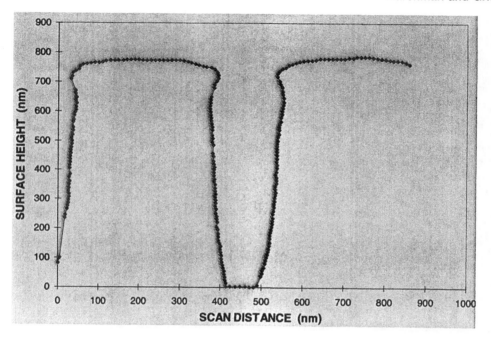

**Figure 18**  Profile of an undercut feature obtained with flared probe and 2-D scan algorithm.

probe shank. In other words, the probe must not flex. Flexing introduces errors into the measurement, in addition to causing instability of the feedback control loop.

The analysis of probe stiffness is identical to that of any cantilevered beam. Imagine a probe tip with uniform circular cross section having radius $R$ and length $L$. Let the elastic modulus of the probe material be $Y$. If a lateral force $F$ is impressed on the probe apex, the probe will deflect a distance $\Delta x$, which obeys the following expression (42):

$$\Delta x = \frac{4FL^3}{3\pi Y R^4}$$

Note that the geometrical factors, $L$ and $R$, have the strongest influence on the deflection. This indicates that efforts to find materials with higher modulus would only produce marginal gains compared to the effects of the probe's size and shape.

## D.  Next-Generation Metrology Probes

The need to precisely control probe shape and achieve the maximum possible stiffness has motivated recent efforts to employ carbon nanotubes as probes. Carbon nanotubes are recently discovered materials made purely of carbon arranged in one or more concentric graphene cylinders and having hollow interiors. This arrangement makes them essentially giant, elongated fullerenes, a relationship made clear by the strategy they share with spheroidal fullerenes (e.g., $C_{60}$—buckminsterfullerene) for closing by incorporating a total of 12 pentagons into their hexagonal network. *Single-wall nanotubes* (SWNTs) come closest of all carbon fibers to the fullerene ideal due to their remarkably high degree of perfection; their poverty of defects and the special nature of carbon–carbon bonding confers upon them material properties, such as strength, stiffness, toughness, and electrical and thermal conductivities, that are far superior to those found in any other type of fiber

(43). There also exists larger-diameter cousins of single-wall (fullerene) nanotubes, typically having 4–20 concentric layers, known as *multi-wall nanotubes* (MWNTs). These may offer extra advantages as SPM probes due to their increased radial diameter, typically between 20 and 100 nm. In addition to having a perfect cylindrical shape, these tubes exhibit an unusually high elastic modulus—in the terapascal range (44). Dai et al. were the first to mount a nanotube on a microcantilever and scan with it (45). The most important characteristic of carbon nanotube tips is their immunity to wear and erosion. Measurement precision would be greatly enhanced if their shape remains unchanged after accidental contact with the surface during scanning. The ability to measure their exact shape and size will provide a new level of accuracy and precision in SPM metrology. The transmission electron micrograph in Figure 19 shows an MWNT that has been mounted on a pyramidal AFM tip apex.

There are two practical impediments to widespread use of these probes. First, techniques for handling these tiny objects are not yet mature enough to allow mass production of mounted tips. Manual mounting of nanotubes to silicon cantilever-based probes and etched-quartz fibers (46) has been performed in the developmental mode, but does not offer the throughput for an adaquate supply in production-mode applications. Optical microscope images taken during different steps of the tip-mounting process for etched optical fibers are shown in Figure 20. One can see from the final step image that alignment of the tube during mounting is critical. Second, single-walled nanotubes that tend to grow with diameters in the 5–20-nm range are too slender to resist flexing. Even with an elastic modulus of one terapascal, a 1-micrometer-long nanotube needs to have a diameter of at least 50 nm to achieve the necessary stiffness. Fortunately the multiwalled nanotube appears to have sufficient diameter to offer adaquate stiffness and resistance to flexing during scanning of high-aspect features. Developments in this field have been rapid, and we expect these difficulties to be overcome in the near future. Presently, metrologic performance is being gauged through studies using manually mounted nanotube tips in different AFM force imaging modes on IC features of interest.

## E.  Throughput

Another very serious issue facing the implementation of scanning probe microscopy into the semiconductor device fabrication line is low throughput. Presently, probe microscopes image too slowly to compete with optical or electron microscopes in terms of speed. Since

**Figure 19**   TEM micrograph of a multiwall nanotube mounted on a pyramidal tip apex.

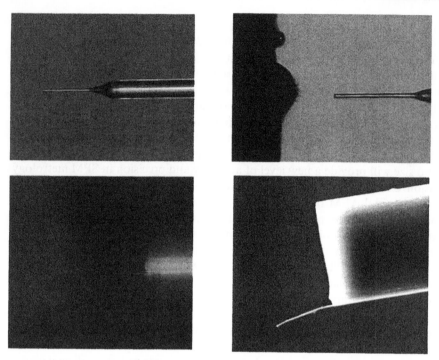

**Figure 20**  Picture sequence for nanotube tip-mounting procedure.

protection of the probe tip is of paramount importance, the system should not be driven faster than its ability to respond to sudden changes in surface height. This is a source of complaint from those used to faster microscopes. In many instances the probe microscope is, however, providing information unavailable from any other tool, so the choice is between slow and never. In addition, when comparing the speed of a probe microscope against cross-sectional SEM, the sample preparation time for the cross sectioning should be taken into consideration along with the fact that the sample has been irretrievably altered.

In additon to imaging time, finding the desired feature is complicated because the physical structure of most probe microscopes excludes high-magnification viewing of the imaging tip on the sample surface. Microcantilever-based SPMs must also contend with the cantilever's blocking, or shadowing, the surface region of interest. Thus the feature must be found by imaging with the SPM at slow speed, with a limited field of view (with the exception of the lateral-force optical fiber microscope by Marchman). In many cases, the majority of tip wear occurs during this step, because larger scan ranges and tip speeds are used in the search phase.

## IV.  ACCURACY AND CALIBRATION

The probe–sample interaction is a source of error for all measuring microscopes, but the interaction of a solid body (stylus) with a sample offers several advantages when high accuracy is needed. The most important advantage arises from the relative insensitivity of a force microscope to sample characteristics such as index of refraction, conductivity, and composition. Optical and electron beam tools are sensitive to these characteristics (47,48), so errors can arise if a reference artifact has a composition different from the sample to be measured. For instance, calibrating an SEM-CD measurement with a metal line may give inaccurate results if the line to be measured consists of photoresist.

A comparison between probe microscope measurements of photoresist lines and scanning electron microscope (SEM) measurements of the same lines was made by Marchman et al. (46). The probe microscope measurements, made with a commercial atomic force microscope (49), agree closely with the cross-sectional SEM measurements, while the plan view measurements deviate from them by up to 0.1 µm, depending on the magnification and how the SEM signal is interpreted. In the calibration of a probe microscope there are two fundamental problems: measuring the behavior of the position sensors and finding the shape of the probe tip.

## A. Position Sensor Calibration

Calibrating position sensors is the easier of the two chores and we will discuss it first. Lateral position sensor calibration is equivalent to a pitch measurement, and an excellent reference material is a grating, a two-dimensional one if possible. Periodic reference materials are available commercially, or they can be reliably fabricated through, for instance, holographic techniques (50). Some samples are self-calibrating, in that they contain periodic structures with a well-known period.

Vertical calibration is more troublesome because most step height standards contain only one step. This makes it impossible to generate a calibration curve unless multiple standards are used. In addition, the step height sometimes differs by orders of magnitude from the height to be measured. Surface roughness measurements often involve height differences of less than a nanometer, though they are sometimes calibrated with step heights of 100 nm or more. The vertical gain of a piezoscanner may be substantially different in the two height ranges.

## B. Tip Shape Calibration

The image produced by a scanning probe microscope is simply a record of the motion of the probe apex as it is scanned across the sample. The image will faithfully represent the sample as long as the apex and the point of interaction coincide. If the interaction point wanders away from the apex, the relationship between image and sample becomes more complicated. An example of this is shown in Figure 21. As the probe moves across the step, the apex follows the dashed line rather than the true step profile. Note that the interaction point stays fixed at the upper edge of the step until the apex reaches the lower plane. The dashed line is, in fact, an inverted image of the probe. Several researchers have analyzed this process (51–55). Villarrubia has conducted the most thorough study, making use of results from mathematical morphology (54,55). We will use his notation. The behavior represented by the dashed line is often called "convolution," which is a misnomer. Convolution is a linear process, while the interaction of the probe with the sample is strongly nonlinear. The correct term for this process is *dilation*.

The mathematical description of dilation uses concepts from set theory. Imagine two objects $A$ and $B$, represented by sets of vectors. The individual vectors in the objects will be denoted $a$ and $b$. The dilation of $A$ by $B$ is given by

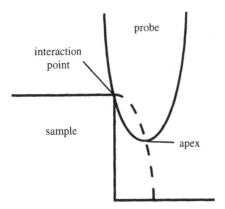

**Figure 21**   Point of interaction and apex trajectories during probe shape mixing.

$$A \oplus B = \bigcup_{\underline{b} \in B} (A + \underline{b})$$

where $A + b$ is the set $A$ translated by the vector $b$. This operation replaces each point of $A$ with an image of $B$ and then combines all of the images to produce an expanded, or dilated, set. In scanning probe microscopy, we work with the following sets: $S$, the sample; $I$, the image; and $T$, the probe tip. In the analysis, one frequently encounters $-T$, the reflection of the probe tip about all three dimensional axes. Villarrubia denotes the reflected tip as $P$. It can be shown that

$$I = S \oplus (-T) = S \oplus P$$

In other words, the image is the sample dilated by the reflected probe tip. We can see this in Fig. 21 where the image of the step edge includes a reflected copy of the probe shape. When actually performing calculations, one must convert this abstract notation into the following expression:

$$i(x) = \max_{x'}[s(x') - t(x' - x)]$$

where $i$, $s$, and $t$ represent the surfaces of the sets.

   If the probe shape is known, then it is often possible to arrive at an estimate of the sample shape substantially better than that represented by the raw image. An example of this analysis is shown in Figure 22. We know that the probe reached the position shown because the dashed line represents the collection of points visited by the probe apex. The sample cannot extend into region $A$ because the sample and the probe are not allowed to overlap. We cannot make the same claim about region $B$, however, because the probe was not capable of occupying that space. The subtraction of region $A$ from the image is an example of *erosion*. The erosion of set $A$ by set $B$ is defined as

$$A - B = \bigcap_{\underline{b} \in B} (A - \underline{b})$$

In our context, it can be shown that the best estimate of the true surface we can obtain from a probe tip $T$ is

$$I - P$$

Clearly, it is important to know the precise shape of $P$.

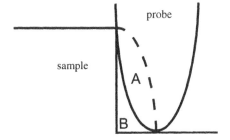

**Figure 22** Sample dilation by the reflected probe tip.

It has been known for many years that scanning probe microscope scans can be used to measure the probe shape. In fact, it is the best way to measure probe shape. Villarrubia has thoroughly analyzed this process, showing the manner in which any image limits the shape of the probe. His argument is founded on the deceptively simple identity

$$(I - P) \oplus P = I$$

If we erode the image with the probe and then dilate the result with the same probe, we get the image back. Villarrubia shows how this expression can be used to find an upper bound for the probe. It is based on the observation that the inverted probe tip must be able to touch every point on the surface of the image without protruding beyond that image. The algorithm implementing this is, however, subtle and complex, so we will not give it here. One of his publications provides full C source code for implementing the algorithm (55).

Some sample shapes reveal more about the probe than others (56,57). Those shapes specially designed to reveal the probe shape are called probe tip characterizers. The characterizer shape depends on the part of the probe that is to be measured. To measure the radius of curvature of the apex, a sample with small spheres of known size might be used. If the sides of the probe are to be imaged, then a tall structure with re-entrant sidewalls should be used. A re-entrant test structure for calibrating cylindrical and flared probes on commercially available CD metrology AFMs is shown in Figure 23. The raw scan data from such a structure is shown in Figure 24a. Of course, this profile is actually a combination of both the probe and characterizer shapes mixed together. If the tip characterization structure is of known shape, it can be subtracted from the raw data in order to produce a mirror image of the actual tip. As a result of this, we can see from Figure 24b that an undercut flared probe was used for acquiring the data. A particle sticking to the probe wall is now visible from the extracted image. It should be noted that the characterizer walls must be more re-entrant than the flared probe's. A simpler and quicker technique used for calibration in on-line metrology applications most often employs etched silicon ridges to determine the bottom width of a probe. This series of ridge structures is referred to as the *Nanoedge*(58). A triangular-shaped profile would appear in the SPM image for a tip of zero width. In fact, a set of trapazoidal lines is obtained, as shown in Figure 25. The width of the trapazoidal tip yields the size of the probe's bottom surface, after the width of the feature ridge (< 2 nm) is taken into account. Therefore an offset in accuracy of only 2 nm is achieved with this technique. Quantitative correction of subsequent CD measurements is achieved by subtracting the bottom-width value from the raw scan data. This method is valid as long as the proximal points are located at the bottom of the probe. Even if this is the case initially, wear and damage can cause this to change. Fortunately, the undercut structure can be used periodically to verify the probe shape and, hence, the validity of the

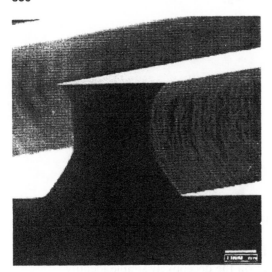

**Figure 23**  Flared silicon ridge (FSR) tip characterization structure.

bottom-width subtraction technique. The AFM scans in Figure 26 show the same structure before and after tip width removal from the image data. As a sanity check, one can overlay the corrected AFM linescan data on top of an SEM cross-sectional image of the feature to verify that the shapes do indeed match (see Figure 16).

## V.  SPM MEASUREMENT PRECISION

An example of the precision gauge study for a CD metrology AFM will be described in this section (59). Currently, the primary AFM commercially available for making dimensional measurements on full wafers in the fab is known as the SXM workstation. This system is the product of more than 14 years of research and development by the IBM Corporation in the area of scanned-probe technology (60). The SXM workstation can be

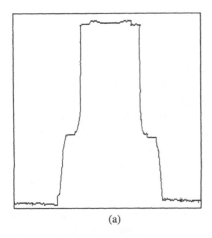
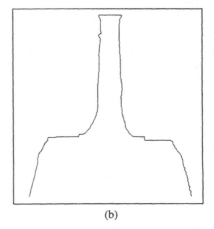

(a)                                                                    (b)

**Figure 24**  (a) Raw data scan of FSR and (b) flared probe image after subtraction.

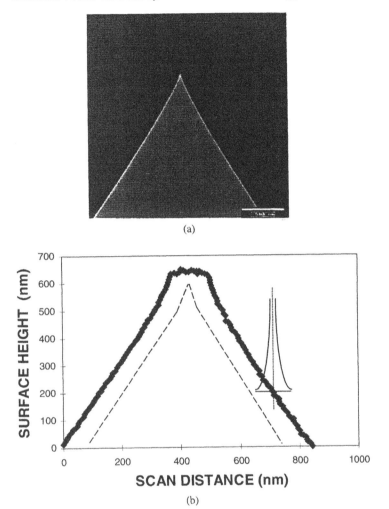

(a)

(b)

**Figure 25** (a) SEM sectional image of the Nanoedge tip characterizer, (b) AFM scan of the Nanoedge structure for tip width estimation.

operated in two modes: standard (1-D) and critical dimension (CD). Standard mode operates with one-dimensional resonance force sensing and simple one-dimensional raster scanning, but CD uses two-dimensional resonant force sensing along with the 2-D surface contour scanning algorithm described earlier (Figure 6). Tip position is obtained accurately in all three dimensions from calibrated capacitive position sensors at each axis. Image distortion due to piezoelectric scanner nonlinearity is minimized by using the capacitive monitors to provide the image data. In the standard mode, precise profiles of sample features can be obtained as long as the half-angle of the conical tip is greater than the structure being scanned. In the CD mode, the tip shape is cylindrical with a flared end. The bottom corners of the boot-shaped CD tip sense the sidewalls of a feature as the tip scans along them. The position of these protrusions at the bottom corners of the tip is key for imaging the foot of a sidewall, which can be imaged if only they remain sharp and are at the lowest part of the probe body. A significant advantage of the CD-mode AFM over other techniques, including optical and SEM, is that the

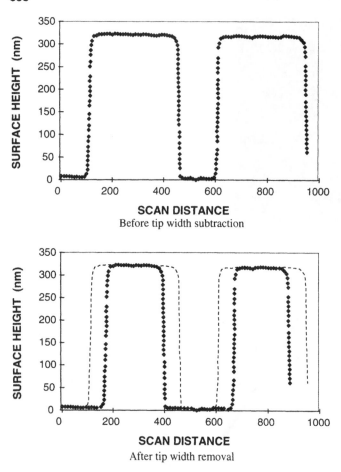

Before tip width subtraction

After tip width removal

**Figure 26**   AFM image dilation from tip shape mixing.

number of data points can be set to increase at feature sidewalls or abrupt changes in surface height.

Initially, screening experiments were performed in order to determine the amount of averaging necessary during each measurement and the relative weighting of different factors in the precision. AFM images are composed of a discrete number of line scans (in the $X$–$Z$ plane)—one at each value of $Y$. For better-precision estimates, each measurement was performed as close to the same location as possible (to within the stage precision) in order to minimize the effects of sample nonuniformity. Another important screening task was to determine how many linescans per image are necessary to provide adaquate spatial averaging of the line-edge roughness in order to reduce the effects of sample variation on the instrument precision estimate. However, it is desirable to require as small of a number of linescans per image as possible due to the relatively long amount of time necessary to acquire AFM data. To determine the minimum required number of scans, alternate lines were successively removed from an image until a noticeable change in the edge-roughness value was observed. The minimum spatial averaging for edge roughness (sample variation) was at least 8 linescans over 2 microns of feature length. Initial repeatability tests performed with different operators indicated that there was no observable effect with the SXM system when run in the fully automatic mode. Once an automation recipe was created, the feature was consistently located and measured with no assistance from the operators.

Gauge precision was obtained by making measurements under different conditions. Determining the most significant factors to vary depends on one's understanding of the physical principles of the instrument's operation. Based on the principles of scanning probe microscopy outlined earlier, the main factors that should be included for this study are the scanning mode, tip shape, and sample loading effects. A diagram illustrating the nested model used in this study is seen in Figure 27. The highest level contains the scanning mode, with the tip next. There were two scanning modes (standard and CD), two tips for each mode, and three to five loading cycles. A loading cycle value consisted of a mean of three or five static measurements. Instead of using a three-level nested model, we found it advantageous to use a separate model for each scanning mode. Due to the three-dimensional nature of AFM data, it was possible to obtain width (at each $Z$), height, and wall angle (left and right) measurements all at the same time. The reproducibility measurements were obtained on nominal 0.25-μm equal lines and spaces that were patterned in oxide on a silicon wafer. Tip width calibration was also performed before each cycle. Precision estimates for the standard and CD modes are shown in Tables 1 and 2, respectively. The nested experimental design allows the use of analysis of variance (ANOVA) for calculating these estimates and actually separating the various components of precision:

$x_{ijk} = k$th measurement of the $j$th cycle on the $i$th tip

$\bar{x}_{ij} = $ average of $n$ measurements in the $k$th cycle on the $i$th tip

$\bar{x}_i = $ average of $b \times n$ measurements on the $i$th tip

$\bar{x} = $ total of average of $a \times b \times n$ measurements taken in the experiment

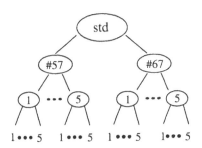

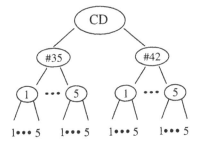

**Figure 27** Nested loop model diagram for ANOVA sampling plan.

**Table 1** Standard-Mode AFM Precision Gauge Studies

**HEIGHT**

| | | | | |
|---|---|---|---|---|
| Tip57 (i=1) | Rep k = 1 | | 2 | 3 |
| Cycle j = 1 | | 318 | 318 | 316 |
| | 2 | 316 | 317 | 315 |
| | 3 | 317 | 316 | 314 |
| Tip67 (i=2) | Rep k = 1 | | 2 | 3 |
| Cycle j = 1 | | 317 | 317 | 317 |
| | 2 | 315 | 315 | 316 |
| | 3 | 315 | 316 | 316 |
| $\sigma_e$ = | | 0.9 | | |
| $\sigma_c$ = | | 0.9 | | |
| $\sigma_t$ = | ~ | 0 | | |
| $\sigma$ = | | 1.2 | | |

**LEFT WALL ANGLE (DEG)**

| | | | | |
|---|---|---|---|---|
| Tip57 (i=1) | Rep k = 1 | | 2 | 3 |
| Cycle j = 1 | | 79.1 | 79.2 | 78.9 |
| | 2 | 78.7 | 78.9 | 79.3 |
| | 3 | 79.3 | 79.3 | 78.9 |
| Tip67 (i=2) | Rep k = 1 | | 2 | 3 |
| Cycle j = 1 | | 76.2 | 76 | 76.3 |
| | 2 | 75.4 | 75.9 | 75.9 |
| | 3 | 75.3 | 75.6 | 75.9 |
| $\sigma_e$ = | | 0.2 | | |
| $\sigma_c$ = | | 0.2 | | |
| $\sigma_t$ = | | 2.3 | | |
| $\sigma$ = | | 2.3 | | |

**RIGHT WALL ANGLE (DEG)**

| | | | |
|---|---|---|---|
| Tip57 (i=1) | Rep k = 1 | | 2 |
| Cycle j = 1 | | 77.7 | 77.7 |
| | 2 | 77.3 | 77.3 |
| | 3 | 77.3 | 77.3 |
| Tip67 (i=2) | Rep k = 1 | | 2 |
| Cycle j = 1 | | 79.4 | 79.7 |
| | 2 | 80.4 | 79.8 |
| | 3 | 80.5 | 79.8 |
| $\sigma_e$ = | | 0.3 | |
| $\sigma_c$ = | | 0.2 | |
| $\sigma_t$ = | | 1.7 | |
| $\sigma$ = | | 1.8 | |

**TOP WIDTH**

| | | | | |
|---|---|---|---|---|
| Tip57 (i=1) | Rep k = 1 | | 2 | 3 |
| Cycle j = 1 | | 170.2 | 171.4 | 174.1 |
| | 2 | 174.7 | 178.3 | 176.0 |
| | 3 | 176.3 | 170.2 | 174.0 |
| Tip67 (i=2) | Rep k = 1 | | 2 | 3 |
| Cycle j = 1 | | 181.5 | 185.3 | 189.2 |
| | 2 | 169.2 | 181.1 | 174.6 |
| | 3 | 181.2 | 179.6 | 187.2 |
| $\sigma_e$ = | | 3.7 | | |
| $\sigma_c$ = | | 3.5 | | |
| $\sigma_t$ = | | 4.4 | | |
| $\sigma$ = | | 6.8 | | |

**MIDDLE WIDTH**

| | | | | |
|---|---|---|---|---|
| Tip57 (i=1) | Rep k = 1 | | 2 | 3 |
| Cycle j = 1 | | 211.2 | 210.7 | 216.6 |
| | 2 | 224.2 | 227.4 | 222.8 |
| | 3 | 220.0 | 222.5 | 225.3 |
| Tip67 (i=2) | Rep k = 1 | | 2 | 3 |
| Cycle j = 1 | | 257.6 | 256.4 | 258.1 |
| | 2 | 257.1 | 252.5 | 254.5 |
| | 3 | 253.2 | 253.8 | 254 |
| $\sigma_e$ = | | 2.2 | | |
| $\sigma_c$ = | | 4.5 | | |
| $\sigma_t$ = | | 24.7 | | |
| $\sigma$ = | | 25.2 | | |

**BOTTOM WIDTH**

| | | | |
|---|---|---|---|
| Tip57 (i=1) | Rep k = 1 | | 2 |
| Cycle j = 1 | | 281.7 | 273.6 |
| | 2 | 288.4 | 290.3 |
| | 3 | 275.0 | 285.1 |
| Tip67 (i=2) | Rep k = 1 | | 2 |
| Cycle j = 1 | | 297.1 | 293.7 |
| | 2 | 297.4 | 292.7 |
| | 3 | 296.4 | 298.2 |
| $\sigma_e$ = | | 5.4 | |
| $\sigma_c$ = | | 3 | |
| $\sigma_t$ = | | 10 | |
| $\sigma$ = | | 11.7 | |

**Table 2** CD-Mode AFM Precision Results

| | | | |
|---|---|---|---|
| Tip35 (i=1) | Rep k = 1 | 2 | 3 |
| Cycle j = 1 | 315.1 | 315.1 | 315.2 |
| 2 | 315.2 | 315.1 | 315.1 |
| 3 | 315.6 | 315.5 | 315 |
| 4 | 315.9 | 315.9 | 315.7 |
| 5 | 316 | 315.2 | 315.7 |
| Tip42 (i=2) | Rep k = 1 | 2 | 3 |
| Cycle j = 1 | 315.7 | 315.5 | 315.3 |
| 2 | 315.4 | 315.8 | 315.5 |
| 3 | 315.3 | 316.2 | 316 |
| 4 | 316.7 | 316.5 | 316.1 |
| 5 | 315.8 | 315.8 | 315.1 |
| $\sigma_e$ = | 0.3 | | |
| $\sigma_c$ = | 0.3 | | |
| $\sigma_t$ = | 0.2 | | |
| $\sigma$ = | 0.5 | | |

| | | | |
|---|---|---|---|
| Tip35 (i=1) | Rep k = 1 | 2 | 3 |
| Cycle j = 1 | 82.9 | 83 | 83.3 |
| 2 | 82.9 | 83.2 | 82.9 |
| 3 | 83.1 | 82.8 | 83.1 |
| 4 | 83.1 | 82.8 | 82.7 |
| 5 | 83.4 | 83 | 83.1 |
| Tip42 (i=2) | Rep k = 1 | 2 | 3 |
| Cycle j = 1 | 82.7 | 83 | 82.9 |
| 2 | 82.8 | 82.8 | 82.8 |
| 3 | 82.7 | 82.9 | 82.8 |
| 4 | 82.7 | 82.8 | 82.9 |
| 5 | 82.8 | 83 | 83.4 |
| $\sigma_e$ = | 0.18 | | |
| $\sigma_c$ = | 0.04 | | |
| $\sigma_t$ = | 0.1 | | |
| $\sigma$ = | 0.21 | | |

| | | | |
|---|---|---|---|
| Tip35 (i=1) | Rep k = 1 | 2 | |
| Cycle j = 1 | 84.9 | 84.9 | 8 |
| 2 | 85.1 | 84.9 | 85. |
| 3 | 84.8 | 84.9 | 84. |
| 4 | 84.9 | 84.8 | 84. |
| 5 | 85 | 84.9 | 8 |
| Tip42 (i=2) | Rep k = 1 | 2 | |
| Cycle j = 1 | 84.7 | 85 | 84. |
| 2 | 85 | 85.1 | 85. |
| 3 | 84.8 | 85 | 8 |
| 4 | 85.1 | 85.1 | 84. |
| 5 | 85 | 84.9 | 84. |
| $\sigma_e$ = | 0.09 | | |
| $\sigma_c$ = | 0.06 | | |
| $\sigma_t$ = | ~ 0 | | |
| $\sigma$ = | 0.11 | | |

**TOP WIDTH**

| | | | |
|---|---|---|---|
| Tip35 (i=1) | Rep k = 1 | 2 | 3 |
| Cycle j = 1 | 180.1 | 181.3 | 179.8 |
| 2 | 181 | 181.7 | 181 |
| 3 | 182.1 | 180.4 | 181 |
| 4 | 181.9 | 178.6 | 181.1 |
| 5 | 178.9 | 181.8 | 179.3 |
| Tip42 (i=2) | Rep k = 1 | 2 | 3 |
| Cycle j = 1 | 178.9 | 178.1 | 180.3 |
| 2 | 179.9 | 179.6 | 178.1 |
| 3 | 180.8 | 179 | 177.3 |
| 4 | 178.7 | 178.4 | 175.9 |
| 5 | 179 | 175.5 | 177.5 |
| $\sigma_e$ = | 1.3 | | |
| $\sigma_c$ = | ~ 0 | | |
| $\sigma_t$ = | 1.5 | | |
| $\sigma$ = | 2 | | |

**MIDDLE WIDTH**

| | | | |
|---|---|---|---|
| Tip35 (i=1) | Rep k = 1 | 2 | 3 |
| Cycle j = 1 | 205.3 | 203.9 | 203.4 |
| 2 | 204.2 | 204.9 | 203.1 |
| 3 | 205.8 | 204.4 | 204.5 |
| 4 | 205.4 | 203.5 | 205.2 |
| 5 | 203.2 | 206.1 | 202.1 |
| Tip42 (i=2) | Rep k = 1 | 2 | 3 |
| Cycle j = 1 | 204.1 | 204 | 205.6 |
| 2 | 205.6 | 204.4 | 204 |
| 3 | 208 | 205.1 | 204 |
| 4 | 204.3 | 205.2 | 204.5 |
| 5 | 205.4 | 200.5 | 202.5 |
| $\sigma_e$ = | 1.4 | | |
| $\sigma_c$ = | 0.4 | | |
| $\sigma_t$ = | 0.2 | | |
| $\sigma$ = | 1.5 | | |

**BOTTOM WIDTH**

| | | | |
|---|---|---|---|
| Tip35 (i=1) | Rep k = 1 | 2 | |
| Cycle j = 1 | 233.5 | 233.9 | 231. |
| 2 | 233.3 | 232.7 | 232. |
| 3 | 233.9 | 233.4 | 233. |
| 4 | 234.5 | 232.5 | 233. |
| 5 | 230.9 | 234.8 | 232. |
| Tip42 (i=2) | Rep k = 1 | 2 | |
| Cycle j = 1 | 233.9 | 232.1 | 235. |
| 2 | 234.7 | 233.4 | 231. |
| 3 | 235.9 | 235.1 | 234. |
| 4 | 232.5 | 232.8 | 229. |
| 5 | 234.1 | 229.4 | 229. |
| $\sigma_e$ = | 1.5 | | |
| $\sigma_c$ = | 0.9 | | |
| $\sigma_t$ = | ~ 0 | | |
| $\sigma$ = | 1.7 | | |

The precision of the AFM in this report is defined as

$$\sigma = \sqrt{\sigma_e^2 + \sigma_c^2 + \sigma_t^2}$$

where $\sigma_e^2$ is the component for error (repeatability), $\sigma_c^2$ is the cycle variance component (reproducibility), and $\sigma_t^2$ is the tip-to-tip variance component. Estimates of these variance components are calculated using the sum of mean squares,

$$MS_e = \frac{\sum_{i=1}^{2} \sum_{j=1}^{b} \sum_{k=1}^{n} (x_{ijk} - \bar{x}_{ij})^2}{ab(n-1)}$$

$$MS_e = \frac{n \sum_{i=1}^{2} \sum_{j=1}^{b} (\bar{x}_{ij} - \bar{x}_i)^2}{a(b-1)}$$

$$MS_e = \frac{bn \sum_{i=1}^{2} (\bar{x}_i - \bar{x})^2}{a-1}$$

and

$$\hat{\sigma}_e = \sqrt{MS_e}$$

$$\hat{\sigma}_c = \sqrt{\frac{MS_c - MS_e}{n}}$$

$$\hat{\sigma}_t = \sqrt{\frac{MS_d - MS_c}{bn}}$$

The caret over each sigma indicates that each is merely an estimate of the variance component. A negative variance component estimate usually indicates a nonsignificant variance component and is set to zero. Variance component estimates can be biased, so the results should be interpreted with caution. The standard mode height precision (see Table 1) is on the single-nanometer level, as one would expect. Height data can be gathered very well with the standard-mode tip. It can also be seen that the error and cycle components of wall-angle variance are quite good. As one would expect, tip variance is the main contributor to the overall imprecision. As described earlier, the angle of the feature sidewall was larger than that of the probe, so imaging of the probe shape occurred instead of imaging of the feature.

The CD-mode height measurements (Table 2) were more precise than in the standard mode and had a variance of $\sigma = 0.5$ nm. The total measurement precision for wall angle was also much better in the CD mode and were a few tenths of a degree. The top-width precision was about 2 nm and showed no clear systematic components. A dramatic improvement in middle- and bottom-width measurements was realized by scanning in the 2-D mode with the flaired boot-shaped probes. Top- and middle-width measurement precision values were on the single-nanometer level. The improvement in CD-mode precision values was due mainly to the ability to image the feature walls with the bottom corners of the flaired probes. This ability essentially eliminated the effect of probe shape on overall precision. The increased number of data points at the feature edges due to the 2-D scanning algorithm also helped to improve the CD measurement precision. Automatic algorithm threshold levels for top- and bottom-width locations can also be set more reliably when there is a greater density of data points at the edges. The AFM top width (AFM TOP) was determined by averaging a band of points centered at the 90% level of each edge. The width of each band was adjustable and set to 5% for our tests (61). The bottom-width (AFM BOT) bands were centered at the 10% level and 5% wide. The middle width (AFM MID) is centered between the top and bottom. Tip width was found prior to each cycle in order to calibrate the linewidth measurements. It was found that nonuniformities of the tip characterization sample produced variations in the reproducibility results. It is very important to note that the precision and accuracy of the AFM critically depend on, and are limited by, the tip calibration procedure.

## VI.  APPLICATIONS TO CRITICAL-DIMENSION METROLOGY

Although conventional scanning electron microscopes (SEMs) have theoretical resolutions on the nanometer scale, the determination of actual edge location in an intensity profile becomes somewhat arbitrary (62). As a result, the measured linewidth, or critical dimension, may vary by as much as 50% of the nominal feature size, depending on which edge detection algorithm, threshold, magnification, or beam parameters are used (63). To make matters worse, the effects on measurement linearity and accuracy are not constant and actually vary with sample type. It is also not practical to produce a calibration standard for every measurement case, even if one had a priori knowledge of which one to use. Despite these limitations, SEMs are still the dominant tool for in-line measurement of submicron features in the cleanroom. This is due primarily to their nanometer-level precision, high throughput, and potential for full automation.

### A.  Evaluation of SEM Measurement Linearity

Linearity describes how well an image corresponds to actual changes in the real feature size. One must determine if the measurements obtained from an image correspond in a direct fashion to reality. CD-SEM linearity is typically determined by comparing measurements of tests features to those obtained by a more accurate reference, the AFM in this case. Typically, SEM width measurements are first plotted against the corresponding reference values. A regression analysis yields the equation for a straight line that best fits the distribution of measured points:

$$Y_1 = \alpha X_1 + \beta + \varepsilon_1$$

where $\alpha$ is the slope defect, $\beta$ is the offset, and $\varepsilon$ is the error component. Therefore, first-order regression analysis can be used to determine the accuracy (offset), magnification calibration (slope defect), and variation (error term). This technique provides a simple means to quantify the linearity and accuracy.

A series of etched poly-silicon lines ranging in width from 50 to 1000 nm will be used in this discussion. The AFM surface rendering of a nominally 60-nm line in the set is shown in Figure 28a. The effect of sample variation on measurement precision was minimized with spatial averaging by performing measurements on each scan (Figure 28b) at successive intervals along the line. CD-SEM measurements of the same structures were then plotted against the AFM reference values, as shown in Figure 29. The degree of linearity and accuracy is given in terms of the slope defect ($\alpha$), goodness of fit ($R^2$), and offset ($\beta$) in the figure. The extremely small sizes of the poly-Si lines do present a challenge for any SEM to resolve. In addition to having a measurement offset of $-56$ nm, the SEM in this example was not able to resolve the 60 nm line. It is important to note that the assumption of *process linearity* is not necessary, because we are comparing the SEM measurements to values obtained from an accurate reference tool. By plotting the SEM

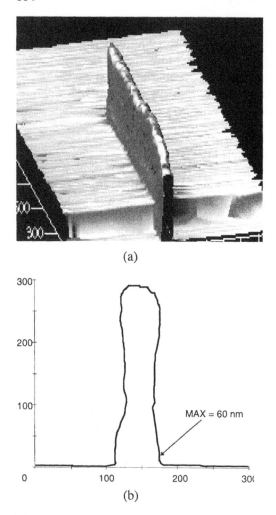

(a)

(b)

**Figure 28** (a) Surface rendering from AFM data of nominally 60-nm poly-silicon line. (b) A single linescan from the image in (a).

widths against actual reference values, we should obtain a linear trend in the data—even if the actual distribution of feature sizes is not.

The relative matching, or tool-induced shift (TIS), between CD-SEMs can be studied by repeating this process for each system. Once the most linear algorithm has been determined, the additive offset needed to make the measurement curves (hopefully of the same shape) overlap with the reference curve must be found. The SEM-to-AFM measurement offsets are illustrated directly in Figure 30. All three SEMs continued to track changes in linewidth down to 50 nm in a linear fashion, but there existed offsets between the three tools. The curves are fairly constant with respect to each other for feature sizes larger than 250 nm. To achieve matching for 300-nm lines, an offset of +6 nm would be added to the measurements from the first SEM (Vendor A, Model 1). Similarly, an offset of −6 nm is needed for the second SEM (Vendor A, Model 2) and +8 nm for the third SEM from Vendor B. It is interesting that the tool from Vendor B actually matched Model 1 of Vendor A better than Model 2 from Vendor A. Even if an SEM may be from the same

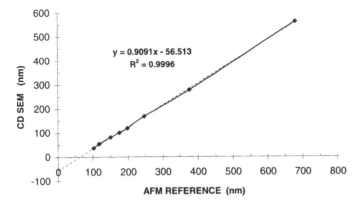

**Figure 29**  CD SEM measurements plotted against corresponding AFM reference values.

vendor, it doesn't mean that it will be any easier to match to the previous model. It will be shown later that matching is greatly improved and simplified if one maintains a homogeneous tool set of the same *model* (hardware configuration). The plot in Figure 30 also demonstrates the ability of CD-SEMs to detect, or resolve, 10-nm changes in width of an isolated poly-silicon line. The dip in all three curves at the 200-nm linewidth is highly suggestive of an error (of about 5 nm) in the AFM tip width calibration at that site. This reinforces the assertion made earlier that AFM performance is essentially limited by the ability to characterize and control the tip shape.

## B.  CD-SEM Matching for Wafer Metrology

The edge detection algorithm that best reproduces the measurement response (i.e., the linearity) of a reference tool can be found from a series of features having a trend in

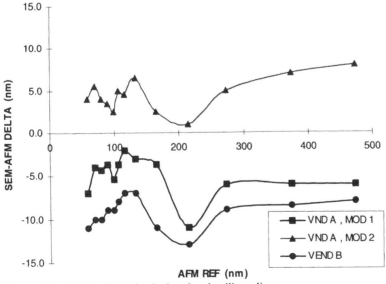

**Figure 30**  SEM offsets for isolated poly-silicon lines.

sizing or profile. This is usually accomplished by changing the coded design size (as was done in the previous section) or stepper focus/exposure conditions about a nominal-size node of interest during lithography. In addition to varying edge profile, the latter technique provides even smaller increments in width variation than can be achieved by the former method of design size coding. In this regard, measurement resolution (the ability to detect small changes in the feature size) is tested even more thoroughly. By varying the lithographic conditions in this way, we are able to test the combined effects of both the edge profile and size on matching. This also provides a way to see how robust the matching between tools will be over a known range of litho-process variation. In this algorithm determination method, the AFM is used to obtain the surface height topographs at each sidewall profile (focus level) condition. AFM topographs at the optimum focus level and at each opposite extreme are shown in Figure 31. This technique can also be implemented on etched layers by varying the etching conditions instead. These surface topographs illustrate a wider range in the process variation than is ever (hopefully) encountered in the production line. Initially, there was great difficulty in finding an SEM edge algorithm that produced the same measurement curve (CD vs. focus level) shape as that of the AFM bottom-width curve. The middle (optimum stepper focus) profile in Figure 31 shows why! The CD-SEM is capable of providing only a top-down view of the feature, so only the maximum portion, indicated by the dotted lines, is able to be seen. This is the widest part of the feature and will be referred to as AFM MAX. As the sidewall is varied, the threshold point in the SEM edge detection algorithm that corresponds to desired width will also change. Since it would not be practical to try to change this threshold in production, a worst-case value of mismatch can be specified over a given range of process variation. The SEM and AFM width measurements are plotted in Figure 32. Indeed, it is possible to find SEM algorithms that follow AFM MAX in shape, but not for AFM BOT. Basically, the AFM MAX-to-BOT difference (unrecoverable error) represents the worst-case SEM contribution to the measurement of pre- to postetch offset. The true extent of this error depends on the isotropy of the etch process. MAX-to-BOT offset will also vary for different material layers, feature sizes, and pattern types (isolated, periodic, holes). The bias could falsely be attributed to material or pattern dependency of SEM charging when it is actually a purely geometric effect.

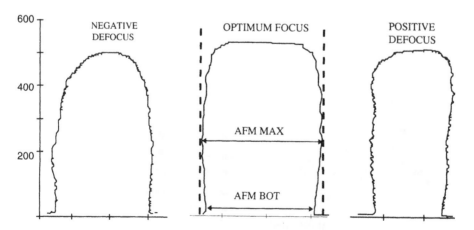

**Figure 31**   Feature profile variation across stepper focus.

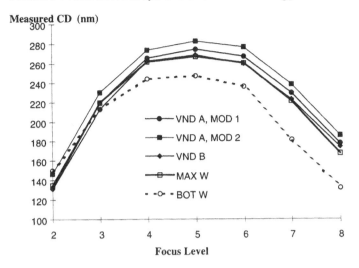

**Figure 32** Comparison of different SEMs and their measurements across profile variation.

One of the more difficult features to image for all microscopes is the hole pattern, also referred to as a contact or via. SEM (left) and AFM (right) top-down images of a nominal 350-nm-diameter hole patterned in deep-UV photoresist on an oxide substrate are shown in Figure 33. Cutaway views of the AFM data are shown in Figure 34. Holes induce more charging effects with the SEM, so signal collection from the bottom is much more difficult than with lines. A larger size and material dependence of matching parameters occurs with holes. On newer-model SEMs, an effective diameter is computed by performing a radial average about several angles (see Figure 33). The hole structure also presents the most difficulty for the AFM reference tool to image. The key issues with AFM analysis of holes are measurement scan location and angular direction, sample induced variations, and tip shape/size effects. The usual method for imaging holes with CD-mode AFM consists of first performing a quick standard-mode overview scan of the hole. The location of the measurement scan is then set by placing the CD-mode image indicator (the long, white rectangular box in Figure 33) at the desired position. Essentially, a small set of CD-mode scans (denoted by 1, 2, 3 in the figure) are taken within a section of the top-down standard image. Only scanning in the horizontal image direction between left and right are possible due to the method of lateral tip-to-sample distance control used in CD-mode AFM. The result is that we can obtain CD imaging only within a horizontal band through the hole center. Centering of this horizontal band through the hole turns out to be a major component of AFM hole-diameter imprecision and error. Sensitivity to centering depends on the hole radius, of course. Radial averaging along different scan angles is also not possible, which can lead to sample variation (i.e., edge roughness or asymmetry) affecting the measurement quality. Although scanning in different directions is possible in the standard mode, diameter measurements cannot be performed, because the tip cross section is elliptical (see the elliptical trace in the AFM image of Figure 33), as seen if one were looking up at the bottom of a probe.

A new method for improving hole-width measurement with the CD-mode AFM has been developed by Marchman (63). This new technique involves imaging an entire hole with the CD-scan-mode line by line and *not* using the standard overview mode for measurement centering. CD-mode width measurements are then made at each image linescan,

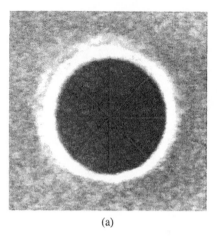

(a)

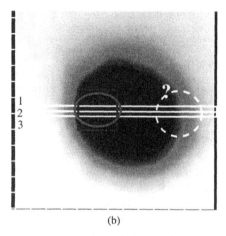

(b)

**Figure 33**  Top-down SEM and AFM views of DUV resist contact on oxide substrate.

except at the walls parallel to the scan direction. The AFM MAX and AFM BOT width measurements are plotted for each scan line number in Figure 35. A polynomial is then fit to the data points, whose maximum indicates the hole diameter after subtraction of the probe width. This technique eliminates issues of centering and spatial averaging, as well as improving the static measurement averaging. Residuals from the polynomial fit can also be used to estimate the combined precision components of sample variation and tool random error. The overall hole measurement reproducibility was reduced to 3 nm using this technique. It should be noted that this technique is still susceptible to errors caused by changes in the hole diameter along different angular directions through the hole. These errors can be corrected somewhat by correlating the horizontal diameter in the SEM image to the AFM value. Then relative changes in hole diameter at different angles can be found from the SEM image, assuming beam stigmation and rotational induced shifts (RIS) in the SEM have been corrected properly (64). As noted earlier, eccentricity in the probe front will introduce errors in hole-diameter measurements in the vertical direction. A more serious issue is starting to arise for sub-200-nm etched-silicon CD probes—they develop a rectangular footprint. This causes the diameter measurement to be less than the actual hole diameter when the point of interaction switches from one probe bottom corner

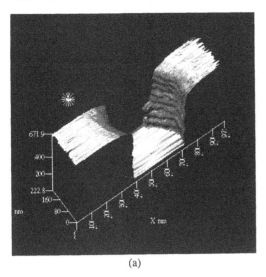

(a)

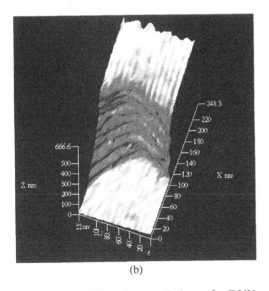

(b)

**Figure 34** AFM surface renderings of a DUV resist hole.

to the other. A pointed probe shape characterization structure can be used to measure the probe size in two dimensions to correct this problem.

Hole-diameter measurements from four SEMs, two "identical" systems of the same model type from each vendor, are plotted against stepper focus in Figure 36. The AFM diameter of the hole (in deep-UV photoresist on oxide substrate) was used to provide the offset necessary to match these four systems about that point.

The last factor in CD-SEM matching variation to consider at this time is that of material composition of the feature. The chart in Figure 37 shows the SEM-to-AFM offset of each tool for different material layers. Measurements where performed at the optimum feature profile on each material combination. This was done in order to isolate the effect on matching due to material type. The dependence of SEM bias on material type is clearly

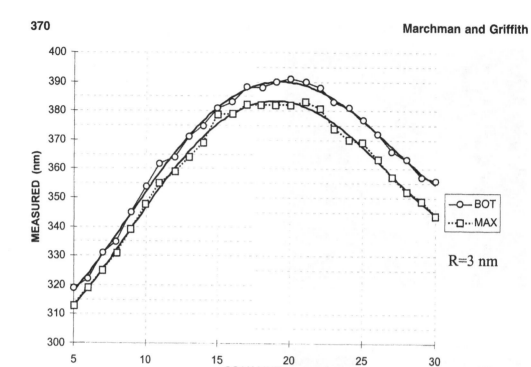

**Figure 35**  Polynomial fitting to determine the contact diameter.

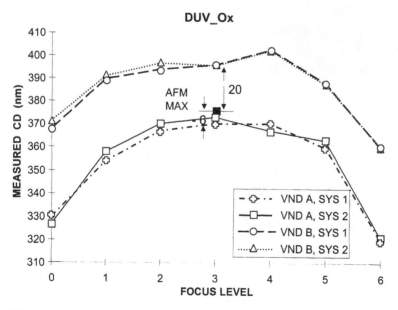

**Figure 36**  Contact hole matching data for CD-SEMs.

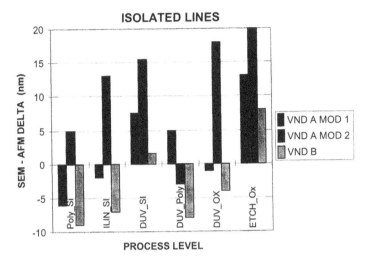

**Figure 37**  Process-level material induced SEM bias shifts.

different for all three SEMs. Material dependence of SEM measurement offset was also studied for dense lines, and different behaviors were observed.

## C.  Photomask Metrology

Modern optical exposure tools use reduction optics to print patterns from photomasks. Mask features are typically four times larger than the printed features, so optical microscopes have sufficed for photomask CD metrology up until the most recent lithography generations. The alternative metrology tools, electron microscopes and scanning force microscopes, have, however, encountered problems, arising from the tendency of the photomask to hold electric charge. SEMs unavoidably inject charge into the sample, often resulting in charge buildup that not only degrades the quality of the image but may also damage the mask through electrostatic discharge.

Scanning force microscopes (SFMs) are less inclined to generate sample charging, but they are, nevertheless, susceptible to charge because of their sensitivity to electrostatic forces. The SFM can confuse electrostatic forces with surface forces, resulting in a scan that does not faithfully reflect the shape of the sample. This effect is especially troublesome to probe microscopes that attempt to scan at extremely tiny forces in a noncontacting mode. The balance-beam force sensor in the surface/interface SNP measures the surface height with the probe in contact with the surface. The repulsive probe–sample forces are high enough to make this tool insensitive to these charging problems. The SNP may, consequently, be operated without special charge suppression measures, such as ionization sources.

Figure 38 shows an image of a phase-shifting mask taken with the surface/interface SNP. The complicated three-dimensional structures on these masks must be held to very tight tolerances for the mask to work properly. The ability to perform the measurement nondestructively is especially important in this situation. Figure 39 shows the precision achievable in photomask scans, taken on a sample different from that of Figure 38 (65).

**Figure 38** Scan of a phase-shifting mask with the surface/interface SNP. The pitch of this structure is 1301 nm, and the spaces bewteen the lines are 388 nm deep.

## VII. CONCLUSION

In order to gain insight into the physical states occurring during the development of IC fabrication processes as well as the monitoring of existing ones, it is now necessary to measure features with nanometer precision and accuracy in all three dimensions. Unfortunately, adequate calibration standards do not exist for submicron features on wafers and masks. The scanning probe microscope has become a good option for providing on-line reference values to higher throughput in-line tools, such as the CD-SEM. The accuracy of SPM metrology is not affected significantly by changes in the material properties, topography, or proximity of other features.

Unfortunately, the probe shape can affect measurement uncertainty in several ways. The radius of curvature of a conical probe must be determined to know the region of wavelength–amplitude space that can been reached. If the width of a cylindrical probe is uncertain, then there is a corresponding uncertainty in the width of each measured object. Durability of the probe tip is especially important. If the probe is changing during a measurement, it will affect the precision of the measurement as well as the acuracy. Finally, the stability of the probe against flexing is important in determining the precision of a measurement. Susceptibility to flexing sets a fundamental limit how deep and narrow a feature may be probed.

The SEM will most likely continue to dominate in-line CD metrology for the next few years due to its nanometer-scale resolution and high throughput. However, a combination of the SEM and SPM in the future may provide both throughput and accuracy. The primary advantage of SPM over using SEM cross sections to provide reference profiles is that of spatial avaraging. Essentially, each slice of the SPM image can be thought of as independent cross sections. As feature sizes shrink, it will be necessary to perform more measurements at each site in order to improve averaging and minimize the effects of increasing edge roughness. A more thorough estimation of the amount of averaging required for each technology node is given in the literature. As fabrication processes are pushed further in order to achieve smaller critical dimen-

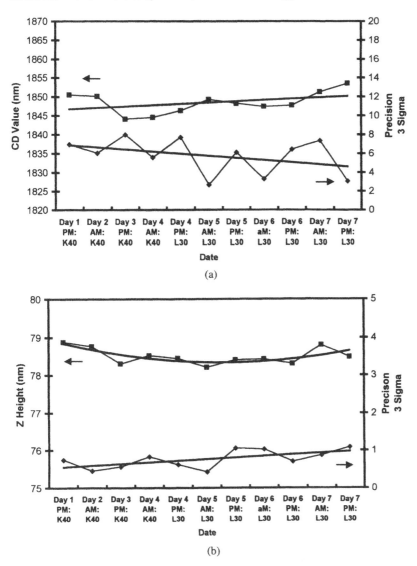

**Figure 39** Measurement data from an optical photomask taken with the surface/interface SNP. (a) CD measurement and (b) height measurement. Two tips, designated K40 and L30, were used to make the measurements over seven days.

sions, the amount of sample variation and roughness as a percentage of nominal feature size will also increase.

## REFERENCES

1. G. Binnig, H. Rorher, C. Gerber, E. Weibel. Phys. Rev. Lett. 49:57–61, 1982.
2. G. Binnig, H. Rorher. Sci. Am. 253(2):50, 1985.
3. J.G. Simmons. J. Appl. Phys. 34:1793, 1963.
4. G. Binnig, C.F. Quate, Ch. Gerber. Phys. Rev. Lett. 56:930–933, 1983.

5. D. Sarid. Scanning Force Microscopy. 2nd ed. New York: Oxford University Press, 1994.
6. T.R. Albrecht, S. Akamine, T.E. Carver, C.F. Quate. J. Vac. Sci. Technol. A 8:3386–3396, 1990.
7. O. Wolter, Th. Bayer, J.J. Greschner. Vac. Sci. Technol. B 9:1353–1357, 1991.
8. G. Meyer, N.M. Amer. Appl. Phys. Lett. 53:1045–1047, 1988.
9. J.P. Spatz, S. Sheiko, M. Moller, R.G. Winkler, P. Reineker, O. Marti. Nanotechnology, 6: 40–44, 1995.
10. Y. Martin, C.C. Williams, H.K. Wickramasinghe. J. Appl. Phys. 61:4723–4729, 1987.
11. Y. Martin, H.K. Wickramasinghe. Appl. Phys. Lett. 64:2498–2500, 1994.
12. H.K. Wickramasinge. Sci. Am. 261(4):98–105, 1989.
13. M. Tortenese, H. Yamada, R. Barrett, C.F. Quate. In: The Proceedings of Transducers '91. IEEE, Pennington, NJ, 1991, Publication No. 91CH2817-5, p. 448.
14. S.C. Minne, S.R. Manalis, C.F. Quate. Appl. Phys. Lett. 67:3918–3920, 1995.
15. S.R. Manalis, S.C. Minne, C.F. Quate. Appl. Phys. Lett. 68:871–873, 1996.
16. G.L. Miller, J.E. Griffith, E.R. Wagner, D.A. Grigg. Rev. Sci. Instrum. 62:705–709, 1991.
17. S.A. Joyce, J.E. Houston. Rev. Sci. Instrum. 62:710–715, 1991.
18. J.E. Griffith, G.L. Miller U.S. Patent 5,307,693, 1994.
19. R.E. Betzig. U.S. Patent 5,254,854, Nov. 4, 1991.
20. E. Betzig, P..L Finn, J.S. Weiner. Appl. Phys. Lett. 60: 2484–2486, 1992.
21. H.M. Marchman. To be published.
22. J.E. Griffith, H.M. Marchman, G.L. Miller. J. Vac. Sci. and Technol. B13:1100–11055, 1995.
23. H. Yamada, T. Fuji, K. Nakayama. Jpn. J. Appl. Phys. 28:2402, 1990.
24. J.E. Griffith, G.L. Miller, C.A. Green, D.A. Grigg, P.E. Russell. J. Vac. Sci. Technol. B8:2023–2027, 1990.
25. R.C. Barret, C.F. Quate. Rev. Sci. Instrum. 62:1391, 1991.
26. D. Nyyssonen. Proc. SPIE Microlithography 1926:324–335, 1993.
27. J.E. Griffith, D.A. Grigg. J. Appl. Phys. 74:R83-R109, 1993.
28. EC Teague. In: C.K. Marrian, ed. The Technology of Proximal Probe Lithography. Vol. IS10. Bellingham, WA: SPIE Institute for Advanced Technologies, 1993, pp 322–363.
29. J. Schneir, T. McWaid. T.V. Vorburger. Proc. SPIE 2196:166, 1994.
30. J. Schneir, T. McWaid, J. Alexander, B.P. Wifley. J. Vac. Sci. Technol. B12:3561-3566, 1994.
31. H.M. Marchman, J.E. Griffith, J.A. Trautman. J. Vac. Sci. Technol. B13:1106–1111, 1994.
32. H.M. Marchman, J.E. Griffith, J.Z.Y. Guo, J. Frackoviak, G.K. Celler. J.Vac. Sci. Technol. B 12:3585–3590, 1994.
33. J.E. Griffith, H.M. Marchman, L.C. Hopkins. J. Vac. Sci. Technol. B12:3567–3570, 1994.
34. H.M. Marchman, J.E. Griffith. Rev. Sci. Instrum. 65:2538–2541, 1994.
35. M. Stedman. J. Microscopy 152:611–618, 1988.
36. M. Stedman, K. Lindsey. Proc. SPIE 1009:56–61, 1988.
37. K.L. Westra, A.W. Mitchell, D.J. Thomson. J. Appl. Phys. 74:3608–3610,1993.
38. J.E. Griffith, H.M. Marchman, G.L. Miller, L.C. Hopkins, M.J. Vasile, S.A. Schwalm. J. Vac. Sci. Technol. B11:2473–2476, 1993.
39. M.J. Vasile, D.A. Grigg, J.E. Griffith, E.A. Fitzgerald, P.E. Russell. J. Vac. Sci. Technol. B9:3569–3572, 1991.
40. H.M. Marchman. U.S. Patent 5,394,500, Feb. 28, 1994.
41. J.E. Griffith, H.M. Marchman, L.C. Hopkins, C. Pierrat, S. Vaidya. Proc. SPIE 2087:107–118, 1993.
42. R.P. Feynman, R.B. Leighton, M. Sands. The Feynman Lectures on Physics. Vol. 2. Reading, MA: Addison-Wesley, 1964.
43. B.I. Yakobson, R.E. Smalley. Am. Scientist 85:324, 1997.
44. M.M.J. Treacy, T.W. Ebbesen, J.M. Gibson. Nature, 381:678–680, 1996.
45. H. Dai, J.H. Hafner, A.G. Rinzler, D.T. Colbert, R.E. Smalley. Nature, 384:147–150, 1996.
46. H.M. Marchman, et al. To be published.
47. D. Nyysonen, R.D. Larrabee. J. Res. Nat. Bur. Stand. 92:187–204, 1987.

48. M.T. Postek, D.C. Joy. J. Res. Nat. Bur. Stand. 92:205-228, 1987.
49. Digital Instruments, Inc. Nanoscope III.
50. E.H. Anderson, V. Boegli, M.L. Schattenburg, D. Kern, H.I. Smith. J. Vac. Sci. Technol. B9:3606–3611,1991.
51. D.J. Keller. Surf. Sci. 253:353–364, 1991.
52. G. Pingali, R. Jain. Proc. SPIE 1823:151–162, 1992.
53. D.J. Keller, F.S. Franke. Surf. Sci. 294:409–419, 1993.
54. J.S. Villarrubia. Surf. Sci. 321:287–300, 1994.
55. J.S. Villarrubia. J. Res. Natl. Inst. Stand. Technol. 102:425–454, 1997.
56. J.E. Griffith, D.A. Grigg, M.J. Vasile, P.E. Russell, E.A. Fitzgerald. J. Vac. Sci. Technol. B9:3586–3589, 1991.
57. D.A. Grigg, P.E. Russell, J.E. Griffith, M.J. Vasile, E.A. Fitzgerald. Ultramicroscopy 42–44:1616–1620, 1992.
58. Nanoedge sample is supplied with the Veeco SXM System and manufactured by IBM MTC, FRG.
59. H.M. Marchman. Proc. SPIE 2725:527–539, 1996.
60. The SXM Workstation is currently marketed by Zygo Corporation.
61. The top location and bandwidth settings were determined by optimizing the measurement repeatability.
62. H.M. Marchman. An overview of CD metrology for advanced CMOS process development. In: International Conference on Characterization and Metrology for ULSI Technology. Gaithersburg, MD: NIST, March 23–27, 1998.
63. H.M. Marchman. J. Vac. Sci. Technol. B 15:2155–2161, 1997; H.M. Marchman. Future Fab International, Aug. 1997.
64. K.M. Monohan, R.A. Forcier, W. Ng, S. Kudallur, H. Sewell, H.M. Marchman, J.E. Schlesinger. Proc. SPIE, 3050:54–67, 1997.
65. Data courtesy of Robert Lipari, Surface/Interface, Inc.

# 16

# Electrical CD Metrology and Related Reference Materials

**Michael W. Cresswell and Richard A. Allen**
*National Institute of Standards and Technology, Gaithersburg, Maryland*

## I. INTRODUCTION

In the fabrication of integrated circuits, the steps of depositing a thin film of conducting material, patterning it photolithographically, and then etching it and stripping the remaining resist are repeated several times as required levels are created. On each occasion, the purpose is to pattern a film into a geometry that is consistent with the design of the circuit. The process control mission is to ensure that each respective set of process steps replicates patterning that meets engineering specifications. In most cases, a measure of this compliance is the closeness of the linewidths of features that are produced in the pattern to their intended "design," or "drawn" widths. Ideally, the linewidths of all features on each level would be sampled after the level is patterned, to provide an indication of whether or not the process is under adequate control. However, such a comprehensive metrology operation is neither economically nor technically feasible. Instead, the as-patterned linewidths of a limited selection of features that constitute a "test pattern," are measured. The test pattern is printed at the same time as the circuitry whose fabrication is being monitored, but at a separate location on the substrate that is reserved exclusively for process-control purposes. An example of a commonly used test pattern is shown in Figure 1 (1).

Usually, test patterns include features that have drawn linewidths matching the minimum of the features being printed in the circuit. These linewidths are typically referred to as the process's *critical dimensions* (CDs). It is the widths of the features in the test pattern that are measured by some means to determine if the respective sequence of patterning steps produces results that comply with engineering specifications. The presumption is that, if the CDs of the line features in the test pattern are found to be replicated within predefined limits, the CDs of the features replicated in the synthesis of the integrated circuit are replicated within those limits. The several common linewidth-metrology techniques in use today are electrical CD (ECD) (discussed in this chapter), scanning electron microscopy (SEM) CD, and scanning probe microscopy (SPM) CD.

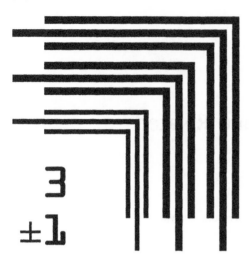

**Figure 1**  Example of SEMI Standard test-pattern cells for isolated features and line-and-space groups.

## II.  ELECTRICAL CRITICAL-DIMENSION MEASUREMENT TECHNIQUE

### A.  Definition of Electrical Critical Dimension

The basic definition of electrical critical dimension (ECD) is the average conductive width of a feature patterned in a thin film whose length is much greater than its width. This operational definition is the property of a conducting feature of most interest to the chip designer, who needs to know the current-carrying capacity of a feature of a given nominal width. Such ECD measurements complement physical CD measurements, e.g., SEM and SPM that support the lithography processes by providing the physical extent of patterned features, measuring both the conductive width and any material on the sidewalls that does not contribute to the electrical path width. This definition above applies to any electrically isolated conductive region, including patterned thin films, e.g., polysilicon and metal, as well as implanted (or diffused) conductive regions, which are junction isolated from the surrounding silicon material.

To better understand the concept behind ECD, consider a conductive film of uniform thickness, or height. There must be no lateral variation in the resistivity of this film, although there may be a uniform variation of the resistivity in the vertical direction. Within this area are patterned both the test patterns from which the ECD are extracted and the integrated circuit. Although the actual features may have any of a number of cross-sectional geometries or exhibit surface roughness, each feature may be represented by an equivalent three-dimensional rectangle that has the same height and length as the actual structure. In this rectangle the electrical properties, the resistivity profile, and total electrical resistance must be identical to those in the patterned feature. This equivalence is illustrated in Figure 2. The width of this rectangle is defined as the ECD of the feature. Notice that any local variability in the lateral dimensions of the feature will produce a measured ECD that is wider or narrower than its physical linewidth at any single point along its length. This will also be observed in the case of lateral variations in the resistivity of the film within the individual feature. In actual practice, any difference between the physical linewidth of the feature and its ECD may be largely attributable to the redis-

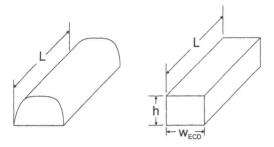

on-wafer feature          equivalent 3D representation

**Figure 2** Equivalent three-dimensional rectangular representation of the feature has the same height, length, resistivity profile, and total electrical resistance as the feature.

tribution of resistivity of the material of the feature near its sidewalls. The differences may have been exacerbated by the lack of a standard definition of the physical linewidth of the feature until very recently (2).

Note that an alternate or supplementary definition of ECD is where the feature is defined with a trapezoidal cross section and the ECD is the half-height width of the trapezoid. This model will give a value for ECD identical to that of the rectangular cross section described earlier.

## B. Electrical Linewidth Test Structures

This section describes the fundamentals of electrical linewidth test structures and related metrology. It introduces the concept of *sheet resistance*, whose measurement is featured in the standard method of ECD determination. Limitations in the design of standard test structures, known as cross-bridge resistors, for linewidth measurement, using Kelvin four-terminal measurements, are addressed, with enhancements for applying reference-length corrections. Multiple reference-segment cross-bridge resistors and a statistical model for facilitating diagnostics and enhancing measurement confidence are described. In the final subsection, an alternative to the cross-bridge resistor, the multibridge resistor, is described. The multibridge resistor has certain advantages over the cross-bridge resistor when determining process limits, such as the effects observed when exceeding the minimum design rules and the effects observed at the onset of dishing in a CMP-planarized conductive layer.

A unique attribute of ECD measurements extracted from electrical linewidth test structures is that the result reflects the electrical linewidth of the feature that, by definition, is the same along its entire length. It is not necessarily a measure of the dimension of the cross section of the feature at some particular point along its length.

## 1. Determination of Electrical Critical Dimension of an Idealized Feature

Figure 3 shows electrical current, $I$, being forced along an idealized uniformly conducting feature of a rectangular cross section of height $h$, width $w$, and *physical* length $L$. For this forced current, there is a voltage drop $V$ through the feature. The ratio of the measured voltage to the forced current, $\langle V/I \rangle$, which in this idealized case is exactly the end-to-end electrical resistance of the conductor, relates to the dimensions of the conducting feature according to the expression

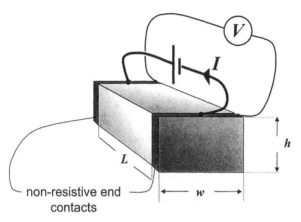

**Figure 3** Electrical current, $I$, being forced along an idealized uniform conductor of rectangular cross section of height $h$, width $w$, and length $L$.

$$\langle V/I \rangle = \frac{\rho \cdot L_E}{w \cdot h} \tag{1}$$

where $L_E$, $w$, and $h$ are the *electrical* length, width, and thickness, respectively, of the conductor and $\rho$ is its resistivity(3). In this idealized case, $L_E$ is equal to the feature's *physical* length $L$. This feature, when incorporated into a test structure for ECD determination, is referred to as the *reference segment*.

For practical reasons, reference segments cannot be electrically "accessed" with ideal, nonresistive contacts. They must be accessed using some form of resistive contacts. To minimize the effects of these contacts, a four-terminal method of contacting the reference segment is used. A pair of electrical test pads is connected to extensions from the end of the reference segment. A second pair of test pads is attached to the endpoints of the reference segment to enable measurement of the voltage drop across the reference segment. These voltage taps are referred to as *Kelvin voltage taps* (4) and this four-terminal procedure is commonly referred to as a *Kelvin measurement of resistance*. An example of a Kelvin measurement configuration is shown in Figure 4. The physical length, $L$, is the center-to-center spacing of the Kelvin voltage taps. When one or more reference segments are connected to test pads to allow for forcing and measuring signals, this construct is referred to as a *test structure*.

## 2. Expression of Electrical Critical Dimension in Terms of Sheet Resistance

In the preceding section, the reference feature is described in terms of resistivity and film thickness. An alternate approach is to express $\langle V/I \rangle$ as the product of its aspect ratio in the horizontal plane, $L_E/w$, and a quantity, $R_S$, which is known as the film's sheet resistance (5). Unlike resistivity $\rho$, the parameter $R_S$ is a characteristic of the film's thickness as well as the material from which it is formed.* After the value of $R_S$ has been separately determined, inspection of Eq. (1) shows that the electrical linewidth $w$ of the reference segment is then, in principle, given by (6).

---

* Although $R_s$ is nominally in the unit of ohms, it is not a resistance per se. Rather, a single square of arbitrary dimensions of the material being characterized would exhibit the resistance $R_s$. Often, $R_s$ is written in terms of the non-SI unit of ohms per square ($\Omega/\square$), to distinguish it from a resistance.

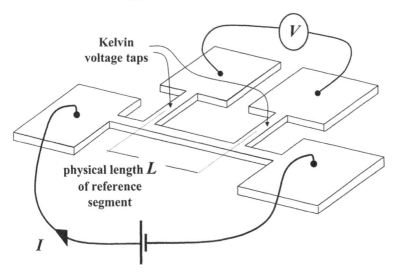

**Figure 4** Example of a four-terminal, or Kelvin, resistance measurement test structure.

$$w = \frac{L_E}{\langle V/I \rangle} \cdot R_S \qquad (2)$$

According to Eq. (2), the electrical linewidth $w$ may be determined from separate measurements of three quantities: these are the $\langle V/I \rangle$ of the reference segment as measured through use of the test pads, the film's sheet resistance, $R_S$, and the reference segment's *electrical* length, $L_E$. Although the latter is generally less than its measurable *physical* length, which, as shown in Fig. 4, is defined by the center-to-center spacing of the Kelvin voltage taps, the formulation in Eq. (2) is generally favored over the alternative one requiring a knowledge of resistivity and film thickness. This is because sheet resistance, unlike the other two quantities, $\rho$ and $h$, can be conveniently extracted from other $\langle V/I \rangle$ measurements made on four-terminal sheet resistors locally co-patterned in the same film as the reference segment. This enables the linewidth of the reference segment to be determined without knowledge of the film's thickness or material resistivity.

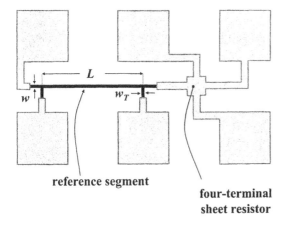

**Figure 5** A standard six-test-pad test structure for the electrical measurement of the width of a conducting feature.

## 3.   Standard Electrical Linewidth Cell

Figure 5 shows a six-probe-pad test structure for the electrical measurement of the ECD of its reference segment. An important element of the test structure shown in Fig. 5 is the incorporation of the four-terminal sheet resistor that was introduced in Sec. II.A. Its presence allows measurement of the sheet resistance of the conducting film and the $\langle V/I \rangle$ of the reference segment with a minimum number of test pads. In general, any such structure, incorporating one or more reference features and/or sheet resistors, is referred to here as an *electrical linewidth cross-bridge resistor* or simply as a *cross-bridge resistor*. Note that the cross-bridge-resistor architecture has been adopted both as ASTM and SEMI Standards for linewidth metrology (7).

In general, these specifications are formulated to minimize errors due to differences between the reference length, which is a physical characteristic of the cross-bridge resistor, and the effective electrical length of the reference segment, which is the quantity that is needed to ascertain electrical linewidth according to Eq. (2). The guidelines generally having the most impact are those that require that the widths of the voltage taps that are used for measuring $\langle V/I \rangle$ be less than 1.2 times the reference-segment width, and that the reference length $L$ typically be greater than 80 μm (8).

Depending on how well actual conditions comply with these requirements, the *electrical linewidth* according to Eq. (2), when applied to $\langle V/I \rangle$ measurements made on cross-bridge resistors having the general layout shown in Fig. 4, will closely approximate the *physical linewidth* when:

The reference segment has a geometrically uniform cross section.

The voltage taps used for $\langle V/I \rangle$ extraction are appropriately narrow relative to the reference length so that the reference length and the electrical length of the reference segment are the same.

Appropriate values of $R_S$ and $L$ are available; e.g., the four-terminal sheet resistor should be located sufficiently close to the reference segment so that the value of sheet resistance extracted from it is applicable to the film material in which the reference segment is patterned.

The cross-bridge resistor is patterned with no geometrical defects, or, if defects are present, they can be detected and eliminated from subsequent data analysis.

The reference segment is made from material that exhibits uniform and linear $\langle I/V \rangle$ characteristics

The Kelvin voltage-measurements have been made by a voltmeter that has an input impedance several orders of magnitude higher than the impedance of the device under test.

The current density of the forcing current is chosen sufficiently small to prevent Joule heating during testing and sufficiently large to be resolved by the voltmeter.

Each of these issues is addressed in detail in the following sections.

## 4.   Sheet-Resistance Metrology

The quality of sheet-resistance metrology has special importance, because any uncertainty in local sheet resistance measurements directly impacts the uncertainties of the accompanying electrical linewidth metrology. Typically, a 1% uncertainty in sheet-resistance measurement contributes a 1% uncertainty to the corresponding electrical linewidth.

A number of papers have followed the pioneering paper by van der Pauw on the subject of the extraction of sheet-resistance from four-terminal sheet resistors (9). These have compared the merits and complexities of using four-terminal sheet resistors having different planar geometries (10–12). The term *planar geometry* means structures patterned in a film of conducting material of uniform thickness, with sidewalls perpendicular to the film's surface, along their entire perimeters. In all cases, an estimate of sheet resistance is obtained through a sequence of $\langle V/I \rangle$ measurements

Two four-terminal sheet-resistor architectures, known as the Greek cross and box cross configurations, are shown in Figure 6. In both cases, voltage $V$ is measured across adjacent terminals $C$ and $D$ when current $I$ is forced between opposite adjacent terminals $A$ and $B$. This "test orientation" provides the first of two complementary $\langle V/I \rangle$ values, $\langle V/I \rangle_1$ and $\langle V/I \rangle_2$, with measurement of both generally necessary for the determination of sheet resistance. The value of $\langle V/I \rangle_2$ is obtained by repeating the current-force/voltage-measure procedure at the complementary test orientation, i.e., when current is forced through terminals $D$ and $A$, and voltage is measured across terminals $B$ and $C$, as shown on the right in Fig. 6 (13). The two complementary $\langle V/I \rangle$ values that are provided by the respective test orientations are used to determine sheet resistance by numerical solution of the relationship (9).

$$\exp\left(\frac{-\pi \cdot \langle V/I \rangle_1}{R_S}\right) + \exp\left(\frac{=\pi \cdot \langle V/I \rangle_2}{R_S}\right) = 1 \tag{3}$$

In those special cases when $\langle V/I \rangle_1$ and $\langle V/I \rangle_2$ are nominally equal, which is anticipated when the resistor is patterned symmetrically with fourfold rotational symmetry, Eq. (3) reduces to the more familiar expression

$$R_S = \frac{\pi}{\log_e 2} I \langle V/I \rangle \tag{4}$$

Once $R_S$ is determined, it may be used to calculate the electrical linewidth of the reference feature according to Eq. (2).

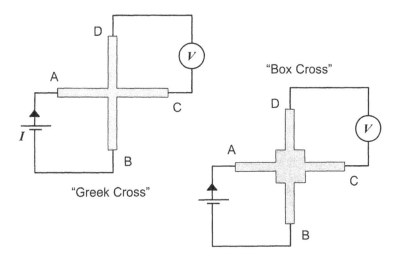

**Figure 6** Two four-terminal sheet-resistor architectures, known as the Greek cross and box cross configurations.

### 5.  Providing for Arbitrarily Wide Voltage Taps and Short Reference Segments

As stated in Sec. II.B.2, the electrical length of the reference segment of a cross-bridge resistor tends to be less than the reference segment's physical length that serves as a metric for ECD measurement. This is because the voltage taps are wider than single point contacts. Thus, at the contacts, part of the current along the reference segment is shunted though the portion of the voltage tap adjacent to the reference feature. In the standard cross-bridge resistor, this is addressed by making the length of the bridge many times the width of the contacts. Since for certain applications it is desirable to produce cross-bridge resistors of arbitrary width, in this section a method is described for measuring and compensating for this effect.

The difference between the physical and electrical lengths of a reference feature introduced by a single nonpoint contact is defined as the parameter $\delta L$ (14), which is incorporated into an appropriate modification to Eq. (2) as

$$w = \frac{(L - \delta L)}{\langle V/I \rangle} \cdot R_S \qquad (5)$$

where the electrical length $L_E$ is defined as the physical length $L$ minus $\delta L$. Since the reference length is defined as the center-to-center tap spacing, each voltage tap of a pair of voltage taps contributes approximately 50% to the effective value $\delta L$, depending on the electrical symmetry of their junctions to the reference segment. Typically, in the geometries used for the cross-bridge resistor, the physical length of a reference segment is greater than its electrical length.

The magnitude of $\delta L$ depends on the width of the voltage taps relative to that of the reference features. However, in a typical thin-film implementation, the impact of $\delta L$ is greater than would be expected from the nominal width of the voltage tap due to lithographic inside-corner rounding (Figure 7) at the junction of the voltage taps to the reference feature (15). Since the widths of the tap and the reference segment are not known a priori and the extent of inside-corner rounding is highly process dependent, $\delta L$ must be determined via measurement rather than quantitatively. Thus, by measuring $\delta L$ and then using its measured value in Eq. (3), the restrictions on the length of the reference feature

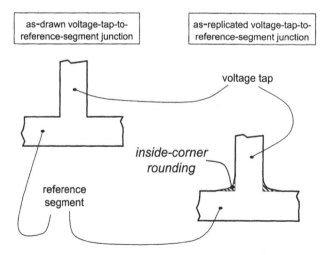

**Figure 7**   The phenomenon of inside-corner rounding enhances the effective $\delta L$.

and the voltage-tap widths, as specified in the standard, may be lifted. In practice, center-to-center voltage-tap spacing can be reduced so long as the $\delta L$ contributed by one of a pair of voltage taps is not influenced by the proximity of other voltage taps. This typically means that the voltage taps that define a particular reference segment may be spaced by as little as several times the drawn voltage-tap linewidth. Consequently, the relative shortness of the reference length that can be allowed reduces the sensitivity of the structure to random patterning defects and increases the spatial resolution of linewidth-measurement mapping. Note that the application of $\delta L$ depends on the assumption that, locally, features such as voltage taps reproduce consistently from location to location.

When cross-bridge resistor test structures are modified to allow compensation for $\delta L$, the new test structure is referred to as the *short bridge resistor*. A simple example of the short bridge resistor is shown in Figure 8. To determine $\delta L$, the two reference segments accessed through probe pad pairs 5 and 6 and 6 and 7 are used. Although these segments have equal physical reference lengths and linewidths, the electrical length of segment $A$ (test pads 5 and 6) is less than that of segment $B$ (test pads 6 and 7), because the former has $n$ ($= 5$ in this case) "dummy taps" distributed along its length. The quantity $\delta L$ is determined by forcing a current $I$ between test pads 1 and 4 and comparing the measured voltage drops $V_A$ and $V_B$, which relate the *electrical* lengths of the two respective segments, according to the ratio

$$\frac{V_A}{V_B} = \frac{L - (n+1) \cdot \delta L}{L - \delta L} \tag{6}$$

An important aspect of the short bridge resistor designs in Fig. 8 is that the dummy tap junctions to the reference segments are required to match physically those of the respective voltage taps. In particular, the dummy taps, *and* the Kelvin voltage taps, of the cross-bridge resistor in the lower part of Fig. 8 actually cross the reference segment and are all tapped from the same side of the reference feature. Use of this geometry, as opposed to simple T-junctions, produces more consistent results when the quality of the lithography and etching is marginal.

Figure 9 shows $\delta L$ measurements made on bridges fabricated in chrome-on-glass. Also shown are the corresponding values of $\delta L$ obtained from current-flow modeling *applied to structures having no inside-corner rounding*. The measured values are 0.3–0.5 μm higher than the modeled ones; this increase is directly attributable to the inside-corner rounding. For example, in Fig. 9 it can be seen that the as-fabricated 1-μm drawn tap

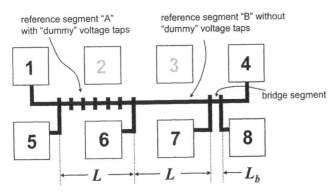

**Figure 8**  Example of test structures patterned in a thin film for enabling $\delta L$ extraction.

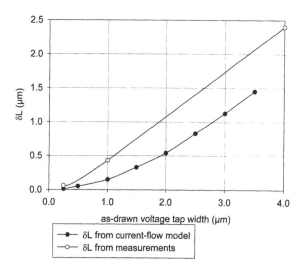

**Figure 9** Comparison of measured and modeled $\delta L$, as determined from Eq. (3), of bridges having drawn linewidths of 1.0 μm and voltage-tap widths ranging from 0.50 μm to 4.0 μm.

widths have the reference-length shortening expected of a 1.8-μm-wide tap with no corner rounding.

A second example of the $\delta L$ effect is shown in Figure 10. In this example, all voltage taps were drawn with a linewidth of 1.0 μm, independent of drawn versus measured electrical linewidth of the reference segments. This data highlights the dependence of $\delta L$ on the width of the feature. Additionally, since these reference segments had a physical reference length of only 8.0 μm, this data shows that $\delta L$ can amount to an appreciable fraction of the reference-length total length.

## 6. Cross-Bridge Test Structures with Multiple Reference-Feature Segments

We have previously shown in Fig. 8 a test structure for extracting $\delta L$ from one reference segment of a cross-bridge resistor and using it in an adjacent reference segment to obtain a value of the linewidth of the latter that is unaffected by errors due to voltage-tap-induced electrical length shortening. We now show an alternative short-bridge-resistor architecture that provides values of $\delta L$ and linewidth from the same reference segments, such reference segments having the same linewidths and common voltage taps.

Figure 11 shows variations of the cross-bridge resistor with multiple reference-length segments connected in series, each segment having a different drawn reference length (16). Several adjacent reference segments share voltage taps. This means that, for example, in the two-segment cross-bridge resistor in the lower part of Fig. 11, both segments use the center tap that serves both for physical-length definition and for voltage measurement. In all the analyses that follow, it is assumed that the junction of each voltage tap in a multireference-segment structure has electrical-physical properties that are sufficiently close to one another that any deviations from this condition do not significantly impact the linewidth-measurement outcome. That is, they all exhibit the same values of $\delta L$.

The cross-bridge resistor in the upper part of Fig. 11 has five reference-length segments but no sheet resistor, and the other has an integral sheet resistor but only two

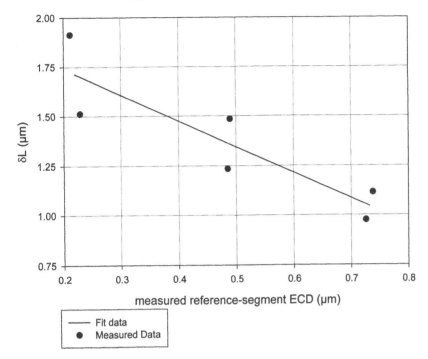

**Figure 10**  Plot of the measured $\delta L$ values, generated by voltage taps having a drawn linewidth of 1.0 μm, versus measured electrical linewidth of reference segments having physical reference lengths of only 8.0 μm.

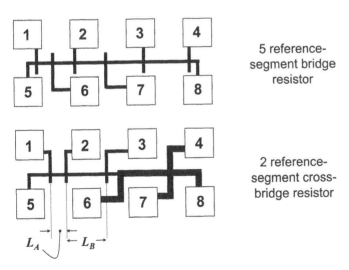

**Figure 11**  Variations of the standard cell cross-bridge resistor for electrical linewidth measurement that feature multiple reference segments connected in series.

reference-length segments. The two cross-bridge resistors shown in Fig. 11 happen to be similar to the ones that were selected for patterning the monocrystalline silicon films that are described later.

The principles of linewidth and $\delta L$ extraction for a two-segment structure are quite straightforward. For example, separate applications of Eq. (3) to the two reference segments of the lower cross-bridge resistor in Fig. 11 yield

$$w = \frac{L_A - \delta L}{\langle V/I \rangle_A} \cdot R_S \tag{7}$$

and

$$w = \frac{L_B - \delta L}{\langle V/I \rangle_B} \cdot R_S \tag{8}$$

As long as $L_A$ is different from $L_B$, Eqs. (7) and (8) may be solved simultaneously to provide values of both $\delta L$ and $w$. Again, note that the cross-bridge resistors in Fig. 11 have crossover voltage taps but no dummy taps. These are not necessary when multiple reference segments are available. Whereas the cross-bridge resistor in the lower part of Fig. 11 has a minimum of two reference segments for the coextraction of ECD and $\delta L$, the five-segment version in the upper part of the figure effectively allows the same extraction with statistical averaging.

For improved statistical averaging and analysis purposes, users may choose to feature architectures that chain more than two reference segments in series, such as that shown in the upper part of Fig. 11. Least squares fitting of the respective measured $\langle V/I \rangle$ values to the appropriate corresponding expression embodying both $\delta L$ and $w$ is achieved by minimizing a quantity $Q$, in this case, given by

$$Q = \sum_{j=1}^{3} \left[ \left( \frac{V}{I} \right)_j - \frac{L_j - \delta L}{w} \cdot R_S \right]^2 \tag{9}$$

Depending on the substrate area that is available in a particular application, there may be benefits, which will be presented in the next section, to incorporating multiple cross-bridge-resistor test structures into the test pattern; each resistor has the same or a different reference-segment drawn linewidth. For example, Figure 12 shows electrical linewidth and $\delta L$ values coextracted from a set of six three-segment test structures having drawn linewidths ranging from 0.5 μm to 1.0 μm and drawn tap widths of 1.0 μm. The constant offset of the measured linewidth from the corresponding drawn linewidth is clearly visible in this data. In the following subsection, we extend this statistical approach from one to many die sites.

## 7. A Statistical Model for $\delta L$ and Electrical Critical Dimension (ECD) Extraction from Multiple Placements of Multireference-Segment Cross-Bridge Resistors

Through use of multiple reference-segment cross-bridge resistors, the user is able to average multiple estimates of the ECD for the same lithographic feature. In the cases when measurements are available from multiple cross-bridge resistors, there exists an opportunity to analyze larger aggregate databases of $\langle V/I \rangle$ measurements. For example, one may identify features in the design with design or mask errors that cause their as-drawn width to be different from the desired value.

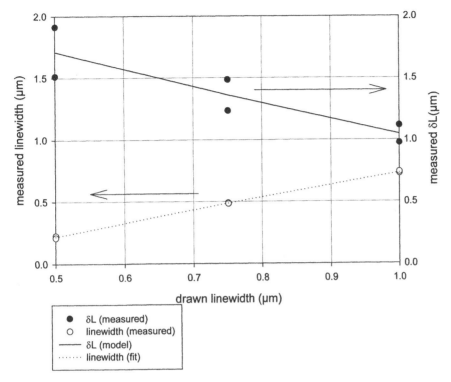

**Figure 12** Electrical linewidths and $\delta L$ values co extracted from a set of six three-segment test structures having drawn linewidths ranging from 0.5 μm to 1.0 μm and drawn tap widths of 1.0 μm.

As in the case described earlier, we have a test pattern with $n$ placements of a cross-bridge resistor, each with $m$ reference segments; within these $n$ placements is a selection of design linewidths within the desired range. However, for this analysis the entire test chip is replicated at $p$ die sites. Reference to Eq. (8) indicates that minimization of the quantity $S_{nmp}$ given by

$$S_{nmp} = \sum_{i=1}^{n} \sum_{j=1}^{m} \sum_{k=1}^{p} \left[ \langle V/I \rangle_{ijk} - \frac{L_j - \delta L_i}{w_{Eik}} \cdot R_{Sik} \right]^2 \tag{10}$$

generates the required electrical linewidths, $w_{Eik}$, of the chained reference segments of the $i$th structure, having the same drawn linewidth, at the $k$th die site. In Eq. (10):

$\langle V/I \rangle_{ijk}$ is the $\langle V/I \rangle$ measurement for the $j$th segment of the $i$th cross-bridge resistor at the $k$th die site.
$L_j$ is the physical reference length of the $j$th segments of the cross-bridge resistors.
$R_{sik}$ is the separately measured local sheet resistance of the $k$th die site.
$\delta L_i$ is the reference-length shortening per voltage tap characteristic of the respective cross-bridge resistors at each die site.

The formulation in Eq. (10) assumes that the applicable values of $\delta L$ are essentially characteristic of the drawn reference-segment and voltage-tap linewidths only and do not materially change from one die site to another for corresponding cross-bridge-resistor placements. Again, this model is considered reasonable, because $\delta L$ is typically less than a 5–10% correction to $w$. However, depending on the application, it could be checked for validity by assigning $\delta L$ a "$k$" subscript in Eq. (10) and calculating $\delta L$ for each die site.

An example of a selection of electrical linewidths calculated by minimizing the quantity $S_{mnp}$ in Eq. (10) is shown in Figure 13. The results in Fig. 13 were obtained from a database of $\langle V/I \rangle$ and local sheet-resistance measurements made on seven die sites each having two placements of cross-bridge resistors with three reference segments for each of 10 drawn reference-segment linewidths ranging from 0.5–3.0 μm. The cross bridges having drawn reference-segment linewidths in the range 0.5–1.0 μm had drawn voltage-tap widths of 1.0 μm, those with drawn reference-segment linewidths in the range 2.0–3.0 μm had drawn voltage-tap widths of 2.0 μm, and the others had equal drawn voltage-tap and reference-segment linewidths. The electrical linewidth results are shown for a random selection of three die sites.

## 8. Split-Cross-Bridge Resistor Test Structure

The cross-bridge-resistor test structure (17) extends the standard cross-bridge-resistor test structure to allow for measurement of both linewidth and line spacing. This structure, shown in Figure 14, adds a split-bridge-resistor segment to the standard cross bridge. This segment, whose total width is equal to that of the bridge, comprises two equal-width lines separated by a space. One of these two lines is tapped to allow for ECD measurement of the feature. The widths of the bridge resistor, $W_b$, and the split resistor, $W_s$, are deter-

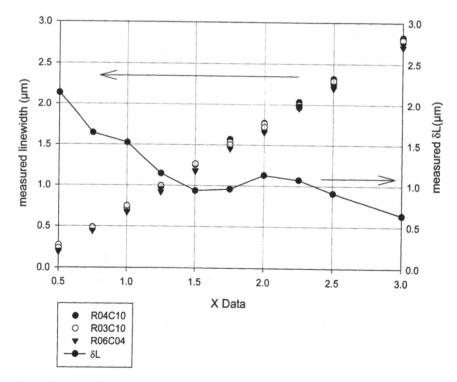

**Figure 13** Electrical linewidths obtained from a database of $\langle V/I \rangle$ and local sheet-resistance measurements made on seven die sites, each having two placements of cross-bridges with three reference segments for each of 10 drawn reference segment linewidths ranging from 0.5 μm to 3.0 μm.

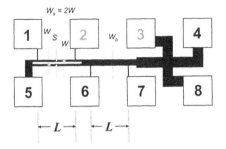

Figure 14 The split-cross-bridge resistor test structure allows for the measurement of line spacing in addition to linewidth.

mined electrically, using Eqs. (4) and (5). The width of the space is determined by subtracting the width of the split bridge from that of the bridge resistor:

$$S = W_b - W_s \tag{11}$$

The pitch is thus

$$P = \frac{W_b - W_s}{2} \tag{12}$$

Note that under conditions of "normal" process, the measured pitch should always be exactly the design pitch. That is, under normal process conditions, where there might be a degree of overetch or underetch, the loss of width of either the conductor or space will be compensated by an equal gain in width in the space or conductor. Thus, the measured pitch indicates whether there is either a catastrophic failure of the test structure (e.g., an open or short in the split) or a failure in the measurement.

## 9. Multibridge-Resistor Test Structure

The multibridge-resistor test structure (18), shown in Figure 15, is often used in industry to calculate electrical linewidth without direct calculation of the sheet resistance. It is based on the assumption that there is a process-specific offset from the design value of every bridge structure and that this offset is independent of linewidth. That is, if the electrical linewidth determined by the multibridge method is $W_M = W_D + \Delta$ where $W_D$ is the design linewidth and $\Delta$ is the difference between the design linewidth and the electrical linewidth, then Eq. (5) becomes

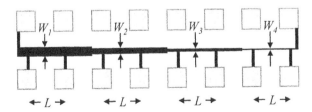

Figure 15 The multibridge-resistor test structure provides the offset of the CD from the design value without the direct calculation of the sheet resistance.

$$R = R_S \frac{L}{W_D + \Delta} \tag{13}$$

Taking the reciprocal of Eq. (13) yields

$$\frac{1}{R} = \frac{1}{R_S L} W_D + \frac{\Delta}{R_S L} \tag{14}$$

If the reciprocals of the measured resistances are plotted against design linewidth, $\Delta$ is given by the intercept divided by the slope of a straight-line fit to the data, and thus the electrical linewidth of each segment can be determined. Deviation of the points from the straight-line fit commonly occur at or below the design rule, where the lithography begins to fail or, for CMP processes, where there are features wider than those at the onset of dishing.

## C.  Common Applications of Electrical Critical-Dimension Metrology

In VLSI fabrication, two applications of ECD metrology dominate. One is process and tool diagnosis, and the other is process control. These are differentiated more by the intended function than by the actual procedure that is applied.

Postprocess diagnosis and/or parameter extraction is usually done at the wafer level, where the features being measured are located in the kerf region. If the ECD test structures were allocated real estate within the chip layout, this measurement could be done after dicing. The level of ECD needed for cost-effective postprocess diagnosis and parameter extraction depends on the process; e.g., if the process yield is marginally acceptable, 100% ECD inspection may be useful. However, the merits of each individual case must be weighed separately.

The second application of ECD metrology to fabrication is in-process CD control, the verification that the CDs of patterned features meet the specific process requirements. Whether this postetch verification is done using ECD, done using an alternate technology, or even done at all is impossible to assess, because such information is usually highly proprietary. This decision is made primarily using economic considerations. That is, the cost of performing the test, which includes the management of contamination that is introduced by the testing process, is compared to the perceived cost of not performing it. Loss of yield is the primary cost of not performing postetch ECD metrology.

As discussed earlier, each reference feature is incorporated into a more complex pattern called an *electrical test structure*. A distinguishing characteristic of electrical test structures is the attachment of probe pads that resemble the bonding pads that are arrayed around integrated circuits. However, instead of having hundreds of test pads, ECD test structures typically have probe-pad counts of $2 \times n$, where $n$ is typically between 4 and 16. (Note that for instances where the ECD structures are to be placed in a narrow scribe region, the structure will be laid out in a $1 \times n$ configuration.) A minimum of six probe pads is required for a single ECD test structure. Increasing the probe-pad count up to $2 \times 16$, normally the largest that is used, enables the linewidths of up to 14 separate features to be incorporated into a single ECD test structure.

Whatever the probe-pad count of the ECD test structure in a particular application, the real estate that is consumed by it is quite substantial and usually dominated by the probe pads, which are typically 80 µm square and located on 160-µm centers. Thus, ECD test structures generally have overall sizes that range from 0.25 mm by 0.4 mm to 0.25 mm by 2.64 mm. Accordingly, when used, they are usually located in the scribe line area.

## D.  Copper–Damascene Implementations

The process cycle of depositing a film of conducting material, patterning it by photolithographic imaging, and then etching it and stripping the remaining resist is not used in copper–damascene implementations. In principle, the same general ECD metrology considerations that have been introduced earlier still apply both to the *analysis of electrical measurements* extracted from the probe pads of a test structure replicated in copper as well as to the *architecture of the test structure* itself. Details of extracting linewidth information are more complex because of the copper features' having barrier layers incorporated into their upper, lower, and sidewall surfaces. A recent article on this subject discloses an ingenious technique for backing out the feature's dimensions from test-structure measurements in conjunction with measurements of interlayer capacitance (19).

An example of the complexities that may be introduced by sidewall-resistivity redistribution of the type featured in copper–damascene processing is shown in Figure 16. In this example, the standard central region of the copatterned extended area, as defined in Sec. II.A, has pure copper disposed over a thinner film of barrier-layer material in contact with its lower surface. The test-pattern feature is modeled as having uniform physical cross-sectional dimensions along its length. The curve in Fig. 16 represents ECD as a function of the resistivity of the side-layer material. If the latter is that of the copper core, then the measured ECD is the same as the physical width of the feature, as one would expect. However, as the sidewall material resistivity becomes progressively higher, characteristic of actual implementations, the feature's ECD decreases significantly from its physical linewidth.

Another application for ECD is determining the onset of dishing in copper–damascene processing. Dishing is a phenomenon related to chemical-mechanical processing, wherein features in one material are removed at a more rapid rate than the surrounding material. This commonly occurs where a soft-metal feature is polished away at a more rapid rate than the harder, surrounding oxide. This effect is more pronounced for wider features, and features below a certain dimension are virtually immune. And ECD test structures can be used to evaluate the onset point for dishing as well as provide an estimate of the effect (18).

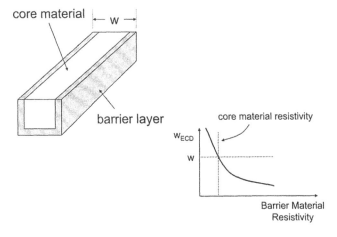

**Figure 16**  ECD of a feature encapsulated in a higher-resistivity-barrier material may diverge significantly from its physical linewidth.

### E.  Other Applications of Electrical Critical-Dimension Metrology

Perhaps the application of electrical CD metrology most closely related to VLSI inte-grated-circuit fabrication is GaAs monolithic microwave integrated circuit manufacture. The preference for gold or gold alloys for gate metal makes liftoff the process of choice in these applications (20). However, there is no special difference in the ECD metrology from that already described for VLSI circuit fabrication, and an approach similar to that used for silicon integrated circuit fabrication has been shown to work quite well (21).

There have been several reports of the linewidths of metal features replicated on *quartz* substrates, e.g., photomasks, that have been measured by ECD metrology (8,22,23). An as-yet-unreported application is the modification of the basic ECD test structure so that it is electrically accessed from outside the pellicle. This will allow a direct comparison between measurements of features on the photomask and on the chip.

Another recent application is that of the certification of a new generation of CD reference materials; this application is described in detail in Sec. III.

Outside the scope of typically integrated circuit CD metrology are applications such as the thin-films used in printed circuit board manufacture; indeed, some of the basic work from which the cross-bridge resistor was derived had its origins in this application (24).

### F.  Equipment Used for Electrical Critical-Dimension Metrology

The purpose of incorporating probe pads into test patterns for linewidth monitoring, as referenced in Sec. II.C, is to enable a device known as a wafer prober to position a matching set of $2 \times n$ probes over the test structure whose features are to be tested. These probes, which are made of tungsten or another hard and conductive material, are mounted in a particular configuration in a probe card. They are simultaneously lowered onto the probe pads with sufficient force to pierce any native oxide or other nonconductive surface that may coat the conducting surface of the test pad. The purpose is to force current through one pair of probe pads of the test structure and to measure voltages using other pairs of probe pads in a preprogrammed sequence. The source and measure-ment instruments, along with instrument control software, as a package are referred to as a process-control test system. Within a process-control test system are a set of instruments for accurately forcing and measuring currents and voltages and a high-performance switch matrix. The switch matrix provides electrically balanced paths from the instruments to the probes that contact the wafer under test, ensuring that the results are independent of the path between the instruments and the wafer. Accurate $\langle V/I \rangle$ measurements require the process control system to include a voltmeter, or multimeter, capable of resolving at least 100 nV. Since such a voltage-measurement resolution exceeds that typically needed for transistor characterization, the capability of the process control system may need to be enhanced in this respect. The voltmeter must have a high input impedance, e.g., 10 G$\Omega$, to avoid errors due to current loss into the inputs of the meter.

During probing of the test structure, the observed $\langle V/I \rangle$ values are typically too small to lend themselves reliably to a two-terminal measurement procedure that lumps the $\langle V/I \rangle$ value of the feature of interest together with the impedances of on-substrate fea-tures, such as interconnect and probe pads, and off-substrate sources, such as probes, the switch matrix, and wires. Thus, a Kelvin measurement, of the kind described in Sec. II.B.1, is preferred.

When performing a four-terminal resistance measurement, it is important that the forced current be sufficient to produce a measurable voltage but not so high as to generate self-induced, or Joule, heating of the feature. Joule heating causes the various $\langle V/I \rangle$ values, which are a function of temperature, to vary in unpredictable ways. That is, in the presence of an unregulated heat source, it is impossible to ensure temperature uniformity during a series of measurements. Therefore, the extracted ECD, which relies on a number of $\langle V/I \rangle$ measurements, will likely be incorrect. For a typical conducting film, there will be a range of acceptable currents that will lead to consistent results. Typically, one will perform a series of measurements of a single feature to determine the approximate range of current densities in which there is no temperature-induced change in resistance. A current density is chosen within this range such that all features can be measured without entering either the region where the voltages are too small to be measured or the region where the structure begins to experience Joule heating.

Since an ECD measurement is very fast, consuming less than 1 second per $\langle V/I \rangle$ measurement, and the measurement procedure easily can be automated to measure a number of sites on a wafer as a group, very large amounts of data are acquired. While the data analysis can be automated to a large degree, it is critical that the data analysis be done in such a manner as to identify and eliminate outliers, such as those due to probing errors or a defective test structure. In most cases, when such faults occur, the magnitude of the result is so far from the expected value that they can easily be identified by the test software.

## G.  Comparison of the Performance of Electrical Critical-Dimension with Alternative Critical-Dimension Metrologies

Other than ECD metrology, the tools that are commonly used for monitoring the line-widths of features replicated in conducting films are SEM, atomic force microscopy (AFM), and scatterometry. The relative merits of these common approaches to CD metrology are compared in Table 1. One set of factors includes capital and operating costs and contamination risk. However, the factors that usually dominate are uncertainty and precision; these will be discussed in the next subsection. If the feature being measured is chemically and physically stable, ECD metrology offers a level of measurement precision superior to that offered by other techniques. It is not possible to devise one universal algorithm for the selection of the optimum metrology to adopt for a particular application. However, it must be recognized that what may be an advantage in one application could possibly be irrelevant to, or even a disadvantage in, another.

Other metrology techniques include transmission optical microscopy, reflective optical microscopy, and transmission electron microscopy. Transmission optical microscopy is used for checking linewidth-control features on printed masks, but it is not usable with optically opaque wafers. Reflection optical microscopy is limited by the wavelength of the light source and available optics to features well in excess of 0.5 μm. Transmission electron microscopy (TEM) is a time-consuming, destructive, and costly metrology tool; TEM is thus limited in use to postprocess diagnostics.

## H.  Precision and Uncertainty

*Precision* and *uncertainty* have different meanings and are generally meaningful in different applications of CD determination. *Precision* refers to either the repeatability of a measurement, i.e., short-term measurement variation, or reproducibility, the long-term measurement variation; for ECD measurements, the repeatability and reproducibility are nearly identical, in contrast to other CD metrology techniques. *Uncertainty*, on the

**Table 1**  Relative Advantages and Disadvantages of Alternative CD Metrology Approaches

| ECD technique | Relative advantages | Relative disadvantages |
|---|---|---|
| Electrical | Highest repeatability<br>Moderate to low cost<br>Nonvacuum<br>High speed<br>Measures sheet resistance as well as ECD<br>No inherent limitations for features scaled down to 0.05 μm<br>Well suited for postprocess diagnostics | Probe contact to probe pads may generate particulates<br>Test-structure real estate requirements<br>Extracts the average conducting width of features over a minimum feature length of 5–10 μm<br>Conductive layers only |
| SEM | Applicable also to nonconducting patterned films such as photo-resist<br>Minimal requirements for test-pattern real estate<br>High linear resolution (can inspect features as short as 1 μm or less)<br>Widely used | High operating and capital costs<br>Charging (particularly when isolated test feature is located on an insulating subfilm)<br>May require extensive sample preparation for postprocess diagnostics<br>Vacuum<br>Performance limitations for features scaled down to 0.05 μm<br>Results have been shown to be operator dependent in some cases. |
| AFM | Qualitatively the same as SEM<br>Low to moderate cost<br>No performance limitations for features scaled down to 0.05 μm | Probe-tip maintenance<br>May require extensive sample preparation for postprocess diagnostics<br>Relatively slow |
| Scatterometry | Noncontact<br>Fast | Test pattern needs to be in the form of a grating—average CDs are reported<br>Not yet widely implemented |

other hand, refers to the scientific "best estimate" of how close the measured value, in this case of the CD, is to the (unknown) "true value." While both precision and uncertainty are important to both usages of ECD, in-process control and postprocess diagnosis, in general, precision is the most critical for process control, where the process engineer is interested in seeing how well the tools are capable of reproducing identical features multiple times. In contrast, for postprocess diagnosis, the uncertainty is more important, since this information is used to determine if the chip will perform to the desired specification. The precision of a test system for measuring the ECD, stated in terms of $3\sigma$, is better than 0.1%. This number represents *published* results for features down to sub-0.5 μm (25,26). There are two ways to determine the uncertainty of a measurement. The first is to compare typical results with those achieved with a sample calibrated traceable to international standards, and the second is to calculate the uncertainties of each element of the measurement and estimate the combined uncertainty. Since there are no CD standards, for electrical measurements, the only technique available is the second. The uncertainty can thus be calculated from the uncertainties due to the measurement instruments and length. The combined uncertainty from these sources depends on several factors, but typically is of the order of 5–20 nm. This number is dependent on a number of design and process factors and is not simply a function of linewidth.

## III. CRITICAL-DIMENSION REFERENCE MATERIALS

By virtue of its superior repeatability and robustness, ECD metrology can be meaningfully linked to a limited selection of hard-to-obtain absolute measurements in a CD traceability path. In the specific implementation described here, the substrates are silicon-on-insulator wafers, and the application is fabrication and certification of a new generation of certified CD reference materials with linewidths in the tens-of-nanometers range.

### A. Methods Divergence

When a measurement instrument is selected, it is not expected that the measurements extracted from this instrument will be characteristically dependent on the instrument type. However, for CD measurement, this is often the case. For example, the linewidth of a feature as measured by AFM will differ from the linewidth of the same feature measured by an electrical parametric tester by a characteristic amount; this difference is called *methods divergence* (27). The observed methods divergence can amount to a significant fraction of the nominal linewidth—thus, the need for a CD reference material with unambiguously known linewidth to provide a link between the multiple rulers providing multiple values of CD.

### B. Physical Requirements for Critical-Dimension Reference Features

Critical-dimension reference features must have both long-term physical stability and ideal geometrical properties. In principle, their cross-sectional profiles may have any shape as long as that shape is uniform along its length. However, the geometrical attribute of the cross-sectional profile, to which the certified linewidth and associated uncertainty values apply, must be specified in order for these values to be useful. This requirement is equivalent to one knowing the reference feature's cross section. For example, if a feature has a cross section known to be rectangular, its nominal linewidth may be stated unambiguously as the width of the rectangle. Alternately, if the reference feature has a symmetrically trapezoidal cross section, its nominal linewidth may be optionally stated to be its upper surface linewidth, or its lower surface linewidth, or its width at some other unambiguously defined height. Whatever feature geometry is chosen, it must have an accompanying description of the exact geometrical attribute of the cross-sectional profile to which its certified linewidth applies.

Other physical properties, which allow applicability of the CD reference material to a wide selection of metrology instruments, are that the material be chemically and physically uniform, electrically conductive, electrically isolated from the substrate, and attached to a chemically homogeneous, microscopically flat surface.

## C. Traceability

*Traceability* in the context of measurement science means that the result of a measurement is traceable to the international definition of the meter as maintained by the Bureau International des Poids et Mésures (BIPM) in France. Normally this traceability is achieved through a thoroughly documented "unbroken chain" of reference to the measurement authority, such as the National Institute of Standards and Technology (NIST) in the United States or Physikalisch-Technische Bundesanstalt (PTB) in Germany. There are several methods of establishing traceability and, sometimes, a measurement is rendered traceable via multiple methods. By far the most commonly used (and best understood) is a method in which an institution acquires standard reference materials from NIST, or from another international authority, and uses these as the institution's primary standards for calibration of the instruments used by that institution.

## D. Elements of the Single-Crystal Critical-Dimension Reference Material Implementation

The previous sections have introduced fundamental requirements for CD reference materials. This section will show how these requirements are met by the single-crystal CD reference material (SCCDRM) implementation. It includes details of the implementation such as substrate preparation, lithographic processing, electrical testing, and CD certification strategy.

The principal elements of the SCCDRM implementation are:

Selection of silicon-on-insulator (SOI) substrates as the starting material
Selection of particular crystallographic orientations of the silicon surface film to ensure essential geometrical characteristics of the reference features.
Reference-feature patterning by lattice-plane selective etching
Certification with electrical CD measurements that are calibrated to silicon lattice counts, where the lattice spacings are traceable to fundamental physical constants

### 1. Silicon-on-Insulator Material

Silicon-on-insulator (SOI) substrates have a surface layer of semiconductor-grade silicon separated from the remaining substrate material by an insulating layer. They resemble silicon wafers and are compatible with regular semiconductor-wafer processing equipment. Integrated circuits are ordinarily formed in the surface layer through the application of the same type of wafer processing that is used for ordinary bulk-silicon wafers. The original motivation for SOI development was the need for radiation-hardened devices for space and military applications. However, for some commercial applications, the higher starting-material expense is offset by higher device speed through the elimination of parasitic capacitances to the bulk substrate. In some cases, the use of SOI materials allows circuits to be designed with a higher device density than would otherwise be possible. Two SOI technologies are in current, widespread use: separation by implantation of oxygen (SIMOX), and bonded and etched-back silicon-on-insulator (BESOI). SIMOX is produced by implanting oxygen into bulk wafers and annealing to produce a buried silicon dioxide layer. BESOI is produced by bonding two thermally oxidized wafers together and thinning one back to the desired thickness.

Generally, either BESOI or SIMOX can be used to produce the CD reference materials described later; however, note that BESOI provides the advantages of arbitrarily thick features and an extremely smooth buried oxide (30).

## 2. Crystallographic Notation

To describe the SCCDRM implementation, an orientation notation convention widely applied in the solid-state physics community is used (28). Namely, specific lattice planes in cubic crystals such as silicon are indicated by parenthesizing the components of their normal vectors. For example, (100), (010), and (001) are three mutually orthogonal planes. The notation 100 means any family of such planes, each having any one index of 1 or −1 and the other two indices being 0. Similarly, [111] is a specific lattice vector, and < 111 > means a family of like directions, including those with any selection of the vector components being negative.

Applicable trigonometry of the (110) plane, containing the [−112] and [1 − 12] vectors, for the case of the cubic lattices, is shown in Figure 17 (13). Although the silicon lattice has a diamond structure, consisting of two interpenetrating face-centered cubic sublattices, the simplified illustration in Fig. 17 is useful here to illustrate the principles of the SCCDRM implementation (29).

One attribute of the crystallography of cubic lattices is that planes of the {111} family intersect those of the {110} orthogonally along lines coinciding with the directions of the < 112 > family of vectors. In Fig. 17(a), for example, the (1 − 11) plane intersects the (110) plane orthogonally, the line of intersection having a [−112] direction. It thus follows that, if line features are drawn in < 112 > directions and are patterned on {110} surfaces, they will have vertical sidewalls that are {111} planes.

## 3. Test-Structure Design

The design of the cross-bridge resistor, from which the measurements presented in the next section were obtained, is illustrated in Figure 18. The layout in Fig. 18 is configured for replication in SOI material having a (110) orientation. For this implementation, feature edges extend in the directions of the [−112] and [1 − 12] lattice vectors. The latter corre-

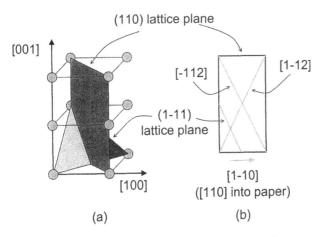

(a)                    (b)

**Figure 17** Applicable trigonometry of the (110) plane, containing the [−112] and [1 − 12] vectors, for the case of the cubic lattices.

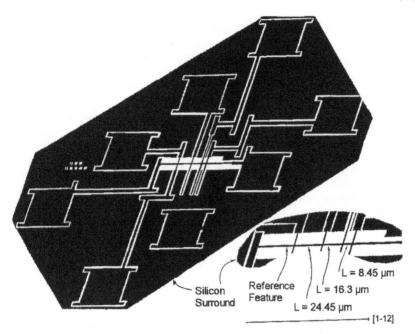

**Figure 18** Three reference-segment cross-bridge-resistor pattern used in the BESOI implementation.

spond to intersections of the vertical sidewall {111} planes at the (110) surface. Four crossover Kelvin voltagetaps define three reference segments. The convex outside corners of the pattern have corner-protect tabs to prevent excessive outside-corner dissolution during pattern transfer from a silicon nitride in situ hard mask used to delineate the features. The orientations shown in Fig. 18 are consistent with the plane of the paper being a (110) plane, the [110] vector being directed into the paper. Alignment of the principal axes of the structure with the indicated lattice vectors ensures vertical, atomically planar, sidewalls of the reference-segment and voltage-tap features.

The perimeter of the cross-bridge resistor is generally defined by the removal of surface-film silicon from regions extending laterally several micrometers around the boundaries of the test structure. However, surface silicon is further removed from regions extending 10 µm from the boundaries of the reference segments. The motivation for not removing all of the surface silicon "surround" that was not incorporated into the cross-bridge resistor was the possibility that it would help mitigate the effects of optical proximity effects during photolithographic exposure. Conversely, leaving the surface-silicon surround offers the possibility of minimizing the adverse effects of oxide charging during reference-segment linewidth measurements by scanning electron microscope beams. The cross bridges were connected to a modified standard probe-pad configuration with a center-to-center periodicity of 160 µm. In Fig 18, the wider lines connected to the pads have a drawn width of 10 µm. For this design, the reference-segment lengths are 8.15 µm, 16.30 µm, and 24.45 µm, respectively. Cross-bridge resistors on the test chip having reference-feature drawn widths ranging from 1 µm to 2 µm have matching voltage-tap drawn widths. Structures having reference-feature drawn widths below 1 µm have 1-µm voltage-tap drawn widths, and those having reference-feature drawn widths above 2 µm have 2-µm voltage-tap drawn widths.

## 4. Fabrication Process

In the microelectromechanical systems (MEMS) technology used, one of two etch solutions, KOH (30) and tetramethyl ammonium hydroxide (TMAH) (31), are used to reveal {111} planes. Since each of these also etches photoresist, a patterned in situ hard mask is needed. This hard mask is typically patterned silicon nitride; an alternate hard mask that may be used for TMAH is silicon dioxide.

An example of a reference-material artifact made by etching a (110)-bulk-silicon surface, having silicon nitride in situ masking and with features aligned in a [112] direction having submicrometer linewidths, is shown in Figure 19. The vertical, {111}, sidewalls generate high levels of contrast for the pitch calibration of electron-beam-linewidth systems (32). Note that the silicon nitride caps on the features, having served for *in situ* masking, had not been removed when this image was recorded. The aspect ratio of their overhang to the height is a measure of the lattice-plane selectivity of the KOH etching.

Another view of the lattice-plane selectivity of the etch is visible in the top-down image of a voltage-tap/reference-segment junction, as replicated in BESOI material, shown in Figure 20. The extent of the {111} planes that occur in the acute angle facets highlights the need to systematically correct for the effects of voltage-tap-induced reference-length shortening, as discussed in Sec. II.B.5. The cross-bridge resistors were fabricated in a process in which the nitride hard mask was patterned using an optical stepper, ordinarily used for a 0.5-µm CMOS process. Reference segments having as-replicated linewidths from 0.15 µm to 0.25 µm were actually derived from 0.35-µm drawn features. Linewidth thinning is attributed to two sources: the first is the small, but finite, etch rate of {111} silicon in KOH, and the second is angular misalignment, of up to approximately 0.15°, of the reference segment with respect to the <112> lattice vectors. The latter has the effect of "snapping" the sidewalls to the lattice <112> directions (33). In so doing, the feature linewidth may be easily reduced by several nanometers for every micrometer of segment length. The effective angular misalignment is typically generated largely by rotational misalignment of the substrate wafer in the stepper tool.

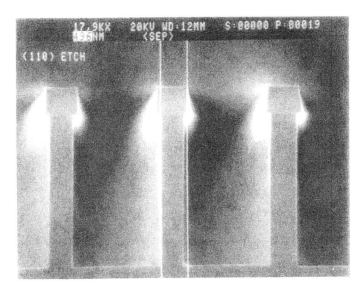

**Figure 19**  A voltage-tap/reference-segment junction as replicated in BESOI material.

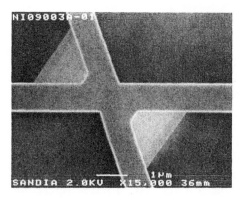

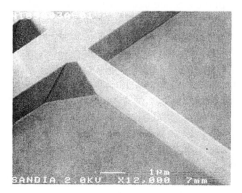

**Figure 20**  Example of a reference-material artifact made by etching a 110 bulk-silicon surface, having silicon-nitride in situ masking and with features with submicrometer pitch aligned in a [112] direction.

A related feature-thinning phenomenon occurs when a feature on the reticle has a patterning defect that becomes transferred to the photoresist in hard-mask patterning. For example, the upper part of Figure 21 shows that a missing absorber defect can result in unintended thinning in pattern transfer to the (110) surface silicon by KOH etching. On the other hand, an extraabsorber defect is generally not transmitted in this particular pattern transfer process, as illustrated in the lower part of the figure.

## E.  Enhancements to the Electrical Critical-Dimension Statistical Model

This section introduces an expression for the $\langle V/I \rangle$ measurements made on reference segments of one of the new multireference-segment cross-bridge resistors that have been

[110] top-down view

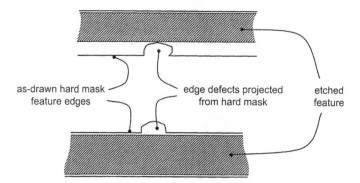

**Figure 21**  Unintended thinning of the replicated feature can occur when the corresponding feature on the reticle has a patterning defect that becomes transferred to the photoresist for hard-mask patterning and subsequently to the hard mask itself.

described in Sec. III.D.3 in terms of the cross-bridge-resistor geometry and sheet resistance. This expression is incorporated into a nonlinear least squares regression model to enable the extraction of the *physical* linewidths of multiple reference segments from extended sets of $\langle V/I \rangle$ measurements made on multiple segments of multiple structures at a particular die site.

## 1. Providing for Reference-Segment Electrical Critical-Dimension Differentiation

For a voltage measured across Kelvin voltage taps separated by a physical length $L_j$, for the $i$th instance of a particular cross-bridge resistor at a die site having sheet resistance $R_S$ and a forced current bridge current $I$, the voltage/current ratio $\langle V/I \rangle_{j,k}$ is given by

$$\langle V/I \rangle_{ij} = \frac{L_j - \delta L_i}{w_{Eij}} \cdot R_S \tag{15}$$

where $w_{Eij}$ and $\delta L_i$ are the corresponding electrical linewidth and voltage-tap induced physical reference-length shortening, respectively. Unlike the formulation in Eq. (10), the electrical linewidths of the respective segments of a particular cross-bridge resistor in Eq. (15) are subscripted by the index $j$. This is because, in the SCCDRM implementation, different segments of the same cross-bridge resistor generally have systematic differences in their physical and electrical linewidths that depend on their lengths. This difference, referred to as *rotational linewidth thinning*, is due to the angular misalignment-induced linewidth thinning introduced in Section III.D.4 (33). The amount of such thinning increases with segment length.

The amount of rotational linewidth thinning of a feature of length $L$ is

$$w_F = \sin \phi \left[ \frac{w_D}{\sin \theta} - \left| \frac{L}{\tan \theta} - \frac{L}{\tan \phi} \right| \right] \tag{16}$$

where $w_D$ is the design width, $\theta$ is the angular misalignment of the mask, and $\phi$, the angle between the (111) planes that intersect the (110) plane orthogonal, is exactly $70.528°$.

## 2. Linewidth Extraction Algorithm

The electrical linewidth, $w_{Eij}$, featured in Eq. (15), may be expressed in terms of $w_{Pi}$ and $\delta w_{\theta j}$ according to

$$w_{Eij} = w_{Pi} - \delta w_{\theta j} \tag{17}$$

where $\delta w_{\theta j}$ is given by Eq. (16). The unknown quantities $w_{Pi}$ and $\theta$ are then extracted by the minimization of a $S_{nmq}$, given by

$$S_{nmq} = \sum_{i=1}^{n} \sum_{j=1}^{m} \sum_{l=1}^{q} \left[ \langle V/I \rangle_{ijl} - \frac{L_j - \delta L_i}{w_{Eij}} \cdot R_{Si} \right]^2 \tag{18}$$

The summations apply to all $\langle V/I \rangle$ measurements that are made at a single particular die site, which, in this example, are attributed one particular value of $\theta$ where there are $q$ repetitions of each $\langle V/I \rangle$ measurement. To minimize the effects of random variation of the tap widths, $\delta L$ in Eq. (18) can be replaced by a value determined from current-flow modeling (34).

After minimization of Eq. (17) generates a complete set of values of $w_{Pi}$ for a particular die site, the corresponding physical segment widths are found by reversing the correction for rotational linewidth narrowing. Namely, physical widths of the respective reference-segments are given by

$$w_{\text{phys},jk} = \sin\theta \left[ \frac{w_{Pi}}{\sin\phi} + \left| \frac{L}{\tan\theta} - \frac{L}{\tan\phi} \right| \right] \qquad (19)$$

### F. Examples of Comparison of Electrically Extracted Physical Critical Dimensions with SEM and AFM-CD Measurements

The measurements were made on structures having the designs described in Sec.. III.D.3. Because of limitations of the lithographic tooling features, drawn linewidths less than 0.35 μm did not replicate reliably. Cross-bridge resistors having only 11 different drawn reference-segment linewidths in the range 0.35–3.0 μm are presented.

Figure 22 shows electrical measurements of the physical linewidths of the 24.45-μm reference segments that were extracted from $\langle V/I \rangle$ measurements made on a die. Also shown are SEM measurements made on the reference segments (35). For the purpose of this comparison, the electrical measurement was corrected for surface depletion, which is a function of the doping and temperature (36). Surface depletion is likely a significant source of observed methods divergence in silicon films, since it leads to a surface region of low conductivity, which would appear to narrow the width of the feature. This correction is not normally used for calibration purposes, since any surface depletion will add a constant offset between the primary calibration measurement, described in the next section, and the ECD measurement.

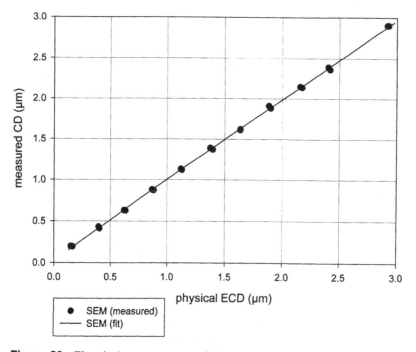

**Figure 22** Electrical measurements of the physical linewidths of the 24.45-μm reference-segments data that were extracted from $\langle V/I \rangle$ measurements that were made on a die that was patterned on a 150-mm BESOI wafer with a surface-film thickness of 1.01 μm.

Note that similar results were seen for a separate die site, which was measured via both ECD metrology and AFM.

## G. High-Resolution Transmission Electron Microscopy and Reference-Feature Linewidth Certification Strategy

High-resolution transmission electron microscopy (HRTEM) enables the viewing and counting of individual silicon lattice planes across lateral extents from the level of several tens of nanometers up to ones in excess of 0.5 μm.

Figure 23 shows a top-down HRTEM micrograph of one of these features. The darker region is the silicon and the lighter region is an oxide, which was used to encapsulate the reference feature to protect during the thinning required for HRTEM imaging. Figure 24 is a narrow region of the line imaged in Fig. 23, along with a higher-magnification view of the left-side region of the silicon. In the enlarged region, the lattice sites can be discerned and counted. Other larger-area images such as these have demonstrated the feasibility of certifying the widths of SCCDRM reference-segment linewidths by HRTEM imaging to within one or two lattice spacings, approximately 1 nm. Lattice counts may be made automatically by high resolution scanning the phase-contrast images, as illustrated in Figure 25.

One strategy for attaching absolute linewidth estimates, with uncertainty levels complying with generally accepted definitions, combines the higher-cost HRTEM imaging with the low-cost robustness and repeatability of electrical linewidth metrology. This

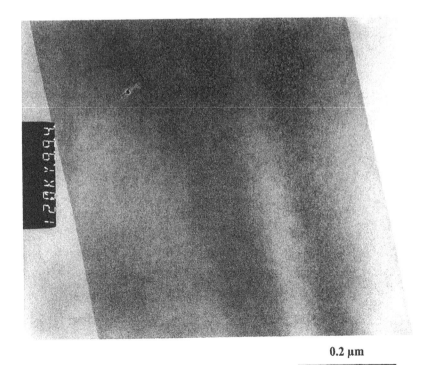

**Figure 23** HRTEM micrograph of a silicon-to-vacuum surface having a maximum extent of 0.35 μm.

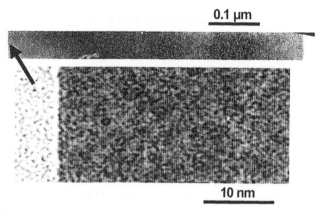

**Figure 24** Enlargement of the edge of the line shown in Fig. 23 showing the lattice planes parallel to the edge, which is encapsulated in silicon dioxide.

strategy is considered to be capable of providing reference materials having nominal line-widths traceable to fundamental constants with acceptable levels of uncertainty and to be manufacturable with acceptable cost. The approach is to test electrically the finished SOI wafer for all reference-segment linewidths by using the methods and statistical techniques described earlier in this section. A selection of reference segments is then additionally submitted to HRTEM imaging. The electrical measurements of all reference segments thereby become correlated to the absolute measurements provided by the limited number of HRTEM images. Determination of the total uncertainty that can be achieved from this approach is the subject of ongoing research.

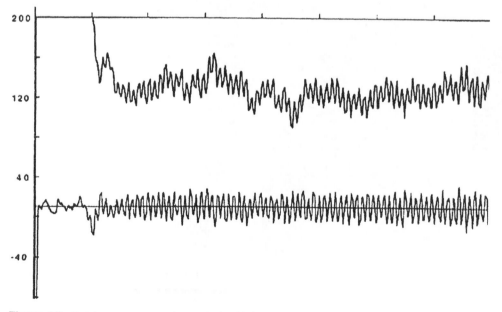

**Figure 25** Lattice counts may be made by high-resolution scanning and digitizing of the phase-contrast images. (Special acknowledgment: Dr. T. J. Headley, Sandia IMRL)

## H.  SCCDRM Implementation with Nonvertical Sidewalls

An alternate approach is to use features patterned in (100) epitaxial silicon material seeded on BESOI (bonded and etched-back SOI) substrates. Reference features aligned with < 110 > lattice vectors that are patterned in the surface of such material are distinguished from those used in the earlier work by their having atomically-planar {111} sidewalls inclined at 54.737° to the substrate surface. Their sidewalls are otherwise rendered as smooth and planar as those studied previously by the chemical polishing provided by the delineating etch. The (110) implementation may offer advantages over, or complement the attributes of, the vertical-sidewall one for selected reference-material applications. As discussed before, analysis of the offsets between the HRTEM and ECD linewidth measurements that are made on the same SOI features is being investigated to establish the feasibility of a traceability path at acceptable cost. Other novel aspects of the (100) SOI implementation that are reported here are the ECD test-structure architecture and the fact that the HRTEM lattice-plane counts are derived from feature cross-sectional, rather than top-down, imaging of the {111} lattice planes.

## IV.  SUMMARY

Electrical-linewidth metrology is a method for measuring a key parameter of semiconductor devices. In contrast to physical metrology techniques, which give an estimate of the spacing between edges of a feature, electrical metrology gives the average conductive-path width of a feature. This conductive-path width assures the designer whether the fabricated features are wide enough to handle the intended current densities. Additionally, the conductive-path width is key in determining the impedance of the feature. Since the actual cross section of the feature cannot be determined by electrical metrology, the ECD is given as an equivalent three-dimensional rectangular representation of the feature, having the same height, length, and total electrical resistance. The feature(s) to be measured electrically are not simply isolated lines, but must be designed in such a way as to allow for electrical contact at specific locations. This is accomplished by incorporating each feature into a more complex pattern called an electrical test structure. A distinguishing feature of electrical test structures is the attachment of probe pads that in some ways resemble the bonding pads that are arrayed around integrated circuits.

   A wafer prober is used to position a set of probes over the probe pads of the test structure whose features are to be tested. In most instances, these probes and probe pads are arrangned in a $1 \times n$ or $2 \times n$ configuration. The probes are made of tungsten, or another hard and conductive material, and are lowered in unison onto the probe pads with sufficient force to pierce any native oxide or other nonconductive surface that may coat the conducting surface of the test pad. Each individual measurement consists of forcing a dc current between a pair of probe pads of the test structure and measuring the resulting voltages present at other pairs of probe pads, in a preprogrammed sequence.

   Two parameters are measured to determine the ECD of a feature: the sheet resistance, which is defined as the resistance of an arbitrarily sized square element of a thin film of uniform thickness and resistivity, and the bridge $\langle V/I \rangle$, sometimes referred to as the "bridge resistance." Although both of these parameters are given in the SI unit of ohms, it is important to remember that they refer to significantly different values in their measurement and extraction. The ECD extraction can be enhanced by determining the electrical length of the bridge, which differs from the physical length by a parameter related to the presence of the voltage taps used to measure the bridge $\langle V/I \rangle$s. From these measurements, the ECD, with its advantages in repeatability, measurement speed, and cost, is extracted.

Other than ECD metrology, the tools that are commonly used for monitoring the linewidths of features replicated in conducting films are scanning electron beam microscopy (SEM), atomic force microscopy, and scatterometry. If the feature being measured is chemically and physically stable, ECD metrology offers a level of measurement precision superior to that offered by other techniques.

The foregoing material has mentioned the use of ECD metrology as a secondary reference for certifying the linewidths of CD reference materials. By virtue of its superior repeatability and robustness, ECD metrology can be meaningfully linked to a limited selection of hard-to-obtain absolute measurements in a CD traceability path. A method for producing CD reference materials was developed that exploits the strengths of ECD measurements. Using this method, samples with uniform, and known, physical geometries are measured electrically and calibrated to the silicon lattice by HRTEM.

## REFERENCES

1. SEMI Standard P19-92. SEMI International Standards 1994. Microlithography Volume (1994).
2. SEMI Standard P35-99. SEMI International Standards 1999. Microlithography Volume (1999).
3. The International Dictionary of Physics and Electronics. New York: Van Nostrand, 1956, p 774.
4. http://www.accuprobe.com/technic/measure.htm.
5. Compilation of Terms. International Standards Program. Semiconductor Equipment and Materials International. 1993, p E–100.
6. M.G. Buehler, S.D. Grant, W.R. Thurber. Bridge and van der Pauw sheet resistors for characterizing the line width of conducting layers. J. Electrochem. Soc. 125(4):650–654 (1978).
7. SEMI Standard P19-92. SEMI International Standards 1994. Microlithography Volume, 83 (1994). See also ASTM 10.04 F1261-89.
8. P. Troccolo, L. Mantalas, R. Allen, L. Linholm. Extending electrical measurements to the 0.5 μm regime. Proc. SPIE Integrated Circuit Metrology, Inspection, and Process Control V., Vol 1464, pp 90–103 (1991).
9. L.J. van der Pauw. A method of measuring specific resistivity and Hall effect of discs of arbitrary shape, Philips Res. Rep., 13:1 (1958).
10. J.M. David, M.G. Buehler. A numerical analysis of various cross sheet resistor structures. Solid-State Electronics 20:539–543 (1977).
11. M.G. Buehler, W.R. Thurber. An experimental study of various cross sheet resistor test structures. J. Electrochem. Soc. 125(4):645–650 (1978).
12. M.I. Newsam, A.J. Walton, M. Fallon. Numerical analysis of the effect of geometry on the performance of the Greek cross structure. Proceedings of the IEEE International Conference on Microelectronic Test Structures (ICMTS 96), March 19–21, 1996, pp 35-38.
13. M.W. Cresswell, J.J. Sniegowski, R.N. Ghoshtagore, R.A. Allen, W.F. Guthrie, A.W. Gurnell, L.W. Linholm, R.G. Dixson, E.C. Teague. Recent development in electrical linewidth and overlay metrology for integrated circuit fabrication processes. Jpn. J. Appl. Phys. 35(1) (12B):6597–6609 (December 1996).
14. R.A. Allen, M.W. Cresswell, L.M. Buck. A new test structure for the electrical measurement of the widths of short features with arbitrarily wide voltage taps. IEEE Electron Device Letters, 13(6):322–324 (June 1992).

15.  R.A. Allen, M.W. Cresswell. Elimination of effects due to patterning imperfections in electrical test structures for submicrometer feature metrology. Solid-State Electronics (3) 435–442 (1992).

16.  M.W. Cresswell, R.A. Allen, W.F. Guthrie, J.J. Sniegowski, R.N. Ghoshtagore, and L.W. Linholm. Electrical linewidth test structures fabricated in mono-crystalline films for reference material applications. IEEE Trans. Semiconductor Manufacturing 11(2), 182–193 (May 1998).

17.  M.G. Buehler, C.W. Hershey. The split-cross-bridge resistor for measuring the sheet resistance, linewidth, and line spacing of conducting layers. IEEE Transactions on Electron Devices ED-33(10):1572–1579 (1986).

18.  L.M. Head, H.A. Schafft. An evaluation of electrical linewidth determination using cross-bridge and multi-bridge test structures. 1999 IEEE IRW Final Report, pp 41–44 (1999).

19.  T. Turner. Metrology: requirements for dual-damascene Cu-linewidth resistivity measurements. Solid-State Technology April 2000, p 89.

20.  Gate metal makes lift-off the process of choice in these applications.

21.  T. O'Keeffe, M.W. Cresswell, L.W. Linholm, D.J. Radack. Evaluation and improvement of E-beam exposure routines by use of microelectronic test structures. Proceedings of the IEEE VLSI Workshop on Test Structures, Long Beach, CA, pp 81–94 (1986).

22.  L.W. Linholm, R.A. Allen, M.W. Cresswell, R.N. Ghoshtagore, S. Mayo, H.A. Schafft, J.A. Kramar, C.T. Teague. Measurement of patterned film linewidth for interconnect characterization. Proc. IEEE 1995 Int. Conference on Microelectronic Test Structures, Vol. 8, pp 23–26 (1995).

23.  R.A. Allen, P. Troccolo, J.C. Owen III, J.E. Potzick, L.W. Linholm. Comparisons of measured linewidths of sub-micrometer lines using optical, electrical, and SEM metrologies. SPIE 1926:34.

24.  See, for example: P.M. Hall. Resistance calculations for thin-film patterns. Thin Solid Films, I: pp 277–295 (1967/1968).

25.  L.W. Linholm, R.A. Allen, M.W. Cresswell. Microelectronic test structures for feature-placement and electrical linewidth metrology. Handbook of Critical Dimension Metrology and Process Control. SPIE Critical Reviews of Optical Science and Technology, Volume CR52, pp 91–118 (1993).

26.  E.E. Chain, M. Griswold. In-line electrical probe for CD metrology. Proc. SPIE Process, Equipment, and Materials Control in Integrated Circuit Manufacturing II, Volume 2876, pp 135–146 (1996).

27.  M.W. Cresswell, J.J. Sniegowski, R.N. Ghoshtagore, R.A. Allen, L.W. Linholm, J.S. Villarrubia. Electrical test structures replicated in silicon-on-insulator material. Proc. SPIE Metrology, Inspection, and Process Control for Microlithography X, Vol. 2725, p 659 (1996).

28.  C. Kittel. Introduction to Solid State Physics. 5th ed. New York: Wiley (1976).

29.  S.K. Ghandi. VLSI Fabrication Principles—Silicon and Gallium Arsenide. Ch. 1. New York: Wiley. (1983).

30.  R.A. Allen, R. N. Ghoshtagore, M. W. Cresswell, J. J. Sniegowski, L. W. Linholm. Comparison of properties of electrical test structures patterned in BESOI and SIMOX films for reference-material applications. Proc. SPIE 3332:124 (1998).

31.  M. Shikida, K. Sato, K. Tokoro, D. Uchikawa. Comparison of anisotropic etching properties between KOH and TMAH solutions. Proc of IEEE Int. MEMS-99 Conf. Orlando, FL, pp. 315–320 (21 Jan. 1999).

32.  Y. Nakayama, K. Toyoda. New sub-micron dimension reference for electron-beam metrology system. Proc. SPIE Integrated Circuit Metrology, Inspection, and Process Control, Vol. 2196, pp 74–78 (1994).

33.  R.A. Allen, L.W. Linholm, M.W. Cresswell, C.H. Ellenwood. A novel method for fabricating CD reference materials with 100-nm linewidths. Proceedings of the 2000 IEEE International Conference on Microelectronic Test Structures, pp 21–24 (2000).

34. W. E. Lee, W. F. Guthrie, M. W. Cresswell, R. A. Allen, J. J. Sniegowski, L. W. Linholm. Reference-length shortening by Kelvin voltage taps in linewidth test structures replicated in mono-crystalline silicon films. Proceedings of the 1997 IEEE International Conference on Microelectronic Test Structures, Vol. 10, pp 35-38 (19xx).

35. M.W. Cresswell, N.M.P. Guillaume, R.A. Allen, W.F. Guthrie, R.N. Ghoshtagore, J.C. Owen, III, Z. Osborne, N. Sullivan, L.W. Linholm. Extraction of sheet resistance from four-terminal sheet resistors replicated in monocrystalline films with nonplanar geometries. IEEE Transactions on Semiconductor Manufacturing, Vol. 12, No. 2, pp 154–165 (1999).

36. S.M. Sze. Physics of Semiconductor Devices. 2nd ed. pp 372-374 (1981).

# 17

# Metrology of Image Placement

**Alexander Starikov**
*Intel Corporation, Santa Clara, California*

## 1. INTRODUCTION

The microelectronics industry has had a long period of remarkable growth. From the time the integrated circuit (IC) was first introduced, the number of transistors per chip has steadily increased while both the size and cost of making a single chip have decreased (1). Rapid evolution in lithography and materials processing were among the primary technologies that enabled the increase of the device count per substrate from 1 to more than 1 billion ($10^9$) and also the mass production of computers. As the customer expectations of system-level IC reliability increased, device-level reliability improved and their failure rates fell (2).

Dimensional metrology of device features on planar substrates has been an important technology supporting manufacture of ICs by microlithography. Although with decreasing device sizes the measurements became ever more difficult to make, the economics demanded that the metrology sampling rates decrease. Even a rough estimate suggests that metrology sampling has already dropped to $< 10^{-8}$ measurements per device per layer in volume production (assumption: DRAM at 250-nm design rules; sampling: 4 measurements per field, 5 fields per wafer, 5 wafers per batch of 25).

Optical microlithography is used in mass production for printing 250-nm features. At this technology level, device design rules stipulate image linewidth control to $< 25$ nm and image placement to $< 75$ nm. Many aspects of microlithography and dimensional control for microlithography have become so advanced that they push the limitations of the available materials, manufacturing tools, and methods.

If the historical rates of change in device size, number, cost, reliability, and so forth were extrapolated for one more decade (3), many implications of a simple dimension scaledown would become difficult to rationalize.

Is there an end to the current trends in microlithography? What other manufacturing technology can we use to make at least 2,000 of such devices for a price of a penny? How do we practice the metrology of image placement? How does it serve the needs of microlithography? What are the known gaps? Can this metrology be done better?

---

Adapted from: A. Starikov, Metrology of Image Placement, Characterization and Metrology for ULSI Technology, 1998 International Conference, AIP Proceedings 449, 1998, pp. 513–535.

This chapter describes the metrology of image placement in microlithography, its current practices, and methods that may be used to answer such questions. It also contains references to many definitive publications in the field—both new and well seasoned.

## A. Metrology and Lithography—A Historical Perspective

Lithography is an old art. Its fundamentals and many strategic solutions have been used for centuries. Implementation, however, has always been a function of applications and available technology. Learning from the great masters of the past helps one to value and understand lithography, enriches one's own skill, and broadens the horizon.

Consider this: The Stone Age ended shortly after humans learned how to put images on any substrate, other than the stone walls of their caves! Rich with information, portable images served as a means to store and move information from one human to another. Our eyes collect a vast amount of data, from where it moves rapidly through the optic nerves to the brain. Establishing how the data is correlated, that is, learning, is a very slow and tedious process. Vagaries of life may stop the knowledge from being passed on through generations. In order to increase the knowledge accumulated by humanity, we embed the information into pictures and symbols, replicate it, and share it with others. Should it be surprising that the greatest improvements in printing technologies have invariably led to the accelerated growth of knowledge and standard of living and to changes of manufacturing methods and the power structure of our society?

Mass production of recorded information by lithography is extremely efficient. Consider an example of wood block lithography, refined by 1764 by Harunobu and his coworkers (4). It relied on a symbiotic social contract among three types of craftsmen. An *Ukiyo-e* artist targeted an emerging class of professional people with a moderate income, produced a design concept, and secured funding from a publisher. In a close collaboration with the artist and printer, an engraver selected from plank-wise-cut cherry wood and carved the shapes for each color into a set of blocks. The printer selected the paper and pigments, mixed consistent batches of paint, and maintained a stable distortion of paper and blocks by adjusting their moisture. Finally, this craftsman spread and shaded the paint, aligned wrinkle-free paper to the block, and made an impression. Water-based paint penetrated the porous paper, resulting in a durable image. When a large print size was required, the standard *oban* format was used in multiples, stitching was reflected in the design and accommodated throughout. A typical run of these beautiful prints could exceed 200 copies.

This technology was as simple as it was effective. A corrected artist's line drawing in brush and black ink was pasted face down on a block, rubbed off, and oiled until it could be seen through. The outlines of the shapes were then incised and engraved. Alignment marks, *kento*, were added in the border area of the first block to be printed in black. Multiple proofs printed from this *kento* block were used in a similar manner, enabling a tight position control of the shapes in each color block and its own alignment targets. Up to 10 color blocks were built this way, perfectly matched to each other. The trial prints and minor adjustments were made as required, until the delicate quilt of pure colors was free of gaps and over printing.

Image size and image placement, their impact on print quality, and the effectiveness of a corrective action were evaluated by visual review. Any image on a block, one printed layer, and each finished lithograph were evaluated much more critically than the most demanding customer ever would. Decisions from the customer perspective were made at all stages of this manufacturing process, passing or rejecting, with immediate corrections as needed. The highest level of image quality and efficiency of this lithography process was achieved by total integration of the whole process, from materials and technology to people.

Many aspects of lithography used to produce *Ukiyo-e* can now also be found in offset printing on paper. Among these are the limited resolution of a human eye and the market economy. When you handle a box of cereals or some other object decorated with offset printing, examine the print quality. Do you, the consumer, need a smaller feature size and tighter overlay? Higher-performance technology is available. But would you, the customer, see any difference in quality and pay more to have it? Not likely. Such considerations are also found in microlithography, and they are likely to remain.

Unlike all the prior forms of lithography, *micro*lithography is used to manufacture *micro*electronics. Integrated circuits manufactured with microlithography enable writing, reading, moving, and analysis of data by computers, rather than by humans. Computer-assisted learning, manufacturing, and control became common. We use simple computers to build more sophisticated ones, fueling remarkable improvements in microlithography, materials, processing, and metrology. Recognizing this unique relationship of microlithography, ICs, and computers in manufacturing, we increasingly rely on computer-assisted learning and control of manufacturing and metrology systems. Using computers, we can gather and use measurements and other information related to product quality, as required for this extreme form of mass production. With computers, we can rapidly analyze the new information and correlate it with our existing knowledge of materials and processes as well as with mathematical models. Once we have learned how to do the job better, we can instruct computers to run our metrology systems, tools, and processes. Once we have taught one computer, teaching many is a snap, and they usually do not blink or deviate from the process of record. This way, we can improve, and keep improving, the dimensional metrology for microlithography, efficiency of manufacturing, and quality of products. Hopefully, this leaves more time for everything else that we like doing!

Over the past decade, conventional optical tools used in metrology of image placement made rapid gains in their precision and accuracy. New methods of error diagnostics and culling and numerical quality feedback for process learning and integration were developed. Heavily computerized metrology of image placement enabled an accelerated metrology and technology learning, fueling the process of integration and more efficient use of other tools and processes in microlithography.

## II. DEVICE DESIGN RULES AND DIMENSIONAL METROLOGY

### A. Linewidth, Centerline, Overlay, and Edge-to-Edge Overlay

To define metrology requirements in microlithography, we use a generally applicable convention (5–8) illustrated in Figure 1. We use the name *feature* to denote the simplest element in one layer of the design of a thin-film device. Feature *linewidth* (*critical dimension,* or CD) in layer $t$ (*reference* or *target*) and level $b$ (*resist, current,* or *bullet*) is denoted $LW_t$ and $LW_b$, respectively. If edge coordinates are denoted as $X_1$, $X_2$, $X_3$, and $X_4$, then

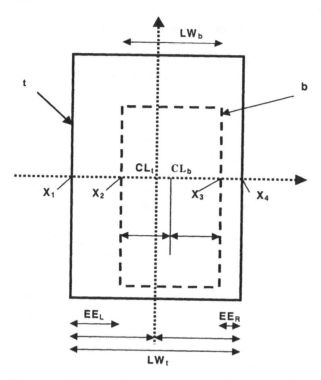

**Figure 1** Linewidth, centerline, overlay, and edge-to-edge overlay.

$$\text{LW}_t = |X_2 - X_3| \quad \text{and} \quad \text{LW}_b = |X_1 - X_4|$$

Likewise, the *centerlines* of the target and bullet features are denoted, respectively, as $\text{CL}_t$ and $\text{CL}_b$,

$$\text{CL}_t = \frac{X_2 + X_3}{2} \quad \text{and} \quad \text{CL}_b = \frac{X_1 + X_4}{2}$$

These are established for two coordinate axes of each layer, typically using a Cartesian coordinate system with axes $X$ and $Y$. The origin and orientation of the coordinate systems are a matter of convention and are not defined in absolute terms. Since microlithography involves patterning very many features in each layer, a set of all centerline coordinates, called *registration* (9,10), is of primary interest. Registration, denoted $\mathbf{R}(X; Y)$, is a vector field made up of centerline vectors of all features in a layer. In microlithography, devices are replicated with some periodicity, and registration is referred to registration grid. Registration is always defined with respect to some agreed upon reference, $\mathbf{R}_r(X; Y)$.

The parameter of primary importance in microlithography is the error in the centerline of one feature in one layer and one feature in another layer, selected as reference. This parameter is called *centerline overlay*, or *overlay* (O/L). Referring to Fig. 1, the overlay of two features whose centerlines are $\text{CL}_b$ and $\text{CL}_t$ is defined as

$$\text{O/L} = \text{CL}_b - \text{CL}_t$$

When referring to all features in bullet and target layers, an overlay vector field is defined (10) as

$$\mathbf{O}/\mathbf{L}(X; Y) = \mathbf{R_b}(X; Y) - \mathbf{R_t}(X; Y).$$

Device design rules require that the linewidth of features in all layers be manufactured and assured within specified tolerances. Linewidth measurements are made for the purposes of both process control and quality assurance. These measurements are made on specialized CD metrology systems.

Image placement of individual features within one layer is important, on its own, only to the extent that it affects, CD. Device design rules generally require that O/L of device features in multiple layers be within specified tolerances with respect to features in other layers. It is at this point, that the *relative* distances in origin, rotation, and registration errors come into play. For the purpose of standardization, the scale used in all coordinate systems is defined in absolute terms. Most often, the metric system of units is used, with *meter* being the standard of length. Measurements of displacement and pitch, closely related to measurements of length, are used to establish the scale underlying the metrology of linewidth and image placement (see Chapter 14, by Postek and Vladár, in this volume and Refs. 11 and 12).

In addition to CD and O/L, design rules specify requirements for edge-to-edge overlay for features of different image layers. Left and right edge-to-edge overlay is denoted in Fig. 1 as $EE_L$ and $EE_R$. In the business of microlithography, as it is practiced today, edge-to-edge overlay is seldom measured directly. In order to control edge-to-edge overlay, CD measurements in each layer are made with CD metrology systems, and centerline layer-to-layer overlay is measured with O/L metrology systems. The data is then mathematically combined to produce the estimates of edge-to-edge overlay.

It would seem that the measurements of linewidths, centerlines, and edge-to-edge overlay are linked by a trivial combination of the edge coordinates $X_1$, $X_2$, $X_3$, $X_4$. However, since CD metrology and O/L metrology are carried out by very different means, this linkage cannot be taken for granted. The profound differences between our practices of CD and O/L metrology and the nonrandom nature of errors in dimensional metrology make it difficult to estimate edge-to-edge overlay. The accurate metrology of both image linewidth and placement is of the essence here.

## B.  Overlay Budgets and Bookkeeping

It is well recognized that overlay in thin-film devices is a result of the image placement errors incurred at many stages of the manufacturing process. Registration errors in each reticle, at alignment and image formation, those driven by processes of image recording and image transfer, are among the contributors. It is important to recognize that the same factors also affect metrology of image placement. One may then conjecture that superior metrology of image placement for microlithography applications may be achieved only by a process (6) similar to continuous quality improvement business processes already used in microelectronics manufacturing (2,13).

As in any manufacturing process, all groups involved in IC manufacturing are governed by social contracts. When microlithography pushes the limits of the tools, materials, and processes, this social contract becomes an essential factor. It predicates the manufacturability and quality of a product, the cost of an entire manufacturing process, and, ultimately, its commercial viability. The optimal use of resources largely determines the commercial success of this business.

In order to achieve the required device O/L, and to do so at a reasonable cost, it is necessary for all the parties affecting O/L to collaborate on O/L budgeting. Using a

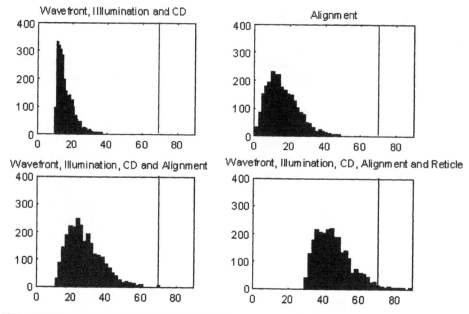

**Figure 2** Components of deep trench to active area edge-to-edge overlay error budget. It is based on characterized masks, lithography system aberrations, field and wafer flatness, focus, and alignment. The feature type-, size-, and orientation-specific CD and placement errors are included (see details in Secs. VI.A, VI.B, and VI.B). (From Ref. 17.)

sensible model that describes components of device O/L (14–17), a portion of the O/L budget is allocated to all contributing sources. Each error source is then controlled individually so that the total error does not exceed the required tolerance. An example of modern edge-to-edge overlay budgeting is shown in Figure 2.

Expectations for O/L metrology are often defined on the basis of a common business practice. For example, for a 256Mb DRAM process with CD of 250 nm it is expected (3) that the O/L budget is under 75 nm. The error allocated to O/L metrology is then stated as 10% of the budget. That is, the metrology error must be < 7.5 nm.

What does an O/L metrology error of < 7.5 nm mean? Can this requirement be met? What is the impact of not meeting this requirement? Answers to these questions are needed in order to make sound decisions on both technology and business aspects.

The ultimate purpose of O/L metrology in microlithography (5,14–16) is to limit losses in device yield, performance, and reliability that are due to O/L from its nominal value of zero. Mathematical models estimating the fraction of good fields in lithography applications address this objective (5,14–17). On the other hand, O/L budgets and metrology requirements are often stated as a sum of the absolute value of the mean error plus three standard deviations:

$$O/L_x = \left|\langle O/L_x \rangle\right| + 3\sigma_{O/L_x} \quad \text{and} \quad O/L_y = \left|\langle O/L_y \rangle\right| + 3\sigma_{O/L_y}$$

It is important to keep in mind that the error of the mean carries a very strong penalty in yield. The average inaccuracy of O/L metrology is added directly to the mean O/L error of devices, because the O/L data is used to remove alignment offsets in a closed control loop (8). Consider the impact of the uncontrolled mean error of metrology used to control a process with ±75-nm control limits. Assuming that the O/L distribution is Gaussian and centered (that is, $1\sigma_{O/L_x} = 25\,\text{nm}$ and $\langle O/L_x \rangle = 0\,\text{nm}$) it is expected that 0.27% of all O/L values are outside the control limits. A mean inaccuracy of just 10 nm would almost double this failure rate. This is why the accuracy of O/L metrology is of such paramount importance.

For evaluating the accuracy of O/L metrology, we will use the concept of *measurement uncertainty* (18). However, unlike the metrology on a single specimen commonly assumed as the environment, in microlithography applications we are concerned with errors in many measurements made to support the microlithography process. We presume that O/L can be measured quite accurately by some means, no matter how slow and expensive. Since we cannot afford doing this in production, we typically incur much larger O/L metrology errors. Therefore, as a practical matter, our goal is to estimate and reduce those errors that cannot be removed by the known and expedient calibrations. In this approach, we first reduce all known errors: imprecision is suppressed by averaging, and any available calibration is applied. Then we estimate the inaccuracy for the population of measured structures, wafers, and so forth. These residual errors are reported as the mean and standard deviation. For a population of residual measurement error $\{E\}$, measurement uncertainty is stated as

$$U = |\langle E \rangle| + 3\sigma_E$$

In addition, we will follow a similar format in estimating the impact of errors from various error sources (19). When estimating the uncertainty due to a particular error mechanism, we suppress all other error sources and report the mean and standard deviation of this particular error. The measurement uncertainty of a population $\{e\}$ of the errors of this particular type is

$$U_e = |\langle e \rangle| + 3\sigma_e$$

Such estimates are established and tracked to gauge the progress made to improve the quality (accuracy) of O/L metrology.

## 1. Applications and Software Tools

In microlithography, it is usually required that the image placement errors in multiple image layers be either matched to each other or matched with respect to a reference. For example, when a new layer is printed on a substrate with a prior (target) image layer, this prior layer is used as reference, and the layer-to-layer errors are minimized by making adjustments in placement of the current (resist, bullet) layer.

Figure 3 graphically illustrates the application environment in step-and-repeat lithography on wafers and separation of overlay errors into intrafield and interfield components. In applications of metrology of image placement in microlithography, it is important to separate the error of image placement into portions that may be removed at their respective sources and the residual error.

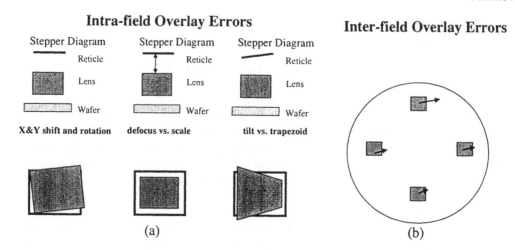

**Figure 3** Common overlay errors in lithography printers: the mismatch of the bullet and target images of one field (a) and multiple fields in one wafer (b). (From Ref. 20.)

The step-and-repeat and step-and-scan systems used in lithography offer a number of adjustments to reduce both intrafield and interfield errors. The O/L errors due to translation, rotation, and mismatched scale occur in any lithography situation; one can get a demonstration just by using a copy machine. While it is easy to keep a sheet of paper flat against the glass of a copier, defocus and tilt are relatively hard to control in microlithography on the processed wafers. When either the reticle or the wafer is defocused and/ or tilted, anisotropic field scale error, trapezoid, or orthogonality distortion may be the result. The intrafield errors that occur due to defocus and tilt are illustrated in Figure 4. These errors can be explained (21) by considering in geometrical optics how a defocused/ tilted wafer (reticle) plane intercepts the ray bundles forming the image. In a telecentric projection lens, the chief rays are parallel with the optical axis, so pure defocus results in no distortion. Tilt of either reticle or wafer, on the other hand, results in anamorphic distortions (Fig. 4). The exact form of this distortion depends on the orientation of the tilt axis with respect to the imaged field. In a case of nontelecentric lens, as most lithography lenses were a decade ago, defocus leads to changes of magnification, and the outcome of tilt is one of the trapezoidal errors.

Keeping in mind the role and place of O/L metrology in microlithography, we need to notice that the raw metrology data will be put through a particular kind of data analysis, aiming to isolate and remove the systematic error sources. The purpose of this analysis, and of the subsequent adjustments and corrective actions, is to reduce O/L error in devices. An example of such analysis (from Ref. 21) is shown in Figure 5, to illustrate a typical error management methodology and conventions.

The systematic component of intrafield O/L error shown in Fig. 5b may be further broken down into a combination of the lower-order terms in a Taylor series expansion in the vicinity of the center (Figure 6, from Ref. 21). These lower-order terms are grouped to represent the individual adjustable parameters ("knobs") of step-and-repeat and step-and-scan projection systems: $X$ and $Y$ translation, field rotation, $X$ and $Y$ magnification (this example illustrates anisotropic magnification), orthogonality, and trapezoid.

A typical (21) expansion showing the O/L vector field ($dx$; $dy$) as a power series of space coordinates ($x$; $y$) is shown next:

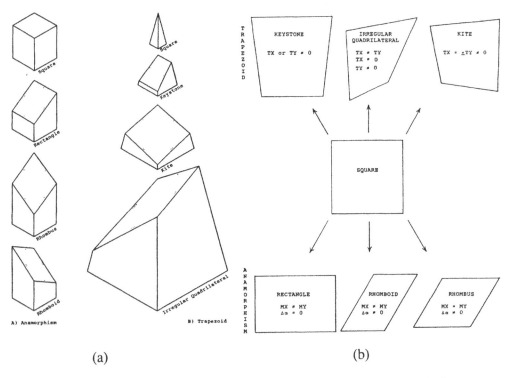

(a)                                                    (b)

**Figure 4** Intersection of a tilted reticle/wafer plane with the chief rays at the field edges.

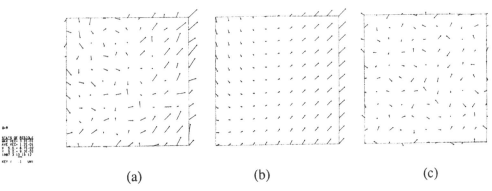

(a)                              (b)                              (c)

**Figure 5** Typical intrafield overlay vector field pattern (a). The tail of each vector is at a measurement site, the vector represents overlay error at that site (greatly increased for display); maximum vector, $X$ and $Y$ standard deviation are displayed. Overlay is shown decomposed into the systematic components (b) and the residual error (c). The systematic components may be adjusted to reduce the residual error.

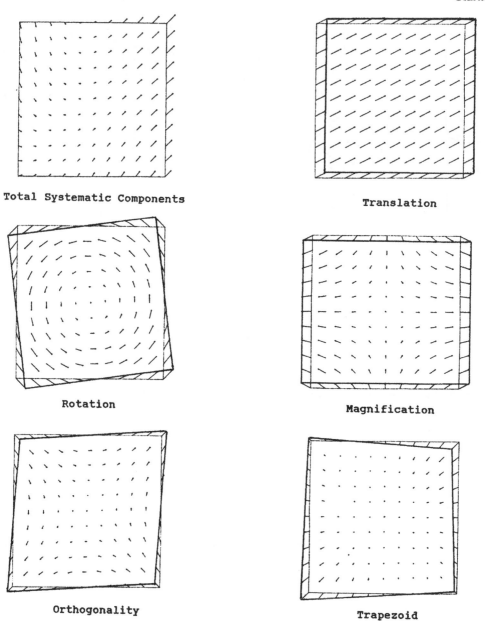

**Figure 6**  Decomposition of the systematic components in Fig. 5.

$$dx = \Delta X - \Delta\Theta * y + M_x * x + \Delta\alpha * \frac{y}{2} + TX * x * y + TY * x^2 + \mathrm{BP} * x * (x^2 + y^2)$$

$$+ \mathrm{D5} * x * (x^2 + y^2)^2 + SX * y^2 + CX * y^3 + \Sigma_N\{XN * |x^N|\} + O_x$$

$$dy = \Delta Y + \Delta\Theta * x + M_y * y + \Delta\alpha * \frac{x}{2} + TY * x * y + TX * y^2 + \mathrm{BP} * y * (x^2 + y^2)$$

$$+ \mathrm{D5} * y * (x^2 + y^2)^2 + SY * x^2 + CY * x^3 + \Sigma_N\{YN * |y^N|\} + O_y$$

where $\Delta X$ and $\Delta Y$ are $X$ and $Y$ translation, $M_x$ and $M_y$ are magnification coefficients, $TX$ and $TY$ are trapezoid coefficients, BP is barrel-pincushion coefficient, D5 is 5th order, $SX$ and $SY$ are coefficients of square distortion, $CX$ and $CY$ are the coefficients of cubic distortion, $\Sigma_N\{\ldots\}$ is a sum of asymmetry terms whose coefficients are $XN$ and $YN$, and, finally, $O_x$ and $O_y$ are the residual terms of the series. Naturally, when this series expansion is truncated at a lesser number of terms, the remaining error becomes a part of the residual terms $O_x$ and $O_y$. Table 1, illustrating these expansion terms with formulae, appropriate dimensional units for each coefficient, and some typical error sources, is included here for reference.

We might also point out here that the advanced optical microlithography of today is pushed far beyond the realm of geometrical optics. Because the intrafield image placement errors in extended optical microlithography are increasing relative to device O/L budget (17), the concept of lens distortion is rapidly loosing its utility (6,17). The subject of the intrafield error of image placement is now in a state of flux (see Secs. VI.A–VI.C and the references cited there).

## C. Physical Definitions: Edge, Linewidth, and Centerline

Various forms of metrology of linewidth and image placement may be traced to a single common root. All of them involve the definition of an edge and the estimation of edge positions in thin-film features. Unfortunately, what constitutes an edge and how to estimate the edge coordinates are among the most fundamental and difficult questions of dimensional metrology (23).

Consider the schematic presentation of a feature in an image formed in a layer of thin film supported by a planar substrate in Figure 7. SEMI Standard definition of linewidth (24), due to Nyyssonen and Larrabee (25), specifies the *linewidth* (LW, CD) as the distance between two opposite line edges in the thin film material with coordinates $X_1$ and $X_2$:

$$\text{LW} = |X_1 - X_2|$$

These edge coordinates are defined along the sidewall of the film at a certain agreed-upon height $H$ from the interface. These definitions of edges and edge coordinates pertain to a single agreed-upon cross section of the feature.

A definition of *centerline*, consistent with its use in microlithography, must be linked to the definition of *linewidth*. Referring to Fig. 7, we define (6–8,19) the *centerline* as an average coordinate of two opposing edges; that is,

$$\text{CL} = \frac{X_1 + X_2}{2}$$

In centerline estimation, it is usually expected that the shape is mirror-symmetrical. When this is the case, the exact value of height $H$ is not important for estimating the centerline. This estimate does not change, although each of two opposite edges may be in error by amount $\delta$ from the edge coordinates defining the linewidth (see more in Sec. III.B.).

Microscopic observations in Figures 8 (from Ref. 6) and 9 reveal significant variability in materials forming the features of interest. As the device linewidths continue to shrink, the variability of the feature sidewall becomes a larger fraction of CD. Any practical attempt to estimate linewidth and centerline must account for edge profile and material variations along a representative sample along the feature length.

**Table 1** Distortion Components

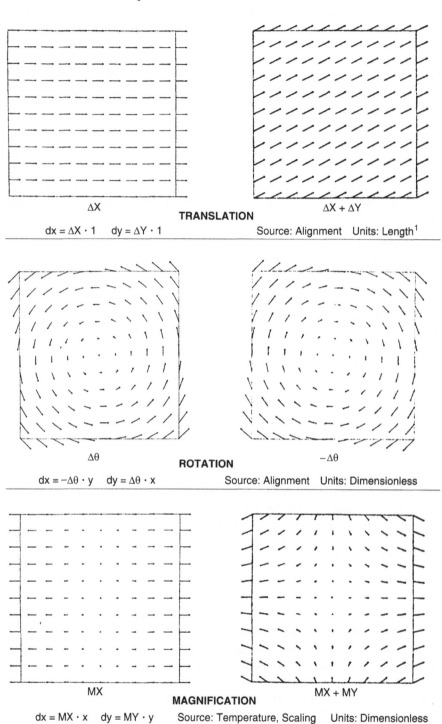

TRANSLATION

$\Delta X$     $\Delta X + \Delta Y$

$dx = \Delta X \cdot 1$    $dy = \Delta Y \cdot 1$     Source: Alignment   Units: Length[1]

ROTATION

$\Delta \theta$     $-\Delta \theta$

$dx = -\Delta \theta \cdot y$    $dy = \Delta \theta \cdot x$     Source: Alignment   Units: Dimensionless

MAGNIFICATION

$MX$     $MX + MY$

$dx = MX \cdot x$    $dy = MY \cdot y$     Source: Temperature, Scaling   Units: Dimensionless

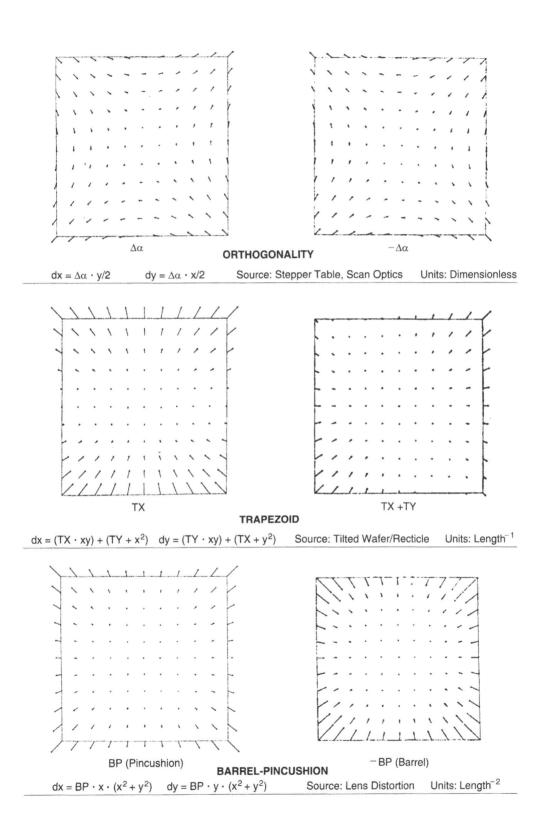

**ORTHOGONALITY**

$\Delta\alpha$          $-\Delta\alpha$

$dx = \Delta\alpha \cdot y/2$      $dy = \Delta\alpha \cdot x/2$      Source: Stepper Table, Scan Optics      Units: Dimensionless

**TRAPEZOID**

TX          TX +TY

$dx = (TX \cdot xy) + (TY + x^2)$    $dy = (TY \cdot xy) + (TX + y^2)$     Source: Tilted Wafer/Recticle     Units: Length$^{-1}$

**BARREL-PINCUSHION**

BP (Pincushion)          $-$BP (Barrel)

$dx = BP \cdot x \cdot (x^2 + y^2)$     $dy = BP \cdot y \cdot (x^2 + y^2)$     Source: Lens Distortion     Units: Length$^{-2}$

**Table 1** (*Continued*)

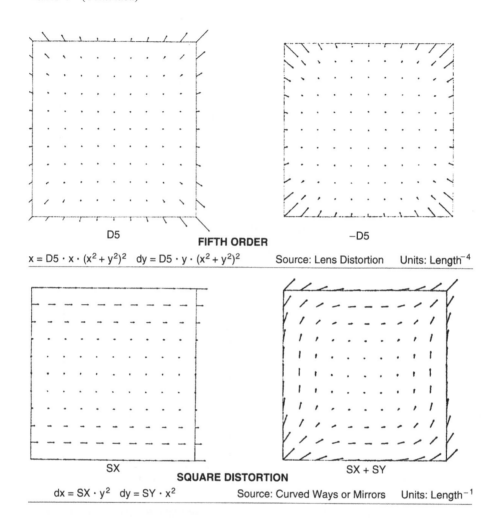

D5                                                      −D5

**FIFTH ORDER**

$x = D5 \cdot x \cdot (x^2 + y^2)^2$   $dy = D5 \cdot y \cdot (x^2 + y^2)^2$      Source: Lens Distortion      Units: Length$^{-4}$

SX                                                      SX + SY

**SQUARE DISTORTION**

$dx = SX \cdot y^2$   $dy = SY \cdot x^2$      Source: Curved Ways or Mirrors      Units: Length$^{-1}$

The problem of edge localization is compounded by the fact that, when localizing edge positions to a few nanometers, even the definition of an "interface" can no longer be taken for granted. Not a single instrument can adequately estimate coordinates of the surface defined by the "interface" of two materials. Given the dimensions of interest and the realistic materials, the simple definitions of LW and CL are difficult to support with the available metrology means (23–27).

Dimensional metrology in microlithography becomes more complicated when the process interactions and asymmetry are present, as illustrated in Fig. 9. As the material interfaces and the function of devices come into consideration, there may be multiple definitions of *linewidth, centerline, overlay,* and *edge-to-edge overlay* applied to the same physical structure.

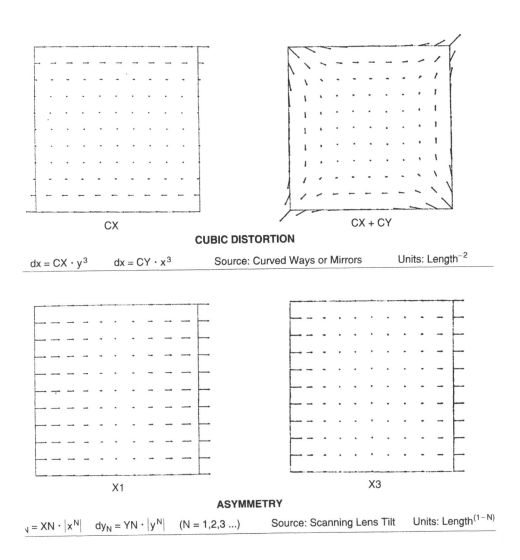

**CUBIC DISTORTION**

$dx = CX \cdot y^3$ $\quad$ $dx = CY \cdot x^3$ $\qquad$ Source: Curved Ways or Mirrors $\qquad$ Units: Length$^{-2}$

**ASYMMETRY**

$_N = XN \cdot |x^N|$ $\quad$ $dy_N = YN \cdot |y^N|$ $\quad$ $(N = 1,2,3 \ldots)$ $\qquad$ Source: Scanning Lens Tilt $\qquad$ Units: Length$^{(1-N)}$

## D. Metrology of Image Placement: Registration, Alignment, and Overlay

There are three basic forms of the metrology of image placement: the metrology of registration (long distance $X/Y$ metrology), the measurement of alignment offset (alignment), and the metrology of overlay. These forms of metrology of image placement are fundamentally related to each other. They are also linked to CD metrology by the definitions of *edge, linewidth*, and *centerline*. Consequently, there are many similarities in both metrology methods and instruments.

Unlike the metrology of linewidth, the spatial resolution of the metrology system is not the primary factor that limits performance. The metrology of image placement can be practiced with the same basic tools from one generation of IC products to another. The

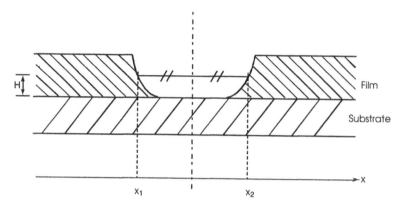

**Figure 7** Physical definitions of edge, linewidth, and centerline.

systems on the market today may look similar to those 10 or 20 years ago. Superior precision and accuracy in image placement are achieved by incremental improvements of all subsystems.

Only brief reviews of $X/Y$ metrology and alignment are presented here. The rest of this chapter treats the conventional optical overlay metrology, concentrating on the commonly used methods, considerations of accuracy, and technology limitations.

### 1. Metrology of Registration

The tools typically used in long-distance $X/Y$ metrology (the metrology of registration) are based on an optical microscope as a position sensor and an interferometric stage as a means of position readout (28–30). The measurements provided by these tools are usually defined as centerline-to-centerline comparisons within an array of essentially similar targets. Related instrument variants are the metrology systems using SEM (scanning electron microscope) (31–33) or AFM (atomic force microscope) (34) as position sensors.

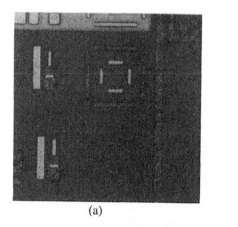

(a)

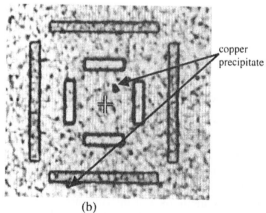

(b)

**Figure 8** Typical bars-in-bars O/L structure. Grainy metal of the bullet mark (final image) and target mark (a). Outer bullet bars of resist (islands) over a blanket AlCu film over polished W stud (substrate) inner target marks (b).

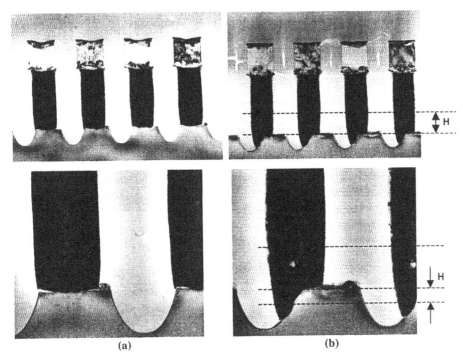

**Figure 9** TEM cross sections of the shallow trench, contact, and metal features illustrating a practical situation where *edge, linewidth, centerline, overlay,* and *edge-to-edge overlay* may and should be defined in multiple ways. Note that, unlike (a), asymmetric contact in (b) makes some definitions of contact linewidth and centerline ambiguous.

In order to test performance and maintain the advanced $X/Y$ metrology systems, their users build and use high-quality stable grid reference materials. Such reference materials, with only limited calibration, are sufficient for the purposes of process control. However, to support a multicompany international business environment, the standard of length is required and the standard of 2-D grid is desirable. The national standards laboratories (11,12,28,32–36) manufacture and/or certify the 1-D scale (length), producing the standards of length widely used today. A certified 2-D grid reference material has recently become available from PTB (32,33).

Self-calibration (37–40) of a 2-D grid is possible with available $X/Y$ metrology systems. To exploit this grid calibration path, a development effort by SEMI (Semiconductor Equipment and Materials International) member companies and NIST (National Institute of Science and Technology) in this area are currently under way. In order to accommodate measurements of a 2-D artifact with the required rotations and translations, the SEMI Task Force had to introduce two nonstandard registration structures. The new 2-D grid artifact features box marks and frame marks but no cross marks (40). Unlike a cross mark, which is the industry standard registration mark for photomasks (41), a box and a frame allow estimation of the centerline at the center of the mark itself. The box mark is the simplest, and, since redundancy is not required for performance enhancement, in a multilaboratory multiuser environment it has a distinct advantage.

The $X/Y$ metrology systems are expected to measure registration errors in arrays of essentially similar features. In this application, a constant additive error of centerline estimation does not affect the accuracy of registration measurement. A typical $X/Y$

metrology system may not be built or maintained to be accurate on dissimilar features. When analysis of the reticle registration error contributions to device overlay and metrology errors (42) involves comparisons of image placement in dissimilar features, inaccuracy of $X/Y$ metrology systems becomes apparent. It may be expected that accurate registration measurements on dissimilar features will soon become a requirement for $X/Y$ metrology systems.

## 2. Alignment

Alignment is closely related to O/L metrology (8). Like O/L metrology, alignment involves estimation of the centerline-to-centerline distance between two dissimilar targets: a wafer target and a reticle target. This distance, called *alignment offset*, is used to mechanically position (10) the wafer with respect to the reticle on a lithography printer. The goal of this operation is that, once the new image is recorded in photoresist over a previously imaged substrate, layer-to-layer overlay be minimized. A good account of recent development in alignment is available (43).

Optical alignment systems and interferometric stages of advanced lithography printers have long been used in stepper self-metrology (44,45). This form of $X/Y$ metrology is now quite common.

## III. EXPECTATIONS AND BASIC REQUIREMENTS IN METROLOGY OF IMAGE PLACEMENT

Microlithography is used to mass produce thin-film devices with an unprecedented level of product uniformity. All process steps must maintain certain essential device properties related to patterning, such as linewidth and image placement. It is generally expected that image placement is tightly controlled in all process steps (6–8,14–17,46,47):

> All sources of registration errors are small and stable (in space and time).
> All lithography systems have their image placement errors closely matched.
> All features designed with the same (redundant) centerline must maintain redundancy.
> Any O/L metrology system may be used to measure product O/L.
> Any illumination wavelength, focus setting, or detection algorithm may be used.
> Any two devices in near proximity must have the same layer-to-layer overlay.
> Any two O/L metrology structures in proximity must yield similar estimates of O/L.
> O/L estimates in developed resist image, after etch or deposition, must be similar.
> Overlay measured with an optical system must be the same as that with SEM or AFM.
> Microscopy and electrical probe–based metrology data must be similar.

When these a priori expectations are not reasonably satisfied, an effort to improve O/L control in manufacture or to improve metrology of image placement is mounted. The reason is very simple: product yield, product performance, and productivity are at stake.

## A. Reproducibility

In order for any metrology to take place, measurements must be *reproducible* (48). By that we mean that metrology must yield somewhat similar or consistent results. In testing reproducibility, a tool is used as expected and permissible and the variability of metrology is evaluated.

Some elements of reproducibility relate to the random events in a measurement: photon noise, flicker and wander of illumination, electronics noise, conversion of analog signal to digital, digital control loops, and so forth. In order to express the impact of random events, many measurements are made over a relatively short period of time, without changes in the state of the instrument. The sample or population standard deviation is then usually reported as the measure of variance. This is referred to as a *static precision* test (48).

In the course of making measurements on multiple targets and wafers, there will also be a small drift in focus, illumination, position of the image with respect to the CCD array, placement of a wafer on the instrument, etc. By exercising many complete cycles of measurements, the contributions to nonreproducibility of both random events and drift in subsystems may be evaluated. This is referred to as a *dynamic precision* test (48). An interlaboratory long-term study of reproducibility may also be conducted along the same lines. Substitution of an operator, creation of a measurement sequence, or preventive maintenance may be a part of such a test.

Occasionally, some unexplained events may grossly corrupt a measurement, resulting in an outlier. Outlier removal (49,50), an inherent part of metrology practices, is based on an a priori expectation of reproducibility. The culling of discrepant data, outliers, is made on the grounds of the statistics, and it is applied to measurements of the same quantity.

One related property is an expectation of the sameness in centerline estimates made on the multiple samples of one feature and selected along its length. This property is expected even before a measurement is made; that is, it constitutes a priori information. Estimates of centerline variation along a feature are used in error diagnostics and data culling (6–8,19,46,47). A similar approach is taken in CD metrology (23,25–27,52). These forms of error diagnostics and culling are commonly available in CD-SEMs, if not readily accessible by the tool user.

In some important applications cases, adequate precision may be difficult to achieve. Considerations in tool-limited precision are outlined in Sec. V.E. The improvement of precision in applications is described in Sec. VI.D.3. Precision limitations of optical O/L metrology are described in Sec. VII.A.1.

## B. Symmetry

As stated in Sec. II.C, our instruments are already hard pressed to resolve the fine structure in feature sidewalls. Often (see Fig. 8b) these instruments are used in applications where a direct detection of edges is impossible. The current technology and business practices in metrology of registration, alignment, and O/L metrology rely on an *implied* symmetry (6–8,19,42,46,47). Symmetry of the target, the position sensor, and the measurement procedure is expected and required for the accurate conventional metrology of image placement.

However, while the concept of symmetry is simple, its implementation and enforcement are not. Consider the human factor. A metrologist cannot quantify, just by looking at an image on a display, that the image asymmetry will result in a 10-nm error (see Figs. 12, 13, 16, 17). A human will have a low detection rate when, in 1% of all measurements, a centerline estimate is affected by a particle near an edge of the target (19). The number of phenomena and parameters that may lead to detectable image asymmetry is large, and every aspect of device manufacture and metrology is a potential culprit. A single deviation may, directly or through interactions, lead to inaccurate metrology. In order to assess such metrology problems, new computer-assisted methods are required.

Symmetry, more than any other property, affects the accuracy of O/L metrology and alignment. A high degree of symmetry, and of tolerance to unavoidable asymmetries, is essential for an O/L metrology tool, image formation, image transfer, and subsequent processing of targets. One of the essential tasks of product development and process integration is to account for all aspects of the O/L error budget, including the error budget of O/L metrology, and to reduce their impact. Much of this chapter is dedicated to improving symmetry as a means to achieving accurate O/L metrology.

Symmetry is the property we expect to be present in the O/L measurement targets and the property that must be preserved by metrology tools. Symmetry is expected before a measurement be made, that is, it is also considered a priori information. Reliance on this form of a priori information (see examples in Figs. 8, 9, 13, 16, 20) permits effective error diagnostics, efficient data culling, and error reduction (6–8,19,42,46,47,52–57).

Unless O/L metrology is attempted with SEM, image asymmetry of a "CD-SEM" image is still given little attention. However, both the image and metrology differences for $X$- vs. $Y$-axis (anisotropy) are already tested (58) and controlled. The more recent development of SEM image analysis for lithography process diagnostics (59) has led to clear demonstrations of its utility in process integration, fueling a rapid proliferation of image analysis capabilities and improved process control in applications (60–68). This serves to show that the computerized analysis of the images used by metrology systems can efficiently produce characterization of the manufacturing process, enabling better process control, higher productivity, and an accelerated rate of technology learning.

As to optical O/L metrology, sources of tool-related asymmetry and their impact on metrology error, as well as methods used for testing accuracy, are described in Sec. V.F.1. Tool sensitivity to process related target asymmetry is in Sec. V.F.2. Typical processes and their impact on target asymmetry and inaccuracy of centerline estimation, error diagnostics, culling, and process integration are described in Sec. VI. The limitations of the optical technology in applications are discussed in Sec. VII.A.2.

## C.  Redundancy

Redundancy is an essential requirement of microlithography (6–8,19,46,47). It is the preservation of redundancy that enables this extreme form of mass production of ICs. A single reticle may contain a billion ($10^9$) individual features. All devices are processed extremely uniformly, resulting in a superb control of feature linewidth and centerline. Metrology of image placement on just a few strategically placed features ensures image placement (6,14,15,17,19,21,69–75) for all other features linked to them by redundancy. The same is true for CD control and the metrology of linewidth.

In microlithography of integrated circuits, registration (and redundancy of the centerline definition) is preserved very well. Redundancy is expected to be present in the O/L measurement marks even before a measurement is made. This form of a priori information about the measurement marks is very useful as a means of error diagnostics and culling (6–8,19,42,46,47,53–55). It is also an effective tool for optimizing the O/L target design and placement on difficult processed layers (6–8,19,46,47,53–55).

Applications of redundancy in metrology of image placement are illustrated in Figs. 13, 14, 16, and 20. The use of redundancy to improve the precision and accuracy of the metrology of image placement is described in Sec. VI. The mechanisms of the intrafield redundancy failure are covered in Secs. VI.A–VI.C.

## IV. SENSORS USED IN THE METROLOGY OF IMAGE PLACEMENT

### A. Optical Systems

Optical instruments are most commonly used today for long-distance $X/Y$ metrology, alignment, and O/L metrology. Optical systems (76,77) may be separated into two large groups: those detecting the image in the spatial coordinate domain, and those detecting the image in the Fourier (spatial frequency) domain. Advantages in simplicity, robustness, and speed have made the latter quite common as an alignment system (44,45,78,79). They are also used in a number of applications in dimensional metrology (80,81). The development of new sensor technologies of this type is ongoing in both academic and industrial groups.

In this chapter, we wish to keep the O/L metrology closely linked to the standard definitions of edge, linewidth, and centerline. Since these definitions are in terms of the spatially localized properties, our review of the metrology of image placement is limited to those systems that detect a small-scale target and produce an image in the spatial coordinate domain. Accordingly, we limit our treatment to various forms of microscopy.

### B. Scanning Electron and Other Small Probe Microscopy

It is often said that the high spatial resolution achievable with a SEM is not required for precise and accurate optical O/L metrology. That is true to some extent: it is possible to achieve high-performance position metrology with low-resolution imaging sensors.

For example, it is possible to build O/L targets composed of bars with equal linewidth defined by isotropic etch in a single thin film. When the primary error mechanisms typical in real applications are suppressed in such a way, metrology errors become small, on the order of 1 nm (82).

However, in order to state the accuracy of optical O/L metrology in real applications, all sources of error must be assessed (83). This may require the use of other metrology systems for sample characterization. Often, an entirely different metrology instrument (84,85) must also be brought in to overcome the resolution- and application-related limits of the conventional optical O/L metrology systems.

Scanning electron microscopes may have the advantage of resolving the fine details of the sidewall in a feature whose CD is being measured. This SEM capability that made them so useful in CD metrology (26,58–68) also makes them useful in O/L metrology (84–89). The different physics of image formation and of interaction with the sample also helps the application of SEM-based O/L metrology for the certification of accuracy.

Using a single tool and edge-detection criterion for both CD and O/L metrology permits linking these forms of metrology, so there is no ambiguity in estimating the edge-to-edge overlay required by the device design rules. These are the basic considerations regarding the merits of SEM-based O/L metrology, as reflected in the *International Technology Roadmap for Semiconductors* (3).

Comprehensive reviews of SEM-based dimensional metrology are available (see Chapter 14, by M. Postek and A. Vladár, in this volume; also see Refs. 90 and 91). SEM-based dimensional metrology has rapidly developed in recent years. At this time, it is the tool of choice in CD metrology. In order to develop a superior CD metrology capability before this one becomes limited, and to gain a better account of the feature sidewalls by a nondestructive technique, other forms of small-probe scanning microscopy are being developed (see Chapter 14 in this volume; also see Refs. 90–93). Evaluations of SEM and other forms of scanning probe microscopy for O/L metrology are ongoing (84–89). If proven accurate and practical in applications, they may become the next O/L metrology tools of choice.

## C. Electrical Probe

Electrical probe-based systems have been successfully applied to the metrology of image placement (94). In evaluations of the image placement capabilities of lithographic printers, they deliver copious amounts of high-quality data at a minimal price. Electrical probe–based metrology has also become indispensable as a means of quality assurance.

Some of the differences in electrical CD measurements vs. microscopy-based measurements may be traced to the spatial and material properties underlying the standard definitions of edge, linewidth, and centerline. Comparative analysis of the errors in O/L metrology (95,96) enabled greater accuracy to be achieved by improving the target symmetry. A comprehensive review of the recent development to enhance performance of this method and a list of essential references is available (see Chapter 16 in this volume).

## V. OPTICAL OVERLAY METROLOGY

In this section, we review the essential elements of the conventional optical metrology of image placement and the errors attributable to metrology systems and metrology structures. It is important to keep in mind that, while our treatment is in the framework of the conventional optical O/L metrology applied to present-day microlithography and thin-film device processing, these metrology methods are quite general.

## A. Overlay Metrology Structures

Many structures are suitable for O/L metrology. Some of these are compatible with at least some available metrology systems. For a given metrology purpose, some of these structures are better than others. This author found many advantages in 1-D O/L metrology structures designed to measure O/L in one axis at a time (6,42); see examples in Figs. 1, 13, 14, 17, and 21. However, this metrology may be slow, and, because various O/L metrology systems are somewhat different, not every one of them can handle them with same ease.

The users of metrology equipment expect some commonality across all available systems. It is in users' best interests to have a choice among the substantially compatible metrology systems. In order to promote system-to-system compatibility, competitiveness, and efficiency of our industry, SEMI described (9) designs of the most commonly used O/L metrology structures.

An O/L metrology system is expected to handle the SEMI standard O/L metrology structures whose designs are illustrated in Figure 10. They comprise a set of the target- and bullet-level features. These features are defined mirror symmetric, concentric, and isotropic and sized to fit in a field of view of a typical optical O/L metrology system.

The simplest possible design is a box-in-box structure, shown in Fig. 10A. It is made up of a single solid square shape for both the target and bullet layers. For a given value of edge-to-edge distance between the target and bullet features, this is the most compact

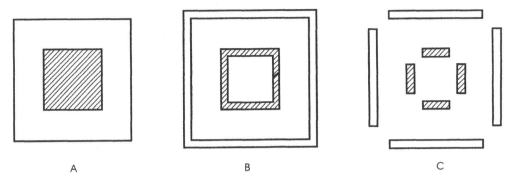

**Figure 10** Common O/L metrology structures: box in box (A), frame in frame (B), and bars in bars (C).

design. A 1-D analog of this structure, bar in bar, is illustrated in Figs. 1, 13a and b, and 17a and b, as well as in devices and SEM O/L metrology structures shown in Figs. 14 and 21. An image of a 1-D or 2-D structure with just two edges is exemplified by the images of the center bar of the target from Figure 13a, shown in Figs. 13c and d. This structure supports both statistics- and symmetry-based diagnostics and culling (6–8,19,42,46,47,53–55).

Both the frame-in-frame design of Fig. 10B and the bars-in-bars design of Fig. 10C have two pairs of edges per axis per layer (see examples in Figs. 8, 16, and 20a; also see the outer bars targets in Figs. 13a and d, 14, and 17a and b). Both designs support symmetry- and statistics-based diagnostics and culling. They can also support redundancy-based diagnostics and culling. In some cases, it may be possible to recover a measurement (19,46,47). In addition, these targets may be used to improve measurement precision by averaging the redundant centerline estimates (19). The differences between the frame-in-frame and bars-in-bars designs are small and process specific. For example, a set of bars on the wafer coated with photoresist or chemical-mechanical polished may lead to less asymmetry and thickness variation in the target and its vicinity.

Although SEMI described the designs of the commonly used O/L metrology structures, the issues of applicability and performance of O/L metrology on targets built from these designs are left to the user (9). Consider the applicability of the common O/L metrology structures, typically designed as wide chrome-on-glass (COG) lines or spaces, at a layer with very small critical dimensions. Since the image placement errors in real lithography systems vary as a function of feature width and polarity (see Sec. VI.B), registration errors recorded by these O/L metrology structures will be different from those of devices. What should the metrologist do? As a member of the team responsible for the device O/L budget, the metrologist needs to learn how much of the relative error of placement will be incurred. When the dissimilar features used in the product move too much with respect to each other, the device O/L budget may be impossible to maintain, and the team will work to reduce these errors. To deal with the remainder, the metrologist may measure O/L on device structures (for SEM-based O/L metrology, see Sec. VIII.B), design productlike O/L metrology targets, or use the conventional ones and occasionally estimate how much error is incurred. A similar dilemma is bought about by the use of a phase-shifting mask (PSM) or optical proximity correction (OPC). When patterned, these mask features may be displaced (see Sec. VI.A) with respect to isolated COG features used in O/L metrology targets. A solution to this problem is found through the assessment of all sources of O/L error and through error control at its source. This is supported by conventional O/L metrology, with an occasional validation of the O/L budget and model by whatever means is required.

The ability to measure O/L precisely and accurately on conventional O/L metrology structures has also been scrutinized. In the recent literature (43), there are numerous studies of how the target design, polarity, and linewidth affect the robustness, precision, and accuracy of alignment and O/L metrology. For example, some users reported (97,98) that a box-in-box structure leads to exacerbated asymmetry and variation in the O/L metrology structures; they concluded that for their applications a frame-in-frame or bars-in-bars structure is better. Other users (99,100) have conducted detailed performance-based comparisons of many possible design choices to pick their best target. It is typical for the users to pursue multiple target designs and placement at a single layer (see examples in Fig. 16d), to reduce the risk of gross problems and to accelerate metrology learning in a product environment.

What is the best O/L measurement structure? The user must establish that in the course of O/L metrology budgeting and metrology integration. The methods of metrology error diagnostics and integration make this task easier; see Sec. VI.D.

## B. Optics

A typical O/L metrology system is a bright-field polychromatic microscope. Illumination bandwidth is usually selected between 400 nm and 700 nm. Both single-band and multiple user-defined illumination bands are available. Stable broadband light sources—W halogen and Xe arc—are common. These partially coherent optical systems use Köhler illumination with filling ratio $\sigma > 0.5$. The primary measurement objective may have a numerical aperture (NA) of 0.5–0.95, depending on the model and the system configuration.

These optical parameters of the position sensor may affect various aspects of system performance. Examples: the width of the illumination band strongly affects the measurement accuracy (98,101,102); a larger filling ratio reduces the sensitivity to grain and to some types of target asymmetry (103) and improves system robustness and accuracy when viewing through clear asymmetric films. With other parameters kept the same, a system with a higher NA yields a better correlation of the O/L data taken in the developed image to those in the final image (104). Consider the differences in the two observations shown in Fig. 14 for options to improve viewing of targets in Figs. 14, 16, 17a and b, and 20a.

Illumination wavelength and bandwidth, numerical aperture, filling ratio, uniformity of the source in the pupil plane, and other parameters of the optical systems are subject to manufacturing tolerances. These variations affect both the job portability and tool-to-tool matching in systems of one design. When these optical parameters differ by design, not just due to tolerances of manufacture, tool-to-tool or another kind matching may be problematic; see Sec. V.F.

The early optical O/L metrology systems were equipped with multiple objectives. That was required for their use for inspection and CD metrology. As the SEM-based CD metrology displaced optics-based CD metrology, optical metrology systems evolved and excelled in O/L metrology. Typically, they now have a single measurement objective, which is designed expressly for O/L metrology. In such objectives, asymmetric aberrations are minimized in the design and manufacture, as well as through image quality–based performance tests and the selection of the best available units. Both the illumination and imaging optics have been simplified, leaving a minimal number of folds and moving parts. They were designed and built for maximum symmetry and long-term stability. In addition, vendors developed and implemented automated performance tests and maintenance and qualification procedures to keep these tools at a top performance level.

The more recent optical O/L metrology systems feature a superior conventional bright-field polychromatic microscope, with options for phase contrast imaging or interference microscopy. When applied to the more difficult process layers (105), such a system may use an intensity, amplitude, and phase image of an O/L metrology target.

## C.  Focus and X/Y Coordinate Linkages

In the course of O/L metrology, it is often necessary to acquire images of the target and bullet portions of a measurement structure at two different focus planes. In doing so, three different types of inaccuracy may be introduced.

Dual focusing may result in a few nanometers of X and Y translation for every micron of movement in Z. In order to reduce this error, manufacturers of metrology systems can improve the hardware or introduce a feedback loop or calibration. Much improvement has been made in this area over the past few years. Correction of this error may also be made at the cost of lower throughput (see Sec. V.F.1 on TIS). To enable further progress in stage technology, NIST is developing a stepped-microcone structure (106).

Focusing is also associated with two error mechanisms that affect the imaging of a sample rather than the mechanics of its positioning. Defocus may change the impact of asymmetry in the imaging system on the measurement error (see Sec. V.F.1 on TIS). Defocus may also change the impact of asymmetry in an O/L measurement structure on the measurement error (see Sec. V.F.2 on WIS). Automated error diagnostics using a through-focus sequence are now widely available.

## D.  Digital Signal Processing

SEMI standard O/L measurement structures fit in a typical field of view of between $30\,\mu m \times 30\,\mu m$ and $50\,\mu m \times 50\,\mu m$. Unlike the earlier systems, which used a scanning slit and a PMT detector, the newer systems use CCD cameras. Because CCD devices are spatially uniform, distortion free, and stable over long periods of time, the introduction of CCD cameras improved both the accuracy and reproducibility of O/L metrology systems.

The image of an O/L metrology or alignment target projected onto CCD is detected by a rectangular array of small detectors (pixels). The signal from each pixel is amplified and converted from analog form into digital (A/D conversion). The estimation of edge or centerline coordinates may be carried out via a variety of digital signal processing (DSP) algorithms. These algorithms may have a significant impact on performance. To properly use a metrology system, a metrologist must understand how a measurement is made. Although the algorithms may be closely guarded by some vendors for competitive reasons, a metrologist must know which algorithms are used and how.

Some of the differences between CD and O/L metrology are evident in their DSP and in how they use calibrations and standards. For example, in CD metrology (23–27) a distance between positions of suitable transitions in an image of a sample is compared to that in an image of a certified reference material (standard). In this way the calibration of measurement can be made to arrive at an accurate CD estimate. In this sense, all CD metrology systems are comparators. Critical-dimension metrology requires an essential similarity of all germane properties of a sample being measured and the standard, as well as similarity of the image of the sample and standard. Centerline estimation in conventional metrology of image placement, on the other hand, does not rely on such comparison, such calibration, and such use of standards. The accuracy of O/L metrology is ensured by maintaining the symmetry of the sample and of the metrology system, including data acquisition and DSP.

In centerline estimation, the image of a sample is compared with a symmetrical replica of itself, not with an image of a standard. The extreme robustness of the metrology of image placement in applications is supported by appropriate DSP. The image contrast, polarity, number of fringes, and fringe separation may vary from one target to another, without necessarily impacting the precision and accuracy of the metrology of image placement. To achieve that, the algorithms used for centerline estimation rely on a priori expectations of symmetry, statistical cohesiveness, redundancy, culling, and various methods of error recovery.

## E. Precision

The precision of optical systems depends on the numerical aperture (NA), the filling ratio ($\sigma$), the center of the illumination band, and the bandwidth; on the spatial sampling rate, the electronics noise, and the digital representation of its intensity; and on the DSP algorithms used in centerline estimation. Hardware-limited precision of new O/L metrology systems has improved from $3\sigma = 30$ nm at the end of the 1980s to $3\sigma = 1$ nm today. That is, when the target quality is not the limiting factor, many commercial systems perform at this level in both static and dynamic precision tests. This is typically achieved on O/L metrology targets formed in photoresist and many etched CVD films.

In addition to properties of the imaging optics, spatial sampling, errors of A/D conversion, and electronics noise (107), the signal-to-noise ratio (SNR) of an image of a real O/L metrology target is a strong function of the applications. In some cases, target- or sample-limited precision may be so poor as to render metrology useless (see Secs. VI.D.3 and VII.A.1.a).

In order to improve the precision on such targets, it is desirable to use an optical imaging system that produces an image with strong signal (peak-to-valley) and high-edge acuity (sharp, high first derivative of normalized intensity). In addition, selecting a larger total combined length of target edges can be used to improve the effective SNR. In this case, a bars-in-bars target may be preferred over a box-in-box target. By using four edges in a bars-in-bars target, rather than two edges of a box-in-box target, it is possible to improve the target-limited precision by a factor of 1.4 or even better (19) (with culling).

## F. Accuracy

The accuracy of O/L metrology was taken for granted until a metrology crisis in the late 1980s, when many users reported systematic errors in excess of 100 nm. Since that time, a wide proliferation of new methods of error diagnostics enabled rapid improvements, changing users' perceptions of reality (108). Nevertheless, large errors are still reported in both alignment and O/L metrology in applications on what is called "difficult layers."

Some sources of error are clearly attributable to the O/L metrology system. They are systematic and do not change as a function of application. Examples of such errors are the asymmetric distortion of an imaging system and image translation in the $X/Y$ plane when refocusing. These errors may be reduced and/or compensated for by calibrations. Vendors of optical O/L metrology systems made much progress in reducing such errors.

However, by far the largest errors still occur in metrology on realistic processed O/L targets. These errors depend on the optical technology used in the tool. For a given tool type, these errors depend on the quality of tool manufacture and setup. They also vary due to the interactions with the adverse, but unavoidable, effects of processing on the O/L measurement structures. Complexity and variability of such effects lead to a situation where it is difficult to foresee all possible error mechanisms and to design a tool configuration that either is insensitive to or compensated for all error sources.

Calibrated realistic (difficult to measure) O/L reference materials are useful in evaluating O/L metrology equipment. Such reference materials and tests of the optical O/L metrology systems are described in Sec. VII.C. However, owing to the variability of the process-related errors, it is impractical to use calibrations as a means to gain accuracy in O/L metrology. Unlike CD metrology, O/L metrology has many attributes of a null test (109). Its accuracy is based not on comparisons to an O/L standard, but on target symmetry. As with any measurements, O/L measurements may be made more accurate by the characterization and removal of error sources. New error diagnostics methods have been developed. They are easily understood, easy to implement, and very effective.

For example, we consider overlay to be a physical property of an O/L metrology structure. Therefore, we hold that the value of O/L is unique. It then follows that our estimates of O/L must not change as a result of a change in the measurement method. Therefore, when the O/L estimates obtained in equivalent, but different, measurement procedures are discrepant, these discrepancies are interpreted as errors of measurement. Once a proper correction is made, it is possible to verify having made an improvement. This process results in measurements that are more accurate. The same approach (109) could be used to build the pyramids in Egypt or to set a 90° angle on a table saw. Highly accurate metrology of image placement can be achieved without calibration to an O/L standard.

## 1. Tool-Induced Shift (TIS)

In order to estimate the impact of tool asymmetry on measurement error, a simple self-consistency test was devised (7). For example, assuming that we are in the wafer coordinates (110), O/L measurements made at 0° and 180° orientation with respect to the tool must be the same, that is, $O/L_{0°} = O/L_{180°}$. A discrepancy in these two estimates of O/L is a measure of O/L metrology error. We call this type of error *tool-induced shift* (TIS) (also see Ref. 111):

$$\text{TIS} = \frac{O/L_{0°} - O/L_{180°}}{2}$$

When an estimate of TIS is available, it is possible to remove this error from the O/L measurement. We define (in a wafer coordinate system) a TIS-corrected O/L estimate as an average of the measurements made at 0° and 180° orientation with respect to the tool:

$$\text{COL} = \frac{O/L_{0°} + O/L_{180°}}{2} = O/L_{0°} - \text{TIS}$$

Many metrologists have demonstrated that this calibration for TIS and reduction of TIS are a very effective means of improving system-to-system matching.

The example (8) illustrated in Figures 11a and b shows that matching between the systems of the same type may be improved from $3\sigma = 54.3$ nm to $3\sigma = 8.7$ nm simply by calibrating for the TIS error of metrology systems. Another example (8), illustrated in

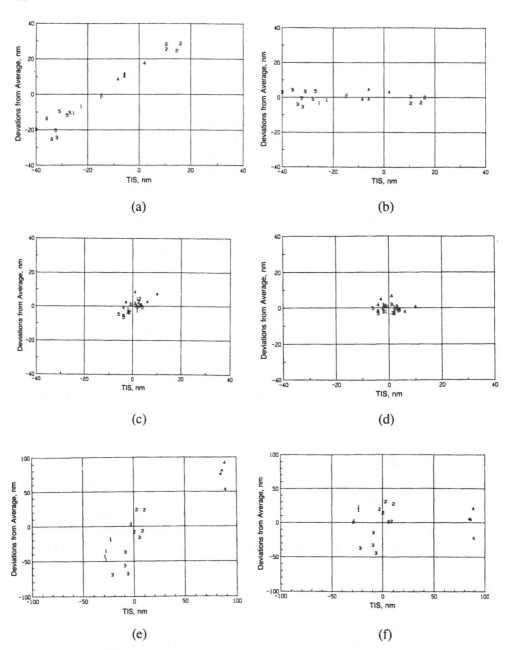

**Figure 11** TIS as the cause of the tool-to-tool mismatch in the systems of the same type (a) and compensation for TIS as a means to improve tool matching (b). TIS as measure of performance: When TIS is small, tools of the same type are well matched (c, d). Mismatch of dissimilar systems on the same sample is only partially correlated with TIS (e), and correcting TIS does not result in matching (f). This suggests that calibrating for TIS is not sufficient to make accurate measurements.

Figs. 11c and d, shows that when the TIS of these systems is small, matching is good even without TIS calibration.

One source of TIS in O/L metrology systems is the lateral stage move when focusing (see Sec. V.C). Asymmetries of the illuminator and imaging optics are the other sources (8). Figures 12a and d illustrate them for the case of a bright field NA = 0.7, $\sigma = 0.7$ monochromatic microscope used to view a 370-nm-deep trench mark under 1-μm resist.

The concepts of TIS as an estimate of errors due to tool asymmetry and of COL as an O/L estimate compensated for this error are simple, and associated procedures are easy to apply. Since its introduction, TIS has become an industry standard measure (108) of tool performance. Automatic procedures to estimate and calibrate for TIS have been implemented on all O/L metrology systems. A number of experimental aids were developed for TIS evaluation, removal by calibration, tool setup, and maintenance (6–8,42,47,106,112–119); one of the earliest aids of this kind is illustrated in Figs. 13a and d, and 17a and b. The users and manufacturers of optical O/L metrology equipment collaborated on reducing this form of error. Experimental assessments and the modeling of metrology error sources brought about a superior understanding of O/L metrology. Systems designed and built exclusively for O/L metrology emerged. As a result, TIS was reduced from as much as 100 nm in 1989 (6–8) to under 5 nm by the mid-1990s.

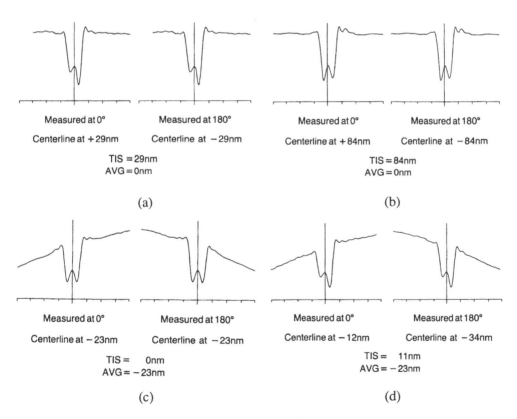

**Figure 12**  Some sources of TIS and WIS: asymmetric illumination (a), coma of imaging objective (b), asymmetric resist over sample (c), asymmetric resist and illumination (d).

Tool-induced shift as a measure of tool performance, the associated tests of accuracy, and the methods to improve accuracy are applicable to alignment and registration metrology. Improvement of alignment accuracy was reported (47,112,119).

As useful as these procedures are, TIS estimates on a few samples do not yield an exhaustive assessment of the error due to tool asymmetries. As many metrologists discovered, a tool that had a given level of TIS in performance testing might fare worse on some new samples. A related observation is that adjusting a tool for the lowest TIS on one layer may result in higher TIS on another layer. The more comprehensive tests of tool asymmetry are introduced in Sec. V.A.3.

Some error mechanisms may result in poor accuracy, even when the tool is symmetric or corrected for TIS. Consider the example (8) of tool-to-tool matching in Figs. 11e and f. Four metrology systems of dissimilar configuration measured the same sample as in Figs. 11a and b. Although corrected for TIS, dissimilar systems did not match as well. That is because some other error, different for each of them, was left uncompensated for (see WIS, next section).

Tool matching is a measure of similarity between the systems. It is a necessary but not sufficient condition for accurate metrology. Since the 1992 upsurge in user confidence (108), it is now widely recognized that low TIS is required for alignment and O/L metrology systems and that TIS is not a panacea for errors in the metrology of image placement.

## 2. Wafer-Induced Shift (WIS)

The notion of wafer-induced shift (WIS) was introduced (7) to account for the errors due to asymmetry in the measurement structures. By WIS, we mean an error that is attributable to some asymmetry borne by the sample. This error would be incurred by a perfectly symmetric O/L metrology system. Such an error is exemplified by a symmetric tool on a symmetric target coated with asymmetric resist film, often resulting in an asymmetric signal, such as is illustrated (8) in Fig. 12c.

The technology and specific configuration of an instrument largely predicate the magnitude of WIS. For a particular type of sample asymmetry, one system type may be generally less liable to err than another. The selection of a detection criterion, focus, and other parameters may also affect this type of error. In addition, for a fixed measure of sample asymmetry, the WIS of a perfectly symmetric system varies as a function of such parameters as film thickness, defocus, and illumination band.

Unlike TIS, errors due to asymmetry in a wafer structure (WIS) may not be easy to estimate. However, physical properties of the O/L measurement structures can be characterized (6,102,120) (see examples in Figs. 17c, 18, and 19e and f). Wafer-induced shift may then be evaluated by modeling the impact of sample asymmetry on viewing and on the resulting metrology error (6–8,121–125); see examples in Figs. 12b and d.

Both TIS and WIS are very simple notions, too simple to account for all sources of inaccuracy. Many errors in O/L metrology defy this classification. Metrologists are urged to dwell less on the labels and more on the substance, to work with explicit physical observations of the error mechanisms.

## 3. Separation of Sample- and Tool Asymmetry-Related Errors

In order to gain accurate O/L metrology, we must identify an occurrence of asymmetry, attribute its source, and either to compensate for it or to reduce its magnitude. To do so, it is necessary to separate (8) asymmetry in the images of target and bullet features of an O/L metrology structure into parts that are attributable to a tool and to a sample.

These goals are addressed by a direct analysis of the target image (42), one level at a time. Consider a Taylor series expansion of the image intensity in the vicinity of the left and right edges of a line:

$$I_L(x) = I(x_L) + I'(x_L)(x - x_L) + \tfrac{1}{2}I''(x_L)(x - x_L)^2 + \cdots$$

$$I_R(x) = I(x_R) + I'(x_R)(x - x_R) + \tfrac{1}{2}I''(x_R)(x - x_R)^2 + \cdots$$

where $I_L(x)$ is image intensity in the vicinity of the left edge whose position is estimated as $x_L$, $I'(x_L)$ is the first derivative with respect to $x$, $I''(x_L)$ is the second derivative, etc. Comparisons of the coefficients in the Taylor series expansion yields a complete set of required measures of asymmetry. For example, the value of

$$\frac{|I(x_L) - I(x_R)|}{I_{MAX}}$$

may be considered a 0th-order measure of asymmetry. This measure of asymmetry in the alignment signal was found useful for error diagnostics and culling (46,47,112). The 1st-order measure of asymmetry may be expressed as

$$ASY = \frac{[1/I'(x_L) + 1/I'(x_R)]\Delta I}{2}$$

where $\Delta I$ is an increment in the image intensity threshold. The physical meaning of this measure is the shift in an apparent position of a centerline, estimated at two image intensity thresholds separated by $\Delta I$. To keep the convention simple, we use signal peak-to-valley or 100% as this increment:

$$ASY = \frac{[1/I'(x_L) + 1/I'(x_R)](I_{MAX} - I_{MIN})}{2}$$

Such measures of image asymmetry (ASY) may be acquired at both 0° and 180° orientations of a wafer with respect to the metrology system. In a manner similar to TIS and COL (but in the coordinate system of the tool), we define a tool-related asymmetry (TASY) and sample-related asymmetry (SASY):

$$TASY = \frac{ASY_{0°} + ASY_{180°}}{2}$$

$$SASY = \frac{ASY_{0°} - ASY_{180°}}{2}$$

Experimental study (42) reveals a strong dependence of variation of TIS on TASY. This experiment was designed so that the observed change of TIS were due solely to the asymmetry of the imaging system, but not to the stage shift incurred when refocusing. The O/L measurement targets illustrated in Figure 13a were printed and processed to ensure the best symmetry. At the same time, the depth of Si etch was selected near 100 nm, 200 nm, etc. to result in either high or low contrast and to change the phase and amplitude content in the mark's image formed by the metrology system. The linewidth of the center bar in 1-D O/L metrology targets was changed from 0.7 µm to 2 µm, while the outside bars were kept at 2 µm. The TIS of the O/L measurements was reported by the metrology system. And ASY and TASY also were estimated from O/L data, using 30% and 70% edge detection thresholds on the center bar to approximate the first derivative. Experiment demonstrated that the variation of TIS is strongly dependent on the variation of TASY; see Fig. 13e.

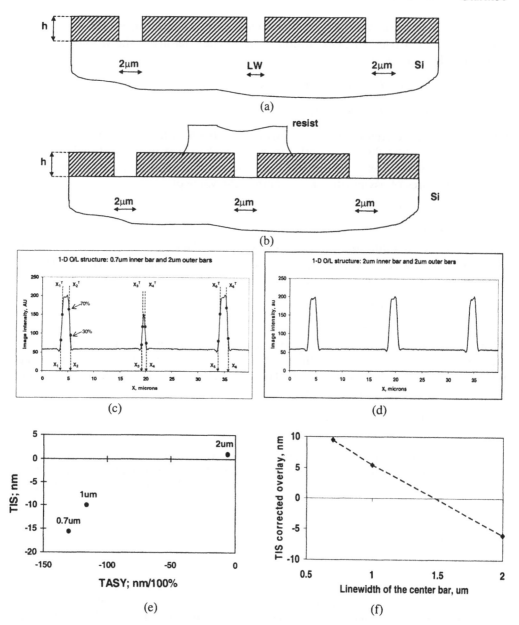

**Figure 13** 1-D O/L metrology mark in (a), with three open bars being 2 μm, cannot lead to large TIS (e), even though the system is asymmetrical, because the images of three bars are similarly asymmetrical (d). When the center bar in (a) is 0.7 μm, this structure does lead to large TIS (e) because imaging and TIS of the inner vs. outer bars are not the same (c). Target (a) is printed in one layer. Position variation (f) of the center bar vs. outer bars, as function of its width, is due to asymmetry in the lithography printer (see Section VI.B.1.).

In many cases, ASY and TASY provide a more useful measure than TIS of the tool performance on a target being measured. When the illuminator is misaligned, a symmetric O/L measurement structure may result in an asymmetric image (see Figs. 12a, and 13c and d). When both the target and bullet portions of an O/L measurement structure are similar, their images may be asymmetric (Fig. 13d). However, two individual nonzero TIS values cancel each other (8) in an evaluation of TIS, resulting in TIS = 0 (Fig. 13e). On the other hand, ASY and TASY may still provide useful measures of the asymmetry of the imaging system.

Furthermore, it appears that once a dependency of TIS on TASY is known, the value of ASY is sufficient to calibrate for TIS during the measurement on symmetric targets. This correction of centerline estimates can be done with ASY estimates derived from the images of O/L metrology structures in both the target and bullet levels, one at a time.

The same approach may be exercised for the detection of the impact of sample asymmetry on the inaccuracy of centerline estimation, that is, on WIS. For example, when a narrow-trench target is observed with a symmetric bright-field microscope, estimates of image asymmetry due to resist asymmetry (SASY) can be correlated with inaccuracy. Once this dependence is known, ASY estimates obtained for a symmetric imaging system may be used to calibrate for WIS. For example, by adjusting the wavelength of light until the estimates of ASY are reduced, the measurement error is also reduced (126). Unlike passive alignment and metrology systems, an active monochromatic bright-field system can be accurate on targets covered with asymmetric resist.

## VI.  METROLOGY ERRORS IN THE MANUFACTURING ENVIRONMENT

Device design rules treat CD and O/L as if they were independent random variables. This may be a matter of tradition and convenience, but CD and O/L are neither random nor independent. To illustrate this point, consider the  metrology test site (8) shown in Figure 14. This test site consists of a device array, SEM and optical O/L metrology structures, isolated lines, and gratings. The last image was printed in 1-μm single-layer resist (SLR) coated over the topographic features of the substrate. The photograph in Fig. 14a was taken in a bright-field microscope with NA = 0.2 set up for coherent illumination with the 10-nm band centered at 546.1 nm. The photograph in Fig. 14b, on the other hand, was made with NA = 0.9 set for incoherent illumination with white light.

One can observe that the image size and image placement in the SLR images vary as a function of substrate topography and reflectivity variations. Exposure monitor structures (127) (EMSs) were used here to evaluate printing conditions in various topographic environments. Consider the image of a narrow line and of an EMS (wider line) printed across an area where oxide was etched through to Si (bright square, 40 μm × 40 μm in the lower right portion of this image). Reflectivity variations in Fig. 14a clearly illustrate that the SLR is only partially planarized. Its thickness is changing over distances comparable with 20 μm. The EMS, whose printed linewidth is about five times more sensitive to exposure dose than the adjacent narrow line, shows linewidth variations typical of reflective and edge notching in SLR. It is also apparent that both linewidth and centerline of EMS vary as a function of position. The same effects are present, though hard to observe with the naked eye, in all images illustrated here.

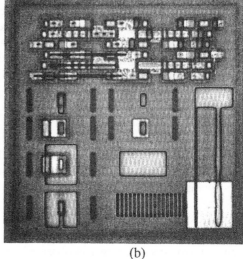

(a)                                                    (b)

**Figure 14**  Metrology test site observed with a bright-field microscope: NA = 0.2 and narrowband illumination at 546 nm (a); same site observed at NA = 0.9 and white light (b).

This example illustrates two important points. First, the same physical phenomena that result in systematic CD variations in thin-film devices may also cause systematic O/L variations. Second, both the device O/L and its metrology may be affected in similar ways.

## A.  Reticle

A reticle written by a modern mask writer may have registration errors with variation of $3\sigma < 10$ nm (at wafer magnification). This control of registration is achieved over large distances. The scale in a reticle image is maintained to within a small fraction of a part per million (ppm, or $10^{-6}$). This capability is supported by $X/Y$ metrology systems that are kept stable and calibrated over long periods of time.

While mask-writing tools produce precise and accurate registration of every feature, they are not perfect. Overlay measurement errors begin accumulating from the moment a reticle is built. For example, when multiple sets of identical O/L measurement structures are measured, O/L metrology data may be systematically discrepant by $3\sigma > 10$ nm due to mask registration errors; these discrepancies are always present in O/L metrology. In addition, when O/L measurement structures have built-in redundancy, redundancy failures due to the same causes may also become a part of the reported measurements (8,19).

A detailed study of dimensional errors in binary conventional reticles is available (128). It reports butting errors at stripe boundaries of over 100 nm (at mask scale). This work suggests that the worst-case image placement errors in reticles are much larger than would be expected, based on conventional sampling. This sampling and an assumption of Gaussian distribution of registration errors do not fully account for the systematic errors of the mask writer. Startling as these gross errors may seem, the extent of butting errors in reticule is consistent with the magnitudes of stable redundancy failures actually observed in the O/L measurement structures (6). Patterning for IC manufacture is compounded by the interactions with realistic circuit design/layout and imperfect lithography systems (129).

The implications of image placement errors in conventional reticles on device O/L and on O/L metrology are severe. Image placement errors at a stripe boundary extended over an array of devices may result in a large number of devices that exceed the O/L control limits. Since the O/L metrology budget is only 10% of permissible device O/L, the impact of such errors on O/L metrology would be much worse.

The centerline of a feature in a conventional reticle is determined by two edges in chrome on quartz (130). In the less aggressive microlithography of the past, image placement of the reticle features correlated well with the placement of the printed image. Therefore, maintaining tight control in the relative placement of such centerlines in reticles was the only requirement related to reticle registration errors.

The introduction of optical proximity correction (OPC) and phase-shift masks (PSMs) changed this paradigm (131,132). The image placement properties of a PSM reticle are strongly affected by the left–right asymmetry of transmission, thickness, roughness, sidewall, and, often, the sizing of the phase shifter. The image placement properties of chrome on quartz reticle with OPC also are affected by the symmetry of the sub-resolution-correcting shapes and adjacent features. The assurance of registration error in such reticles now requires the control of multiple parameters and increased interaction with the customer.

## 1. The Impact of Optical Proximity Correction and Phase-Shift Masks on Image Placement and Its Metrology

In the new lithography paradigm (131), complex transmission (amplitude and phase) of a reticle is modified in order to improve some aspects of the entire patterning process. To achieve enhancements, the properties of OPC and PSM reticles are inextricably linked to those of the imaging system, image recording and transfer processes, as well as the measures of patterning performance. Some examples of this kind are described in this section, and more are given in Secs. VI.B, VI.C, and VI.D. The impact of this new lithography paradigm on image placement errors, and on metrology of image placement, is profound.

Consider those cases of COG (chrome-on-glass) reticle patterns, with or without OPC, where good feature fidelity at nominal dose and focus is accompanied by lateral displacement when defocus occurs. Although the reticle and projection system may have no errors, defocus and interference of light are enough to introduce large image placement errors in modern aggressive ($k_1 < 0.5$) optical microlithography.

Let us also note here that for some structures in COG reticles (designed as a test of asymmetry), the impact of asymmetric illumination or coma on printed features may be quite large. In one report, $0.2\lambda$ of coma in a lens with NA = 0.28 using G-line (436-nm) illumination was shown to lead up to 300 nm of lateral shift (133). In a more recent test (134) of a KrF (248-nm) system with NA = 0.55, a lateral shift of over 35 nm was observed in a target of similar design.

Consider also the example of strong PSM (altPSM) used to improve CD control at the gate level of CMOS devices. While CD control improves on most device structures, asymmetric layouts in PSM may exacerbate the effects of coma in projection optics and also result in gross CD mismatch in "left–right" device pairs (135).

The mismatch of transmission or phase-shift error, such as due to manufacturing tolerance, may result in significant errors of image placement, especially with defocus. A large phase error results in image displacement so severe that it warrants the use of a PSM as a focus monitor (136). Image placement of a pattern printed with 90°/270° altPSM changes (137) by about 100 nm as a result of 1-μm focus move in a 248-nm projection system with NA = 0.5.

A pragmatic application of altPSM requires careful attention to phase shifter errors. Both experiment and models are used today to estimate the impact on image placement in the next generations of devices (138–144). Such learning, done well ahead of time, allows our industry to anticipate the need and to drive the efficient development of PSM manufacturing capabilities. The required control of the phase shifter may be estimated (144) for the required image placement performance, so it may be made to this specification, and assured, during PSM manufacture, see Figure 15a.

What does this imply for the metrology of image placement? First, considerations of the accuracy and utility of the metrology of image placement in microlithography applications is becoming strongly linked to the type of features used in the measurements. Metrologists will be increasingly more involved in characterization, attributing to proper error sources and accounting for all sorts of image placement errors—not just errors of the metrology of image placement. Second, there is going to be a strong drive to replace "typical" (SEMI standard) O/L metrology structures with those that are more representative of the product. When this is not practical, they may be replaced with those that are less affected by the many sources of image placement error, with supplemental metrology of devices or devicelike structures. Third, demand for O/L metrology on devicelike O/L metrology structures and on real devices will increase, driving refinements of small-probe microscopy (SEM, AFM) for such applications. These considerations will come out often in the subsequent sections of this chapter.

## B. Image Formation

Registration errors found within a reticle are modified in the course of image formation on a lithographic printer. Registration errors in the aerial image differ from those in the reticle (at wafer scale). The simplest form of this error in image placement is a local translation varying slowly in space, known as *distortion*. Lens distortion is a well-known error of the optical projecting imaging systems, and it is tightly controlled.

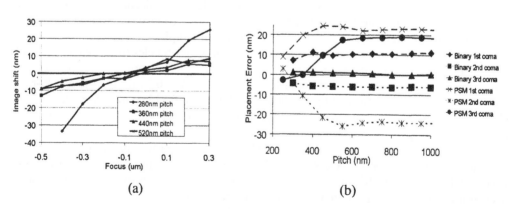

(a)                                                            (b)

**Figure 15** Experimental study of image placement errors in 80-nm lines of varying pitch due to 170° PSM phase for a lithography system with λ = 193 nm and NA = 0.6 (a). Model-based prediction of image placement errors, for 100-nm lines printed on the same system, due to coma (b): PSM vs. COG reticle

However, lens distortion alone is neither complete nor adequate to account for image placement errors at imaging. Indeed, these systems are used to print a wide range of images over a variety of exposure–defocus conditions. In such an environment, the image placement error of devices varies as a function of feature type, size, and polarity, as well of as focus and exposure conditions.

It is well known that asymmetric aberrations of the imaging optics may result in a large lateral shift (145) of the image of a pinhole or a narrow slit, even when the conventional measures of image quality are high. It is also well known that the aberrations and setup of the condenser (6,8,146–151) may lead to considerable and variable errors of image placement. While such errors of image placement may have been relatively small in the earlier generations of IC products, their fraction in device O/L naturally increased in the heavily enhanced ($k_1 < 0.5$) optical microlithography. A recent assessment (17) of device O/L (see Figs. 2 and 21) suggests that these errors are already a significant part of a 50-nm O/L budget.

By now, it has become apparent that the application of altPSM to the manufacture of advanced ICs is strongly contingent upon the reduction of the image placement error due to asymmetries of the lithography systems. As shown in Fig. 15b (from Ref. 144), for a lithography system with NA = 0.6 at $\lambda = 193$ nm used for patterning 100-nm lines, image placement in presence of $0.02\lambda$ coma may more than double for PSM over COG. Clearly, image placement errors as large as 20 nm are not compatible with the sub-50-nm O/L budget of the next generation of ICs (3). Perhaps even more disconcerting are the concurrent CD errors of this magnitude, such as "left vs. right" in device pairs and ends of array (135). In order for such CD errors not to consume the entire CD control budget (3), asymmetric aberrations of lithography systems must be reduced.

The full impact of asymmetric aberrations and of illuminator misalignment on image placement and on its metrology in aggressively enhanced microlithography (6,17,134–144, 146–151) is only beginning to be felt.

## 1. Overlay Metrology and Process Integration

In today's lithography environment, device CD is under 250 nm, but a typical alignment or O/L metrology target may be as wide as $2\,\mu m$. Owing to the differences of their printing, there may be considerable discrepancies of image placement between the O/L metrology structures and the devices. Consider an example (6) of a lithography system with NA = 0.5, $\lambda = 248$ nm and $\sigma = 0.5$ whose illuminator is misaligned (off center) by 6% of full NA. When the image of COG reticle is defocused by $+/-500$nm, the image placement discrepancies between devices and O/L metrology structures approach $+/- 25$ nm. This error exceeds the total O/L metrology budget of $+/- 7.5$ nm!

Are the errors of image placement (such as are illustrated in Figs. 2, 14–16, and 21) due to reticles, or incurred while printing them, a part of the O/L metrology error budget? They should not be. However, such errors are present throughout the entire IC manufacturing process. As IC makers tighten the O/L control, these errors keep surfacing in the course of O/L metrology for the first time. Metrologists are the first to observe them. Consequently, until they are attributed to their source and the proper owner is established, they are often considered a part of the metrology error.

Conventional optical O/L metrology tooling has come a long way. These metrology systems can be precise to <5 nm ($3\sigma$). Inaccuracy attributable to its manufacture and setup (TIS) has been reduced to 5 nm, and even that may be reduced by calibration. Based on these statements, it would seem that the available O/L metrology is better than required for manufacturing of products with an O/L budget of less than 75 nm and allowed O/L metrology error of under 7.5 nm. Yet the error sources associated with device manufacturing come into play, and the actual metrology on product often falls short of the system-limited performance levels.

The real solution here is for the metrology community to interact with the other groups on O/L budgeting. In an efficient metrology business process, an error can be attributed and owner assigned so that it is removed where it makes most sense. In this process, the conventional optical O/L metrology systems with their diagnostics capabilities are rapidly becoming the engine of technology characterization and the catalyst for process integration. This approach is very useful in the metrology of image placement for microlithography. To a considerable extent, this chapter is as much about metrology of image placement as it is about process integration.

## C.  Image Recording

As observed in Fig. 14, the process of image recording may lead to errors of image placement when the SLR (or the substrate below it) is asymmetric. The accuracy of measurements made on O/L metrology structures located in proximity to substrate topography may be affected severely. When that happens, a metrologist may rely on a priori information and identify errors due to printing and viewing.

Once the errors are identified and attributed, the magnitude and/or frequency of errors may be reduced. To achieve that efficiently, O/L metrology error diagnostics have been developed (6–8,19,46,47) and feedback automated (19). Numeric estimates of metrology errors have become an essential part of an efficient O/L metrology business process.

Consider the errors of image recording (6–8,104) illustrated in Figures. 16a and b. The initial estimates of these errors were up to 250 nm, in either developed or etched image. Having redesigned the O/L structures and their placement in the kerf (scribe), an assessment was made of mark-related error. Automated error diagnostics (19) produced required error estimates, with a minimal loss of tool throughput. Figures 16e and g and 16f and h present a summary of this evaluation for the O/L metrology structures in Figs. 16c and d. The photograph of Fig. 16c, made with a low-NA bright-field monochromatic microscope, suggest that SLR standing-wave effects due to adjacent scribe and chip structures are readily observable. They are seen much reduced for the O/L mark shown in Fig. 16d. The numeric estimates of the O/L metrology errors were reported as redundancy failure for the target- (substrate) and bullet- (resist) level bars. These errors, analyzed with KPLOT (21), are shown in Figs. 16e and g and 16f and h.

For the bullet (resist) portion of the O/L metrology structure in Fig. 16c, the estimates of measurement uncertainty solely due to redundancy failure are

$$U_{\text{RED}}^{X} = \left| \langle \text{RED}_{X} \rangle \right| + 3\sigma_{\text{RED}} = 20 \, \text{nm}$$

$$U_{\text{RED}}^{Y} = \left| \langle \text{RED}_{Y} \rangle \right| + 3\sigma_{\text{RED}} = 96 \, \text{nm}$$

These errors are largely due to the SLR effects at printing, but also due to viewing of bullet-level trenches in asymmetric resist. These errors were triggered by the adjacent topography in the chip (above the target), as seen in Figs. 16c and e. In this example, the long-distance interaction of resist planarization with adjacent die structures was not accounted for too well.

A much better result may be achieved by placing an O/L metrology mark in a different "neighborhood," separating it from the nearest "neighbors," or by designing defenses against the detrimental effects of long-distance process interactions. Indeed, by better placing a similar O/L metrology structure in the same kerf, these errors were reduced; see Figs. 16d and f:

$$U_{RED}^X = |\langle RED_X \rangle| + 3\sigma_{RED} = 11\,nm$$

$$U_{RED}^Y = |\langle RED_Y \rangle| + 3\sigma_{RED} = 15\,nm$$

The largest remaining component of redundancy error, $|\langle RED_Y \rangle| = 9\,nm$, is likely due to mask making. Since this error is constant and the source is difficult to eliminate, an efficient solution here is calibration.

Estimates of measurement uncertainty due solely to redundancy failures in the target (deep-trench, substrate) portion of the O/L structures shown in Fig. 16c are (see Fig. 16g):

$$U_{RED}^X = |\langle RED_X \rangle| + 3\sigma_{RED} = 29\,nm$$

$$U_{RED}^Y = |\langle RED_Y \rangle| + 3\sigma_{RED} = 18\,nm$$

and for the strucures in Fig. 16d (see Fig. 16h) are:

$$U_{RED}^X = |\langle RED_X \rangle| + 3\sigma_{RED} = 27\,nm$$

$$U_{RED}^Y = |\langle RED_Y \rangle| + 3\sigma_{RED} = 27\,nm$$

Their magnitudes and spatial properties are comparable, and the errors appear to be unrelated to the adjacent topography. These observations are consistent with the process flow: The substrate structures were printed and polished at the earliest stages of build sequence.

Automated error diagnostics and feedback of quality estimates are very useful. They help to highlight the errors in O/L metrology and to identify the error mechanisms. They also provide the relevant input for many other groups affected by or affecting the O/L error. When corrective actions are taken, their utility can be evaluated and progress monitored. A systematic use of error diagnostics and numeric quality measures enables metrology integration as a business process.

## D.  Semiconductor Processing

Semiconductor processes are an endless source of the most challenging alignment and O/L metrology error mechanisms. To those involved in the integration, the goals of device processing and metrology may sometimes be seen as quite divergent. One might even say that the goal of IC processing is not to build good alignment and O/L metrology targets, but to build good devices. That it is, but...

Consider an example of an advanced BEOL (back end of the line) process. The best CD control and electric properties of metal interconnections are achieved when chemical-mechanical polishing (CMP) results in the global planarization of contacts and dielectric, that is, in a flat top surface. However, once the metal is sputtered and the resist is spun on, conventional alignment systems and O/L metrology systems will fail. After all, they are

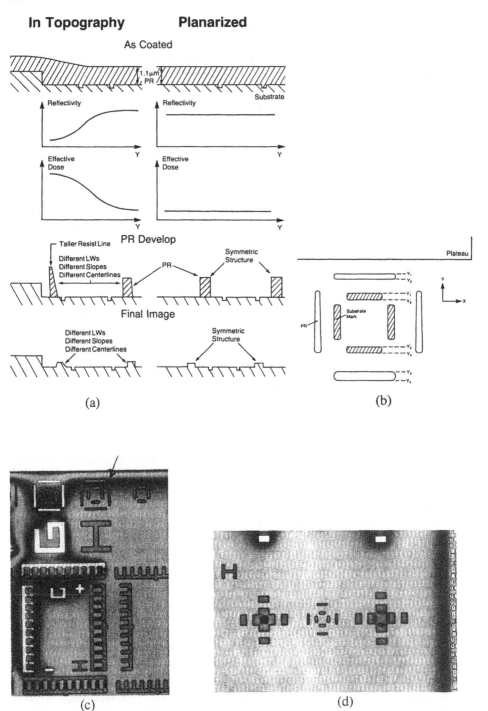

**Figure 16** Evolution of the bullet structure (resist bars) when printed in single-layer resist (SLR) near topography: partially planarized vs. fully planarized (a). O/L Metrology structures near topography observed at low NA and monochromatic light: (c) nonplanar SLR vs. (d) planar SLR. Definition of redundancy failure (b): $RED_Y = (Y_1 + Y_4)/2 - (Y_2 + Y_3)/2$. Redundancy error estimates in resist bars (e, f) and substrate bars (g, h).

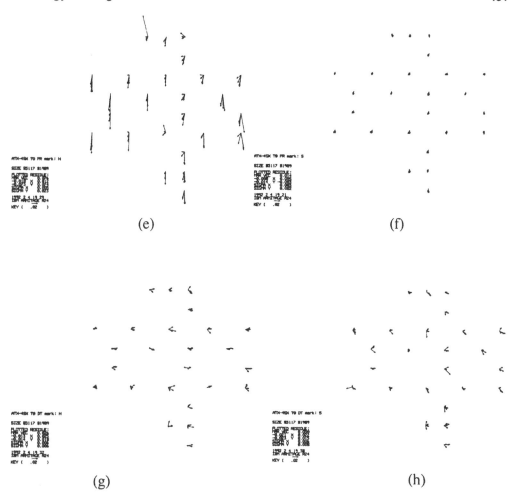

**Figure 16** (*continued*)

optical systems, and, for them, a flat metal film is a mirror through which a substrate target cannot be observed.

The ultimate goal of IC manufacturing is to make high-quality devices at reasonable cost so that the people in metrology and processing can make their living.

When the social contract of the various groups involved in IC manufacture is seen this way, a solution is always found. To arrive at the most efficient solution, a metrologist needs to know about processing, and the people in processing need to know about metrology. The quantitative feedback of metrology error diagnostics provides the relevant measures of quality. This provides the subject and the language for their dialog, and helps to quickly implement the most efficient corrective actions. The result is an effective business process, enabling its practitioners to achieve device performance at the least possible cost.

## 1. Asymmetry in the Resist Coatings

The accuracy of alignment systems in the presence of resist asymmetry was the most recognizable process related alignment issue of the 1980s. This problem was given much attention in the lithography community (43). One of these accounts (148,149) reports the alignment errors that are radial on a wafer and amount to $3\sigma > 0.6\,\mu m$. It was established that the longer the spin time, the larger the errors. When the spin time was reduced from 10 s to 2 s, alignment improved two-to-three-fold. Unfortunately, both resist thickness uniformity and CD control deteriorated. The basic problem: a bright-field monochromatic alignment system may be severely affected by the spun-on resist. The high sensitivity of some alignment and O/L metrology systems to asymmetric resist flow across the targets was recognized as a serious problem. These systems were either modified to reduce such sensitivity or withdrawn from the market. As to the process, a spin-on deposition of photoresist is still universally practiced today.

When photoresist is spun onto a patterned wafer, centrifugal and capillary forces govern its flow (154,155). An alignment or O/L metrology structure in the form of a trench or an island presents a disruption of this flow. Consequently, the thickness of wet resist upstream (on the side closer to center of the wafer) is different from that downstream (away from the center). The evaporation of resist solvent results in a somewhat isotropic resist shrinkage (154) to about 25% of its wet thickness and helps to reduce asymmetry, but only over a short distance. Reflow of wet and baked resist also modifies the initial asymmetry in its profile.

The characterization of asymmetry in resist coverage over metrology structures was undertaken in an effort to better understand and solve the problem (6). The top surface of resist flow profiles was measured with a Tencor Alpha-Step 200 contact profilometer. A set of surface relief profiles formed by 1-μm novolak resist spun across 370-nm-deep trenches is shown in Figure 17c. These trenches are located about 2 inches from the center of the wafer. Trench widths range from 1 μm to 15 μm; the direction of resist flow is from right to left. The test structures, data collection, and processing afforded a reduction of asymmetries due to the shape of the diamond stylus and the direction of scanning. Measured resist profiles were leveled, referenced to the top of the substrate and centered with respect to the trenches below. The left–right asymmetry readily seen in the photoresist profiles of Fig. 17c is the reason for the inaccuracy in centerline estimation (6–8,42,121–126).

A qualitative assessment of the problem may be gained by reviewing the image formed by a low-NA bright-field monochromatic microscope (Figs. 17a and b). This microscope configuration is similar to some early alignment and O/L metrology systems. Owing to interference of light in a thin film of resist, the image contrast is strongly dependent on resist thickness. Variability of the image contrast, antireflection, and left–right asymmetry in the images of alignment targets resulted in outright failures, poor signal-to-noise ratio, degraded precision, and gross errors of centerline estimation.

A quantitative assessment of the inaccuracy of centerline estimation is possible with a good physical model of the optical metrology system and light interaction with the sample. This approach to optical CD metrology (23,25,83,130,156,157) was pioneered in the 1980s by Larrabee, Nyyssonen, Kirk, and Potzick at NBS (National Bureau of Standards, now NIST). An essential part of optical dimensional metrology infrastructure at NIST, this approach was used in generating the standard reference materials of line-widths.

The same approach was successfully applied in studies of alignment (6–8,121–126). The modeling of alignment errors due to resist asymmetry (and other sources of error) proved extremely useful for understanding the causes of errors of centerline estimation in conventional optical alignment. Analyses were also done for conventional optical O/L metrology systems (6–8). An example illustrating the notion of WIS, as error of measurement due to asymmetric resist coating, is shown Fig. 12c.

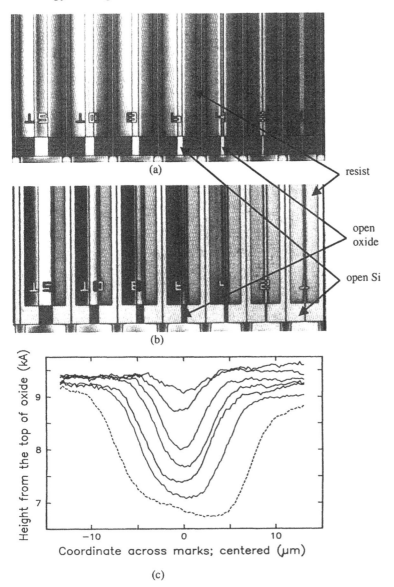

resist

open oxide

open Si

(a)

(b)

(c)

**Figure 17** Resist flow across trenches in oxide (a) and islands of oxide (b) results in asymmetric images of these substrate marks as viewed in a bright-field monochromatic optical microscope. Resist profiles over trenches are shown in (c).

The modeling of metrology error sources became indispensable in the development of high-performance alignment and O/L metrology systems. A majority of the recently introduced alignment and O/L metrology systems underwent exhaustive trials by performance modeling, which by now have become an inherent part of design evaluation. As a result, vendors rapidly evolved their conventional optical systems and, more recently, those with interferometric or phase imaging. These models have also become available to the metrology community. They have served to heighten user and vendor awareness of the integration issues. Many users have reported (43) successes in process integration. A great deal more has been done, but has not been published, for competitive reasons.

## 2. Asymmetry in Sputtered Metal

Sputtering of metals is commonly used in IC manufacturing to deposit metals for device interconnection or metallization layers in BEOL (158,159). For example, a W plug may be formed by first etching a hole in the interlevel dielectric (ILD) and then sputtering metal until the via is filled. On those layers where the linewidth of via (or contact) openings is large and the aspect ratio (depth/CD) is small, bias sputtering is used. Given the size of a metal target and sputtering chamber with respect to the size of a wafer, metal incident upon the surface travels predominantly from the center. In this situation, a target perpendicular to the radial direction is covered with an asymmetric (160) metal film.

An example (from Ref. 160) of W sputtered over a 1-μm via with a 0.55 aspect ratio is illustrated in Figure 18. The cross-sectional SEM microphotograph in Fig. 18b depicts an asymmetric relief in the top surface of tungsten. The V-shaped groove of W fill is apparently shifted with respect to the centerline of the opening as defined by its sidewalls. A physical model illustrated in Fig. 18a, SIMulation by BAllistic Deposition (SIMBAD), produces a good account of both gross and fine structure of deposited W.

This process has some similarities with spin-on deposition of resist. It also produces an asymmetric film over targets, leading to inaccurate centerline estimation. Unlike asymmetric photoresist and other clear coatings, an asymmetric metal film over the targets has the potential to limit the applicability of optical metrology systems. Indeed, the only information from which a centerline estimate may be derived in such applications is associated with an asymmetric top layer—the targets are not observable. Solutions for this problem may involve either an additional processing step to remove the metal over targets or the development of sputtering systems that produce highly symmetric sidewall coverage. A process integration effort to reduce metrology errors to an acceptable level—without recurring costs at manufacture—is a more efficient solution.

Image placement error in metal-to-via alignment was radial across the wafer (161), with $3\sigma = 0.9\,\mu m$. Accuracy was improved by selecting a target for which the sputtering process leads to the smallest measurement error (and target asymmetry). This is an example of how critical process integration is for alignment and O/L metrology.

Automated error diagnostics and numerical quality feedback enable such a solution most efficiently. The early examples (53–55) of process integration for a similar optical

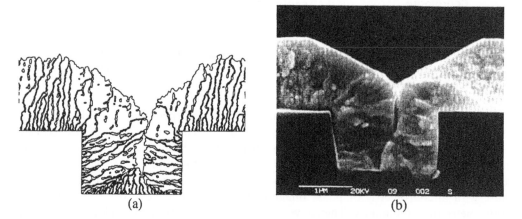

|            (a)            |            (b)            |

**Figure 18**  SIMBAD model of W sputtering over a contact opening (a) and SEM cross section (b) of a feature processed under the same conditions as assumed in the model.

alignment system on similar BEOL layers show that automated metrology error diagnostics and feedback can be very useful.

Modern BEOL requires (158,159,162) filling of vias with CD < 300 nm and aspect ratios of more than 3:1. In order to improve metallization in such environments, collimated sputtering may be used. However, alignment and O/L metrology on layers with collimated sputtering is a much more complicated problem. Not only is there a radial effect but also a collimator "ripple" across the entire wafer surface and an associated rapid spatial variation in the asymmetry of sidewall coverage. The utility of automated error diagnostics and feedback in such applications is expected to be even higher.

## 3. Granularity in Sputtered Metal

Sputtering often results in severe granularity of the metal and the formation of seams and voids when filling small openings (Figs. 8, 9, and 18). Copper precipitate forming on the top surface of AlCu metal, near an edge of a target feature, may also contribute to poor precision and reproducibility. BEOL metrology is severely impacted by W CMP (preceding AlCu deposition; see the example in Fig. 8b). Because the planarization angle is small (Figures 19e and f), the incident light is deflected by only $2°$–$3°$ and is collected by even a moderate-NA objective. This leads to a weak dark fringe formed in a bright-field image ("no contrast"). At worst, as the surface relief over W studs becomes less than 30 nm, even the phase contrast in the image may become so low that neither target capture nor measurement precision meets requirements.

Increasing the sampling size from within an image of an O/L metrology structure, increasing its size, may help considerably. On the other hand, statistics-based error diagnostics and culling are even more effective when dealing with fast spatial variability in the target (and in its image). Rejecting or culling (6–8,19,46,47) a small portion of the image of the O/L measurement structure allows one to recover a measurement, rather than rejecting it, and improves precision and reproducibility. This practice is different from the statistics, based outlier rejection (49–51)—validation or culling of data within an O/L metrology structure, relying on an a priori expectation and a known measure of the sameness of the structure along its length—and is entirely appropriate within a metrology integration business process. The same approach is commonly used in CD metrology.

(A note of caution: Indiscriminate culling outside a proper metrology integration process may obscure the frequency and magnitude of the real anomalies in the materials. When the materials' properties are inconsistent with design rule expectations, they may lead to device failures.)

Automated error diagnostics, culling, and feedback are very effective alignment and O/L metrology for microlithography (6–8,19,46,47,53–55). This method provides 100% quality assurance of metrology data. The performance of the conventional O/L metrology and alignment systems is improved. A reduced O/L-related rework rate and increased effective throughput of metrology tooling are among the significant gains. In addition, the automated quality feedback afforded by such error diagnostics fuels the systematic and efficient process integration.

## 4. Chemical-Mechanical Polishing (CMP)

Chemical-mechanical polishing (CMP) is a rapidly growing technology. Pioneered at IBM (154,155), it combines the processes forming metal studs and interlevel dielectric (ILD) with a process where chemical erosion activated by mechanical abrasion results in a layer of nearly coplanar dielectric and metal. This provides relief for the shrinking depth of

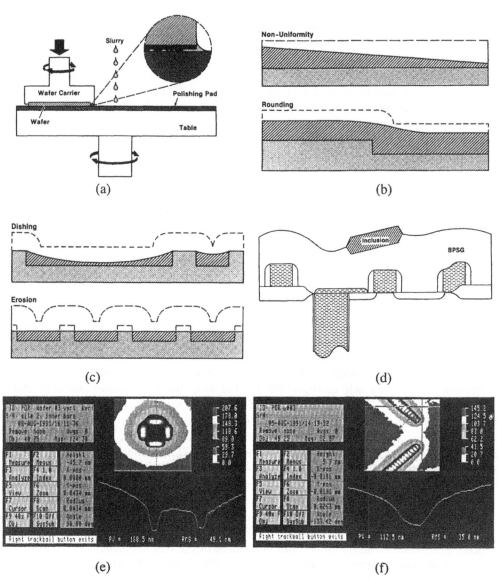

**Figure 19** Illustrations of CMP process (159): wafer moves relative to polishing pad (a), long-distance CMP effects on nonuniformity and rounding of wafer surface (b), dishing and erosion (c); debris and inclusions (d) may result in scratches. Observations (6) of long-distance CMP effects on asymmetry of the top surface and variation of film thickness around an O/L metrology structure (e) and an alignment mark (f).

focus in optical microlithography and enables superior metallurgy and higher productivity of IC manufacturing. Chemical-mechanical polishing is now being rapidly introduced worldwide for state-of-the-art BEOL layers. In order to apply this technology successfully, process integration on an unprecedented scale is required.

Chemical-mechanical polishing plus sputtered metals commonly used in BEOL result in diverse error mechanisms that profoundly affect alignment and O/L metrology. Pattern recognition, precision, and accuracy are affected. Modern BEOL is a veritable wonderland of lithography and metrology errors. It provides a metrologist with stimuli to learn every aspect of modern lithography and IC processing.

A paper aptly titled *Integration of Chemical-Mechanical Polishing into CMOS Integrated Circuit Manufacturing* (163) describes the CMP process and some typical side effects (Figs. 19a and d). The polishing slurry of abrasive and chemically active materials is poured onto the surface of a stiff but pliable polishing pad (Fig. 19a). The pad rotates with respect to the wafer. A wafer held by a wafer carrier rotates about its center. The center of the wafer also rotates with respect to the polishing pad. As the device side of the wafer is pressed against the pad and slurry, abrasion and chemical reactions result in a preferential removal of either metal or ILD. With careful integration of the CMP process, the resulting improved planarity of processed wafers makes it possible to overcome the limited depth of focus of optical lithography systems and to manufacture advanced BEOL with multiple metal and contact levels (158,159,163).

One of the important aspects of CMP integration has to do with CMP interfering with centerline estimation on processed targets. The long-range nonuniformity and rounding (Fig. 19b), dishing, and erosion (Fig. 19c) are quite common. These effects are a function of the local pattern density over a range as large as 1 mm. They are affected by the rapid changes in pad condition, and they vary across a wafer and wafer to wafer. Asymmetry and film thickness variations in alignment and O/L metrology targets lead to failures of pattern recognition, lack of robustness, and inaccurate alignment and O/L metrology. Particles in the slurry may also result in scratches across targets. Studies of W CMP impact on alignment targets (Fig. 19f) and O/L metrology targets (Fig. 19e) were routinely conducted on samples covered by metal film. Since the metal was normally deposited in the manufacturing process, application of an optical noncontact scanning interference profilometer is a natural technique to use (the Maxim*3D® system manufactured by Zygo Corp. was used here).

IBM began introducing various types of CMP, Damascene interconnections, and related processes in the early 1980s. As these new processes moved from research into manufacturing lines, the integration issues had to be addressed. Many new methods for alignment and O/L metrology developed at IBM were born out of the need to support these new processes. Overlay metrology became a catalyst for process integration, enabling manufacturability. Practicing O/L metrology as a business process yielded significant competitive advantages. As CMP processes proliferate among IC manufacturers, the new metrology methods become recognized as the enabler of their manufacturability.

To assess the impact of asymmetric ILD and film thickness variation on the accuracy of alignment and O/L metrology, consider a target processed with W CMP illustrated in Figure 20a. Owing to the long-distance effects of CMP and to the nonuniform distribution of W-filled areas, the $SiO_2$ ILD may become asymmetric in the vicinity of the target (see Fig. 19e). Viewing of the W-filled target portion of the O/L metrology structure is affected by the asymmetry of the clear ILD film.

The potential impact of an asymmetric W surface and of surrounding ILD film on centerline estimation in W stud structures was assessed (164) using the O/L metrology structures in Fig. 20a. A single O/L measurement structure was measured in nine fields across the entire diameter of an 8-inch wafer. Several repeated measurements were made on each field. The wafer was rotated 180° and remeasured in the same way. After averaging and correcting for TIS, the estimated redundancy failure in the W target was computed from two measurements: one made with the inner edges of the bars, and one with the outer edges. Redundancy error was estimated as

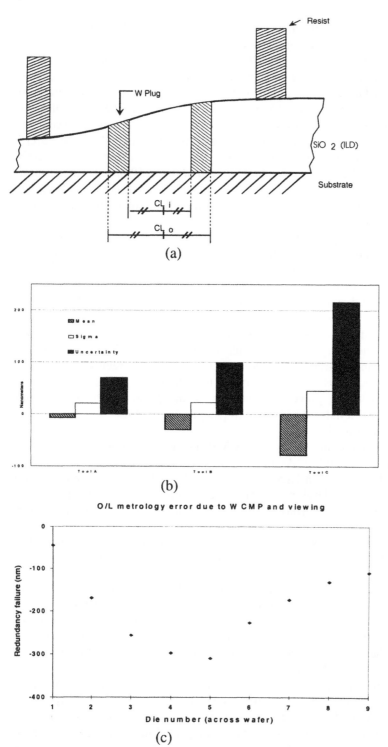

**Figure 20** O/L metrology structure with target-level W plugs processed with CMP (a). Redundancy errors incurred by three O/L metrology systems in viewing the target-level structures are summarized in (b). Variation of redundancy failures in one target is shown as a function of position across the wafer (c).

$$RED_W = CL_o - CL_i$$

where $CL_o$ is the centerline of the outer edges and $CL_i$ is the centerline of the inner edges. These estimates were then corrected for the error of mask making and imaging, producing calibrated estimates $RED'_W$. The mean calibrated error $\langle RED'_W \rangle$ and standard deviation $\sigma_{RED'_W}$ for one set of nine targets was produced. The measurement uncertainty, due solely to the apparent redundancy failure in the inner W bars, was estimated as

$$U_{RED'_W} = |\langle RED'_W \rangle| + 3\sigma_{RED'W}$$

The parameter $RED'_W$ is an estimate of the difference of the inaccuracies in the centerline estimation for inner vs. outer edges of the same target. The result is strongly affected by the sensitivity of instruments to asymmetric clear films (WIS) typically encountered in IC manufacture. When this sensitivity is high, the appropriate selection of several setup parameters and DSP options becomes an important issue.

This experiment was set up by the applications engineers of three O/L metrology tool vendor companies (designated here as A, B, and C) and run under comparable conditions at their respective demo facilities. For each system, the target-limited precision and TIS were much higher than would be typical on resist or etched targets. However, the systematic redundancy failure incurred at viewing W targets embedded in asymmetric ILD (i.e., WIS) was by far the largest error. Unlike TIS and imprecision, WIS presently cannot be removed by tool calibration and averaging.

The results of this study are summarized in Fig. 20b. Although this is an evaluation of possibly the worst case for O/L metrology, it is difficult to believe that the systems with hardware-limited precision $3\sigma < 5$ nm could produce data with $> 70$ nm measurement uncertainty. The error of viewing is alone at $4\sigma$! However, it is true that processing effects can lead to a very large systematic inaccuracy of centerline estimation. For example, redundancy failures observed for system C are shown in Fig. 20c. This error can be as large as 300 nm, and it varies as much across a single wafer. This observation is consistent with a slow and nearly radial across water variation of ILD thickness commonly observed in CMP.

This example shows that conventional optical O/L metrology can be successfully practiced on CMP layers. With careful attention to metrology integration and with automated error diagnostics used to drive it as a business process, O/L metrology error on CMP layers can be brought down to a level consistent with a 75-nm O/L budget. Some of the important elements of this process that have actually been used in 256-Mb technology are:

1.  O/L budget, with error allocation to major sources and O/L metrology
2.  Physical characterization of tool- and process-related error mechanisms
3.  Models of optical instruments and interaction with samples
4.  Automated error O/L metrology diagnostics, culling, and feedback
5.  O/L structures designed and placed to reduce the impact of CMP
6.  Process integration driving over a threefold error reduction for each design release
7.  Account for limits in conventional metrology; have a working alternative
8.  SEM-based O/L metrology[*] for O/L marks with AlCu over W plug
9.  SEM O/L metrology structures in kerf/scribe

---

[*] A. Starikov, M. G. Rosenfield, 1992, unpublished.

The first (165) "all good" 256-Mb chips were made, in fact, with optical microlithography supported by optical alignment systems and optical O/L metrology systems. And CMP was used at both the front end and back end critical layers.

Manufacturers of metrology equipment made refinements in their hardware and software to reduce metrology errors in processed layers. In addition, all alignment and O/L metrology systems on the market today make at least some use of error diagnostics and culling. The number of successful cases of process integration is many and growing (6–8,19,46,53–56,166–174).

## VII.  OVERLAY METROLOGY AS A BUSINESS PROCESS

As stated in Sec. I and described in detail in Sec. VI, microlithography is an extreme form of mass production. It puts unique requirements on dimensional metrology. When metrology errors were frequent and gross, a metrologist could spot them. However, as the magnitude and frequency became small, comprehensive automated metrology error diagnostics and culling became indispensable. Practical applications of overlay metrology for microlithography demand automated error diagnostics.

Methods of error diagnostics were described in Sec. III and illustrated in applications to optical O/L metrology systems in Sec. IV and in the manufacturing environment in Sec. V. Error diagnostics support both reactive and proactive modes of use.

When O/L metrology error diagnostics detect a problem with a measurement, culling of data is made in response to a problem that has already occurred. This reactive use of O/L metrology error diagnostics is very effective at reducing the impact of the problem, for example, in terms of error magnitude and frequency. Automated error diagnostics also provide a "paper trail" of metrology quality and play an important role in product quality assurance.

However, the most significant impact of metrology error diagnostics is achieved in proactive uses for process integration. Here, the O/L metrology error diagnostics serve as a numeric quality feedback. This feedback fuels a metrology integration process in which the actions are taken to reduce the magnitude and incidence of the problem itself. These actions may be directed at changing something about the metrology system or the way it is used. They may also be directed outside the metrology itself, changing the environment in which metrology is conducted. The outcome is an efficient metrology that is optimal for its environment. That is the goal of the O/L metrology business process.

### A.  Limitations of Conventional Technology

It is a good practice of a metrology business process to address the fundamental technology limitations. This reduces the risk of some day being unable to support manufacturing. With some foresight, alternatives can be developed in a steady businesslike environment (such as outlined at the end of Sec. VI.D) rather than in a crisis.

### 1.  Target-Limited Precision

The precision of centerline estimation is a function of the width of an edge response, the strength of the signal, the noise level, the sampling density, the A/D conversion, the algorithm used, and so forth. However, in some process environments, the signal strength may be low and the noise high. In other words, the signal-to-noise ratio (SNR) may be low. Therefore, the target-limited precision may be poor and, in extreme cases, render metrology useless.

Examples of applications where the target quality may strongly impact precision are targets with small or slowly varying amplitude and phase (height). For grainy materials, sampling considerations compound these difficulties.

For example, conventional optical systems may produce a poor signal on targets processed with CMP and covered by metal. The reason for the poor signal is clear from the physical characterization (6) illustrated in Figs. 19e and f: CMP often results in surface relief under 30 nm and sidewall angles of less than 2°. When viewed with bright-field optical systems whose NA is greater than 0.5, samples like this do not strongly reflect the specular component outside the NA, so the amplitude contrast is low. On the other hand, they resolve well enough so that the phase contrast also is low. Various forms of interference microscopy, including the computerized interference microscopy used in these studies, are able to produce phase (directly related to profile), amplitude, and conventional intensity image. Using phase information in the images they collect, the newer alignment and O/L metrology systems can better cope with such targets (105, 167–174).

Although their applicability is extended, these systems are still limited by the same basic causes and, in some cases, may be rendered useless. These limitations are usually overcome by practicing metrology as a business process. In this approach, putting the increased demands and resources into new metrology tooling is seen as just one of many available options. The most efficient one is found through technology learning and optimization.

## 2. Target Asymmetry

When a symmetric target is built on, coated with, or surrounded by asymmetric films, these films may make it either difficult or impossible to estimate the centerline.

When these films are transparent, accurate metrology may be (at least in principle) expected from an optical metrology system. Over the years, the accuracy of O/L metrology in the presence of asymmetric clear films has improved. A user may reasonably demand that a system used in O/L metrology for microlithography be insensitive to asymmetry in clear films, such as photoresist or interlayer dielectric. Realistic calibrated O/L reference materials of this kind and performance evaluation of optical O/L metrology systems are described in Sec. VII.C.1.

The case of absorbing films over targets, on the other hand, represents a potential fundamental limitation to the applicability of optical systems: they simply cannot detect the target directly. A metrologist has to be clear about the available options:

1. When an absorbing film over a symmetric target is also symmetric, accurate metrology may be expected to take place.
2. When an absorbing film over a symmetric target is asymmetric, accurate O/L metrology with an optical system cannot be expected.

If a user desires to employ optical systems, the opaque film must be rendered symmetric or removed. (A word of caution: Symmetry of the top surface is required, but not sufficient, for accuracy. Top surface relief may have shifted from the target below. Cross-checking metrology data and alternative metrology serve to validate such assumptions.)

In order for a conventional O/L measurement to be accurate, the target must be fundamentally symmetric; that is, its sidewalls must have mirror symmetry. The same must be true for absorbing films above the target. If these are not symmetric but this asymmetry is detectable to the tool, then the state of the symmetry (and resulting accuracy) can be improved. For that goal, estimates of asymmetry in images of O/L metrology structures are used as the error diagnostics, to cull some data or to modify the process, improving the symmetry of the image. As long as a metrology system can detect asymmetry, its utility can be extended with both reactive and proactive corrections.

*a.   Failure to Detect Asymmetry*

When an O/L metrology system is unable to detect asymmetry of the targets, technology employed in such a system is not suitable for stand-alone applications. The problem here may be the fundamental inability to detect or resolve the feature sidewalls or simply the lack of error diagnostics. However, without the estimates of inaccuracy and systematic corrective measures that they fuel, O/L metrology could not be a quality process. For this metrology to be useful in the manufacture of ICs, assurances of target symmetry must come from elsewhere.

Methodologies used to manufacture high-quality ICs may be also used to manufacture high-quality O/L metrology structures. Physical characterization, modeling, and statistical process control are among them. In O/L metrology, as in IC manufacturing, the first step is to achieve and maintain high levels of process control in the O/L measurement structures. The next step is to reduce those effects in manufacturing processes that result in asymmetry (and inaccuracy). Finally, the accuracy of O/L metrology is directly ascertained by using alternative metrologies that overcome the limitations of the metrology systems used in production. The scanning electron microscope (26,84–89) supports one such alternative.

## B.   SEM-Based Overlay Metrology

Conventional optical systems are predominantly used today for O/L metrology. However, a number of applications issues may limit the precision and accuracy of these optical systems, necessitating the use of a different instrument. Often (84–89) it is possible to use a scanning electron microscope (SEM).

In order to reduce the asymmetries inherent in the use of left-to-right scanning and off-axis detectors, a special SEM system is required. A low-voltage scanning electron microscope with an annular microchannel plate (MCP) detector and frame-rate image detection were used early on (84,85). Symmetric images of the specially designed O/L metrology structures and IC structures, some modified for O/L metrology with SEM, were recorded and saved for DSP. The precision and linearity in SEM-based O/L metrology were estimated to be competitive with those of optical O/L metrology systems. With appropriate tool setup, the TIS observed in a SEM is very small.

This SEM was used for O/L metrology in IC devices (84). A critical parameter in bipolar technology, emitter-to-base O/L, was measured in transistors, as shown in Fig. 14a. This measurement was made in developed photoresist and in the final image. Unlike the adjacent O/L metrology structures for SEM and optical metrology, the layout of this transistor is asymmetric. It was found that SEM O/L measurements in developed images of transistors differed from those in O/L metrology marks by as much as 20 nm. Optical O/L metrology on O/L measurement structures designed for optical metrology was discrepant by about as much, as compared to SEM O/L metrology made on SEM O/L structures. After etch, this level of correlation deteriorated, suggesting that the image transfer by reactive ion etch (RIE) in an asymmetric environment may have shifted the image sidewalls.

The newer SEM systems typically have an immersion lens that does not have asymmetry introduced by the detector. The left–right scanning may still be an issue, but it can be mitigated by the use of lower beam current and by accumulating image at frame rate (see Fig. 22 in Chapter 14 in this volume).

The capability to measure O/L in devices is very useful in both process control and quality assurance. In addition, it allows one to assess metrology errors due to differences in imaging and image transfer between the device features and O/L measurement marks (8). Suitably designed and sized topographic features may be measured by an optical O/L metrology system and another O/L metrology system using SEM or some other small-probe microscopy. This way, the errors associated with the resolution limits and process sensitivities in optical O/L metrology systems may be addressed (84–89).

The need for SEM-based O/L metrology has been significantly elevated by an increased use of image enhancements (see Secs. VI.A and VI.B). As the device size decreases and the O/L budget is tightened, the image placement is significantly affected by imperfections in OPC and PSM reticles, lens, and condenser aberrations. The notion that lens distortion remains constant is also broken down when illumination is redefined for CD control and by variations of dose and focus, especially for dissimilar feature types and sizes. In this interactive lithography, the control of image placement errors requires O/L metrology in devices. Some aspects of this new environment are illustrated in Figure 21.

There also are some cases when IC processing and the choice of materials severely affect the precision and/or accuracy of optical O/L metrology, limiting its utility. On the other hand, SEM-based O/L metrology may be unaffected. One important case is metrology in developed resist image for blanket metal over a W CMP contact target. In this case, SEM-based O/L metrology is not only useful to ensure the accuracy of the optical systems; O/L metrology may fail entirely because of the lack of contrast. On the other hand, SEM produces strong atomic weight contrast with retro-reflected electrons. Since the SEM image is not severely affected by asymmetry in the top of the contacts (see Figs. 19e and f and 20a), it enables accurate SEM O/L metrology in CMP layers. A part of a comprehensive metrology solution (see Sec. VI.D.4), it may be used for in-line O/L metrology if the conventional metrology becomes inadequate.

In order to practice conventional optical O/L metrology, its accuracy must be ascertained. For this purpose, SEM-based O/L metrology has been developed and proven useful (84–89). The requirement for SEM-based O/L metrology capability is an important part of the *National Technology Roadmap for Semiconductors* (3).

## C. Standards, Calibration, and Ensurance of Accuracy

Linewidth is a distance from an edge to the opposite edge. The accurate measurement of this distance is based, in part, on the standard of length. Scale calibration to the standard of length is a fundamental requirement of linewidth (CD) metrology. Unlike CD metrology, conventional optical O/L metrology is implemented as a differential measurement in two concentric features. The quantity being measured, centerline-to-centerline separation of these features or overlay, is only a small fraction of the critical dimension. Consequently, the scale calibration requirements of O/L metrology are much less stringent.

Indeed, consider a product with CD = 250 nm, a CD tolerance of 25 nm, and a metrology error of 2.5 nm. To accurately measure this CD, the error due to scale must be a small part of the metrology error. If the scale error may be up to 10% of the CD metrology error budget, scale must be calibrated to better than 0.25 nm/250 nm = 0.001 = 0.1%. On

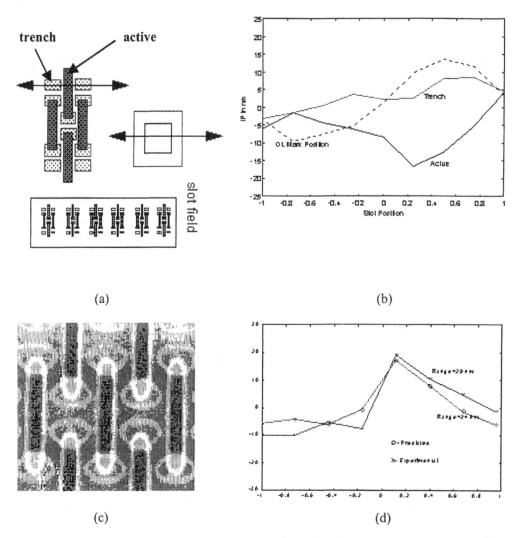

**Figure 21**  Deep trench to active overlay metrology and image placement errors in devices and in O/L metrology marks. The overlay measurement mark is large, the deep trench and active area shapes are small and different (a). When all are printed at multiple positions in a slot field, their image placement is variably affected by aberrations; model prediction is shown in (b). To confirm a predicted discrepancy between the O/L metrology mark and the device overlay, the device O/L is measured with SEM (c) and compared with that in O/L metrology marks (d).

the other hand, a typical O/L tolerance is 75 nm and metrology error is 7.5 nm. When measuring O/L on the concentric SEMI standard targets, the centerline-to-centerline distance is under 75 nm. If the error scale may be up to 10% of O/L the metrology error, then the scale must be calibrated to 0.75 nm/75 nm = 0.01 or 1%. That is, in O/L metrology, the calibration of scale to the standard of length is 10 times less critical.

There is another large difference between the metrology of image placement and the metrology of linewidth. The metrology of linewidth relies on a comparison to a standard of linewidth and on a calibration through the range of linewidth measurements. For this comparison-based metrology to work, all germane properties of a standard and the feature being measured must be similar. The metrology of image placement, on the other hand, is made accurate by maintaining various forms of symmetry. Therefore, comparison to an O/L standard is not required.

As explained earlier, owing to their limited resolution of feature sidewalls, conventional optical metrology systems cannot be self-sufficient. In order to produce statements of accuracy, SEM or other forms of small-probe microscopy have to be used.

The SEM-based metrology of O/L is feasible on devices, small features, and difficult layers (84–89,115). When conventional optical O/L metrology is too limited to be practical, SEM-based O/L metrology may be used to replace it. In cases where it is possible to simultaneously measure CD and O/L without much loss of throughput, SEM-based metrology may become a tool of choice. Among the examples are emitter level in bipolar technology, gate level in CMOS, and contact and metal levels in BEOL.

## 1. Overlay Reference Materials

As a practical matter in O/L metrology, the user needs to know from metrology, not from device yield, if the measurements are accurate. The user knows that accurate O/L metrology can be gained by ensuring the symmetry of the metrology systems and targets. A wealth of experience has already been gained about how to use a priori information, transposition techniques, and available adjustments in O/L metrology. As with other metrology forms (109) before it, the accuracy of O/L metrology has been greatly improved by these methods.

There is also a strong motivation for developing the difficult-to-measure certified O/L reference materials. They offer the convenience and expediency of ensuring that, on a typical and difficult sample, a system is accurate to a stated level. Also, because an error mechanism may not be discovered in an application, even after all the known checks have been made, such materials can be used to support a statement that the error is less than a stated value.

Unlike the case of CD metrology, O/L reference materials cannot readily be used to calibrate metrology systems. However, O/L reference materials would be useful in selecting an O/L metrology system that is more immune to the unavoidable side effects of common IC processing. There is a considerable need for the O/L reference materials for evaluating the sensitivity of a tool to various process-related error mechanisms. What is required is a certified difficult-to-measure, but certifiable, O/L metrology structure. Such structures may be used to evaluate how well this type of O/L measurement system will perform in the applications.

In order to ascertain accuracy, $X/Y$ metrology, SEM metrology, and optical O/L metrology were used. A pair of one critical and one noncritical $5X$ reticles was built to manufacture them. The O/L metrology targets in the critical level reticle were measured on a Nikon-3i metrology system. Evaluation of $X/Y$ metrology capability on small distance measurements suggested that the error was under 5 nm (at wafer scale).

A library of O/L metrology reference materials, known as TV-LITE, has been built on 5-inch ultraflat Si wafers. These wafers were oxidized in small batches, multiple measurements were made to ensure that wafer-to-wafer and across-wafer film thickness variation was small. The thickness of $SiO_2$ was selected and manufactured so that the images of etched structures had a high- and low-amplitude contrast (42,164). Reference wafers were exposed on a $5X$ I-line lithography system. Linewidth measurements were made within batches and batch to batch so that the printed targets would be as similar as possible. The important step, etching, was made with every measure taken to minimize the radial asymmetry in etched $SiO_2$. Mark designs are illustrated in Fig. 13a and b (also see Figs. 17a and b). Images of the high contrast samples, without resist overcoat, are shown in Figs. 13c and d. Then SEM and optical metrology were conducted in order to estimate sidewall asymmetry. Asymmetry in the structures of the critical layer was found to be below measurable.

Overlay metrology on multiple structures and wafers was conducted. Multiple measurements on each site were made and averaged to reduce imprecision. These measurements were made at both 0° and 180° and corrected for TIS. The resulting measurements were found to be within a total uncertainty of 4.5 nm from the calibration values for the high-contrast materials and 5.6 nm for the low-contrast materials.

These wafers were then spin-coated with 1-µm-grade novolak resist, baked, and exposed again. The noncritical image cleared the reference structures from resist, but not over the selected test structures; see Fig. 13b. Resist asymmetry, a known source of measurement inaccuracy, was readily observable. These reference materials were used to evaluate the accuracy of the available O/L measurement systems.

Applications engineers of vendors A, B, and C set up their systems according to their best judgment. In order to suppress the imprecision, multiple dynamic repeats were made on each structure. In addition, measurements at 0° and 180° orientations were used to assess and calibrate for TIS.

All systems easily measured the wafers with high-amplitude contrast. After averaging, correcting for TIS, and subtracting the calibration values, the mean residual error, variance, and uncertainty were produced (Figure 22a). All three conventional optical systems demonstrated the required uncertainty of < 7.5 nm. However, for applications where averaging and calibration for TIS are not practical and one repeat and wafer orientation are used, the error would almost double.

Metrology on the wafers with low contrast was conducted in the same way, producing the estimates of mean residual error, variance, and uncertainty shown in Fig. 22b. On these samples, the combination of resist asymmetry and low contrast resulted in higher imprecision, TIS, and residual error. The uncertainty of all three systems was better than 12 nm. A more realistic estimate based on one repeat and orientation would exceed 20 nm. Uncertainty due to TIS was over 13 nm.

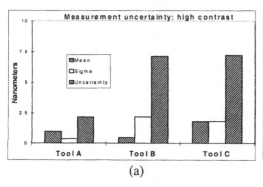
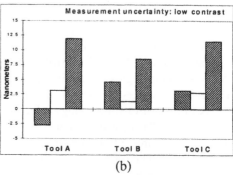

(a) (b)

**Figure 22** Test of O/L metrology accuracy on calibrated O/L metrology reference material: high-contrast samples (a) and low-contrast samples (b).

These experiments demonstrate the utility of certified O/L reference materials for the evaluation of measurement capability. A metrology requirement of 7.5 nm may be met, at least in some cases. In order to gain this level of metrology capability across all process layers; further improvements in tools and process integration are required.

## VII. CONCLUSIONS

At the end of the 1980s, conventional optical O/L metrology systems had a precision of 30 nm and a TIS as large as 100 nm. Several technology strategists of the time had serious reservations about the feasibility of optical microlithography for IC products with 250-nm critical dimensions and an O/L budget of 75 nm. In memory products, the inability to improve overlay was perceived to be a limiting factor, and no major innovation was foreseen to accelerate improvements.

The metrology of image placement with conventional optical systems has advanced much faster than was perceived possible. They still are the preferred O/L metrology system, with a precision and TIS of less than 5 nm. With available enhancements, these systems are able to detect faint targets, diagnose metrology errors, and cull erroneous data. Automated error diagnostics provide the quality feedback required for the accelerated learning of technology and process integration. Similar techniques are becoming widely used in alignment and stepper self-metrology.

Mathematical models of manufacturing processes, manufacturing tools, and metrology systems have been developed. The accuracy of optical O/L metrology has been established on realistic calibrated O/L reference materials. Errors of O/L metrology can be automatically diagnosed and reported, on both test and product samples.

Alternative O/L metrology based on SEM has been developed. Capabilities to measure O/L in devices, to verify accuracy, and to complement the optical O/L metrology systems on difficult layers have been demonstrated.

It appears that, as a whole, the metrology of image placement is very healthy (3) and can continue to advance at a fast pace.

## ACKNOWLEDGMENTS

I am very grateful to Dr. Kevin Monahan and Dr. Michael Postek for stimulating dialogs on dimensional metrology. I owe to Mr. James Potzick and Dr. Robert Larrabee many good insights on the practical realities of metrology. It is due to the keen interest of Dr. Alain Diebold, the editor of this volume, that O/L metrology is covered as required for applications in heavily enhanced optical microlithography.

This chapter is based largely on a body of original work produced at the IBM East Fishkill and the IBM Yorktown Heights Research Center between 1985 and 1994. I gratefully acknowledge both the permission of the IBM Corporation to publish additional portions of that work for the first time and the assistance of Dr. Timothy Brunner in making it possible.

I was fortunate in having had many an opportunity to learn about microlithography and alignment from Drs. Joseph Kirk, John Armitage, and Douglas Goodman. The atmosphere created by Joe attracted top-notch researchers. My interactions with Drs. Diana Nyysssonen and Christopher Kirk on metrology and modeling of optical instruments were very stimulating and fruitful.

Dr. A. Rosenbluth of IBM Research and Mr. S. Knight of IBM Burlington, VT, supported my 1986 study of resist flow. Dr. A. Lopata and Mr. W. Muth were the first to join in on the studies of O/L accuracy; they coined the terms TIS/WIS and shared the new methods with vendors of metrology tools. Dr. C. Progler extended these methods in his work on Suss and SVGL alignment systems. Mr. D. Samuels and Mr. R. Fair of IBM East Fishkill and Drs. T. Brunner and M. Rosenfield of IBM Research were instrumental in building and characterizing TV-LITE. Mr. K. Tallman of IBM East Fishkill helped me to appreciate the intricacies and usage of device design rules. Drs. H. Landis of IBM Burlington, VT, and W. Guthrie of IBM East Fishkill shared their knowledge of materials and processes used in BEOL. The development of error diagnostics for the process integration of alignment and O/L metrology in BEOL was conducted with ATX-4 and -4SX Task Forces. The process integration of O/L metrology in the 256-Mb DRAM joint venture of IBM, Siemens, and Toshiba (TRIAD) was supported by Drs. A. Gutmann of Siemens and K. Okumura of Toshiba.

The development of optical O/L metrology, error diagnostics, and data culling was supported over the years by metrology vendors. I would like to thank BioRad Micromeasurements, Ltd., KLA Instruments, Inc., and Optical Specialties, Inc., for their collaboration. I would like to acknowledge the assistance of Mrs. J. Gueft of BioRad in the preparation and initial runs of the studies of accuracy.

Figure 3 is a contribution from Dr. A. Diebold of SEMATECH. Dr. J. P. Kirk of IBM Corp. contributed his original illustrations for Figs. 4–6 and Table 1. The TEM cross sections in Fig. 9 were produced by J. Z. Duan and C. Matos of Intel Corp, Santa Clara, CA. Figure 15 was contributed by Dr. R. Schenker of Intel, Portland, OR. The illustrations of sputtered W over target in Fig. 18 are courtesy of Mr. T. A. Janacek of Alberta Microelectronic Centre, Edmonton, Alberta, Canada. Dr. H. Landis of IBM Corp. provided the illustrations of CMP process in Figs. 19a and d. Dr. C. J. Progler of IBM East Fishkill contributed Figs. 2 and 21.

## REFERENCES

1. GE Moore. Lithography and the future of Moore's law. Proc. SPIE 2439:2–17, 1995.
2. DL Crook. Evolution of VLSI reliability engineering. 28th Annual Proceedings on Reliability Physics, New Orleans, LA, 1990, pp 2–11.
3. International Technology Roadmap for Semiconductors, 1999 Edition, Lithography. International Technology Roadmap for Semiconductors, 1999 Edition, Metrology.
4. See, for example: EF Strange. Hiroshige's woodblock prints. Mineola, NY: Dover.
5. RM Booth Jr, KA Tallman, TJ Wiltshire, PL Yee. A statistical approach to quality control of non-normal lithography overlay distributions. IBM J. Res. Development 36(5):835–844, 1992.
6. A Starikov. Overlay in Subhalf-Micron Optical Lithography, Short Course. SEMICON/ WEST '93, San Francisco, CA, 1993.
7. DJ Coleman, PJ Larson, AD Lopata, WA Muth, A Starikov. On the accuracy of overlay measurements: tool and mark asymmetry effects. Proc. SPIE 1261:139–161, 1990.
8. A Starikov, DJ Coleman, PJ Larson, AD Lopata, WA Muth. Accuracy of overlay measurements: tool and mark asymmetry effects. Optical Engineering 31(6):1298–1310, 1992.
9. Specification for Overlay-Metrology Test Patterns for Integrated-Circuit Manufacture. Mountain View, CA: Semiconductor Equipment and Materials International, 1996 (SEMI P28-96).
10. Specification for Overlay Capabilities of Wafer Steppers. SEMI P18-92.

11. MT Postek, AE Vladár, SN Jones, WJ Keery. Interlaboratory study on the lithographically produced scanning electron microscope magnification standard prototype. J. Res. National Bureau Standards 98:447–467, 1993.

12. MT Postek, AE Vladár, SN Jones, WJ Keery. Report on the NIST Low Accelerating Voltage SEM Magnification Standard Interlaboratory Study. Proc. SPIE 1926:268–268, 1993.

13. LH Breaux, D Kolar, RS Coleman. Pareto charts for defect analysis with correlation of in-line defects to failed bitmap data. In: K. Monahan, ed. Handbook of Critical Dimension Metrology and Process Control. SPIE CR52:332–358, 1993.

14. WH Arnold. Overlay simulator for wafer steppers. Proc. SPIE 922:94–105, 1988.

15. WH Arnold, J Greeneich, Impact of stepper overlay on advanced design rules. Proceedings of OCG Interface '93, San Diego, CA, 1993, pp 87–99.

16. N Magome, H Kawai. Total overlay analysis for designing future aligner. Proc. SPIE 2440:902–912, 1995.

17. C Progler, S Bukofsky, D Wheeler. Method to budget and optimize total device overlay. Proc. SPIE 3679:193–207, 1999.

18. RD Larrabee. Report on a Workshop for Improving Relationships Between Users and Suppliers of Microlithography Metrology Tools, NISTR 5193. National Institute of Standards and Technology, Gaithersburg, MD, 1993.

19. NH Goodwin, A Starikov, G Robertson. Application of mark diagnostics to overlay metrology. Proc. SPIE 1673: 277–294, 1992.

20. AC Diebold. In-line metrology. In: Y Nishi, R Doering, eds. Handbook of Semiconductor Manufacturing Technology. New York: Marcel Dekker, 2000, Chap 24, pp 745–795.

21. JD Armitage Jr, JP Kirk. Analysis of registration overlay errors. Proc. SPIE 921:207–222, 1988. Data analysis package KPLOT described in this paper is still available, free of charge, from Dr. Joseph P. Kirk, IBM Microelectronics Div., Hopewell Junction, NY 12533.

22. Applications SW for analysis of image placement errors is available from New Vision Systems, Inc., and Yield Dynamics, Inc. Most metrology and lithography system vendors either license such SW packages for distribution with their tools or supply packages of their own. Many IC manufacturers developed their own optimization strategies and SW.

23. RD Larrabee. Submicrometer optical linewidth metrology. SPIE 775:46–50, 1987.

24. Specification for metrology pattern cells for integrated circuit manufacture, SEMI P19-92. Also see the updates and extensions: SEMI P35-0200, Terminology for Microlithography Metrology and SEMI Draft Document #2649, Specification for Electrical Critical Dimension Test Structures and SEMI 2860, Specification for Dimensional Metrology Pattern Test Cells for Integrated Circuit Fabrication.

25. D Nyyssonen, RD Larrabee. Submicrometer linewidth metrology in the optical microscope. J. Res. National Bureau Standards 92(3):187–204, 1987.

26. MT Postek, DC Joy. Submicrometer microelectronics dimensional metrology. Scanning Electron Microscopy 92(3):205–228, 1987.

27. RD Larrabee, MK Postek. Parameters characterizing the measurement of a critical dimension. In: K Monahan, ed. Handbook of Critical Dimension Metrology and Process Control. SPIE CR52:2–24, 1993.

28. JS Beers, WB Penzes. NIST Length scale interferometer measurement assurance. NSTIR 4998, 1992.

29. H Feindt, D Sofronijevic. Ultraprecise mask metrology: development and practical results of a new measuring machine. Proc. SPIE 1138:151–157, 1989.

30. T Ototake, E Matsubara, H Keiichi. Advanced mask metrology system for up to 4-Gbit DRAM. Proc. SPIE, 3096:433–443, 1997.

31. DP Paul. Scanning electron microscope (SEM) dimensional measurement tool. Proc. SPIE 632:222–231, 1986.

32. H Bosse, W Hässler-Grohne. A new electron microscope system for pattern placement metrology. 17th Annual BACUS Symposium on Photomask Technology and Management. Proc. SPIE 3236:160–169, 1997.

33. W Hässler-Grohne, H Bosse. Current activities in mask metrology at the PTB. Proc. SPIE 3412:386–394, 1998.

34. J Kramar, E Amatucci, DE Gilsinn, J-S J Jun, WB Penzes, F Scire, EC Teague, JS Villarubia. Toward nanometer accuracy measurements. Proc. SPIE 3677:1017–1028, 1999.

35. Technical descriptions of existing and proposed standard reference materials for semiconductor manufacturing metrology. NIST, 1996.

36. Physikalisch-Technische Bundesanstalt, Bundesallee 100, D-38116 Braunschweig, Germany.

37. MR Raugh. Absolute two-dimensional sub-micron metrology for electron beam lithography. Precision Engineering 7(1):3–13, 1985.

38. M Raugh. Self-Consistency and Transitivity in Self-Calibration Procedures. Technical Report CSL-TR-91-484. Computer Systems Laboratory, Stanford University, Stanford, CA, 1991, pp 1–36.

39. MT Takac, J Ye, MR Raugh, RFW Pease, NC Berglund, G Owen. Self-calibration in two-dimensions: the experiment. Proc. SPIE 2725:130–146, 1996.

40. RM Silver, T Doiron, W Penzes, E Komegay, S Fox, M Takac, S Rathjen, D Owens. Two-dimensional calibration artifact and measurement procedure. SPIE Proc. 3677:123–138, 1999.

41. Specification for registration marks for photomasks. SEMI P6-88.

42. A Starikov. Beyond TIS: separability and attribution of asymmetry-related errors in center-line estimation. Presentation at SPIE Microlithography Symposium, San Jose, CA, 1994.

43. SPIE Proceedings on Microlithography and SPIE Proceedings on Integrated Circuit Metrology, Inspection and Process Control. Proceedings of KTI (OCG) Microlithography Seminars.

44. MA van den Brink, CG de Mol, HF Linders, S Wittekoek. Matching management of multiple wafer steppers using a stable standard and a matching simulator. Proc. SPIE 1087:218–232, 1989.

45. MA van den Brink, H Franken, S Wittekoek, T Fahner. Automatic on-line wafer stepper calibration system. Proc. SPIE 1261:298–314, 1990.

46. A Starikov, AD Lopata, CJ Progler. Use of *a priori* information in centerline estimation. Proc. KTI Microlithography Seminar, San Jose, CA, 1991, pp 277–294.

47. CJ Progler, A Chen, E Hughlett. Alignment signal failure detection and recovery in real time. J. Vac. Sci. Technol. B11(6):2164–2174, 1993.

48. CD Metrology Procedures, SEMI P24-94.

49. C Eisenhart. Realistic evaluation of the precision and accuracy of instrument calibration systems. J. Res. National Bureau Standards—C, Engineering Instrumentation 67C(2):161–187, 1963.

50. F Proschan. Rejection of outlying observations. Am. J. Physics 21(7):520–525, 1953.

51. JR Taylor. An Introduction to Error Analysis (The Study of Uncertainties in Physical Measurements). New York: Oxford University Press, 1982.

52. CM Nelson, SC Palmateer, AR Forte, SG Cann, S Deneault, TM Lyszczarz. Metrology methods for the quantification of edge roughness II. SPIE Proc. 3677:53–61, 1999.

53. RS Hershel. Alignment in step-and-repeat wafer printing. Proc. SPIE 174:54–62, 1979.

54. MS Wanta, JC Love, J Stamp, RK Rodriguez. Characterizing new darkfield alignment target designs. Proc. KTI Microelectronics Seminar INTERFACE '87, San Diego, CA, 1987, pp 169–181.

55. GE Flores, WW Flack. Evaluation of a silicon trench alignment target strategy. Proc. SPIE 1264:227–241, 1990.

56. P Dirksen, CA Juffermans, A Leeuwestein, C Mutsaers, TA Nuijs, R Pellens, R Wolters, J Gemen. Effects of processing on the overlay performance of a wafer stepper. Proc. SPIE 3050:102–113, 1997.

57. T Kanda, K Mishima, E Murakami, H Ina. Alignment sensor corrections for tool-induced shift (TIS). Proc. SPIE 3051:846–855, 1997.

58. KM Monahan, F Askary, RC Elliott, RA Forcier, R Quattrini, BL Sheumaker, JC Yee, HM Marchman, RD Bennett, SD Carlson, H Sewell, DC McCafferty. Benchmarking multimode CD-SEM metrology to 180 nm. Proc. SPIE 2725:480–493, 1996.

59. JM McIntosh, BC Kane, JB Bindell, CB Vartuli. Approach to CD-SEM metrology utilizing the full wavefront signal. Proc. SPIE 3332:51–60, 1998.

60. DM Goldstein, B Choo, B Singh. Using CD SEM linescan and image correlation to flag lithographic process drift, at KLA-Tencor Corp. Yield Management Solutions Seminar, YMSS 7/99, SEMICON/West '99, San Francisco, CA, 1999.

61. JY Yang, IM Dudley. Application of CD-SEM edge-width measurement to contact-hole process monitoring and development. Proc. SPIE 3882:104–111, 1999.

62. M Menaker. CD-SEM Precision—improved procedure & analysis, Proc. SPIE 3882:272–279, 1999.

63. DG Sutherland, A Veldman, ZA Osborne. Contact hole characterization by SEM waveform analysis. Proc. SPIE 3882:309–314, 1999.

64. E Solecky, R Cornell. CD-SEM edge width applications and analysis. Proc. SPIE 3882:315–323, 1999.

65. JM McIntosh, BC Kane, E Houge, CB Vartuli, X Mei. Dimensional metrology system of shape and scale in pattern transfer. Proc. SPIE 3998:206–217, 2000.

66. J Allgair, G Chen, SJ Marples, DM Goodstein, JD Miller, F Santos. Feature integrity monitoring for process control using a CD SEM. Proc. SPIE 3998:227–231, 2000.

67. B Su, R Oshana, M Menaker, Y Barak, X Shi. Shape Control Using Sidewall Imaging. Proc. SPIE 3998:232–238, 2000.

68. B Choo, T. Riley, B Schulz, B. Singh. Automated process control monitor for 0.18-μm technology and beyond, Proc. SPIE 3998:218–226, 2000.

69. EA McFadden, CP Ausschnitt. Computer-aided engineering workstation for registration control. Proc. SPIE 1087:255–265, 1989.

70. LC Mantalas, HJ Levinson. Semiconductor process control. In: K Monahan, ed. Handbook of Critical Dimension Metrology and Process Control. SPIE CR52:230–266, 1993.

71. TE Zavecz. Machine models and registration. In: K Monahan, ed. Handbook of Critical Dimension Metrology and Process Control. SPIE CR52:134–159, 1993.

72. NT Sullivan. Semiconductor pattern overlay. In: K Monahan, ed. Handbook of Critical Dimension Metrology and Process Control. SPIE CR52:160–188, 1993.

73. SM Ashkenaz. Sampling considerations for semiconductor manufacture. In: K Monahan, ed. Handbook of Critical Dimension Metrology and Process Control. SPIE CR52:316–331, 1993.

74. HJ Levinson, ME Preil, PJ Lord. Minimization of total overlay errors on product wafers using an advanced optimization scheme. Proc. SPIE 3051:362–373, 1997.

75. JC Pellegrini, ZR Hatab, JM Bush, TR Glass. Supersparse overlay sampling plans: an evaluation of methods and algorithms for optimizing overlay quality control and metrology tool throughput. Proc. SPIE 3677:72–82, 1999.

76. JW Goodman. Introduction to Fourier Optics. New York: McGraw-Hill, 1968.

77. GO Reynolds, JB DeVelis, GB Parrent Jr, BJ Thompson. The New Physical Optics Notebook: Tutorials in Fourier Optics, PM-01. SPIE Press, 1989.

78. S Wittekoek, J Van der Werf. Phase gratings as waferstepper alignment marks for all process layers. Proc. SPIE 538:24–31, 1985.

79. K Ota, N Magome, K Nishi. New alignment sensors for wafer stepper. Proc. SPIE 1463:304–314.

80. SSH Naqvi, SH Zaidi, SRJ Brueck, JR McNeil. Diffractive techniques for lithographic process monitoring and control. J. Vac. Sci. Technol. B 12(6):3600–3606, 1994.

81. CJ Raymond, SW Farrer, S Sucher. Scatterometry of the measurement of metal features. Proc. SPIE 3998:135–146, 2000.

82. HS Besser. Characterization of a one-layer overlay standard. Proc. SPIE 1673:381–391, 1992.

83. JE Potzick. Accuracy in integrated circuit dimensional measurements. In: K. Monahan, ed. Handbook of Critical Dimension Metrology and Process Control. SPIE CR52:120–132, 1993.

84. MG Rosenfield, A Starikov. Overlay measurement using the low voltage scanning electron microscope. Microelectronic Engineering 17:439–444, 1992.

85. MG Rosenfield. Overlay measurements using the scanning electron microscope: accuracy and precision. Proc. SPIE 1673:157–165, 1992.

86. J Allgair, M Schippers, B Smith, RC Elliott, JD Miller, JC Robinson. Characterization of overlay tolerance requirements for via to metal alignment. Proc SPIE 3677:239–247, 1999.

87. VC Jai Prakash, CJ Gould. Comparison of optical, SEM, and AFM overlay measurement. Proc. SPIE 3677:229–238, 1999.

88. T Koike, T Ikeda, H Abe, F Komatsu. Investigation of scanning electron microscope overlay metrology. Jpn. J. Appl. Phys. 38:7159–7165, 1999.

89. CJ Gould, FG Goodwin, WR Roberts. Overlay measurement: hidden error. Proc. SPIE 3998:400–415, 2000.

90. MT Postek. Scanning electron microscope metrology. In: K Monahan, ed. Handbook of Critical Dimension Metrology and Process Control. SPIE CR52:46–90, 1993.

91. H Marchman. An overview of CD metrology for advanced CMOS process development. In: DG Seiler, AC Diebold, WM Bullis, TJ Shaffner, R McDonald, EJ Walters, eds. Characterization and Metrology for ULSI Manufacture, AIP Conference Proceedings 449, AIP, Woodsbury, NY, 1998, pp 491–501.

92. D Nyyssonen, L Landstein, E Coombs. Application of a 2-D atomic force microscope system to metrology. Proc. SPIE 1556:79–87, 1992.

93. JE Griffith, DA Grigg. Dimensional metrology with scanning probe microscopes. J. Appl. Phys., 74:R83–R109, 1993.

94. TF Hasan, SU Katzman, DS Perloff. Automated electrical measurements of registration errors in step-and-repeat optical lithography systems. IEEE Trans. Electron. Devices ED-17(12):2304–2312, 1980.

95. LW Linholm, RA Allen, MW Cresswell. Microelectronic test structures for feature placement and electrical linewidth metrology. In: K. Monahan, ed. Handbook of Critical Dimension Metrology and Process Control. SPIE CR52:91–118, 1993.

96. MW Cresswell, WB Penzes, RA Allen, LW Linholm, CH Ellenwood, EC Teague. Electrical test structure for overlay metrology referenced to absolute length standards. Proc. SPIE 2196:512–521, 1994.

97. BF Plambeck, N Knoll, P Lord. Characterization of chemical-mechanical polished overlay targets using coherence probe microscopy. Proc. SPIE 2439:298–308, 1995.

98. J-H Yeo, J-L Nam, SH Oh, JT Moon, Y-B Koh, NP Smith, AM Smout. Improved overlay reading on MLR structures. Proc. SPIE 2725:345–354, 1996.

99. SD Hsu, MV Dusa, J Vlassak, C Harker, M Zimmerman. Overlay target design characterization and optimization for tungsten CMP. Proc. SPIE 3332:360–370, 1998.

100. D-FS Hsu, JK Saw, DR Busath. Characterization and optimization of overlay target design for shallow-trench isolation (STI) process. Proc. SPIE 3677:217–228, 1999.

101. J-S Han, H Kim, J-L Nam, M-S Han, S-K Lim, SD Yanowitz, NP Smith, AM Smout. Effects of illumination wavelength on the accuracy of optical overlay metrology. Proc. SPIE 3051:417–425, 1997.

102. S Kuniyoshi, T Terasawa, T Kurosaki, T Kimura. Contrast improvement of alignment signals from resist coated patterns. J. Vac. Sci. Technol. B 5(2):555–560, 1987.

103. Y Oshida, M Shiba, A Yoshizaki. Relative alignment by direct wafer detection utilizing rocking illumination of Ar ion laser. Proc. SPIE 633:72–78, 1986.

104. A Starikov. On the accuracy of overlay measurements: mark design. IEEE/DARPA Microlithography Workshop, New Orleans, Jan. 1990.

105. N Sullivan, J Shin. Overlay metrology: the systematic, the random and the ugly. In: DG Seiler, AC Diebold, WM Bullis, TJ Shaffner, R McDonald, EJ Walters, eds. Characterization and Metrology for ULSI Manufacture. AIP Conference Proceedings 449, AIP, Woodsbury, NY, 1998, pp 502–512.

106. RM Silver, JE Potzick, F Scire, CJ Evans, M McGlauflin, E Kornegay, RD Larrabee. Method to characterize overlay tool misalignments and distortions. Proc. SPIE 3050:143–155, 1997.

107. N Bobroff. Position measurement with a resolution and noise limited instrument. Rev. Sci. Instr. 57(6):1152–1157, 1986.

108. J Hutcheson. Demand for Overlay Measurement Tools, Executive Advisory, VLSI Research, Inc., San Jose, CA, 1992.

109. CJ Evans, RJ Hocken, WT Estler. Self-calibration, redundancy, error separation, and "Absolute Testing." Annals CIRP 45(2):617–634, 1996.

110. This convention is intuitive: The value of O/L is unique and it is linked with the O/L metrology structure on the wafer. However, it is inconvenient to rotate the metrology, so the wafer is rotated. Most O/L metrology systems make use of the tool coordinate system, not the wafer coordinate system.

111. Alternative definition of TIS, based on four measurement orientations of a wafer with respect to the tool: S Stalnaker, RA Jackson, FA Modawar. Automated misregistration metrology for advanced devices manufacturing. Proceedings of KTI Microlithography Seminar INTERFACE '90, San Diego, CA, 1990, pp 137–141.

112. CJ Progler, AC Chen, TA Gunther, P Kaiser, KA Cooper, RE Hughlett. Overlay Performance of X-ray Steppers in IBM Advance Lithography Facility (ALF). Publication No. 231. Waterbury Center, VT: Karl Suss America, 1993.

113. A Kawai, K Fujiwara, T Kouichirou, H Nagata. Characterization of automatic overlay measurement technique for sub-half micron devices. Proc. SPIE 1464:267–277, 1991.

114. RM Silver, JE Potzick, RD Larrabee. Overlay measurements and standards. Proc. SPIE 2439:262–272, 1995.

115. MMW Cresswell, RN Ghoshtagore, RA Allen, LW Linholm, S Mayo, JR Lowney, CA Reber, SJ Everist. Electrical Overlay Metrology. 3/99 NIST report to SEMATECH, Napa, CA.

116. NP Smith, RW Gale. Advances in optical metrology for the 1990s. Proc. SPIE 1261:104–113, 1990

117. PM Troccolo, NP Smith, T Zantow. Tool and mark design factors that influence optical overlay measurement errors. Proc. SPIE 1673:148–156, 1992.

118. V Nagaswami, W Geerts. Overlay control in submicron environment. KTI Microelectronics Seminar INTERFACE '89, San Diego, CA, 1989, pp 89–106.

119. T Kanda, K Mishima, E Murakami, H Ina. Alignment sensor corrections for tool-induced shift (TIS). Proc. SPIE 3051:845–855, 1997.

120. A Mathai, J Schneir. High-resolution profilometry for improved overlay measurements of CMP-processed layers. Proc. SPIE 3332:182–191, 1998.

121. CP Kirk. Theoretical models for the optical alignment of wafer steppers. Proc. SPIE 772:134–141, 1987.

122. GM Gallatin, JC Webster, EC Kintner, F Wu. Modeling the images of alignment marks under photoresist. Proc. SPIE 772:193–201, 1987.

123. N Bobroff, A Rosenbluth. Alignment errors from resist coating topography. J. Vac. Sci. Technol. B 6(1):403–408, 1988.

124. C- M Yuan, AJ Strojwas. Modeling of optical alignment and metrology schemes used in integrated circuit manufacturing. Proc. SPIE 1264:203–218, 1990.

125. AK Wong, T Doi, DD Dunn, AR Neureuther. Experimental and simulation studies of alignment marks. Proc. SPIE 1463:315–323, 1991.

126. A Starikov. Accuracy of alignment and O/L metrology systems by means of tunable source and handling of signal. US Pat. 5,276,337, 1994.

127. A Starikov. Exposure Monitor Structure. Proc. SPIE 1261:315–324, 1990.

128. JN Wiley, LS Zubrick, SJ Schuda. Comprehensive detection of defects on reduction reticles. Proc. SPIE 2196:219–233, 1994.

129. X Quyang, T. Deeter, CN Berglund, RFW Pease, MA McCord. High-Spatial-Frequency MOS Transistor Gate Length Variations in SRAM Circuits. ICMTS 00-25, Proceedings of the 2000 International Conference on Microelectronic Test Structures, Monterey, CA, 2000.

130. J Potzick. Automated calibration of optical photomask linewidth standards at the national institute of standards and technology. Proc. SPIE 1087:165–177, 1989.

131. MD Levenson. Extending the lifetime of optical lithography technologies with wavefront engineering. Jpn. J. Appl. Phys. 33:6765–6773, 1994.

132. J Potzick. Photomask metrology in the era of neolithography. Proc. SPIE 3236:284–292, 1997.

133. Structures for Test of Asymmetry in Optical Imaging Systems, IBM TDB, Oct. 1990, 90A 62756/FI8890364, Pub. No. 5, pp 114–115.

134. T Saito, H Watanabe, Y Okuda. Evaluation of coma aberration in projection lens by various measurements. Proc. SPIE 3334:297–308, 1998.

135. J Finders, AM Mulders, J Krist, D Flagello, P Luerhrmann, M Maenhoudt, T Marshchner, P de Bisschop, K Ronse. Sub-0.25 micron lithography applying illumination pupil filtering (quadrupole) on a DUV step-and-repeat systems, Olin Microlithography Seminar Interface '98, San Diego, CA, 1998.

136. RD Mih, A Martin, T Brunner, D Long, D Brown. Using the focus monitor test mask to characterize lithographic performance. Proc. SPIE 2440:657–666, 1995.

137. T Brunner. Impact of lens aberrations on optical lithography. Olin Microlithography Seminar, INTERFACE '96, 1996. Also see: IBM J. Res. Development, 41:57–67, 1997.

138. DS Flagello, J de Klerk, G Davies, R Rogoff, B Geh, M Arnz, U Wegmann, M Kraemer. Towards a comprehensive control of full-field image quality in optical photolithography. Proc. SPIE 3051:672–685, 1997.

139. P Dirksen, CA Juffermans, RJ Pellens, M Maenhoudt, P Debisschop. Novel aberration monitor for optical lithography. Proc. SPIE 3679:77–86, 1999.

140. S Nakao, K Tsujita, W Wakamiya. Quantitative measurement of the ray shift aspect of coma aberration with zero-crossing method. Jpn. J. Appl. Phys. 37:6698–6702, 1998.

141. H Fukuda, K Hayano, S Shirai. Determination of high-order lens aberration using phase/ amplitude linear algebra. J. Vac. Sc. Technol. B 17(6):3318–3321, 1999.

142. T Terasawa. Subwavelength lithography (PSM, OPC). Proceedings of the ASP-DAC 2000, 2000, pp 295–299.

143. S Bukofsky, C Progler. Interaction of pattern orientation and lens quality on CD and overlay errors. Proc. SPIE, 4000:315–325, 2000.

144. R Schenker, H Kirchauer, A Stivers, E Tejnil, Alt-PSM for 0.10-μm and 0.13-μm poly patterning,. Proc SPIE 4000:112–120, 2000.

145. M Born, E Wolf. Principles of Optics. 6th ed. New York: Pergamon Press, 1980.

146. DW Peters. The effects of an incorrect condenser lens setup on reduction lens printing capabilities. Kodak Microelectronics Seminar, INTERFACE '85, San Diego, CA, 1985, pp 66–72.

147. DS Goodman, AE Rosenbluth. Condenser aberrations in Köhler illumination. Proc. SPIE 922:108–134, 1988.

148. TA Brunner. Pattern-dependent overlay error in optical step and repeat projection lithography. Microelectronic Eng. 8:13–24, 1988.

149. C-S Lee, JS Kim, IB Hur, YM Ham, SH Choi, YS Seo, SM Ashkenaz. Overlay and lens distortion in a modified illumination stepper. Proc. SPIE 2197:2–8, 1994.

150. T Saito, H Watanabe, Y Okuda. Effect of variable sigma aperture on lens distortion and its pattern size dependence. Proc. SPIE 2725:414–423, 1996.

151. T Saito, H Watanabe, Y Okuda. Overlay errors of fine patterns by lens aberration using modified illumination, Proc. SPIE 3051:866–889, 1997.

152. Y Kishida. Solution of problems associated with reduction projection stepper (in Japanese). Semiconductor World 7:122–127, 1984.

153. KA Chievers. A modified photoresist spin process for a field by field alignment system, Kodak Microelectronics Seminar INTERFACE '84, 1984, pp 66–72.

154. DB LaVergne, DC Hoffer. Modeling planarization with polymers. Proc. SPIE 539:115–122, 1985.

155. LM Manske, DB Graves. Origins of asymmetry in spin-cast films over topography. Proc. SPIE 1463:414–422, 1991.

156. D Nyyssonen. Theory of optical edge detection and imaging of thick layers. JOSA 72:1425–1436, 1982.

157. CP Kirk, D Nyyssonen. Modeling the optical microscope images for the purpose of linewidth measurement. Proc. SPIE 538:179–187, 1985.

158. KH Brown, DA Grose, RC Lange, TH Ning, PA Totta. Advancing the state of the art in high-performance logic and array technology. IBM J. Res. Development 36(5):821–828, 1992.

159. WL Guthrie, WJ Patrick, E Levine, HC Jones, EA Mehter, TF Houghton, GT Chiu, MA Fury. A four-level VLSI bipolar metallization design with chemical-mechanical planarization. IBM J. Res. Development 36(5):845–857, 1992.

160. SK Dew, T Smy, RN Tait, MJ Brett. Modeling bias sputter planarization of metal films using a ballistic deposition simulation. J. Vac. Sci. Technol. A 9(3):519–523, 1991.

161. C Lambson, A Awtrey. Alignment mark optimization for a multi-layer-metal process. Proc. of KTI Microlithography Seminar, INTERFACE '91, San Jose, CA, 1991, pp 37–52.

162. SK Dew, D Lin, MJ Brett. Spatial and angular nonuniformities from collimated sputtering. J. Vac. Sci. Technol. B 11(4):1281–1286, 1993.

163. H Landis, P Burke, W Cote, W Hill, C Hoffman, C Kaanta, Koburger, W Lange, M Leach, S Luce. Integration of chemical-mechanical polishing into CMOS integrated circuit manufacturing. Thin Solid Films 220:1–7, 1992.

164. A Starikov. SRDC/TRIAD overlay metrology development. IBM internal document, East Fishkill, NY, June 24,1994.

165. IBM, Siemens and Toshiba alliance announces smallest fully-functional 256Mb DRAM chip. Press Release, June 6, 1995.

166. D Meunier, B Plambeck, P Lord, N Knoll, The implementation of coherence probe microscopy in a process using chemical mechanical polishing. OCG Microlithography Seminar, INTERFACE '95, San Diego, CA, 1995, pp 155–169.

167. B Plambeck, N. Knoll, P Lord. Characterization of chemical-mechanical polished overlay targets using coherence probe microscopy. Proc. SPIE 2439:298–308, 1995.

168. AW Yanof, W Wilson, R Elias, JN Helbert, C Harker. Improving metrology signal-to-noise on grainy overlay features. Proc. SPIE 3050:114–122, 1997.

169. E Rouchouze, J-M Darracq, J Gemen. CMP-compatible alignment strategy. Proc. SPIE 3050:282–292, 1997.

170. PR Anderson, RJ Monteverde. Strategies for characterizing and optimizing overlay on extremely difficult layers. Proc. SPIE 2196:383–388, 1994.

171. JH Neijzen, RD Morton, P Dirksen, HJ Megens, F Bornebroek. Improved wafer stepper alignment using an enhanced phase grating alignment system. Proc. SPIE 3677:382–394, 1999.

172. F Bornebroek, J Burghoorn, JS Greeneich, HJ Megens, D Satriasaputra, G Simons, S Stalnaker, B Koek. Overlay performance in advanced processes. Proc. SPIE 4000:520–531, 2000.

173. G Rivera, P Lam, J Plauth, A Dunbar, M Philips, G Miraglia, L Rozzoni, E Castellana. Overlay performance on tungsten CMP layers using the ATHENA alignment system. Proc. SPIE 3998:428–440, 2000.

174. NT Sullivan. Critical issues in overlay metrology. Characterization and Metrology for ULSI Technology, 2000 International Conference, AIP Proceedings, Woodsbury, NY, 2000 (in press).

# 18

# Scatterometry for Semiconductor Metrology

**Christopher J. Raymond**
*Accent Optical Technologies, Albuquerque, New Mexico*

## I. INTRODUCTION

The production of a microelectronic circuit is a complex operation, performed by dedicated, expensive machines that coat, expose, bake, develop, etch, strip, deposit, planarize, and implant. Each of these process steps must be executed by design in order for the final product to function properly. For controlling a process, it is important to have accurate quantitative data that relates to the physical composition of the device being made, but the complexity of the operations coupled with the small dimensions of devices today makes this a difficult procedure. In the best of all worlds, the number of process measurements would be kept to a minimum, and the machines used would ensure that the measurements are accurate, repeatable, and inexpensive.

To make matters more difficult for the process engineer, but to the delight of consumers, computers are only getting faster. The most straightforward way to accomplish this is to reduce the dimensions of the features (transistors, gates, etc.) on the chip itself. This perpetual shrinking places an increasing demand on present metrology techniques in terms of the accuracy and precision of metrology measurements. Much has been done to improve current metrology techniques in order to accommodate measurement requirements for advanced products. For example, more complex scanning electron microscopes (SEMs) can image in a range of different detection modes (1), thus providing the process engineer with different ways to look at their product. Likewise, in contrast to conventional two-data-point (psi and delta) film measurement tools, advanced ellipsometers can operate over a range of wavelengths and/or several incident angles (2). Yet despite these improvements to these technologies, metrology remains a serious concern given today's aggressive production standards.

Consider, for example, the manufacture of a device using 0.18-µm features, a geometry in production at the time of this publication. Typical process tolerances for linewidths are specified at 10% of the nominal linewidth (18 nm), but metrology tools must be capable of measurement precision even smaller than this. As a general rule of thumb, the semiconductor industry would like metrology precision, or the repeatability of a given measurement, to be 10% of the tolerance, or 1% of the nominal linewidth. The measurement precision requirement on these devices, then, is a mere 1.8 nm, a stringent requirement to say the least.

But measurement precision is just one facet of the metrology problem. Metrology measurements must also be accurate, in the sense that the information they provide truthfully reflects the actual dimensions of the device being measured. Absolute accuracy can only be assessed when a measurement is made on some standard, i.e., on a sample whose dimensions are known with complete certainty. While many thin-film standards exist, and in the case of patterned features pitch or period standards are available (3), there are no standards currently available for linewidth measurements, which ironically are one of the most common measurements made in the industry. Given the lack of a standard, the best that can be done to quantify the accuracy of some CD measurement technology is to compare its results to those made from another technology. For example, SEM-CD results are often compared to those made by atomic force microscopes or to electrical linewidth measurements. As we will see in Sec. IV, scatterometry measurements of linewidths have also often been compared to SEM data. The measurements from these various technologies are often consistent with one another, but they do not always agree (in other words, they tend to differ systematically). So until a linewidth standard emerges, the question of which technology is the most accurate will remain unanswered.

In addition to precision and accuracy, given that consumer demand for appliances that use microelectronic devices has never been higher, throughput in production fabs is a great concern. This is especially true when it comes to the time required for metrology measurements, which, from a purely economic perspective, do not add any direct value to a product (although this point is often debated). Also, due to the difficulty in making microelectronic devices, it is expected that at some point metrology tools that measure every wafer (and multiple sites on each wafer) will be needed to meet the stringent requirements for linewidth control (4). A metrology that is capable of performing measurements rapidly will have a competitive advantage over slower methods.

Bearing all these considerations in mind, the perfect metrology, then, is one that is capable of meeting or exceeding industry standards for measurement precision (and some measure of accuracy) while being versatile enough to measure several different process parameters. For on-line applications, it should also be able to perform these measurements in a rapid manner. Finally, of course, this perfect metrology should be cheap.

Scatterometry is an optical metrology technique that fulfills many of these needs and thereby offers advantages over existing metrologies. Unlike traditional scatterometry, which analyzes diffuse scatter from random defects and particles in order to quantify the size and frequency of these random features, the scatterometry described here refers to a diffraction-based measurement where the scattering features are periodic, such as line/ space photoresist gratings or arrays of contact holes. This method could also have been called *diffractometry* or *diffraction reflectometry*, but because of its similarities with the measurements made by traditional scatterometers, it has come to be known by the same name. Scatterometry has many attributes common to other optical metrologies; it is simple and inexpensive, and as we will see its measurements are highly repeatable (good precision) and consistent with other methods (accurate). Diffraction-based metrology has already proven useful in a variety of microlithographic applications, including alignment (5), overlay (6), and temperature measurement (7).

In general terms *scatterometry* can be defined as the measurement and characterization of light diffracted from periodic structures. The scattered or diffracted light pattern, often referred to as a *signature*, can be used to characterize the details of the grating shape itself. For a periodic device (such as a series of lines and spaces in resist), the scattered light consists of distinct diffraction orders at angular locations specified by the following well-known grating equation:

$$\sin \theta_i + \sin \theta_n = n\frac{\lambda}{d} \qquad (1)$$

where $\theta_i$ is the angle of incidence, $\theta_n$ is the angular location of the $n$th diffraction order, $\lambda$ is the wavelength of incident light, and $d$ is the spatial period (pitch) of the structure. Due to complex interactions between the incident light and the material itself, the fraction of incident power diffracted into any order is sensitive to the shape and dimensional parameters of the diffracting structure and may be used to characterize that structure itself (8,9). In addition to the period of the structure, which can be determined quite easily, the thickness of the photoresist, the width of the resist line, and the thicknesses of several underlying film layers can also be measured by analyzing the scatter pattern.

The scatterometric analysis can best be illustrated in two parts. First, in what is known as the forward problem, the diffracted light "fingerprint" or "signature" is measured using a scatterometer. Scatterometers are simple instruments, consisting essentially of an illumination source (such as a HeNe laser), some focusing optics, perhaps a rotation stage, and a detector. Details of the forward problem and various types of scatterometers will be discussed in Sec. II. The second step, the inverse problem, consists of using the scatter signature in some manner to determine the shape of the periodic structure. This involves the use of a model that defines some relationship between the scatter signatures and the shape of the features associated with those signatures. For many applications a theoretical diffraction model can be used, where the complete stack information is fed to the model and the signature data is generated by allowing each grating shape parameter to vary in discrete steps over a certain range. Another solution of the inverse problem is to use empirical data to train a prediction model, such as a multivariate statistical regression. Both methods have been used with success and will be discussed in Sec. III.

## A. History of Scatterometry

The earliest application of diffraction for semiconductor metrology purposes appears to have been that of Kleinknecht and Meier in 1978 (10). In their work, diffraction from a photoresist grating was used to monitor the etch rate of an underlying $SiO_2$ layer. Using Fraunhofer diffraction theory, they showed that the first diffraction order could be used to determine when an etch process had undercut the photoresist. In later work, the same authors used a similar technique to measure linewidths on photomasks (11). However, due to limitations in the scalar diffraction model, both of these applications were limited to specific grating geometries.

More advanced diffraction models—which would prove to be useful for the theoretical solution to the inverse problem—come in many forms (see Ref. 12 for examples). One model that has received a considerable amount of attention is known as rigorous coupled-wave theory (RCWT). Although RCWT had been used in other applications for a number of years (13), it was the method proposed by Moharam and Gaylord in the early 1980s that proved to be most useful with respect to diffraction theory (14). It is in one such example of their work where the first "2-$\Theta$" scatter signature (diffraction efficiency as a function of angle) appears (8). It is known as a 2-$\Theta$ signature because of the two theta variables present in Eq. (1). 2-$\Theta$ techniques in particular would prove to be useful ways for collecting diffraction data, and, as we will see, they exhibit good measurement sensitivity.

The first application of RCWT for grating metrology purposes involved the measurement of linewidths on chrome-on-glass photomask gratings (15). Performed at the University of New Mexico, this work used a library of diffraction data as a "lookup table" as a solution for the inverse problem. Around the same time, but on a different research horizon (also at UNM), angle-resolved scatter (ARS) measurements were maturing as a means for characterizing microstructure surface roughness (16). It is important to note that the experimental arrangement for ARS measurements enabled the incident angles and detection angles to be varied quite easily, a feature that would prove to be important for diffraction-based scatter measurements.

Thus it was the research marriage of the experimental arrangement for ARS measurements to the theoretical foundation of RCWT that led to the development of several "flavors" of scatterometry at UNM. The flexibility of the ARS arrangement meant that a variety of diffraction data could easily be measured, while the power and accuracy of RCWT allowed a wide variety of diffracting structures to be modeled. Original scatterometry applications on large-period samples measured the entire diffraction-order envelope at a fixed angle and used this as a means for determining the dimensions of the scattering structure (17,18). However, for short-pitch structures ($\lambda \sim$ period), at most a few diffraction orders exist, and the need for a new type of analysis of the diffraction data arose. The measurement of diffraction efficiency as a function of incident angle—a 2-$\Theta$ scatter signature—was proposed as a means for metrology for short-pitch structures by McNeil and Naqvi in 1993 (19,20). Since then, angle-resolved diffraction signatures have been used successfully in a number of applications (21–27), the details of some of which will be discussed in Sec. IV.

Although 2-$\Theta$ techniques have received the most attention in the literature, other forms of scatterometry have had a number of successful applications as well. The diffraction characteristics of latent image (predevelop) photoresist gratings have been used to monitor both focus (28) and exposure (29) dose during lithography. A "dome" scatterometer, which monitors diffraction in two dimensions, has been used to measure the trench depth of 16-MB DRAMs (30). More recently, a novel in situ scatterometric sensor was developed and used to monitor the postexposure bake (PEB) process for chemically amplified resist samples (31). The success of scatterometry in applications such as these continues to make it an attractive subject of ongoing research. The details of these other forms of scatterometers, and their respective applications, will be summarized in various sections of this chapter.

## II. THE FORWARD PROBLEM

In the forward problem in the simplest sense we are concerned with obtaining a scatter signature. By signature, we mean that the diffraction data must be measured in such a way that it brings insight into the physical composition of the features that interact with the incident light. A variety of scatterometer configurations have been explored and published in the literature in the past decade. In this section we explore these different configurations.

### A. Fixed-Angle Scatterometers

A fixed-angle scatterometer is one in which one of the theta variables present in Eq. (1) is fixed. Figure 1 illustrates a basic fixed-angle scatterometer, where the incident angle is denoted by $\theta_i$ and the measurement angle for the $n$th diffraction order is denoted by $\theta_n$. In the simplest case both angles may be fixed. This means that for some incident angle $\theta_i$ the detector monitors the diffraction intensity of some diffraction order $n$ located at angle $\theta_n$.

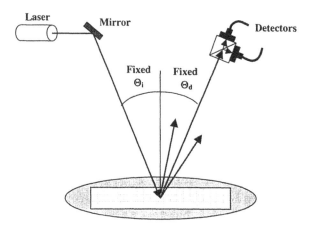

**Figure 1** Fixed-angle scatterometer.

This type of scatterometer generates just a single data point (or two points if both the S and P polarizations are measured), but despite its simplicity can yield very useful data. Hickman et al. used this configuration to monitor the 1st order diffraction efficiency of predeveloped photoresist gratings (29). This type of grating is known as a *latent image* grating because there is no surface relief present in the resist, and the presence of a grating is due to a weak spatial modulation in the index of refraction. By looking at the data from different exposure dose sites they observed that the diffraction efficiency increases with increasing dose. This is because longer doses induce larger differences between the refractive index in the bright and dark regions of the image, which in essence produces a "stronger" grating and thus a larger diffraction efficiency. So, because dose and 1st-order diffraction efficiency are correlated, this form of scatterometer has useful applications for dose monitoring, and it might be used as an endpoint detector for dose.

Milner et al. extended this work by investigating the effects of focus on diffraction efficiency, also for latent image gratings (28). They noted that, for a fixed dose, the 1st order diffraction efficiency peaks at optimum focus. This is because at optimum focus the latent image has the best edge acuity; i.e., the edges of the image are all sharply defined because they are in focus. Physically this means the index of refraction transition between the bright and dark regions is also sharp, and therefore the grating acts more efficiently. The peak in the 1st-order intensity is due to this effect. Based on these two examples, it should be clear that even the simplest scatterometer arrangement can be useful for process-monitoring applications.

An alternative to fixing both angles is to allow the $\theta_d$ variable to change in the configuration shown in Fig. 1. The best application of this is to measure several diffraction orders at their respective $\theta_d$ for a fixed $\theta_i$. This is known as an "envelope" scan or an "order" scan because the entire envelope of diffraction orders is measured. This type of scatterometer is useful for measuring large-pitch gratings because, as Eq. (1) illustrates, they generate lots of diffraction orders. Krukar used this scatterometer configuration for the measurement of 32-μm-pitch etched gratings (32), and was able to characterize the etch depth, linewidth, and sidewall angle of these samples.

Finally, single-angle scatterometers have also proven useful for the measurement of time-varying semiconductor processes, most notably for monitoring the postexposure bake (PEB) process associated with chemically amplified resists. The PEB process is used to complete the chemical alteration of the photoresist prior to development. The duration of the bake can be adjusted to "fine-tune" the chemical composition of the resist. Such fine-tuning can alter the final (postdevelop) CD without affecting the resist profile, and in some manner can compensate for process variations encountered during previous steps. The grating at this point is a weakly diffracting, latent image grating but still has measurable diffraction characteristics. By monitoring the diffraction during PEB, the process can be controlled to produce a desired final CD.

Various fixed-angle diffractive techniques have been explored for PEB process control (33,34). Perhaps the most thorough investigation involved the design and use of a "scatterometric PEB sensor," which had a very compact design and was installed directly to a postexposure bake station at SEMATECH's microelectronics fabrication facility in Austin, Texas (31). The sensor monitored the time-dependent 0th- and 1st-order diffraction characteristics from 800-nm-pitch gratings (nominal 350-nm CDs), which at the time were production environment geometries. Empirical diffraction data was used to train multivariate statistical prediction models (PLS and PCR) that were eventually used for CD determination. One of the main conclusions of this work had to do with the importance of monitoring both the 0th- and 1st-order diffraction characteristics in order to decouple thickness variations from CD changes. If only the 0th-order data was used, the measurements were inferior to those that incorporated both the 0th- and 1st-order data.

## B. Variable-Angle Scatterometers

Most of the applications of scatterometry to date have involved the use of a 2-$\Theta$ scatterometer, which is an extension of the fixed angle apparatus and can be seen in Figure 2. Some incident light, such as the He-Ne laser beam ($\lambda = 633$ nm) shown in the figure, is incident upon a sample after passing through some optical scanning system (which simply steers and focuses the beam). The incident light could also be a laser at a different wavelength, or might even be a variable broad spectral source (35–37). By some manner, be it mirrors or lenses, the incident beam is scanned through a series of discrete angles denoted by $\theta_i$. Likewise, using the grating Eq. (1), the detector of the scatterometer is able to follow and measure any single diffraction order as the incident angle is varied. This

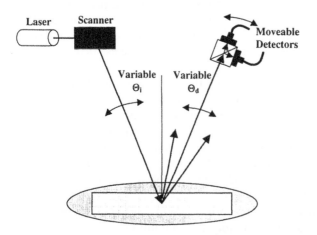

**Figure 2** Variable-angle (2-$\Theta$) scatterometer.

measurement angle is just denoted as $\theta_d$ since it tracks a single diffraction order. The tandem motion of the incident beam and the detector allows the 2-$\Theta$ signature to be measured.

In principle a 2-$\Theta$ scatterometer can measure a signature for any diffraction order and over any range of angles between 0 and 90. Since the incident light can also be decomposed into two orthogonal polarizations—something we will see in Sec. III—it can also measure these polarizations independently. Assuming the signature is measured in discrete 1-degree steps, the maximum number of data points that can be realized for a single diffraction order is 182 (91 angles for two polarizations). In practice, however, the top end of the angle range is typically limited to $\sim$ 50 degrees. Furthermore, most applications have made use of just a single order, the 0th- or specular diffraction order, measured in both polarizations. In general, this makes for an adequate signature (comprised of $\sim$ 102 data points).

Another variation of the basic 2-$\Theta$ scatterometer is one that incorporates measurements of the polarization state of the diffraction orders. In the same way that ellipsometers make measurements of the "psi" and "delta" of the polarized light reflected from a sample, a 2-$\Theta$ scatterometer can make similar measurements of the diffraction orders. Thus both the intensity and phase of the diffracted orders are in effect being recorded. This additional phase data has been shown to increase the overall sensitivity of the measurement (38–41) (meaning the measurement resolution is smaller). But since intensity measurements alone have demonstrated requisite sensitivity in a number of applications, for the moment it appears polarization-resolved measurements may not be necessary. As device geometries continue to shrink, however, this additional sensitivity will likely become more attractive.

Because they are inexpensive, maintenance free, and robust, HeNe lasers are typically the illumination sources used in 2-$\Theta$ scatterometers. The use of shorter-wavelength sources has been explored by some authors (38,42–44), with the general conclusion that sensitivity is enhanced when these sources are used. But there are only a few practical shorter-wavelength laser sources available (442 nm, 325 nm) that can take advantage of this enhancement, and even at 325 nm the sensitivity improvement is at best an order of magnitude and is considerably sample dependent. So, as is the case with the polarization-resolved measurements, for the time being it appears HeNe sources will remain dominant until a serious demand for greater measurement sensitivity emerges.

## C. Dome Scatterometers

The final scatterometer configuration we shall examine is known as a dome scatterometer, so named because it measures a (typically) large number of fixed-angle diffraction orders simultaneously by projecting them onto a diffuse hemispherical "dome." Dome scatterometers are useful for measuring doubly periodic devices, like contact holes and memory arrays. The double periodicity of the features produces a two-dimensional diffraction pattern (unlike 2-$\Theta$ diffraction, where all the orders lie in one plane).

A figure of a basic dome scatterometer appears in Figure 3. An image of the diffraction orders is captured by a CCD camera and is then downloaded to a computer for analysis. Note that the intensity of the orders is the salient data point for dome scatterometry applications—the position of the orders will yield only the periodicity of the features. Therefore, a CCD with a large dynamic range should be used in order to optimize the resolution with which the order intensities can be measured. It is also important that the diffuse reflectance properties of the dome be uniform.

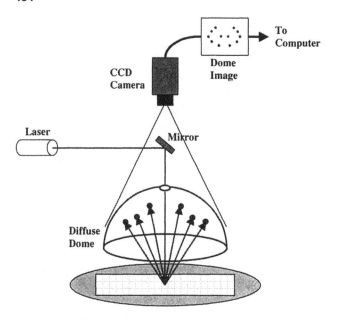

**Figure 3**  Dome scatterometer.

Dome scatterometry has been used by Hatab et al. for the measurement of the depth and diameter of DRAM memory cells (30). By training empirical prediction models (this procedure will be discussed in Sec. III.B) from dome signatures of samples whose dimensions had been characterized by a SEM, the dome scatterometer was able to perform measurements on unknown samples. Typical measurement results, for both the depth and diameter, agreed with subsequent SEM measurements to within $\sim 1\%$.

## III.  THE INVERSE PROBLEM

As was stated earlier, the "inverse problem" is the second half of the scatterometry method. It is at this point that the scatter signature is used in some manner to back out the grating parameters. There are a number of different approaches to the solution of the inverse problem, and each has its own merits, depending on the application being pursued. In this section we explore these different approaches, beginning with a discussion of the models that can be used to generate diffraction signatures. From that point we will discuss the different ways these models can be implemented and compare their relative merits.

Throughout this section, an emphasis will be placed on the use of pure a priori theoretical modeling methods coupled with library search techniques, since this method is currently the most widely used.

### A.  Theoretical Models

One manner in which the data can be compared is through the use of a theoretical diffraction model. The data from the model can be used as a library for the comparison process, or, as we will see later, it may be used to train a regression or statistical model. Let's examine some theoretical models.

## 1. Rigorous Coupled-Wave Theory

Despite its imposing name, rigorous coupled-wave theory (RCWT) is simply a mathematical mechanism that allows for the direct solution of the electromagnetic fields diffracted by a grating. At its very heart lie Maxwell's equations, which are used in a variety of optical inspection techniques, most notably ellipsometers for the measurement of thin films. By utilizing Maxwell's equations in vector differential form, and by applying boundary conditions at all interfaces, RCWT calculates the field strengths in all regions of the grating. The details of RCWT have been discussed in several references (8,12,14,20). We present a brief overview of the method here.

Consider the grating stack illustrated in Figure 4, which consists of four regions: the incident region (typically air), the grating itself, a region of underlying planar layers (which may or may not exist), and finally the substrate. For simplicity we assume the grating has a perfectly square profile—information on how complex gratings are modeled will be discussed later. We need to derive expressions for the fields in each region, apply the appropriate boundary conditions for the interfaces between each region, and solve for the reflected and transmitted diffraction-order amplitudes. Using the geometry described in Fig. 4 we shall derive the field expressions, beginning with region 1 and working our way down the stack.

In incident region 1, the incident light is composed of two orthogonal polarizations, known as S and P, or TE and TM, respectively. The S (TE) polarization refers to an orientation where the electric field is transverse to the plane of incidence, whereas the P (TM) polarization case refers to the orientation where the electric field is parallel to the plane of incidence, meaning the magnetic field is transverse. By defining unit vectors $\hat{S}$ and $\hat{P}$ in the direction of the S and P polarizations, respectively, and using the geometry illustrated in Fig. 4, the incident field $E_{1i}$ can be represented mathematically as

$$\bar{E}_{1i}(x, y, z) = \begin{bmatrix} \hat{S}_0 \sin \phi \\ \hat{P}_0 \cos \phi \end{bmatrix} e^{-j(k_{x0}x + k_y y + k_{1,z0}z)} \tag{2}$$

where the various $k_{x0}$, $k_y$, and $k_{1,z0}$ components are the wave vector magnitudes for the propagating fields and are as defined later.

But there are also scattered fields present in region 1, manifested in the reflected diffraction orders. These scattered fields propagate in the $-z$ direction and can be represented by a summation of the individual orders:

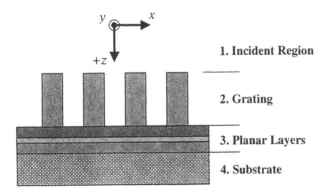

**Figure 4** Geometry of the grating stack used by RCWT.

$$\bar{E}_{1s}(x, y, z) = \sum_{n=-\infty}^{\infty} \begin{bmatrix} \hat{S}_n R_{sn} \\ \hat{P}_n R_{pn} \end{bmatrix} e^{-j(k_{xn}x + k_y y - k_{1,zn}z)} \tag{3}$$

where $R_{sn}$ and $R_{pn}$ are the reflection coefficients for the $n$th diffracted order and the $x$ and $y$ wave vector components are defined as

$$k_{xn} = \frac{2\pi}{\lambda} \left[ n_1 \sin\theta \cos\phi - \frac{n\lambda}{d} \right] \tag{4}$$

$$k_y = \frac{2\pi}{\lambda} [n_1 \sin\theta \sin\phi] \tag{5}$$

where $\theta$ is the angle of incidence, $\phi$ is the polarization angle, and $n_1$ the index of refraction of the incident medium. When $(k_0 n_1)^2 \geq (k_{xn}^2 + k_y^2)$, the $z$ component of the wave vector will be real (propagating waves) and can be expressed by

$$k_{1,zn} = \sqrt{(k_0 n_1)^2 - k_{xn}^2 - k_y^2} \tag{6}$$

Conversely, when $(k_0 n_1)^2 \leq (k_{xn}^2 + k_y^2)$, the $z$ component is imaginary (evanescent waves) and is written as

$$k_{1,zn} = -j\sqrt{k_{xn}^2 + k_y^2 - (k_0 n_1)^2} \tag{7}$$

Thus the total fields present in region 1 is the sum of Eqs. (2) and (3).

Next we must consider the fields in grating region 2. The index of refraction is periodic in the $x$ direction in this region, and hence the permittivity may be expressed as a Fourier series expansion:

$$\varepsilon_2(x) = \varepsilon_0 \sum_{l=-\infty}^{\infty} B_l e^{j(\frac{2\pi l x}{d})} \tag{8}$$

Likewise, the fields present in this region may also be expressed in a similar Fourier series expansion, with coefficients that are a function of $z$. The expression for the electric fields is

$$\bar{E}_2(x, y, z) = \sum_{n=-\infty}^{\infty} \begin{pmatrix} \hat{x} C_{xn}(z) \\ \hat{y} C_{yn}(z) \\ \hat{z} C_{zn}(z) \end{pmatrix} e^{-j(k_{xn}x + k_y y)} \tag{9}$$

where once again $n$ is an index for the diffraction orders. The magnetic fields are represented by a similar relation. Within the grating region the laws of electrodynamics must still apply, meaning Maxwell's equations must still hold true. Specifically, the curl equations can be applied, which results in a set of second-order partial differential equations. So-called "state space" techniques can be used to solve this set of equations and allow, for example, the $x$ component of the electric field in region 2 to be written as

$$E_{2x}(x, y, z) = \sum_{n=-\infty}^{\infty} \sum_{l=-\infty}^{\infty} C_{xl} W_{xnl} e^{\lambda_{xl}z} e^{-j(k_{xn}x + k_y y)} \tag{10}$$

where $W_{xnl}$ and $\lambda_{xl}$ are the eigenvectors and eigenvalues, respectively, of the solution obtained using the state space techniques. The magnetic fields expressions are similar.

Finally, we must consider the fields in any underlying planar layers (region 3) that might be present, as well as the substrate (region 4). The fields in a single planar layer can be expressed as the sum of plane waves traveling in both the $+z$ and $-z$ directions. For example, the $x$ component can be represented by

$$E_{3x}(x, y, z) = \sum_{n=-\infty}^{\infty} (X_{3n} e^{-jk_{3,zn}z} + Y_{3n} e^{jk_{3,zn}z}) e^{-j(k_{xn}x + k_y y)} \tag{11}$$

where $X_{3n}$ and $Y_{3n}$ are the unknown amplitudes for the $n$th diffraction order. Note that for multiple planar layers an expression equivalent to Eq. (11) (but with different coefficients) would be needed for each layer. Likewise, in the substrate region, the fields are represented in the same manner, but since there are no interfaces beneath the substrate, the waves can only propagate in the $+z$ (downward) direction. Therefore the electric fields ($x$ component) in the substrate can be represented by

$$E_{4x}(x, y, z) = \sum_{n=-\infty}^{\infty} X_{4n} e^{-jk_{4,zn}z} e^{-j(k_{xn}x + k_y y)} \tag{12}$$

where, as before, $X_{4n}$ is the unknown amplitude of the transmitted $n$th diffraction order.

With expressions for the fields in each of regions in hand, the final step for the solution of the diffracted order intensities is to match boundary conditions at all interfaces. Maxwell's equations state that the tangential component of the fields be continuous across an interface, which, when applied to these various equations, will result in a system of linear equations that can be solved for the complex field amplitudes in region 1 (reflected orders) and region 4 (transmitted orders).

As we have seen in Eqs. (2) through (12), the fields in the various regions are expressed as infinite summations over the diffraction orders. The series summations can be shifted to start at zero by redefining the summation index as

$$M = 2n + 1 \tag{13}$$

where $M$ is known as the number of modes retained in the calculation. The factor of 2 in Eq. (13) accounts for complementary positive and negative diffraction orders, while the $+1$ term accounts for the specular, or 0th diffraction order. Of course, for a practical implementation these summations must be finite, and ideally we would like to use as few modes $M$ as possible, for this translates into faster modeling time. Fortunately the RCWT solution for the fields reaches a rapid convergence. The convergent criterion—the point at which successive modes retained in the calculation are insignificant—must be carefully chosen and should be less than the noise level for the instrument performing the measurement of the forward problem. The use of $M$ modes beyond the convergent point, then, would be useless, since the signature differences would be within the measurement noise.

The number of modes at which a solution becomes convergent is application specific, but depends strongly on grating pitch and the optical properties of the stack. To a lesser extent it will also depend on the geometry of the stack. In a nutshell, large-pitch gratings with highly absorbing materials (like metals) and tall lines (the grating height) require a higher number of modes. The P polarization also typically requires more modes than its S counterpart to reach convergence.

The computation speed for generating a signature using RCWT is most strongly dependent on the eigen solution needed for Eq. (10) and on the matching of boundary conditions for the final solution. The computation time for the eigen solution is proportional to $M^3$, while the linear system solution for the boundary conditions is proportional to $M^2$. Bearing these proportionalities in mind, in particular the cube relationship, the importance of judicious mode selection as outlined in the previous paragraph should be apparent. As an example, going from a calculation that requires 5 modes to one that requires 10 does not double the computation time, as one might expect. In fact, it increases by a factor of $2^3 = 8$, almost an order of magnitude larger!

## 2. Other Diffraction Models

Although RCWT appears to have received the most attention in the literature with respect to scatterometry applications, there are other ways to model vector diffraction. Despite the differences in models, it is important to note that they all start with the fundamental Maxwell's equations. From there, it is a matter of how Maxwell's equations are implemented and solved that make the approaches different.

In the broadest sense, diffraction models can be categorized into differential methods and integral methods. Whereas RCWT is concerned with solving a set of differential equations, integral methods seek solutions using equations in integral form. A typical integral approach will begin by deriving expressions for unknown induced currents on the surface of a grating (45). The current densities can then be solved for by using numerical integration techniques. With the current densities in hand, the fields can be computed using traditional radiation integrals. As was the case with RCWT, integral methods will also have analysis methods that must be carried out to some convergent point. Still, for grating measurement applications some methods have shown pronounced differences in comparisons to RCWT solutions (46). This issue aside, the choice of using a differential versus integral method will depend on the numerical analysis software available as well as the speed with which they perform on the desired application.

The Chandezon method is another "flavor" of a differential diffraction model that utilizes a coordinate transform to make the ensuing computations more tractable (47). The method introduces a conformal coordinate system that maps corrugated surfaces to planar surfaces, thus making the matching of boundary conditions straightforward. This conformal mapping technique also transforms Maxwell's equations from the Fourier domain into a matrix eigenvalue problem, the solution of which is readily obtainable using numerical analysis methods. The strength of the Chandezon method is in applications that require complex profiles to be modeled, especially those that possess conformal layers, such as films deposited over gratings.

The user of a scatterometer should generally not have to concern herself with the type of diffraction model being used. Aside from an interest in the speed with which a model can perform *convergent* computations (and how convergence is defined), the model should be thought of as a mechanism for generating scatter signatures for comparison to measurement data. In the best of all worlds, a suite of several methods (integral and differential) would be available in a "supermodel," and the best individual model (probably the quickest) for a given application would be used automatically.

## 3. Grating Parameterization

Regardless of the diffraction model used, the theoretical data will only be as useful as the degree to which it reflects accurately the physical composition of the stack being measured in the forward problem. To that end, it is important to have a method of describing the stack in a manner that allows flexibility for the various features that might exist while minimizing the number of parameters overall. This would include provisions for uniform (unpatterned) film layers in the stack as well as a means for describing the grating profile itself.

If the grating profile is nominally square, parameters such as thickness and linewidth are readily implemented into diffraction models. But when the grating sidewalls deviate from vertical, or corner rounding is apparent, the square-profile assumption may not provide an adequate means to parameterize the profile. Fortunately, diffraction models like RCWT can accommodate arbitrary grating profiles by slicing each profile into many layers of thin rectangular slabs and then running the theoretical model over all the layers while applying the appropriate boundary conditions at each interface. First proposed by Peng et al. (48), this slicing technique approximates a nonsquare profile by a number of thin rectangular gratings stacked one upon the other. Figure 5 illustrates the slices that approximate a nonsquare profile. The number of slices, and the thickness of each slice, can also be established by way of some convergence criterion. The technique has been used in the past to study the diffraction characteristics of dielectric (49), photoresist (8,35), and etched (23,25,35) gratings. Although profile models can be readily integrated into diffraction models, the additional profile parameters, as well as the slicing that must be performed, make for longer computation times per signature and larger libraries overall. Therefore, care must be taken to minimize the number of parameters in the profile model.

But even with a slicing technique in place, a profile model must be chosen that adequately reflects profiles associated with a real process. Given the various processes that occur in the production of a microelectronic device, and the different materials that might be encountered for any process, the variety of line profiles that one might encounter seems limitless. Fortunately there is one parameterized profile that reflects a broad spectrum of such profiles (35). Figure 6 depicts this profile model and examples of profiles it can model. In essence it is a trapezoid with some linewidth, height, and sidewall, with the added provision that the corners of the trapezoid are allowed to be rounded. The corner rounding is expressed mathematically as a segment of a circle with some given radius. As the figure illustrates, this scheme describes a variety of profiles using only five parameters: height, linewidth, sidewall, top corner radius, and bottom corner radius. Asymmetric profiles can also be accommodated and will result in additional sidewall and corner radii parameters. In the simplest of this rounded trapezoid model, with no corner rounding and vertical sidewalls, the profile is rectangular. When the sidewall deviates from vertical, the profile becomes a trapezoid, a common shape seen in etched lines. Note that overhanging sidewalls are readily accommodated as well. Finally, with rounding added to the corners, the profile is able to describe even the most complex resist shapes.

Apart from the grating profile, the scatterometry user must also consider the influence of other layers on the overall diffraction signature. This might include unpatterned,

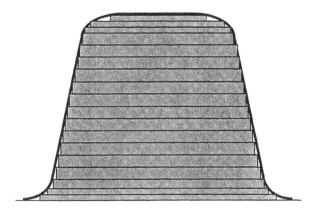

**Figure 5** Slicing technique used to accommodate arbitrary profile shapes.

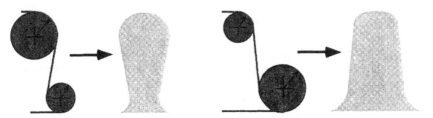

**Figure 6** "Rounded trapezoid" profile model and examples of profiles that can be generated from this model.

uniform layers below (and possibly above) the grating region. Unless their thicknesses are known reliably, they should also be included as free parameters in the diffraction model. In certain circumstances film layers may be omitted from the model, but this is only when they lie beneath an opaque layer that fully attenuates the incident light. In this case no light reaches the layer of interest, and its absence in the model will not affect the diffraction signature. A general rule of thumb for determining the thickness at which a layer of some material is fully opaque is to calculate the classical "skin depth" of the material, defined mathematically as

$$\delta = \frac{\lambda_0}{2\pi k} \tag{14}$$

where $\lambda_0$ is the incident (free space) wavelength of the scatterometer and $k$ is the absorption (the imaginary part of the index of refraction) of the material at the same wavelength.

With a good profile model and knowledge of all uniform layers that need to be included in the diffraction signature computations, the scatterometer user is ready to build a parameter space, or library. In general, there are two things to consider when building a library. First, are all parameters that are expected to vary accounted for? For example, if resist thickness is left fixed in the model (no iterations), but it does in fact vary, the other match parameters will be compromised. Second, is the overall size of the library (number of signatures, or discrete parameter combinations) manageable in terms of the time necessary to compute all the signatures? When one builds a library it is very easy to allow lots of parameters to vary over a large range with a small resolution. The danger in this, however, is that libraries can get very large very quickly, and may take a long time to compute.

As an example, consider a simple patterned resist grating, beneath which is a bottom antireflective coating (BARC) film layer followed by a silicon substrate, as seen in Figure 7. Of primary interest is the linewidth or CD and the sidewall angle of the lines. Assume rounding of the corners is insignificant and not included here. But the thicknesses of the resist and the BARC layer may vary and therefore must be included as unknown parameters. In total, then, there are four parameters allowed to vary: CD, sidewall, resist height, and BARC thickness. The process engineer should have specifications for the nominal dimensions for each of these parameters, as well as some idea of their expected variation ($+/-10\%$ is a good rule of thumb, but will depend in the application). All that is left to do is to determine an adequate resolution for each of the parameters, also commonly referred to as the *step size* for the parameter iterations. Table 1 summarizes an example for a nominal 180-nm process for this stack, and shows the different library sizes that result from using different parameter resolutions. There are three cases illustrated here, which result in library sizes of 78,625, 364,854, and 1,062,369 signatures.

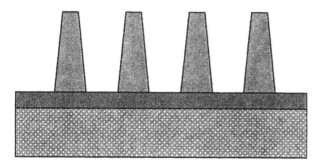

**Figure 7**  The resist on BARC on silicon substrate stack.

While any of these libraries might be adequate for the process engineer, with today's computing power the third case—a library exceeding 1 million signatures—would probably take a full day to compute. Depending on the importance of the application and the need to use such small parameter resolutions, this may be an acceptable consequence for the process engineer (bearing in mind the library only needs to be generated once). As a sidenote, all three of the library designs illustrated here allow for overhanging ($> 90°$) sidewalls, something that is not usually observed for typical resist processes. Therefore, any of these libraries could be made about half the sizes noted here by not including parameter sidewalls beyond $90°$. This is just one example of how a good understanding of a process can be very beneficial in the design of a library.

## 4.  Model Accuracy

An important consideration for any model-based metrology, such as scatterometry or ellipsometry, is the accuracy of the model with regard to the data it generates. As we discussed in Sec. III.A.3, diffraction models are based on the fundamental Maxwell's

**Table 1**  Library Sizes as a Function of Different Parameter Resolutions

| Parameter name | Nominal dimension | Lowest iteration ($\sim -10\%$) | Highest iteration ($\sim +10\%$) | Resolution (step size) | Number of iterations | Library size (no. of signatures) |
|---|---|---|---|---|---|---|
| Linewidth | 180 nm | 162 nm | 198 nm | 1 nm | 37 | $37 \times 17 \times$ |
| Sidewall | 88° | 80° | 96° | 1° | 17 | $25 \times 5 =$ |
| Resist thickness | 6000 Å | 5400 Å | 6600 Å | 50 Å | 25 | **78,625** |
| BARC Thickness | 800 Å | 700 Å | 900 Å | 50 Å | 5 | |
| | | | | | | |
| Linewidth | 180 nm | 162 nm | 198 nm | 0.5 nm | 73 | $73 \times 17 \times$ |
| Sidewall | 88° | 80° | 96° | 1° | 17 | $49 \times 6 =$ |
| Resist thickness | 6000 Å | 5400 Å | 6600 Å | 25 Å | 49 | **364,854** |
| BARC Thickness | 800 Å | 700 Å | 900 Å | 40 Å | 6 | |
| | | | | | | |
| Linewidth | 180 nm | 162 nm | 198 nm | 0.5 nm | 73 | $73 \times 33 \times$ |
| Sidewall | 88° | 80° | 96° | 0.5° | 33 | $49 \times 9 =$ |
| Resist thickness | 6000 Å | 5400 Å | 6600 Å | 25 Å | 49 | **1,062,369** |
| BARC Thickness | 800 Å | 700 Å | 900 Å | 25 Å | 9 | |

equations, which have been in use for more than 100 years now and are the basis for modern electrodynamics. Therefore, the models themselves are entirely accurate, in the sense that they generate correct results for a given set of input conditions. But what if the input conditions—things such as the stack composition—are not accurate? What if the stack composition implemented in the model does not reflect the actual stack being measured?

This is often a concern with regard to the optical constants (the index of refraction, $n$, and absorption, $k$) of the materials being used. For some materials this is not a concern. Simple oxides such as $SiO_2$ have well-known and consistent optical properties. But for a material such as poly-silicon, for example, the optical properties of the material may depend on the operating conditions of the deposition chamber. Unfortunately, there is no way to quantify in advance the degree to which optical constant variations will affect scatterometry measurements. This can be done only for applications where measurement data is already available; by generating libraries derived from different optical constants, and matching the data to the different libraries, differences can be investigated. It is important to note, however, that scatterometry is not the only technology that might encounter this problem. In fact, no technology that uses a model will be immune to this problem. All that the process engineer can do is keep this consideration in mind and try to use the most accurate and up-to-date information about their materials as possible.

## B. Empirical Models

Up to this point we have explored the various *theoretical* diffraction models available to the scatterometer user and the ways such models are utilized. We have also limited our discussions to examples where the diffracting features were periodic in one dimension only, such as photoresist lines and spaces. But these features do not represent real microelectronic devices, which have much more sophisticated geometries in all dimensions and are very time-consuming to model (at the time of this publication anyway)! So in the absence of theoretical data, can this technology be applied to real devices?

Provided they are fabricated in a periodic fashion, they can be measured using scatterometry, but due to the complexity of their shapes a theoretical model cannot be used. Instead, we must rely on empirical data for the solution to the inverse problem. An empirical model needs two components: (1) trusted measurement data (presumably from some other measurement technology) and (2) scatter signatures from those measurement sites. The model generator then looks for statistical consistencies between the signature and parameter data, with the aim of using those consistencies to make measurements on future unknown samples. Common empirical models include multivariate statistical algorithms such as partial least squares (PLS) or principal component regression (PCR). Neural networks are another example of an empirical model; they have been used by several investigators for the generation of prediction models using scatterometry data (32,50–51).

Dome scatterometers, as discussed in Sec. II.C, are good scatterometers to use in tandem with empirical models. The complexity of multidimensional periodic features produces a large number of diffraction orders. This, in fact, is why they are not amenable to modeling—the mode number required for a convergent solution would be tremendously large! But as we discussed earlier, the dome can simultaneously image these orders, and an empirical model can take this large amount of data and process it in such a way that it will give excellent measurement ability. As discussed earlier, Hatab et al. (30) have used a dome scatterometer and empirically trained models (PLS and PCR) for the measurement of DRAM cells.

In principle, any scatterometer can be used with an empirical model, provided the amount of data it provides is amenable to the algorithms used by the model. Scatterometers that measure only a few points, for example, would not be suitable for many of these techniques. But 2-$\Theta$ methods provide enough data and have been used with success, as we will see in Sec. IV.

Finally, provided general computing power continues to improve, there will come a day when even the most complex periodic features will be modeled theoretically and in as little time as if they were lines and spaces today. But whether or not this will put empirical models out of business remains to be seen.

## C. Model Regression Versus Library Search

With raw measurement data in the form of scatter signatures and our model available, the final step in the scatterometry process is to make some measurements by comparing the two. We have already discussed the similarities between scatterometry and ellipsometry in terms of the various optical measurements that can be performed. Yet while the two methods are similar, it is in the final part of the inverse problem that they differ the most. The thin-film reflectance models used by ellipsometers are very simple compared to those for patterned materials, even in the case where polarization-resolved measurements are being performed on multilayer film stacks. So, given the simplicity of uniform-film models, it stands to reason that they run much faster. This allows ellipsometers to fit their measurement data to their models by way of a regression in real time.

Apart from empirical models, scatterometer models are more complicated and, in general, cannot perform enough iterations quickly enough to perform a regression (at least for measurements that must be performed in a few seconds). Therefore, the usual approach for the final part of the inverse problem is to generate a library of scatter signatures ahead of time. Then, when the raw signatures are measured, the library is searched for the best match to the data. A common metric used as the best match criterion is the root mean square error (RMSE) or mean square error (MSE), which are both a point-by-point (angle) comparison between the theoretical and measured data. The parameters of the modeled signature that has the minimum RMSE or MSE are taken to be the parameters of the unknown signature.

Computer calculation times are a fraction of what they were 10 years ago, and at some point they will be fast enough to run diffraction models in real or near-real time. At that point it is almost certain that scatterometers will use similar regression schemes for the solution to the inverse problem. In fact for a few specific scatterometry applications— those that are computationally quick (low modes)—regressions may already be tractable approaches. Bischoff et al. (50) have implemented several regression schemes using diffraction models and have had promising results. By training a neural network from RCWT data, their regression to the neural network ran quickly and gave better measurement results than those that used a PLS regression.

Despite the need to generate lots of signature data a priori, there are advantages to performing full library searches over regressions. For starters, the solution to the regression method may depend on the initial starting point (in terms of parameters) in the model, making it quite possible that the final solution is a local, not global, minimum (2). But in the case of the library search, because an RMSE is calculated for every signature, the global minimum will always be identified. Also, with the fully generated library in hand, it is easy to check the sensitivity of the measurement to the various parameters (by studying the signature changes) and even to enhance the sensitivity in local regions by adding or removing data points.

Finally, in addition to identifying the global minimum, a full library search can also retain the best $N$ number of matches. The parameters of these $N$ matches in tandem with the degree to which they match (their root mean square errors) can be used as a metric for the measurement uncertainty, a useful feature for the process engineer (23).

## IV. REVIEW OF APPLICATIONS

Up until this point we have had a thorough discussion of the various types of scatterometers (the forward problem) and the manner in which data from these instruments is used to make dimensional measurements (the inverse problem). We can now put both halves of the process together and review significant applications of the technology.

As the comprehensive list of references at the end of this chapter will indicate, in the past decade scatterometry has played a ubiquitous role in semiconductor metrology in general. Indeed, the number of applications has been so diverse, and their collective data so extensive, that to summarize them all would be beyond the scope of this chapter. Furthermore, given the incredible pace of production in a modern fab, process engineers will only be interested in a "turnkey" solution to their metrology problems. Bearing in mind these considerations, in our review of applications in this section data will be presented only for applications for which a commercial instrument is available. The intent here is to present ideas for metrology solutions to the process engineer. To this author's knowledge, at present there is only one commercial scatterometer being manufactured; it is an angle-resolved (2-$\Theta$) system from Accent Optical Technologies, known as the CDS-200. But despite the fact that there is only a single system available, as we shall see it has a broad range of applications.

### A. Photoresist Metrology

The most common types of measurements performed in the semiconductor industry are probably linewidth measurements in developed photoresist. Because the pattern in the photoresist forms the mask for any subsequent processing, it is important to have full knowledge of the pattern. There have been several applications of scatterometry for photoresist metrology. Most have used the 2-$\Theta$ technique.

As we saw in our discussion of the forward and inverse problems, 2-$\Theta$ scatterometry is amenable to a number of different experimental arrangements and analysis methods. One of the earliest applications investigated the effects of different angle ranges, diffraction orders, and model comparison techniques (21). Two angular ranges were used; one from 2 to 60 degrees, the other from 2 to 42 degrees (the 0- and 1-degree angles cannot be measured because the detector occluded the incident beam in this experimental arrangement). The smaller range was investigated because it corresponded to the angular range covered by a proposed commercial scatterometer, now currently available from Accent Optical Technologies. Also, the measurements in this research concentrated on the use of 0th-order data for the analyses, though the use of the $\pm$1st orders was explored as well. As we will see, the results indicated the 0th-order signatures alone contained sufficient information about the unknown structure to yield accurate characterization of the diffracting structure, even when the smaller angle range was used. For the solution to the inverse problem, two approaches were taken: the use of a partial least squares (PLS) regression model (trained from RCWT data), and the library search method (minimum mean square error, or "MMS") discussed in Sec. III.

The samples in the study were grating structures of 1-μm pitch and nominal 0.5-μm lines, exposed and developed in Shipley SNR89131 photoresist (a negative resist). The substrate was bare silicon. Because the process produced clean vertical sidewalls and the resist thickness was well known, it was only necessary to keep one parameter free in the final parameter space: linewidth, which was set in the model to range from 350 to 750 nm in 1-nm, steps.

The effects of using different angle ranges and analysis algorithms can be seen in Figures 8 and 9. For comparison purposes SEM measurements are included on the plots. Figure 8 indicates very good agreement between both top-down and cross-sectional SEM and scatterometric measurements when the MMS search algorithm is used. The measurements are actually enveloped by the top-down and cross-sectional SEM measurements over the entire exposure range. The PLS algorithm results for this same wafer can be seen in Fig. 9; the results are not as consistent as those using MMS (particularly at low exposure energy), but they are still consistent with the SEM measurements. With regard to angular range, the 60- and 42-degree data sets for either analysis algorithm yield approximately the same linewidth. With regard to the use of different diffraction orders, Figure 10 shows the results of using 1st-order data to predict linewidths from the same sample as used for Figs. 8 and 9. Although the PLS results differ from the SEM results at higher linewidths, the MMS results are consistent with the SEM results for the entire range.

In this application, the samples, and thus the model, were limited to variations in linewidth only. But in a real production environment, variations in other aspects of the process (e.g., resist and/or film thicknesses) will affect the scatter signature and influence

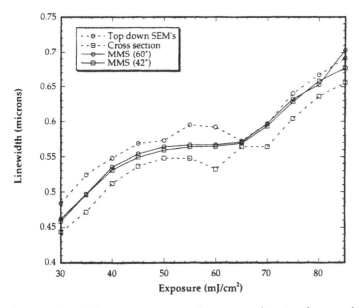

**Figure 8**  MMS scatterometry results compared to top-down and cross-sectional SEM measurements. (From Ref. 21.)

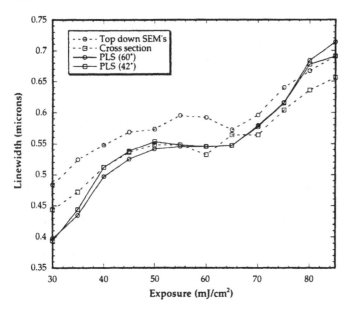

**Figure 9** PLS scatterometry results compared to top-down and cross-sectional SEM measurements. (From Ref. 21.)

the measurement if left unaccounted for. Since the diffraction model can account for film thicknesses, even when they are underlying, 2-$\Theta$ CD measurements should not be affected by these changes. Ideally, the technique would be able to quantify these film thicknesses while at the same time performing CD measurements.

Research has been performed to investigate the ability of 2-$\Theta$ scatterometry for simultaneously measuring linewidths and underlying film layers (22). A large wafer set, 25 in total, was used. Each possessed several underlying film layers whose thicknesses were

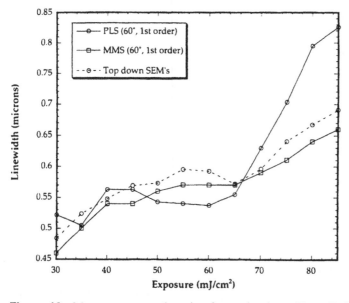

**Figure 10** Measurement results using first-order data. (From Ref. 21.)

intentionally varied within a specified process latitude. These variations were chosen to simulate real process drift. Thus it was necessary to include the additional unknown thickness parameters in the theoretical model; from a single scatter signature, four process parameters were determined.

On each wafer from this application, there was an $SiO_2$ film followed by a poly-Si layer, an antireflection coating (ARC), and finally the APEX-E resist structures. The poly-Si layer was intentionally varied among the wafers at thicknesses of 2300, 2500, and 2700 Å, which represent the centers and extremes of a real process window. Furthermore, because the resist and ARC thicknesses could vary from wafer to wafer, they were also included in the theoretical model. The expected oxide variation, however, was small enough that this thickness was left fixed in the model. In addition to all the film thickness variations, a $9 \times 9$ focus/exposure (F/E) matrix was printed on each wafer, which required a broad range of linewidths in the model. Furthermore, each F/E location comprised two different line/space gratings. The first had nominal 0.35-μm resist lines with a 0.8-μm period. The second had 0.25-μm lines with a 0.75-μm period.

For the diffraction measurements the standard 2-Θ arrangement discussed earlier was used. Based on the success of the previous research, only the smaller angular range (from 2 to 42 degrees) and the 0th order were used in these measurements. Similarly, only the library search (MMS) algorithm was used for identifying the unknown parameters.

Figures 11 and 12 depict the linewidth measurements as a function of exposure for two wafers in this study. On the figures, measurement results for three different metrology techniques (scatterometry, top-down SEM, and cross-sectional SEM) are shown. As is evidenced on the plots, the correlation between scatterometry and cross-sectional SEM measurements is excellent over the entire linewidth range. The average difference between scatterometry and cross-sectional SEM measurements for the nominal 0.35-μm lines is −1.7 nm; for the nominal 0.25-μm lines it is −7.3 nm.

The remaining three parameters were all thickness measurements; Tables 2–4 depict these results. Included in the tables are comparisons to ellipsometer measurements. For resist (grating) thickness (Table 2), all measurements show an average 71.2-nm or 65.2-nm difference between the scatterometry results and those calculated from ellipsometric data.

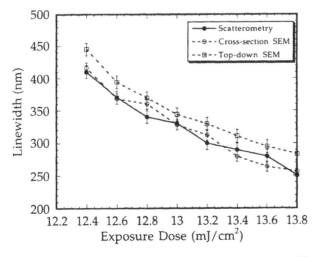

**Figure 11**  Nominal 0.35-μm linewidth measurements. (From Ref. 22.)

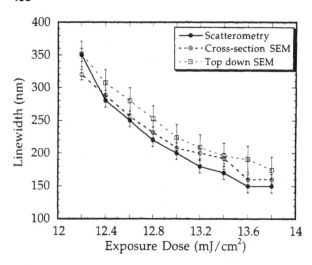

**Figure 12**  Nominal 0.25-μm linewidth measurements. (From Ref. 22.)

However, it is important to note that the ellipsometer measurements were taken immediately after the resist was spun on the wafers. Since APEX-E is a chemically amplified resist, after lithography it undergoes a postexposure bake (PEB), a process known to shrink the resist thickness by about 10%. The difference between the two measurements was hypothesized to be due to this shrinkage. This was later confirmed to be true by a cross-sectional SEM.

Table 3 shows the ARC thickness measurement results for a series of five different wafers from this study. Because the ARC was very thin and had an index ($n = 1.63$) close to that of the resist ($n = 1.603$) at the 633-nm wavelength used by the scatterometer, it was more difficult for scatterometry to determine this particular thickness. Despite this, agreement between the ellipsometer and scatterometer measurements for both gratings was good, with a bias of 6.4 nm for the 0.25-μm lines and 12.8 nm for the 0.35-μm lines.

The poly-Si thickness measurements across the same series of five wafers, seen in Table 4, also show good agreement with ellipsometric values. Unlike the resist and ARC thicknesses, which were essentially the same from wafer to wafer, there were three different nominal poly-Si thicknesses deposited across the five wafers. As is seen in the table, the 2-Θ scatterometer was able to determine these different thicknesses.

**Table 2**  Comparison of Scatterometer and Ellipsometer Resist Thickness Measurements

|  | Nominal 0.25-μm lines | | Nominal 0.35-μm lines | |
|---|---|---|---|---|
| Sample | Ellipsometry | Scatterometry | Ellipsometry | Scatterometry |
| Wafer 2 | 782 nm | 700 nm | 782 nm | 720 nm |
| Wafer 7 | 782 nm | 710 nm | 782 nm | 720 nm |
| Wafer 12 | 781 nm | 710 nm | 781 nm | 720 nm |
| Wafer 17 | 781 nm | 710 nm | 781 nm | 710 nm |
| Wafer 22 | 780 nm | 720 nm | 780 nm | 710 nm |
| Avg. bias | 71.2 nm | | 65.2 nm | |
| 1σ of bias | 7.8 nm | | 4.8 nm | |

**Table 3** Comparison of ARC Thickness Measurements Using Scatterometry and Ellipsometry

| Sample | Nominal 0.25-μm lines | | Nominal 0.35-μm lines | |
|---|---|---|---|---|
| | Ellipsometry | Scatterometry | Ellipsometry | Scatterometry |
| Wafer 2 | 68.7 nm | 80.0 nm | 68.7 nm | 77.5 nm |
| Wafer 7 | 68.0 nm | 65.0 nm | 68.0 nm | 85.0 nm |
| Wafer 12 | 67.8 nm | 75.0 nm | 67.8 nm | 77.5 nm |
| Wafer 17 | 67.5 nm | 65.0 nm | 67.5 nm | 77.5 nm |
| Wafer 22 | 66.7 nm | 75.0 nm | 66.7 nm | 85.0 nm |
| Avg. bias | 6.4 nm | | 12.8 nm | |
| 1σ of bias | 3.7 nm | | 4.4 nm | |

As we have discussed earlier, measurement precision is also a very important aspect of any metrology tool. The precision of an instrument relates to the variation of measured values when a measurement is performed repeatedly. The most judicious definition of precision is one that accounts for measurement variation due to both short-term and long-term effects (52). For short-term repeatability, also known as *static* repeatability or simply repeatability, the sample is left in place while the repeated measurements are performed. To determine long-term repeatability, which is more generally known as *reproducibility* but may also be referred to as *dynamic* repeatability, the sample is removed from the tool and remeasured some time later, with the repeated measurements being performed over several days. Collectively these two components define tool measurement precision, which is expressed mathematically as

$$\sigma_{precision} = \sqrt{\sigma^2_{repeatability} + \sigma^2_{reproducibility}} \tag{15}$$

Thus precision as defined here reflects both standard deviations, $\sigma_{repeatability}$ and $\sigma_{reproducibility}$, of the repeated measurements and may be expressed in multiples ($2\sigma$, $3\sigma$, etc.) of the base figure.

For the multiparameter data set we have been discussing, the straight (short-term) repeatability was not calculated. Instead, the reproducibility was assessed. Ten consecutive measurements were made on each of the two grating samples having the nominal under-

**Table 4** Comparison of Poly-Si Thickness Measurements

| Sample | Nominal 0.25-μm lines | | Nominal 0.35-μm lines | |
|---|---|---|---|---|
| | Ellipsometry | Scatterometry | Ellipsometry | Scatterometry |
| Wafer 2 | 247 nm | 230 nm | 248 nm | 235 nm |
| Wafer 7 | 292 nm | 260 nm | 290 nm | 280 nm |
| Wafer 12 | 248 nm | 230 nm | 248 nm | 230 nm |
| Wafer 17 | 290 nm | 260 nm | 292 nm | 280 nm |
| Wafer 22 | 267 nm | 250 nm | 265 nm | 250 nm |
| Avg. bias | 23.4 nm | | 13.6 nm | |
| 1σ of bias | 7.6 nm | | 3.0 nm | |

lying film thicknesses. Between each scan the wafer was removed, replaced, and repositioned so that the laser spot was centered on the grating. All four parameters for each of the gratings were determined for every scan using the MMS algorithm. The average and standard deviation for each parameter were then calculated.

The results ($3\sigma$) for the reproducibility study are summarized in Table 5. For both gratings and all four parameters the reproducibility ($1\sigma$) is less than 1 nanometer. In particular, the $3\sigma$ reproducibility of linewidth measurements is 0.75 nm for nominal 0.25-μm lines and 1.08 nm for nominal 0.35-μm lines. For the remaining parameters the $3\sigma$ value is still good (subnanometer in some cases). The ARC thickness reproducibility is said to be less than 0.5 nm, because this parameter did not vary for any of the measurements, and 0.5 nm was the step size used in this library. Overall, the reproducibility for all parameters is well within the 1% tolerance specification required for a production-based metrology technique and exceeds the most recent SEM repeatability figures (53).

Finally, an extensive resist measurement gauge study has been performed by Baum and Farrer at Texas Instruments (54). In this investigation a CDS-200 2-$\Theta$ scatterometer was used to characterize photoresist gratings (linewidth and sidewall) in a wide variety of nominal dimensions, including different line-to-space-aspect ratios and grating pitches. The intent was to see whether scatterometry measurements could be affected by these different packing ratios and also to correlate this large number of measurements to CD-SEM measurements from the same sites. In total, nearly 30,000 measurements were performed.

The linewidth results from this study are neatly summarized in Figure 13. As is evidenced in the figure, the scatterometry results are along the same line as those made by the SEM. The few points that lie outside the line were later determined to be sites at extremes in the process window and whose profiles were probably quite degraded (C. Baum, personal communication, 1999). In addition, this investigation also encompassed an extensive full-precision study of the tool, the results of which are summarized in Table 6. As is typical for scatterometry measurements, the $3\sigma$-precision results are subnanometer for all parameters measured.

## B. Etched Grating Metrology

Because etched features represent the final dimensions of a microelectronic device, they are important features to measure. Although measurements are often performed on resist lines as a means for determining critical dimensions, the resulting etched feature may not always have the same CD as the resist line. Complicated resist shapes, particularly a foot at the bottom of the resist, as well as variations in the etch process itself will all contribute to the final etched critical dimension. Thus measurements of the final etched structures reflect the cumulative result of etch and lithography processes.

As we saw in our review of the history of scatterometry (Sec. I.B), etched features were the first samples investigated by scatterometry. More recently, 2-$\Theta$ techniques have

**Table 5**  Reproducibility of the Multiparameter Measurements

| Sample size | Linewidth | Resist height | ARC height | Poly-Si height |
|---|---|---|---|---|
| 0.25 μm | 0.75 nm | 1.29 nm | < 0.5 nm | 0.72 nm |
| 0.35 μm | 1.08 nm | 0.78 nm | < 0.5 nm | 0.78 nm |

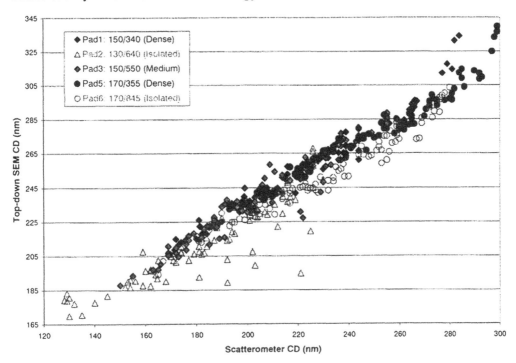

**Figure 13** Scatterometry resist CD measurements compared to SEM data. (From Ref. 54.)

been applied to a sample set of etched features with more modern geometries. We begin our review of applications of scatterometry to etched features by considering an investigation sponsored by SEMATECH (55). In this study, which ran parallel to the SEMATECH resist study we reviewed in Sec. IV.A, five wafers contained a series of etched poly-Si gratings with sub-half-micron CDs. In order to test the robustness of scatterometry to poly-Si variations, the thickness of the poly-Si was varied across the five wafers. Two had nominal thicknesses of 230 nm, two had nominal thicknesses of 270 nm, and one had a thickness of 250 nm. These thicknesses were chosen to mimic real process drift encountered in fabricating a gate layer. In addition to the poly-Si thickness variations, a $9 \times 9$ focus/exposure (F/E) matrix was printed on each wafer. Different exposure doses yield different etched linewidths, providing a broad range of CDs to be measured. Furthermore, each F/E location comprised two different line/space etched gratings. The first had nominal 0.35-μm etched lines with a 0.8-μm period; the second had nominal 0.25-μm etched lines with a 0.75-μm period.

**Table 6** Precision Data (all Figures 3σ) for Resist Stack Measurements

| Parameter | Repeatability | Reproducibility | Precision |
|---|---|---|---|
| ARC thickness | 1.0 Å | 0.7 Å | 1.2 Å |
| Resist thickness | 4.0 Å | 3.8 Å | 5.6 Å |
| Linewidth | 0.45 nm | 0.2 nm | 0.47 nm |
| Sidewall | 0.02° | 0.01° | 0.02° |

The inverse problem was once again solved by using RCWT to generate a library a priori and matching the measured signatures to the library. In the library, initially only two parameters were left as unknowns—linewidth and thickness. Measurement results using this two-parameter model—which assumed the grating profile was square—were less than satisfactory. In particular, the mean square errors for the best matches were higher than desired. It was thought that the square-profile assumption was not valid, and for the second measurement attempt a trapezoidal profile model was used in the theoretical model. A trapezoidal profile allows the overall etch shape to be parameterized by a single value, the sidewall angle. The measurements did improve considerably with the addition of the third parameter.

Results for the 0.25-µm linewidth (CD) measurements as a function of exposure dose for the optimum focus row can be seen in Figure 14. Also shown in the figure are results from AFM and SEM measurements. The measurements are consistent with the top-down SEM and AFM results, with a slight systematic offset. In comparison to the cross-sectional SEM, however, the results are excellent (the average difference between the two is 14.7 nm), indicating that the systematic offset with respect to the top-down SEM and AFM are artifacts from those metrologies and not from scatterometry. It should be noted that a significant sidewall was measured on these samples, and it is not known how the top-down SEM interprets a linewidth dimension in light of this. The scatterometry and AFM results are taken in the middle of the line. With respect to the AFM measurements, because the output is a convolution of the AFM tip shape with that of the line being scanned, it is very important to have the tip shape accurately characterized (56). Ironically, this is usually accomplished with a SEM. At the time these measurements were performed, this particular AFM (a Veeco Instruments Dektak SXM) had been recently acquired, and it was not known how well the tip had been characterized.

Figure 15 depicts the 0.35-µm linewidth results for a sample on which cross-sectional SEM measurements, top-down SEM measurements, and two sets of AFM measurements were also performed. As was the case for the 0.25-µm features, the cross-sectional and scatterometry measurements agree well (average difference = 12.8 nm). The two consecutive AFM scans were performed in order to assess the uncertainty present in this type of measurement. The average difference between the two scans was 5.5 nm.

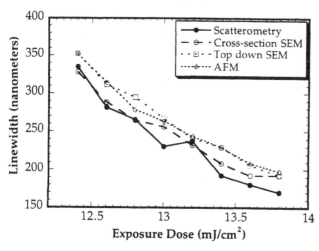

**Figure 14**  Measurement results for nominal 0.25-µm etched CDs. (From Ref. 55.)

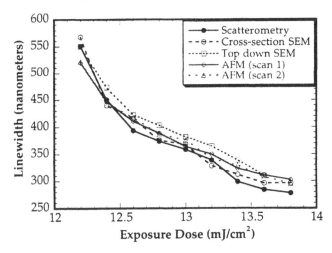

**Figure 15** Measurement results for nominal 0.35-μm etched CDs. (From Ref. 55.)

Measurement results for the poly-Si thickness can be seen in Table 7. Included in this table are thickness measurements made with an ellipsometer as the wafers were being fabricated (prior to patterning with photoresist and etching). The values cited in the table are averages taken at several points across the wafer. For the scatterometer measurements, the average was determined from measurements made on both gratings along the optimum-focus row (nine locations), for a total of 18 measurements; for the ellipsometer measurements, the average was obtained from five measurements made across the same optimum-focus row. As is illustrated in the table, the two measurement techniques agree very well across three wafers with the three different nominal poly-Si thicknesses. The ellipsometer measurements are consistently higher than the scatterometry measurements; overall the average difference between the two techniques (for all three of the wafers) is 14 nm.

Finally, the sidewall-angle measurement results for two of the wafers, both gratings, can be seen in Table 8. Measurement values are cited for the optimum-exposure-dose location (13.0 mJ) as well as the adjacent exposure sites above and below the optimum dose. Cross-sectional SEM sidewall measurements, which were measured from the SEM photograph, are shown for each location in comparison to the scatterometry sidewall-angle results. Most locations are significantly trapezoidal, with angles typically between 80 and 83°.

As part of this study the repeatability and reproducibility of the 2-Θ scatterometer for etched CD measurements were assessed by making 20 2-Θ measurements on the nominal 0.25-μm lines on one of the etched samples. For the first 10 scans the wafer was kept mounted on the scatterometer for all scans (static/repeatability measurements); for the second 10 scans the wafer was removed, replaced, and manually repositioned

**Table 7** Poly-Si Thickness Measurement Results

| Wafer number | Scatterometry | Ellipsometry | Difference |
|---|---|---|---|
| 14 | 236 nm | 247 nm | 11 nm |
| 19 | 273 nm | 290 nm | 17 nm |
| 24 | 251 nm | 265 nm | 14 nm |

**Table 8** Sidewall-Angle Measurement Results

| | Exposure dose (mJ) | | | | | |
|---|---|---|---|---|---|---|
| | 12.8 mJ | | 13.0 mJ | | 13.2 mJ | |
| Wafer no. feature size | Scattering | SEM | Scattering | SEM | Scattering | SEM |
| 19/0.25 μm | 82° | 82° | 90° | 81° | 80° | 82° |
| 14/0.35 μm | 83° | 83° | 80° | 81° | 80° | 80° |

between scans (dynamic/reproducibility measurements). For each series of scans, the average and standard deviation for each parameter were then calculated.

The results for the precision study can be seen in Table 9. For static measurements, both the linewidth and thickness measurement repeatability ($3\sigma$) is well under 1 nanometer (0.63 nm for both). This is true as well for dynamic measurements of the linewidth—the $3\sigma$ reproducibility is 0.78 nm. For dynamic measurements of the poly-Si thickness, the $3\sigma$ reproducibility is slightly higher, at 1.02 nm. The sidewall-angle measurement is demonstrating zero variation because, even with a 0.5° step size in the library, every one of the 20 scans performed predicts the same sidewall angle.

More recently several investigators at Texas Instruments have also been involved in using 2-$\Theta$ scatterometry for metrology of etched gratings. Bushman and Farrer (25) used the technique to characterize a variety of gratings with different pitches across six wafers. In total more than 1200 measurements were made, including full-profile measurements. They concluded that, in comparison to other metrologies, "scatterometry has the most potential of all the metrology techniques for use as a process monitor and control sensor for etch because of its speed and potential for making in-situ measurements."

The most extensive treatment of 2-$\Theta$ etched CD measurements was performed by Baum et al. (23) (also at TI). In this research nearly 3000 nonrepeated measurements were performed on gate-level polysilicon gratings on a thin oxide underlayer. A variety of grating pitches (from 450 nm to 1500 nm) and grating linewidths (from approximately 80 nm to 280 nm) were measured and compared to duplicate measurements obtained from a top-down SEM. The large range of linewidths resulted in overlapping measurements for each pitch, so linearity with pitch could be evaluated. The results from the scatterometer are in excellent agreement with the SEM in nearly all cases. Figure 16 summarizes the linearity between the two methods for a large amount of data.

## C. Photomask Metrology

Of critical importance to the overall quality of a finished microelectronic device is the accurate knowledge of the structures on the photomask from which the device is imaged. Because the mask represents the very first step in the process, any errors present on the

**Table 9** Precision Results for the Etched Samples ($3\sigma$, nominal 0.25-μm CDs)

| Precision type | Etched linewidth | Poly-Si thickness | Sidewall angle |
|---|---|---|---|
| Static/repeatability | 0.63 nm | 0.63 nm | 0.0° |
| Dynamic/reproducibility | 0.78 nm | 1.02 nm | 0.0° |

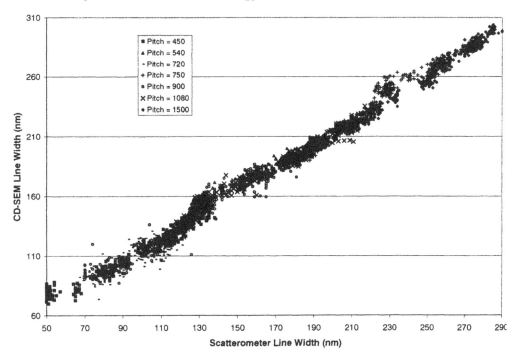

**Figure 16** Linearity of ~ 3000 SEM and scatterometry etched CD measurements. (From Ref. 23.)

mask are inevitably transferred to the product. The first diffractive technique geared for the measurement of photomask linewidths appears to have been that of Kleinknecht and Meier in 1980 (11). Though their measurements agreed well with SEM data, approximations in their model put limitations on the breadth of potential applications; in particular, application was restricted to grating periods greater than the illumination wavelength.

In 1992 Naqvi et al. used RCWT and an early version of a 2-Θ scatterometer for linewidth measurements of chrome-on-glass photomask gratings (15). The rigorous model did not impose any limitations on the scope of the application. Measurements were performed on six different gratings, and all agreed well with measurements performed by other, conventional metrology techniques.

Gaspar-Wilson et al. performed the most recent study of scatterometry for mask metrology, and they extended the breadth of the application to include the characterization of phase-shifting masks (PSMs) (24,58). This type of mask, which relies on interference effects for contrast enhancement in the image, is becoming more prevalent in the semiconductor industry because it extends existing lithography tools to smaller geometries. Due to their complexity, however, they require more stringent metrology requirements. In this work the 2-Θ technique was used in both reflective and transmissive modes to measure the 0th, 1st, and 2nd diffracted orders from a variety of phase-shift gratings. For the most part, the use of the 0th order alone was sufficient for accurate characterization, although full-profile characterization for a more complex mask required the use of all three orders. The transmissive and reflective measurements were consistent with one another, indicating that either could be used independently. Numerous comparisons to AFM measurements were made, and all showed a high degree of agreement. Figures 17 and 18, for example, illustrate the linearity between the techniques for linewidth and depth measurements.

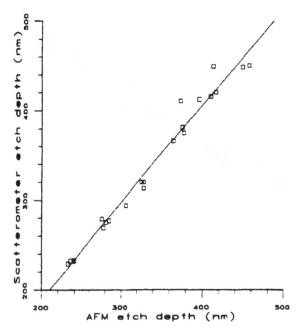

**Figure 17**   Linearity of AFM and scatterometry photomask etch-depth measurements. (From Ref. 58.)

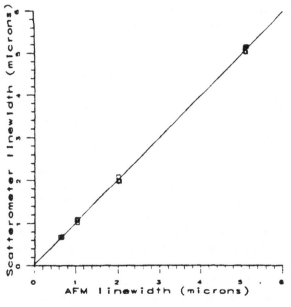

**Figure 18**   Linearity of AFM and scatterometry photomask linewidth measurements. (From Ref. 58.)

## D. Grating Profile Metrology

In addition to the primary parameters of interest associated with gratings, such as thickness and linewidth, more complex structural shapes, such as rounded corners and nonrectangular profiles, may be detected and characterized by 2-Θ scatterometry. Such irregular shapes are common in the production of submicron structures of photoresist, since even slight changes in focus and/or exposure conditions can have a significant effect on the profile once it is developed. They also are common in etched materials, since etch chemistries do not affect all materials in the same manner. The measurement of such profiles is important because, if left unaccounted for, they could have an effect on other scatterometric dimensional measurements. In addition, nonvertical sidewalls and the presence of a rounded lower corner on the resist (known as a "foot") may affect the manner in which any underlying materials are processed.

As we discussed in Sec. III.A.3, grating profiles can be measured by implementing some grating profile model, such as the one seen in Figure 6, and then slicing the profile in the manner illustrated in Figure 5 so that the diffraction model can compute 2-Θ signatures. This profile model was used to quantify resist profiles that had both top and bottom corner rounding as well as slightly nonvertical sidewalls (57). Because it was well known, the overall thickness of the resist was left fixed, allowing the profile to be parameterized with four variables: top radius, top CD, bottom radius, and bottom CD.

The profile model was applied to 12 exposure sites from one wafer. The measurement process for the profiles was done in two steps. First, a square profile model was used to determine the approximate linewidth of each site. If the profile features are subtle, the resulting linewidth measurement will be a good approximation of the profile. In fact, even when the profile features are more prominent, results from using a square-profile model will reflect an average of the entire profile (9). Then, using this approximate linewidth, the profile model was used to generate an additional library. Experimental measurements indicated that the use of the profile model improved the match between theory and experiment for all 12 sites. Figure 19 depicts one of the experimental signatures from this data in comparison to the best-match modeled signatures from the square and profile

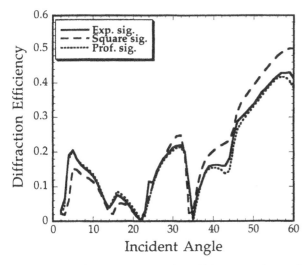

**Figure 19**  Theory (square and profile) vs. experiment for a photoresist grating. (From Ref. 35.)

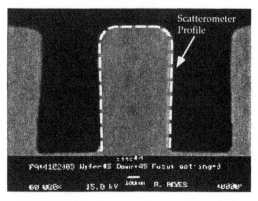

Top CD = 300 nm
Top Radius = 140 nm
Bottom CD = 615 nm
Bottom Radius = 40 nm
Sidewall Angle = 92°

Square CD = 550 nm
Scatterometry CD = 541 nm
SEM CD = 536 nm

**Figure 20**  Scatterometry profile results in comparison to a cross-sectional SEM image. (From Ref. 57.)

models. The use of the profile model clearly improves agreement between theory and experiment. The physical cross section of this site, in comparison to a SEM image, can be seen in Figure 20. The two profiles agree very well, including such features as the dimensions of the top corner radius (140 nm) and overhanging sidewalls (92°).

  Similar profile measurement results were obtained for the remaining locations on this wafer. Overall, with respect to all 12 locations, if the square-profile model was used in the diffraction model, the difference between the SEM and scatterometry CD measurements was a respectable 19.3 nm. However, if the profile model was used, the difference improved to 10.1 nm.

  Characterization of etched feature profiles using 2-$\Theta$ scatterometry has also been considered (57). Most prior work involving etched samples incorporated sidewall angle into the diffraction models—vertical sidewalls are not that common for these types of gratings. But for one set of samples reported in the literature, the addition of another profile parameter improved the matches to the model considerably. Figure 21 depicts the shape used in this modified trapezoid, or "stovepipe," model, in which the trapezoidal sidewall stops short of a square top section. This shape can be implemented with four parameters: the top square region, the overall height of the grating, and the height and sidewall angle of the trapezoidal region.

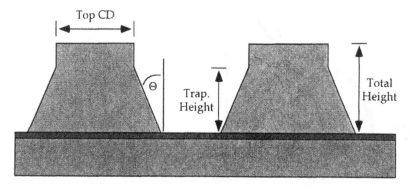

**Figure 21**  "Stovepipe" model used on the etched poly-Si profiles. (From Ref. 35.)

The stovepipe-profile model was applied to eight locations on a single etched wafer. The original trapezoid model improved the agreement to experimental data considerably (compared to using a simple square-profile model), and the use of the stovepipe model improved the match further still. An actual measured profile for one location on this wafer can be seen superimposed on the SEM photograph in Figure 22. The two profiles agree very well, both visually and dimensionally. The other sites yielded similar results, with good agreement in comparison to SEM photographs.

## V. FUTURE DIRECTIONS

In this chapter we have seen that applications of scatterometry within the semiconductor industry have been numerous and widespread. But given the nature of the technology, and the wide range of processes that exisit in any fab, scatterometry is bound to see new applications as the industry requires them. Indeed, in many ways the applications discussed here have only been the tip of the iceberg.

Of foremost importance for the future of scatterometry is requisite measurement sensitivity for the dimensions that will be encountered as the industry moves to smaller geometries. We have already discussed various ways in which the measurement sensitivity can be enhanced (38–44), but such improvements may not be needed for several years. The ultimate resolution of a scatterometer system is a function of two components: (1) the noise level of the scatterometer and (2) the signature sensitivity for the features being modeled. On the scatterometer side, typical reflectance measurements are capable of resolving $\sim 0.1\%$ differences. This 0.1% figure can be thought of as the "noise floor" of the system. On the model side, if two signatures for some parameter difference are distinguishable (the signatures separate beyond the tool noise), then the parameter change (resolution) is also taken to be distinguishable. Therefore, for two signatures to be considered

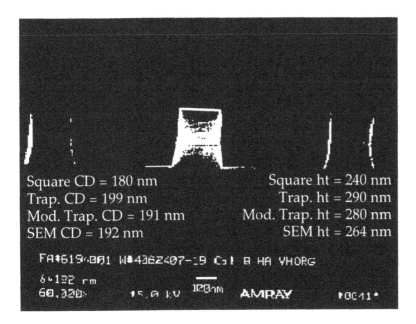

**Figure 22** Scatterometry and SEM profiles for an etched grating. (From Ref. 57.)

resolvable, they must differ by more than 0.1%. Although this signature separation will vary with the application, for purposes of assessing HeNe illumination scatterometry on small features we can model the generic lithography stack seen in Figure 7 (resist on BARC on silicon) and relate this to the tool noise level. For a 1:1 line/space aspect ratio, the inspection of 100-nm CDs will have gratings with a 200-nm pitch and, in accordance with the 1% rule, will require a measurement resolution of 1 nm. RCWT shows that 1-nm CD changes to this stack result in $\sim 0.2\%$ signature changes, so 1-nm signature differences for this stack are resolvable using a variable-angle HeNe scatterometer. Likewise, the same exercise can be performed for a nominal 70-nm CD stack with a 140-nm pitch. In this case the signatures need to be resolvable at the 0.7-nm level, and RCWT indicates the signatures differ by $\sim 0.15\%$ for these 0.7-nm parameter changes. Thus these signatures are also resolvable. It is not until CDs reach the $\sim 50$-nm point (in a 100-nm pitch) that the signature differences are about equal to measurement noise and the technology reaches a theoretical lower limit. Of course, optimized grating design, by way of pitch and height geometries or certain materials, as well as the incorporation of phase measurements may extend the method to even smaller resolutions.

Beyond CD metrology applications, the ability of diffraction-based instruments to perform overlay measurements has been explored, but this has not moved much beyond the literature. Both projection moiré techniques and interferometric diffraction methods have been proposed (59). Although both show promise, a pilot study in conjunction with a representative of the semiconductor industry would be an excellent proving ground for this application.

The advent of parallel computing coupled with advances to diffraction models will only broaden the extent of scatterometry applications. At present, modeling is moving from well-explored, one-dimensional grating applications to two-dimensional, real-device patterns. In the past, as has been reviewed here, empirical models have been used for characterizing doubly periodic patterns. It is only a matter of time before computation power advances to the point where these devices can be rigorously modeled.

In our description of the various flavors of scatterometers in Sec. II, one might have noticed that, fundamentally, the differences between ellipsometers, surface roughness/particle scanners, and the scatterometers discussed in this chapter are very slight. All rely on optical reflectance measurements and may use magnitude information, phase information, or both. All rely on models to derive a measurement, either theoretical or empirical. All can be configured for single-wavelength, fixed-angle operation or may use variable wavelengths and/or angles. In short, these instruments typify the nature of optical measurements today, particularly in the semiconductor industry. But tomorrow, it seems likely that one will simply purchase one optical inspection tool, complete with a variety of illumination sources, detection modes, and models. Then, by way of selecting an appropriate recipe, the user will configure this one tool to perform measurements of film thicknesses, CDs or lateral dimensions, or surface roughness/particle counts.

## VI. SUMMARY

In the history of semiconductor manufacturing, metrology has never been more of a concern than it is today. Complex processes, aggressive production standards, and strong demand have created a production environment that relies on metrology more and more. Especially attractive are those inspection technologies that are fast, accurate, and precise.

In this chapter we have reviewed the details of scatterometry, a promising alternative to existing metrologies. We began with descriptions of the various forms of scatterometers used to measure a scatter signature (known as the forward problem). We then illustrated the different ways the scatter data is analyzed, either theoretically or empirically, in order to determine specific process or metrology parameters (the inverse problem). Finally, we reviewed a number of significant applications of the technology in order to illustrate the strengths of the method compared to other inspection tools and to summarize the more pertinent measurements for which it can be used. In all instances scatterometry measurement results have compared favorably to other inspection technologies (the best way to assess measurement accuracy). In addition, the precision (measurement repeatability and reproducibility) of the technology has been shown to be excellent and to exceed that of any other method. For these reasons, it seems likely that the future of scatterometry—both in the fab and in the lab—remains bright.

## ACKNOWLEDGMENTS

The author wishes to thank Bob McNeil and Sohail Naqvi, who were original pioneers in this field and who are still making contributions to its development. In addition, I thank my colleagues who have made their own contributions to scatterometry: Steve Prins, Steve Farrer, Mike Murnane, Ziad Hatab, Shoaib Zaidi, Babar Minhas, Steve Coulombe, Petre Logofatu, Susan Wilson, Scott Wilson, Richard Krukar, Jimmy Hosch, and Chris Baum.

## REFERENCES

1. KM Monahan. Proc SPIE 2439:325–333, 1995.
2. JA Woollam, B Johs, CM Herzinger, J Hilfiker, R Synowicki, CL Bungay. Proc SPIE CR72:3–58, 1999.
3. NF Zhang, MT Postek, RD Larrabee, L Carroll, WJ Keery. Proc SPIE 2725:147–158, 1996.
4. Semiconductor Industry Association. The National Technology Roadmap for Semiconductors. 1994.
5. S. Wittekoek, H. Linders. Proc SPIE 565:22–28, 1985.
6. R Pforr, S. Wittekoek, R van Den Bosch, L van Den Hove, R Jonckheere, T Fahner, R Seltmann. Proc SPIE 1674:594–608, 1992.
7. MP Lang. M.S. thesis, University of New Mexico, Albuquerque, NM, 1992.
8. MG Moharam, TK Gaylord, GT Sincerbox, H Werlich, B. Yung. Applied Optics 23(18):3214–3220, 1984.
9. MR Murnane, CJ Raymond, ZR Hatab, SSH Naqvi, JR McNeil. Proc SPIE 2196:47–59, 1994.
10. HP Kleinknecht, H Meier. J Electrochem Soc: Solid-State Sci Tech 125(5):798–803, 1978.
11. HP Kleinknecht, H Meier. Applied Optics 19(4):525–533, 1980.
12. R Petit. Electromagnetic theory of gratings. In: Topics in Current Physics. Berlin: Springer-Verlag, 1980.
13. H Kogelnik. Bell Syst Tech J 48:2909–2947, 1969.
14. MG Moharam, TK Gaylord. J Opt Soc Amer 71(7):811–818, 1981.
15. SSH Naqvi, SM Gaspar, KC Hickman, JR McNeil. Applied Optics 31(10):1377–1384, 1992.
16. RD Jacobson, JR McNeil. Applied Optics 31(10):1426–1435, 1992.
17. KP Bishop, SM Gaspar, LM Milner, SSH Naqvi, JR McNeil. Proc SPIE 1545:64–73, 1991.
18. SSH Naqvi, JR McNeil, RH Krukar. J Opt Soc Am A 11(9):2485–2493, 1994.
19. JR McNeil, SSH Naqvi, SM Gaspar, KC Hickman, KP Bishop, LM Milner, RH Krukar, GA Petersen. Microlithography World 1(15):16–22, 1992.

20. SSH Naqvi, JR McNeil, RH Krukar, KP Bishop. Microlithography World 2(3): 1993.
21. CJ Raymond, MR Murnane, SSH Naqvi, JR McNeil. J Vac Sci Tech B 13(4):1484–1495, 1995.
22. CJ Raymond, MR Murnane, SL Prins, SSH Naqvi, JR McNeil. J Vac Sci Tech B 15(2):361–368, 1997.
23. C Baum, R Soper, S Farrer, JL Shohet. Proc SPIE 3677:148–158, 1999.
24. SM Gaspar-Wilson, SSH Naqvi, JR McNeil, H Marchman, B Johs, R French, F Kalk. Proc SPIE 2439:479–494, 1995.
25. S Bushman, S Farrer. Proc SPIE 3213:79–90, 1997.
26. J Bischoff, H Truckenbrodt, JJ Bauer. Proc SPIE 3099:212–222, 1997.
27. SA Coulombe, BK Minhas, CJ Raymond, SSH Naqvi, JR McNeil. J Vac Sci Tech B 16(1):80–87, 1998.
28. LM Milner, KP Bishop, SSH Naqvi, JR McNeil. J Vac Sci Tech B 11(4):1258–1266, 1993.
29. KC Hickman, SSH Naqvi, JR McNeil. J Vac Sci Tech B 10(5):2259–2266, 1992.
30. ZR Hatab, JR McNeil, SSH Naqvi. J Vac Sci Tech B 13(2):174–182, 1995.
31. SL Prins. Monitoring the Post-Exposure Bake Process for Chemically Amplified Resists Using a Scatterometric Sensor. PhD dissertation, University of New Mexico, Albuquerque, NM, 1996.
32. RH Krukar. A Methodology for the Use of Diffracted Scatter Analysis to Measure the Critical Dimensions of Periodic Structures. PhD dissertation, University of New Mexico, Albuquerque, NM, 1993.
33. JL Sturtevant, SJ Holmes, TG van Kessel, PC Hobbs, JC Shaw, RR Jackson. Proc SPIE 1926:106–114, 1993.
34. R Soper. Linewidth prediction based on real-time post-exposure bake monitoring. Unpublished internal SEMATECH report, Austin, TX, 1995.
35. CJ Raymond. Measurement and Efficient Analysis of Semiconductor Materials Using 2-Θ Scatterometry. PhD dissertation, University of New Mexico, Albuquerque, NM, 1997.
36. JR McNeil, S Wilson, RH Krukar. US Patent 5867276, filed 3/97, awarded 2/99.
37. X Niu, NH Jakaatdar, J Boa, CJ spanos, SK Yedur. Proc SPIE 3677:159-168, 1999.
38. SA Coulombe. Analysis of Dielectric Grating Anomalies for Improving Scatterometer Linewidth Measurement Sensitivity. PhD dissertation, University of New Mexico, Albuquerque, NM, 1999.
39. BK Minhas, SA Coulombe, SSH Naqvi, JR McNeil. Applied Optics 37:5112–5115, 1998.
40. SA Coulombe, PC Logofatu, BK Minhas, SSH Naqvi, JR McNeil. Proc SPIE 3332:282–293, 1998.
41. PC Logofatu, SA Coulombe, BK Minhas, JR McNeil. J Opt Soc Amer A 16(5):1108–1114, 1999.
42. SA Coulombe, BK Minhas, CJ Raymond, SSH Naqvi, JR McNeil. J Vac Sci Tech B 16(1):80–87, 1998.
43. E Yoon, CA Green, RA Gottscho, TR Hayes, KP Giapis. J Vac Sci Tech B 10:2230–2233, 1992.
44. BK Minhas, SL Prins, SSH Naqvi, JR McNeil. Proc SPIE 2725:729–739, 1996.
45. CA Balanis. Advanced Engineering Electrodynamics. New York: Wiley, 1989, pp 670–712.
46. BH Kleeman, J Bischoff, AK Wong. Proc SPIE 2726:334–347, 1996.
47. L Li, J Chandezon, G Granet, JP Plumey. Applied Optics 38(2):304–313, 1999.
48. ST Peng, T Tamir, HL Bertoni. IEEE Trans Microwave Theory and Techniques MTT-23(1):123–133, 1975.
49. MG Moharam, TK Gaylord. J Opt Soc Amer 72:1385–1392, 1982.
50. J Bischoff, JJ Bauer, U Haak, L Hutschenreuther, H Truckenbrodt. Proc SPIE 3332:526–537, 1998.
51. IJ Kallioniemi, JV Saarinen. Proc SPIE 3743:33–40, 1999.
52. S Eastman. SEMATECH Technical Transfer document 94112638A-XRF, 1995.
53. T Yamaguchi, S Kawata, S Suzuki, T Sato, Y Sato. Japanese J App Phys I 32(12B):6277–6280, 1993.

54. C Baum, S Farrer. Resist Line Width and Profile Measurement Using Scatterometry. SEMATECH AEC-APC Conference, Vail, CO, 1999.
55. CJ Raymond, JR McNeil, SSH Naqvi. Proc SPIE 2725, 1996.
56. JE Griffith, DA Grigg. J Appl Phys 74(9):R83–R109, 1993.
57. CJ Raymond, SSH Naqvi, JR McNeil. Proc SPIE 3050:476–486, 1997.
58. SMG Wilson. Light Scatter Metrology of Phase Shifting Photomasks. PhD dissertation, University of New Mexico, Albuquerque, NM, 1995.
59. SSH Naqvi, S Zaidi, S Brueck, JR McNeil. J Vac Sci Tech B 12(6):83–109, 1994.

# 19

# Unpatterned Wafer Defect Detection

**Po-Fu Huang, Yuri S. Uritsky, and C.R. Brundle**
*Applied Materials, Inc., Santa Clara, California*

## I. INTRODUCTION

As device design rules continue to shrink and die sizes grow, the control of particulate contamination on wafer surfaces has become more and more important in semiconductor manufacturing. It has been found that defects caused by particles adhering to the wafer surface were responsible for more than 80% of the yield loss of very-large-scale integrated circuits (VLSIs)(1). Although particulate contaminants could be introduced at any point during the wafer manufacturing and fabricating processes, particles generated within process equipment are the most frequent cause. Not only mechanical operations (e.g., valve movement, wafer handling, shaft rotating, pumping, and venting) but also the wafer processing operation (e.g., chemical and physical reactions) can produce particles.

Since the production of a device involves numerous processes and takes many days, it would be too late to look for the defects and their sources at the end of the process. Currently, defect metrology is carried out at critical process steps using both blanket (unpatterned) and patterned wafers. There is a dispute about whether defect control using blanket wafers is necessary. The opposing argument states that, in addition to the problem associated with the individual process step that can be identified using either blanket or patterned wafers, metrology on blanket wafers does not reveal any of the problems related to the integration of the processes and, therefore, should be eliminated. There are several additional factors to be considered, however. Inspection speed, cost, and sensitivity (the smallest size of particle detectable) are all better on blanket wafers. In addition, though a problem may be originally identified on a patterned (production) wafer, to isolate which tool and the root cause within that tool requires many partition tests using blanket wafers (either Si monitor wafers, for mechanical tests, or full film wafers, for process tests) to be performed. Finally, the specifications for particle adders in a tool/process and the statistical monitoring (baselining) of that specification are carried out on blanket wafers. Therefore, to ensure high yield in mass production, blanket wafer monitoring is absolutely necessary and will not go away any time soon.

The light-scattering-based technique for defect detection has been used for inspection of unpatterned wafers for more than two decades. For a perfectly smooth and flat surface, the light reflects specularly. In the presence of a surface contaminant or surface roughness, the specular light is disturbed, and, as a result, scattered light is produced. For a given incident light source, the amount of light scattered by the surface and the contaminant varies with the physical and optical properties of the surface and the contaminant, such as the size, shape, orientation, refractive index of the surface contaminant, and the thickness and refractive index of the substrate.

In the early days, skilled operators visually inspected wafer surfaces for particles in a darkened room. During such inspection, parallel beams of intensive light illuminated the wafer surface at an oblique angle, and the light scattered from any surface defect was observed with the naked eye. With this technique, experts are actually able to detect and count particles of sizes down to about 0.3 μm on a polished bare silicon wafer. However, the reliability, repeatability, and accuracy of these measurements are inadequate, because the concentration level of the inspector may vary from time to time, and the involvement of subjective judgment makes comparison between inspectors difficult. It is obviously also very slow and costly, requiring skill and experience.

During the last 20 years, great effort has been directed toward developing laser-based wafer-surface scanners as a metrology. The first commercial laser-scanning instruments appeared in the early 1980s. The amount of light scattered by a defect is related to its optical size. Current established automated laser-scanning systems are capable of detecting particles with minimum detection size, MDS, as small as 0.1 μm in optical diameter on a smooth surface (note that this does not mean they can provide accurate statistics down to this level). Very recently, new systems with MDS values of 0.06–0.08 μm on a polished bare silicon wafer have become available. These wafer-surface scanners provide information such as the size, number, and location of each particle detected on the wafer surface. The semiconductor manufacturing industry uses this information (size, counts, and location) for product qualification, manufacturing tool monitoring, and the first step in defect root-cause analyses. As the critical dimensions of semiconductor devices continue to shrink, the size of "killer defects" also decreases. The ability to accurately and repeatably detect, locate, and size such small particles on a wafer surface with these instruments is, therefore, of paramount importance.

## II. PRINCIPLES OF LIGHT SCATTERING

In a surface scanner, a narrow beam of light illuminates and is scanned over the surface. For a perfectly smooth and flat surface with no surface contaminant, the light reflects specularly (i.e., the angle of reflected light equals the angle of incident light, Figure 1a). In the presence of a surface defect, which can be a particle on the surface, surface roughness, or subsurface imperfection, the specular light is disturbed and, as a result, a portion of the incident light is scattered away from the specular direction (Fig. 1b and c). In the specular direction, an observer sees the region causing the scattering as a dark object on the bright background (bright field). In contrast, away from the specular direction the observer sees a bright object on the dark background (dark field). Both bright-field and dark-field techniques are used for wafer inspection, but the latter gives a better detection sensitivity for small defects. A photomultiplier tube (PMT) is used to collect the scattered light in the dark field. The amount of light scattered from the surface imperfections, which depends on the physical and optical properties of these imperfections, is measured as the light scattering cross section, $C_{sca}$, measured in square micrometers.

The study of light scattering by particles can be dated back to the 19th century. The general theory of light scattering by aerosols was developed in 1908 by Gustav Mie (2). It gives the intensity of light ($I$) scattered at any angle, θ, by a sphere with a given size parameter (ratio of the perimeter of the sphere to the wavelength of the incident light

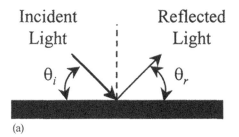

Incident Light    Reflected Light

$\theta_i$    $\theta_r$

(a)

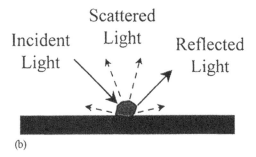

Scattered
Incident    Light
Light                Reflected
Light

(b)

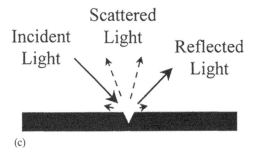

Scattered
Incident    Light
Light                Reflected
Light

(c)

**Figure 1**    Light scattering from (a) a perfect surface, (b) a surface with particle, and (c) a surface with a pit.

$(\pi d/\lambda)$ and complex refractive index $(m)$ that is illuminated by light of intensity $I_0$ (W/cm²) and wavelength $\lambda$. Basically, the relationship between the light-scattering cross section and the particle size can be divided into three regimes according to the particle size $(d)$ relative to the wavelength of incident light $(\lambda)$. For particles much smaller than the wavelength of incident light (i.e., $d < 0.1\lambda$), Rayleigh-scattering theory applies. In this case the light-scattering cross section for light from a particle with a complex refractive index of $m$ is

$$C_{sca} = \frac{2\pi^5}{3} \frac{d^6}{\lambda^4} \left| \frac{m^2 - 1}{m^2 + 2} \right|^2 \qquad \text{for } d \ll \lambda \tag{1}$$

For particles larger than $0.1\lambda$, the Mie equations must be used to determine the angular distribution of scattered light. For a sphere suspended freely in a homogeneous medium, the intensity of the unpolarized scattered light at a distance $R$ in the direction $\theta$ from the center of the sphere is then given by

$$I(\theta) = \frac{I_0\lambda^2(i_1 + i_2)}{8\pi^2 R^2} \qquad \text{for } d \sim \lambda \tag{2}$$

where $I_0$ is the intensity of the light source and $i_1$ and $i_2$ are the Mie intensity parameters for scattered light with perpendicular and parallel polarization, respectively.

For particles that are much larger than the wavelength of the incident light, simple geometric optical analysis applies, and the scattering cross section equals approximately twice the geometric cross section; i.e.,

$$C_{sca} \approx \frac{\pi d^2}{2} \qquad \text{for } d \gg \lambda \qquad (3)$$

In the case of a particle adhering on the surface of a wafer, the scattered field becomes extremely complicated due to the complex boundary conditions of the surface. The amount of light scattered from the surface depends not only on the size, shape, refractive index (i.e., composition), and orientation of the particle, but also on the profile (e.g., thickness, surface roughness) and refractive index (i.e., composition) of the surface. The light-scattering system for a surface has been solved exactly by Fresnel. Approximations that combine these two theories to solve the complex system composed of a particle and a surface have been determined by researchers (3,4). Details on these developments are beyond the scope of this review.

For a real wafer, the surface is neither perfectly smooth nor without any surface defects. Some of the examples of surface imperfection that diminish the yield are particles, crystal-originated particles (COPs), microroughness, ions, heavy metals, organic or inorganic layers, and subsurface defects. The light scattered from each of these imperfections is different. The signal generated by general surface scatter (e.g., roughness, surface residues) is not at a discrete position, as is the case for a particle, but rather is a low-frequency signal, observed across the effected regions of the wafer. This low-frequency signal is often referred to as *haze*. Figure 2 illustrates a signal collected along a scan line (this is part of the "review mode" operation of the scanner—a detected defect is examined in more detail). The signal from a discrete defect sits on top of the background haze. The MDS is limited by the *variation* of this haze background ("noise"), which is statistical, plus effects of varying microroughness. It corresponds to a signal-to-niose ratio ($S/N$) of

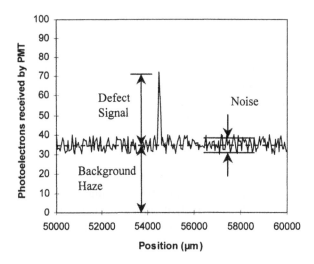

**Figure 2** Signal, background, and noise: light-scattering signal in a scan across a light point defect (LPD) (review mode).

about unity. However, a ratio of 3 is recommended (although a ratio of 5 is more conservative) and is standard industry practice, to separate the defect signal from the background haze with the high certainty needed by industry.

The "haze" value is defined as the ratio of the scattered light, collected by the PMT, to the incident light and is expressed in parts per million (ppm). This haze value ($\gamma$) is a measure of the surface quality, such as uniformity, and can be related to the RMS surface roughness ($\sigma$) by the surface reflectivity ($R_0$) and wavelength of the incident light ($\lambda$), with the assumption that nearly all of the scattered light is collected (5):

$$\sigma = \frac{\lambda}{4\pi} \sqrt{\frac{\gamma}{R_0}} \qquad (4)$$

A ray of light behaves like an electromagnetic wave and can be polarized perpendicularly (S) or parallel (P) to the incident plane, which is the plane through the direction of propagation of light and the surface normal. Figure 3 plots reflectivity against the thickness of an overlayer on a substrate (in this case oxide on Si) for S and P polarization. On a bare surface, or one with only a few angstroms of native oxide, S polarization has a maximum in reflectivity and therefore a minimum in scattering. P polarization behaves in the opposite manner. Therefore for rough bare surfaces, the S polarization channel can be used to suppress substrate scattering and hence to enhance sensitivity to particle defects. For specific overlayer thickness, destructive interference between surface and overlayer/substrate reflectivity reverses the S and P behavior (see Fig. 3).

## III. CONFIGURATION AND OPERATING PRINCIPLES OF LIGHT-SCATTERING PARTICLE SCANNERS

There is a variety of commercial particle scanners available, but only a few are in common use. There is an intrinsic difference in requirements for detection of defects on patterned versus unpatterned wafers. For patterned wafers it is of paramount importance to recognize defects that are violations of the pattern itself. To do this requires complex image

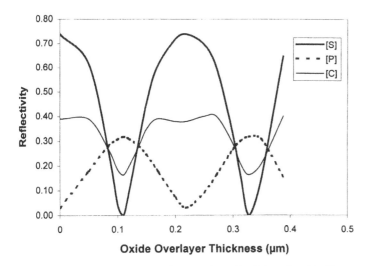

**Figure 3** Effect of polarization on the reflectivity of oxide films.

recognition, which is mostly a software and algorithm issue. Defect survey tools designed to do this compare die to die patterns and look for differences. These differences might be opens, shorts, or dimensional variations in the pattern, or they may be added defects, such as resist residue and particles. Their real strength is in the die-to-die comparison and not necessarily in their sensitivity down to very small sizes (impossible anyway in highly patterned product wafer surfaces). They can be, and are, also used for blanket wafers and in some cases have some advantages (e.g., improved $X$, $Y$ location accuracy and possibly higher "capture rate"—see later). These tools, however, tend to be two to five times as expensive as the blanket wafer scanners, have a larger footprint, and be much slower. The tools designed specifically for blanket wafers concentrate more on speed and the best sensitivity possible. In this section we describe the basic configuration, operating principles, and capability limitations of the most popular models in use today. How well some of them do in determining the parameters of importance (sensitivity, repeatability, capture rate, false count rate, $X$, $Y$ accuracy) is discussed later.

## A. KLA-Tencor Surfscans

### 1. 6200 Series

The KLA-Tencor Surfscan 6200 series (Figure 4a), which replaced the older 4000 and 5000 series, illuminates the wafer surface using a normal incident laser beam. The incident laser beam is circularly polarized and is swept across the wafer by an oscillating mirror (170 Hz) in the $X$ direction, while the wafer is being held and transported by a vacuum puck in the $Y$ direction. The combination of these two motions provides a raster scan of the entire wafer surface. With the illuminated spot and detector on its focal lines, a cylindrical mirror of elliptical cross section is placed above the wafer to redirect scattered light to the detector. The collection angle of the 6200 series is limited by the physical build of this mirror and is illustrated in Fig. 4b. After loading, each wafer undergoes two scanning cycles. The prescan (fast) is for haze measurement and edge detection, while the second scan (slow) is for surface defect detection and edge/notch determination. Any defects scattering light above a defined threshold of intensity are categorized as light point defects (LPDs). The 6200 series is designed for smooth surfaces, such as bare silicon, epitaxial silicon, oxides, and nitrides. For rough surfaces, such as polymers and metals, the sensitivity of the 6200 degrades severely. The reason for this can be traced to the scattering geometry. Normal incidence, and the detector's collecting over a wide solid angle, constitutes an efficient arrangement for collecting scattering from the surface roughness, thereby decreasing the signal/noise ratio, i.e., the particle scattering/surface background scattering.

### 2. KLA-Tencor 6400 Series

Unlike the 6200 series, the 6400 series uses so-called double dark-field technology (Figure 5a), a low incident angle ($70°$ from the surface normal), and a low angle of collection (fixed low-/side-angle collection optics). Both the incident laser beam and collected light can be polarized perpendicularly (S) and in parallel (P). In addition, the collected light can be left as is (i.e., unpolarized, U) and the incident can be circularly polarized (C). An optimal arrangement of incidence-collection polarization helps to suppress the haze signals due to an imperfect surface of the substrate and leads to a better signal-to-noise ratio for defect detection. As a result, the 6400 series works better than 6200 series on rough surfaces, such as metal films. For example, the P-P polarization is the most sensitive to the surface

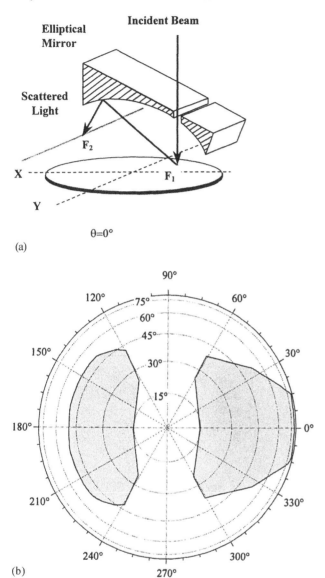

**Figure 4**  Tencor Surfscan 6200: (a) configuration, (b) angular response (shaded area).

roughness and is used for smooth surfaces, such as bare silicon wafer. The C-U polarization is for oxide, nitride, and other moderate surfaces. The S-S polarization is the least sensitive to surface roughness and is suitable for rough surfaces, such as metal. However, because of the fixed low-/side-angle collection optics, the collection angle varies depending on the position of the laser spot with respect to the detector (Fig. 5b). As a result, a sphere located on the right-hand side of the wafer could appear twice the size as when located on the left-hand side of the wafer. Furthermore, the grazing-incident angle exacerbates the variation of the size and shape of the laser spot as the beam sweeps across the wafer. The end result of these configurational differences from the 6200 series is that (a) on smooth

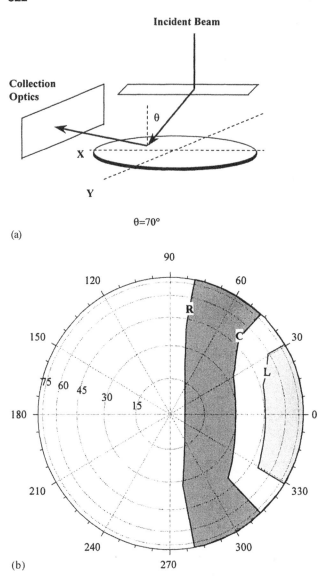

(a)

(b)

**Figure 5** Tencor Surfscan 6400: (a) configuration, (b) angular responses for a laser spot on the far side (enclosed by the line L), at the center (line C), and on the near side (line R) of a 200-mm wafer.

surfaces the sensitivity (i.e., minimum detectable size) is poorer, (b) on rough surfaces the sensitivity is better, and (c) the sizing accuracy is much more uncertain and dependent on the orientation of the defect (unless it is spherical).

## 3. KLA-Tencor SP1 Series

The SP1 series was designed for handling 300-mm wafers in addition to 200-mm wafers. To minimize the distance of wafer transportation during the scanning process, the laser beam is stationary while the wafer rotates and translates under the laser beam (Figure 6a).

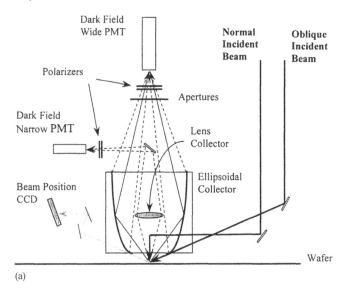

(a)

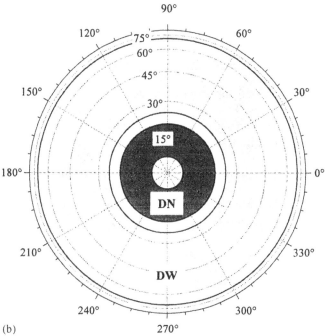

(b)

**Figure 6** KLA-Tencor Surfscan SP1: (a) configuration, (b) angular response (5°-20° for narrow channel; 25°-70° for wide channel).

The basic version of the SP1 (so-called SP1-classic) is equipped with a normal laser beam. Similar to the 6200 series, this is mainly for smooth surface inspection in dark field. Differential interference contrast (DIC) technology is used in the bright-field detection. Any surface feature with appreciable slope change can be detected as a phase point defect (PPD). This bright-field DIC channel can detect large defects, about 10 μm and larger, such as mounds, dimples, and steps. An additional oblique beam (70° from the surface normal) is added to the TBI (triple beam illumination) version for inspection of rough surfaces (i.e., the SP1-TBI combines the capabilities of the 6200 and 6400). Two sets of collection optics (channels) are used to collect scattered light in the dark field. The dark-field narrow channel (DNN for normal beam and DNO for oblique beam) collects scattered light from 5° to 20° from the surface normal using a lens collector. The dark-field wide channel (DWN for normal beam and DWO for oblique beam) equipped with an ellipsoidal mirror collects scattered light from 25° to 70° from the surface normal. These different angular configurations offer the *possibility* of distinction between different *types* of defect (e.g., a signature for a shape or particle/scratch distinction). This area of application is not well developed, but we report our efforts in this direction later. Since the normal beam and oblique beam share the same collection optics, they cannot be used at the same time. Also, since both the lens collector and the ellipsoidal mirror are axisymmetrical, the collection angle (Fig. 6b) covers the complete circle. Therefore, the SP1 is not sensitive to the *orientation* of defects (unlike the 6400).

## B. Other Inspection Tools

### 1. Applied Materials WF-7xx Series

The Applied Materials (Orbot) WF-7xx series is intended primarily as a patterned-wafer inspection tool. Four PMTs are located at 45° compared to the wafer axis to filter out signals scattered from edges of dies, and 10–20° up from the wafer surface to increase the signal-to-noise ratio (Fig. 7). Five objectives (1.6x, 3x, 5x, 10x, and 20x) can be used for fast inspection and slow inspection with good sensitivity. Scattered light is collected by these four PMTs and then converted to gray-level values at each pixel location via the perspective dark-field imaging (PDI) technique. Pixels with alarms are identified in each channel. Ranking is then assigned to each alarm location based upon information from the four channels. The alarm locations are grouped together and identified as defects. A defect map is then generated. Although the WF-7xx series is used primarily for patterned-wafer inspection, they can be used for unpatterned wafers as well. The pixel size ranges

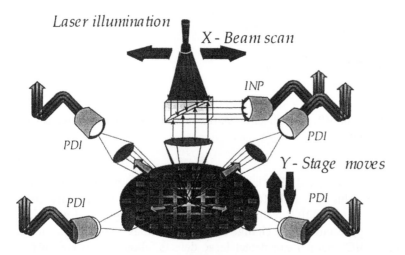

**Figure 7** Optical configuration of WF7xx patterned-wafer inspection tool.

from 0.5 μm to 7 μm, depending on the selection of objectives. It is claimed that sub-0.1-μm defects are detectable in favorable cases.

## 2. KLA AIT

The AIT is a patterned-wafer defect detection tool employing low-angle incident light and two PMT detectors at a low collection angle (Figure 8). It has tunable S, P, circular polarization of incident and collected light to allow several combinations of polarization settings. The AIT also has a tunable aperture and programmable spatial filter for capability on patterned wafers. It appears that the AIT is being used in industry for both patterned and unpatterned defect detection. This is not because of any advantage over the 6000 or SP1 series for sensitivity or accuracy in unpatterned wafers, but because it is fast enough and versatile enough to do both jobs fairly well.

## 3. KLA 2100

The 2100 series has been the Cadillac of scanners for a number of years for patterned-wafer inspection. Unlike all the other tools, it is a true optical imaging system (not scattering). It was very expensive (maybe four or five times the cost of the 6000 tools), but gave accurate image shape and size information, with close to 100% capture rate (i.e., it didn't miss defects), but *only* down to about 0.2–0.3 μm. Because it scans the whole wafer in full imaging mode, it is also slow. In addition to its use for patterned wafers, it is used as a reference, or verification, tool to check the reliability and accuracy of the bare wafer scanners.

## IV. REQUIREMENTS FOR A RELIABLE MEASUREMENT

Wafer inspection is the first and crucial step toward defect reduction. When a particle/defect excursion is observed during a semiconductor manufacturing process, the following questions are asked: How many defects are on the surface? Are they real? How big are they? What are they? Where do they come from? How can they be eliminated? Wafer inspection provides the foundation for any subsequent root-cause analysis. Therefore, the

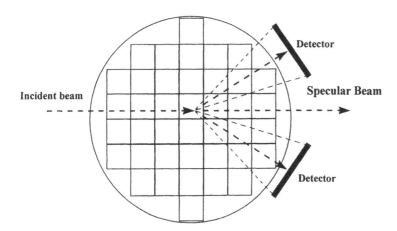

**Figure 8** Optical configuration of AIT patterned-wafer inspection tool.

success of defect reduction relies on an initial reliable wafer inspection by light scattering. There are five requirements to be considered for a reliable measurement.

## A. Variation of Measurements

First of all, the variation of the measurements made by a well-trained operator under a well-controlled environment must be acceptable. The counting variance, the variation of defect counts, has been used as an indication of the reliability of a wafer inspection tool. However, since the counting variance is in direct proportion to the average number of defect counts, it is not a fair measure of the reliability of the measurement. The coefficient of variation (CoV), which takes repeatability and reproducibility into account, is a better measure of the uncertainty of the measurements of a given inspection tool. The coefficient of variation is defined as

$$CoV = \frac{\sqrt{\sigma_{RPT}^2 + \sigma_{RPD}^2}}{2(\mu_{RPT} + \mu_{RPD})} \tag{5}$$

where $\mu_{RPT}$, $\sigma_{RPT}$, and $\mu_{RPD}$, $\sigma_{RPD}$ are the average count and corresponding standard deviations of the repeatability test and reproducibility test, respectively. The repeatability is normally determined by the accepted industry approach of measuring a given test wafer continuously 30 times without any interruption. The reproducibility test can be carried out either in the short term or the long term. The short-term reproducibility is determined by measuring the test wafer 30 times continuously, loading/unloading the wafer between scans, while the long-term reproducibility is obtained by measuring the test wafer regularly once every day for 30 days.

Due to their well-defined spherical shape and commercial availability in calibrated sizes down to very small values (0.06 μm), polystyrene latex (PSL) spheres have been used to specify the performance of wafer inspection tools. However, real-world defects are hardly spherical and exhibit very different light-scattering behavior than PSL spheres. A low variation of measurements of PSL spheres, therefore, does not guarantee the same result for measurements of real defects. Custom testing is needed for characterizing the performance of the tool for typical defects of interest. This is done whenever the metrology tool performance becomes an issue.

## B. Sizing

Defects must be sized in a reproducible and repeatable manner, since it is the total number of defects larger than a certain size that is most commonly used as a specification in a wafer process. Such a number is meaningless if the cut size (threshold) of the metrology tool used and the surface scanner cannot be precisely stated. The sizing is achieved by calibrating the light-scattering response (e.g., the cross section, in μm²) of known size defects. Since PSL spheres can be obtained in known graduated sizes over the particle size ranges of interest, they have become the standard for this calibration.

Since real defects are neither polystyrene latex nor (usually) spheres, it must be kept in mind that the reported size of a real defect using this industry standard calibration approach represents a "PSL equivalent size" and does not give real sizes. We will refer to this again several times, but note here that it is *currently* considered more important to the industry that the sizing be repeatable, with a good precision, than that it be *accurate*; that is, the goal is that for a given, non-PSL, defect, everyone gets the same "PSL equivalent size," independent of which specific scanner or model of scanner is used. A careful calibration procedure is essential if this is to be achieved, and it is often *not* achieved.

The experimental calibration procedure involves at least three steps: the deposition of known-size PSL spheres; the measurement of the light-scattering cross section for these spheres, and the construction of the calibration curve. Prior to deposition, the wafer to be used for calibration must be scanned to ensure that it is representative of the wafers for which the eventual calibration is going to be used. It should have essentially the same haze and noise response (which is going to define the lower size limit for *reliable* particle size detection) and the same light-scattering response for a given-size PSL sphere. This last point simply means it must be of the same substrate material; a calibration for particles on 2000-Å oxide must be done using PSL spheres on a 2000-Å oxide wafer (preferably one from the same process recipe in the same tool, to ensure no material or microroughness differences).

Deposition of the PSL spheres is done with a commercial particle deposition system, PDS (e.g., the MSP2300D), which can sort sizes (via a differential mobility analyzer, DMA), filter out any contaminant particles that will be of a different size, and eliminate doublets or triplets. The PSL spheres can be deposited either in spots or on the full wafer. Spot detection can give a much higher deposition density. This high density is necessary for the smallest PSL spheres in order to distinguish them from the haze background of the wafer. The minimum detectable size (MDS) can be determined by the appearance of the deposited spot containing that size PSL sphere, in a scanned wafer map. This occurs at an $S/N$ ratio of about unity. This does *not* mean you can use the tool, practically, down to that level. Remember, an $S/N$ ratio of 3–5 (depending on which model) is recommended to *reliably* distinguish particle LPDs from haze LPDs with low "false counts" (see later). This MDS is also referred to as the *noise equivalent size*. The spot diameter in the deposition varies, depending on the size of PSL spheres and the settings of the PDS, but on the MSP2300D it is always possible to have at least eight deposited spots on one 200-mm wafer.

The disadvantage of using spot deposition is that the wafer may have a nonuniform haze level (quite common), which could affect the results in any given spot. Full wafer deposition averages out such localization effects, giving a more representative result. It is, therefore, better to use full wafer deposition whenever possible, i.e., for all but the smallest PSL spheres used in the calibration.

Owing to the spherical shape of the PSL spheres, destructive interference occurs at specific sizes for a given laser wavelength (488 nm is the value used in most scanners) and substrate, giving rise to dips in the calibration curve. These data points should be excluded from the calibration. For the same reason, the same set of PSL sphere sizes should be used for all calibrations. Both these points are illustrated in Figure 9. A set of PSL sphere sizes corresponding to the upper curve is obviously different from that constructed from the set of PSL sphere sizes corresponding to the lower curve.

Figure 10 shows Tencor 6200 calibration curves for a bare Si substrate and for a 2000-Å oxide film. The bare calibration goes down to 0.1 μm (MDS), but the 2000-Å oxide goes only to 0.2 μm, because of the rougher oxide surface. The curves are different because of the effect of substrate material on the scattering cross sections. If the bare Si calibration were used in error for the 2000-Å oxide it would result in defects being undersized in "PSL equivalent" by 50–300%, depending on the size. Another way of expressing this is that if a spec was required to be set at 0.3 μm for the oxide, then using the bare Si curve would create an actual threshold of 0.415 μm. All particles between 0.3 and 0.415 μm would be missed, giving a lower (wrong) total count. Given the importance of these specs, it is

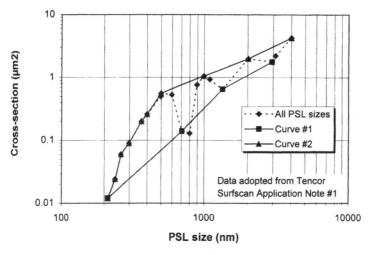

**Figure 9** Calibration curves for bare silicon on the Surfscan 4500.

absolutely essential that the proper calibration curve for the surface scanner be used and that, for each change in material, thickness of material, or even change in process recipe, a new calibration be done and a new recipe set. Without this, total particle counts will be under- or overestimated, and, at the limit of sensitivity, haze will be confused with the real particle counts.

None of the foregoing procedures, which are aimed at giving *reproducible* results, addresses the issue that real defects are not PSL spheres, so their *true sizes* differ from their "PSL equivalent" sizes. There are, in principle, three ways of relating the "PSL equivalent size" to true size, but in practice only one is used: that is to *review* the defects by SEM (or optical microscopy if the defect is large enough). From our long experience of doing this on the 6200 series (see Sec. III.A.1 and Chapter 20, Brundle and Uritsky), we can give some rough guides for that tool. If the defect is a particle in the 0.2–1-μm range, 3-dimensional in shape, and with no strong microroughness on it, the "PSL equivalent" size is often correct within a factor of 2. For particles smaller than 0.2 μm or much larger than 1 μm,

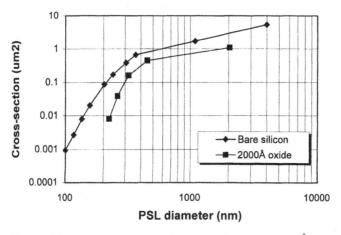

**Figure 10** Calibration curves for bare silicon and 2000-Å oxide substrates on the 6200.

the "PSL equivalent" size may be off by a factor of 2–10. For particles that are very 2-diminsional, the correspondence can be worse. Finally, for defects that are *not* particles (e.g., pits, scratches, COPs, bumps), no correlation at all should be expected.

The other possible methods for determining actual sizes are (a) either to construct calibration curves using real particle deposition of known sizes, such as Al, $Al_2O_3$, W, and $SiO_2$, or to back-construct them from compilations of data comparing PSL equivalent sizes to SEM determined true sizes, or (b) to model the light-scattering behavior from fundamental principles. Progress has been made in this area, but there is no data incorporating the actual optical configurations of commercial scanners that would allow translation of the theory (already verified by scatterometry for some situations) into practical use.

Industry, of course, is well aware that real sizes are not being determined. As stated earlier, it is considered more important that measurements be repeatable and verifiable. To this end, specs usually also define the model scanner (wisely not wanting to deal with the issues of cross-calibration among different models). Obviously there is no real connection between the ITRS and size requirements (which presumably refer to the effects of real defect sizes!) and industry practice.

## C.   Capture Rate

An absolute definition of *capture rate* would specify 100% when all of the particles of interest are detected. Now obviously, if you are working close to the MDS there is a sizable probability of missing a significant fraction. (The very term *minimum detection size* implies calibration with PSL spheres and, therefore, a discussion on capture rates of PSL spheres.) One moves significantly higher than the MDS threshold (i.e., to the $S/N$ ratio of 3:1 suggested by KLA/Tencor for the 6200) to avoid such statistical variations; i.e., you move to a higher bin size to create a high level of precision. In this situation, what reasons are there for the capture rate of PSL spheres to be less than 100%? Other than the small percentage of variability discussed in Sec. V.B., which is basically due to counting electronics, there should be only one fundamental reason that is significant: The area defined by the laser spot size and the raster of the laser beam means PSL spheres that are close enough together will be counted as one LPD event, reducing the capture rate. The amount below 100% depends on the distribution density of the PSL spheres. In actual use we will see later that there seem to be instrumental factors not under the control of the user that also affect capture rate significantly.

How is the capture rate determined? A common procedure used to be to use the KLA 2100 imaging scanner as the benchmark, i.e., with the assumption it has a 100% capture rate. This works only for *real* sizes larger than 0.2 μm. At smaller sizes there is no practical way other than relying on the PDS to know how many PSL spheres are deposited into a spot and then to check this with SEM direct searching and compare the total found to that found by the scanner.

Once one moves away from PSL spheres, *capture rate* has a significantly different meaning and is rarely an absolute term, since *real* defects may scatter very differently from PSL spheres. For instance, a physical 1-μm defect with a 0.1 "PSL equivalent size" in one model scanner may have a very different value in another model because of strong angular effects not present for PSL spheres. The practical definition of *capture rate*, then, becomes comparative rather than absolute. For example, CMP microscratches of a certain character (see later) are detected effectively in a particular SP1-TBI mode, but only to about

5% of that number in the 6200. One would then say that the *capture rate* of the 6200 was 5% (relative to SP1-TBI).

This again gets us onto dangerous political grounds. Should we be giving numbers in terms of "approved PSL equivalent" specifications, or should we try to establish their physical meaning? Probably both. In this case, it is advisable when working with real defects of a known type, which can *only* come after defect review (see Chapter 20), to get a feel for the capture rate for that particular defect for the scanner used, by both comparing to other scanners and verifying by SEM/EDX. In general one might reasonably expect that a demonstrated higher capture rate for PSL spheres for one scanner versus another might lead to a higher capture rate for real defects; but this always needs verifying.

## D. False Counts

False counts is a way of checking, when there is doubt, on whether LPDs represent real defects. What causes false counts? Remember the recipe has been set with the detection threshold at a level ($S/N = 3{:}1$, for instance) such that there is confidence there are no false counts in the smallest bin used (the threshold defines that size). Given this, the only way a significant number of false counts can occur (industry likes there to be less than 5%) is if the haze and/or noise level, $N$, increases. This can happen if, for the particular wafer in question, the surface roughness has increased (either across the whole wafer or in patches). All this is saying is that the recipe used is now inappropriate for this wafer (or patch on the wafer), and there will be doubt about the validity of the number of defects detected in the smallest bin size. Such doubt usually arises when an apparent particle count rises without *any* changes having occurred in the process being monitored.

The first step in establishing whether there *are* false counts is to revisit LPDs in the smallest bin size in review mode (e.g., Fig. 2), and establish whether each $S/N$ is greater than 3. If $S/N$ is lower than 3, this LPD must be revisited by optical imaging or SEM review to establish if it is genuine (optical imaging will be appropriate only if we're talking about a large minimum bin size). If nothing is detectable, the LPD is considered a false count (or a *nuisance defect*). If the number of false counts is found to be too high, the recipe for the scanner has to be changed, increasing the threshold, and the wafer rescanned. Now the minimum bin size being measured will be larger, but the number of LPDs detected in it will be reproducible.

## E. Defect Mapping

Defect mapping is important at three different levels, with increasing requirements for accuracy. First, the general location of defects on the wafer (uniform distribution, center, edge, near the gate valve, etc.) can give a lot of information on the possible cause. Second, map-to-map comparison is required to decide what particles are *adders* in a particular process step (pre- and postmeasurement). Finally, to review defects in a SEM, or any other analytical tool, based on a light-scattering defect file requires $X$, $Y$ coordinates of sufficient accuracy to be able to refind the particles.

*Targeting error* is the difference between the $X$, $Y$ coordinate values, with respect to some known frame of reference. For the general distribution of defects on a wafer, this is of no importance. For map-to-map comparison, system software is used to compare pre- and postcoordinates and declare whether there is a match (i.e., it is not an adder). However, if the targeting error is outside the value set for considering it a match (often the case), a wrong conclusion is reached. Here we are talking of errors of up to a few hundred micrometers compared to search match routines of 100 μm. For defect review by SEM/EDX, high accuracy is crucial. To reliably detect a 0.1-μm defect in SEM without a lot of time-searching requires an accuracy of ±15 μm in coordinates. None of the blanket wafer scanners approach this accuracy.

The targeting error is a combination of the following sources of error.

*Random error due to spatial sampling of the scattered light.* This type of error arises from the digital nature of the measurement process. As the laser spot sweeps across the wafer surface, particles and other surface defects scatter light away from the beam. This scattered light signal is present at all times that the laser spot is on the wafer surface. In order to process the scattered light signals efficiently, the signal is digitized in discrete steps along the scan direction. Unless the defect is directly under the center of the laser spot at the time a sample is made, there will be error in the defined coordinates of this defect. Depending on the laser spot size and the sampling steps, this type of error can be as much as ±50 μm.

*Error due to the lead screw nonuniformity.* It is assumed that the wafer translation under the laser beam is linear in speed. However, random and systematic errors exist, due to the imperfection of the lead screw, and will be integrated over the travel distance of the lead nut. The contribution of this type of error depends on the wafer diameter and the tolerance of peak-to-peak error of the lead screw.

*Error due to the sweep-to-sweep alignment (6200 and 6400 series).* The 6200 and 6400 series use a raster-scanned laser spot to illuminate the surface contaminant. In order to keep the total scan time as short as possible, data is collected on sweeps moving from right to left across the wafer and from left to right on the next consecutive pass. To align between consecutive sweeps, a set of two high-scattering ceramic pins is used to turn the sampling clock on and off. Random errors of as much as twice the size of the sampling step could occur if sweeps are misaligned.

*Error due to the edge/notch detection.* When the start scan function of the 6200 or 6400 series is initiated, the wafer undergoes a prescan for edge detection as well as the haze measurement. The edge information is gathered by a detector below the wafer. The edge information is stored every 12 edge points detected. The distance between successive edge points could be as much as few hundred microns. Since curve fitting is not used to remove the skew between sweeps, a systematic error on edge detection results. Similar systematic errors exist for the SP1 series. The center of the wafer is defined by the intersection of two perpendicular bisectors from tangent lines to the leftmost edge and the bottom-most edge. Once the center of the wafer is found, the notch location is searched for. The notch location error can be significantly affected by the shape of the notch, which can be quite variable. The edge information (the center of the wafer) and the notch location (the orientation of the wafer) are what tie the defect map to the physical wafer.

*Error due to the alignment of the laser spot to the center of rotation.* This type of systematic error applies only for a tool with a stationary laser beam, such as the SP1 series.

*Error due to the alignment of the center of the wafer to the rotation center of the stage.* This type of systematic error arises only for a tool with rotational stage, such as the SP1 series.

The final mapping error is a combination of all of the types of errors just described and can be characterized by a first-order error and a second-order error. The first-order error is the offset of defects after alignment of the coordinate systems. The second-order error (or point-to-point error) is the error of the distance between two defects after correction of the misalignment of the coordinate systems. The first-order error has not received as much attention as its counterpart. However, the success of map-to-map comparison depends not only on point-to-point accuracy but also on the first-order mapping accuracy. If a defect map has a first-order error of as much as 700 µm, which it can have at the edge of a wafer using the SP1, the software routines for map-to-map comparison will fail, even if the second-order point-to-point accuracy is very good.

## V.  CHARACTERIZATION OF KLA/TENCOR PARTICLE SCANNER PERFORMANCES

In this section, practical applications for the use of KLA/Tencor scanners (6200, 6400, SP1, SP1-TBI) are given. The purpose of these studies is to evaluate the performance of the Surfscans and establish a better understanding of these tools so as to better interpret particle measurement data, and to establish their most reliable mode of usage.

### A.  Capture Rate and Sizing of PSL Spheres

The counting performance of these instruments under either low or high throughput was characterized using a bare silicon wafer with PSL spheres (0.155 µm). A total of about 1000 spheres were deposited on a 200-mm bare silicon wafer using a MSP2300D particle deposition system. The wafer was measured five times using each tool. Since this is with a bare Si wafer, normal practice often is to rely on the generic, KLA/Tencor-supplied, calibrations. Doing this the average determined size of these PSL spheres was 0.151 µm for the SFS 6200 and 6400 and 0.156 µm for the SP1 series. The differences from the known true 0.155 µm represents a difference of the system electronics (e.g., a change in the gain of the PMT) compared to the "standard" for the generic calibrations. This could be a drift or simply an incorrect setting.

The capture rates for the different instruments, assuming exactly 1000 particles were deposited (unknown), are shown in Figure 11. Clearly, in the low-throughput mode the 6200, SP1, and SP1-TBI do a good job. The 6400 captures about 10% less than the other tools. The reason is unknown, but it should be taken into account when comparing results from different tools. On tools 6200A and 6400, switching to high-throughput mode reduced the capture rates by about 5%, but this made no significant difference in the other tools. The higher-throughput mode involves faster rastering and an automatic (factory-set) compensation for this. Clearly this compensation is slightly incorrect for these particular tools, but it is OK for these particular 6200B, SP1, and SP1-TBI tools.

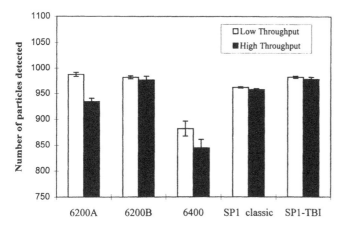

**Figure 11**  Capture of 0.155-μm PSL spheres on bare silicon wafer.

## B.  Repeatability

To understand the measurement variation of the KLA-Tencor Surfscan 6200 and SP1-TBI on real-world particles, a 200-mm (for SFS 6200) and a 300-mm (for SP1-TBI) bare silicon wafer contaminated by environmental particles was used for the repeatability and reproducibility test on two SFS 6200 and one SP1-TBI. The repeatability test was done by scanning the test wafer 30 times continuously without any interruption. The reproducibility test was 30 continuous measurements, with loading/unloading of wafer between measurements. Both high-throughput and low-throughput modes were used in this study to determine the variation due to the throughput settings. The results (Tables 1 and 2) indicate that about 4% of the LPD counts were due to the measurement variation of the instruments, which is acceptable. No contamination trend was observed during the course of this experiment. Although the throughput setting had no apparent effect on the measurement variation, it significantly affected the number of defects captured. Whereas, for the PSL spheres of a specific size in Sec. V.A there was a maximum 5% effect, here both 6200s showed a 20–25% capture rate loss at high throughput (Table 1). The greater effect is probably because there is a distribution of PSL equivalent sizes present, many being close to the threshold level. Inadequate electronic compensation in high throughput pushes these to a lower PSL equivalent size, and many fall below the detection limit.

For the SP1-TBI, more than half of the total particles are missed in high-throughput mode (Table 2)! Figure 12 shows the size distribution, as measured in low- and high-

**Table 1**  Average LPD Counts, Repeatability, and Reproducibility for Two SFS 6200s

| Throughput | 6200 | Repeatability test | | Reproducibility test | | |
| | | Mean LPD | 1σ | Mean LPD | 1σ | CoV (%) |
|---|---|---|---|---|---|---|
| High | A | 1962 | 20 | 2101 | 67 | 3.5 |
| | B | 2113 | 17 | 2118 | 23 | 1.4 |
| Low | A | 2346 | 50 | 2267 | 17 | 2.3 |
| | B | 2505 | 71 | 2353 | 39 | 3.3 |

200-mm water, environmental particles, 30 scans

**Table 2**  Average LPD Counts, Repeatability, and Reproducibility for a Surfscan SP1-TB1

| Throughput | Channel[a] | Repeatability test | | Reproducibility test | | |
|---|---|---|---|---|---|---|
| | | Mean LPD | 1σ | Mean LPD | 1σ | CoV (%) |
| High | DCN | 224 | 6 | 227 | 6 | 3.7 |
| | DNN | 146 | 3 | 147 | 3 | 2.7 |
| | DWN | 169 | 6 | 172 | 5 | 4.7 |
| | DCO | 221 | 3 | 223 | 4 | 2.4 |
| | DNO | 151 | 4 | 154 | 3 | 3.2 |
| | DWO | 163 | 4 | 165 | 3 | 3.0 |
| Low | DCN | 514 | 7 | 521 | 8 | 2.0 |
| | DNN | 331 | 8 | 347 | 9 | 3.5 |
| | DWN | 338 | 6 | 346 | 9 | 3.2 |
| | DCO | 503 | 11 | 497 | 8 | 2.6 |
| | DNO | 313 | 7 | 319 | 10 | 4.0 |
| | DWO | 318 | 8 | 323 | 12 | 4.5 |

300-mm wafer, environmental particles, 30 scans.
[a] DCN: Dark-field composite channel with normal incident; DNN: Dark-field narrow channel with normal incident; DWN: Dark-field wide channel with normal incident; DCO: Dark-field composite channel with oblique incident; DNO: Dark-field narrow channel with oblique incident; DWO: Dark-field wide channel with oblique incident.

throughput modes. Clearly most of the particles in the lowest bin sizes are lost. The high-throughput mode involves scanning about seven times faster, with an automatic gain compensation. From Sec. V.A., this clearly worked for PSL spheres of a fixed (0.155-μm) size, but it does not work here. Again, PSL equivalent size is being pushed into bins below the cutoff threshold. This dramatic change in behavior—0.155 μm PSL spheres in high-/ low-throughput modes where no capture rate loss occurs, compared to a 50% loss for real environmental contaminants—points to the need to be careful of assuming apparently well-defined PSL calibrations map to real particle behavior. Use of the low-throughput setting is strongly recommended for measurement using the SP1-TBI. Several other SP1-TBI users have also come to this conclusion.

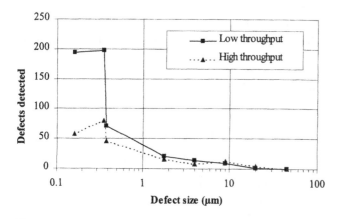

**Figure 12**  Size distribution reported by the SP1-TBI under high and low throughput.

## C. *X/Y* Positioning Accuracy

### 1. Electron-Beam Etched Standard XY Calibration Wafer

The standard $XY$ calibration wafer (200 mm) contains electron-beam-etched pits (2 μm in diameter) arranged in groups of 5 × 5 matrix formation (with 1-mm spacing in the $x$ and/or $y$ directions), as shown in Figure 13. This wafer was scanned five times on each Surfscan. The center pits from each 5 × 5 matrix group were selected using MasterScan v1.2 software (MicroTherm, LLC) to determine the $XY$ positioning accuracy and sizing accuracy of all Surfscans. The wafer was then measured by a SEM (JEOL 848) to confirm the size and location of the pits. Based on measurements with the etch-pit wafer, the SEM stage itself appears accurate to within ±10 μm. Since the actual position of all etch pits was known to within ±10 μm, correction of the coordinate system alignment (1st-order correction) for each tool could be done using a computer program for the comparison of the real positions versus scanner-recorded $X$, $Y$ positions. Such a global alignment showed that the 1st-order correction errors of the tools varied from ±50 μm to ± 700 μm (the SP1-TBI). Once this correction was made, the remaining 2nd-order point-to-point correction ranges could be determined. They are summarized in Table 3. They ranged from 30 μm to 130 μm, with the 6200B and the SP1-TBI being best at around 30–40 μm. There was also clear evidence from the study that (a) accuracy in the 6000 series is a function of wafer-loading orientation (6), and (b) for the SP1-classic, accuracy depended strongly on the location of the defect on the wafer. This latter aspect is discussed in more detail in the next section, where it is also found for PSL spheres.

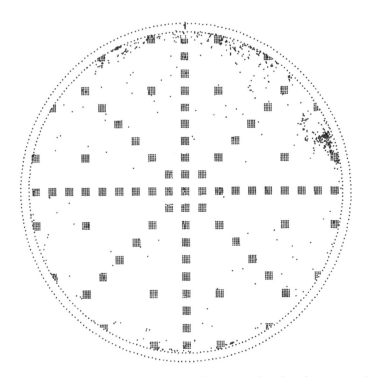

**Figure 13**   Electron-beam-etched $XY$ standard wafer of 5 × 5 matrix formation.

**Table 3** Second-Order Positioning Error of Surfscans

| Surfscan | Electron-beam-etched pits (2 µm) | PSL (0.72 µm) |
|---|---|---|
| 6200A | 93 ± 73 | 81 ± 51 |
| 6200B | 33 ± 20 | 53 ± 26 |
| 6220 | 59 ± 41 | — |
| 6400 | 133 ± 177 | 60 ± 32 |
| SP1-classic | 104 ± 60 | 98 ± 55 |
| SP1-TBI (normal) | 40 ± 20 | 39 ± 21 |
| SP1-TBI (oblique) | 38 ± 14 | 38 ± 22 |

### 2. Bare Silicon Wafer Loaded with 0.72-µm PSL Spheres

Polystyrene latex spheres of 0.72-µm diameter were deposited on a bare silicon wafer to determine the positioning accuracy of all Surfscans of interest. The wafer was scanned three times on each Surfscan. Actual defect coordinates were measured using a JEOL 848 SEM to within about ±10 µm. $XY$ data from each scan was then manipulated to compensate for the effect of coordinate system misalignment (1st-order correction). Figure 14 shows the 2nd-order positioning errors of each Surfscan as a function of the distance of individual PSL spheres from the wafer center. The data shows that positioning of the SP1-classic had a strong dependency on defect location. Table 3 lists the average positioning error and its spread measured by each tool after optimized map-to-map coordinate system alignment. The SP1-TBI gave the most accurate position among all Surfscans tested and showed no difference between pits and PSL spheres. The 6400 positioning for PSL spheres (60 ± 32 µm) was more accurate than that of electron-beam-etched pits (133 ± 177 µm). The suspected reason is the different distribution of PSL spheres (random) compared to the regular patterning of the etch pits. A larger percentage of the etch pits is nearer the edges, which will have larger errors. Therefore the average error is larger. For robust SEM review of small defects, a defect mapping accuracy of ±15 µm is required. None of these instruments is able to provide such a defect mapping accuracy.

This problem can be corrected using an automatic high-accuracy optical defect review system (MicroMark 5000, MicroTherm, LLC) prior to SEM review (7), to update the very inaccurate scanner $X$, $Y$ files. The MicroMark 5000 works by using essentially the same laser-based dark-field scattering as the scanners to refind the LPDs. As each one is found, an entry in a new $X$, $Y$ file is made, now accurate to ±5 µm because the MicroMark stage is very accurate and the effective pixel size is very small. The tool can update at up to 1000 LPD defects an hour. It can also be used to place laser fiducial marks anywhere on the wafer, including "crosshairs" around specific LPD sites, if needed. Both the updating to ±5-µm accuracy and the laser-marking capability are essential to our ability to find and review very small defects (see Chapter 20).

### 3. SP1 Defect Mapping Error Due to Wafer Loading Orientation

The mapping error of the SP1 series arises mainly from the uncertainty in determining the center and orientation of the wafer, the misalignment of the wafer center and the rotation center, the misalignment of the rotation center and the laser spot, and the error contributed by the lead screw pitch. To determine the effect of wafer loading orientation on the

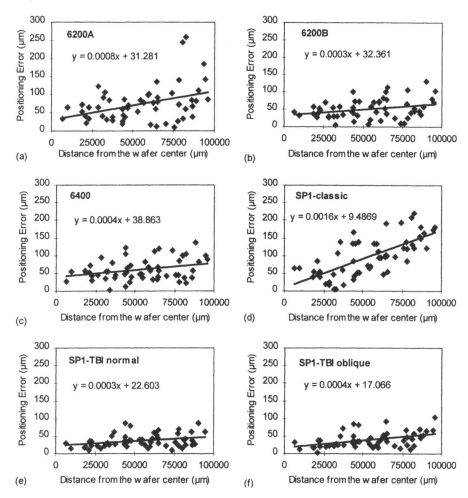

**Figure 14** $X/Y$ positioning error of defects detected by the Surfscan 6200, 6220, 6400, SP1-classic, and SP1-TBI as function of the distance from the wafer center. Data was obtained using the same bare silicon wafer loaded with 0.72-μm PSL spheres.

mapping accuracy, an $XY$ standard wafer (which is different from the one used in previous work) was loaded into the wafer cassette with an arbitrary orientation. A dummy scan was carried out at high throughput, and the wafer was unloaded, with notch up ($0°$). The wafer was then reloaded and unloaded with the notch rotated $30°$. This was repeated six times, rotating each time. A second set of measurements was repeated on another day, but in this set the robot was allowed to "initialize" the system first; that is, the robot runs through a procedure to optimize alignment of the stage, etc. The coordinate files in both sets of measurements were then corrected for coordinate misalignment (i.e., the 1st-order correction was eliminated) to bring them into alignment. After this was done, a consistent error ($20 \pm 10$ μm) was found on all maps (Figure 15a). This indicates that wafer orientation does not affect the point-to-point (i.e., the second-order) accuracy. However, the size of the first-order errors clearly showed a strong dependence on the wafer loading orientation (Fig. 15b). Data also showed that the mapping error was least when the wafer was loaded with notch up ($0°$). Figure 16 shows the $XY$ positions of the geometric center of all marks

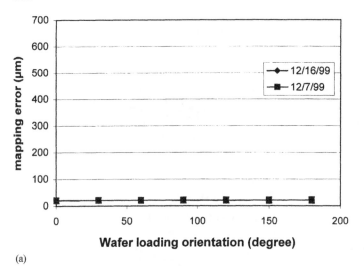

(a)

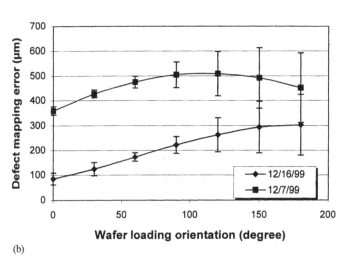

(b)

**Figure 15**   (a) Second-order defect mapping error of SP1-TBI. (b) First-order defect mapping error of SP1-TBI.

for various wafer loading orientations. Notice that the geometric centers move counter-clockwise around a center point and that the rotation angle of these geometric centers is similar to the increment of the loading orientation ($\sim 30°$). This indicates that the first-order mapping error was likely dominated by the misalignment of the wafer center to the rotation center of the stage. The calculated centers of rotation were (99989, 99864) and (99964, 99823) for measurements done on 12/7/99 and 12/16/99, respectively. A variation range of about 50 μm in radius was observed on these calculated rotation centers. The angular offset of the patterns of the geometric centers suggests that the alignment of the wafer center to the rotation center depends on the initial wafer loading orientation as well.

Figure 17 shows the total rotation needed to correct the misalignment of the coordinate system for each map. The similar trend shown for the two sets of measurements suggests that the notch measurements were repeatable. The consistent offset ($\sim 0.035°$) was a result of a change in misalignment (improvement) of the rotation center by the robot

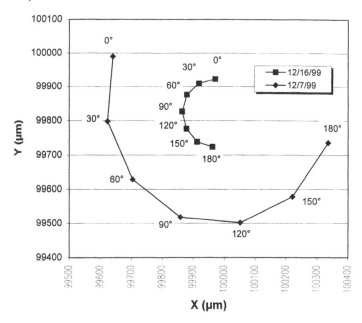

**Figure 16**  Position of the geometric centers of all electron-beam-etched pits on SP1 maps.

initialization in the second run. Additional measurements were repeated several times for the same loading orientation. No significant difference was found from measurements done at the same time for a given loading orientation. The variation of total rotation of measurements with the same orientation indicates the uncertainty of the notch measurement. This is equivalent to about $17\,\mu m$ at the edge of a 200-mm wafer loaded with notch up (0°).

To summarize, although the point-to-point accuracy was not affected by the orientation of the wafer loaded into the cassette, the first-order mapping accuracy of the SP1-TBI showed a strong dependency on the wafer loading orientation. Our data clearly shows

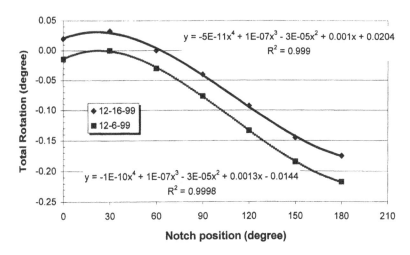

**Figure 17**  Total rotation needed to correct misalignment of the SP1 coordinate system (0° denotes the wafer was loaded with notch up).

that loading the wafer with notch up gives a more accurate defect map, as does the act of initializing with the robot. The accuracy level of these maps varies with the initial position of the wafer with respect to the stage rotation center.

## D.  Sizing Accuracy

The sizing performance of a surface inspection tool is generally specified by the manufacturer based on polystyrene latex (PSL) spheres on bare silicon wafers. However, real-world defects on the surface of the wafer are usually not perfect spheres. They could be irregular chunks, flakes, bumps, voids, pits, or scratches. We measured the size of electron-beam-etched pits using all Surfscans of interest. For the 6XY0 series, the reported size for these pits was consistent across the wafer surface (Figure 18h), though the 6400 significantly underestimates the size by 70%. Pit sizes reported by the SP1 series with normal illumination, SP1-classic (Fig. 18e) and SP1-TBI DCN (Fig. 18f), and DNN (Fig. 18g) were strongly dependent on the location of the pits with respect to the wafer center (i.e., the rotation center). Similar to the 6400, the SP1-TBI with oblique illumination (Figure 18f, DCO) underestimates the pit size by 70%. Table 4 summarizes the averaged LPD sizes for electron-beam-etched pits measured by all Surfscans of interest. For comparison, sizes for 0.72-μm and 0.155-μm PSL spheres measured by these Surfscans are also listed in Table 4. In contrast to the case of pits, the 6400 and SP1-TBI oblique overestimated the PSL spheres by 40% and 30%, respectively. This is due to the oblique incidence, but the magnitude of the discrepancy will depend strongly on the type and size of the defect.

The strong variation of size with radius found for etch pits with the SP1 and SP1-TBI may be connected to the fact that the rotational speed of the wafer is changed as it translates under the laser beam (to attempt to keep the dwell time per area constant).

## E.  Discrimination of Particles and Chemical-Mechanical Planarization Microscratches on Oxide Substrate

Discrimination and classification of different kinds of defects on unpatterned wafers is normally performed by analytical tools such as optical, electron, scanning tunneling, and atomic force microscopes. Restricted by the nature of these analytical tools, such measurements are extremely time consuming. Therefore, such defect classification methods do not meet the requirement of massive manufacturing of semiconductor industry. Recent development of the light-scattering inspection tools with high lateral resolution, sizing capability, and two or three dark-field channels for light scattered into different solid angles provides a possible alternative approach for defect discrimination and classification.

The KLA-Tencor Surfscan SP1 is equipped with two dark-field channels for collecting light scattered in narrow angles (5–20°) and in wide angles (25–70°). Both narrow (DN) and wide (DW) channels were calibrated with standard PSL spheres. For a given PSL sphere, the sizes predicted by both channels must then be the same, giving a size ratio (DN/DW) of unity. However, for a real surface defect that tends to scatter light preferably into one of the channels, the size ratio will deviate from the unity. Therefore, the size ratio can be used as a measure for discrimination between particles and surface defects such as crystal-originated particles (COPs) and microscratches. Discrimination of COPs based on the size ratio has been reported elsewhere (8). In this section, the focus is on the discrimination of chemical-mechanical planarization (CMP) microscratches and particles on an unpatterned wafer.

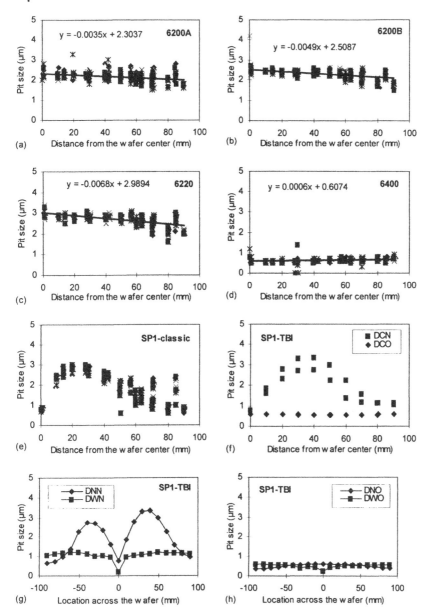

**Figure 18** Electron-beam-etched pit size reported by Surfscan 6200, 6220, 6400, SP1-classic, and SP1-TBI as function of the distance from the wafer center. Data was obtained using the same *XY* standard wafer for all Surfscans.

Such CMP utilizes a chemically active and abrasive slurry, composed of a solid–liquid suspension of submicron particles in an oxidizing solution. Filters are used to control the particle size of the abrasive component of the slurry. Over time and with agitation, the colloids tend to agglomerate and form aggregates that are sufficiently large to scratch the wafer surface during polishing. These micron-scale scratches (microscratches) are often missed because of their poor light-scattering nature. The fundamental difference in the light-scattering behavior between particles and microscratches can be

**Table 4**  LPD Size Reported by Surfscans

| Surfscan | Electron-beam-etched pits | PSL (0.72 µm) | PSL (0.155 µm) |
|---|---|---|---|
| 6200A | $2.4 \pm 0.4$ | $0.63 \pm 0.07$ | $0.155 \pm 0.03$ |
| 6200B | $2.3 \pm 0.5$ | $0.62 \pm 0.05$ | $0.151 \pm 0.02$ |
| 6220 | $2.8 \pm 0.5$ | — | — |
| 6400 | $0.6 \pm 0.1$ | $1.0 \pm 0.1$ | $0.150 \pm 0.05$ |
| SP1-classic | $1.8 \pm 1.1$ | $0.74 \pm 0.05$ | $0.156 \pm 0.003$ |
| SP1-TBI (normal) | $2.0 \pm 0.9$ | $0.71 \pm 0.05$ | $0.156 \pm 0.03$ |
| SP1-TBI (oblique) | $0.6 \pm 0.1$ | $0.92 \pm 0.04$ | — |

illustrated by numerical simulation using DDSURF/DDSUB (9). The light-scattering patterns for a particle and a scratch on bare silicon substrate are very different (Figures 19 and 20). As a gross test of the capability of the DN/DW ratio to distinguish particles from pits/scratches, this ratio was compared for two wafers: bare Si with 0.72-µm PSLs and bare Si with etched pits (one of the $X$, $Y$ standard wafers). The DN/DW size ratio was plotted against frequency of occurrence in both cases. For particles the ratio was $1 \pm 0.5$ (Figure 21); for the etch pits it was $7 \pm 3$ (Figure 22), a very clean distinction.

For a CMP oxide wafer (18,000-Å oxide) we had found that measurements on the SP1 gave very different results from those on the 6200. A large number of defects in the

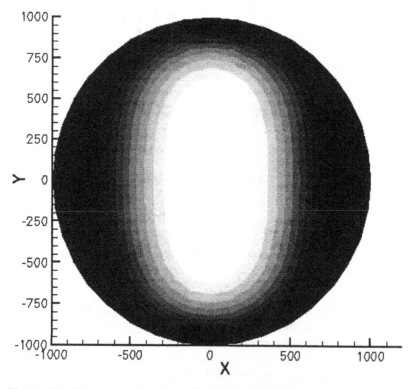

**Figure 19**  Light-scattering pattern for a 0.2-µm PSL sphere on silicon substrate. The gray scale is corresponding to the intensity of the scattered light. The white is hot and the dark is cool.

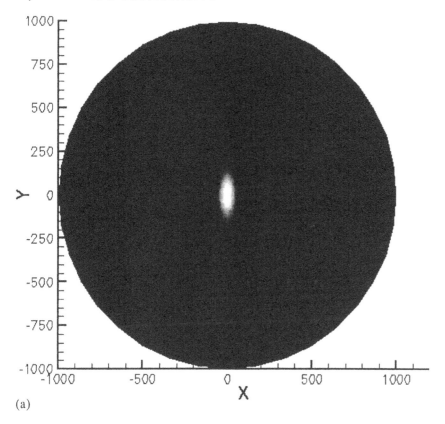

(a)

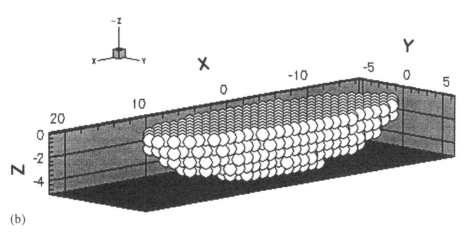

(b)

**Figure 20** (a) Light-scattering pattern for a microscratch on silicon substrate, (b) model of a microscratch used to generate the light-scattering pattern shown in (a). Notice that the surface is at $z = 0$, and the negative $z$ is the depth into the surface.

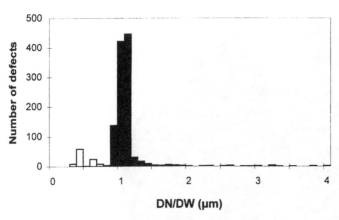

**Figure 21**  DN/DW ratio for 0.72-μm PSL spheres.

0.2-μm range were detected by the SP1 (sum of both narrow and wide) but missed by the 6200. We suspected the SP1 was detecting microscratches that the 6200 was missing. The wafer was re-examined on the SP1-TBI under normal illumination with both narrow and wide channels set at a 0.2-μm threshold. The DN/DW size ratio was plotted against frequency of occurrence (Figure 23). Two lobes appear with the separating value being at a ratio of about 2.5. The anticipation would be that the lobe with a ratio below 2.5 represents particles and that with the ratio above 2.5 represents microscratches. To verify this, 65 defects in the smallest (0.2-μm) bin size were reviewed by the Ultrapointe Confocal Laser microscope. All were real defects, and 57 of the 65 were, indeed, microscratches. The physical size (6 μm in length) observed by a confocal microscope (Ultrapointe) far exceeds the PSL equivalent size of 0.2 μm.

To summarize this section, then, it is clear that by using all the available channels in the SP1-TBI (different angular scattering) it is possible to derive signatures of defects that are very different physically, e.g., particles versus pits or microscratches. It remains to be seen whether the distinction is sufficient to be useful for particles with less dramatic difference, though we have also been able to easily distinguish small COPs from very small particles this way. Other scanners, with availability of multiple channels, such as the Applied Excite, or ADE tool, can also perform this type of distinction.

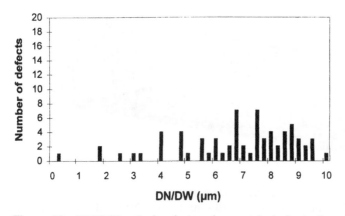

**Figure 22**  DN/DW ratio for electron-beam-etched pits (∼ 2 μm in diameter).

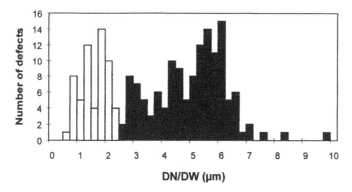

**Figure 23** DN/DW ratio for CMP microscratches.

## VI. CONCLUSIONS

In this chapter we have tried to give a summary of the principles behind using light-scattering for particle scanners, a description of the important parameters and operations in practically using scanners, and a discussion of the caveats to be aware of. It is very easy to get data using particle scanners and just as easy to misinterpret that data without expertise and experience in how the scanners actually work. Owing to the fact that particle requirements in the industry are at the limit of current scanner capability (e.g., sizing accuracy, minimum size detectability, accuracy/reproducibility in particle counting) it is very important they be operated with a full knowledge of the issues involved.

We have also presented the results of some of our efforts into delineating, in a more quantitative manner, some of the important characteristics and limitations for the particular set of scanners we use (6200, 6400, SP1, SP1-TBI).

In the future we expect several developments in the use of particle scanners. First, there is a push toward integrating scanners into processing tools. Since such a scanner has to address only the particular process in hand, it is not necessary that such a tool be at the forefront of all *general* capabilities. It can be thought of more as a rough metrology check; when a problem is flagged, a higher-level stand-alone tool comes into play. Second, there should be greater activity toward obtaining a particle "signature" from the design and use of scanners with multiple channels (normal, oblique, incidence, different angular regions of detection) to make use of the difference in scattering patterns from different types of defects. Third, it is likely that there will still be a push to better sensitivity (i.e., lower particle size detection). This, however, is ultimately limited by the microroughness of the surface and so will be restricted primarily to supersmooth Si monitor wafers. The issue of improved targeting accuracy will be driven by customers' increased need to subsequently review defects in SEMs or other analytical tools. Either the scanner targeting accuracy must improve (primarily elimination of 1st-order errors) or on-board dark-field microscopes have to be added to the SEMs (e.g., as in the Applied SEMVision) or a stand-alone optical bench (such as MicroMark 5000) must be used to update the scanner files to a ±5-μm accuracy in a fast, automated manner.

Finally the issue of always working in "PSL equivalent sizes" must be addressed. Everyone is aware that for real defects, real sizes are not provided by scanners and the error can be very different for different types of defects. The push to greater sensitivity, then, is more in line with "we want to detect smaller" rather than any sensible discussion of *what* sizes are important to detect.

## ACKNOWLEDGMENTS

The authors would like to acknowledge stimulating discussions with many of our colleagues, in particular, Pat Kinney of MicroTherm, and with Professor Dan Hireleman and his group (Arizona State and now Purdue University).

## REFERENCES

1. T Hattori. In: KL Mittal, ed. Particles on Surfaces: Detection, Adhesion, and Removal. New York: Marcel Dekker, 1995, pp 201–217.
2. H van de Hulst. Light Scattering by Small Particles. New York: Dover, 1981.
3. K Nahm, W Wolfe. Applied Optics 26:2995–2999, 1987.
4. PA Bobbert, J Vleigler. Physica 137A:213, 1986.
5. HE Bennett. Scattering characteristics of optical materials. Optical Engineering 17(5): 1978.
6. Y Uritsky, H Lee. In: DN Schmidt, ed. Contamination Control and Defect Reduction in Semiconductor Manufacturing III. 1994, pp 154–163.
7. P-F Huang, YS Uritsky, PD Kinney, CR Brundle. Enhanced sub-micron particle root cause analysis on unpatterned 200 mm wafers. Submitted to SEMICON WEST 99 Conference: Symposium on Contamination-Free Manufacturing for Semiconductor Processing, 1999.
8. F Passek, R Schmolk, H Piontek, A Luger, P Wagner. Microelectronic Engineering 45:191–196, 1999.
9. T Quinteros, B Nebeker, R Berglind. Light Scattering from 2-D surfaces—A New Numerical Tool. DDA 99—Final Report. TR 350, 1999.

# 20

# Particle and Defect Characterization

**C. R. Brundle and Yuri S. Uritsky**
*Applied Materials, Inc., Santa Clara, California*

## I. INTRODUCTION

### A. Definition of Particle Characterization and Defect Characterization

Within the subject area of this chapter, *particle characterization* refers to procedures for establishing the nature of the particle material sufficiently well so that its origin and root cause can be determined. This type of work nearly always involves analysis on full wafers (either 150 mm, 200 mm, and, now, 300 mm). Occasionally, during detailed detective work to establish root cause, particle removal from chamber hardware may also be examined to attempt a match to the offending wafer-based species.

Concerning size characterization, specialists are rarely asked to deal with anything larger than a few tens of microns, since the state of the semiconductor processing equipment industry is such that these should really be considered "boulders." The exception to this is a class of "particles" that actually turns out, on examination, to be thin-film patches—i.e., large in lateral dimension but very thin (down to tens of angstroms in some cases). At the other end of the scale, the smallest particles we are asked to deal with *should* be decided by the IRTS requirements—i.e., the size that will cause yield loss for the design rule concerned. Currently this is around 0.1 µm in the most stringent cases. In practice this is not the situation. Light-scattering (LS) tools (see Chapter 19) are the industry standard method for detecting particles (so-called "particle scanners"), and anything genuinely detectable (i.e., giving a signal distinguishable from the substrate background) is likely to be considered a particle violation. Owing to the variation between different types of LS tools, variability in the manner in which they can be (and are) used, variation of the substrates to be considered (smooth, rough, Si, oxide, metals), and finally the fact that sizing calibration in LS is based on polystyrene latex spheres (PSLs), whereas real particles are a variety of shapes and material, the true size of the "smallest" particles being considered (i.e. those at the threshold of distinction from the substrate background) varies from 0.05 µm to as large as microns, depending on the specific circumstances. A *typical range* of sizes (smallest to largest) encountered currently in fulfilling requests for full-wafer particle analysis is 0.15–5 µm. This range will move lower as IRTS specifications change and, more practically important, if LS detection tools improve detection limits. Since the performance of scanners is so variable, one might say that the smallest genuine size one has to deal with is zero, since one objective of characterization is simply to determine whether a detected particle is real or a "nuisance" defect representing surface rough-ness that the scanner is unable to distinguish from a real defect (see Chapter 19).

The *number* of particles on a wafer (within some LS-determined nominal size range) available for characterization may vary from a few (less than 10) to 1000s. Clearly, if there are 1000s, it is not practical to analyze every particle. The analyst must devise a protocol, which will depend on the total number, the spatial pattern (if any), the objective of the analysis, and the time availability for analysis (usually the gating factor). If this is a new problem with no previous particle analysis experience, the protocol can be established only while the analysis is being done. A common approach is to analyze a sufficient number of particles to ensure one has identified the statistically significant types present. A less ambitious goal, but maybe sufficient, depending on objectives, may be to identify the majority species.

So far we have considered only particles, which, in the overall scheme, form just one class of physical defect that may lead to yield reductions for fully processed wafers (the important parameter!). If the problem has persisted to where detailed characterization is requested, however, particles are likely to be the dominant class (see later for a discussion on the distinction between *characterization* and *classification*). Scratches, pits, and other forms of surface imperfections are another important class, either simply because they are detected by LS "particle scanners" and are, therefore, de facto, a problem or because they really impact yield. A third class is any kind of pattern violation on patterned wafers (shorts, opens, linewidth variations, incomplete vias, etc.). Though obviously of extreme importance in the fab, and sometimes caused by particle contamination during previous processing steps, this kind of "final defect" will not be discussed further in this chapter, for two reasons. First, it is an area in which the authors have less experience; second, the situation where a foreign particle (either in the layer of interest or buried in the layers below) is directly responsible for the pattern violation represents only a small fraction of pattern violations.

## B. Significance of Particle Characterization

Our experience, being in an equipment manufacturing company rather than a fab envir-onment, is obviously biased toward solutions of problems at the individual process tool level rather than overall yield level. Inside our own industry, particle characterization, leading toward determination of the particle source, is important in three distinct phases of the business: (1) development of new processes and new hardware kits/chambers, (2) meeting customer specifications during demos and β-site operations, and (3) responding rapidly to serious excursion failures associated with installed base tools. A fourth area that is becoming important but is barely attempted yet is in baselining tools, which are in spec, to provide a knowledge base for solving future excursion problems or improving baseline performance.

For the first phase, though the presence of particle problems during initial process development and hardware development may seem to be separable issues, they often are not. The chemistry and physics of the processing procedures can, and usually do, affect hardware to a greater or lesser degree (not surprising, since they mostly involve deposition or etching under harsh conditions), resulting in particle (and other) failures through their synergistic interaction. Regarding phase 3, our involvement, since we are not part of a fab team (or even in a fab company), is usually at the point where a particular tool or chamber has been identified as the cause of a problem but routine empirical approaches to elim-inating it (e.g., a wet clean, swapping parts) have not worked.

## C. Defect Characterization: Relationships to Defect Detection, Defect Review, and Defect Classification

Unfortunately, these terms are used somewhat interchangeably, with arbitrary distinctions, and therefore they can mean different things to different people.

*Detection* involves the initial survey, or inspection, of the whole wafer in some manner, preferably very quickly, since such wafer inspections may be required for thousands of wafers in short time periods during normal fab operation. The industry method of choice is light scattering (able to scan a wafer in seconds) or sometimes scanned optical imaging (much longer), with pattern recognition for patterned wafers. The objective is to give the number of defects above a certain nominal threshold size (nominal because of the limitations of the metrology calibration (see Chapter 19).

*Review* means, in its broadest sense, establishing more about individual detected "hits" than just counting them. In a very real sense, defect survey tools already have some review capabilities. Light-scattering scanners attempt to provide size binning, based on LS intensity and calibration, along with detection. The stored data on each individual "hit" can also be pulled up and reviewed after the fact. A narrower, but more common, use of the term *review*, as in *defect review tool* (DRT), is to physically revisit an already detected defect and examine it further in some way. Image-based survey/detection tools, or LS tools with an imaging mode, can also meet this definition of a DRT and supply additional information—i.e., topography/shape, *if* the defect is large enough. More commonly, however, tools are designed to do either the best job of detection (survey/inspection tools) *or* review (DRT). Such DRTs do not have to scan whole wafers, but simply read coordinate files from the prior survey tool and drive to the appropriate defects. Basically they are swapping speed for some other parameter (e.g., spatial resolution, quality of information). In the fab, DRTs in practical use are either optical (image) review or SEM review. Other methods may develop into fab-compatible DRTs, such as AFM or Auger (see later), but these are not viable currently. Optical is by far the most used currently in fab environments (ease, speed), but, because of the limits of optical resolution, SEM DRT is becoming essential in more and more cases. For particles on unpatterned wafers, no useful image/topography information is usually achievable below $\sim 0.3\,\mu m$ using optical review. Obviously, the purpose of review is to get more information—size distribution, shapes (from imaging)—or even simply to verify that "detected" defects really exist and are not an artifact of the survey tool (e.g., substrate roughness confused with particle presence, so-called "nuisance defects"). As well as providing better spatial resolution, SEM DRTs may also provide element classification (by EDX), maybe even automatically for routine cases. At this level of information the distinction between "defect review" and "defect characterization" becomes very blurred (but the *expectation* for review is that it will handle wafers quickly, and maybe automatically, with a *minimum of human expertise needed*).

A reasonable operational distinction might be that review leads to defect classification—size, shape, particle or scratch, element A present or absent—whereas particle characterization involves enough analytical firepower and specialized expertise to establish *what* the particle is (e.g., a piece of amorphous carbon as opposed to diamond, for instance), with the objective of determining *where* it came from (and preferably *why*), i.e., root cause.

## D. Patterned and Unpatterned Wafers

The authors' experience is largely with unpatterned (blanket) wafers. It is important to understand distinctions and where each type of wafer analysis belongs. There are several types of blanket wafers that typically get submitted for particle characterization work.

### 1. Silicon Monitor Wafers

These are used extensively and routinely in two ways. The first way is to check the level of particle production while exercising only tool hardware (e.g., loadlock entry, robot operation, gate valve operation). Such tests and the wafers themselves are referred to as *mechanical*. The second way is to check for particle production during simulations of some part of the process in the tool chamber (e.g., running gas sequences, heater on/off, plasma or RF on/off). The simulation may or may not be representative of the real process. Often, the objective is to stress the system to see if the particle count changes. The wafers used are usually prime wafers, because one wants to start with a substrate as smooth and clean as possible so that small numbers of small particles can be reliably detected and subsequently analyzed (see Chapter 19).

### 2. Monitor Wafers Other Than Silicon

Sometimes monitor wafers other than Si are used (particularly oxide wafers). If used for mechanical tests, the reason would have to be expedience, unless one was specifically looking for Si particles, in which case a non-Si substrate might be needed. Usually the reason is that a process simulation is needed and a Si substrate is inappropriate for the simulation. Whatever the reason, not using a prime Si wafer will result in a larger minimum detectable size.

### 3. Full Film Wafers

These are actual production wafers of the particular process step, e.g., W or TiN CVD deposition (though the process may have been modified specifically for a particle test). For the examples given, then, we would be looking for particles on, or in, a film of W or TiN—hence the term *in-film* particles or on *in-film* monitor wafers.

It is typical to receive requests for particle characterization on one or more of the three types of blanket wafers described earlier when particle excursions have been identified at the individual tool and chamber level or during development. Patterned wafers, which involve integration of lithography, deposition, etching, and CMP, are a completely different issue. They are very often subjected to defect review to classify pattern defects (shorts, opens, design rule violations, scratches, particles), but they are less likely to be submitted for particle characterization. Once a pattern defect had been classified it may be considered necessary to investigate the root cause, usually if one cannot tell which part of the integration (or which one of the possibly many process steps) is leading to the eventual pattern defect. To do this traditionally requires cross-sectioning plus SEM and TEM, to get to the offending layer, which of course destroys the wafer. With the advent of the FIB/SEM dual-beam tool, it is now possible to do some of this work on-wafer.

## E. Techniques Used for Particle Characterization

Keeping in mind our working definition of characterization, the predominant technique used for particle characterization over the past 10 years has been the combination of EDX with SEM. This will likely remain so for the foreseeable future, though use of other techniques with complementary attributes will become more and more an additional requirement—i.e., a suite of techniques working together is needed (1–4).

## 1. SEM/EDX

If you can find the particle (the key to success, discussed in Sec. II), a modern field-emission SEM will be able to establish its shape, size, and topography for sizes well below that required in the ITRS for the foreseeable future (the spatial resolution of the SEM should be on the order of a few nanometers). By tilting or using a SEM with multiple detectors at different angles (e.g. the Applied SEMVision), it is also possible to get a reasonable 3D representation. Energy-dispersive x-ray emission (EDX) works by having a solid-state x-ray detector inside the SEM to detect x-rays emitted from the material under the electron beam. The detector resolves the energies of the x-rays, leading to identification of the elements present. In an instrument designed for state-of-the-art ana-lytical work, the EDX approach allows detection of all elements except H, He, Li, and Be, and can distinguish between all the elements it detects, using the appropriate x-ray lines for analysis. Most SEM/EDX systems designed for fab use, however, are designed to be DRTs and in doing so compromise the parameters necessary for full analytical capability (thin window or windowless detector; ability to go to 30-kV excitation energy; energy resolution of detector) in favor of other parameters appropriate to DRT (speed, automa-tion, "expert-free" usage). Such instruments, or the mode in which they are used, therefore lead to more limited characterization capabilities.

With enough expertise and available standards it is sometimes possible to establish particle composition semiquantitatively, from peak intensities, using a good analytical SEM/EDX (5). It is nearly always possible to state whether an element is there as a minor or a major component. Sometimes this is sufficient, along with the morphological information from the SEM image, to characterize the particle. When it is not, the reasons may be as follows.

1. *EDX is not a small-volume technique on the scale we are concerned with.* For a 1-μm particle (assumed spherical for the sake of discussion), less than half of the emitted x-rays may come from the particle (material dependant) at 20-kV excitation energy. The rest originate from the substrate because the electron (e-beam) passes through the particles and spreads out in the substrate below it. Obviously, there is therefore a problem of distinguishing the particle signal from the substrate signal. At 0.1-μm size, less than 1% may originate from the particle. This excitation-volume problem can be partly overcome by reducing the excitation voltage, which reduces the penetration depth, but then some x-ray lines needed for the analysis may not be excited. There are other variations possible (wavelength dispersive x-ray emission, WDX, and microcalorimetry x-ray energy determination—see Sec. III.A) that, in principle, can get around these problems, but they are not yet available in commercial full-wafer SEM tools. Despite this analyzed-volume drawback to EDX, it is still quite possible in many cases to arrive at a definitive analysis for particles of 0.1-μm size (see Sec. II).
2. *EDX lacks chemical-state information.* Elemental presence and maybe compo-sition can be determined, but no information on chemical bonding is provided. This can be important in characterization, particularly for organic-based mate-rial. Detection of carbon, for instance, does not get you very far in a root-cause

determination, since there are usually multiple possible sources. A distinction between amorphous C, graphite, and pump oil may be necessary, for instance—i.e., what is the chemistry of the C detected?

3. *Electron-beam effects.* If the substrate is insulating, severe charging can occur, destroying the SEM image resolution and making it difficult, in EDX, to know if the beam is really on the particle. For oxide wafers this typically occurs for oxide thicknesses greater than 500 nm, but this depends critically on many factors. If the particle itself is insulating, it may fly off the surface (bad) or show enhanced contrast (good). For small particles ($< 0.5\,\mu m$) on a noninsulating surface, charging is rarely an issue, since it bleeds away into the surface. Focusing the beam into a small area results in high beam current densities, which can damage the material being studied (melting, evaporation, compositional change) and preclude successful analysis. Organic materials are particularly susceptible.

One has to be aware of all these effects when using SEM/EDX, or any other e-beam technique.

## 2. Optical Methods

The reason for adopting alternate or additional methods to SEM/EDX is either to overcome one or more of the three deficiencies of SEM/ EDX (too large a sampling volume, insufficient chemistry information, e-beam damage) or that a faster or less expert-based approach is applicable. Optical methods are certainly faster and experimentally easier (no vacuum required, for instance). Interpretation is not necessarily easier, however, and, of course, spatial resolution is far inferior to SEM. Ultrapointe markets a scanning laser (visible wavelength) confocal microscope that accepts full wafers and reads imported light-scattering files. It is usually considered a review tool (DRT) to follow defect detection by an LS scanner, but its confocal plus extensive image-processing capabilities do allow more extensive characterization usage (6). Image shape information is useful only down to about $1\,\mu m$, but one can *detect* features much smaller (sub-$0.1\,\mu m$ in some cases) and, from the confocal mode, establish the *height* of the defect relative to the wafer surface (about $0.3$-$\mu m$ height resolution). Thus one can determine whether the feature is on, in, or below the surface. From the reflective properties it is possible to distinguish, for instance, metallic from dielectric particles and maybe greater detail in specific cases where prior "fingerprinting" standards are available.

The major characterization capability, however, results from combining an optical microscope of this type with a Raman spectrometer (making it Raman microscopy (6)). Raman is a vibrational technique. The incident laser light excites vibrations in the material. The details of the vibrational frequencies indicate what chemical groups are present and, for solids, also give information on the stress and the phase of the material. Adding a Raman spectrometer detection system to the Ultrapointe therefore provides a way to get spatially resolved chemical-state information down to the $0.5$-$\mu m$ level. If the defect/particle is a strong Raman scatterer, it is often possible to get information on smaller sizes (we have done this down to $0.2\,\mu m$), though now the background wafer signal may dominate. On the other hand, there are many materials where no vibrational frequencies useful for analytical purposes are excited, so *no* information is obtained at any size. Owing to the strong material variability, a Raman expert is essential, and the most productive usage is in a "chemical fingerprinting" mode using a library of known spectra. The depth probed by Raman is also very variable, depending on the optical properties of the material. For highly reflective materials it may be as little as a few tenths of a micron; for optically transparent material it may be as much as 10 µm. This range can be an advantage or a disadvantage, depending on the situation; e.g., a thin defect (100 nm) on a strong Raman-scattering substrate may give poor defect/substrate discrimination, but a large defect buried under or in a transparent layer will be accessible to Raman analysis (provided you can find it optically!).

Finally, in a Raman setup, visible fluorescence is often produced in addition to Raman scattering. This may obliterate the Raman spectrum. Photoluminescence radiation is sometimes extremely specific and intense and can be used for characterization. A good example is the distinction of anodization $Al_2O_3$ from ceramic $Al_2O_3$ (5). The latter always contains trace Cr, which fluoresces strongly at a specific visible wavelength, identifying the $Al_2O_3$ as coming from a ceramic source (see Sec. III.B).

Another type of vibrational spectroscopy is infrared (IR) absorbtion, usually performed in a Fourier transform IR (FTIR) spectrometer. It provides information of the same general type as Raman, but the instrumentation is completely different. Instead of an incident, single-wavelength laser, a multiwavelength lamp (white light) is used. When coupled to microscope optics, this restricts the resolution capability to 10–20 µm. For this reason (insufficient spatial resolution), there are no commercial full-wafer FTIR DRTs. Infrared analysis is usually done "off-line" (6), and often on wafer pieces. It does have some advantages over Raman, however. There is a much larger library of known spectra available, there is no interfering fluorescence, and the sampling depth is generally shorter, meaning that generally thinner defects are detectable than for Raman.

## 3. Scanning Auger Microscopy (SAM)

Auger spectroscopy has been used in surface science for over 30 years. The Auger process (named after the discoverer of the phenomena, Pierre Auger) is the atomic decay process complementary to x-ray emission, following initial excitation by an electron beam. Instead of an x-ray's being emitted, an electron (the Auger electron) is emitted. Its energy is related specifically to the electron energy levels involved in the process and therefore, as with x-ray emission, is characteristic of the atom concerned. X-ray and Auger emission *always* occur together; it is just a question of which decay process dominates, and this is completely dependant on which core levels of which atoms are involved. The difference in the analytical capability of Auger versus EDX stems from the following two factors.

1. Auger spectroscopy is a far more surface-sensitive technique . Escape depths range only from less than a nanometer to a few nanometers (dependant on the KE of the Auger electron and the material).
2. There is the potential for chemical-state information in Auger spectroscopy from the small "chemical shifts" observable in Auger electron energies (if the energy analyzer used is capable of resolving them).

Scanning Auger microscopy, SAM, is achieved by rastering the incident e-beam over the surface, as in SEM, while synchronously detecting the Auger electron of choice via an electron energy analyzer. In imaging mode a SEM image is always obtainable along with an Auger image—they both come from electrons emitted from the surface. But since the number of low-energy secondaries emitted (the SEM image) is several orders of magnitude greater than the number of Auger electrons emitted, SAM requires higher beam currents (tens of nanoamps instead of a fraction of a nanoamp) and is *much* slower than the SEM/

EDX analysis combination. Since SAM is *so* surface sensitive, clean, UHV conditions are required so that one doesn't simply observe a hydrocarbon contamination layer.

It is because of the requirement for an electron energy analyzer and UHV technology, and the slower nature of the technique, that commercial wafer-level SAM defect review/characterization tools have only recently been introduced (7). In the defect characterization arena, one would want one as an alternative to SEM/ EDX in the following situations:

1. A need to analyze the elemental composition of particles in the smaller size range, where EDX is ineffective because of its large volume probe depth (i.e., taking advantage of the surface sensitivity of Auger)
2. Situations where "particles" turn out to be thin-film patches (surface sensitivity again)
3. A need for some chemical-state information

Since an EDX detector can easily be added to a SAM, it is possible to get SEM, EDX, and Auger information from the same tool. This is, in fact, what we do in our PHI SMART 300 (300 mm SAM). But since SAMs are designed to maximize Auger performance, not SEM, the SEM resolution capabilities are inferior (dark-space resolution of 15 nm instead of 2–3 nm).

## 4. Secondary Ion Mass Spectroscopy

Secondary ion mass spectroscopy (SIMS) is a destructive analytical technique in which material is removed from a surface by ion beam sputtering and the resultant positive and negative ions are mass analyzed in a mass spectrometer (1,2). From the ions produced it is possible to say what elements were present in the material removed and sometimes in what molecular or chemical form (particularly organic/polymeric compounds). Quantification is notoriously difficult, requiring standards of the same, or very similar, material. Time-of-flight SIMS (TOF-SIMS) is a particular version of SIMS (there are many versions), having high sensitivity and the ability to go to very high amu values at high mass resolution. It uses an FIB source, so it is capable of spatial resolution below 0.1 μm. But even with the good sensitivity of TOF-SIMS, the size of a particle from which sufficient diagnostic information can be obtained, before the particle is sputtered away, is limited. We have demonstrated usefulness on Al and $Al_2O_3$ particles down to 0.2 μm. A different version of SIMS, quadrupole SIMS (i.e., the mass analysis is done using a quadrupole mass spectrometer), is much simpler (and cheaper) but has far lower mass range, resolution, and sensitivity. Quadrupole SIMS analyzers have been added as analytical attachments to FIB/SEM DRTs, but this has not proved very generally useful because of (a) the high sputter rate needed to get sensitivity (i.e., even large particles are gone quickly) and (b) the limited level of mass spectral information.

## 5. Atomic Force Microscopy (AFM)

In AFM a surface is physically profiled using the attractive or repulsive force changes as a function of distance between the surface and a (very close) tip; it is capable of height resolution of subangstrom level and lateral resolution of 10 Å. Whereas 300-mm-wafer-capable AFM tools are available, they are usually configured primarily for determining film roughness at the nano level, rather than for profiling defects. The latter can be done, however, provided an accurate $X$, $Y$ stage and a means of reading LS files are included. It is still often difficult to find defects because of the high magnification, and therefore limited field of view, of AFM when small defects are involved (targeting issue—see Sec. II). Once these are found, very detailed topographic information can be provided. AFM is particularly useful in detecting bumps and distinguishing them from pits or scratches, even down to extremely small dimensions (nanometers), which is a problem for SEM, since no material contrast is present when no foreign material is present. Sometimes shallow features (bumps or dips) are completely undetectable by SEM yet are easily observed and quantified in AFM (2).

### 6. Electron Spectroscopy for Chemical Analysis (ESCA)

Also known as XPS, x-ray photoemission, ESCA is a standard method in surface science for obtaining chemical state information in the outermost $\sim 20$ Å of material. The method involves ejecting electrons, by x-ray impact, whose binding energies, BE, are characteristic of the atom from which they came. If the energies of the ejected electrons are determined, the atomic composition can be determined. The technique is very similar to Auger, except for two important distinctions:

1. The level of chemical-state information available from small "chemical shifts" in the photoelectron energies is much better than in Auger.
2. You cannot easily focus an x-ray beam down to a submicron spot as you can an e-beam.

Owing to the latter point, there are no XPS laboratory commercial tools (let alone fab-compatible DRTs!) for defect/particle work. Non-full-wafer instruments exist with capabilities down to the $\sim 10$-μm range. It is possible to achieve sufficient spatial resolution in XPS using soft-x-ray synchrotron facilities at National Labs to make the technique usable down into the submicron range (8). Though a very powerful analytical technique, we will not discuss it further here because of the lack of full wafer instrumentation.

## II. CHARACTERIZATION BY SEM/EDX

### A. Finding Particles from Light-Scattering Files

In general, a two-stage process is used to analyze particles on any wafer: LS particle scanner detection to provide information about particle size and spatial distribution, followed by high-magnification scanning electron microscopy (SEM) coupled to EDX in order to analyze for chemical composition and morphology (9). Because the orientation of the wafer in the SEM coordinate system differs from the orientation in the particle scanner coordinate system, a coordinate transformation is required to map the particles from the reference frame of the scanner to that of the SEM. The fundamental problem in locating particles lies in developing an efficient methodology (10) for establishing the coordinate system accurately.

The accuracy to which particle positions are known in the SEM is of paramount importance for locating small particles. For a 0.1-μm particle, the minimum magnification required on the SEM is approximately 2000×. This high magnification is derived from several factors: the 0.2-mm resolving limit of the human eye, the faint contrast between the surface of the wafer and the particle, and the ability of the operator to focus the SEM. At 2000× magnification, the field of view on the SEM is correspondingly small. For a typical 100 × 100-mm CRT, a 2000× magnification translates to a field of view of only

50 μm × 50 μm, meaning that the accuracy of particle positions must be known to within a radius of 25 μm for them to be seen on the SEM screen.

Achieving such accuracy is a challenging task on any wafer but especially on 200- and 300-mm notched blanket wafers due to a lack of suitable reference marks. Patterned wafers are significantly easier to process, because pattern edges provide convenient and relatively long reference lines with respect to which the wafer can be oriented.

In case of blanket notched wafer (standard for most 200-mm wafers), alignment depends solely upon the 1 × 1-mm notch (11). In addition to difficulties caused by the notch dimensions, another problem is a variability of notch shape from vendor to vendor. Figure 1 shows two distinctly different but common notch shapes. The shape of the so-called "rounded" notch (Fig. 1b) requires a more sophisticated coordinate transformation procedure than notch with well-defined line segments (Fig. 1a). The laser scanner and the SEM *must* use the notch to establish the coordinate system of the wafer in its reference frame, for it is the only alignment feature available. Obviously, aligning the wafer via the notch is prone to rotational error. For 200/300-mm wafers, any errors in rotational alignment are magnified by the large radius; for example, a 1-degree error in aligning the notch translates to positioning error of 1700 μm (error $= r\Delta\theta$) for particles near the edge of the 200-mm wafer.

## B. Reduction of Targeting Error

The difference between the predicted positions in the scanner file and the positions finally observed on the SEM stage is known as the *targeting error*. Since the SEM's $x/y$ stage is accurate to within a few micrometers, nearly all of the targeting error is caused by inaccuracies pertaining to the particle map. According to its manufacturer (12), the scanner (in our case the Tencor 6200 series) measures particle positions with a resolution defined by a rectangular cell of $10 \times 26$-μm, but these positions are referred to the less accurately determined coordinate system. The coordinate system is developed based on the measurements of the wafer's edge during wafer prescan (C.T. Larson, personal communication, 1992). For this purpose a coarser grid is used ($120 \times 26$-μm data cells). The coordinate system is then aligned to the user-selectable coordinate system origin and notch position. Utilization of the coarser grid affects especially the accuracy of determination of the notch

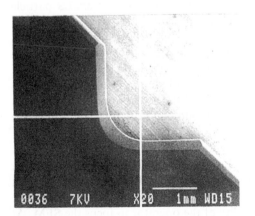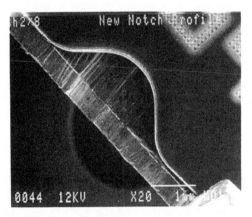

**Figure 1** SEM images of two common notch shapes: (a) well-defined line segments, (b) rounded shape.

position. Considering that the notch is only 1 mm deep, the 120-μm sampling could completely skip over the important notch minimum point. That means that the uncertainties produced by particle-to-particle resolution limits (point-to-point errors, or second-order errors) are expected to be smaller than the error introduced by this coordinate system misalignment (1st-order error) based on notch and wafer center position.

Thus, a key step in reducing targeting error is to eliminate the coordinate system misalignment. For this, the coordinate transformation is conducted in a multiple-step process. A transformation is accomplished by measuring the wafer orientation in the SEM frame and transforming the scanner map data to the experimentally determined SEM orientation. Since the scanner uses the center and notch points to establish its coordinate system, the center and notch points must also be used in the SEM reference frame to establish the SEM coordinate system. The key difference here is that the accuracy of the SEM is much greater than that of the scanner.

To locate the notch point in the SEM, two approaches are used (13). In the first approach, shown in Figure 2a, the notch is determined by the intersection of the two line segments making up the "V" of the notch. Two points are taken from each line segment (V1, V2 for one line segment; H1, H2 for the other line segment), and the lines formed by the points intersect to yield the notch location. Each point is determined using high magnification ($\sim 1000\times$) so that accuracy is ensured. In the second approach, shown in Fig. 2b, a circle is fitted to the lower curvature of the notch using a least squares algorithm. The center of the circle is taken as the notch point. The curve-fitting approach is particularly well suited for "rounded" notch shapes such as shown in Fig. 1b. These notches do not have the sufficiently well-defined line segments necessary to determine the notch intersection point. Oftentimes, both approaches will be used and an average notch point will be computed.

To locate the center in the SEM, a least squares fit is applied to four circumference points on the wafer edge, one from each "quadrant" of the wafer, as shown in Figure 3. Again, high magnification ($\sim 1000\times$) is used to ensure accuracy.

After this initial transformation, the targeting errors are still large, apparently due to errors in alignment of the scanner coordinate system to the actual wafer. This alignment

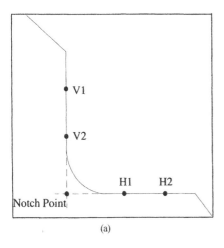 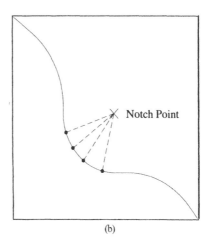

(a)                                                              (b)

**Figure 2** Locating the notch point under the SEM by (a) intersecting the two notch line segments and by (b) locating the center of the least squares circle fit to the notch curvature.

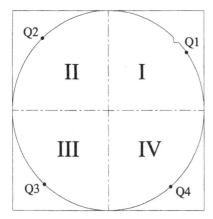

**Figure 3** Locating the center of the wafer by a least squares circle fit to four points along the circumference of the wafer. One point is taken from each "quadrant" of the wafer.

error can be eliminated once several particles (termed *reference particles*) are located with the SEM. By comparing the coordinates of two particles in the SEM and scanner frames, a new coordinate transformation is made. The new transformation effectively eliminates the influence of the uncertainties of the scanner coordinate system's alignment.

Further improvements in accuracy are obtained by using more than two reference particles and averaging the coordinate transformation parameters (14). In addition, realizing that scanner errors are both random and systematic, statistics can be applied to reduce the systematic portion of the error. The approach is as follows. Once two reference particles are found in the SEM, *actual* coordinate transformation parameters may then be computed between the scanner and the SEM using the two reference particles instead of the center and notch. The difference between the coordinate transformation parameters obtained using the two reference particles and those obtained using the center and notch are saved into a log file for each wafer analyzed. For each subsequent new wafer analyzed, the average difference in the coordinate transformation parameters is applied to the center and notch parameters before the transformation mathematics are performed. This has the effect of improving the center and notch coordinate transformation parameters by reducing systematic errors. As a result, after processing about 12–15 wafers scanned by the same scanner, the error in locating reference particles decreases to consistently below ~100 μm.

## C. Wafer Marking/File Updating

In spite of the various techniques for improving positioning accuracy, final targeting accuracy measured on a SEM's stage has never been better than about 40–50 μm for blanket wafers using blanket wafer scanners (KLA/Tencor 6200 or SP1 series). Implementation of the techniques to ensure consistency between the defect map coordinate system and the SEM stage's coordinate system allows minimizing and in some cases even eliminating the systematic error, but these methods cannot eliminate the random error that exists in the defect map coordinates (15). A recent study indicates that random error of the defect maps of the today's most popular unpatterned wafer scanners ranges from about ±30 μm to over ±200 μm (16). The random error makes submicron defects difficult (and often impossible) to locate by SEM, especially when the defect does not generate sufficient imaging contrast due to low secondary electron emission (such as low $Z$

organic-based and/or low topographical features, for example, COPs). We observe some cases in which defects that are easily detected by wafer scanners provide very little or no contrast under SEM (12,18). It is also likely that a class of defects exists in which the reverse is true. This problem becomes much worse when an analyzed wafer is even slightly warped (usual case) and low imaging contrast is combined with poor focusing capabilities (there is a complete lack of features on which to focus in the case of a blanket wafer). These issues are at the heart of problems facing blanket wafer SEM-based defect analysis.

In order to optimize the particle analysis, a new technique for SEM-based defect characterization on unpatterned wafers has been developed. This technique, embodied in the MicroMark[TM] 5000 (MicroTherm, LLC), employs optical redetection on an accurate stage to produce high-accuracy defect maps. Precision surface marking is used (when necessary) to simplify wafer alignment on the SEM's stage. The wafer preparation methodology and some experimental data are described next.

Before SEM review, and after the wafer is scanned using a wafer scanner (i.e. KLA/ Tencor 6x00, SP1), the wafer is processed by the MicroMark[TM] 5000 system (MM5000). The MM5000 revisits defect sites using the wafer scanner's defect map. At each defect site, the system records an image of the surface using a laser dark-field optical technique. The image is analyzed to locate the defect's location in the field of view. From this, and from the MM5000's stage position, the defect's coordinate is calculated relative to the wafer's coordinate system. If required, the system can create tiny fiducial marks to simplify defect map alignment to the SEM's stage. The MM5000 employs a 300-mm-capable air-bearing $xy$ stage for high-speed and high-accuracy operation. The system outputs a standard defect map file (in TFF format) with coordinates that are accurate to within $\pm 5\,\mu m$ over the entire wafer. The system is fully automatic and performs defect map updating at a rate that exceeds 1,000 defects per hour.

Optical redetection of defects by the MM5000 has several advantages over attempted direct SEM-based detection. First, since the contrast mechanism of the MM5000 system is the same as that of the wafer scanner (light scattering), the system detects the same defects with high probability. Second, the contrast is high enough using light scattering that small defects can be detected at low magnification (large field of view), making the detection insensitive to defect map inaccuracies. This enables the system to operate robustly despite inaccuracies in the original defect maps. The redetection process also provides a means for filtering out false counts from the wafer scanner and the possibility of preselecting defects for SEM review based on optical characteristics and creating a new file containing only these defects.

Figure 4 shows a SEM image of laser marks created by the MM5000 around a small defect. The marks are made at precisely controlled $xy$ coordinates, and an entry is made in the defect file at the coordinate. These marks ensure a robust method for high accuracy, since they are easily located by SEM and they have precisely known coordinates relative to the MM5000's updated defect map file. A mark pattern is created around the defect to indicate precisely the defect's location on the wafer. By marking individual defects, the problem of locating them in subsequent analytical tools is nearly eliminated, even if the wafer must be cut up for analysis by a small-sample analytical tool. This method, termed mark-assisted defect analysis (MADA) has been described in detail elsewhere (17). In the example shown in Fig. 4, a defect, invisible via low-magnification SEM, is located within the laser mark "crosshairs." Increasing magnification on the region defined by the "crosshairs" brings the defect immediately into view.

The efficiency of the MADA has been determined using the following experiment. SEM defect review tests were performed with and without the MicroMark[TM] 5000 process

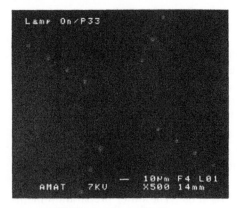 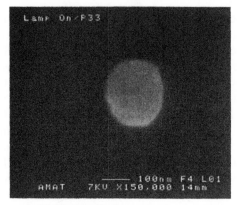

SEM Image of Laser Marks                         SEM Image of Defect
Intersecting at Defect Position                  Found Using MicroMark™ 5000

**Figure 4**  Example of the use of wafer-marking small defects.

on a defect map produced by the industry-standard wafer scanner system (KLA/Tencor SP1). The measurements were made using a state-of-the-art SEM defect review tool (FEI DualBeam™ SEM) that provides wafer-positioning accuracy better than $\pm 2\,\mu m$. Tests with the MicroMark™ 5000 defect file provided defect characterization yields exceeding 90% for defects in the 0.1–0.2-μm range. The defects were located by SEM without searching, and the combined error of stage and defect file averaged $\pm 4.2\,\mu m$. Tests using only the Tencor SP1 defect file resulted in unsatisfactory defect navigation results. (a) Difficulty was experienced in locating suitably small defects for good 2nd-order correction, and (b) subsequent defect characterization yields were under 50% in the 0.1–0.2-μm range, with an average defect map error of $\pm 35\,\mu m$ (18).

## D.  Use and Limitations of Classification and Characterization of Defects by SEM Alone

Much of the recent interest in automatic defect classification, ADC, in optical review and SEM review of defects concerns image recognition and image classification. This is useful for pattern violations on patterned wafers (opens, shorts, dimension violations) and for large defects where the objective is to provide classification of a descriptive nature, such as round, long, thin, large, small, on surface, in-film. For particles on bare wafers this rarely provides a root cause, however, and very different materials can give very similar shapes. It is usually necessary to have a materials composition analysis.

There are, however, occasionally cases where an image *is* distinctive enough to assign a root cause based on previous experience with the same defect. For example, Figure 5 shows the SEM image of particles often found as a particle failure mode in dielectric CVD of Si oxide. They are small spheres of $SiO_2$, caused by a gas-phase nucleation due to process parameter instability. They occur as single particles, Fig. 5a, and in clusters, Fig 5b. They are also usually evenly distributed over the wafer and have a narrow range of sizes, as recorded by the scanner bin split (0.2–0.3 μm). Of course the first time we saw these particles it *was* necessary to confirm, by EDX, that they were $SiO_x$. Note that in this EDX analysis, small $SiO_2$ particles on an $SiO_2$ film actually represent one of the most difficult cases to deal with, requiring very careful work. Now, it is probably no longer

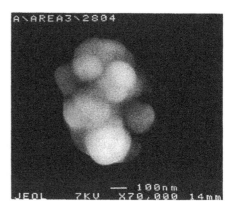

**Figure 5** SEM images of typical gas-phase nucleation particles ($SiO_2$) formed during dielectric CVD process: (a) small individual sphere, (b) agglomerate of small, rounded particles.

necessary. But it takes only about 30 seconds to record the EDX (if you have one on your SEM!) to check that no elements other than Si and O are present. Silicon nitride particles can be formed in an exactly parallel manner during nitride CVD.

A second case of a very distinctive shape is shown in Figure 6. The images represent TiN flakes that have come from TiN having been deposited on a particular piece of chamber hardware, the RF coil, during a TiN PVD process. Eventually, sufficient stress is built up so that the film peels off in flakes that curl up under the stress ("potato chips"). If these shapes of particles are seen during TiN PVD, it's a good bet that the foregoing explanation is correct, but it is still a good idea to check by EDX that they are of TiN composition, since it is conceivable that some other film deposit could flake off in a similar manner. On the other hand, if there are several morphologies of TiN particle present, this distinctive "potato chip" shape is a pointer to those that come from the RF coil, as opposed to other regions of the chamber hardware.

The third example, Figure 7, shows the real danger of relying on images alone for classification. The spherical particles imaged were both found on tungsten silicide wafers. Any automated image analysis would undoubtedly say they are the same. But in fact the particle in Fig. 7a is made of carbon, whereas the one in Fig. 7b is made of tungsten and silicon, and they originate from different sources. Note also the large discrepancy of Tencor scanner sizing (in parentheses at the top of the images). Because of the difference in material, the "PSL equivalent" sizes are very different, even though the true sizes are similar.

### E. SEM/EDX Necessary and Sufficient to Establish Characterization

The last example is obviously a case where EDX is needed, since it indicated two separate sources of particles. To decide *what* those sources were took detailed partitioning experiments in collaboration with the process engineers. There are many other similar cases. Figure 8 shows the SEM image and EDX of a long, thin particle. The presence of Ni, P, O is sufficient to tell us it originated from degradation of a Ni-electroplated coating on a hardware part, once we had established, in close collaboration with the designers and process engineers, that such parts existed in the chamber and examined the EDX of that part. Note that the shape of the particle is quite similar to that of the stainless

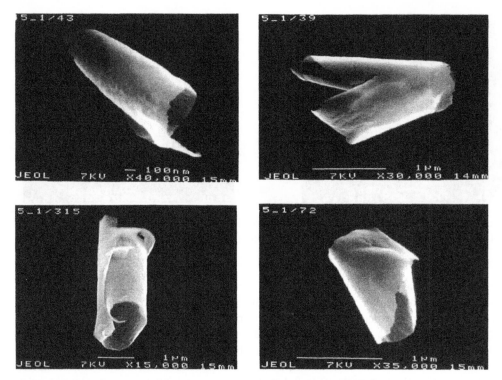

**Figure 6**   SEM images of particles originated from an RF coil (TiN PVD process).

steel particle shown in Figure 9, but its composition and origin are unconnected! The determination of stainless steel in the particle in Fig. 9 is a classical example of EDX being positively able to identify the nature of a particle (from the Fe, Ni, and Cr intensities). But since there might be many sources of stainless steel in chamber hardware, it still takes a lot of collaborative activity and partitioning tests involving the hardware designers, process engineers, and defect analysts to find the particular culprit. In this case it was an "illegal"

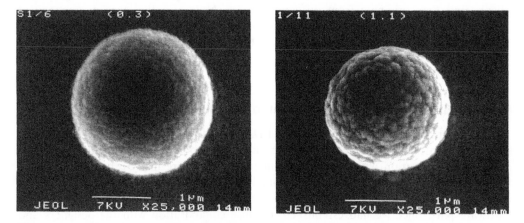

**Figure 7**   SEM images of quasi-spherical particles consisting of (a) carbon, (b) tungsten/silicon (by EDX). Spheres were detected on a tungsten silicide deposition test wafer.

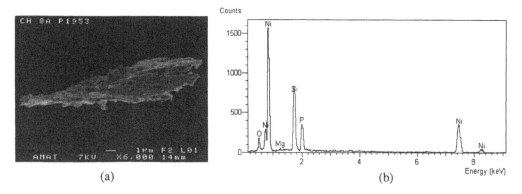

(a)                                              (b)

**Figure 8** (a) SEM image and (b) EDX spectrum obtained from a particle originated from degraded electroplated film.

stainless steel screw. Note that both the stainless steel and Ni particles were found on the same wafer, though they obviously came from different hardware sources. In both cases, process attack was involved, however.

Perhaps the best-known example of root-cause determination directly from SEM/EDX is that shown in Figure 10. The SEM image is of a "particle," which is actually a collection of many smaller particles, originating from plasma degradation of a viton O-ring (9). The signature of the viton is the presence of Mg in the EDX spectrum, in addition to C, O, and F. Mg is not used anywhere in the chamber hardware, whereas, of course, C, O, and F can come from many sources. In the viton O-ring it comes from the magnesium carbonate used as a "filler" in the elastomer. Other types of O-rings use different fillers with different signatures. Again, it is necessary to know the EDX of the different types of O-rings (the standard spectra for this problem) to be able to confirm that it is *viton* elastomer debris. Also, this particular example points to the dangers of any kind of automated analysis. Since the "particle" is actually a collection of particles, it is highly heterogeneous and not all regions show the presence of Mg (see Fig. 10). In this case it was necessary to "search" the particle for the Mg signature.

Figure 11 shows particles arising from a Si-etch process using Br-containing plasma chemistry. When process parameters are not optimized, fragmented residual

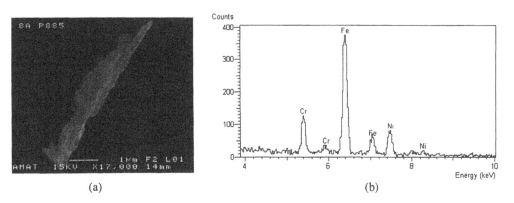

(a)                                              (b)

**Figure 9** SEM image and (b) EDX spectrum obtained from stainless steel particle (W CVD process chamber).

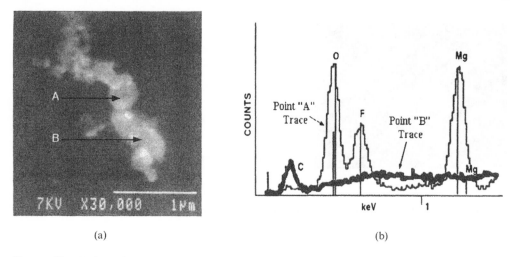

(a)                                                        (b)

**Figure 10** (a) SEM image and (b) EDX spectra of O-ring-generated particle.

species conglomerate, forming distinct, easily recognizable, particles. In SEM under low magnification they are shaped like different geometrical figures (cones, squares, triangles, etc.) (Fig. 11). There is also one more feature that helps to unambiguously identify them. This kind of particle is not stable under a focused electron beam. An e-beam of a relatively low (5–7-kV) voltage in the raster mode corresponding to ∼10–15 K magnification produces severe damage or completely destroys them (Fig. 11b).

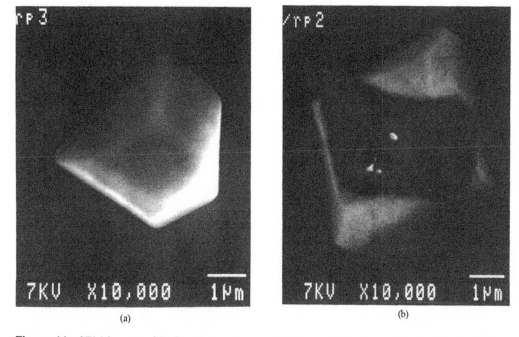

(a)                                                        (b)

**Figure 11** SEM images of Br-based gas-phase nucleation particles formed in the Si-etch process: (a) "as is", (b) damaged by an e-beam.

Nevertheless, to prove these come from an instability of the process chemistry, the presence of the Br must be identified by EDX. This is also tricky, since the e-beam-damaging effects rapidly desorb the Br!

Figures 12A and 12B shows a case where defects are of different morphology but of the same composition. The figures show Si/O particles detected on test wafers during a W deposition process. The test wafer (bare Si) has been run through the chamber when only one gas line was open (SiH$_4$ gas line). Particles scavenged by the wafer were then analyzed using SEM/EDX. Two particle groups were detected based on their "look" (morphology). The first (Fig. 12A) was very dispersed porous particles consisting of Si and O (ratio the same as for a SiO$_2$ standard using 4-kV voltage). The second particle group (Fig. 12B) also consisted of Si and O (also close to 1:2 Si/O ratio), but morphologically they looked absolutely different: they were much less dispersed and porous and appeared as agglomerated small, rounded primaries. If SEM and image classification alone had been used here, the conclusion would have been two sources of defect when, in fact there is only one, an oxygen leak in the SiH$_4$ line.

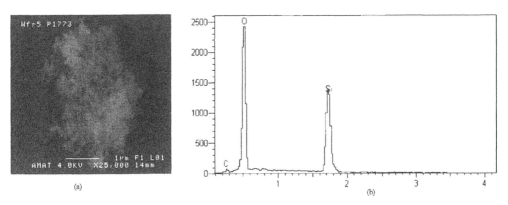

**Figure 12A**  (a) SEM image and (b) 4-kV EDX taken from porous SiO$_2$ particles (W CVD process).

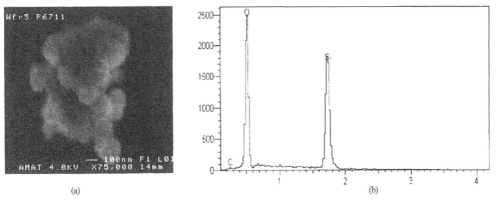

**Figure 12B**  (a) SEM image and (b) 4-kV EDX spectrum of dense and rounded SiO$_2$ particles (W CVD process).

## F.  Cases Where More Subtle Aspects of EDX Analysis Are Needed to Obtain a Characterization/Root Cause

In the previous examples the interpretation of the EDX spectrum itself was fairly straight-forward, once it been obtained. We also regularly come across cases where a more advanced approach is needed. High-voltage excitation may be required to access the x-ray lines providing the definitive signature; comparison of low and high voltage may be needed to determine what is on top and what is underneath (or inside); spectral overlaps may inhibit analysis; and very careful determination of whether a particular element is really in the particle, or actually in the substrate, may be needed. It is also true that in many of these cases EDX may not be the optimum analytical technique, even though with skill it can do the job. Specific examples are given next.

### 1.  Barium Strontium Titanate CVD Particles

The CVD process for producing films of barium strontium titanate, BST, has three separate inlet gas lines by which the Ti-based, Ba/Sr-based, and solvent precursors are delivered into the BST process chamber. During development of the process, severe particle problems occurred (not unusual at that stage of any process). After initial tests ruling out hardware issues, the particle characterization task was to clarify which line, or lines, generated what particles.

The major Ti and Ba lines ($TIK_\alpha$ and $BAL_\alpha$) overlap in EDX. They are separated by only 46 eV, whereas the standard Si solid-state detector has a resolution of only 130 eV. We use the highest-resolution detector available (Ge), which still cannot separate the $TIK_\alpha$ and $BAL_\alpha$ lines, but can just resolve the $TIK_\beta$ and $BAL_\beta$ lines nearby. Working at 20 kV to a maximize $S/N$ ratio, it is just possible to detect $TIK_\beta$ as a shoulder on the $BAL_\beta$ peak, Figure 13a and b. Analysis of many particles on many wafers, during the development of the process, led to the conclusion that there are *two* groups of Ba/Ti particles present. One, Fig. 13c and d, has Ti as the dominant component. Note the morphologies are also different in the two groups. The two groups were eventually found to correspond to particles being generated from the different gas lines. More efficient filtering was implemented, leading to reduction of the particle levels to meet specification.

### 2.  WCVD: Apparent Haze Appearance on Bare Si Monitor Wafers

The WCVD deposition process has two main kinds of particle testing: (a) after W deposition (i.e., in-film particles) and (b) so-called "gas-on" test, or "mechanical" test, on a bare Si wafer run through the W deposition chamber with flow of Ar gas to scavenge particles after $NF_3$ plasma clean. This process is well established, and extended particle statistics/characterization data have been presented elsewhere (9,19). In that work the light-scattering inspection of gas-on test wafers reported particle composition corresponding to the end of the lifetime of chamber parts (Al-based particles) and/or by-product species (W/Ti based). The case reported here was different—light scattering showed increased haze in the central regions of the bare Si monitor wafers, i.e., something that method could not easily distinguish from surface roughness.

SEM, Figure 14a, showed that there were real particles present, though they were very small (typically 30 nm) and at high density. Low-voltage EDX (4 kV) was used to examine a high-density region while restricting the probing volume. No foreign element was observed, but of course only x-ray lines well below 4 kV could be generated. Going to 15 kV and making the beam raster size as small as possible, so as to focus it on an

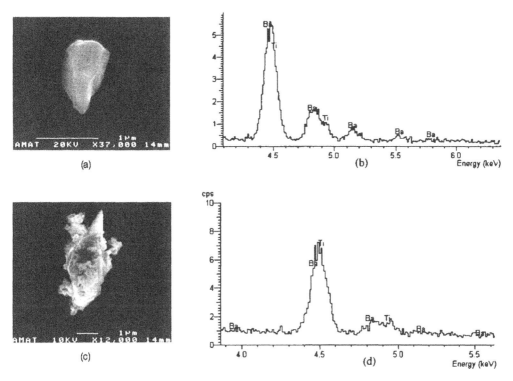

**Figure 13** SEM images and 20-KeV EDX spectra obtain from Ba/Ti-based particles. The presence of Ti in the upper particle (a and b) is deduced from the shoulder on the right side of the BaL$_\beta$ peak. This particle is considered as Ba dominant. The lower particle (c and d) also shows a double shoulder in the BaL$_\beta$/TiK$_\beta$ region, but this one is considered as Ti dominant.

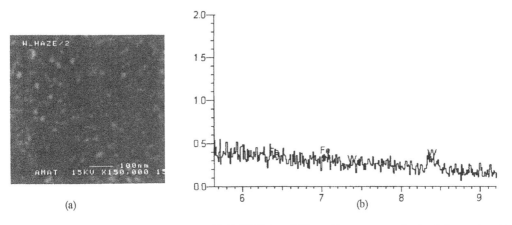

**Figure 14** (a) SEM image and (b) 15-KeV EDX (see WL peak) spectrum obtained from "haze" particles.

individual tiny particle instead of the general high-particle-density area, revealed that the particles were W, Fig. 14b, where this minor, high-energy WL x-ray line at 8.4 kV was consistently (but weakly) observed. The major WM line, which is accessible using 4 kV of energy, overlaps the Si substrate line and at these signal strengths had no chance to be observed. The finding of W indicated that a W-containing vapor, probably $WF_x$, was not being properly evacuated from the chamber. Adjusting the process window fixed the problem.

### 3.  Aluminum-Based Particles on an Oxide Wafer

This example represents a particle problem occurring in reactive plasma cleaning, where $NF_3$ chemistry is used to clean the chamber components. $NF_3$ is very aggressive, and inappropriate usage often generates particle problems from hardware attack.

For EDX, this case is one of the most difficult, because the suspected parts are made of similar materials. The first was the chamber dome, made of ceramic (Al/O), and second was the chamber heater made of Al-alloy. The monitor wafer used to scavenge particles is 200-nm thermal oxide. The EDX spectra of the majority of particles showed the presence of O, F, Al, and Si. The F peak is the result of the penetration of fluorine radicals into hardware materials due to too harsh plasma. The question is: Does the particle contain Al and O, as well as F (in which case the ceramic dome is to blame)? Or does it just contain Al and F (in which case it is the Al alloy to blame)?

The representative particle and corresponding EDX spectra taken from the particle (solid line) and background (dashed line) are shown in Figure 15a and b. The intensity of O from the particle is very low (O : F ratio of about 1 : 6, when the corrected O background is subtracted). Ceramic parts attacked by $NF_3$ always show higher O : F ratios (typically 1:1), leading us to conclude that these particles come from Al Alloy degraded by $NF_3$. These findings were supported by EDX obtained from ceramic templates located within the processing chamber for long RF hours, where such low O : F ratios were never observed.

### 4.  Focused Ion Beam Application

Initially FIB was used for modification of ICs and microelectronic devices and for TEM sample preparation. With improvements in performance, FIB has become one of the most effective instruments for the root-cause analysis of particle defects, because of its ability to

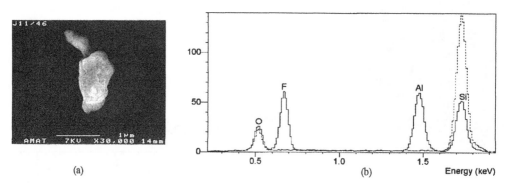

**Figure 15**  (a) SEM particle image and (b) 7-KeV EDX spectra: solid line—from the particle; dashed line—background.

cut into particles, or into the film containing the particle, with 0.1-μm resolution. The following examples show the value of FIB in understanding the mechanism of defect formation. Two specific problems are chosen to illustrate this.

During oxide deposition a particle spike occurred. Based on light-scattering data, the majority of the particles were caused by the so-called "gas-phase nucleation" problem. Particles were spread evenly on a test wafer, and the majority were 0.2–0.3 microns in size. Subsequent SEM/EDX confirmed these observations, since particles representing the major trend were 0.2 μm in size and of nearly spherical shape. By EDX, there was no difference between the elemental composition of such particles and background (500-nm oxide). This particle type represented about 70% of all particles detected. The rest (~30%) were of larger size and often like an agglomeration of tiny primaries (Figure 16a). Low-voltage EDX (4 KeV) did not show any additional to background elements. At this point one might reasonably conclude that all particles were SiO$_2$ from gas-phase nucleation, some individually, some agglomerated. This was exactly the case discussed in Sec. II.D. However, the use of higher voltage (7 KeV, Fig. 16b) revealed Al and F. This result indicated that *hardware* shed particles, probably, due to NF$_3$ plasma chamber clean. To

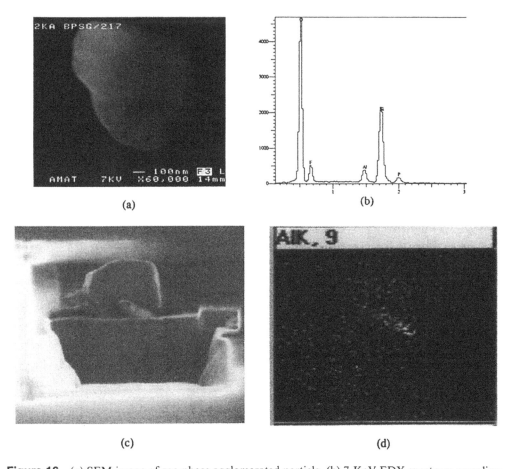

(a)

(b)

(c)

(d)

**Figure 16** (a) SEM image of gas-phase agglomerated particle. (b) 7-KeV EDX spectrum revealing the F and Al. (c) FIB cross section right through the particle. (d) AlK x-ray map confirming that the seed is Al-based material.

confirm this assumption, the process engineers requested that we find the detailed morphology and composition and whether the Al/F was on, in, or under the 500-nm oxide film, or encased in a $SiO_2$ particle. If located at the interface, it requires different corrective actions than if a seed is present inside of a gas-phase nucleation particle.

Using FIB, this particle was iteratively cross-sectioned, finally revealing that a tiny seed sits on the surface of bare Si wafer, landing prior to oxide deposition (Fig. 16c). Analysis by EDX confirmed that it consists primarily of Al (Fig. 16d) and trace F (F map is not shown).

In a second example, a severe particle spike (more than 5000 particles) had been detected on TiN wafers. The TiN film (100 nm) had been deposited on a 300-nm oxide film produced by the PE-TEOS process. The defect was shaped as individual knobs about 0.1–0.15 microns in diameter (Figure 17a and b). Also, agglomerated ones were located (Fig. 17c) with sizes in some cases close to 1 micron. It was suspected that the cause of this problem was associated with contamination present either at (a) the Si/oxide interface, or (b) at the oxide/TiN interface. The cross-sectioning of large defects (Figure 18a and b) showed that the TiN film could be excluded from consideration, since the bump clearly goes through the TEOS layer to the substrate, Fig. 18b. Further FIB revealed that the initial bare Si wafer was contaminated, probably with carbon-based species. At high temperature the contaminant outgases, causing a bubble that eventually gets decorated by the TiN deposition. Thus the source of the problem was contamination at the oxide/Si interface, and the TiN PVD process was cleared from blame.

## G.   Automated SEM/EDX DRT and Automatic Defect Classification

What we have described so far is the use of SEM/EDX in the analytical laboratory to solve defect (particle) issues on unpatterned wafers. We have close and rapid communication with the process engineers (or customer engineers for field issues), who "own" the problem, and often we are able to give turnaround in hours. However, often the problem is brought to the lab's attention only after some period of trying to solve the problem empirically (partition experiments, changing parts and process recipes). In the fab this approach is not desirable, because instant turnaround is the goal, to avoid scrap product and downtime. As SEM gradually replaces optical for defect review in the fab, it is, therefore, natural that it be viewed as very important to move toward automated SEM and ADC and away from "expert" intervention.

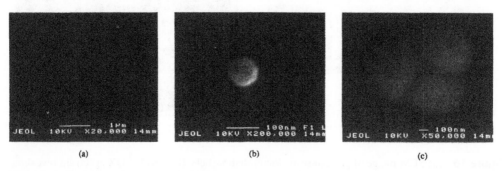

**Figure 17** (a, b) SEM images of individual small knobs under different magnification. (c) Agglomerated defect.

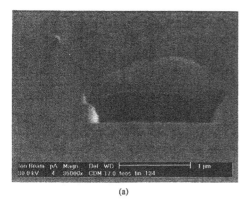
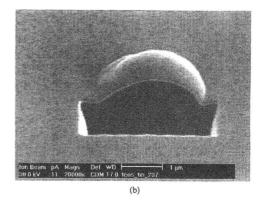

(a)                                    (b)

**Figure 18** (a, b) SEM images of FIB cut through the middle of agglomerated defects. The cross sections show that oxide film piles up above the surface of the bare Si wafer due to initial wafer contamination.

A number of SEM companies make automated SEMs for full-wafer DRT use in fabs. They usually do not have the full flexibility of analytical lab SEMs, the most common difference being a restricted voltage range. The idea is simple. The SEM, working off the light-scattering defect files, drives automatically to the predicted defect site, takes an image, and moves on to the next site. There is a large difference in the capability to do this, and the usefulness of the data if it can do it, for unpatterned compared to patterned wafers. For patterned wafers, the pattern itself, and the observation of pattern violations, is key information. This can be done automatically from the SEM image with appropriate recipe writing (which must always be done for a new pattern, new film, and new defects and which can take much time), and classification into shorts, opens, CD variations, etc., can be achieved by manually examining the output. For particles, and particularly for particles on unpatterned wafers, the usefulness of doing this automatically in purely SEM mode is greatly reduced. First, on unpatterned wafers it is much harder to just "drive to the defect" and image it (the "targeting" issue). Even with on-board optical microscopes (e.g., Applied SEMVision), small, but real, particles can be missed and therefore incorrectly classified as "false counts" for the original scanner. Second, since there is no pattern, most of the useful classification disappears. The ones that remain viable—large, small, round, long, on top of film, under film, etc.—generally not only do not give enough information, but also can be positively misleading. As we saw earlier, it is primarily materials characterization that is needed. Particles can have different materials composition and origin but very similar shapes, and they can also have very different shapes but the same composition and origin.

Owing to the necessity for materials analysis, automated SEM DRTs do, usually, have EDX options. This slows down the whole process, since an EDX spectrum takes longer than an image. Second, as demonstrated in the previous sections, a correct (and therefore useful, as opposed to misleading) interpretation requires the EDX to be taken in the correct way, and what *is* the correct way can *only* be established as each new defect case comes up. (E.g., what size is the particle? And what voltage should be used? Is more than one voltage needed? Is there spectral overlap to be dealt with that requires better statistics and longer data acquisitions? Is a careful comparison on and off the particle needed?) Current automated SEM DRT is not at this stage. Usually one standard spectrum at one voltage is acquired, with no assurance that it is even dominantly representative of the particle rather than the background.

Automatic defect classification has been a debated topic for the past few years. For SEM this involves "simply" pattern recognition, and it can, and does, work for patterned wafers. The protocol then is that the automated SEM produces the images, and, instead of a human doing the classification into opens, shorts, etc., an ADC program can be used to automatically bin into defined classes. Except for exceptionally simple classification, this still requires establishing a sufficient database for the wafer level/pattern/types of defects at hand in order to establish the bins and a working ADC recipe (which can then be self-learning as new data is added). Obviously, then, ADC is best suited for fab-like situations where established processes/tools are being constantly monitored for variation from base-line conditions.

If automated acquisitions of EDX on particle is difficult, ADC of the EDX data is even more so. For it to have a chance of working at all requires that the data itself not be misleading. Given that the current level of automated acquisition is going to produce data that can be misleading in many cases, ADC on this data will have problems. At current levels of sophistication it has to be restricted to the following situations.

1. Either particles are so large that there is no doubt that the signal comes entirely from the particles, or the ADC classification scheme is immune to variable contributions from the wafer.
2. Either the mere identification of an element, or set of elements, as being present is sufficient to bin the defect into a useful classification (more likely a useful characterization), or the difference in relative strength of different EDX lines for different classification is pronounced enough for the ADC to be able to make a decision.
3. Spectral overlaps in the data must not be an issue.

From these constraints, it should be obvious that EDX ADC is *never* a viable proposition for the type of particle root-cause analysis we are involved in. Almost every case for us is a "new" case, and far more time would be spent trying to acquire enough quality data to write reliable ADC recipes than we spend solving the problem with exper-tise. This does not mean that EDX ADC can never be successful. For tool-level situations that are thoroughly baselined and a set of particle defects established that conform to constraints 1–3, it is quite possible to write ADC recipes so that automated data collection, plus that ADC, can track the statistical changes in the baseline particle behavior. This has been demonstrated using the SEMVision DRT. The questions are: How many tools have been EDX baselined sufficiently well to adopt the procedure (not many!)? And what fraction of real problems will conform to constraints 1–3 (not known!)? Anything outside conforming to constraints 1–3 will require procedures corresponding more to what is thought of as an "expert system," rather than ADC.

## III. ADVANCED FULL-WAFER PARTICLE CHARACTERIZATION TECHNIQUES

### A. Incorporation of Add-Ons into Commercial Full-Wafer SEM/EDX Tools

Originally EDX was coupled to SEM because it uses the same electron beam column to generate the x-ray signal and because the solid-state detector is small and fits inside the SEM chamber, near to the sample. Similarly there are several other electron-beam-based techniques that can be added on to commercial SEMs and that, in principle, can be used for full-wafer particle characterization. In practice, however, none have yet been very successful in such an add-on mode, so they are only briefly discussed here.

## 1. Wavelength Dispersive X-Ray Spectroscopy (WDX)

Wavelength dispersive x-ray spectroscopy, WDX, differs from EDX only in the manner in which the x-rays are detected and the energy analyzed. Instead of a Si or Ge solid-state detector, conventional crystal spectrometers are used to disperse the x-ray wavelengths. The advantage is higher spectral resolution by a factor of 10. This resolves all of the EDX spectral overlap issues. It also allows x-ray analysis to be done by restricting the e-beam excitation to low voltage and using only the lower-energy region of emitted x-rays, which always overlap in EDX. Thus the analysis volume can be greatly reduced, improving the particle/wafer background separation of signal. The serious disadvantages are an enormous loss of signal strength (use of crystal spectrometers, plus their great distance away from the particle—*outside* the SEM chamber through a viewing port) plus the fact that *several* crystals have to be used to cover the needed energy range. No viable commercial add-on has been available until recently (20). It remains to be seen whether it has the ease of use and signal strength for full-wafer small-particle work.

## 2. Microcalorimetry X-Ray Spectroscopy

Microcalorimetry is another way of improving EDX resolution by an order of magnitude (21). A detector is used, such as a low-temperature superconducting junction that is sensitive to the very small amounts of heat imparted by the x-ray photon impact (proportional to x-ray energy). Originally developed for astrophysics, full low-T-physics liquid-He-cooling technology is required to work at the necessary 50 mK. Resolution, with sufficient sensitivity, has been demonstrated to be capable of even chemical shift information on the x-ray line. The question is whether working at 50 mK can be successfully adopted by the semiconductor manufacturing industry.

## 3. Auger Spectroscopy

Electron energy analyzers have been added to SEMs and TEMs in research before, and a few years ago we tried a simple, compact add-on Auger analyzer (the Pierre from Surface/Interface) on a 200-mm wafer commercial SEM (JEOL 6600F). Count rates were low (performance was affected by the magnetic field of the SEM lenses). But, more significantly, since the SEM is not clean UHV, hydrocarbons crack under the e-beam on the surface so that all one can see in the surface-sensitive Auger spectrum is carbon. It became quickly obvious that a stand-alone full-wafer Auger system would be far more effective (and far more expensive!)—see Sec. III.B.

## 4. Electron Backscattering Diffraction

Electron diffraction is a method of getting structural-phase information (22), just like x-ray diffraction. It has been successfully used in the SEM by interpreting the diffraction patterns cast in backscattering onto a phosphor screen (the setup is available commercially as an add-on), but not for particles on full wafers. We have investigated its use on cut-up sections of wafers (19). Obviously it will give information on the particle only if it is crystalline. We have found that the patterns from small particles are generally too weak to be useful—mainly because small particles give a small signal and partly because, even if crystalline, a particle tends to have an amorphous outer sheath.

The conclusion from this section is that the use of these techniques, in conjunction with a commercial SEM/ EDX, is premature for defect review work. We expect, however, that eventually successful add-on versions will appear. The closest at this moment is WDX.

## B. Stand-Alone Full-Wafer Defect Review Techniques Other Than SEM/EDX

Outside of the purely optical review category, there are a few full-wafer review techniques (200 mm or 300 mm) available in addition to SEM/EDX. These are Raman spectroscopy, scanning Auger spectroscopy (SAM), atomic force microscopy (AFM), and secondary ion mass spectroscopy (SIMS) (20). We have the first three available in our laboratory, and the fourth, SIMS, is available through outside analytical laboratories. None of these has yet gained any foothold for use inside the fab, though we except this will happen gradually. The principle of operation of the methods and their advantages and drawbacks, were discussed in Sec. I, so here we just concentrate on presenting a few examples where the root-cause analysis answers could not be obtained by SEM/ EDX alone.

### 1. Raman Spectroscopy

Figure 19 shows the Raman spectra of two carbon-based particles found on the same Si monitor wafer after wafer-handling tests. The Raman spectrometer is attached to our 300-mm Ultrapointe confocal laser microscope DRT (6). The Ultrapointe uses 488-nm blue light and can find defects based on scanner files rather easily, because, unlike SEM, it is a high-contrast optical tool using the same wavelength as the particle scanners (i.e., fewer targeting problems). Also, since it is an optical imaging tool, we are not usually using it for particles below ~0.3 μm. In Raman mode, we are able routinely to get spectra down to about 0.7-μm size, sometimes lower, depending on Raman cross section. The laser spot size (~1 μm) is the limiting factor. In Fig. 19 the particle sizes were 0.3 μm. The EDX technique cannot distinguish the two types, diamond and graphite, which are very easily distinguished by Raman. Knowledge of the wafer-handling hardware materials, plus partitioning experiments by the engineers, allowed their root causes to be determined.

Figure 20 shows the photoluminescence spectra (obtained via Raman spectrometer, but in a different wavelength range from the Raman vibrational lines) of two ~0.7-μm

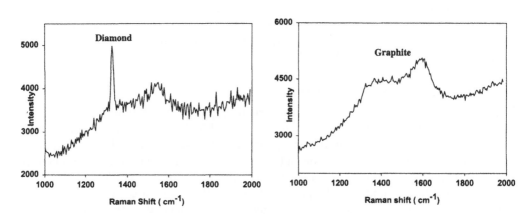

**Figure 19** Raman spectra of small (~ 0.3-μm) particles found on a bare Si monitor wafer.

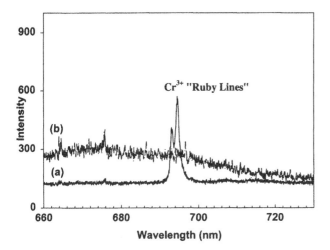

**Figure 20** Photoluminescence from ~ 0.7-μm $Al_2O_3$ particles: (a) from ceramic; (b) from anodized Al.

$Al_2O_3$ particles. One is from a ceramic source, i.e., crystalline α-alumina; the other is from anodized Al. The former shows the $Cr^{3+}$ "ruby lines." The latter possesses no $Cr^{3+}$. We use this distinction, which is very sensitive, routinely to separate sources of $Al_2O_3$ (5).

### 2. Atomic Force Microscopy (AFM)

Figure 21 shows AFM images of two Epi defects found on the same wafer. One is a symmetrical mound, 12 μm in diameter and 0.7 μm high. The other is an $Al_2O_3$ flake (previously identified as such from EDX) 1 μm across but only 250 Å high. Both are difficult to image in SEM (especially the first one), in one case because of lack of materials contrast and in the other because of the small $z$ change and flat surface of the particle. In the AFM they are very easy to precisely define, in terms of geometries, *once found.* To do the latter required laser-marking with the MicroMark 5000, however, since in AFM the time taken to search, if there is a targeting issue, is prohibitive. Seiko solved this problem years ago by adding an on-board dark-field microscope to their AFM to first find defects before going into AFM mode (23). An equivalent system has just been made commercially available in the United States (24). An alternative approach for AFM is to add an AFM objective to an existing optical review station. Several commercial options are available here also.

Note that AFM provides topography information *only,* though it can do this down to the angstrom level in the $Z$ (height) direction. Note also that the two Epi examples shown represent worst-case scenarios for direct sizing by the original light-scattering scanner. It estimated the 12-μm defect to be 300 μm, and the 1-μm defect to be 0.1 μm!

### 3. Scanning Auger Microscopy (SAM)

Figure 22 shows Auger images, and spectra, typical of a type of "particle" we have been finding in a variety of situations over the past few years. The data was taken on our PHI SMART 300, a 300-mm-capable SAM DRT. Auger, being surface sensitive and chemistry sensitive, has no difficulty in establishing the "particle" as a thin-film (< 50 Å) patch

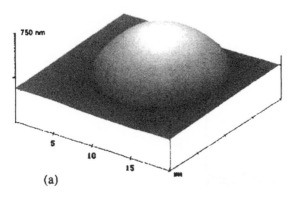

(a)

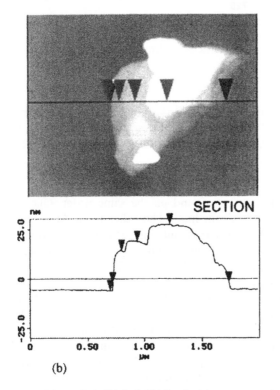

(b)

**Figure 21** AFM (Digital 7000) of: (a) a symmetrical mound in an Epi layer (Tencor greatly overestimates size); (b) an $Al_2O_3$ thin flake on an Epi layer (Tencor greatly underestimates size).

of carbon or hydrocarbon material (7). EDX is incapable of distinguishing this patch from the wafer, because it is too thin to give a detectable EDX signal. In this particular case no other elements are present, but in other situations we have observed C and F together, i.e. a fluorocarbon, and C, O, and F together, which turned out to be a fluoroether (confirmed by SIMS—see Sec. III.B.4). The sources are greases, oils, lubricants, etc.—either legally or illegally in the processing tools. Very often they are observed initially by light scattering *only* if there is an inorganic seed at the center, which has a much larger scattering cross-section. Since they form by a 2-D nucleation

**Figure 22** "Fleur-de-lis" patch found on monitor wafer. Auger images and spectra show it contains carbon only (probably hydrocarbon from oil contamination).

phenomenon, it is not surprising to find a seed at the center—rather like a snowflake, for instance, which is one of the names that has been given to these types of defect (others are: fleur-de-lis; dendrite, fractal, and our original description, area/seed). However, for every one that has a seed, there may be 10–10,000 that do not. So the true defect count of this type is 10–10,000 times larger than actually suggested by light scattering. The key issue is whether they cause yield loss.

Figure 23 shows the composite Auger image of a composite Al/Ti particle, found on a monitor wafer in a W etchback chamber. In the original work (25) we established it got there as a side product of the etchback process, $TiF_x$, attacking Al metal hardware metal (not oxide, because the particle has Al metal in it—shown by detailed Al Auger spectra). The most striking finding, however, was that the monitor wafer was covered with a monolayer of Ti. This means the $TiF_x$ vapor hangs around in the process

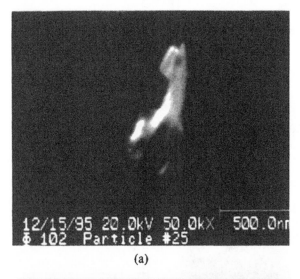

(a)

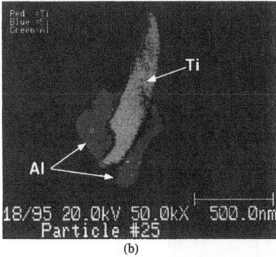

(b)

**Figure 23**  A Ti/Al composite particle found after a W etchback process: (a) SEM image; (b) composite SAM map of defect.

chamber and reacts with any fresh surface put into it, depositing this thin Ti film on it (25). Again, the issue is whether it matters.

## 4. SIMS

C. A. Evans and Associates have a 200-mm TOF-SIMS instrument available as an analytical service. Figure 24 is the F image from one of those oil contamination situations discussed in the previous section. In this case it comes from an experimental process chamber. Similar O- and C-based SIMS images were also obtained. The whole surface of the wafer was covered with the contamination, the "snowflake" being just thicker regions (detected by light scattering). Figure 24 shows a portion of the complete SIMS negative ion spectra from one of the patches. The peaks marked in the unknown are a definitive signature not just of a fluoroether, but, in this case, a very specific one—Krytox

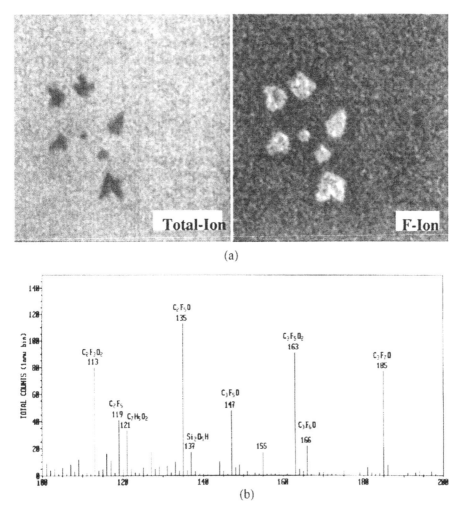

(a)

(b)

**Figure 24** TOF-SIMS of "snowflake particles" found on a monitor wafer: (a) total ion image and F ion images; (b) SIMS spectrum from a snowflake particle in the 100–200-amu range.

(12). Knowing this, the root cause was easily tracked down to a compressor being used in conjunction with the experimental chamber.

## IV.  CONCLUSIONS AND THE FUTURE

We believe that SEM/EDX will remain the heart of particle characterization for full-wafer work for the foreseeable future (particularly coupled with FIB), but then several other full-wafer methods will gradually come into play. Scanning Auger microscopy is already doing so and making significant contributions to root-cause analysis. Raman spectroscopy is extremely useful and, since it is a technique not requiring vacuum and is coupled to the familiar area of optical DRT, we anticipate several commercial tools to come on the market. It is, however, ultimately limited by the size of particle it can handle.

Regarding automation of SEM/EDX work and the possible use of ADC for characterization of particle defects on unpatterned wafers, it is clear that a lot of development, both in the methodology of using the tool and in providing schemes for handling the EDX data, is needed before it can move very far away from the expert user. The basic reason is that, once one moves away from defect related to pattern violations on patterned wafers, materials analysis is key to the characterization of defects, not an image, and this is much more difficult conceptually to automate.

## ACKNOWLEDGMENTS

We acknowledge many discussions and input from our colleagues at Applied Materials, Inc., particularly our customers, the process engineers owning most of the problems we have used as examples. The rest of the DTCL analytical staff also has contributed, and we want to specifically thank Richard Savoy for much of the SEM/EDX work, Giuseppina Conti for the Raman examples, and Edward Principe for SAM work.

## REFERENCES

1.  Encyclopedia of Materials Characterization. Edited by CR Brundle, CA Evans, Jr., S Wilson. Butterworth-Heinemann and Manning, 1992.
2.  CR Brundle, Y Uritsky, P Kinney, W Huber, A Green. Microscopy and spectroscopy characterization of small defects on 2-mm wafers. Proceedings of the American Institute of Physics #449 on Characterization and Metrology for ULSI Technology: 1998 International Conference, Edited by DG Seiler, AC Diebold, WM Bullis, TJ Shaffner, R McDonald, EJ Walters, Woodbury, New York, 1998, pp 677–690.
3.  PD Kinney, Y Uritsky, EL Principe, R Savoy, I Mowat, L McCaig. Evaluation of single semiconductor defects using multiple microanalysis techniques. Proceedings of the Material Research Society Symposium on Electron Microscopy of Semiconducting Materials and ULSI Devices, Edited by C Hayzelden, C Hetherington, F Ross, Volume 523, San Francisco, 1998, pp 65-70.
4.  YS Uritsky, G Conti, CR Brundle, R Savoy, PD Kinney. Chemical characterization of semiconductor defects and particles on the wafer surface by multiple microanalysis techniques. Proceedings of Microscopy and Microanalysis 1999, Volume 5, Supplement 2, Edited by GW Bailey, WG Jerome, S McKernan, JF Mansfield, RL Price. Spriger, 1999, Portland, OR, pp 742–743.

5. CR Brundle, Y. Uritsky, J Tony Pan. Micro, July/August 1995, pp 43–56.

6. G Conti, YS Uritsky, CR Brundle, J Xu. Semiconductor defects and thin film characterization by high-resolution images and by vibrational spectroscopy. Proceedings of Microscopy and Microanalysis 1999, Volume 5, Supplement 2, Edited by GW Bailey, WG Jerome, S McKernan, JF Mansfield, RL Price. Spriger, 1999, Portland, OR, pp 48–49.

7. EL Principe, Y Uritsky, CR Brundle, DF Paul, DG Watson, KD Childs. You never know what you'll find on a wafer! Proceedings of the Semicon-West Conference, July 20, San Francisco.

8. F Gozzo, B Triplett, H Fujimoto, R ynzunza, P Coon, C Ayre, PD Kinney, YS Uritsky, G Ackermann, A Johnson, H Padmore, T Renner, B Sheridan, W Steele, Z Hussain. Chemical analysis of particles and semiconductor microstructures by synchrotron radiation soft x-rays photoemission spectromicroscopy. Proceedings of the Material Research Society Symposium on Application of Synchrotron Radiation Techniques to Material Science IV, Volume 524, San Francisco, 1998, pp 227–232.

9. Y Uritsky, V Rana, S Ghanayem, S Wu. Microcontamination. 12(5):25–29, 1994.

10. YS Uritsky, HQ Lee. U.S. Patent #5,381,4. Particle Analysis of Notched Wafers. Jan.10, 1995.

11. SEMI Standards for Silicon Wafers, M1-92:13, 1992.

12. SurfScan 62, Wafer Surface Analysis System, User Manual, 1991.

13. Y Uritsky, H Lee. Particle analysis on 2-mm notched unpatterned wafers. Proceedings of the Electrochemical Society Symposium on Contamination Control and Defect Reduction in Semiconductor Manufacturing III, Volume 94-9, Edited by DN Schmidt, 1994, pp 154–163.

14. Y Uritsky, HQ Lee, PD Kinney, K-H Ahn. U.S. Patent #5,267,17. Method of Particle Analysis on a Mirror Wafer, Nov. 30, 1993.

15. M-P Cai, Y Uritsky, PD Kinney. Optimization of wafer surface particle position map prior to viewing with an electron microscope. In: WM Bullis, DG Seiler, AC Diebold, eds. Semiconductor Characterization Present Status and Future Needs. Woodbury, New York: AIP Press, 1996, pp 243–247.

16. P-F Huang, YS Uritsky, PD Kinney, CR Brundle. Defect detecting, positioning, and sizing by KLA-Tencor Surfscans (62, 64, SP1-classic, and SP1-TBI). Proceedings of the Electrochemical Society Symposium on Analytical and Diagnostic Techniques for Semiconductor Materials, Devices, and Processes, Edited by BO Kolbesen, C Claeys, P Stallhofer, F Tardif, J Benton, T Shaffner, D Schroder, K Kishino, P Rai-Choudhury. Volume 99–16:450–457, 1999.

17. P Kinney, Y Uritsky. Locating defects on wafers for analysis by SEM/EDX, AFM, and other microanalysis techniques. Proceedings of the Society of Photo-Optical Instrumentation Engineers (SPIE) Conference on In-line Characterization Techniques for Performance and Yield Enhancement in Microelectronic Manufacturing II, Edited by S Ajuri, TZ Hossain. SPIE 3509, 1998, pp 204–209.

18. C Roudin, PD Kinney, Y Uritsky. New sample preparation method for improved defect characterization yield on bare wafers. Proceedings of the Society of Photo-Optical Instrumentation Engineers (SPIE) Conference on In-Line Methods and Monitors for Process and Yield Improvement, Edited by S Ajuria, JF Jakubczak. SPIE Volume 3884, 1999, pp 166–173.

19. Y Uritsky, S Ghanayem, V Rana, R Savoy, S Yang. Tungsten in-film defect characterization. Proceedings of the American Institute of Physics #449 on Characterization and Metrology for ULSI Technology: 1998 International Conference, Edited by DG Seiler, AC Diebold, WM Bullis, TJ Shaffner, R McDonald, EJ Walters. Woodbury, New York, 1998, pp 805–809.

20. NC Barbi, MA Henness. MaxRay Parallel Beam Spectrometer, Noran Instruments, Rev A, August 1998.

21. DA Wollman, KD Irwin, GC Hilton, LL Dulcie, DE Newbury, JM Martinis. High-resolution, energy-dispersive microcalorimeter spectrometer for X-ray microanalysis. J of Microscopy 188(3): 196–223, 1997.

22. DT Dingly, C Alabaster, R Coville. Inst Phys Conf, Ser 98, 1989, pp. 451–454.

23. N Fujino, I Karino, K Kuramoto, M Ohomori, M Yasutake, S Wakiama. First observation of 0.1 micron size particles on Si wafers using Atomic Force Microscopy and Optical Scattering. J Electrochem Soc 143 (12) 4125–4128, 1996.

24.  LM Ge, MG Heaton. Characterizing wafer defects with an integrated AFM-DF review system. Micro, February 20, pp. 57–63.
25.  Y Uritsky, L Chen, S Zhang, S Wilson, A Mak, CR Brundle. J. Vac. Sci. Technol., A 15(3):1319–1327, 1997.

# 21
# Calibration of Particle Detection Systems

**John C. Stover**
*ADE Corporation, Westwood, Massachusetts*

## I. INTRODUCTION

If you have read some of the preceding chapters, then you are well aware that contemporary "particle scanners" work by scanning a focused laser spot over the wafer while measuring the flashes of scattered light that occur when the traveling spot encounters a surface feature or defect. Feature location is determined by knowing where the laser spot is on the wafer. It is desirable to know feature location to tens of micrometers, and this is not always an easy measurement. But a far more difficult problem is determining specific feature characteristics. Even learning the average feature diameter is a serious problem, because different features (pits, mounds, particles of different materials, etc.) all scatter differently. The scattered light changes in intensity, direction, and polarization as a function of feature characteristics. Similar changes occur with source wavelength, polarization, and incident angle. In some cases, such as surface roughness, we have a pretty good understanding of the relationship between how the surface is rough and how the light scatters. Unfortunately, we are still learning how many other surface defects scatter. But even if we knew all of the specific particulars, it would still be difficult problem, because the scanner tries to solve "the inverse problem." That is: From a limited amount of scatter data, what is the defect? This is opposed to: How does a given defect scatter? And to make the problem a little more difficult, each year, as line widths are reduced, the surface features considered "defects" get smaller, which makes the list of critical surface features longer.

This situation makes even defining the term *scanner calibration* difficult. Is a scanner calibrated when it correctly reads scattered power into a detector? Probably not, because, in general, scanners do not report scattered power. Scattered power per unit incident power (in parts per million, or PPM) is often used, and a "calibrated scanner" could correctly report this value. Certainly the definitions used to quantify scattered light, given in the next section, are used in scanner calibration. However, scanner users don't care about scattered-power measurements or power ratios. They expect their (expensive) instrumentation to tell them something that will be useful in solving their production yield problems. Their questions are more like: How big is that thing? How many are there? And even, what the heck was it? Over a decade ago, the scanner industry took a different approach to calibration to answer these questions. As scanners evolve, so must the calibration process.

The early scanners consisted of one scatter (or dark-field) detector. With only one scatter signal it is impossible to learn much about defect characteristics; however, by measuring standard sources of scatter, it is possible to "calibrate" a scanner in terms of that known feature. This allows comparisons to be made between instruments and gives a rough estimate of defect size. Polystyrene latex spheres (or PSLs) were chosen for this standard, and their use has been critical to scanner development. With only one scatter signal available, no real analysis was possible, and it made sense to report defect size in "PSL equivalents." Modern scanners now employ several detectors, and this complicates reporting PSL equivalents because now one defect will generally be assigned a different PSL equivalent size by each dark-field measurement. Comparing results from scanners employing different optical designs (so that they are literally measuring different things) is a calibration-related concern, and this is also addressed. The chapter concludes by looking at the future of scanner calibration, which (quite naturally) requires defect identification prior to defect sizing. This is going to be a long, hard road to travel. But the measurement and modeling tools are being generated that can make significant progress on this rather open-ended problem.

## II. TERMS FOR QUANTIFYING SCATTERED LIGHT

Scatter signals are quantified in watts of scattered light per unit solid angle. In order to make the results more meaningful, these signals are usually normalized, in some fashion, by the light incident on the sample. There are three normalization methods commonly employed, and all of them are defined here.

If the feature in question is uniformly distributed across the illuminated spot on the sample (such as surface roughness), then it makes sense to normalize the scatter signal (in watts per steradian) by the incident power. This simple ratio, which has units of inverse steradians, was commonly referred to as *the scattering function*. Although this term is occasionally still found in the literature, it has been generally replaced by the closely related *bidirectional reflectance distribution function* (or BRDF), which is defined (1–3) as the differential ratio of the sample radiance to its irradiance. After some simplifying assumptions are made, this reduces to the original scattering function, with the addition of a cosine of the polar scattering angle in the denominator. Defined in this manner, the BRDF has become the standard way to report angle-resolved scatter from features that uniformly fill the illuminated spot. We have NIST to thank for this basic definition (1) and ASTM to thank for a standard method of measurement (2):

$$\text{BRDF} = \frac{P_s/\Omega}{P_i \cos \theta_s} \tag{1}$$

The scatter function is often referred to as the "cosine-corrected BRDF" and is simply equal to the BRDF multiplied by the cosine of the polar scattering angle. Figure 1 gives the geometry for the situation and defines the polar and azimuthal angles ($\theta_s$ and $\phi_s$) as well as the solid collection angle ($\Omega$).

Integration of the scatter signal (or in effect the BRDF) over much of the scattering hemisphere allows calculation of the *total integrated scatter*, or TIS (3,4). This integration is usually carried in such a way that areas near both the incident beam and the reflected specular beam are excluded. In the most common situation, the beam is incident on the sample at normal (or near normal), and the integration is carried from small values of $\theta$s to almost 90 degrees. If the fraction of light scattered from the specular reflection is small

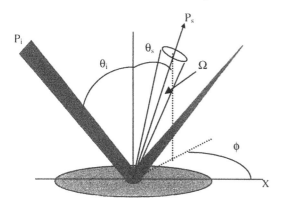

**Figure 1**  Scatter analysis uses standard spherical coordinates to define terms. (Courtesy of The Scatter Works, Inc.)

and if the scatter is caused by surface roughness, then it can be related to the RMS surface roughness of the reflecting surface. In this case, it makes sense to normalize the scatter measurement by the total reflected power, $P_r$ (or nominally the reflected specular beam for a smooth surface). As a ratio of powers, the TIS is a dimensionless quantity. This is done because reductions in scatter caused by low reflectance should not influence the roughness calculation. The pertinent relationships are as follows, where $\sigma$ is the RMS roughness and $\lambda$ is the light wavelength:

$$\text{TIS} = \frac{P_s}{P_r} \cong \left(\frac{4\pi\sigma}{\lambda}\right)^2 \tag{2}$$

Of course all scatter measurements are integrations over a detector aperture, but the TIS designation is reserved for situations where the attempt is to gather as much scattered light as possible, whereas "angle-resolved" designs are created to gain information from the distribution of the scattered light. Early-generation (single-detector) scanners were basically TIS systems, while the newer designs employ multiple detectors.

Scatter from discrete features, such as particles and pits, which do not completely fill the illuminated spot, must be treated differently. This is because changes in spot size, with no corresponding change in total incident power, will change the intensity (watts/unit area) at the feature, and thus the scatter signal (and BRDF) will change without any relation to changes in the feature being measured. Clearly this is unacceptable if the object is to characterize the defect with quantified scatter. The solution is to define another quantification term, known as the *differential scattering cross section* (or DSC), where the normalization is the incident intensity at the feature (3). The units for DSC are area/steradian. Because this was not done in terms of radiometric units at the time it was defined, the cosine of the polar scattering angle is not in the definition. The same geometrical definitions found in Fig. 1 also apply for the DSC:

$$\text{DSC} = \frac{P_s/\Omega}{I_i} \tag{3}$$

These three scatter parameters, the BRDF, the TIS, and the DSC, are obviously functions of measurement system variables such as geometry, scatter direction (both in and out of the incident plane), incident wavelength, and polarization, as well as feature characteristics. This is a key point, because it means that different scanner designs will naturally give different BRDF, TIS, and DSC results for the same defect. Correlation between different scanner designs cannot be performed using these units. It is the dependence of the scatter signal on these system parameters that makes scatter models useful for optimizing instrument designs. It is their dependence on feature characteristics that makes scatter measurement a useful metrology tool.

The terms TIS, BRDF, and DSC are absolute, not relative, terms. For a given instrument configuration, the DSC of given particle on a given substrate in a given direction is a fixed value, which can be repeatedly measured and even accurately calculated from models. The same is true for TIS and BRDF values associated with surface roughness of known statistics. Scatter-measuring instruments, like scanners or lab scatterometers, can be calibrated in terms of these quantities. But as has already been pointed out, the user of an industrial scanner will almost always be more interested in characterizing defects than in the resulting scatter values. The calibration challenge is to use methods that ensure appropriate measurement of the basic scatter quantities but relate the results in a manner that is meaningful to the user.

## III. POLYSTYRENE LATEX SPHERES AS REFERENCE STANDARDS

Polystyrene latex spheres (or PSLs) were chosen as a scattering standard for scanners because of their spherical shape, well-known refractive index (5), and ready availability. The big markets for PSLs are for testing filters and particle counters, where very accurate sizing (tighter than 10–20%) is not critical. Only a very small fraction of the PSLs produced end up being used as scattering standards in the semiconductor industry. Because the semiconductor industry represents such a small market segment, there has never been a strong motivation for PSL manufacturers to provide more accurate sizing of the spheres. The PSLs are available in sizes ranging from tens of nanometers to tens of micrometers. In general, the larger spheres, which can be seen in optical microscopes, are more accurately sized than the smaller ones. In a given batch, the size distribution is generally quite narrow (on the order of 1–2%), but there are exceptions.

Scanners have traditionally been calibrated by depositing PSLs of a single size on wafers in full or half (coverage) depositions. The nominal size of the sphere is used to calibrate the scanner response. By applying a wide coverage on the wafer of identical PSLs, a check may be made on the uniformity of scanner response. Modern particle deposition systems provide the capability of making spot depositions, which gives the advantage that several different PSL sizes can be placed on a single wafer, although spot depositions alone do not provide a uniformity check.

Scanners are calibrated using the measured response from a range of PSL sizes that span the dynamic range of the instrument. A relationship between detector response and PSL diameter is produced by curve fitting (least squares, piecewise linear, etc.) the data. This allows detector responses within the range for any defect to be converted into "PSL, equivalent sizes." For single-detector scanners, this is about all that can be done; however, more sophisticated scanners developed in the mid-1990s have two or more detectors and are capable of making additional judgments about defect characteristics.

For example, take a system with two scatter detectors, both of which are calibrated in PSL equivalents. They record signals from a defect and both report about the same PSL equivalent size. Then one could reasonably assume that they have measured a PSL or a particle with a very similar refractive index (perhaps silicon oxide or some other dielectric).

If they report quite different PSL equivalent sizes, then one can assume that the defect scatters differently than a PSL and is not "PSL like." In fact, exactly this kind of approach has been used to develop scanners capable of discriminating between surface particles and surface pits. Learning ways to provide additional defect information is currently the subject of research in the scanner industry and associated universities. Before these issues are more carefully examined, we need to learn a little more about PSL spheres and how they have been used (and misused) over the last two decades.

When we use PSL spheres as scattering standards for our scanners we rely on two physical attributes, the sphere diameter and the material index. Shape and orientation will also strongly influence scatter from particles larger than about a fifth of a wavelength (6), but these are not issues for the spherical PSLs.

In a 1996 paper by Mulholland et al. (7), the issue of uncertainty in PSL diameter was investigated. They used a differential mobility analyzer to carefully size PSLs at the National Institute of Standards and Technology (NIST). In this process, charged particles of different diameters flowing in an air stream are drawn sideways toward a charged plate with a narrow slit in it. The deviation induced in a particle path by the electric field varies with air resistance, and that depends on particle diameter. The result is that only a narrow range of particle diameters (independent of material) are able to slip through the slit. By carefully controlling airflow rate, voltage, and system geometry, accurate particle sizes can be determined. They reported the ratio of manufacturers' diameter to measured diameter and plotted the result as a function of measured diameter for several commercial sources of particles. Very few of the commercially advertised diameters fell within even 5% of the measured NIST sizes. Instead, diameters were scattered all over the chart, as far away as 25% from the measured diameters.

Put yourself in the place of an engineer charged with designing a scanner that can get down to 100-nm PSL sensitivity. Without accurate sizing information, which "100-nm PSLs" would you choose to calibrate your equipment and prove its sensitivity? The larger ones of course! Thus, with no particular intent or malice, the industry scanners have been calibrated with PSLs that are often more than 10% larger than their indicated sizes. Of course if everyone is doing it, then it probably won't matter, right? Unfortunately, this is not the case. Even if everyone uses the same wrong sizes, there are problems. But before we examine the effect of diameter variations on scatter levels and related scanner capabilities, we need to have a quick look at PSL index values and variations.

Variations in the PSL index will also cause changes in scattering levels. In a paper by Mulholland et al. (5), the measured index of PSLs was investigated. Of course index varies with wavelength, and there is a measured uncertainty. Using the data from this paper, this issue was considered in the PSL sizing work done for SEMATECH (8) in 1999. The dispersion relationship and the estimated uncertainty (about ±0.01) are shown in Figure 2. The actual effect of this uncertainty should be investigated for each particular scanner design. The example in the next section is a typical, and the result is that index uncertainty does not impose the difficulty that size uncertainty does.

## IV. RELATIVE SCATTERING CHARACTERISTICS OF POLYSTYRENE LATEX SPHERES

PSL contamination is not a problem in production lines. Instead, particles of silicon, silicon oxide, and a few metals are found. Since we use PSLs to calibrate our scanners,

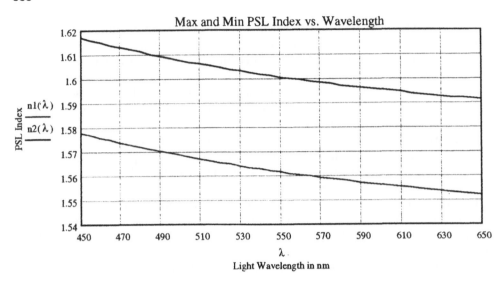

**Figure 2** Maximum and minimum variations of PSL refractive index are plotted as a function of wavelength. (Calculated from the work in Ref. 8. Courtesy of The Scatter Works, Inc.)

it is worthwhile spending a little time comparing the relative scattering characteristics of various particle materials.

Figure 3 shows a plot of the incident-plane DSC for a number of different PSL diameters, all plotted for an incident angle of 45 degrees from a P-polarized source of 488-nm light. (As of 1999 many scanners use this wavelength, although shorter ones will undoubtedly be used as shorter-wavelength lasers with longer lifetimes are developed.)

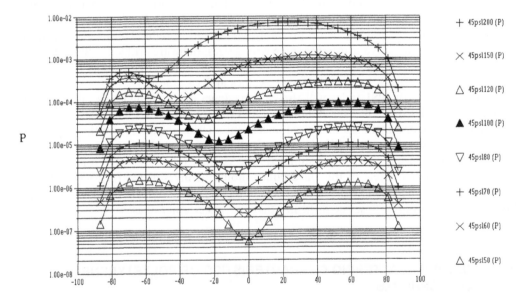

**Figure 3** Modeled DSC response for several different PSL diameters in the incident plane for a 488-nm P-polarized source incident at 45 degrees. The diameters vary from 50 to 200 nm. Notice that the dip moves left as the PSL diameter increases. (Courtesy of ADE Corporation.)

These plots were created with a scattering model that has proved to be a very accurate (6,9). Notice that larger spheres scatter more (as expected), but the curves change shape as the diameter changes. Another point is that a nominal increase in PSL diameter makes a dramatic increase in scattered light. A little study of this figure shows that a rule of thumb might be an 80% increase in scatter for a 10% increase in diameter. This makes scatter an extraordinarily sensitive measure of particle diameter, which is what we want, but it also means that care must be taken to get the calibration diameters correct. Remember that Fig. 3 shows only the incident-plane response. In fact the full hemispherical response looks a bit more like a donut. Larger particles tend to make slightly misshapen "donuts." Figure 4 gives the results for the same PSL sizes using an S-polarized source. The prominent dip in the response is gone, but the sensitivity to small changes in particle diameter is still there.

Of course it is not practical to even attempt outlining results for all of the possible wavelength and incident-angle combinations, but three more charts will help. First, in Figure 5 we see the complex refractive index (at a wavelength of 488 nm) for several particle materials that are commonly encountered during scanner measurements. They may be loosely separated into four groups. These are dielectric (lower left, where the PSLs are found), semiconductors (lower right), gray metals (upper right), and shiny metals (upper left). We will see that the index variations between these classes can have dramatic effects on the scatter patterns of identically sized particles. Conversely, particles of the same diameter and roughly the same index can be expected to scatter about the same.

To see these index-induced changes, examine Figure 6, which is similar to Fig. 3, except that here the DSC for spherical silicon particles is modeled instead of PSLs. Obviously, particle material plays a key role, for both level and shape variations with size are quite different. Notice that in general the (high-index) silicon response is larger. In

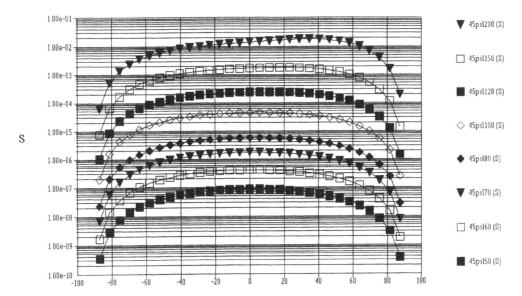

**Figure 4** Modeled DSC response for several different PSL diameters in the incident plane for a 488-nm S-polarized source incident at 45 degrees. The diameters vary from 50 to 200 nm. Notice the absence of the dip found in the P-polarized response of Fig. 3. (Courtesy of ADE Corporation.)

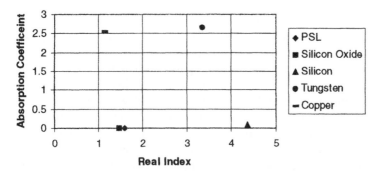

**Figure 5** Complex refractive index ($N = n + i_k$) for a few materials commonly found in semiconductor manufacturing. The indexes easily group into dielectrics, semiconductors, gray metals, and shiny metals. Thus nickel and iron will be located close to tungsten and silver close to copper.

Figure 7, the DSC responses for 100-nm spherical particles are compared from each material group. The differences are dramatic even on the log scale. In general, PSLs (and thus the dielectric class) scatter a lot less light than the other materials.

Modeling can be used to examine the importance of the small uncertainty in PSL index. This is done in Figure 8 by comparing the DSC response of two 100-nm PSLs with indices of 1.59 and 1.60 to the response of a 99-nm PSL with an index of 1.59. The same scattering geometry is used as for the other charts. The difference in most scattering directions is small and for this situation is roughly equivalent to a 1-nm change in diameter. Of course this may be different for other geometries and thus should be checked for specific scanner geometries.

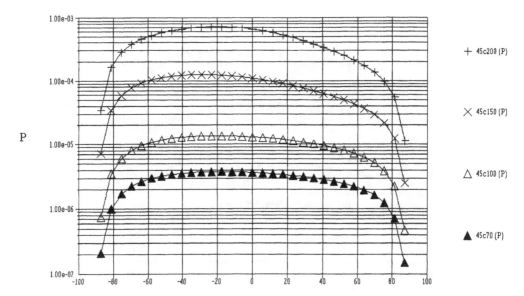

**Figure 6** Modeled DSC response of different-sized silicon particles in the incident plane for a P-polarized source incident at 45 degrees. The diameters are 40, 60, 80, and 100 nm. Notice that the shapes change in a different manner than the PSL responses of Fig. 3. (Courtesy of ADE Corporation.)

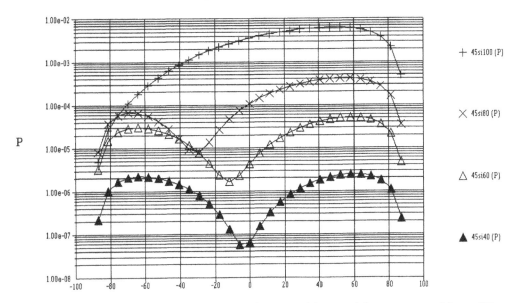

**Figure 7** Modeled DSC responses of four different particle materials are compared for a 488-nm P-polarized source incident at 45 degrees. The DSCs differ in amplitude and shape, which implies that PSL equivalent measurements are not an accurate measure of diameter for generic particles. (Courtesy of ADE Corporation.)

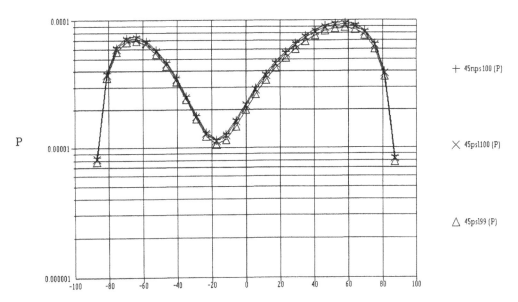

**Figure 8** Variations in the PSL DSC caused by small changes in particle index and diameter are compared for a 488-nm P-polarized source incident at 45 degrees. An index change of 0.01 is roughly equivalent to a diameter variation of 1 nm. (Courtesy of ADE Corporation.)

Reviewing Figs. 1–8 we can reach some conclusions worth emphasizing about the use of PSLs as calibration standards. First, we see that the bottom line for PSLs is that they are made of the right stuff, but size does matter. Second, we realize that using PSL equivalent signals is not going to be a very accurate method of sizing real particles (other than dielectrics) and that, in general, metal and semiconductor particles are smaller than their PSL equivalent diameters. In addition, although the issue of other real defects (like surface pits) has not yet been raised, it seems clear that there is no reason to believe that PSL equivalent diameters will be an accurate indicator of pit size.

## V.  MAKING USE OF POLYSTYRENE LATEX SPHERE CALIBRATION FOR OTHER SURFACE FEATURES

Multiple-detector scanners offer additional capabilities and complications. For example, based on the model data of the last section one would expect such a scanner, calibrated with PSLs, to report different PSL equivalent values when a particle of a different material is detected. Following the reasoning from the single-detector era, the scanner is reporting multiple PSL equivalent sizes for one defect. Which one is right? None of them are right, of course! We can pretty much conclude that they are all incorrect. We know this because different results from multiple detectors indicate that the defect is not a PSL, and because we know that PSL equivalent values for real defects are incorrect. However, the situation offers the possibility of providing limited defect identification, if enough information can be gathered by the scanner.

Although there are questions about accuracy, modern scanners are able to tell the difference between surface pits and surface particles. To see how this is generally done, examine Figure 9, which shows the DSC response for a series of conical pits using the

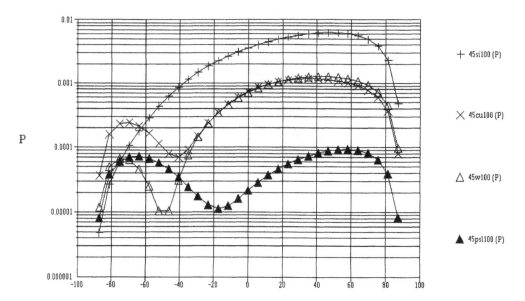

**Figure 9**  Modeled DSCs for conical pits of different diameters are compared for the same P-polarized 488-nm source of the previous figures. The diameters are 70, 100, 150, and 200 nm. Notice the absence of a central dip in the response. (Courtesy of ADE Corporation.)

same P source incident at 45 degrees. Unlike the particle responses of Figs. 3, 6, and 7, the pits have DSCs without the prominent dip near surface normal. Generally pit/particle discrimination techniques have been used that make use of the dip that appears in the particle DSC for a high-incident-angle P-polarized source but is absent for the pit DSC. Given that this capability exists, there are several obvious questions. How strongly does the pit/particle identification process depend on calibration? Given that a defect is identified as a pit, what is its average diameter? Of course there are many similar questions related to other defect identification and sizing capabilities that the next-generation scanners are expected to have. It will be shown in this section that the ability to answer these types of questions depends critically on the accuracy with which scanners are calibrated.

Consider a hypothetical scanner based on the geometry of the P-polarized 488-nm source, incident at 45 degrees, that was used in the figures of the last section. This scanner is to be used in a production situation where the only defects are conical surface pits and spherical silicon particles. Calibration is carried out with PSLs. The scanner design employs detectors at the 0 and +30-degree locations in the plane of incidence. One object of this exercise is to see what effect poor calibration will have on using this scanner to discriminate between PSLs, surface pits, and silicon particles. Now the data of Figs. 3, 6, and 9 are used to obtain the PSL equivalent values for these two defects by interpolating between the PSL values in the figures. The results are shown in Figure 10, where the PSL equivalent ratios of the 0-degree detector over the 30-degree detector is plotted against defect diameter for the PSLs, silicon particles, and pits. Never mind that in a real situation we would also have to worry about other particle materials or that real systems would require integration over the detector aperture. The point is that for this simple system, we can clearly separate pits, PSLs and silicon particles using this ratio. When the ratio is larger than 1.0, the defect is a pit. For ratios close to 1.0, a PSL (or dielectric) has been detected; for ratios less than 1.0, the source of scatter is a silicon particle. The object is to see if this system is effected by calibration errors.

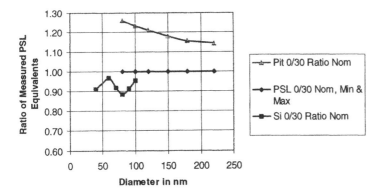

**Figure 10**  Modeled signals for the 0-degree to 30-degree detectors are ratioed to perform defect identification for the hypothetical scanner discussed in the text. In this situation, only calibration PSLs, conical pits, and silicon particles are allowed. The ratio can be used to identify these three scatter sources. Pits ratio larger than unity and silicon particles less than unity. The PSL ratio is always unity, because both detectors were calibrated with PSLs. Other defects give nonunity ratios, because they have DSCs with different-intensity distributions than PSLs, as shown in Figs. 4, 7, and 8. (Courtesy of ADE Corporation.)

To check for calibration effects, we assume that the PSLs used to calibrate the scanner may vary from their assumed true values by as much as 10 nm in either direction, which, based on the previously referenced PSL sizing study (7), is a perfectly reasonable assumption. Figure 11 shows the previous identification results, along with maximum and minimum ratio values found after allowing for the 10-nm uncertainty in the calibration PSLs. The PSLs are still at a ratio of 1.0, because whatever actual size is used, the scanner gets the same signal when that same particle is encountered again. But both the pits and the silicon have changed to include a band of values. This is because their DSCs are shaped differently than the PSLs. The problem is that the bands now overlap. So if the original (factory) settings are based on one set of calibration PSLs but another set is used in the field, the system will suddenly have discrimination problems. The separation ratio of 1.0 is no longer correct. For this model scanner to be able to use the factory-installed pit/particle separation algorithm, they will all have to be calibrated to more accurate PSL standards. When the example of Fig. 11 was run with a 2-nm PSL uncertainty (not shown), feature identification was again possible. The next section discusses calibration requirements for defect sizing.

## VI. SCANNER CALIBRATION, MODEL-BASED STANDARDS, AND DEFECT SIZING

It should also be obvious that once the scanner has identified the defect as a pit or a silicon particle, a relationship can be generated between the PSL equivalent information and the defect size. This relationship is found from the scattering models for these defects and the PSLs. The pit model is used here as an example of what is becoming known as a *model-based standard*. This is probably not a very accurate name, because these models are not really standards in the sense of a physical reference material or a written document, but this is what they are being called. Using this model, an estimate of average pit diameter can

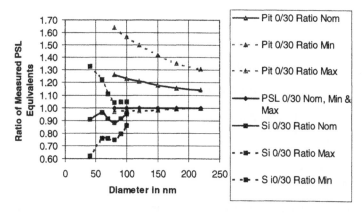

**Figure 11**  If the PSL calibration particles are incorrectly sized by as much as ±10 nm, then the PSL equivalent values with other defects becomes distorted. Here the PSL ratio stays at unity, under the assumption that the (incorrectly sized) calibration PSLs are measured, but the ratios for the pits and silicon particles change to the point where the defect identification rule cannot be trusted. Not shown is a result for a ±2-nm PSL variation in which the defect identification still works. (Courtesy of The Scatter Works, Inc.)

be obtained from the measured PSL equivalent data. This is only an estimate, because pit shape will also change the DSC response, but it is better than just reporting PSL equivalents, and it is far more practical than attempting to produce identical sets of physical reference standards to do scanner-pit calibrations. This model-based approach to defect sizing will be used for many different types of wafer defects.

But there is a problem. Whether the models report DSC values or PSL equivalent values, the models can only be used under the assumption that the PSLs are accurately sized. Accurate sizing of the calibration PSLs is also required for defect sizing. Figure 12 shows the pit-sizing estimate that would be obtained from the 30-degree-detector data, along with the deviations in estimated pit diameter that would be induced by the same 10-nm uncertainty in the PSL calibration. Clearly the same reasoning applies to the silicon particles.

Modern scanners are employing defect-scattering models and PSL calibration spheres. The models are used in tool-specific ways and are generally proprietary to the manufacturer. In effect, they are as model specific as the hardware used in individual products. Of course they have to be lab tested, but they are a part of the scanner system. The user is purchasing the model results as well as the tool hardware.

The PSL calibration standards should not be unique to the scanner in question. Absolute calibration values are required for these systems to function the way the industry needs. The entire international industry benefits from the common use of accurately sized PSL standards. When discussing calibration with your scanner manufacturer it is not unreasonable to request the use of accurate PSL standards. Research into PSL characteristics has been going on at NIST for several years; NIST is in the business of providing measurement traceability, so an obvious solution is to use NIST-traceable PSLs for scanner calibration. SEMATECH provides NIST-traceable PSLs deposited on wafers, and other commercial sources can be expected to appear.

## VII.   CALIBRATION ISSUES FOR HAZE AND SURFACE ROUGHNESS

*Haze* is the industry name for the background scatter measured by particle scanners. It is caused by a combination of surface roughness, bulk scatter from films, Rayleigh scatter from air molecules close to the wafer, and stray light generated by the measurement system itself. If the system is well designed, this measured signal is dominated by the first two

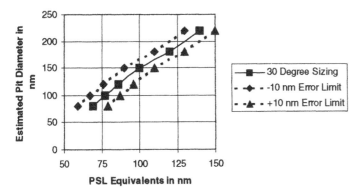

**Figure 12**   PSL equivalent sizes for the hypothetical scanner are converted to pit diameters by using the data of Fig. 9. (Courtesy of ADE Corporation.)

causes (which are wafer effects) rather than by extraneous stray light. Thus, on one hand, if carefully measured, it can be an indication of wafer surface quality, and on the other, it is a troublesome source of noise that limits small defect detection. Both of these effects are reviewed here relative to calibration.

## A.  Haze as a Measured Signal

The relationship between surface roughness and scattered light has been exhaustively studied (3,10–12) and will not be reviewed here, other than to point out that it has been proved that for bare silicon wafers there is a direct correspondence between the measured BRDF and the surface power spectral density (or PSD) function. The PSD is a measure of roughness power as a function of roughness (or spatial) frequency. When the BRDF is integrated over a large solid angle, an integrated scatter signal is obtained. The haze measured by that detector is defined as the ratio of the measured scattered power to the power incident on the wafer, which is similar to TIS. Because the BRDF varies dramatically with incident angle, source polarization, and detector position, the measured haze is a strong function of how the measurement is taken as well as surface quality. Thus for a given instrument, measured haze can produce relative results between wafers, but comparing measurements between instruments of different design is virtually meaningless. It follows that specifying minimum haze resolution for an instrument also has little meaning. Further complicating attempts to bring significance to haze measurements is the fact that in many scanners the numerical values (usually expressed in parts per million) are incorrectly reported, because neither the scattered power nor the incident power are measured in calibrated units. These issues are more completely discussed in SEMI Standard #2730 (13).

There are at least three obvious methods to obtain haze in calibrated units. One is to use a sample with a known BRDF, measure the scattered power with the appropriate detector, and then, using the definition of BRDF, solve for the value of the $P_i\Omega$ product. White diffuse samples are available with well-characterized BRDFs, and the procedure for this type of calibration, which has been in use since the 1970s, is well documented in ASTM Standard E1392-90 (14). The BRDF of such samples has only a small dependence on measurement conditions. The downside for this approach is that the haze is many orders of magnitude larger than that associated with wafers. A second approach is to use a NIST-traceable PSL with a known DSC. This is a little more trouble, for then the beam spot size must be known. For a Gaussian spot, the conversion factor from incident intensity to incident power is the area factor $\pi R_o^2$, where $R_o$ is the $e^{-2}$ Gaussian radius. Not knowing the PSL diameter can be a big source of error in related scatter measurements as pointed out in earlier sections. The most apparent method would be the use of a silicon wafer, with a known haze value, as a physical reference; however, this is difficult, because haze depends on the measurement technique (geometry, wavelength, etc.) in addition to wafer surface conditions. A unique way around this problem has been suggested (15) that employs the use of a wafer with a known PSD, which is a property of the wafer only. Given this information it is possible to calculate the BRDF associated with a given measurement situation and then to integrate over the appropriate aperture to get the haze. Manufacturing surfaces with appropriate PSDs is a challenge. In addition to being robust, the surface should be uniform (statistically) and isotropic. Anisotropic surfaces can create diffraction peaks (or "hot spots") in the BRDF that can distort the measurement and make it change dramatically if the wafer is rotated. The haze from such a standard needs to be significantly higher than the background scanner noise associated with Rayleigh air scatter and electronic noise.

## B. Haze as a Source of Background Noise

Most scanners employ high-pass filtering to reduce problems associated with the haze signal-masking defect signals. This works because haze is a slowly changing signal, while the flash of light from a discrete defect often takes place in only a fraction of a microsecond. Unfortunately, haze is also responsible for a high-frequency background signal (created by individual electron interactions in the detector) called *shot noise*. The RMS noise current from this well-known effect can be evaluated in terms of the electronic charge ($q$), the bandwidth (BW), and the total detector current ($I$) as

$$I_{RMS} = (2qI\text{BW})^{1/2} \tag{4}$$

Scanner designers increase sensitivity by reducing throughput (accomplished by adjusting BW or measurement speed) and by increasing incident power, which works because the signal is proportional to $P_i$, while the noise varies approximately as its square root.

The point is that if the system is correctly calibrated, then minimum PSL sensitivity can be expressed in terms of wafer haze, measured in ppm, for each dark-field detector. Using measured haze to specify scanner PSL sensitivity for a given throughput is a very useful application of haze specifications; however, once again it requires that the dark-field detectors be accurately calibrated using NIST-traceable PSLs to be meaningful.

## VIII.  COMPARISONS BETWEEN CALIBRATED SCANNERS

So now if we carefully calibrate two different-model scanners and then measure wafers with a combination of PSLs and real-world pits and particles, the result should be excellent comparison between the scanners, right? Maybe, but in general, no! What happens?

Consider the silicon particle DSCs of Fig. 6, and assume that the two scanners have detectors at −30 and −10 degrees for one and +10 and +30 degrees for the other. Clearly they will measure different PSL equivalents for the same silicon particles. Thus if they bin defects in PSL equivalent sizes, the two scans of the same wafer will be different. Only if both scanners can perform defect identification and sizing can they be expected to produce results that compare well. On the other hand, PSLs (and other dielectrics) will measure the same on both scanners.

It should be obvious that measurement correlation between scanners can be performed by doing defect identification in one scanner and then using an accurate model to predict the PSL equivalent response to the same defect in the second scanner. Multiple-channel scanners with defect identification capabilities can be expected to provide the data necessary for correlation if they are accurately calibrated.

## IX.  SUMMARY

The following summarizes the main points made in this chapter.

The techniques for appropriately quantifying scatter light are well understood.

Size does matter for PSL standards, especially now as the industry moves from single-detector to multiple-detector scanners used for defect identification and sizing.

PSLs are made of the right stuff. For many scattering geometries the uncertainty in index is on the order of an equivalent change of 1 nm in PSL diameter.

PSL equivalent diameters are roughly the same as the measured diameters for small dielectric particles, but generally oversize metal and semiconductor particles.

The PSLs used as calibration standards for most of the 1980s and '90s have been off by as much as 25% and tend to be larger than indicated. This situation, which was an annoying irritation during the era of single-detector scanners, is a limiting issue for multiple-detector scanners.

In order for multiple-detector scanners to optimally perform defect identification and sizing, accurate PSL calibration standards are required. It is not enough that all users agree on the same incorrect standards. Like the difference between religion and politics, it is a situation where absolute values are required.

NIST traceability is a logical requirement for PSL calibration standards.

Even when scanners are correctly calibrated, different models will not give the same PSL equivalent measurement for real defects. Only if they are sophisticated enough to both identify and size the measured defects can close comparisons be expected.

Unlike PSLs, with fixed diameters independent of the scanner geometry, measured haze is a property of the wafer and the measurement situation. This makes haze measurements, haze calibration, and haze standards very tricky topics.

## REFERENCES

1. FE Nicodemus, JC Richmond, JJ Hsia, I Ginsberg, T Limperis. Geometric considerations and nomenclature for reflectance. U.S. Dept. of Commerce, NBS Monograph 160, 1977.
2. ASTM Standard #E1392-90. Standard Practice for Angle Resolved Optical Scatter Measurements on Specular or Diffuse Surfaces. 1991.
3. JC Stover. Optical Scattering: Measurement and Analysis. 2nd ed. Bellingham, WA: SPIE Press, 1995.
4. HE Bennett, JO Porteus. Relation between surface roughness and specular reflectance at normal incidence. J. Opt. Soc. Am. 51:123, 1961.
5. GW Mulholland, AW Hartman, GG Henbree, E Marx, TR Lettieri. Development of a one-micrometer diameter particle size standard reference material. J. Res. NBS 90(1):3, 1985.
6. CA Scheer, JC Stover, VI Ivakhnenko. Comparison of models and measurements of scatter from surface-bound particles. SPIE Proc. 3275:102, 1997.
7. GW Mulholland, N Bryner, W Liggett, BW Scheer, RK Goodall. Selection of calibration particles for scanning surface inspection systems. SPIE Proc. 2862:104, 1996.
8. V Sankaran, JC Stover. Advanced particle sizing technique for development of high-accuracy scanner calibration standards. Technology Transfer #99083800B-TR, 1999.
9. YA Eremin, NV Orlov. Simulation of light scattering from particle upon wafer surface. Appl. Opt. 3533:6599–6605, 1996.

10. EL Church, HA Jenkinson, JM Zavada. Relationship between surface scattering and micro-topographic features. Opt. Eng. 18(2):125, 1979.

11. WM Bullis. Microroughness of silicon wafers. Semiconductor Silicon, Proc. Silicon Material, Science the Technology, 1994.

12. JC Stover, ML Bernt, EC Church, PZ Takacs. Measurement and analysis of scatter from silicon wafers. SPIE Proc. 2260:21, 1994.

13. SEMI Document 2730. Guide for Developing Specifications for Silicon Wafer Surface Features Detected by Automated Inspection. San Jose, CA, 1998.

14. ASTM Standard #E 1392-90. Standard Practice for Angle Resolved Scatter Measurements on Specular or Diffuse Surfaces. 1991.

15. BW Scheer. Development of a physical haze and microroughness standard. SPIE Proc. 2862:78, 1996.

# 22
# In Situ Metrology

**Gabriel G. Barna**
*Texas Instruments, Inc., Dallas, Texas*

**Bradley Van Eck**
*International SEMATECH, Austin, Texas*

**Jimmy W. Hosch**
*Verity Instruments, Inc., Carrollton, Texas*

## I. INTRODUCTION

Since the early 1960s, semiconductor manufacturing has historically relied on statistical process control (SPC) for maintaining processes within prescribed specification limits. This is fundamentally a passive activity based on the principle that the process parameters—the hardware settings—are held invariant over long periods of time. Then SPC tracks certain unique, individual metrics of this process—typically some wafer-state parameter—and declares the process to be out of control when the established control limits are exceeded with a specified statistical significance. While this approach has established benefits, it suffers from (a) its myopic view of the processing domain—looking at one or only a few parameters, and (b) its delayed recognition of a problem situation—looking at metrics generated only once in a while or with a significant time delay relative to the rate of processing of wafers.

In the early 2000s, while semiconductor manufacturing continues to pursue the ever-tightening specifications due to the well-known problems associated with decreasing feature size and increased wafer size, it is clear that both these constraints have to be removed in order to stay competitive in the field. Specific requirements are that:

Processing anomalies be determined by examining a much wider domain of parameters

Processing anomalies be detected in shorter timeframes, within wafer or at least wafer to wafer

Processing emphasis be focused on decreasing the variance of the wafer-state parameters instead of controlling the variance of the setpoints

Advanced process control (APC) is the current paradigm that attempts to solve these three specific problems. Under this methodology, the fault detection and classification (FDC) component addresses the first two requirements, and model-based process control (MBPC) addresses the last one. In contrast to the SPC methodology, APC is a closed-loop, interactive method where the processing of every wafer is closely monitored in a time scale that is much more relevant (within wafer or wafer to wafer) to the manufacturing process. When a problem is detected, the controller can determine whether to adjust the process parameters (for small deviations from the normal operating conditions) or to stop the misprocessing of subsequent wafers (for major deviations from the standard operating conditions).

The APC paradigm is a major shift in operational methods and requires a complex, flexible architecture to be in place to execute the foregoing requirements. A schematic representation of this architecture is provided in Figure 1. Briefly, this system starts with the processing tool and sets of in situ sensors and ex situ metrology tools to provide data on the performance of the tool. When the performance exceeds some predefined specifications, actions can be taken either to terminate the processing, or to reoptimize the settings of the equipment parameters via the model tuner and the pertinent process models. The purpose of this brief APC overview is to provide the context for this chapter on in situ metrology. In situ sensors are becoming more widespread, for they are one of the key components of this APC paradigm.

The goal of this chapter is to detail the fundamentals of in situ process-state and wafer-state sensors in OEM (original equipment manufacturer) tools and their use in APC, because this is the path that semiconductor manufacturing now has to aggressively pursue. This message is clearly articulated in the 1997 version of the NTRS, which states (1): "To enable this mode of operation (APC), key sensors are required for critical equipment, process, and wafer state parameters. It is essential that these sensors have excellent stability, reliability, reproducibility, and ease of use to provide high quality data with the statistical significance needed to support integrated manufacturing."

Hence, this chapter will provide information for the in situ sensors that are *commercially available* (i.e., not test-bed prototypes) and are currently being used, or soon will be, in OEM tools for the measurement and control of process-state and wafer-state properties. In situ sensors are those that monitor the process state of the tool or the state of the wafer

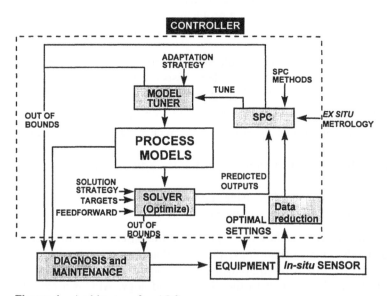

**Figure 1**   Architecture for APC.

during the processing of each wafer. For the sake of completeness, in-line sensors will also be included, for some of the sensor technologies are applicable only in this format. In-line sensors measure wafer state in some location close to the processing, such as a cool-down station in a deposition reactor or a metrology module on a lithography track system. Metrology tools that are currently used off-line, but are seen to be moving toward a simpler in situ or in-line sensor embodiment (e.g., ellipsometer, FTIR film thickness) will be included. Sensors used for tool development (e.g., intrusive RF probes, spatially resolved spectroscopy) are not included. Sensors used for gas flow and pressure control are also omitted, because these are well-established technologies. For each sensor described, the following information will be included based on input from the sensor manufacturer:

The fundamental operating principle behind each sensor, with greater detail for the less common ones
Practical issues in the use and interfacing of these sensors

When a number of manufacturers exist for a given sensor, references will be provided to several manufacturers, although there is no claim that this list will be totally inclusive. The sensors included in this chapter are ones that provide most of the features of an ideal in situ sensor. These features are: low cost, reliability, ease of integration into the processing tool, with sensitivity to equipment and process variations over a broad range of processing conditions. The highest-level sorting will be by the major process-state (temperature, gas-phase composition, plasma properties, etc.) and wafer-state (film thickness, thickness uniformity, resist thickness and profile, etc.) sensors. The major focus is on the technology behind each sensor. Applications will be described only when they are not necessarily obvious from the nature of the sensor. Any particular application example is not intended to promote that particular brand of sensor, but (1) it may be the only available sensor based on that technology, or (2) the specifics may be required to provide a proper explanation for the use of that type of sensor.

## II. PROCESS-STATE SENSORS

Sensors exist for monitoring both the process state of a particular tool and the wafer state of the processed wafer. The wafer state is of course the critical parameter to be controlled; hence measurement of the appropriate wafer-state property is clearly the most effective means for monitoring and controlling a manufacturing process. However, this is not always possible, due to:

Lack of an appropriate sensor (technology limitation)
Lack of integration of appropriate sensors into processing tools (cost, reliability limitations)

In these cases, the alternative is to monitor the process state of the manufacturing tool. In many cases, this is an easier task achieved with less expensive sensors. Nonintrusive radio frequency (RF) sensors can be connected to the RF input lines, or the tuner, of an RF-powered processing tool. A range of optical techniques exists that require only an optical access to the processing chamber. Historically, the most predominant use of such process-state sensors has been for endpoint determination. This is generally performed by the continuous measurement of an appropriate signal (e.g., intensity at a specific wavelength) during the processing of a wafer, looking for a change in the magnitude of the signal. Aside from endpoint determination, the availability of process-state sensors in OEM processing tools is generally paced by integration issues (electrical, optical, cost, reliability). In addition, there is generally a lack of the necessary models that relate these process-state measurements to the critical wafer-state properties. Due especially to this limitation, many of the process-state sensors are typically employed for fault detection. This is the simplest use of such sensors, where the output is monitored in a univariate or multivariate statistical method, to determine deviations from the "normal" processing conditions. This methodology has a significant payback to manufacturing yield by an early determination of operational anomalies and hence the decreased misprocessing of wafers. Using process-state sensors for process control requires much more rigorous models between the sensor signal(s) and the wafer-state parameter. The following is a description of the sensors that have been, or soon will be, integrated into OEM processing tools for use in FDC or MBPC.

## A. Temperature

The measurement and control of wafer temperature and its uniformity across the wafer are critical in a number of processing tools, such as RTP, CVD, PVD and EPI, used for film growth and annealing. Historically, the most commonly used temperature measurement techniques are thermocouples and infrared pyrometry. Infrared pyrometry is based on analysis of the optical emission from a hot surface. It is dependent on two main variables: field of view of the detector, and the optical properties of the material, such as refractive indices and emissivity. While useful only above 450°C, due to the low emissivity of semi-conductors in infrared, pyrometry has been commercialized and is widely utilized in SC manufacturing tools. A newer technique is diffuse reflection spectroscopy (DRS), which provides a noncontact, in situ optical method for determining the temperature of semi-conducting substrates. The technique is based on the optical properties of semiconductors, specifically that the absorption coefficient rapidly increases for photon energies near the bandgap of the material. Hence a semiconducting wafer goes from being opaque to being transparent in a spectral region corresponding to its bandgap energy. A temperature change of the semiconductor is accompanied by a change in the bandgap, which is then reflected as a shift of this absorption edge. Recently, an acoustic technology has been developed. The advantages and disadvantages of these four techniques are presented in Table 1.

Thermocouples are sometimes used for temperature measurement in processing tools. Since they have to be located remotely from the wafer, temperature errors of more than 100°C are possible, with no means for monitoring the temperature distribution across the wafer. Hence, this sensor is not widely used in SC manufacturing tools; it will be omitted from this discussion.

### 1. Pyrometry

Precise wafer temperature measurement and tight temperature control during processing continue to be required, because temperature is the most important process parameter for most deposition and annealing processes performed at elevated temperature (2). As device features become smaller, tighter control of thermal conditions is required for successful device fabrication.

**Table 1** Pros and Cons of Four Temperature Measurement Techniques

| Technique | Advantages | Disadvantages |
|---|---|---|
| Thermocouple | • Easy to use<br>• Low cost | • Cannot be used in hostile environment<br>• Requires mechanical contact with sample<br>• Sensitivity depends on placement |
| Pyrometer | • All optical, noninvasive<br>• Requires a single optical port | • Unknown or variable wafer back-side emissivity<br>• limited temperature range<br>• sensitive to all sources of light in environment |
| Diffuse reflectance spectroscopy | • Optical, noninvasive<br>• Directly measures substrate temperature<br>• Insensitive to background radiation<br>• Can be applied to a wide range of optical access geometries<br>• Wafer temperature mapping capability | • Requires two optical ports<br>• Relatively weak signal level |
| Acoustic thermometry | • Sample emissivity not a factor<br>• Wide temperature range | • Intrusive to the reaction chamber<br>• Physical contact required |

*Source*: Ref. 8.

Optical temperature measurement is historically the primary method for in situ wafer temperature sensing. Known as pyrometry, optical fiber thermometry, or radiation thermometry, it uses the wafer's thermal emission to determine temperature. The optical fibers (sapphire and quartz) (or a lens) are mounted on an optically transparent window on the processing tool and collect the emitted light from, in most cases, the back side of the wafer. The collected light is then directed to a photo detector, where the light is converted into an electrical signal.

All pyrometric measurements are based on the Planck equation, written in 1900, which describes a black-body emitter. This equation basically expresses the fact that if the amount of light emitted is known and measured at a given wavelength, then temperature can be calculated.

As a consequence of this phenomenon, all pyrometers are made of the following four basic components:

Collection optics for the emitted radiation
Light detector
Amplifiers
Signal processing

There are thousands of pyrometer designs and patents. A thorough description of the theory and the many designs, as well as the most recent changes in this field, are well summarized in recent books and publications (3–6).

The two largest problems and limitations with most pyrometric measurements are the unknown emissivity of the sample—which must be known to account for the deviations from black-body behavior—and stray background light. In addition, the measurement suffers from a number of potential errors from a variety of sources. While the errors are often small, they are interactive and vary with time. The following is a summary of these sources of error, roughly in order of importance.

1. Wafer emissivity—worst case is coated back sides, with the wafer supported on pins.
2. Background light—worst case is RTP and high-energy plasma reactors.
3. Wafer transmission—worst at low temperatures and longer wavelengths.
4. Calibration—has to be done reproducibly and to a traceable standard.
5. Access to the wafer—retrofits are difficult for integrating the sensor into the chamber.

The following problems will become much more important as the previous problems are minimized by chamber and pyrometer design.

6. Pyrometer detector drift—electronics (amplifiers) and photodetectors drift over time.
7. Dirt on collection optics—deposition and outgassing coat the fiber or lens.
8. Changes in alignment—moving the sensor slightly can cause an error by looking at a different place on the wafer or by changing the effective emissivity of the environment.
9. Changes in the view angle—changes the effective emissivity and hence the measured temperature.
10. Changes in wavelength-selective filters—oxidation over years will change the filters.

Careful design of the *entire pyrometer environment,* not just the pyrometer itself, can serve to minimize these problems. Nearly all single-wafer equipment manufacturers now have pyrometry options for their tools. Properly designed sensor systems in single-wafer tools can accurately measure wafers quickly (1/20 of a second) from about 250°C to 1200°C with resolution to 0.05°C.

The continually increasing process requirements will continue to push the pyrometric measurement limits to lower temperatures for plasma-assisted chemical vapor deposition (CVD), cobalt silicide RTP, and plasma etch. The need for lower temperatures will mandate more efficient fiber optics. There will be a continued improvement in real time emissivity measurement. Physical properties other than thermal radiance will also be used for measuring wafer temperatures. Repeatability will have to improve. The problem of unknown or changing emissivity is being addressed with the implementation of the ripple technique (7), which takes advantage of the modulation in background light to measure wafer emissivity real time. This method can measure emissivity to ±0.001 at a rate of 20 times per second.

## 2. Diffuse Reflectance Spectroscopy

### a. *Theory of Operation*

Semiconductor physics provides a method for the direct measurement of substrate temperature, based on the principle that the bandgap in semiconductors is temperature dependent (8). This dependence can be described by a Varshni equation (9),

$$E_g(T) = E_g(T=0) - \frac{\alpha T^2}{\beta + T} \tag{1}$$

where $\alpha$ and $\beta$ are empirically determined constants. The behavior of the bandgap is reflected in the absorption properties of the material. If a wafer is illuminated with a broadband light source, photons with energy greater than the bandgap energy are absorbed as they pass through the material. The wafer is transparent to lower-energy (longer-wavelength) photons. The transition region where the material goes from being transparent to opaque occurs over a relatively narrow energy (or wavelength) range. Thus, a plot of light signal level passing through the material as a function of wavelength or energy (its spectrum) yields a steplike absorption edge, as shown in Figure 2. The position of the absorption edge shifts in energy or wavelength as the substrate temperature changes. The magnitude of the shift is material dependent.

### b.  Bandedge Measurement Techniques

There are several different techniques for measuring the optical absorption of semiconductors in situ: transmission, specular reflection, and diffuse reflection.Transmission measurements observe light that passes through the substrate. This technique has the advantage of providing a high signal-to-noise ratio. The main drawbacks are that to implement this approach one must have optical access to both sides of the sample and one is limited to single-point temperature monitoring.

Reflection measurements from a sample can be separated into two components: specular reflection and diffuse reflection. Specularly reflected light is the familiar "angle of incidence equals angle of reflection" component. The diffusely scattered portion is light that is reflected from the sample over $2\pi$ steradians and carries with it color and texture information about the sample. The main differences between diffuse and specular reflection are:

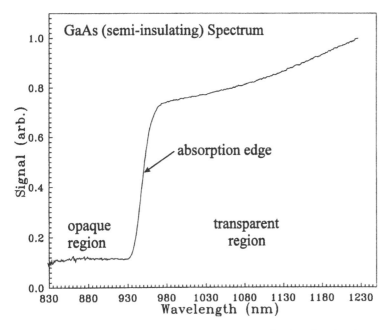

**Figure 2**   Spectrum of 350-micron-thick GaAs wafer showing the absorption edge where the wafer goes from being opaque to being transparent. (From Ref. 8.)

1. Diffuse reflection can be seen over a wide range of angles, while specular reflection can be observed from only a narrow region in space.
2. Diffuse reflection is much weaker than specular reflection.
3. Both diffuse and specular reflection carry with them color and texture information. However, it is much easier to read this information from diffuse reflection, because it is not buried under a large background signal level.

*c. Bandedge Thermometry Limitations*

There are two limitations to bandedge temperature measurement techniques: (1) free carrier absorption in the material, and (2) coating the substrates with opaque material. In semiconductors, there are a number of different phenomena that contribute to light absorption in the material. One of these is the so-called *free carrier absorption*. This absorption term is related to the number of carriers excited into the conduction band from the valence band. Free carriers are thermally excited and the free carrier absorption increases with sample temperature. This absorption occurs over a broad wavelength range and the material becomes more opaque as the temperature rises. The substrate bandedge feature decreases in intensity until, above a material-dependent threshold temperature, it can no longer be seen. In general, the smaller the bandgap, the lower the temperature at which the substrate will become opaque. For silicon, the upper temperature limit for diffuse reflectance spectroscopy (DRS) temperature measurement is approximately 600°C, while for gallium arsenide the upper limit is estimated to be above 800°C.

The second limitation arises when the substrate is covered with an opaque material, such as a metal. In this situation, light cannot penetrate the metal layer and consequently no bandedge spectra can be recovered. Two remedies for this situation are: leaving an open area on the substrate for temperature monitoring purposes, and viewing the substrate from the nonmetallized side.

*d. Temperature Measurement: Practice and Applications*

The initial application of bandedge thermometry (10) measured the heater radiation that passed through a substrate to deduce the sample temperature. This approach suffered from a decreasing signal-to-noise ratio as the heater temperature dropped. As in pyrometry, it was susceptible to contamination from other light sources present in the processing environment. Later on (11–13) two key innovations were introduced to the original transmission technique: (1) the use of an external modulated light source to illuminate the sample, and (2) collection of the diffusely reflected source light to measure the bandedge. The DRS technique is shown schematically in Figure 3. Part of the light is specularly reflected while part is transmitted through the sample. At the back surface, some of the transmitted light is diffusely scattered back toward the front of the sample. Collection optics are placed in a nonspecular position, and the captured diffusely scattered light is then analyzed. The net result is a transmission-type measurement from only one side of the wafer.

A DRS temperature monitor (14) is shown schematically in Figure 4. The system consists of a main DRS module housing the monochromator, power supplies, and lock-in amplifier electronics. The light source and collection optics attach to the exterior of the processing chamber. Data collection and analysis is controlled by a stand-alone computer. The system can output analog voltages for direct substrate temperature control and can communicate with other devices using an RS-232 interface. The device is capable of 1-second updates with point-to-point temperature reproducibility of ±0.2°C. The system can read silicon, gallium arsenide, and indium phosphide substrates from below room temperature to 600°C, > 800°C, and > 700°C, respectively.

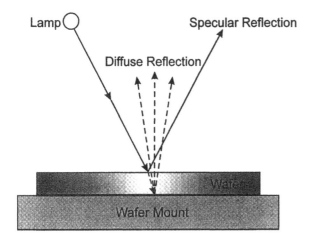

**Figure 3** Schematic showing the DRS measurement technique. (From Ref. 8.)

The DRS technology has been applied to both compound semiconductor and silicon processing. In molecular beam epitaxy (MBE) and its related technologies, such as chemical beam epitaxy (CBE), material is grown layer by layer by opening shutters to molecular sources. The quality of the layers depends, in part, on the temperature and temperature uniformity of the substrate. In typical growth environments the wafer temperature is controlled by a combination of thermocouple readings and pyrometer readings. The DRS technique can monitor and control the temperature of a GaAs wafer in a CBE tool to well within $\pm 1°C$ (15). Even though bandedge thermometry has a fixed upper-temperature limit of $\sim 600°C$ for silicon, this technique is still applicable to several silicon-processing steps such as silicon etching (16), wafer cleaning, and wafer ashing.

## 3. Acoustic Wafer Temperature Sensor

Recently, a new technology has been developed for real-time wafer temperature measurement in semiconductor processing tools (17). This product (18) is based on state-of-the-art

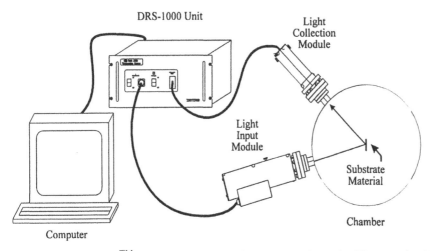

**Figure 4** DRS 1000$^{TM}$ temperature monitoring system schematic. (From Ref. 14.)

acoustic thermometry technologies developed at Stanford University. This sensor fills the need for real-time wafer temperature measurement independent of wafer emissivity, especially in the sub-600°C process regime. It is compatible with plasma processes and mechanical or electrostatic wafer-clamping arrangements.

The velocity of acoustic waves in silicon is a very linear function of temperature. Acoustic thermometry accurately measures velocity of an acoustic wave on the silicon wafer to determine the temperature of the wafer. The acoustic thermometer determines the velocity by very accurately measuring a delay between two points at a known distance.

In its simplest implementation, the acoustic thermometer contacts the wafer with two pins, as shown in Figure 5. One pin is a transmitter and the other a receiver. Both pins have a piezoelectric transducer mounted to their lower end. Both piezoelectric transducers can turn an electrical excitation into an acoustic excitation, and vice versa. The tops of the two pins touch the wafer, to allow the transfer of acoustic energy between the wafer and pins. The pins can be of any inert, relatively stiff material, such as quartz (fused silica) or alumina.

An electrical excitation pulse excites the transmitter pin's transducer to initiate the measurement. This excites an acoustic wave that propagates up the transmitter pin. When the wave reaches the top of the pin, two things happen to the acoustic energy: Most of the energy reflects back down the pin. The reflected energy gives rise to an electrical echo signal at the transmitter's transducer. A small amount of the acoustic energy enters the silicon as an acoustic wave, which propagates out from the transmitter pin. When the acoustic wave reaches the receiver pin, a small portion of its energy excites an acoustic wave that propagates down that pin. That wave produces the electrical signal at the receiver pin's transducer. The echo signal corresponds to propagation up a pin and down a pin. The received signal corresponds to propagation up a pin, across the wafer, and down a pin. If the pins are identical, the delay between the received and echo signals corresponds to propagation on the wafer over the known distance between the transmitter and receiver pins. The ratio of the distance to the delay is the velocity of the acoustic wave in the wafer, and is thus an indicator of wafer temperature.

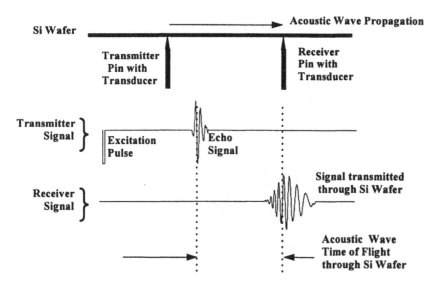

**Figure 5** Geometry and electrical signal for acoustic thermometry. (From Ref. 17.)

The temperature sensor is the wafer itself. Since the acoustic thermometer measures the velocity of propagation between the two pins, it measures an average temperature of the wafer along the propagation path between the pins and not just the temperatures at the pins. Additional pins can yield average temperatures over multiple zones on the wafer. The acoustic wave extends across the whole thickness of the wafer. Thus the measured temperature is an average over the thickness of the wafer as well. It is not particularly sensitive to the lower surface of the wafer, even though the pins touch only the lower surface. This sensor can be designed to operate as the lift-pin mechanism in single-wafer processing tools. The sensor does not influence or contaminate the wafer environment. The pin assembly has to be integrated into the OEM tool for the successful deployment of this technology.

## B.  Gas-Phase Reactant Concentration

Most semiconductor manufacturing processes are chemical in nature. Typically, there is a chemical reaction between a gas-phase chemical species (or mixture) with the surface layer of the silicon wafer. These reactions can be kinetically controlled, hence the interest in wafer temperature. But they can also be controlled by the composition and concentration of the gas-phase reactants. These parameters therefore have to be monitored and controlled in order to provide consistent, reproducible reactions and wafer-state properties. Processing tools always control the primary components of these gas-phase mixtures (flow rate of the gases, pressure). However, there is still a clear need for measuring the composition and/or the concentration of individual species to detect:

Changes in the chemistry as a function of the chemical reaction with the wafer, i.e., endpoint for etch

Changes in the chemistry due to spurious conditions or faults (e.g., leaks, wrong gas)

Rapid changes or a slow drift in the gas-phase chemical composition due to reactor chamber effects such as cleaning, residue formation on the walls, wear of consumable parts within the reactor

A large group of in situ sensors for gas-phase monitoring are spectroscopic in nature, for these are truly nonintrusive sensors. They require optical access through single or sometimes opposing windows, which are made of materials that can provide a vacuum seal and are transparent to the wavelengths being employed. The sensors analyze the composition of the gas phase via absorption or emission methods, as appropriate for a given process. The spectral range is from the UV through the IR, depending upon the nature of the information required. Mass spectroscopy, in the commonly used residual gas analysis (RGA) mode, provides another class of in situ sensors used for monitoring gas composition. These are based on the analysis of the masses of the species (specifically, m/e) entrained in the gas flow. Sampling can be performed via a pinhole orifice to the processing chamber or by sampling the effluent from the reactor. A typical installation requires differential pumping of the RGA, although some of the recent systems do not have this requirement in selected low-pressure applications.

## 1.  Optical Emission Spectroscopy

Optical emission spectroscopy (OES) is based on (19) monitoring the light emitted from a plasma during wafer processing and is used to gain information about the state of the tool and the process. It exploits the fact that an excited plasma emits light at discrete wavelengths that are characteristic of the chemical species present in the plasma. The intensity of the light at a particular wavelength is generally proportional to both the concentration of the associated chemical species and the degree of plasma excitation.

An OES system consists of a viewport to the plasma chamber, an optical coupling system, an optical detector incorporating some means of isolating the wavelength of interest, and a computer or processor to acquire and analyze the spectral image. The viewport is either a window in the reactor or a direct optical feedthrough into the chamber. The OES technique requires a direct view of the portion of the plasma immediately above the wafer, but not of the wafer itself, so the placement of the viewport is not too restrictive. If ultraviolet wavelengths are to be monitored, the window must be of fused silica and not ordinary glass. A number of OES sensor systems are commercially available (e.g., Refs. 20 and 21), and most OEM plasma tools come with their own on-board OES systems. The typical configuration is shown in Figure 6. The optical components and the other associated concerns with OES systems are described in the next sections.

*a. Fixed-Wavelength Systems*

There are several types of optical detectors for OES. Simple systems use fixed-bandpass filters for wavelength discrimination. These are stacks of dielectric films, and they have a bandpass of typically 1–10 nm and a peak transmission of about 50%. The light that is passed by the filter is converted to an electrical signal either by a photodiode or by a photomultiplier tube (PMT). Advantages of these systems are low cost and high optical throughput; disadvantages are the limited spectral information and the mechanical complexity involved in changing the wavelength being monitored.

*b. Monochromators and Spectrographs*

More flexibility is afforded by systems that incorporate a monochromator. A monochromator consists of a narrow entrance slit, a diffraction grating, and an exit slit. Light falling on the grating is dispersed into a spectrum, the diffraction angle being dependent upon wavelength. The light is reimaged onto another slit, which provides wavelength discrimination. By turning the grating, any wavelength within the range of the instrument can be selected. Changing the width of the entrance and exit slits changes the bandpass of the system. Automated systems in which the wavelength can be altered automatically under computer control are often used. The throughput of a monochromator is much lower than that of a bandpass filter; hence PMTs are normally used for light detection in these systems.

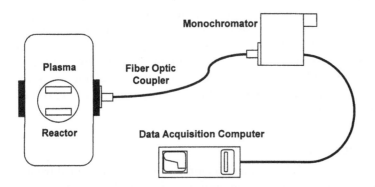

**Figure 6**   Optical emission sensor configuration. (From Ref. 19.)

A variant of the monochromator is the spectrograph. It uses a fixed grating and a solid-state detector array instead of an exit slit and PMT. The advantage of a spectrograph over a monochromator is that many wavelengths can be monitored at once. This is significant for situations where information has to be based on an entire spectral scan, not from only a single spectral peak (an example is found in Sec. III.B.1.e). In the typical installation, light is transferred from the viewport on the tool to the detector head via an optical coupler. Where access space is available, devices mount directly onto the chamber, and no coupler is needed. In other cases, all that is required is to mount the detector against the window on the chamber. More typically, however, space or field-of-view considerations require the use of an optical coupling system. Optical fiber bundles are typically used to bring light into monochromators and spectrographs. This permits locating the detector at a convenient place away from the reactor. Attention to details such as $f$-number matching is required to prevent unnecessary loss of sensitivity. Suppliers of OES equipment can generally provide couplers that are matched to their equipment.

In the typical installation in a semiconductor production environment, an OES system usually consists of a detector head interfaced to a PC running a Windows or NT environment. The computer contains an A/D card for digitization of the analog signal from the detector, and runs an application that performs various functions: processing, data display, endpoint detection, and communication with the process controller.

*c. CCD Array Spectrometers*

A Charge-Coupled Device (CCD) array spectrometer is a spectrograph that consists of a slit, a dispersion grating, and a CCD array detector. Light is typically delivered to the slit by a fiber-optic cable between the spectrometer and a fixture on the window of the reactor. A large number of CCD array spectrometers exist in the commercial market. They can be sorted into two main categories based on the characteristics of the CCD array and the form factor of the final product.

The first category is characterized (22) by the "spectrometer-on-a-card" format, meant primarily as a portable diagnostic tool for use with existing PCs. At least two such systems are currently available (23,24). This means the spectrometer—say, with a 2048-pixel CCD array—and all associated electronics can fit inside the PC, plugging into an expansion slot with an external fiber-optic connection to the light source. This has some obvious benefits:

Compact size—a substantial benefit in the high-cost cleanroom space
Portability—can be used as a roaming diagnostic tool
Multiple spectrometers on a single card—simultaneous spectral analysis in multi-chamber tools, or spatially resolved analysis (24) in a single chamber
Low cost—cost not a barrier to the use of this tool

While there are benefits to these systems, there are some limitations that need to be understood. Specifically, the CCD array typically used in these systems has four inherent deficiencies (relative to PMTs), which, if not adequately offset, will prevent the CCD array from surpassing or even matching PMT system performance. The four inherent CCD array deficiencies (relative to PMTs) are:

1. Small detector pixel height (limited aspect ratio, e.g., 1:1), which limits the sensitivity of certain CCD array devices
2. Absence of inherent CCD array device "gain" (unity gain), which further limits the sensitivity
3. Poor UV response of certain CCD array devices

4.  Limited spectral resolution, due to typical CCD array configuration into very-short-focal-length optical spectrographs, which exhibit more limited wavelength dispersion than is sometimes the case for PMTs and which generally also exhibit severe, uncompensated, internal spectral imaging aberrations normally inherent with very-short-focal-length optical spectrograph designs.

Fortunately, solutions exist that provide offsetting factors for each of these CCD array deficiencies. A brief description of these solutions provides background for understanding the key characteristics of CCD array spectrometers.

Concerning the problem of small detector pixel height, offsetting factors include greatly reduced CCD dark current and noise (especially with small pixel areas), availability of selected array devices having greater than 1:1 (h:w) pixel aspect ratio (e.g., 20:1), and availability of one-dimensional (vertical), internal, secondary, light-concentrating optics with certain CCD array spectrographs. A relatively "tall" spectral image is thereby height focused onto a much shorter array pixel, thus concentrating the light and increasing the signal, without increasing the dark current or affecting spectral resolution (image width).

Concerning the problem of absence of inherent CCD array device gain (unity gain) relative to high-gain PMTs, offsetting CCD array sensitivity factors include the natural integrating properties of array pixels and an inherent CCD array quantum efficiency that far exceeds that of photomultiplier tubes. Collectively, these offsetting factors are so effective that CCD arrays can be rendered sufficiently sensitive to achieve a "full well" device charge count (saturation) for prominent spectral features within the range of 400-ms (or less) exposure time, even with the dimmest of plasma-etching experiments. When the light level is quite high, CCD array exposure times may typically be as low as 10 ms or even less. The high light level allows, for example, 20 or even 40 separate (10-ms) exposures to be digitally filtered and signal averaged (coaddition) for each of 2048 array pixels. Digitally filtering and signal-averaging this many exposures provides a major statistical enhancement of the SNR (signal-to-noise ratio). In addition, data from several adjacent-wavelength pixels may optionally be binned (software summation) in real time for even more SNR enhancement, in cases where spectral resolution is not critical.

Concerning the problem of poor UV response, offsetting factors exist, in the form of fluorophore coatings applied directly to the detector pixels. Satisfactory UV response is thereby achieved.

The problem of limited spectral resolution is one of the most basic problems in using CCD array systems. At most, CCD arrays are only about 1 inch long (e.g., 21–28 mm). This means the entire spectrum must be compressed to fit the 28-mm array length, which limits the spectrograhic wavelength dispersion that may be employed. There is an additional resolution and spectral range tradeoff in the choice of gratings. The total wavelength coverage interval of a CCD array is determined by array dimensions and by the spectrograph focal length and grating ruling density, which together establish the wavelength dispersion. For an array of fixed dimensions, and a spectrograph of fixed focal length, coarsely ruled gratings (600 grooves/mm) provide less dispersion, and hence lower resolution, but a larger total wavelength coverage interval. Finely ruled gratings (1200 or 2400 grooves/mm) provide more dispersion and higher resolution but a smaller total wavelength coverage interval. Centering of a given wavelength range is specified by the user and is fixed at the factory by adjusting the grating angle.

The second category of spectrographs is characterized by high-performance CCD arrays, with applications aimed at stand-alone use (PC or laptop not necessarily required) or integration into OEM processing tools. These are based (19) on research-grade CCD spectrographs that are available with performance that equals or exceeds that of PMT-based systems. For maximum sensitivity, they employ cooled, back-illuminated CCD area arrays. The CCD is operated in a line-binning mode so that light from the entire vertical extent of the slit is collected. These devices have peak quantum efficiencies of greater than 90%, and over 40% throughout the typical spectral range of interest (200–950 nm), compared with a peak value of 20% typical of a PMT. This means that about one photoelectron is created for every two photons to reach the detector. The best such devices have readout noise of only a few electrons, so that the signal-to-noise performance approaches the theoretical limit determined by the photon shot noise. However, the traditional size, cost, and complexity of these instruments make them impractical for use for routine monitoring and process control.

Nonetheless, many of these features are beginning to appear in instruments priced near or even below $10K. Spectrographs in this price range are available (20,25,26) that employ a cooled, back-illuminated CCD with a 3-mm slit height and whose sensitivity matches or exceeds that of currently available PMT-based monochromators. If cost is the prevailing factor, lower cost can be achieved by using front-illuminated CCDs. Performance suffers, since the quantum efficiency is reduced by a factor of 2, and the spectral response of these devices cuts off below 400 nm. Nonetheless, this is a cost-effective approach for less demanding applications.

The issues of size and complexity are being addressed as well. One approach is to integrate the optical head together with the data-acquisition and process-control functions into a single unit (20) with a small footprint and an on-board digital signal processor (DSP) for data analysis. Such a system can be integrated with the host computer for OEM applications or can be connected to a laptop for roaming applications.

Another advantage of such high-performance arrays is that they can be utilized as an imaging spectrograph, where the entrance slit is divided into multiple sections that can couple to different chambers. The resulting spectra can be read independently from the CCD. In this way multiple spectra can be run on the same instrument.

*d. Calibration, Interface, and Upkeep Issues*

Implementing an OES analysis for a new process requires some expertise on the part of the process engineer. First, the spectral line or lines to be monitored must be chosen based upon a fundamental understanding of the spectral signature. Routine acquisition and signal analysis can then be performed by the sensor. The practical issue of the etching of the window or optical feedthrough (or deposition on these components) has to be handled by cleaning or replacing these components. Some OEM vendors address these issues by heating, or recessing, the windows (to slow down deposition) or by installing a honeycomb-like structure over the window (to cut down the deposition or etch on the major cross-sectional area of the window).

*e. Routine Application: Endpoint Detection*

By far the most widespread use of OES sensors is for endpoint detection. Historically, such sensors have been routinely used in plasma etch reactors for decades, since the process state (plasma) is a rich source of useful information. The fundamental principle behind endpoint detection is that as the etch proceeds from one layer (the primary layer being etched) to the underlying layer (the substrate), the gas phase composition of the plasma changes. For example, when etching a typical TiN/Al/TiN stack on an oxide substrate with a Cl-containing chemistry, there is a significant decrease in the AlCl product species, with a corresponding increase in the Cl reactant species, as the etch transitions from the bulk Al to the TiN and oxide layers. So a continuous monitoring of the 261-nm Al emission line intensity will show a decrease during the time when the Al film disappears. Traditional endpoint detection techniques have historically relied on numerical methods such as threshold crossing, first-derivative, or other combinatorial algorithms, which are manually devised to conform to the characteristics of a family of endpoint shapes and can be tuned to declare endpoint for a typically anticipated signal change. Besides the endpoint indication—which is by far the most commonly generated information from this data— the slope of the endpoint signal (at the endpoint) can be used as an indicator of the nonuniformity of the etch process (27).

From the sensor point of view, endpoint detection has been well established for the last 15–20 years. The majority of OEM plasma etch tools have endpoint-detection hardware integrated into the tool. Historically, these were simple, inexpensive photodetectors that viewed the plasma emission through an appropriately selected optical bandpass filter and an optically transparent window in the side of the reactor. In the newer-generation tools, this optical signal is obtained by the use of a short-focal-length-grating monochromator that can be manually scanned to the correct wavelength for the specific process. These have the advantage of the readily variable wavelength selection, the higher spectral resolution required to optically separate closely overlapping spectral lines, and a higher sensitivity of the photomultiplier (PMT) detector (vs. the photodiode detectors).

*f. Emerging Application: Endpoint Detection for Low Exposed Areas*

For etch processes where the material etched is a significant percentage of the wafer surface area (e.g., metal etch), there is a large change in the plasma chemistry when this material is etched off; hence the endpoint signal is very strong and easily detected. The latest challenge in low-exposed-area endpoint detection is for processes such as oxide etch in contact holes where the exposed area of oxide is under 1% of the wafer area. This drives a very small chemistry change at endpoint, which in turn generates a very small change in the large emission signal intensity. The following two subsections provide details on the newly emerging software methods for detecting endpoint based on single-wavelength endpoint curves and on full spectral data.

*Neural network endpoint detection.* The shape of the typical single-wavelength endpoint curve is the natural by-product of the processing tool characteristics, the product design, and the influence of the tools and processes that precede the endpoint-defined tool step. As such, the endpoint shape of a given process exhibits a statistical variation derived from the numerous preceding processes that affect the state of the wafer supplied to the tool requiring endpoint control. It then becomes the challenge of the process engineer, working in conjunction with the endpoint controller system, to effect a practical control algorithm. This algorithm has to be unique enough to recognize the endpoint, but also general enough to comprehend the pattern variability that is a consequence of the tool change (namely, window absorbance) and the product mix. This challenge can be imposing, requiring a lengthy empirical evaluation of numerous endpoint data files in an attempt to achieve the correct numerical recipe that accurately and reliably declares endpoint for the full suite of endpoint pattern variations.

One approach (19) to this problem is a neural-network-based endpoint detection algorithm (28). It utilizes a fast-training neural-network pattern recognition scheme to determine the endpoint signature. Unlike traditional feed-forward neural networks, which require many pattern samples to build an effective network, the methodology employed with this approach minimizes the number of representative sample data files required for training to typically fewer than 10. The process engineer is not burdened with numerical recipe optimization. The following simple three-step procedure outlines the technique.

1.  Acquire representative data files exhibiting a full range of endpoint patterns; new patterns can be later introduced into this data set.
2.  Tag the endpoint patterns in the collected data files; i.e., identify the region in each data set that contains the endpoint.
3.  Train the network—an automatic procedure completed in a few minutes.

This technology has been successfully demonstrated and used in etching oxide with as low as 0.1% open area. Ultimate limits for any specific application are tied to a number of variables. These include the type of tool, process, optical detector, and appropriate selection of emission wavelength(s) to monitor. As a caveat, it is important to note that this technique must "see" the evolution of a distinguishable pattern in order to learn the shape and correctly identify its occurrence as endpoint. The shape may be subtle and complex, but it must be identifiable for successful results.

Another useful feature of this type of endpoint detector is its ability to recognize complex, unusual, nonmonotonic patterns. Since this technique employs a pattern recognizer, it is not limited by complex signal variations that can prove daunting for numerical recipes.

*Evolving-window factor analysis of full spectra.* With only one wavelength being monitored, random plasma fluctuations or detector/amplifier noise excursions can obscure the small intensity change that would otherwise serve to characterize endpoint in oxide etching of wafers exhibiting open areas of 1% or less. An alternate solution to this problem is to use the full spectrum available from the plasma. It is clear that significantly more useful data can be had if one measures the intensities in a broad spectral range vs. at a single wavelength of a chosen species. Since the spectral data still has to be obtained at a fast-enough rate to detect the endpoint, this drives the use of CCD array detectors. With the necessary dispersion optics, these CCD arrays can simultaneously measure the spectral intensities across a broad spectral range at a resolution determined by a number of factors described in Sec. II.B.1.c. This sensor change generates a new dimension in the endpoint data, the spectral dimension (vs. a single wavelength). This, in turn, necessitates the development of algorithms that can make good use of this additional information.

In one particular case (23), an evolving-window factor analysis (EWFA) algorithm is employed to obtain the endpoint information from the multivariate spectral data. This EWFA is a variant of the more classical evolving factor analysis (EFA) technique used for the analysis of ordered—in this case, by time—multivariate data. The EFA technique follows the singular values (factors) of a data matrix as new rows (samples) are added. In a manner similar to PCA analysis, EFA determines how many factors are included in the data matrix and then plots these against the ordered variable (time). Such algorithms are well defined and routinely available (29). The EWFA method considers a moving window of data samples (as in Figure 7), for computational ease. The resulting data is a time-series plot of the appearance, or disappearance, of certain factors in the data set. So looking at the multiple spectral lines provides an increased endpoint signal sensitivity.

A typical representation of this EWFA analysis is given in Figure 8, which shows two of the factors. Factor 3 (Value 3) shows the temporal nature of the rotating magnetic field in this processing tool. Factor 4 (Value 4) shows the endpoint signals from the etching of a four-layer oxide stack; the four endpoint signals are clearly identified in spite of the other temporal variations in the process (the rotating magnetic field).

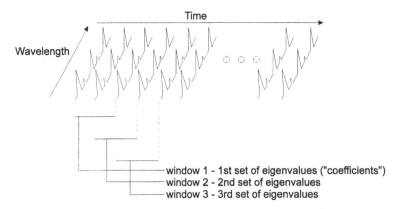

**Figure 7**   EWFA data matrix. (From Ref. 22.)

Automated endpoint detection in oxide etching has been shown to work with this technique down to 0.1% open area.

## 2.  Fourier Transform Infrared Spectroscopy

Infrared spectroscopy, in the mid-IR range of 1–20 μm, can provide a wealth of information about gas properties, including species temperature, composition, and concentration. Its application to gas-phase analysis in semiconductor (SC) manufacturing tools has been more limited than the use of visible spectroscopy for analyzing the optical emission of plasma processes. But there are some useful applications based on infrared spectroscopy, so the technology will be described.

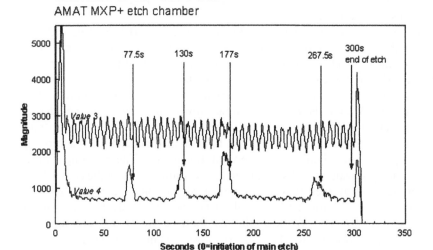

**Figure 8**   EWFA endpoint signal on Value 4. (From Ref. 22.)

*a. Theory of Operation*

Complete spectra from wavelengths of 1.5 to 25µm can be obtained in fractions of a second using a Fourier transform infrared (FTIR) spectrometer. The core of an FTIR is typically a Michelson interferometer, consisting of a beam splitter and two mirrors, one of which moves (30). As shown in Figure 9, incoming radiation in a parallel beam impinges on the beam splitter and is split roughly in half into beams directed at the mirrors. The reflected light recombines at the beam splitter to form the outgoing radiation. If the mirrors are equidistant from the beam splitter, then the radiation recombines constructively. If the paths differ by one-quarter wavelength, then the beams combine destructively. Because the moving mirror travels at constant velocity, the radiation is amplitude modulated, with each frequency being modulated at a unique frequency that is proportional to the velocity of the moving mirror and inversely proportional to the wavelength. Thus, radiation with twice the wavelength is modulated at half the frequency. The key requirements for such an FTIR spectrometer are that they be vibration immune, rugged, permanently aligned, and thermally stable. Another key issue for accurate quantitative analysis is detector linearity. Mercury cadmium telluride (MCT) detectors are high-sensitivity infrared detectors but are notoriously nonlinear. Detector correction methods are required to linearize the response. All these requirements have been addressed, making FTIR a commercially available sensor (31) for possible use in SC manufacturing.

*b. Exhaust Gas Monitoring Applications*

Most of the sensors described in this chapter are used for sensing process- or wafer-state properties during or after a given manufacturing process. However, FTIR lends itself well to another important aspect of SC manufacturing: fault detection based on exhaust gas monitoring from a reactor.

In one particular study (32) on a high-density plasma reactor, the spectrometer was positioned above the exhaust duct of the etch chamber. The IR beam was directed to a set

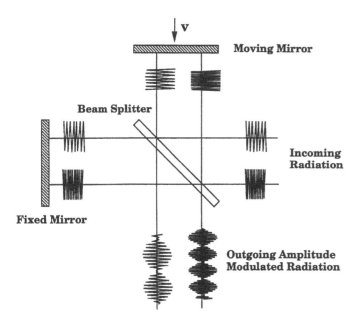

**Figure 9** Modulation of radiation by a moving mirror interferometer. (From Ref. 30.)

of focusing and steering mirrors and into a multipass mirror assembly enclosed in an exhaust-line tee. This tee was placed between the turbo and the mechanical pumps. The multipass cell generated 20 passes through the tee to provide a 5-m path length. The exhaust gas passed through this in-line gas cell, and spectra were collected at 1-cm$^{-1}$ resolution.

The data obtained in this study suggests that FTIR measurements can provide:

1. Exhaust gas monitoring, after the turbo pump, providing a reproducible and rapid measurement of a rich variety of compounds produced during the wafer etch
2. Identification of the mix of compounds that can be used to interpret an etching sequence or the cleaning of a reactor by a reactive plasma
3. Identification for the effects of incorrect chucking, incorrect plasma power, air leaks, and low-pressure gas feed
4. Data for use in fault detection, for a reliable and automated fault detection and classification system

The FTIR technique can also be used for the analysis of the efficiency of large-scale, volatile organic compound (VOC) abatement systems.

### 3. Mass Spectroscopy/Residual Gas Analyzer

In addition to the optical methods previously described, gases can also be analyzed by mass spectroscopy of the molecular species and their fragmented parts. The in situ mass spectrometric sensor for gas analysis is commonly known as a residual gas analyzer (RGA).

*a. Conventional Residual Gas Analyzer*

Quadrupole devices are by far the most widely used in semiconductor manufacturing applications, typically for the maintenance and troubleshooting of process tools (33). Leak checking, testing for gas contamination or moisture, and periodic residual gas analysis for tool qualification have been the main uses of RGA. In other applications, RGAs are used to establish a correlation between a wafer's quality and its measured contamination in the process chamber. Recently RGAs have been used for in situ monitoring to prevent accidental scrap and to reduce wafer-to-wafer variability. To be effective, RGAs have to be able to directly monitor both tool baseline pressures and process chemistries in a nonintrusive fashion.

*Theory of operation.* Conventional RGAs operate by sampling the gases of interest through an orifice between the container for the gases (the processing chamber or exhaust duct in an SC manufacturing tool) and the RGA analyzer (shown in Figure 10). In a conventional RGA, the pressure must be reduced below typical processing chamber pressures prior to ionization. This requires differential pumping and sampling of the process gases, making conventional RGAs a relatively bulky and expensive package. The following brief description of the three basic components of a quadrupole mass spectrometer analyzer—the ionizer, the mass filter, and the detector—are provided to facilitate the understanding of the sensors based on this technology.

*Ionizer.* Gas ionization is usually achieved using an electron-impact process. Electrons are emitted from a hot filament (2200°C) using an electric current. Few metals have a low-enough work function to supply currents in the milliamp range at

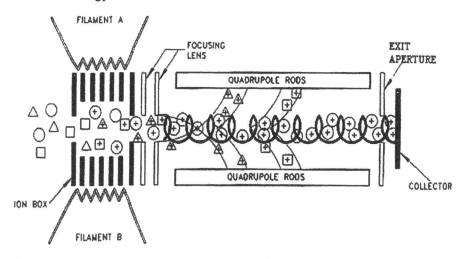

**Figure 10** Quadrupole residual gas analyzer. (From Ref. 33.)

such temperatures. Filaments are usually coated with materials with better thermoemission properties. Typical coatings are thoria ($ThO_2$) and yttria ($Y_2O_3$) and typical base metals are tungsten, iridium, and rhenium. Electrons are then accelerated to acquire an energy in the 30–70-eV range, which corresponds to the highest ionization cross sections for several gases. The ionization occurs in an enclosed area called the ion source. There are many types of sources, but the major distinction is between open and closed sources. The higher the pressure in the source, the greater is the sensitivity to minor constituents. The sensitivity is the minimum detectable pressure relative to the maximum number of ions produced in the source. A closed ion source has small apertures to introduce the sample gas from the process environment, to allow the electrons to enter the source, and to extract the ions into the mass filter. With the use of an auxiliary pump, the filaments and the mass filter and the detector are kept at a much lower pressure than the source. In addition to greater sensitivity, the advantages associated with closed sources are: (1) prolonging the filament lifetime in the presence of corrosive gases and (2) enabling electron multipliers to be used as ion detectors. However, the high complexity and cost associated with the apertures' precision alignment and the required high-vacuum pump make closed-source-type instruments very expensive.

*Mass filter.* The ions are extracted from the source and are focused into the entrance aperture of the mass filter with an energy $V_z$. The mass filter is the cavity enclosed by the four parallel quadrupole rods arranged in a square configuration (see Figure 11). Typical diameter and length of the cylindrical rods are at least 6 and 100 mm, respectively. The species moving through the filter are singly or multiply charged atoms or molecules. Filtering is the common term for selecting ions with a particular mass-to-charge ratio that possess a stable trajectory enabling them to reach the detector while all other ions (with unstable trajectories) are filtered out. Filtering is accomplished by subjecting the ions to lateral forces generated by the combination of dc and RF voltages on the rods. The filtered mass and the mass resolution are given by:

TYPICAL QUADRUPOLE

**Figure 11**  Array concept. (From Ref. 33.)

$$m = \frac{7 \times 10^6 V}{f^2 r_0^2} \tag{2}$$

$$\Delta m = \frac{4 \times 10^9 V_z}{f^2 l^2} \tag{3}$$

where $V$ is the amplitude of the RF voltage, $f$ is the RF frequency, $r_0$ is the radius of the inscribed circle, $l$ is the length of the mass filter, and $V_z$ is the ion energy.

*Detector.*  The filtered ions are accelerated at an exit aperture to reach the detector. Two detection techniques are generally used: Faraday cups and electron multipliers. Faraday cups are in the shape of cavities in which collected ions and any secondary electrons are trapped to generate a current. The current is then converted to a voltage using a sensitive electrometer circuit. The limit of detection of these devices is determined by the ability to make sensitive electrometers. Fundamental limitations associated with Johnson noise in resistors and the noise in the semiconductor junctions determine the lowest detectable current. Alternatively, there are techniques for multiplying the current in vacuum by using a continuous dynode electron multiplier. This is shaped as a curved glass tube, with the inside coating made of a high-resistivity surface ($PbO$-$Bi_2O_3$ glass) with a high secondary electron emission coefficient. A high voltage (3 kV typically) is applied between the ends of the tube. When filtered ions strike the active surface, a shower of electrons is produced and accelerated toward the opposite wall of the surface. Each electron leads to the emission of more electrons, and the process is repeated along the length of the tube causing an avalanche of electrons. A multiplication or gain up to $10^7$ can be achieved. However, the ability to emit electrons decreases with time. The time scale depends on the total number of electrons emitted, which in turn depends on the number of incident ions. At high pressures, large numbers of ions strike the surface, causing a high rate of depletion and hence a shorter lifetime. Another important phenomenon related to the operation at high pressures is the "positive feedback." As the number of positive ions increases inside the tube, the gain can be drastically reduced, since ions, accelerated in the opposite direction, interfere with the electron multiplication process. These phenomena limit the practical use of electron multipliers to the low-pressure ($< 10^{-5}$torr) range.

*b.  Sensor-Type Residual Gas Analyzers*

*Component choices.*  A recent key development in RGA technology is the evolution of sensor-type RGAs. These have miniaturized quadrupoles that allow mass-filter operation at nearly three orders of magnitude higher pressure, thereby not requiring differential pumping of the sensor for many applications. The basis of these new systems is the substantially shortened quadropoles that provide a short path for ions to travel to the detector, thus minimizing the effect of collisional losses at the higher pressures. Their small size allows them to be mounted at several strategic and/ or convenient locations without increasing the footprint. This represents a major breakthrough with regard to the sensor size, cost, and ease of use. A number of tool manufacturers provide such sensors (e.g., Refs. 34 and 35). But any miniaturization attempt has to be carried out without sacrificing mass spectrometry performances of conventional RGAs in term of sensitivity, mass resolution, and mass range.

The optimal use of these sensors requires an understanding of the interactions between the pressure range, the detection technique, and the required sensitivity. At low pressures (below $10^{-5}$ torr), which conveniently coincides with the optimum operating pressure of the high-gain electron multipliers, high sensitivity to low-partial-pressure contaminants is readily achieved. This provides capability for sensitive determination of background moisture levels and low-level leaks in the vacuum system.

But with these smaller sensors currently available, the shorter path length of the ions allows RGA mass filters to operate at pressures in the millitorr range, which also enables the direct monitoring of many semiconductor manufacturing processes. However, these pressures are too high for the efficient operation of the electron multiplier detector. So one solution is to return to the use of a pressure-throttling device (orifice) and a high-vacuum pump. Aside from the cost and size penalties of this approach, there are more serious considerations that have to do with the retained gases, or lack thereof, on the analyzer chamber walls. This leads to measurements that do not necessarily reflect the gas composition of the process chamber but reflect more on the state of the analyzer. This is very noticeable when the pressure is very low in the analyzer chamber. The lower the pressure in the analyzer chamber, the lower the required pressure of the high-vacuum pump, since species not pumped will provide a background measurement that must be taken into account and may be variable with time and temperature. Another solution is to use a Faraday cup detector at these millitorr pressures, but this sacrifices sensitivity due to the lower gain inherent in these detectors. The sensitivity is further decreased by the geometrical aspects of these miniature sensors, since the reduction of the rod size, and hence of the diameter of the inscribed circle between the four rods, results in a smaller acceptance area for the ionized species to reach the detector.

A recent solution to this sensitivity issue at higher pressures has been the development of an array detector. In this configuration, an array of miniature quadrupoles compensates for the loss of sensitivity. The mass resolution and the mass range are maintained by increasing the RF frequency, as seen in the Eqs. (2) and (3). While the array concept was introduced several decades ago, volume production has only recently been enabled by the new fabrication technologies used to handle microparts. This sensor (36) comprises a $4 \times 4$ array of identical cylindrical rods (1-mm diameter) arranged in a gridlike pattern, where the cavities between the rods form a $3 \times 3$ array of miniature quadrupole mass spectrometers. The length of the rods is only 10 mm, which enables the operation of the sensor at higher pressures (10 mtorr). It occupies less than 4-cm$^3$ total volume. The manufacturing method uses glass-to-metal technology to seal the rods and electrical pins. This technology provides lower manufacturing cost and physically identical sensors that are simple to calibrate. The replacement cost of these sensors is low enough to consider them to be consumables.

*Calibration and lifetime.* Calibration of an RGA sensor is performed against capacitance manometers and needs no field calibration other than for fault detection. The data is displayed in torr or other acceptable pressure units and can be directly compared with the process pressure gauge. There are recommended practices published by the American Vacuum Society (AVS) for calibrating the low-pressure devices, but the miniature high-pressure RGAs were developed after those recommended practices were established.

At the high operating temperatures, RGA filaments react strongly with the ambient gases. This interaction leads to different failure mechanisms, depending on the pressure and the chemical nature of these gases. Tungsten and rhenium filaments are volatile in oxidizing atmospheres (such as oxygen), while thoria- and yttria-coated iridium filaments are volatile in reducing atmospheres (such as hydrogen). Corrosive gases such as chlorine and fluorine reduce filament lifetime drastically. In fact, lifetime is inversely proportional to the pressure in the $10^{-5}$–$10^{-2}$ torr range. This corrosion-limited lifetime favors systems that have readily and inexpensively replaceable detectors.

*Sensor interface to original equipment manufacturer tools.* Since both the process tools and the RGA sensors are vacuum devices, the pressure connections to the vacuum should be made with good vacuum practice in mind. Lower-pressure operation increasingly requires the use of large-diameter short lines made with materials and processes that provide low retentivity. Wherever possible, the source of the sensor should be in good pneumatic communication with the gases to be measured. Care should be taken to avoid condensation in the RGA of species from the process. This may require that the sensor and its mounting be heated externally. Temperatures as high as 150°C are common in these cases. Higher temperatures may require separating portions of the system from the sensor, which may deteriorate the system performance and add to the cost.

Quadrupole devices operate internally at very high RF voltages and therefore may radiate in the process chamber or to the ambient outside the chamber. Good grounding practices, such as those used with process plasma RF generators, are important.

Compared to other instruments on the process tool, RGAs generate large amounts of 3-D data (mass, pressure, time) in a very short time. Invariably, the data from these devices is transmitted on command by a serial link. RS232, RS485, and proprietary protocols are all in use. Since multiple devices are commonly bused together, data systems such as a "Sensor Bus" will normally be used. Efficient use of the data by the tool controller represents a major challenge on the way to a full integration of these sensors into OEM tools.

## 4. Acoustic Composition Measurement

Although generally not well recognized, the composition of a *known* binary mixture of two gases can be determined by measuring the speed of sound in the mixture (37). Very high sensitivity ($\sim$1 ppm) is available when a high-molecular-weight precursor is diluted in a low-molecular-weight carrier gas. Acoustic gas-composition measurement is inherently stable, and consequently accuracy is maintained over the long term. There are no components that wear, and the energy levels imparted to the gases are very low and do not induce any unintended reactions. These features make this technique ideal for many MOCVD and CVD processes. This technique is not readily applicable if the individual gas species are unknown or if more than two species are present, because many combinations of different gas species may be blended to produce the same speed of sound. This lack of uniqueness does not pose a problem for blending gases since a cascaded arrangement of sensors and controllers may be used to add one gas at a time. Each successive transducer will use the information for the blend from the previous instrument as one of its component gas' thermodynamic constants. The most obvious application is to determine the gas composition flowing through the tubes that supply mixed gases to a reactor. However, there may be additional value to sampling gases from within the reactor chamber or its outlet. This dual-transducer arrangement can provide information on the efficiency, stability, and "health" of the ongoing process.

*a. Theory of Operation*

The speed of sound, *C*, in a pure gas is related to the gases' fundamental thermodynamic properties as follows (38):

$$C = \sqrt{\frac{\gamma R T}{M}} \tag{4}$$

where $\gamma$ is the specific heat ratio, $C_p/C_v$, *R* is the universal gas constant, *T* is the Kelvin temperature, and *M* is the molecular weight. The same equation form holds precisely for a mixture of gases when appropriate values for $\gamma$ and *M* are calculated based on the relative abundance of the individual species. Likewise, it is only an algebraic exercise to solve the resulting equations for the relative concentration of a mixture when the speed of sound is known or measured (39).

*b. Sensor Configurations*

Building a composition-measuring instrument using this fundamental thermal physics has been accomplished in two distinct ways. The first implementation measures the transit time for an ultrasonic ($\sim$15 kHz) pulse through the gas (40). This time-of-flight implementation requires only a high-resolution timer to measure the time between when a sound pulse is generated and its arrival at a receiver a distance *L* away. The second implementation measures the resonant frequency of a small chamber filled with the target gas mixture (39), as in Figure 12. All wetted components of this chamber are fabricated

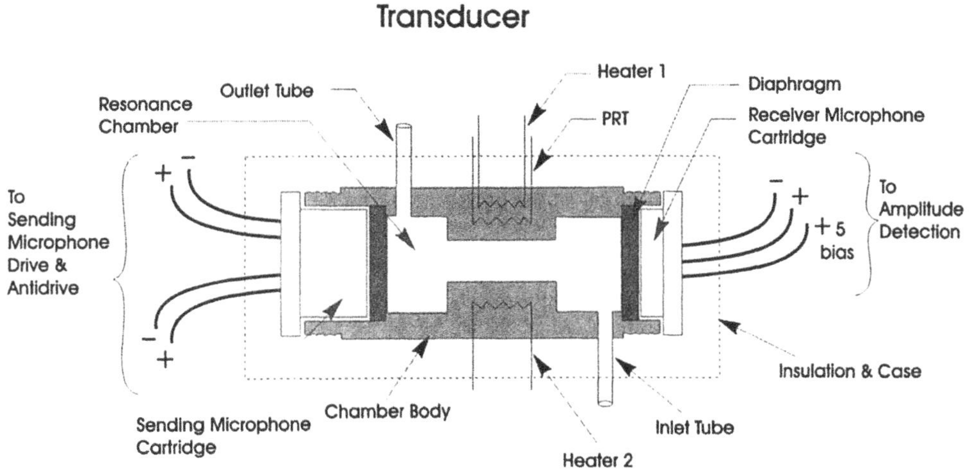

**Figure 12** Cross section of a transducer for a low-frequency-resonance type of acoustic gas analyzer. (From Ref. 37.)

from high-purity and electropolished stainless steel and inconel. A precisely controlled frequency generator is used to stimulate the gas at one end of the chamber, and the intensity of the transmitted sound is measured at the opposite end. Algorithms are designed to maintain the applied frequency at the gas chamber's resonance frequency. The chamber is precisely temperature controlled (to less than $\pm 0.03°C$) and is carefully shaped to resonate in the fundamental mode at a low audio frequency (0.3–4.6 KHz). Low-frequency operation allows the use of metal diaphragms, which avoids process gas contact with the acoustic generators and receivers. Low-frequency sound is also more efficiently propagated through the chamber than ultrasonic frequencies, resulting in useful operation at pressures as low as 70 torr. Since the chamber's length is fixed and its temperature is carefully controlled, the speed of sound, $C$, is simply related to the resonant frequency, $F$, as $C = 2FL$, where $L$ is the effective distance between the sending and receiving elements. It is possible to resolve a gas-filled chamber's resonant frequency, and therefore the speed of sound of the gas, to less than 1 part in 50,000 using the resonant technique. Even though the frequency-generation method employed can generate frequencies only 0.1 Hz apart, even greater resolution may be achieved by measuring the amplitude at several frequencies around resonance and then curve-fitting the instrument's response to the theoretical shape of the resonance peak. There is little effect on the gas-supply dynamics because either implementation adds little additional volume ($< 25$ cc) to the reactor's delivery system.

A typical installation of an acoustic composition measuring and control system (41) is shown in Figure 13. It consists of two major components. First, a transducer is inserted directly into the reactor's supply line. Second, an electronic control console is used for controlling the sensor's temperature, determining the speed of sound, computing the

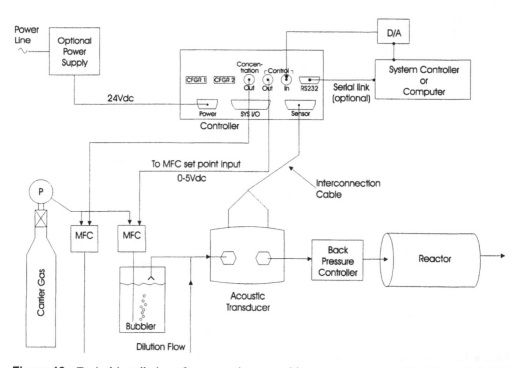

**Figure 13** Typical installation of an acoustic composition measurement system. (From Ref. 37.)

composition, generating feedback control voltages for mass flow sensors and analog and digital transmission of relevant process control data. The gas delivery tube is cut and reconnected so that the gas mixture passes through the transducer. The transducer's temperature may be set to match the transport needs of the materials or to match heated delivery tubes. This installation demonstrates simultaneous control of both the bubbler flow and the dilution flow, thus maintaining constant composition at constant total flow.

*c.  Sensor Sensitivity, Stability, and Calibration*

Figure 14 demonstrates how the speed of sound varies with relative composition for some common gas pairs. The steep slope at low concentrations of a high-molecular-weight gas in a light carrier allows concentrations as small as 1 ppm to be measured. The sensitivity of these techniques is strongly influenced by the difference in mass between the species and the particular range of compositions. The technique is most sensitive for low concentrations (less than 5 mole percent) of a high-molecular-weight species in a light gas. Even when the molecular weight differences are small, e.g., $O_3$ in $O_2$ or $N_2$ in Ar, it is generally easy to discern compositions differing by 0.1% or less for all concentrations.

Acoustic analysis is stable and highly reproducible over long periods of time. Reproducibility can be further improved with daily calibration. Calibration is simple if the installation of the sensor permits pure carrier gas to flow through the sensor. Calibration is the renormalization of the instrument's effective path length, $L$. This is easily accomplished at the point of installation by measuring a known pure gas, generally the carrier gas. The calibration process is brief, requiring only that the sensor and its supply lines are sufficiently flushed to dilute any remaining precursor so that it causes no measurable effect on the speed of sound. This condition is readily observable because the speed of sound will asymptotically stabilize once the other gases are sufficiently flushed.

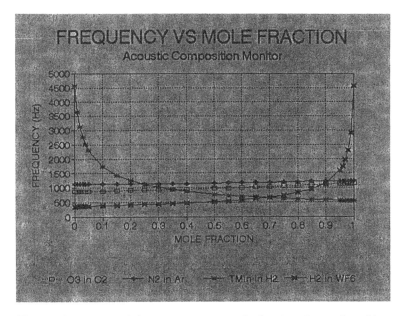

**Figure 14**  Graph of frequency versus mole fraction for various binary gas mixtures. (From Ref. 37.)

Lower temperatures and higher pressures are more easily measured with acoustic techniques, because acoustic transmission is dependent on media density. At low gas density it is difficult to transmit sufficient sound intensity to overcome the parasitic sounds transmitted through the sensor structure itself or to distinguish the signal from the other sounds in the environment. These problems have presently limited the successful operation of this technology to situations where the supply gas already exceeds or can be compressed to pressures over 70 torr.

*d.   Sensor Integration*

Integration of this sensor with the reactor may be either analog or digitally based. These sensors' electronics implementation is dominantly digital, but the composition signal is also available on a precision analog output that allows the concentration to be interpreted by precise analog readout. The preferred interfacing is digital, for the transducer's inherent precision and dynamic range exceeds that normally carried as an analog process control signal. A digital interface allows a graphic display to be used to advantage to uncover process flaws such as improper reactor gas-line switching and to readily view the concentration instabilities that indicate source depletion. It is also straightforward to save a detailed record of each process run when utilizing a digital interface.

The future direction for the development of this instrumentation is to enhance operation from the present operating limit of 65°C and allow its use at temperatures in excess of 120°C. Another productive path will be to learn how to apply these simple and robust transducers in networked arrangements. It is felt that they will have measurement capabilities beyond the present two-gas limit and may reliably provide information on reactor efficiency and perhaps, by inference, on wafer properties such as the rate of film growth.

## C.   Radio Frequency Properties

A significant number of etch and deposition tools are RF powered. For these tools, the process state includes the characteristics of the RF conditions, such as the delivered power or the current, voltage, and phase information—at the fundamental frequency and the higher harmonics. The major benefit of RF sensors is that they are readily available, relatively inexpensive, and readily couple into the RF circuit. Their major drawback is that the accuracy of the measured parameters is a complex function of the placement and impedance matching of the sensor in the RF path. Specifically, any RF sensor of fixed impedance, placed between the matching network and the dynamically changing impedance of the plasma chamber, will generate measurements of questionable accuracy. Calibrating the sensor for variable load impedances can mitigate this issue. But viewed pragmatically, this complexity is the reason why commercial plasma tools are not yet controlled by postmatch RF sensors. If accuracy is not the primary requirement, these sensors can generate useful data for endpoint determination or fault detection.

### 1.   Sensor Technologies

An RF sensor (42) is a device that produces output signal(s) that are of a definite and defined relationship to the electrical energy present in or passing through the sensor. To allow for the placement of sensing elements into controlled and reproducible electromagnetic field conditions, RF sensors are typically designed and built around transmission-line structures.

Minimum system disruption by the RF sensor is important to guarantee that the load seen by the source network is the same with or without the RF sensor. In other words, the measurement device should not significantly change the load it is trying to measure. A typical RF sensor will produce the following insertion disruptions:

1. Capacitance to ground
2. Series inductance
3. Series resistance or insertion loss

A small capacitance to ground is essential in preventing the sensor from increasing the reactance of the load as seen by the source network. The series inductance is generally designed in combination with the capacitance to ground to produce the characteristic operating impedance of the sensor (usually 50 ohms for RF applications). A small series resistance allows the sensor to have low insertion loss (i.e., dissipating power in the sensor instead of the load). Having a small value for the series resistance is crucial to maintaining a high $Q$ (quality factor) of the load network and allowing for high system efficiencies. The following describe the two major types of RF sensors: directional couplers and voltage/current sensors.

The directional coupler is the most common type of RF sensor. It is generally used to measure the forward and reverse power at the generator output by directionally sensing the RF components at the sensor. These values are generally accurate, because the 50-ohm sensor is connected to the stable 50-ohm input to the matching network. Commercially available directional couplers are typically rated in terms of the forward and reverse power the sensing element can withstand and usually have a specific coupling coefficient (e.g., $-30$ dB) over a specified frequency bandwidth and a characteristic impedance (typically 50 ohms). Directional couplers are available from a number of vendors (43,44).

Voltage/current sensors are the second most common type of RF sensor. This sensor's operation relies upon the electrical principles of capacitance and mutual inductance. A capacitor is formed when two conductors are placed in parallel to one another and separated by a certain distance. If the output of the capacitor is connected to a suitable shaping network, the voltage drop across the created capacitor can be controlled to produce an output signal with repeatable attenuation. The ideal capacitive structure, when built into a transmission line, is as wide and as short as possible. This allows for maximum voltage coupling (by maximizing the capacitance) and minimum current coupling (by minimizing the mutual inductance) into the sensor device network. A mutual inductor is easily formed when two conductors are placed in parallel with each other. An ac current traveling in one conductor will produce an ac current in the other conductor traveling $180°$ out of phase with it. The ideal inductive structure, when built into a transmission line, is as long and as thin as possible. This allows for maximum current coupling (by maximizing the mutual inductance) and minimum voltage coupling (by minimizing the capacitance) into the sensor device network. It is important to be aware of a possible current contamination in the voltage sensor and voltage contamination in the current sensor. If this occurs, the dynamic impedance range and maximum sensor accuracy will be sacrificed.

One important thing to recognize when using voltage and current sensors is that each independent sensor must look at the same position on the coaxial line. Also consider that the sensor and the associated cables and electronics have been specifically calibrated as a unit; hence no component can be arbitrarily changed without recalibration. If these rules are not followed, the standing-wave ratio seen by each sensor will be different, allowing for errors in the produced signals.

Voltage and current sensors are typically rated in terms of how many amps or volts the sensor can tolerate. This rating is similar to the transformer rating specified in terms of $VA$ (volts times amps). Also commercially available are $VI$ sensors (45–48).

## 2. Measurement Technologies

A measurement technology is necessary to process the outputs of the RF sensor. In the past, measurement techniques have been typically analog-based signal processing. Since the advent of the digital signal processor (DSP), more and more measurement techniques have migrated to the digital world. For any type of measurement technique to perform well, it must have the following minimum characteristics:

Reproducible results—stable vs. time and environmental conditions
Wide frequency range
Wide sensitivity range
Impedance-independent accuracy
$\pm 180°$ phase measurement capability
Flexible calibration and calculation algorithms

Having a measurement technique with reproducible results is a must for any sensor system. Day-to-day reproducibility allows for maximum reliability of the sensor, while unit-to-unit reproducibility allows for data interpretation to be consistent for each unit purchased. An excellent unit-to-unit reproducibility is absolutely necessary if a sensor system is to be used in a manufacturing environment. Inherent in reproducibility is low drift. Low drift over time in a sensor system's readings is necessary for day-to-day and measurement-to-measurement reproducibility. Also, because of the large temperature ranges produced by many of the new plasma processes, low temperature drift is necessary to maintain maximum accuracy.

Many single-frequency sensor systems are available on the market today, but a sensor system with a measurement technology that performs over a wide frequency range allows the user to look at harmonics (for single-frequency processes) and mixing products (for multiple-frequency processes) without incurring additional cost. Hence, a sensor system with a wide frequency range has the lowest cost of ownership.

Especially if the sensor is used over a wide frequency range, a wide range of sensitivity is required. The magnitudes of the signals at the fundamental vs. the upper harmonics can be significantly different, hence requiring a large dynamic range in the sensor sensitivity.

Some sensor systems have accuracy specifications that depend upon the impedance of the load. For maximum reproducible accuracy, a sensor system that uses a measurement technology with impedance-independent accuracy must be employed. The most important values to be measured are the fundamental electrical parameters of $|V|$, $|I|$, and $\angle Z$ (the phase angle of the load or the phase angle between the voltage and the current). These three parameters are the building blocks of all other electrical parameters (such as power, impedance, reflection coefficient). Some sensor system vendors specify their accuracy in terms of nonelemental parameters; in this case a little algebra is necessary to transform the specifications to the elemental parameters.

Passive loads (formed with capacitors, inductors, and resistors) can result in impedance phase angles only in the $\pm 90°$ range, while active loads can produce any phase angle over the $\pm 180°$ range. Due to the complicated physical processes that govern electron and ion transport in a plasma, the resulting electrical impedance produced by the plasma is active. Hence, to allow for proper measurement of a plasma load, the sensor system must be capable of determining phase angles in the $\pm180°$ range—all possible phase angles. Another consideration for a sensor system is its upgrade path. Typical analog techniques process the sensor signals with circuitry. Due to the fact that any technology improvement requires circuit redesign, analog processing does not allow for a low-cost upgrade path for technology improvements. Hence, the lowest cost of ownership in a sensor design is achieved with a digital technique that allows for signal processing upgrades with new versions of software.

## 3. Signal Processing

Once the sensor signal is obtained (see Sec. II.C.2), it has to be processed to derive the parameters of interest. In some cases, signal processing requires the down-conversion of the RF signals to a lower frequency that is more easily digitized. Once in the digital domain, DSP algorithms provide a very efficient and flexible way to process these sensor signals. In contrast to available analog signal-processing methods, digital signal processing is done completely with software, not hardware. Hence, the flexibility of calculation and calibration algorithms is very high. Any improvements to sensor or calculation technology can be implemented in software, drastically reducing the design cycle for improvements in the signal-processing technology. Another important advantage of having a DSP-based embedded system in the design is completely self-contained operation. Additional hardware is not necessary to support operation of the unit because all calibration information can be stored in DSP nonvolatile memory. In addition, the DSP can allow for user-selectable high-speed filtering of data.

An RF sensor system should be able to extract the following data at the frequency of interest:

| | | |
|---|---|---|
| $\lvert V \rvert$ | RMS voltage | volts |
| $\lvert I \rvert$ | RMS current | amps |
| $\lvert Z \rvert$ | Impedance magnitude of load | watts |
| $\theta$ | Phase Angle of Load | degrees or radians |
| $P_D$ | Delivered (load) power | watts |
| $P_F$ | Forward power | watts |
| $P_R$ | Reverse power | watts |
| $P_{RE}$ | Reactive (imaginary) power | var |
| $\Gamma$ | Reflection coefficient | no unit |

Due to the mathematical relationships among these nine parameters, the RF sensor system must be able to directly measure three of the nine parameters to properly calculate the remaining six. The accuracy with which each of these three fundamental parameters is measured determines the accuracy to which the other six parameters can be calculated. A broadband RF sensor system will allow the user to extract data at harmonics to more thoroughly characterize the behavior of the RF plasma and RF system.

## 4. Sensor Installation and Use

The two typical installations of an RF sensor system are shown in Figures 15 and 16. As demonstrated, the RF sensor can be mounted either before or after the matching network. One thing to realize is that any such 50-ohm sensor will perturb the $V/I/\theta$ values that existed in a non-50-ohm path without the sensor in place. Impedance mismatch between

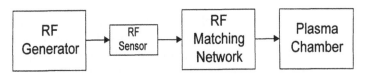

**Figure 15**   RF sensor mounting, prematch. (From Ref. 42.)

the sensor and the point where it is inserted will generate RF reflections, thereby influencing the RF environment. This does not negate their utility, but one needs to consider that the measurement itself changes the RF environment. The second issue is that, whether the sensor is located pre- or postmatch, it reads the instantaneous $V/I/\theta$ values at that point along the transmission path. These values are indicative of the standing wave characteristics at that point in the transmission path. However, these values will be influenced by the plasma properties, which is the primary reason for the use of these sensors for endpoint or fault detection. The changing impedance of the plasma creates changes in the standing wave characteristics along the transmission path, most dramatically between the tuner and the plasma. Hence these sensors, located either pre- or postmatch, will see changes in the plasma. One benefit for locating sensors prematch is the relative ease of mounting the sensor with standard coaxial coupling, assuming that a useful signal can be obtained in this location.

The analysis and interpretation of the available sensor data requires that one comprehend that the sensor measures the instantaneous $V/I/\theta$ values at one specific location in the RF transmission path. What happens at another location (namely, the plasma) can be inferred by correlation (i.e., a change in the standard measured values) or by means of a full RF-circuit model. Such models are generally very difficult to generate; hence the majority of the RF sensor data analysis is performed by the simpler correlative method.

## 5.  Applications of a Radio Frequency Sensor System

In spite of the previously described limitations, RF sensors can be gainfully utilized in a number of applications. In some cases, they are relatively "easy" and inexpensive add-on sensors, and they have shown benefits in applications where accuracy is not a key parameter (as long as sensor reproducibility persists). The following sections describe examples of these applications.

### a.  Etching Endpoint Detection, Deposition Thickness

For this application, the RF sensor system can be mounted before or after the RF matching network. Even in a prematch location, an RF sensor system with enough sensitivity can detect the small variation in plasma impedance that depicts an etching endpoint or accompanies a deposition process. Using $VI$ sensors in a prematch configuration on a

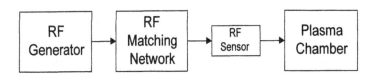

**Figure 16**   RF sensor mounting, postmatch. (From Ref. 42.)

plasma etcher, endpoint signals have been seen in the higher harmonics (49). For an oxide deposition, a capacitance change will also be seen by the RF sensor. The value of the observed capacitance could be correlated to the thickness of the deposited film. Similar information can be obtained from sensors placed postmatch.

*b. Harmonic Signature Analysis*

Due to the nonlinear characteristics of the plasma, the pure sinewave from the RF generator will be turned into a harmonic-rich waveform by the plasma. The number of RF harmonics present, as well as the characteristics of each, will depend on the plasma chamber geometry, the type of energy source for the plasma chamber, and the type of process being run.

Proper use of this technique would create a "harmonic fingerprint" of the process when it is running well. Future process fingerprints would be compared, probably by a multivariate numerical technique, to the master fingerprint at regular intervals. Any significant change between the two would indicate a process shift, allowing the chamber to be taken off-line before a complete lot of wafers is destroyed. If enough data is taken, a database could be created allowing proper interpretation of the bad harmonic fingerprint. Examples of anomalies expected to be found by this technique are a broken wafer, a misplaced wafer, and a dirty chamber (i.e., chamber cleaning required).

It is wise not to limit the harmonic range. Experiments have indicated (49) that higher harmonics (tenth through the twentieth) can contain stronger process correlation than lower harmonics as well as different information. In a particular RF investigation on a cluster-tool polysilicon etcher chamber B, it was found that the seventh harmonic, or 94.92 MHz, had a good etch endpoint trace (EPT). Looking at the higher harmonics, the thirteenth harmonic, or 176.28 MHz, showed a strong reaction to chamber D's RF power cycle. This indicated that the thirteenth harmonic should not be used to characterize chamber B. At the sixteenth harmonic, or 216.96 MHz, an endpoint signal was found with much better EPT characteristics than the seventh harmonic. No other harmonics, up to the twentieth, produced any usable EPT information.

*c. Measurement of Power Delivered to Plasma*

The more difficult application is the accurate measurement and control of power delivered to the plasma. A typical RF system will regulate the output power of the RF generator to very high accuracy. Unfortunately, every RF delivery system has losses. In most cases, the losses change as a function of generator power due to plasma impedance changes. Also, the losses in an RF delivery system may increase as the system ages (e.g., wear of the mechanical components in the tuner). This means that the actual power delivered to the plasma is always less than the output power of the generator and may change from wafer to wafer and lot to lot. *A properly designed and calibrated* RF sensor connected between the matching network and the plasma chamber allows for measurement of power delivered to the plasma. With valid measurement of the true delivered RF power, the RF sensor system can be used to compensate for the losses just described and decrease wafer-to-wafer processing variations. This approach could provide a better tuning algorithm by using the impedance information from the RF sensor to correctly set the capacitor values of the RF matching network, although additional feedback will be required to prevent excessive reflected power from damaging the generator. This all leads toward the implementation of a feedback loop from the postmatch sensor to the RF generator. The goal is to provide more consistent and accurate RF power delivered to the plasma in commercial plasma tools.

## D. Wall-Deposition Sensor

Process-state sensors have classically focused on determining the species concentrations, pressure, gas flow, and RF power characteristics in the processing environment during the processing of each wafer. In the recent quest for more control over the reproducibility of processing tools, attention has been focused on the deposits generated on the internal surfaces of processing tools. Such deposits are formed in both deposition and etch tools, typically at a greater rate in deposition systems. These deposits have several detrimental effects:

> They provide a slowly changing chamber-wall state, which can influence the process chemistry.
> At some point, the deposit can start to flake off the walls and potentially contaminate the wafer.

The last point drives the mechanical and plasma cleaning of tools, in hopes of preventing this source of particle contamination. Without an appropriate sensor, the frequency of these cleaning cycles is based on empirical rules, with an incorrect guess risking either wafer contamination or expensive, unnecessary downtime for tool cleaning.

A sensor has recently been developed (17) for real-time, noninvasive monitoring of chamber wall deposits in etch and deposition process tools. This sensor (50) can be used in an R&D or production environment for optimizing clean cycles and reducing particle contamination. It operates on the principle of acoustic reflectometry (Figure 17). A piezoelectric transducer is attached to the outer wall surface of a process chamber. A short electric pulse applied to the electrodes of the piezoelectric transducer excites an acoustic wave that propagates from the outer wall of the chamber toward the inner wall. If the inner wall is bare (with no deposits), the acoustic wave is reflected as an echo from the boundary between the inner wall and the process environment. This echo propagates to the outer wall, where the transducer converts it into a detectable electrical signal. The transit time of the roundtrip by the acoustic wave is the fundamental measurement. When a film is deposited on the inner wall, any change in the thickness of the deposit causes a

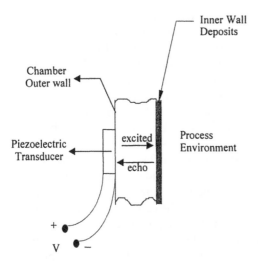

**Figure 17** Cross-sectional view of a chamber with the piezoelectric transducer attached to the outer wall.

proportional change in the transit time. A change in temperature also changes the transit time, but in a very predictable manner. Hence the sensor acoustically monitors, in real time, changes in the average temperature along the cross section of the chamber wall, with a real-time temperature compensation applied to the measurement just described. The sensor system consists of a personal computer that houses the electronics and the transducer module. The transducer module is attached to the outer wall (or window) at the chosen location on the process chamber. The transducer module is a cylinder 2 in. in diameter and 2 in. in height. The primary use of this sensor is for determining and optimizing the chemistry, frequency, and duration of clean cycles for etch and deposition tools.

## III.  WAFER-STATE SENSORS

As stated previously, process-state sensors have been used predominantly for endpoint determination and fault detection and, in some recent cases, for dynamic process control. But clearly, wafer-state sensors provide more direct information for all these tasks. Such wafer-state sensors are slowly being integrated into processing tools, paced by issues of customer pull, sensor reliability, and cost of integration. The following is a description of the wafer-state sensors that have been, or are currently, overcoming these barriers and are being integrated into OEM tools.

### A.  Film Thickness and Uniformity

The thickness of optically transparent thin films (silicon, dielectrics, resists) on a reflective substrate is measured via the analysis of the interaction of electromagnetic radiation with such a film or film stack. These methods rely on single-wavelength (laser) or spectral (white light) sources, impinging on the sample at normal incidence (interferometry and reflectometry) or at some angle off-normal (reflectometry, ellipsometry). The wavelength range is from the UV through the IR. The interaction of the light with the material can be detected through a polarization change (ellipsometry), a change in the phase (interferometry), or a change in the reflected amplitude (reflectometry). Optical models are used to extract the different physical parameters of the films (e.g., thickness) from the known optical indices of the individual layers. These techniques are well-established standard methods for off-line film thickness measurement, and hence the methods will be described only briefly. The emphasis will be on the deployment of these techniques as sensors in OEM tools.

### 1.  Optical Sensors

The spectral reflectivity of transparent thin films on reflective substrate materials is modulated by optical interference. The effect of the interference on the measured spectrum is a function of the film and substrate refractive indices. If the dispersion components of the refractive indices are known over the wavelength range, the thickness of the surface film can be found using a Fourier transform technique. For thin layers ($< 100$ nm), the method of spectral fitting is very effective. Once the film thickness has been found, a theoretical reflectance spectrum can be determined and superimposed on the measured spectrum. This ensures a very high level of reliability for the film thickness measurement.

### a.   Reflectometry Technique

*Theory of operation.*   The thickness of films on a silicon wafer is measured by means of spectrophotometry, utilizing the theory of interference in thin films (51). The basic procedure is to measure the spectral reflectance of the desired sample. The spectral data is then interpreted to determine the thickness of the top layer of the measured stack. The actual reflectance $R_{act}(\lambda)$ is measured and fitted to $R_{theor}(\lambda)$ to find the thickness ($d$) of the last layer. $R_{theor}(\lambda)$ is calculated according to the specific optical model of the measured stack. The "goodness of fit" parameter measures the difference between the measured and the theoretical results and is used as a criterion of correct interpretation. Figure 18 shows a graph of $R_{theor}(\lambda)$ for a layer of 10,000-Å $SiO_2$ on Si substrate. The fitting algorithm used for data processing has to treat several issues, such as: spectral calibration, noise filtering, recognition of characterizing points (minima, maxima, etc.), and calculating a first-order approximation for the thickness and the final fine fitting.

*Optical overview.*   The optical path of one specific reflectometer (52) is shown in Figure 19. In this case, the specular reflection is monitored at the incident angle normal to the wafer surface, and the radiation source is in the visible range. Briefly, the light emitted from the lamp (11) travels through an optical fiber (10) until reaching a condenser lens (9). The light beam then reaches a beam splitter (3), where it is split; half of the light passes through the beam splitter, while the other half is reflected downwards, focused by a tube lens (2) and an objective lens (1) onto the target (wafer). After being reflected by the target, the light beam travels back through the objective lens (1), tube lens (2), and beam splitter (3) until it reaches a "pinhole" mirror (4). From there, the light is sent in two directions:

1.   A portion of the light (the image of the wafer surface) is reflected from the "pinhole" mirror (4), focused by a relay lens (5) onto a CCD camera (8), where it is processed and sent to the monitor for viewing by the operator.
2.   The light that passes through the "pinhole" is also focused by a relay lens (6) and then reflected by a flat mirror toward the spectrophotometer (7), which measures the spectrum of the desired point. This information is then digitized and processed by the computer for the computation of film thickness.

This spectrophotometer also includes an autofocusing sensor for dynamic focusing on the wafer surface during the movement of the optical head over the wafer.

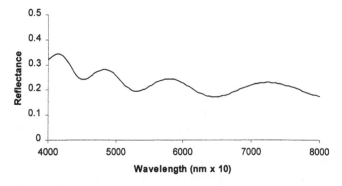

**Figure 18**   Reflectance of $SiO_2$ on Si in water. (From Ref. 51.)

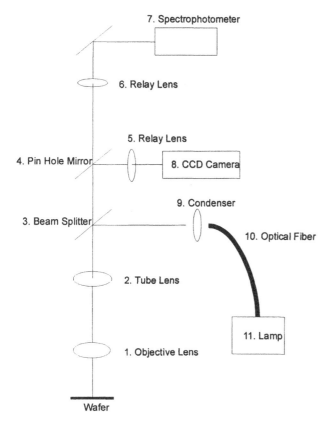

**Figure 19** Optical path of light beam in NovaScan 210. (From Ref. 51.)

*System integration, in-line measurement.* While this chapter is focused on in situ metrology, there are some well-established in-line measurement techniques that are worth including because they provide routine and useful information for APC (specifically, wafer-to-wafer control). Two embodiments of in-line reflectometry for film thickness measurements will be described in this section, one for use in CMP and the other for epi film growth.

Reflectometry is used to monitor and control film thickness in CMP operations. When CMP is used to planarize and remove part of a blanket film, such as in oxide CMP, there is no detectable endpoint, since no new films are exposed. The only way to monitor and control such a process is by means of a sensor that measures the thickness of the film. This is a very difficult task for a slurry-covered wafer that is in motion, hence the measurement is performed in-line in the rinse station of the CMP tool.

A commercially available reflectometry-based sensor (e.g., Ref. 52) is currently being used for CMP tool monitoring. Its primary benefits are as follows:

Provides thickness measurement data for every product wafer, required for rapid feedback control of the CMP process

Performs measurements in parallel to the processing of the next wafer, hence not affecting system throughput unless a very large number of measurements is required

In-water measurement capability obviates need to clean and dry wafers before measurements

Eliminates additional cleanroom space and labor required for off-line measurements

Only one component, the measurement unit, has to be integrated into the polisher. The compact size of this unit, with a footprint only ~40% larger than the wafer, enables easy integration into the process equipment. Two such implementations in commercial CMP tools are represented in Figures 20 and 21.

Two different delivery system principles are applied for the integration of the measurement system into OEM tools. In one case (Fig. 20) the wafer handler transfers wafers down from the wafer-loading station to the water tub of the measuring unit and back. In another configuration (Fig. 21), the measurement unit replaces the unload water track of the polisher. It receives the wafer, performs the measurement process, and delivers the wafer to the unload cassette. In both cases, the wafer is wet during the measurement.

A second commercially available implementation of reflectometry (in this case using an IR source and nonnormal incidence) is the use of FTIR measurement of epi thickness. The in-line measurement of epi thickness has been achieved by the integration of a compact FTIR spectrometer (53) to an Applied Materials Epi Centura cluster tool, as shown in Figure 22. The cool-down chamber top plate is modified to install a $CaF_2$ IR-transparent window, and the FTIR and transfer optics are bolted to the top plate. The IR beam from the FTIR is focused to a 5-mm spot on the wafer surface, and the specular reflection is collected and focused onto a thermoelectrically cooled mercury cadmium telluride (MCT) detector. Reflectance spectra can be collected in less than 1 s. Reference spectra are obtained using a bare silicon wafer surface mounted within the cool-down chamber. Epi thickness measurements are made after processing, while the wafers are temporarily parked in the cluster tool's cool-down chamber, without interrupting or delaying the wafer flow.

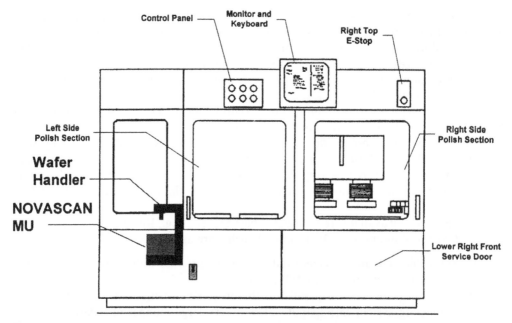

**Figure 20**  NovaScan system integrated in Strasbaugh model 6DS-SP planarizer. (From Ref. 51.)

**Figure 21**   NovaScan in IPEC 372M and 472 polisher. (From Ref. 51.)


A simulated reflectance spectrum is computed from parametric models for the doping profile, the dielectric functions (DFs) of the epi film and substrate, and a multilayer reflectance model. The models for the wavelength-dependent complex DFs include dispersion and absorption due to free carriers, phonons, impurities, and interband transitions. The models are tailored to the unique optical and electronic properties of each material. The reflectance model computes the infrared reflectance of films with multilayered and graded compositional profiles using a transfer matrix formalism (54,55). The model parameters are iteratively adjusted to fit the measured spectrum.

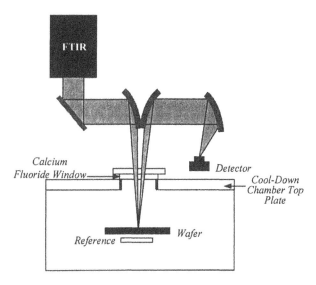

**Figure 22**   Configuration of On-Line Technologies, Inc., FTIR on Applied Materials' Centura 5200. (From Ref. 52.)

Gauge tests demonstrate the relative accuracy of this first-principle analysis of epi layer thickness to be in the range of 0.5–2 nm (5–20 Å). Comparison to destructive SIMS and SRP measurements shows the absolute accuracy to be within the accuracy of these standard measurements.

*b.   Interferometric Technique*

*Theory of operation.*   Interferometry is a well-established technique for the optical measurement of thin, optically transparent films. Some of the light impinging on such a thin film reflects from the top of the film and some from the bottom of the film. The light reflected from the bottom travels farther, and the difference in path length results in a difference in phase. After reflection, the light following the two paths recombines and interferes, with the resulting light intensity a periodic function of the film thickness. The change in film thickness for one interferometric cycle is $\lambda/2n\cos\theta$, where $\lambda$ is the observation wavelength, *n is the index of refraction of the film, and* $\theta$ *is the angle of refraction within the film.*

*Full-wafer imaging sensor.*   The full-wafer imaging (FWI) sensor (56) is a novel sensor developed in the early 1990s (57) based on this interferometric technique. It uses an imaging detector to make spatially resolved measurements of the light reflected from the wafer surface during etching or deposition processes. This sensor takes advantage of the fact that the reflectivity of a thin film on the wafer surface is generally a function of the thickness of the film. By quantifying the changes in reflectivity as the film thickness changes, the FWI sensor determines spatially resolved etching or deposition rate, rate uniformity, spatially resolved endpoint, endpoint uniformity, and selectivity. These measurements are performed on every wafer, providing both real-time endpoint and run-by-run data for process monitoring and control.

The operation of this particular sensor relies on a number of optical phenomena:

*Optical emission*: Optical emission from the plasma is the preferred light source for FWI sensors because it is simpler than using an external light source and it allows direct detection of optical emission endpoint. If plasma light is not available, an external light source can be added. A narrow-bandpass filter is used to select the measurement wavelength. Different wavelengths are best suited to different types of process conditions, the most important characteristics being the intensity of the plasma optical emission as a function of wavelength, the film thickness, and the film's index of refraction. In general, a shorter wavelength gives better rate resolution, but cannot be used in certain situations; e.g., a 0.3-μm-thick layer of amorphous silicon is typically opaque in the blue but transparent in the red.

*Interferometry for transparent thin films*: In practice, during an etching or deposition process, the intensity of light reflected from the wafer surface varies periodically in time. The interferometric signal is nearly periodic in time in most processes because the process rate is nearly constant, even though the signal is strictly periodic in film thickness rather than in time.

*Interferometry for trench etching*: In trench etching, the interference is between light reflected from the top of the substrate or mask and light reflected from the bottom of the trench. A coherent light source, e.g., a laser, must be used because the interference is between two spatially distinct positions. Etching rate is calculated using the same types of techniques discussed earlier for thin-film interferometry. Endpoint time is predicted by dividing the desired trench depth by the measured etching rate.

*Reflectometry for nontransparent films*: Light impinging on a nontransparent film reflects only from the top of the film, so there is no interference. However, the reflectivity of the nontransparent film that is being etched is different from the reflectivity of the underlying material. Thus the intensity of reflected light changes at endpoint. This method is typically applied to endpoint detection in metal etching.

From a system viewpoint, the FWI sensor requires a high data-acquisition rate and uses computationally intensive analyses. So the typical configuration consists of a high-end PC, advanced software, and one or more independent CCD-based sensor heads interfaced to the computer via the PCI bus. Each sensor head records images of a wafer during processing, with each of the few hundred thousand pixels of the CCD acting as an independent detector. The full images provide visual information about the wafer and the process, while the signals from thousands of detectors provide quantitative determination of endpoint, etching or deposition rate, and uniformity. The simultaneous use of thousands of independent detectors greatly enhances accuracy and reliability through the use of statistical methods. The FWI sensor can be connected to the sensor bus by adding a card to the PC. Connecting the sensor head directly to the sensor bus is not practical, due to the high data rate and large amount of computation.

Figure 23 shows a schematic diagram of the FWI sensor head installation. The sensor head is mounted directly onto a semiconductor etching or deposition tool on a window that provides a view of the wafer during processing. A top-down view is not necessary, but mounting the sensor nearly parallel to the wafer surface is undesirable because it greatly reduces spatial resolution, one of the technique's principal benefits. For both interferometry and reflectometry, spatially resolved results are determined by applying the same calculation method to hundreds or thousands of locations distributed across the wafer surface. These results are used to generate full-wafer maps and/or to calculate statistics for the entire wafer, such as average and uniformity.

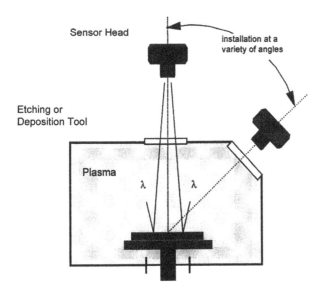

**Figure 23** Full-wafer imaging sensor head mounting. (From Ref. 56.)

Several methods can be used to find the etching or deposition rate from the periodic interference signal. The simplest way is to count peaks, but this is accurate only if there is a large number of interferometric cycles, which is not common in most semiconductor processes. For example, a 0.3-μm-thick layer of polysilicon contains only 3.8 interferometric cycles. The accuracy of simple peak counting is one-half of a cycle, which is only 13% in this example. The accuracy can be improved somewhat by interpolating between peaks, but the accuracy is still fairly low. In addition, false peaks caused by noise in the signal often plague peak-counting methods.

A more accurate way to determine rate is to multiply the change in film thickness per interferometric cycle by the frequency (number of cycles per second). The simplest way to find the desired frequency is to use a fast Fourier transform (FFT) to convert from the time domain to the frequency domain. A local maximum in the signal versus frequency then specifies the frequency to be used in the rate calculation. Accuracy can be further increased by starting with an FFT to provide an initial guess of the frequency and then fitting the signal versus time to an empirical function that models the physical signal. This combined method is more accurate than the FFT alone if there are few cycles, if the interferometric signal is not a pure sinewave, or if the signal-to-noise ratio is low—all of which occur commonly in semiconductor processing. In either method, a frequency window can be used to designate which maximum in the FFT is used to calculate rate. This is a useful way to measure selectivity in etching processes where two materials, e.g., the mask and the film of interest, are etching simultaneously.

For transparent thin films, endpoint can be detected or predicted. The detection method relies on the fact that the periodic modulation of reflected light intensity ceases at endpoint. Endpoint is found by detecting the deviation of the observed signal from an interferometric model. The prediction method uses the measured rate and the desired thickness change to predict the endpoint time. Prediction is the only available endpoint method for deposition processes. It is also necessary in those etching processes where the film is not completely removed.

The FWI technique has been used on devices with feature sizes down to 0.1 μm, aspect ratios up to 50:1, percent open area as low as 5%, film thickness greater than 2 μm, and substrate sizes larger than 300 mm. Endpoint, etching or deposition rate, and uniformity can be monitored for a variety of transparent thin films, including: polysilicon, amorphous silicon, silicon dioxide, silicon nitride, and photoresist. For nontransparent materials, such as aluminum, tungsten silicide, chrome, and tantalum, rate cannot be measured directly, but spatially resolved etching endpoint and thus endpoint uniformity have been determined.

Examples of the use of FWI are shown in the following figures. Figure 24 is an example of the signals from three different CCD pixels recorded during a polysilicon gate etching process. Each pixel imaged a small, distinct region; an image of the wafer is included to indicate the position of these pixels. Two of the pixels are on the wafer and display a periodic signal due to the change in thickness of the thin film. The pixel off the wafer shows the optical emission signal, which rises at endpoint. Analysis of the periodic signal is used to determine rate and/or endpoint, while analysis of the optical emission signal is used independently to detect the average endpoint time for the entire wafer.

Figure 25 is an example of a full-wafer etching-rate-uniformity surface plot. The plot was generated from rate calculations at 4000 locations on a rectangular grid covering the wafer. Two trends are evident. First, the etching rate at the center of the wafer is lower than at the edge. Second, variations within each die are visible as a regular array of peaks and valleys in the etching-rate surface plot. The deepest of these valleys go all the way to

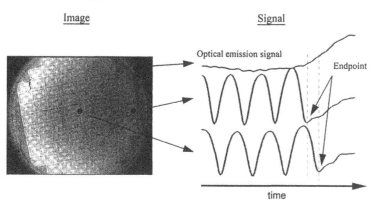

**Figure 24** Signal from three positions: two on the wafer and one off the wafer. (From Ref. 56.)

zero and correspond to areas of pure photoresist mask, which did not etch appreciably in this high-selectivity process.

Figure 26 is an example where an FWI sensor was used to automatically monitor every product wafer. Results for each wafer were determined and displayed while the next wafer was being loaded into the processing chamber. The figure shows the etching rate and uniformity for four consecutive product wafer lots. The process was stable (no large fluctuations in rate or uniformity) but not very uniform (7%, 1$\sigma$). Furthermore, pattern-dependent etching is clearly evident. At the beginning of each lot, several bare silicon warm-up wafers and one blanket (not patterned) wafer were run, and then the patterned product wafers were run. The blanket wafers etched about 10% slower and

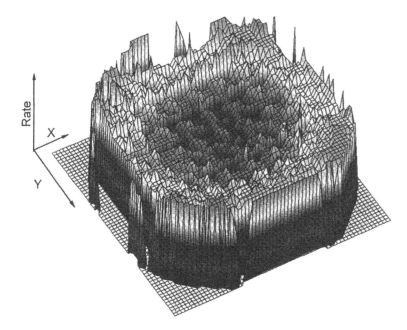

**Figure 25** Full-wafer etching-rate map. Average = 2765 Å/min, uniformity = 3.9% 1 − $\sigma$. (From Ref. 56.)

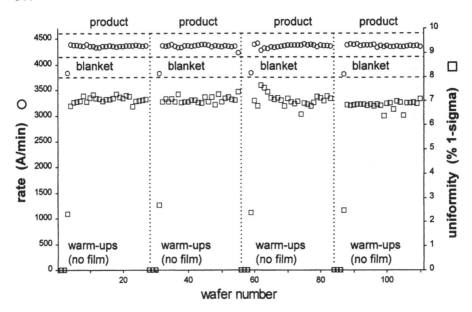

**Figure 26**  Rate and uniformity for four product wafer lots. (From Ref. 56.)

much more uniformly than the product wafers. The difference between the blanket and product wafers demonstrates the need to use real product wafers to monitor a process.

Sensor calibration has been achieved by a comparison between FWI sensors and ex situ film-thickness metrology instruments. The agreement is generally good, even though the two systems do not measure exactly the same thing. The FWI measures dynamic changes in film thickness, while the ex situ instruments measure static film thickness. It is typical to take the thickness-before minus thickness-after measured ex situ and divide this by the total processing time to get the ex situ rate and uniformity values that are compared to the rate and uniformity measured in situ by the FWI.

Integration to the processing tool is required to obtain the benefits provided by an FWI sensor. There are two main technical issues. First, a window that provides a view of the wafer during processing is required. Between wet cleans, this window must remain transparent enough that the wafer stays visible. Second, communication between the tool's software and the FWI sensor's software is useful to identify the process and wafer/lot and to synchronize data acquisition. Both of these technical needs must be met whether the FWI runs on a separate computer or on a subsystem of the tool controller.

The FWI sensor provides different benefits to different users. In R&D, it provides immediate feedback and detailed information that speeds up process or equipment development and process transfer from tool to tool. In IC production, every product wafer can be monitored by the FWI sensor so that each wafer serves as a test wafer for the next. This means that fewer test, monitor, qualification, and pilot wafers are required—a significant savings in a high-volume fab. Also, fewer wafers are destroyed before faults or excursions are detected, and data is provided for statistical process control (SPC) of the process.

## 2. Current Sensor for Film Removal in Chemical-Mechanical Polishing

There are two distinct measurement and control problems in CMP. The first is the measurement of the thickness of a layer during the blanket thinning of that layer by CMP. Since there is no endpoint to this process, the only way to control it is via a pre- and postmeasurement of the specific layer thickness. This has already been discussed in Sec. III.A.1.a. The second is the determination of the endpoint when polishing a metal layer on an oxide substrate. This is an easier problem that has been solved by monitoring the current to the motor that rotates the wafer carrier. Most systems rotate the carrier at constant RPM. This requires the current supplied to the motor to increase or decrease depending on the amount of drag. Fortunately, this drag changes fairly reproducibly as the system polishes through the metal and reaches the oxide interface. A variety of other factors also influence the total motor current, hence there is considerable noise to this signal. However, with proper signal conditioning the endpoint can be detected. In one such endpoint system (58), the motor current signal is amplified, and normalized, and high-frequency components are removed by digital filtering. Proprietary software is then used to call endpoint from the resultant trace.

Other film stacks, such as poly/oxide, poly/nitride, and oxide/nitride, may find this technique useful as well, but this will need to be tested and will likely require different signal-conditioning and endpoint software algorithms. The motor current signal also appears to be useful to diagnose other tool parameters. Correlation between deviations of "known good" data traces and specific tool problems may allow the user to diagnose tool failures and to signal preventative maintenance. This sensor typically comes integrated into the tool by the OEM supplier.

## 3. Photoacoustic Metrology

Section III.A.1 described optical methods for the measurement of optically transparent films. There is also a need for a simple measurement of metal film thickness. This section (59) describes an impulsive stimulated thermal scattering (ISTS) method for noncontact measurement of metal film thickness in semiconductor manufacturing and process control. The method, based on an all-optical photoacoustic technique, determines thickness and uniformity of exposed or buried metal films in multilayer stacks, with repeatability at the angstrom level. It can also be used to monitor chemical-mechanical polishing (CMP) processes and to profile thin metal films near the edge of a wafer. The method is being investigated for use in monitoring both the concentration and depth of ions, including low-energy low-dose boron ions implanted into silicon wafers. While currently this technology is implemented in an off-line tool, it has the potential to be developed as an in situ sensor for measuring properties of both metal films and ion-implanted wafers.

### a. Photoacoustic Measurement Technique

The photoacoustic measurement method used in this tool (60) is illustrated schematically in the inset to Figure 27. Two excitation laser pulses having a duration of about 500 picoseconds are overlapped at the sample to form an optical interference pattern containing alternating "light" (constructive interference) and "dark" (destructive interference) regions. Optical absorption of radiation in the light regions leads to sudden heating and thermal expansion (box 1). This launches acoustic waves whose wavelength and orientation match those of the interference pattern, resulting in a time-dependent surface "ripple" that oscillates at the acoustic wave frequency (61). A probe laser beam irradiates the surface ripple and is diffracted to form a signal beam that is modulated by the oscillating

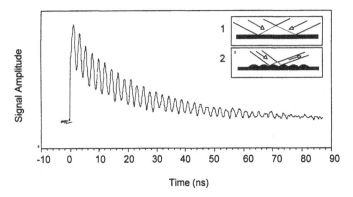

**Figure 27** Signal waveform measured from a 1-micron aluminum film. (From Ref. 59.)

surface ripple (box 2). (The displacement of the surface is grossly exaggerated for purposes of illustration.) The signal beam is then detected and digitized in real time, resulting in a signal waveform such as the one in Fig. 27. With this method, data is measured in real time with very high signal-to-noise ratios: the data shown was collected from a 1-micron aluminum film in about 1 second.

The acoustic wave that is excited and monitored in these measurements is a waveguide or "drumhead" mode whose velocity is a sensitive function of the film thickness. The film thickness is calculated from the measured acoustic frequency, the spatial period of the interference pattern (i.e., the acoustic wavelength), and the mechanical properties (i.e., density and sound velocity) of the sample. The thickness determined in this manner correlates directly to traditional techniques, such as four-point probe measurement and SEM thickness determination. Moreover, the acoustic wavelength that is excited in the film can be rapidly changed in an automated fashion. Data collected at several different acoustic wavelengths can be used to determine sample properties in addition to film thickness. In particular, thermal diffusivities and the viscoelastic properties of the sample can be measured.

A modified form of the optical technique used to determine film thickness can be used to monitor the concentration of ions implanted in semiconducting materials. In this case, the waveform of the diffracted signal depends on the concentration and energy of the implanted ions. Ion concentration and depth can be separately determined from parameters of the measured signal.

### b. Hardware Configuration

The photoacoustic hardware is a small-scale optical system housed in a casting measuring approximately $50 \times 50 \times 10$ cm. The optical system uses two solid-state lasers: a Nd:YAG microchip laser generates the 500-picosecond excitation pulses, and a diode probe laser generates the probe beam that measures the surface ripple. A compact optical system delivers these beams to a sample with a working distance of 80 mm. The spot size for the measurement is $25 \times 100$ microns. For each laser pulse, the optical signal is converted by a fast photodetector to an electrical waveform that is digitized by a high-speed A/D converter. The digitized signal is further processed by a computer to extract the acoustic frequency and other waveform parameters. A thickness algorithm calculates the film thickness from the measured acoustic frequency, the selected acoustic wavelength, and the mechanical properties of the sample.

Integrated metrology requires in situ or in-line monitors that can attach directly to cluster tools and monitor film properties of a wafer in, or emerging from, the process chamber. This photoacoustic measurement technology fulfills many of the requirements for such a sensor. As already described, it is compact and fast, does not require moving parts, and has the long working distance necessary for optical measurement through a viewing port. While currently this technology exists as an off-line metrology tool, it is easily adapted as an in-line sensor. With appropriate optical access to the processing chamber, it is possible to evolve this methodology to an in situ sensor for measuring properties of both metal films and ion-implanted wafers.

### c. Applications

The principal application of this technology is for the measurement of metal film thickness in single-layer and multilayer structures. Figure 28 shows 49-point contour maps of a 5000-Å tungsten film deposited directly on silicon; the map on the left was measured nondestructively with the InSite 300 in about 1 minute, while the map on the right was measured destructively with a four-point probe in about 4 minutes. The contours of the maps are nearly identical, both showing thickness variations of about 500 Å across the surface of the film.

This tool can also measure the thickness of one or more layers in a multilayer structure, such as a 1000-Å TiW film buried beneath a 2000-Å aluminum film. In this case, the system is "tuned" to explicitly measure the relatively dense buried film (TiW has a density of about 13,000 $kg/m^3$, compared to 2700 $kg/m^3$ for aluminum). This tuning is done by first initiating a low-frequency acoustic wave that is sensitive to changes in the TiW thickness but relatively insensitive to changes in the aluminum thickness. This data is processed to generate the TiW contour map. The system then initiates a relatively high-frequency acoustic wave that is sensitive to the combined thickness changes in the TiW/aluminum structure. A contour map of the outer aluminum film can be generated from this combined data.

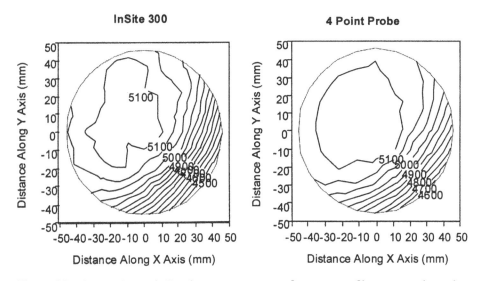

**Figure 28** Comparison of 49-point contour maps of a tungsten film measured nondestructively using the InSite 300 (left) and destructively using a four-point probe (right). (From Ref. 59.)

Full characterization of the uniformity of deposited metal films requires measurement to the edge of the film. This is particularly important for monitoring sputtering and CVD tools, which are often configured to deposit a film blanketing the entire wafer, except in an "edge-exclusion zone" along the outer perimeter of the wafer. For films deposited by a process with such an edge-exclusion zone, the thickness drops from its nominal value to zero within a few hundred microns from the wafer edge. It is important to verify that edge specifications are met, because devices bordering the edge-exclusion zone can represent close to 10% of the total number of devices on a 200-mm wafer or 7% for a 300-mm wafer. Besides the contact issue, probe techniques are limited in this regard by their probe spacing and electrical issues near the edge of the wafer. The small spot size used in this methodology makes it possible to profile this narrow edge-exclusion zone.

This technique has also been applied to the measurement of the thickness of an Al film during CMP. In one particular application, 49-point contour maps were generated from data measured prior to polishing and following 30-second intervals of polishing. Prior to polishing, the film had no distinct pattern. The CMP process imparted a "bull's-eye" contour to the film that is evident after about 60 s and becomes more pronounced as the polishing process continues. The data also indicates that the average removal rate is not constant, varying from ca. 60 angstroms/second during the first 30 seconds to ca. 120 angstroms/second in the final 30-second interval. Measurement at the center and edge of a wafer can be performed in a few seconds, making this approach attractive for real-time monitoring of CMP removal rates.

A variation of ISTS, called impulsive stimulated scattering (ISS), has been used to measure both the concentration and energy of ions implanted in silicon wafers. It is an optical technique that initiates and detects both electronic and acoustic responses in the implanted semiconductor lattice. In these measurements, the signal waveform shows features that vary with the concentration of the implanted ions and a separate parameter that varies with the depth of the implanted ions. Although results at this early phase are preliminary, ISS has effectively measured arsenic, phosphorous, and boron ions implanted at energies ranging from 3 keV to 3 MeV and concentrations ranging from $1e^{11}\,cm^{-2}$ to $1e^{16}\,cm^{-2}$. The method appears to be particularly effective for measuring shallow-junction boron implants made at low energies and/or low concentrations. It measures samples rapidly (less than 2 seconds) and remotely (12-cm working distance), making in situ measurements a real possibility.

## B.　Resist Profile

Measurement for the control of lithography has classically relied on off-line metrology techniques such as scanning electron microscopy (SEM), and more recently on atomic force microscopy (AFM). The SEMs are not applicable to in situ measurements. Due to its very small field of view and slow scan rates, AFM is also not likely to become even an in-line sensor for routine feature-size measurements. Scatterometry is the only current technology that is capable of evolving into an in-line sensor for feature-size measurements.

### 1.　Scatterometer

#### a.　Theory of Operation

Scatterometry, a complex optical technique for critical-dimension metrology, evolved from R&D work at the University of New Mexico (62). It provides critical-dimension information for a lithographic or etched pattern. Scatterometry is based (63) on the analysis of light scattered, or diffracted, from a periodic sample, such as resist lines in a grating. This light pattern, often referred to as a *signature*, can be used to identify the shape and spatial features of the scattering structure itself. For periodic patterns, the scattered light consists of distinct diffraction orders at angular locations specified by the grating equation:

$$\sin \theta_i + \sin \theta_n = \frac{n\lambda}{d} \tag{5}$$

where $\theta_i$ is the angle of incidence, $\theta_n$ is the angular location of the nth diffraction order, $\lambda$ is the wavelength of incident light, and $d$ is the spatial period (pitch) of the structure. The fraction of incident power diffracted into any order is very sensitive to the shape and dimensional parameters of the diffracting structure, and thus may be used to characterize that structure itself (64). In addition to the period of the structure, which can be determined quite easily, the thickness of the photoresist, the width of the resist line, and the thicknesses of several underlying film layers can also be measured by analyzing the scatter pattern. Figure 29 shows the basic sensor configuration for an early version of the bench-top "2-$\Theta$" scatterometer, named appropriately because it has two rotating axes.

The scatterometric analysis can best be defined in two steps. First, in what is referred to as the *forward problem*, the diffracted light "fingerprint", or "signature," is measured using a scatterometer. A He-Ne laser beam is incident upon a sample after traveling through a spatial filter and some focusing optics. The wafer is mounted on a stage that permits it to rotate. Because the beam itself is fixed, this rotation changes the angle of incidence on the wafer. Using the grating equation, the detector arm is able to track any diffraction order as the angle of incidence is varied. Thus, the intensity of a particular diffraction order is measured as a function of incident angle (this is known as a *scatter signature*). Figure 30 illustrates this technique and shows the resulting trace of the zeroth-order intensity vs. the incident angle (65).

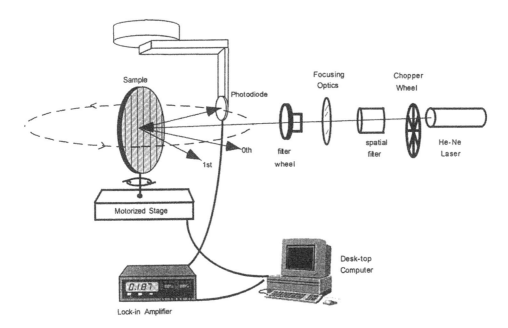

**Figure 29** A "2-$\Theta$" scatterometer. (From Ref. 63.)

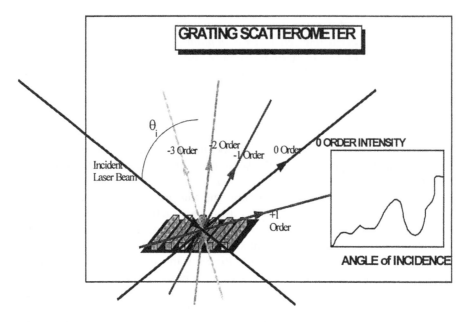

**Figure 30**  Diffracted orders for a grating scatterometer. (From Ref. 65.)

In the second step, known as the *inverse problem*, the scatter signature is used to determine the shape of the lines of the periodic structure that diffracts the incident light. To solve this problem, the grating shape is parameterized (66), and a parameter space is defined by allowing each grating parameter to vary over a certain range. A diffraction model (67) is used to generate a library of theoretical scatter signatures for all combinations of parameters. A pattern-matching algorithm is used to match the experimental diffraction signature (from the forward problem) with the library of theoretically generated signatures (from the inverse problem). The parameters of the theoretical signature that match most closely with the experimental signature are taken to be the parameters of the unknown sample. One algorithm that can be used to select the closest match between theoretical and measured traces is based on minimizing the mean squared error (MSE), which is given by

$$\text{MSE} = \frac{\frac{1}{N}\sum_{i=0}^{N}(x_i - \hat{x}_i)^2}{\frac{1}{N}\sum_{i=0}^{N}(x_i)^2} \tag{6}$$

where $N$ is the number of angle measurements, $x_i$ is the measured reference trace, and $\hat{x}$ is the candidate trace from the theoretical library. It should be noted that because the technique relies on a theoretical model, calibration is not necessary.

Figure 31 depicts an example of an experimental signature in comparison to theoretical data, and illustrates the sensitivity of the technique for linewidth measurements. In the figure the two theoretical scatter signatures correspond to two linewidths that differ by 10 nm. The difference between the two signatures is quite noticeable. The experimental data for this sample—a 1-μm-pitch photoresist grating with nominal 0.5-μm lines—matches most closely with the 564-nm linewidth. Thus the signatures provide a useful means for characterizing the diffracting features.

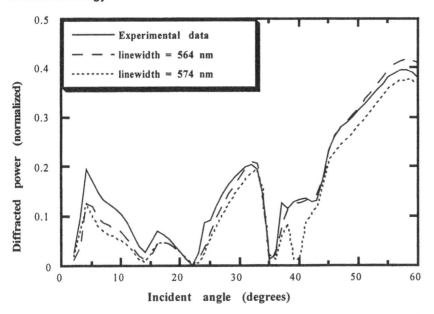

**Figure 31** Scatter signatures for two linewidths differing by 10 nm. (From Ref. 63.)

*b. Sensor Integration*

Changing the angle of incidence provides a simple way to analyze the scatter pattern and thus gather information about the scattering feature. However, it is not necessary to mount the wafer on a rotation stage to vary the incident angle. Instead, the laser/detector assembly might be mounted along a circular track in such a manner that allows it to rotate, as demonstrated in Figures 32 and 33. While in 1998 this tool existed only in an "off-line" configuration (68), evolution of this sensor to an in-line and in situ configuration is in progress. Figure 32 depicts one arrangement where the assembly is mounted to the side of a track for routine monitoring. In Figure 33, the same system is used for inspection through a window in a load-lock station. It is worth noting that the integration of scatterometry into OEM tools is not paced by mechanical and optical problems, as is the case for many other wafer-state sensors. The acceptance and implementation of scatterometry is paced by the availability of data that provides a proof of principle for the effective use of this sensor to SC manufacturing.

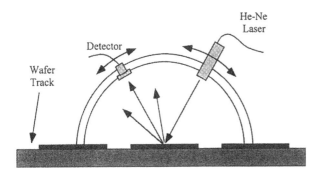

**Figure 32** A scatterometer mounted on a track. (From Ref. 63.)

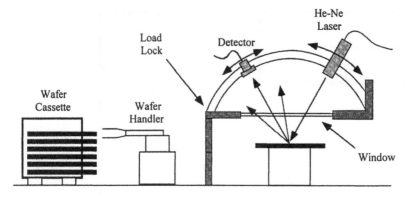

**Figure 33** A scatterometer mounted on a loadlock station. (From Ref. 63.)

*c. Gauge Study*

An extensive gauge study was sponsored by SEMATECH in 1994 that included both resist and etched samples (69,70). Figure 34 shows some of the results of this particular study, the measurement of nominal 0.25-μm linewidths in developed photoresist. As is evidenced in the figure, the measurements are consistent with those performed by both SEM techniques. Figure 35 depicts etched poly-Si CD measurement results for this same study. Once again the scatterometry CD measurements are consistent with other metrology instruments, including AFM measurements that were performed on these samples. Due in part to its simplicity, the repeatability of the technique is typically less than 1 nm (3σ), making it attractive for the aggressive production standards set forth in the SIA roadmap.

Although there are many potential applications for scatterometry, the most widespread to date have been for critical-dimension (CD) measurements on a variety of sample types. As of early 1998, the majority of applications have been in R&D for the measure-

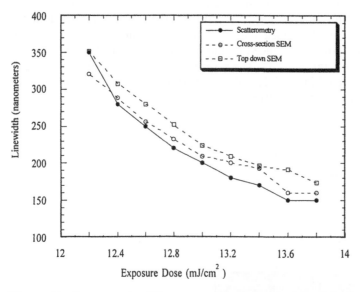

**Figure 34** Scatterometry CD measurements in comparison to both top-down and cross-sectional SEM measurements of the same samples. (From Ref. 63.)

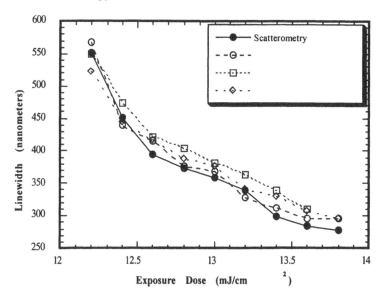

**Figure 35** Scatterometry CD measurements in comparison to AFM, top-down SEM, and cross-sectional SEM measurements of the same samples. (From Ref. 63.)

ment of resist and etched structures. These geometries can be quite complex, such as in the case of a resist/BARC/poly/oxide etch, and scatterometry can provide a good characterization of these geometries from test grating structures. Another possible application is to use the scattering signature from a periodic pattern from the actual device (e.g., DRAM memory) and look at these in terms of a characteristic signature for fault detection. However, the use of scatterometry in manufacturing environments is just beginning, and more data is needed before the technique sees widespread use.

## IV. MEASUREMENT TECHNIQUES FOR POTENTIAL SENSORS

A number of metrology tools exist based on sensing methods that are currently implemented on large, complex hardware. These methods are currently used for ex situ measurements. However, adaptations of these techniques can become implemented as in situ tools in OEM processing tools. This evolution will be paced by our abilities to generate less expensive and more compact versions of these tools, and by the need to implement such methods as in situ sensors for fault detection or model-based process control. Since some of these methods are likely to find their way into OEM processing tools within the next 3–5 years, a brief description of these methodologies is warranted.

## A. Ellipsometry

Ellipsometry, in its single-wavelength, dual-wavelength, or spectral embodiments, is a well-established technique for the measurement of film thickness. The fundamentals of ellipsometry are described in Chapter 2. So the following emphasizes the in situ aspects of this sensor.

## 1.  Theory of Operation

Ellipsometry is the science of the measurement of the state of polarization of light (19). The polarization of light, in the usual conventions, corresponds to the spatial orientation of the E-field part of the electromagnetic wave. Since light is a transverse wave, the polarization is two dimensional; there are two independent orientations of the polarization in a vector space sense. The linear bases are usually referred to as the P and S components. For light that reflects from a surface, or is transmitted through an object, the P polarization lies in the plane of reflection (or transmission), and the S polarization is perpendicular to the plane of reflection. In addition to specifying the amount of P-type and S-type components, the phase difference between them is also important. The phase lag is a measure of the difference in the time origin of the two (P and S) electric field vibrations.

For the case where there is a nonzero phase lag the E-field vector traces out a curve in space. The projection of this curve onto a plane that is normal to the direction of propagation of the light is generally an ellipse, thus the origin of the name *ellipsometry*. A special case of the ellipse is a circle. There are two unique circular polarizations referred to as left-handed and right-handed, depending on whether the P-polarization component leads or lags the S-polarization component. At times it is convenient to use the circular basis set in place of the linear basis set. The full state of polarization of light requires the specification of a coordinate system and four numbers, which includes the amount of unpolarized light. Unpolarized light implies that the polarization is randomly and rapidly changing in time. This naturally occurs for a light source that consists of a large number of independent, random emitters.

The reflectivity of light from a surface depends on the state of incident polarization. For example, at a specific angle for many materials, the P-polarization has no reflection, whereas the S-polarization does. The angle at which this occurs is the *Brewster angle* of the material; it is a function of the index of refraction of the material. Using fully polarized light, the index of refraction of a bulk material may be readily measured by finding the Brewster angle.

Figure 36 illustrates the difference in reflection for P- and S-polarization as a function of incidence angle for 0.5-micron light (green light). The Brewster angle at about 76° is the point where the P-polarization reflectivity goes to 0. At normal incidence and at grazing incidence there is generally much less information about the surface that can be derived from measurements in the change of polarization.

By taking ratios, all the information about the change in the state of polarization may be determined by specifying two numbers, known as the ellipsometric parameters $\psi$ and $\Delta$. Psi ($\psi$) is usually given as the angle whose tangent is the ratio of the magnitudes of the P- and S-components of the reflected light. Delta ($\Delta$) is the relative phase between the P- and S-components. By measuring $\psi$ and $\Delta$ as a function of incident angle, or as a function of wavelength at a fixed incident angle, much information may be determined about the reflecting surface including the effect of thin films (thickness and composition). Figure 37 shows the reflectivity of the P- and S-polarization from a sample of silicon dioxide (220 nm thick) layer on silicon substrate as a function of wavelength. This example illustrates the type of data from which the thin-film thickness and material composition is inferred by regressing on the ellipsometric equations that are defined by this system.

## 2.  System Integration

When the wavelength of the incident light varies (by using a white light source) for a fixed incident angle, the term *spectral ellipsometry* (SE) is used. For a fixed wavelength (laser

**P and S Reflection for Si**

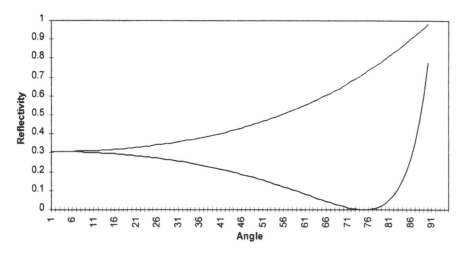

**Figure 36**  Reflection from silicon for 0.5-micron light for P- and S-polarizations. (From Ref. 19.)

source) with variable incident angle, the term *variable-angle ellipsometry* (VAE) is used. An instrument that varies both angle and wavelength is a *variable-angle spectral ellipsometer* (VASE). Figure 38 is a schematic representation of a spectral ellipsometer. The white light source is a broadband emitter such as a Xenon arch discharge. The fixed polarizer passes a specific linear polarization component. The polarization modulator is a device that changes the polarization in a known manner such as a photoelastic polarization modulator. In some instruments the function of these two devices is replaced with a polarization element that is mechanically rotated. The analyzer selects out a specific state of polarization of the reflected light. Since the state of the incident polarization is well defined, the

**P and S Reflectivity for SiO2 on Si**

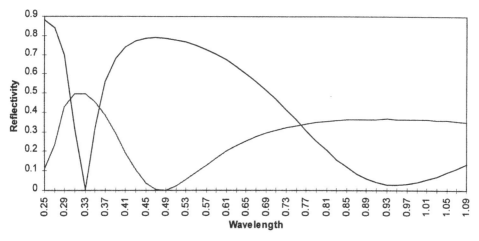

**Figure 37**  Reflection from $SiO_2$ layer on Si for P- and S-polarizations as a function of wavelength. (From Ref. 19.)

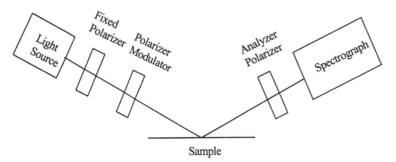

**Figure 38** Schematic representation of a spectral ellipsometer. (From Ref. 19.)

effects of reflection from the sample can be determined. The spectrograph analyzes the white light source into a number of spectral components.

The use of spectral ellipsometers for in situ monitoring and control has been limited by the cost of these units. An additional constraint has been the complexity of the integration of the optics (two opposing windows required) into standard OEM processing tools. The cost issue is slowly improving through the development of lower-cost ellipsometers. When warranted, the optical complexity can be overcome (71). As processing complexity and the inherent cost of misprocessing 200–450-mm wafers continues to increase, spectral ellipsometers will likely find their way into OEM tools for in situ monitoring and control of thin-film growth and composition in real time.

## B. Epi Resistivity and Thickness

In the process of depositing an epi layer of silicon, the resistivity and thickness of the epi layer are the two parameters of greatest interest (72). Traditional methods of monitoring epi layer resistivity measures either the average resistivity of the layer, as is the case of a four-point probe, or the resistivity as a function of depth into the epi layer, as is the case with a Hg probe or *CV* Schottky diode. These traditional methods are all destructive, because probe marks, contamination due to Hg, and metal dots all contaminate the wafer.

A new technique has recently been developed (73) that performs a nondestructive measurement of epi layer resistivity and profile. The technique used is conceptually quite similar to Hg probe or *CV* Schottky measurements. While the technique was introduced as a stand-alone metrology tool, an in-line version incorporated into the cooling station of an epi reactor is an obvious extension of the technology.

### 1. Theory of Operation

Both the *CV* Schottky diode and Hg probe techniques place an electrode on the surface of the semiconductor and then measure the depletion width by looking at the capacitance across the depletion width. They vary the depletion width by varying the voltage on the electrode and measure the capacitance of the depletion width at each electrode voltage. Similarly, this technique positions an electrode near the semiconductor surface, although in this case it does not touch the wafer. It then measures the depletion width for each of multiple voltages on the electrode.

The technique used to position the electrode near the semiconductor surface, but not touching it, is similar to the air bearing effect used in computer hard disk drives, and it is shown in Figure 39. A disk whose bottom surface is made of porous, inert material is pushed toward the wafer surface by air pressure above the porous surface. As air escapes through the porous surface, a cushion of air forms on the wafer, and the air cushion acts like a spring and prevents the porous surface from touching the wafer. The porosity and air pressure are designed such that the disk floats approximately two microns above the wafer surface. A stainless steel bellows acts to constrain the pressurized air and to raise the porous disk when the air pressure is reduced. Note that if the air pressure fails, the disk moves up, rather than falling down and damaging the wafer. Similarly, an electrical failure would not damage the wafer surface. The mechanism is simple, for no electrical or computer feedback of any kind is required. It is analogous to suspending an object between two springs of different spring constants. The porous disk has a hole in the center, and a sensor element is mounted in the hole to prevent the pressurized air from escaping. The sensor consists of an electrode that is 1 mm in diameter. The electrode is made of a material that is electrically conductive and optically transparent. The electrode protrudes from the bottom of the porous disk such that during the measurement it is located about one-half micron above the wafer surface.

A block diagram of the measurement system is shown in Figure 40. As with Hg probe and $CV$ Schottky measurements, depletion width is measured by looking at the capacitance of the depletion layer. The system actually measures the capacitance from the wafer chuck to the electrode, which is the series combination of three capacitances: the capacitance from the wafer chuck to the wafer, in series with the capacitance of the depletion layer, and in series with the capacitance of the air gap. The capacitance of the wafer chuck to the wafer can be ignored, for the area of the wafer is so much larger than the area of the electrode. Even with a 200 mm wafer, the ratio of the areas is more than 22,000:1. And although the effective separations of the capacitor plates may be unequal, it is reasonable to consider the wafer-chuck-to-wafer-capacitance as a short circuit. The capacitance of the air gap cannot be treated so easily, but because there is some electrode voltage at which the semiconductor surface is in accumulation and the capacitance of the depletion width is infinite, the capacitance of the air gap can be measured. Assuming that the actual air gap does not vary with changing electrode voltage, the capacitance of the air gap is the measured capacitance at its maximum value. Subtracting the capacitance of the air gap from the measured capacitance provides the capacitance of the depletion width. If the air bearing does not have infinite stiffness and the air gap changes as a result of the varying electrostatic attractive force created during the measurement, then it is possible to model the behavior and calculate the air gap capacitance at any electrode voltage.

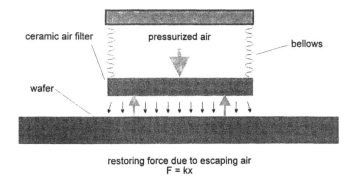

**Figure 39**  Air bearing mechanism. (From Ref. 72.)

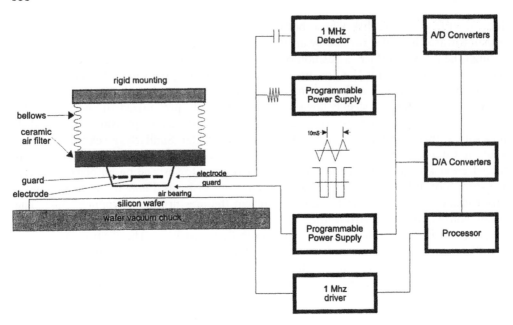

**Figure 40** Block diagram of complete air bearing system. (From Ref. 72.)

At every step in electrode voltage, the capacitance is measured and the charge on the electrode is calculated as the integral of $C\ dV$. The relevant equations necessary to compute the profile of $N_{sc}$ as a function of depth, $W$, are as follows:

$$W = \varepsilon_s \varepsilon_0 A \left[ \frac{1}{C_{total}} - \frac{1}{C_{air}} \right]$$

$$dQ = C_{meas}\ dV \tag{7}$$

$$N_{sc}(W) = \frac{dQ}{qA\ dW}$$

where $A$ is the area of the electrode, $\varepsilon$ refers to dielectric constant, and $Q$ is the elementary charge. Unlike in traditional Hg probe or $CV$ Schottky measurements, the electrode voltage in this system varies rapidly. A full sweep from accumulation to deep depletion is done in about 10 milliseconds, and data from multiple sweeps is averaged in order to reduce the effect of noise. The fast sweep period also serves to reduce inaccuracies due to interface states and carrier generation.

The system displays either plots of resistivity versus depth or $N_{sc}$ versus depth. Resistivity is obtained by converting according to the ASTM standard. A typical profile produced by the system is shown in Figure 41 . Repeatability and reproducibility are quite reasonable compared to other techniques. Resistivity of a single wafer measured at 8-hour intervals over a 3-day period showed a measurement error of 0.75% ($1\sigma$).

## 2.  Calibration and Performance Range

The system is calibrated by presenting it with one or more wafers of known resistivity. The system then creates a piece-wise linear calibration curve so that there is complete agreement at the calibration points; between calibration points, the system interpolates to

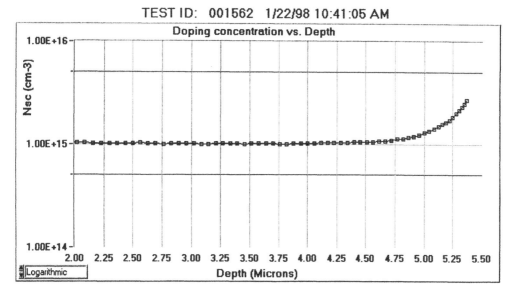

**Figure 41** Epimet measurement profile. (From Ref. 72.)

obtain good calibration. The more calibration points there are, the better the performance across the entire range of calibrated values.

Doping concentration profile can be generated within depletion depth. Figure 42 shows the maximum epi layer depth (for p-type silicon) that the system can deplete. As with a mercury probe or $CV$ Schottky diode, the maximum depletion is a function of resistivity, for a given applied voltage. For reference, Fig. 42 also shows the maximum depth that can be measured by mercury probe using 3-V and 10-V bias voltages. Development is under way to increase the maximum layer depth that can be measured, i.e., to move the line "up" on the graph.

## V. USE OF IN SITU METROLOGY IN SEMICONDUCTOR MANUFACTURING

Semiconductor manufacturing has historically relied on statistical process control (SPC) for maintaining processes within prescribed specification limits. This passive activity, where action is taken only after a process variation limit has been exceeded, is no longer adequate as device geometries shrink below 250 nm. Active control of the process is required to keep the wafer properties within the ever-decreasing specification limits. To achieve this tighter control, a new paradigm, called advanced process control (APC), is emerging. All APC activities require timely observability of the process. Timely observability is the major driving force for the implementation of in situ sensors in OEM processing tools. These sensors determine the process, wafer, or machine states during the sequential processing of wafers and hence provide the necessary information for APC. This section will describe the two major components of APC, fault detection and classification (FDC) and model-based process control (MBPC). The use of metrology-based sensors and multivariate analysis for endpoint detection will be covered at the end of the section.

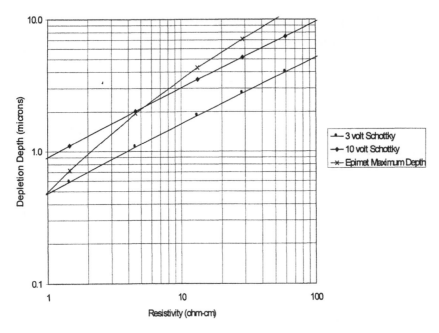

**Figure 42.** Epimet operating range, p-type epi. (From Ref. 72)

## A. Fault Detection and Classification (FDC)

The goal of FDC is to answer two questions:

1. Is the process and process tool running as expected?
2. If not, what is the likely source of the problem?

Fault detection and classification determines anomalous processing conditions by univariate and multivariate (single and multiple signals) analysis methods. These conditions can be intermittent in nature or can result from a gradual drift in the processing tool. The first task (74) of FDC is *fault detection*, i.e., during the processing of each wafer, to determine if the sensor signatures indicate a "nonnormal" tool state and process state. The determination is made using a previously made model that correlates the tool sensor readings to the normal tool state and process state. Creating the model is a bigger problem than might first be envisioned, since the normal state is not static, but slowly changes through time. For example, a plasma etch chamber performance changes as the chamber cleanliness degrades with the processing of each wafer. To detect all faults, the FDC system must be sensitive to small excursions from the normal operation of the process tool at any time during its scheduled maintenance cycle. To prevent false alarms, the FDC model must be tolerant of the expected drifts in the tool during its maintenance cycle. The upkeep of the FDC model by an already-overburdened equipment engineer must be minimal. Data from successive wafers is analyzed against the FDC model, and appropriate statistics indicate whether the wafer was processed in the normal way or whether some aspect of the process was anomalous.

Multivariate analysis methods being used today for fault detection are often grouped under the broad heading of *chemometrics* (75). Chemometric methods do not rely on first-principle interpretations of the sensor data. Instead, chemometric techniques generate a few metrics to describe the composite behavior of all of the sensor readings being tracked (76). Multivariate FDC methods are proving more sensitive to tool and process variance than traditional (SPC) process tracking and univariate fault detection (77). Traditional univariate SPC techniques monitor each sensor reading independent of any other sensors being monitored. Multivariate FDC methods use the inputs from many sensors, each contributing a slightly different view of the process tool. Unlike SPC methods, the multivariate FDC analysis methods are enhanced by the autocorrelation and cross-correlation among the many sensor readings. The result is enhanced FDC sensitivity to small deviations from normal processing. This fault detection sensitivity enhancement, derived from applying multivariate techniques to many sensors monitoring the same tool and process, is similar to the measurement accuracy improvement derived from applying the central limit theorem to multiple measurements of the same object.

The FDC methods strive to detect a fault at the earliest time that is cost effective. Wafer-to-wafer fault detection (the minimum requirement for costly 300-mm wafers) must collect all of the needed information, perform calculations, and pronounce the process OK or abnormal between the time the wafer finishes processing and the time the next wafer is committed to the process. A number of wafer fabs are implementing this level of FDC today (77–79). Real-time fault detection strives to detect faults while the process is running and to stop the process fast enough to prevent wafer damage or tool damage. Real-time FDC, which actually saves wafers, is achievable today for specific failures, such as absence of photoresist on the wafer about to be etched (80). Integrating a real-time FDC system so it can gracefully interdict the processing sequence is a challenge on most process tools today.

Once a fault is detected, the next task is *fault classification*, i.e., providing useful information for identifying the source of the fault. Depending on the FDC algorithm, the model can be used in a reverse sense to establish what sensor signal(s) generated the fault. Even if not very specific (e.g., the pressure reading, throttle valve position, and a mass flow control signal were the largest contributors to the excursion from normal), such information is very valuable during diagnosis of the cause of the fault that must precede repair. The confidence the equipment engineer has in the FDC system to detect real faults *and* assist in diagnosing the problem is critical to the long-term use of the system in production. Immediately after installing an FDC system on a process tool in production, many faults are typically identified. By immediately "fixing" the identified problems, the number of FDC alarms from the tool dwindle to near zero within a couple of weeks, after which the tool will run flawlessly for a considerable length of time (77).

Notice that an FDC alarm only indicates that the conditions during the processing of a given wafer were anomalous. This does not necessarily indicate that the wafer has been misprocessed. There is growing evidence that the multivariate FDC methods, based on the historical relationships of multiple sensors on a semiconductor processing tool, are more sensitive to subtle deviations from normal processing than the few in-line wafer-state measurements typically available in production following the process step (77). Wafer fab experience has demonstrated the correlation between tool performance, measured with a multivariate FDC system, and wafer yield and device performance at final test. Recent demonstration projects have shown that some tool variance on some processes, measured by a multivariate FDC metric, can be restated in units of lost yield points at final test. The goal of that work is to identify the results of specific types of tool variation. Tool variation can fall into three broad categories: (1) has no effect on yield, (2) has a negative effect on yield, (3) has a positive effect on yield or device performance. Different responses are appropriate for each category. The goal is to know exactly what tool variance is unacceptable and justifies halting production to make the repair (from private communication with John Colt, IBM, MS 975E, 1000 River Rd., Essex Junction, VT 05452).

By tracking the trends of multivariate FDC metrics, one can determine that the tool condition is deteriorating, even though no wafer-state measurement at the process step can detect a fault in the wafer. This predictive capability has been called *prognosis* (81). Prognosis of an impending fault predicts that a failure condition is *about to happen*, with enough lead-time to (1) prevent misprocessing a wafer and (2) schedule preventative maintenance before misprocessing conditions occur. As these methods improve, it will be possible to move toward need-based tool maintenance, to replace many time-based preventative maintenance activities.

## B. Model-Based Process Control

The goal of MBPC is to keep the output of the process close to the target value by making small adjustments to the control "knobs" (process input variables) of the process. Small adjustments to the nominal process-variable settings are determined by using a model of the process that describes the relationship between process-variable settings and the resulting output of the process. Process models are usually empirically derived simple linear models (thickness = rate × time + offset). The rule of thumb is that the model should be no more complex than needed to predict process performance with the accuracy needed for the control goals.

The accuracy of the process model must be repeatedly verified at time intervals shorter than the time constant of process drift. Comparing the measurement of the actual results just achieved on the wafer against the model prediction for the process-variable settings used verifies the accuracy of the model. When the difference between the actual value and the predicted value is statistically significant, the process model must be updated or tuned to make it accurately predict the current process tool performance (see Fig. 1). Wafer-state sensors provide the needed measurements of actual results on the wafer just processed. The precision of MBPC is limited to the reproducibility and sensitivity of the wafer-state measurements.

The response time needed from the MBPC system is dictated by the natural drift rate of the process. Due to a lack of real-time in situ wafer-state sensors, very few opportunities for real-time MBPC exist today. A rapid thermal process (RTP) that uses MBPC tends to rely on optical modeling of the process chamber. Wafer-temperature sensors that can accurately track fast process-temperature trajectories are not available (82). Most process tools are stable enough over time that wafer-to-wafer MBPC, using in-line sensors, is quite adequate to deal with the slow drift of the tool. For example, film thickness measurement, performed in the load lock, is useful for controlling deposition tools (83). The longer lag time of after-the-process overlay measurements from stand-alone metrology tools are adequate for lot-to-lot MBPC of lithography overlay that displays a very slow random walk drift off target (84).

Like any other control method, MBPC is limited by the length of the measurement lag time, precision, and stability of accuracy offset of wafer-state measurements used in feedback or feed-forward control loops. Today, MBPC is limited by the availability of in situ and in-line sensors to deliver the needed sensitivity, reproducibility, and short lag time of wafer-state measurements. By shortening the lag time and improving the reproducibility of wafer-state measurements, in situ metrology will improve process control using MBPC methods. Using in situ sensors, MBPC requires physical integration, electrical integration, and software control integration. That is difficult on legacy tools. In situ sensors for

MBPC will probably always require the OEM to build the sensor into the tool at the design stage for the hardware and the control system.

## C. Relationships Between Fault Detection and Classification and Model-Based Process Control

The two parts of APC—FDC and MBPC—are complementary but different. This section will deal with those issues.

### 1. Fault Detection and Classification Implementation Should Precede Model-Based Process Control Implementation

The authors offer the following reasoning in support of this recommendation:

A. Today, the implementation of FDC *is not* limited by the availability of appropriate sensors. The sensors already used by the single-in-single-out (SISO) controllers on the process tool are usually adequate for providing the multidimensional timely observability of the tool needed for multivariate FDC. And FDC implementation *is* limited by the difficulty of (1) accessing the sensor information in the process tool and accessing sensors added to the process tool, (2) coordinating FDC with the needed process information, and (3) tracking the FDC results for each wafer separately. The demonstrated benefits of wafer-level tracking systems will be enhanced when FDC results for each wafer in each process tool are also added to the information being tracked (85).

B. Ensuring that the process tool is running as expected (using FDC) is the first requirement for MBPC to be successful. The process models used in MBPC are empirically developed over a narrow operating range, compared to the wide range of possible process parameter settings on the tool. Those process models are valid only when the process tool is operating in that narrowly defined operating region, i.e., operating as expected. Not having FDC operational, forces one to *assume* that the process and tool are running as expected. Operating a process under that assumption effectively uses product wafers as fault detection sensors. The failure of product wafers at a downstream parametric test indicates that one or more upstream processes are not running as expected. It is getting very costly to use product wafers in this manner.

It is important to understand that MBPC will only work on a process tool that is "in control," from an SPC point of view. One way to guarantee that the process tool is operating as expected is to implement an FDC system before implementing MBPC. The FDC system will quickly shut the system down when hardware breaks. Without an operating FDC system, the MBPC system (and the SPC tracking system used today) can be blind to faults outside its narrow vision of the process that are provided by the one or two wafer-state measurements in the feedback or feed-forward loop. Without an operational FDC system, MBPC runs the risk of attempting to correct for hardware failures, which is rarely successful.

C. Like an SPC-quality tracking system, MBPC relies on feedback from one or two quantitative measurements of only the most important wafer attribute(s) affected by the process. The assumption is: If one measured parameter is on target, then all of the other unmeasured (or unmeasurable) wafer attributes are also within specification. Although that assumption is frequently correct, when it fails, the cost in the number of scrapped wafers can be high. By monitoring 30–50 tool-state and process-state sensors throughout the process tool, a multivariate FDC system greatly improves the probability that the assumption is true.

D. The longer a tool failure goes undetected, the more expensive the error becomes. An undetected tool fault causes expensive processing to continue on material that will be scrapped. Adding to that cost is the lost revenue from salable wafers that could have been processed. Without the diagnostic aids provided by FDC, the time to diagnose the problem will further extend the tool downtime. SEMATECH's cost of ownership (COO) modeling software has been enhanced to quantify the cost of delayed fault detection. The reduced COO from the early detection of a tool failure using an FDC system rapidly pays for the FDC system (86).

E. Current university research into the cost drivers in wafer fabs has shown similar results (87–89):

> Preliminary academic research has shown that integrating measurement tools into semi-conductor process tools follows the economics of information transfer and violates traditional incentive structures suggested by the economics of specialization. Anyone creating a business model for integrating measurement tools into semiconductor process tools must take these findings into account.
>
> One of the first things the researchers noticed was that the greatest savings to the chipmaker was a rapid localization of (the source of a yield) problem. Knowing in which process step was the source of the trouble could drop the wafer fab's potential loss rate from a few hundred thousand dollars per hour to a few hundred dollars per hour. Thus the value of owning yield analysis technologies is very high. A recently developed model quantifies the value of ownership of yield analysis technologies. It strongly suggests that in situ metrology has a high return on investment.

## 2. Different Individuals in a Wafer Fab Are Interested in the Outcomes of Fault Detection and Classification and Model-Based Process Control

A. FDC helps the equipment engineer who is charged with maintaining the performance state of a process tool so that it will produce good product for the maximum length of time and lowest maintenance cost.

B. MBPC helps the process engineer who is charged with the design and maintenance of the robust process to produce product within the specification at the lowest production cost.

## 3. Different Kinds of Metrology Information Is Needed for Fault Detection and Classification and Model-Based Process Control

A. Univariate FDC that is aimed at specific types of failures can often be achieved with one sensor input. For example, univariate FDC systems can shut down a plasma etch within seconds of igniting the plasma by monitoring for an OES intensity level associated with etching photo resist.

B. Multivariate FDC that is sensitive to *any* kind of tool or process failure requires readings from an array of process-state and tool-state sensors. They may directly and/or indirectly sense what is happening in every aspect of the process tool. Process-state and tool-state sensors are predominantly used for FDC. These are the sensors already installed on the process tool and used by the SISO controllers for each process parameter and tool subsystem. The process tool often has limit checks (univariate FDC) on the parameters controlled by these separate SISO control loops. Simple limit failures are detected by most process tools and cause a tool shutdown. Multivariate analysis of the correlation between *these same sensor readings* provides an additional dimension to FDC. Making full use of the information provided by the existing sensors on process tools to implement multivariate FDC techniques is more cost effective than seeking additional sensors for specific failures.

C. Wafer-state sensors can also be used for FDC, when they are available and provide the information in a timely manner for the real-time or wafer-to-wafer FDC analysis. However, the added cost of a wafer-state sensor cannot be justified by the small incremental improvement in FDC that uses the sensors already on the tool. The goal of preventing scrap is not achieved when a wafer-state sensor detects a failure—even if it is detected in real time. For that reason, process-state and tool-state sensors tend to dominate the mix of sensors monitored for univariate and multivariate FDC.

D. MBPC requires quantitative wafer-state sensors, which provide the measurements to feed back or to feed forward to the controller. The capability of the MBPC is directly dependent on the delivery of accurate measurement of the feature being controlled. In addition, the measurement must be delivered in a time shorter than the drift rate of the process being controlled. The in situ sensors described in this chapter can fulfill some of these measurement requirements today. The availability and capability of in situ sensors will pace the adoption of MBPC, since the tool control achievable with MBPC, can never exceed the ability to quickly measure the product being produced.

In theory, it is possible to use process-state measurements to synthesize virtual sensor readings if in situ wafer-state sensors are not available. A virtual sensor is an empirical model that quantitatively correlates the behavior of many process- and tool-state sensors with the actual size of the features produced on the wafer surface. In reality, the poor precision of virtual sensor measurements severely limits their usefulness for process control.

## 4. The Kind of Acceptable Measurement from a Sensor Is Dependent on the Intended Application of the Sensor: FDC vs. MBPC

A. *MBPC*—Wafer-state sensors that provide short lag time between the process and the measurement result are needed for MBPC. These sensors must provide a high-precision correlation to the physical characteristic(s) of the wafers being produced in the process. This indirect method must produce measurements in engineering units. And MBPC demands a quantitative measurement value of the actual size of the wafer feature just produced by the process that can be quantitatively compared to the target value. For example, the film thickness deposited must be measured in angstroms, or the etched line width must be measured in nanometers. The precision, or reproducibility, of the measurement establishes the ultimate limit of the control that can be achieved.

B. *FDC*—Sensors that have a stable and repeatable correlation to process performance are adequate for the multiple sensor readings needed to detect subtle excursions from a normal process tool or process operation. Any sensor that is stable over time and quantitatively responds to changes in the process or tool conditions is worth considering as a process monitor for multivariate FDC. The output of the sensor need not be linear or calibrated in engineering units of measure.

To be most effective at detecting any kind of fault—even failures that have never been seen before—FDC needs the inputs from a wide range of sensors. Sensors that exhibit strong correlation but measure slightly different things are desirable, especially if the sensor already exists on the tool and the cost of including it in the FDC system is low.

It is worth mentioning that random inclusion of any and all tool sensors is not appropriate for FDC. A real example of including an inappropriate sensor occurred on a process tool that had an air-lock-open sensor. The sensor would activate each time an operator would put a new cassette of wafers in the load lock of the process tool, even when another wafer was processing. The sensor was included in the list of sensors for the FDC system to monitor during wafer processing. The result was the FDC system misinterpreted the occasional "air lock open" event as a statistically significant excursion from the normal processing of the wafer currently in the chamber. In fact, the sensor had nothing to do with processing the wafer and should not have been included in the FDC system.

## 5. Who Will Implement Fault Detection and Classification?

Exactly who will implement FDC systems is closely tied to three factors: (1) the persons who own the problem that FDC addresses, (2) the persons who own the data needed for the system, and (3) the persons who possess the appropriate environment for developing the system.

High-volume-production wafer-fab-equipment engineers are primarily responsible for keeping process tools running as expected. Equipment engineers are motivated to implement FDC because it is a cost-effective tool for achieving that primary goal. Equipment engineers are the logical implementers of FDC because they have ownership of both the problem and the information needed to implement an FDC solution. The essential data needed to implement FDC that equipment engineers can provide are (1) time-series readings from sensors inside the process tool, (2) process recipe information, and (3) a way to associate the data with individual wafers. Finally, only a high-volume wafer fab has process tools that run identical or at least similar product wafers frequently enough to establish how the process and tool operate normally—the critical first task in implementing an FDC system.

Equipment engineers are installing the first FDC systems in wafer fabs with the support of the wafer fab managers. In wafer fabs that have developed FDC expertise, the fab managers are pushing for rapid fan-out of FDC to more process tools. It is safe to assume those fab managers believe they receive an acceptable return on their investment in FDC because it is addressing a problem that is high on their priority list.

One can logically argue that the OEMs *could* be responsible for implementing FDC, since every one of the sensors and all the information needed is under their direct control. That is, they own the sensors and all of the required information. However, the OEMs have generally not taken on the task of adding FDC to their tools. A possible reason for this is that FDC never rises high enough on their priority list before they are reassigned to design the next process and build the next process tool. In addition, all of the tools available to an OEM are in the state of rapid change that characterizes research and development. The OEM does not have the stable process environment of a high-volume manufacturing fab needed to identify the normal state of the process and process tool, the first step in developing a FDC system. Until the current business model changes, it is reasonable to expect that the OEMs will continue not to develop or implement FDC on their tools (86–89).

It is reasonable, however, to request that the OEMs provide easier access to sensor signals in their tools in order to make easier the wafer fab equipment engineer's job of implementing FDC on the tool after it is installed in the wafer fab. The OEM could provide the synchronizing signals that equipment engineers need when they add sensors to the tool to enhance FDC. Standards are needed for these and other issues about interfacing add-on sensors to installed process tools. The announcement that major OEMs will adopt the sensor bus SEMI-standard is a step in the right direction.

Standards will make it easier for the semiconductor manufacturers to describe what they need when writing equipment specifications for purchase orders (Private communication with John Pace, IBM).

## 6. Who Will Implement Model-Based Process Control?

Exactly who will implement MBPC systems is closely tied to the same three factors: (1) the persons who own the problem that MBPC addresses, (2) the persons who own the data needed for the system, and (3) the persons who possess the appropriate environment for developing the system.

High-volume-production wafer-fab process engineers are primarily responsible for development and maintenance of a process that will produce a product within specification for extended periods of time at minimal cost. Process engineers will implement MBPC systems when they provide the most cost-effective solution to achieve that goal. The major stumbling block has been the integration of the MBPC system with the existing manufacturing execution system (MES). This issue is covered in the next section, on software. Process engineers are the logical implementers of MBPC because they have ownership of both the problem and the information needed to implement an MBPC solution. First, the process engineer is responsible for the process model, ideally developed by the OEM during process development, that describes the product variance as a function of process input settings. Second, the process engineer has access to the stand-alone metrology tool or in situ sensor wafer-state measurements of the actual size of the wafer feature produced by the process. Most importantly, the process engineer has the authority and access to make slight process input adjustments to keep the product on the specified target. Process engineers, working with control engineers, are installing the first lot-to-lot MBPC systems in wafer fabs today. Those wafer fabs that have developed expertise implementing lot-to-lot MBPC are actively extending the control method to more tools in their fabs. Again, it is safe to assume that those fab managers believe they receive an acceptable return on their investment in MBPC that is addressing a problem high on their priority list.

And MBPC will not be implemented by OEMs until in situ sensors are available to provide the needed wafer-state measurements. The process tool OEMs recognize that using MBPC makes the capability of a process tool totally dependent on the consistent accuracy and lag time of the wafer-state measurement that is fed back to the controller. It is doubtful that an OEM will ever place the performance of his tool in the hands of a downstream metrology tool, over which he has no control. For this reason, implementation of MBPC will likely continue to be the responsibility of the wafer fabs, who have ownership of (and control over) every information component used in MBPC. Capable in situ and in-line wafer-state sensors may change this current assignment of responsibility for MBPC. After the equipment OEM demonstrates that an in situ or in-line sensor, mounted on his tool, can reliably deliver the wafer-state measurement, wafer-to-wafer MBPC will be an option for the equipment OEM. The OEMs will implement MBPC only after they demonstrate it to be the technically acceptable and economically preferred solution for meeting stricter process specifications.

## D. Metrology-Based Endpoint Sensors

Endpoint detection enables a semiconductor process to extend or shorten process time to automatically accommodate variation in the incoming wafer. The control of ending many processes has long relied on process-state sensors whose readings change when the endpoint is reached. Optical emission spectroscopy (OES) and RF sensors have been used for endpoint detection of plasma etch processes that remove a thin film, exposing a different material at "endpoint" (90–95). The change in spindle motor current for CMP processes detects the polishing force change at the discontinuity of materials at endpoint (91). Endpoint sensors must meet the same requirements for sensitivity to small change and reproducibility demanded of FDC sensors. Because relative signal strength is typically monitored to detect the endpoint, no requirement exists for absolute sensor accuracy. The relative change in intensity of a wavelength of light emitted by the plasma indicates endpoint is reached.

New processes today, such as etch of damascene vias, deep trench isolation, and chemical-mechanical polish, are presenting a new kind of endpoint problem. In these processes, no stopping layer is ever reached, so no signal change is produced at endpoint. Instead, the process is to be stopped when a specified depth or thickness is reached in the material being removed or deposited by the process. Endpoint detection requires real-time, in situ quantitative measurements of the film thickness or trench depth. Reflectometry, interferometry, ellipsometry, and other techniques are being pressed into service as real-time, in situ sensors capable of supplying the quantitative metrology needed by these endpoint problems (90,96).

Metrology methods, traditionally provided by stand-alone measurement equipment, are being forced to become smaller, more specific, and less expensive so they can be implemented as in situ sensors for endpoint detection. In most cases these sensors will be designed into the process tool by the equipment OEM rather than retrofitted by the end user in a wafer fab. Limits on the footprint of process tools ultimately drive the requirement for small size. The limited demands of endpoint detection on a specific process drive the OEM's demand for "just enough" metrology accuracy and versatility to do the job. Because the OEM will have to provide sustained maintenance of the sensor over the life of the process tool, the end user's purchase price for the new endpoint sensor will be 3–4 times the purchase price from the sensor manufacturer. Containment of process tool cost drives the demand for low cost for these new metrology-based endpoint sensors.

## E. Spectral Sensors for Conventional Endpoint Detection

Traditionally, only one sensor has been used to generate the signal used for process endpoint detection. For example, the time-series intensity at one or two wavelengths emitted by a plasma etcher has provided the OES endpoint signal. Driven by feature size reduction, endpoint problems are more challenging. For example, contact etch removes material from a few tiny contact holes in the resist surface on the wafer. The open contact holes represent less than 1% (and shrinking) of the total wafer surface. The signal-to-noise ratio of one- or two-wavelength endpoint detection schemes is no longer sufficient to reliably detect endpoint. Several trends are developing to solve these problems.

The use of sensors that produce many measurements (spectrum) is expanding the amount of data available for the endpoint task rather than single measurements. Three examples include: (1) CCD-based spectrographs instead of monochromators for optical OES endpoint, (2) mass spectrometers (residual gas analyzers—RGAs) to track the concentration changes of multiple plasma species at endpoint, and (3) RF sensors that measure the endpoint change of voltage, current, and phase at multiple harmonics of the fundamental RF power frequency.

The volume of raw data delivered by these new spectral sensors requires more sophisticated data management and analysis methods than those used with single intensity readings (e.g., the intensity of one wavelength from a monochromator). A CCD-based spectrograph, for example, delivers, several times a second, the intensity of over 1000 wavelengths emitted from a plasma reactor. Simply extending the traditional methods to selecting many wavelengths to monitor for endpoint detection quickly becomes unwieldy, even for an experienced process engineer. Chemometrics multivariate analysis methods, similar to those used in FDC, are showing great promise for endpoint applications (90,92,93). These chemometrics methods will initially be applied to the spectrum from just one sensor at a time. There is every reason to expect that combining the signals of multiple spectral sensors, each contributing endpoint information, will deliver improved endpoint detection just as multivariate FDC, using many sensors, has improved fault detection.

## VI. SOFTWARE FOR IN SITU METROLOGY

Sensors are an absolute prerequisite for providing the benefits of the APC paradigm to semiconductor manufacturing operations. Stand-alone metrology tools that measure the wafer state after the entire lot of wafers complete processing are adequate for lot-to-lot MBPC. For wafer-to-wafer or real-time MBPC, the wafer-state measurement must be provided with a short time lag. In-line sensors, which measure the wafer immediately before or after processing, are needed for wafer-to-wafer MBPC. In situ sensors, which measure the wafer during processing, are required for real-time MBPC. All FDC applications, on the other hand, require the sensors to be on the tool and to provide data while the process is running.

Timely measurements from appropriate sensors are necessary, but not sufficient, for implementing APC. Extensive software is also required to turn the sensor data into useful information for APC decisions. Software is required for data collection, for data analysis for FDC, and for the control systems that perform MBPC.

### A. Data-Collection Software

The two major sources of data used in APC applications are signals from the processing tool and from add-on sensors connected to the tool. The former is generally collected through the SECS (Semiconductor Equipment Communications Standard) interface available on the tool. The SECS protocol enables the user to configure bidirectional communications between tools and data-collection systems. This standard is a means for an independent manufacturer to produce equipment and/or hosts that can be connected without requiring specific knowledge of each other. There are two components to SECS. The SEMI Equipment Communications Standard E4 (SECS-I) defines the physical communication interface for the exchange of messages between semiconductor processing equipment (manufacturing, metrology, assembly, and packaging) and a host that is a computer or network of computers. This standard describes the physical connector, signal levels, data rate, and logical protocols required to exchange messages between the host and equipment over a serial point-to-point data path. This standard does not define the data contained within the message. The second component is the software standard such as SEMI Equipment Communications Standard E5 (SECS-II). SECS-II is a message content standard that determines the meaning of the messages. While SECS-I has solved the hardware interface issues, SECS-II implementation contains enough tool-specific latitude to make interfacing to individual tools a time-consuming task. The GEM standard, built upon the SEMI E5 (SECS-II) standard, specifies the behavior of a generic equipment model that contains a minimal set of basic functions that any type of semiconductor manufacturing equipment should support. The SECS-II standard provides the definition of messages and related data items exchanged between host and equipment. The GEM standard defines which SECS-II messages should be used in what situations, and what the resulting activity should be. Brookside software (97) is a commonly used package for obtaining machine data from the SECS port on plasma etchers. More recently, network-based equipment server software products are demonstrating easy access to the sensor data inside process tools (98). These equipment servers run on a PC—either added on to the process tool or built in by the OEM. As a result, FDC analysis could be run local to the tool using the data that is extracted from the sensors onboard the process tool.

Add-on sensors are most often sophisticated enough (OES, RGA) that they are run by a dedicated PC. This PC also collects the raw sensor data and transfers it to the analysis software package.

## B. FDC Analysis Software

Once a machine and sensor data are properly formatted and available on a routine basis, univariate or multivariate analysis of individual or multiple signals achieves FDC. For plasma etching, univariate analysis has been performed on the endpoint signal for a long time (97,99,100), primarily due to its simplicity and effectiveness in detecting processing faults. The basis of these analyses is to examine the shape of a signal and to use an algorithm to detect a significant variation between the signal from the currently analyzed wafer relative to an accepted reference trace from a good wafer. In the mid-1990s, multivariate FDC became feasible, enabled by a number of software vendors (101–106) with capabilities for analyzing machine and sensor data. The capabilities of these vendors to provide multivariate FDC was evaluated in a SEMATECH FDC benchmarking study (107). These analyses determined the correlation between the various time-series signals coming from the tool sensors. Models are generated from a number of good wafers that represent the normal variability of the individual signals. The major issues are the choice of the signals, the choice of the good wafers, and the methods used for the analysis of this large volume of time-series data. An important feature of any such multivariate FDC technique is that the method be robust to the long-term steady drift in the sensor signals (which are not associated with the processing faults). At the same time, it needs to stay sensitive to small but significant signal variations at any particular time (108). The availability of pertinent information from the factory computer-integrated manufacturing (CIM) system (e.g., lot number, wafer number, log point) is a critical component for both univariate and multivariate FDC. These form the basis for sorting the machine and sensor signals into necessary groups for analysis. A major driver for determining where to run the analysis is the network latency period. Real-time analyses are generally performed on a local computer, while wafer-to-wafer or lot-to-lot analyses can be performed on the CIM network.

## C. Model-Based Process Control Software

Model-based process control is based on models of the wafer-state properties as a function of the input parameters. With such models and routinely available feed-forward or feedback data from the factory CIM system, MBPC can be performed to keep the output parameters of interest under tight control. The numerical basis for this can be quite straightforward. It is now enabled by commercial software that performs these calculations (109,110). Full automation of such models, where the MBPC software obtains the necessary information from the factory automation and downloads the new recipe to the tool, is a complex and generally lengthy integration task. Such integration is generally reserved for situations where benefits of MBPC have been proven on a local, off-line basis.

## D. Advanced Process Control Factory Integration

The integration of APC, both FDC and MBPC, into the MIS systems of wafer fabs has been a major impediment to the implementation of these systems in wafer fabs. Starting in the 1980s, SEMATECH sponsored programs to address the general problem of integrating software within wafer fabs. The SEMI CIM Framework, in wide use today, is an outgrowth of those SEMATECH programs. The information-handling and software-integration problems specific to APC is addressed by the Advanced Process Control (APC) Framework, a subsection of the CIM Framework. The first version of the APC Framework was created and implemented in a $2\frac{1}{2}$-year $10 million Advanced Process Control Framework Initiative project involving SEMATECH, Advanced Micro Devices (AMD), Honeywell, and ObjectSpace. The project concluded in Q3 1998 and was supported by the National Institute of Standards and Technology. The APC Framework has been commercialized and enhanced by ObjectSpace before being sold to KLA-Tencor.

The APC Framework provides an architecture for implementing both FDC and MBPC solutions, primarily in the run-to-run or lot-to-lot time frames. The APC Framework was developed specifically to work with the current generation of factory and equipment control systems, while using SEMI CIM Framework–compliant system technologies that also integrate seamlessly with the emerging generation of more open, distributed standards–based systems. For these reasons, it is useful primarily to the wafer fab faced with integrating an APC solution into their existing MIS system. By using the APC Framework, the integration work is made easier and has to be done only once. After one APC system has been installed on one tool in the factory, fan-out to other similar tools and other types of tools is much easier because of the reusable interfaces provided by the APC Framework. A great advantage of the APC Framework is that it is designed to utilize existing off-the-shelf software components to accomplish the desired APC objectives. Several of the major semiconductor wafer manufacturers are implementing the APC Framework to demonstrate its utility in speeding up the implementation of sensor-based FDC and MBPC solutions.

## E. Sensor Integration on Process Tools

The cost of scrapping a 300-mm wafer is projected to be extremely high. The cost of losing even one wafer will drive APC to shorter time frames. Wafer-to-wafer FDC will be asked to move to prognosis so that no wafer is ever misprocessed due to a tool failure. Lot-to-lot MBPC will be pushed toward wafer-to-wafer and even real-time control. As those time frames shorten, the demand will grow for in situ and in-line sensors to enable rapid response APC. A key element for implementing in situ and in-line sensors is the integration of the sensor into the process tool. The sensor manufacturers have come together to form the Integrated Measurement Association (IMA), to address the sensor integration issues (112). The first activities of the IMA are focusing on the further development and adoption of standards for integrating sensors into process tools.

One IMA task force is enhancing the SEMI Sensor Bus Standard by adding end-point sensor characteristics to the common device model for sensors. A second task force is working with their European counterparts to develop sensor integration standards for both mechanical, electrical, and software interfaces between the process tool and sensors.

## VII. FUTURE WORK

In situ sensors for process monitoring and control will grow in the future. Three issues will play a role in that future development.

1. Today, isolated success with FDC has been reported and attempts made to justify the cost of installing FDC within the wafer fab that did the work. A compelling business model that clearly shows the benefits of using in situ metrology is needed. The business model should address the financial needs and rewards of the sensor manufacturers, process equipment OEMs, and wafer fab. To justify FDC costs, a method is needed for measuring the financial value of rapid problem identification and resolution. It is difficult to quantify the value today of identifying problems early with an FDC system and repairing the cause before significant wafer loss occurs.

The cost of ownership (COO) model can be used to justify MBPC. Case studies are needed to compare the economic effectiveness of MBPC vs. traditional methods for achieving the tighter control of processes over longer time periods being demanded for future processes.

2. In-line measurements using stand-alone measurement equipment will always be needed in wafer fabs. The superior measurement capabilities of stand-alone metrology equipment is needed for process development and problem solving. In high-volume semiconductor manufacturing, process monitoring and control using the metrology sensors on the process tool will slowly replace the stand-alone metrology tools used today for those tasks. That transition will be driven by the demand for measurement time delays shorter than can be achieved using stand-alone metrology tools. Both FDC and MBPC demand short measurement lag time.

Sensors designed to be installed on the process tool for process control will be designed into the tool and process. The sensor will have to perform the measurement inside the boundaries of mechanical fit, electrical and communication interfaces, and cost. In situ sensors will never be used if they exceed the OEM's cost target.

3. The application of sensors that generate a spectral response will continue to grow. This trend is driven by the demand to measure smaller features. Spectral measurements provide a wealth of information that single-point measurements lack. Examples of spectral sensors include spectral ellipsometry, multiangle scatterometry, optical spectrographs, and FTIR. In quantitative measurement applications, the actual spectrum measured is compared with the closest-matching spectrum generated from first principles. When a close match between experimental and theoretical spectra is achieved, the model parameter values are assumed to describe the physical object being measured. New spectral sensors can also be used qualitatively for tracking tool and process health and some control

applications. Further development in both first-principles modeling and empirical multi-variate modeling is needed in addition to the development of new spectral sensors.

In summary, the development and implementation of in situ metrology for semiconductor manufacturing faces significant challenges in the future. The challenges start with technically how to measure the smaller features of interest with the required precision and accuracy. Using existing metrology methods "smarter" and developing altogether new measurement methods will be required. The second challenge is to reduce the lag time between manufacturing the part and receiving the measurement. Placing the sensor closer to the process chamber is the logic that dictates in situ and in-line metrology be used. Wider use of MBPC methods will be needed to take full advantage of timely measurements provided by in situ metrology. Finally, easier access to the sensor information is needed by the end user, who is motivated to monitor the process and tool using FDC.

## VIII.  WEB LINKS

Table 2 lists web links to major companies manufacturing sensors for semiconductor manufacturing.

**Table 2**  Web Links to Major Companies Manufacturing Sensors for Semiconductor Manufacturing

| Sensor company | Website | Product |
|---|---|---|
| Advanced Energy | http://www.advanced-energy.com/ | RF systems, sensors |
| Applied Energy Sys. | http://www.aenergysys.com/ | |
| Accent Optical Technologies, Inc. | http://www.accentopto.com/ | Scatterometer, FT-IR |
| Digital Instruments | http://www.di.com/ | AFM for CMP (offline) |
| ENI | http://www.enipower.com/ | RF probes |
| Ferran Scientific | http://www.ferran.com/main.html | RGA |
| Ircon | http://www.designinfo.com/ircon/html/ | Noncontact IR thermometers |
| KLA-Tencor | http://hrweb.kla.com/ | Stress, thin-film measurement, wafer inspection, metrology |
| Leybold Inficon, Inc. | http://www.inficon.com/ | RGA, leak detectors, full-wafer interferometry, QCM |
| Lucas Labs | http://www.LucasLabs.com/ | Plasma diagnostics |
| Luxtron | http://www.luxtron.com/ | Endpoint for plasma and CMP |
| Ocean Optics | http://www.oceanoptics.com/homepage.asp | Spectrometer-on-a-card |
| Nova Measuring Instruments | http://www.nova.co.il/ | Spectrophotometric integrated thickness monitors for CMP |
| On-Line Technologies, Inc. | http://www.online-ftir.com/ | FT-IR Spectrometer for wafer-state and gas analysis |
| Panametrics | http://www.industry.net/panametrics/ | Moisture, $O_2$ analyzers |
| Princeton Instruments | http://www.prinst.com/ | Imaging, spectroscopy |
| Quantum Logic Corp. | http://www.quantumlogic.com/ | Pyrometry |
| Rudolph Technologies, Inc. | http://www.rudolphtech.com/home/ | Ellipsometer |
| SemiTest | http://www.semitest.com/ | Epi layer resistivity |
| Scientific Systems | http://www.scientificsystems.physics.dcu.ie/ | Langmuir probe, plasma impedance |
| SC Technology | http://www.sctec.com/sctinfo.htm | Reflectometry for coaters |
| Sigma Instruments | http://www.sig-inst.com/ | QCM |
| Sofie Instruments | http://www.microphotonics.com/sofie.html | Interferometer, OES, ellipsometer |
| Sopra | http://www.sopra-sa.com/ | Spectroscopic ellipsometry |
| Spectra International | http://spectra-rga.com/ | RGA |
| Spectral Instruments | http://www.specinst.com/ | Spectrophotometers, CCD cameras |
| Therma-Wave | http://www.thermawave.com/ | Thin-film and implant metrology |
| Thermionics | http://www.thermionics.com/ | Diffuse reflectance spectroscopy |
| Verity Instruments | http://www.verityinst.com/ | OES, spectral ellipsometer, reflectometer |

## REFERENCES

1. National Technology Roadmap for Semiconductors. 1997 Edition, p 183.
2. Adapted from input by C Schietinger, Luxtron Corp., Santa Clara, CA.
3. F Roozeboom. Advances in Rapid Thermal and Integrated Processing. NATO Series. Kluwer Academic Publishers, Dordrecht, The Netherlands, 1996.
4. C Schietinger, B Adams. A review of wafer temperature measurement using optical fibers and ripple pyrometry. RTP '97, New Orleans, Louisiana, Sept. 3–5, 1997.
5. C Schietinger, E Jensen. Wafer temperature measurements: status utilizing optical fibers. Mat. Res. Soc. Symp. Proc. 429:283–290, 1996.
6. DP DeWitt, GD Nutter. Theory and Practice of Radiation Thermometry. Wiley, New York, 1988.
7. C Schietinger, B Adams, C Yarling. Ripple technique: a novel non-contact wafer emissivity and temperature method for RTP. Mat. Res. Soc. Symp. Proc. 224:23–31, 1991.
8. Adapted from input by Jim Booth, Thermionics Northwest, Port Townsend, WA 98368. http://www.thermionics.com/
9. YP Varshni. Physica 34:149, 1967.
10. ES Hellman, JS Harris Jr. J Crystal Growth 81:38, 1986.
11. MK Weilmeier, KM Colbow, T Tiedje, T van Burren, Li Xu. Can. J. Phys. 69:422, 1991.
12. SR Johnson, C Lavoie, T Tiedje, JA MacKenzie. J. Vac. Sci. Technol. B11:1007, 1993.
13. SR Johnson, C Lavoie, E Nodwell, MK Nissen, T Tiedje, JA MacKenzie. J. Vac. Sci. Technol. B12:1225, 1994.
14. DRS 1000™ Temperature Monitor, Thermionics Northwest, Inc., Port Townsend, WA 98368. http://www.thermionics.com/
15. Zhongze Wang, Siu L. Kwan, TP Pearsall, JL Booth, BT Beard, SR Johnson. J. Vac. Sci. Technol. B15:116, 1997.
16. JL Booth, BT Beard, JE Stevens, MG Blain, TL Meisenheimer. J. Vac. Sci. Technol. A14:2356, 1996.
17. Adapted from input by Sensys Instruments, Sunnyvale, CA.
18. WTS 100 Wafer Temperature Sensor, Sensys Instruments, Sunnyvale, CA.
19. Adapted from input by Mike Whelan, Verity Instruments, Carrollton, TX. http://www.verityinst.com/
20. Model SD1024 Smart Detector Spectrograph, Verity Instruments, Carrollton, TX 75007. http://www.verityinst.com/
21. Model 1015 DS, Luxtron Corporation, Santa Clara, CA. http://www.luxtron.com/
22. Courtesy of Bob Fry, Cetac Technologies Inc., Omaha, NE, 68107.
23. EP-2000 Spectrometer. Cetac Technologies Inc., Omaha, NE, 68107.
24. PC1000 Miniature Fiber Optic Spectrometer, Ocean Optics, Inc., Dunedin, FL, 34698. http://www.oceanoptics.com/homepage.asp
25. Princeton Instruments, Inc., Trenton, NJ 08619. http://www.prinst.com/
26. Spectral Instruments, Tucson, AZ 85754. http:///www.specinst.com/
27. PK Mozumder, GG Barna. Statistical feedback control of a plasma etch process. IEEE Trans. Semiconductor Manufacturing. 7(1): 1994.
28. Neural Endpointer, Verity Instruments, Carrollton, TX. http://www.verityinst.com/
29. Barry M Wise, Neal Gallagher. PLS_Toolbox. Eigenvector Technologies, Manson, WA 98831.
30. Adapted from input by Peter Solomon, On-Line Technologies, East Hartford, CT 06108.
31. On-Line 2100 FT-IR spectrometer, On-Line Technologies, East Hartford, CT 06108.
32. PR Solomon, PA Rosenthal, CM Nelson, ML Spartz, J Mott, R Mundt, A Perry. A fault detection system employing FT-IR exhaust gas monitoring. Presented at SEMATECH Advanced Process & Equipment Control Program, Proceedings (Supplement), September 20–24, 1997), Lake Tahoe, Nevada, pp 290–294.

33. Adapted from input by RJ Ferran and S Boumsellek. Ferran Scientific Inc., San Diego, CA 92121. http://www.ferran.com/

34. Transpector XPR, Leybold Inficon, East Syracuse, NY 13057. http://www.inficon.com/

35. Micropole™ Sensor System, Ferran Scientific, San Diego, CA 92121. http://www.ferran.com/main.html

36. RJ Ferran, S Boumsellek. High-pressure effects in miniature arrays of quadrupole analyzers for residual gas analysis from $10^{-9}$ to $10^{-2}$ torr. JVST A14:1258, 1996.

37. Adapted from input by CA Gogol, Leybold Inficon, East Syracuse, NY. http://www.inficon.com/

38. CA Tallman. Acoustic gas analyzer. ISA Transactions 17(1):97–104, 1977.

39. A Wajid, C Gogol, C Hurd, M Hetzel, A Spina, R Lum. M McDonald, RJ Capic. A high-speed high-sensitivity acoustic cell for in-line continuous monitoring of MOCVD precursor gases. J. Crystal Growth 170:237–241, 1997.

40. JP Stagg. Reagent concentration measurements in metal organic vapour phase epitaxy (MOVPE) using an ultrasonic cell. Chemtronics 3:44–49, 1988.

41. Leybold Inficon's Composer, Acoustic Gas Composition Controller.

42. Adapted from input by Kevin S Gerrish, ENI Technology, Inc., Rochester, NY 14623.

43. Bird Electronic Corporation, Solon, OH 44139.

44. Applied Energy Systems, Inc., Malvern, PA 19355. http://www.aenergysys.com

45. ENI Technology, Inc., Rochester, NY 14623.

46. Fourth State Technology Division of Advanced Energy Industries, Austin TX 78754.

47. Comdel, Inc., Gloucester, MA 01930.

48. Advanced Energy Industries Inc., Fort Collins, CO 80525.

49. D Buck. Texas Instruments, personal communication.

50. CSS 100 Chamber-Wall State Sensor, Sensys Instruments, Sunnyvale, CA 94089.

51. Adapted from input by Ran Kipper, Nova Measuring Instruments Ltd., Weizman Scientific Part, P.O.B. 266, Rehovoth 76100, Israel.

52. NovaScan, Nova Measuring Instruments Ltd., Weizman Scientific Park, P.O.B. 266, Rehovoth 76100, Israel.

53. Model 2100 Process FT-IR, On-Line Technologies, East Hartford, CT.

54. T Buffeteau, B Desbat. Applied Spectroscopy. 43:1027–1032, 1989.

55. F Abeles. Advanced Optical Techniques. North-Holland, Amsterdam, 1967, Chap 5, p 143.

56. Adapted from input by William T Conner, Leybold Inficon Inc., East Syracuse, NY. http://www.inficon.com.

57. T Dalton, WT Conner, H Sawin, JECS, 141:1893, (1994).

58. Model 2450 Endpoint Controller, Luxtron Corporation, Santa Clara, CA 95051. http://www.luxtron.com/Index.html.

59. Adapted from input by John Hanselman, Active Impulse Systems, Natick, MA 01760.

60. InSite 300, Active Impulse Systems, Natick, MA 01760.

61. JA Rogers, M Fuchs, MJ Banet, JB Hanselman, R Logan, KA Nelson. Appl. Phys. Lett. 71: 1997.

62. JR McNeil et al. Scatterometry applied to microelectronic processing. Microlithography World 1(15):16–22, 1992.

63. Adapted from input by Christopher Raymond, Accent Optical Technologies, Inc., Mountain View, CA 94043. http://www.accentopto.com

64. CJ Raymond, SSH Naqvi, JR McNeil. Resist and etched line profile characterization using scatterometry. SPIE Microlithography. Proc. SPIE 3050:476–486, 1997.

65. S Bushman, S Farrer. Scatterometry measurements for process monitoring of polysilicon gate etch. Proc. SPIE. 3213:79–90, 1997.

66. SSH Naqvi et al. Etch depth estimation of large-period silicon gratings with multivariate calibration of rigorously simulated diffraction profiles. J. Opt. Soc. Am. A 11(9):2485–2493, 1994.

67. SSH Naqvi, JR McNeil, RH Krukar, KP Bishop. Scatterometry and simulation of diffraction-based metrology. Microlithography World 2(3): ,1993.

68. CDS-2 Scatterometer, Bio-Rad Laboratories, Albuquerque, NM. http://www.bio-rad.com/

69. CJ Raymond, JR McNeil, SSH Naqvi. Scatterometry for CD measurements of etched structures. Integrated Circuit Metrology, Inspection and Process Control X, Proc. SPIE 2725: ,1996.

70. CJ Raymond et al. Multi-parameter grating metrology using optical scatterometry. JVST B 15(2): ,1997.

71. G Barna, LM Loewenstein, KJ Brankner, SW Butler, PK Mozumder, JA Stefani, SA Henck, P Chapados, D Buck, S Maung, S Sazena, A Unruh. Sensor integration into plasma etch reactors of a developmental pilot line. J. Vac. Sci. Technol. B 12(4):2860, 1994.

72. Adapted from input by Charlie Kohn, SemiTest, Inc., Billerica, MA 01821.

73. Epimet, SemiTest, Inc., Billerica, MA 01821.

74. GG Barna. APC in the semiconductor industry, history and near-term prognosis. SEMI/IEEE ASMC 96 Workshop, Cambridge, MA, Nov. 14, 1996, pp 364–369.

75. The North American/International Chemometrics Society (NAmICS). www.iac.tuwien.ac.at/NAmICS/WWW/welcome.html

76. NB Gallagher. A comparison of factor analysis methods for endpoint and fault detection. AEC/APC Symposium XI, Vail, CO, Sept. 11–16, 1999, Proceedings vol. 2, pp 779–797.

77. R Bunkofske. IBM. How to use real-time process monitoring to improve yield and reduce manufacturing costs. Tutorial, AEC/APC Symposium X, October 11–16, 1998, Vail, CO.

78. D White. DMOS-5, Texas Instruments, Inc., Dallas, TX.

79. A Toprac. Fab-25, Advanced Micro Devices, Austin, TX.

80. P. Hanish. Real-time processing faults detected in dry etch: a case study. AEC/APC Symposium X, Vail, CO, October 11–16, 1998, Proceedings, vol. 1, pp 369–380.

81. A Ison, D Zhao, C Spanos. In-situ plasma etch fault diagnosis using real-time and high-resolution sensor signals. AEC/APC Symposium X, Vail, CO, October 11–16, 1998, Proceedings, vol. 1, pp 389–403.

82. D de Roover, A Emami-Naeini, JL Ebert, RL Kosut. Trade-offs in model-based temperature control of fast-ramp RTP systems. AEC/APC Symposium XI, Vail, CO, Sept 11–16, 1999, Proceedings vol. 2, pp 799–810.

83. PA Rosenthal, O Bosch, W Zhang, PR Solomon. Impact of advanced process control on the epitaxial silicon process. AEC/APC Symposium XI, Vail, CO, Sept. 11–16, 1999, Proceedings vol. 2, pp 525–541.

84. CA Bode, AJ Toprac. Run-to-run control of photolithography overlay. AEC/APC Symposium XI, Vail, CO, Sept. 11–16, 1999, Proceedings vol. 2, pp 715–724.

85. GM Scher. Wafer tracing comes of age. Semiconductor Int. May 1991, p 126.

86. Darren Dance, Wright Williams, and Kelly, Austin, TX.

87. C Weber, E von Hippel. Patterns in rapid problem solving under conditions of high loss rates: the case of semiconductor yield management. (unpublished) MIT working paper, July 31, 1999. Scheduled for presentation at ASMC 2000.

88. C Weber, V Sankaran, K Tobin, G Scher. A yield management strategy for semiconductor manufacturing based on information theory. PICMET—1999, Portland, OR, pp 533–539.

89. C Weber, V Sankaran, G Scher, K Tobin. Quantifying the value of ownership of yield analysis technologies. Proc. IEEE/SEMI/ASMC, Boson, MA, Sept. 8, 1999, pp 64–70.

90. Verity Instruments, Carrollton, TX 75010.

91. Luxtron Corporation, Santa Clara, CA 95051-0941.

92. Cetac, Omaha, NE 68107.

93. Leybold Inficon, Watertown, MA 2472.

94. Peak Sensor Systems, Albuquerque, NM 87106.

95. Fourth State Technology Division of Advanced Energy.

96. Nova Measuring Instruments, Ltd., Israel.

97. Bookside Software, San Carlos, CA, http://www.brooksidesoftware.com/

98. Symphony Systems, Campbell, CA, http://www.symphony-sys.com/
99. GG Barna. Automatic problem detection and documentation in a plasma etch reactor. IEEE Transactions Semiconductor Manufacturing 5(1):56, 1992.
100. BBN Domain Corporation, Cambridge, MA 02140.
101. Perception Technologies, Albany, CA 94706.
102. Triant Technologies, Nanaimo, BC, Canada V9S 1G5.
103. Umetrics, Kinnelon, NJ 07405.
104. Brooks Automation, Richmond, BC, Canada V6V 2X3.
105. Pattern Associates, Inc., Evanston, IL 60201.
106. Real Time Performance, Sunnyvale, CA 94086.
107. SEMATECH Fault Detection and Classification Workshop, Feb. 18–19, 1997.
108. GG Barna. Procedures for implementating sensor-based fault detection and classification (FDC) for advanced process control (APC). Sematech Technical Transfer Document #97013235A-XFR, Oct. 10, 1997.
109. MiTex Solutions, Inc., 363 Robyn Drive, Canton, MI 48187-3959. http://www.mitexsolutions.com/
110. Adventa Control Technologies, Inc., 3001 East Plano Parkway, Plano, TX. http://www.adventact.com/
111. Catalyst APC Software, Control Solutions Division, KLA-Tencor, Suite 300, 811 Barton Springs Road, Austin, TX 78704-1163. http://www.tencor.com/controlsolutions/index.html
112. Integrated Measurement Association. http://www.integratedmeasurement.com/

# 23

# Metrology Data Management and Information Systems

**Kenneth W. Tobin, Jr.**
*Oak Ridge National Laboratory, Oak Ridge, Tennessee*

**Leonard Neiberg**
*Intel Corporation, Portland, Oregon*

## I.  INTRODUCTION TO SEMICONDUCTOR YIELD MANAGEMENT

*Semiconductor device yield* can be defined as the ratio of functioning chips shipped to the total number of chips manufactured. *Yield management* can be defined as the management and analysis of data and information from semiconductor process and inspection equipment for the purpose of rapid yield learning coupled with the identification and isolation of the sources of yield loss. The worldwide semiconductor market was expected to experience chip sales of $144 billion in 1999, increasing to $234 billion by 2002 (1). Small improvements in semiconductor device yield of tenths of a percent can save the industry hundreds of millions of dollars annually in lost products, product re-work, energy consumption, and the reduction of waste streams.

Semiconductor manufacturers invest billions of dollars in process equipment, and they are interested in obtaining as rapid a return on their investment as can be achieved. Rapid yield learning is thus becoming an increasingly important source of competitive advantage. The sooner an integrated circuit device yields, the sooner the manufacturer can generate a revenue stream. Conversely, rapid identification of the source of yield loss can restore a revenue stream and prevent the destruction of material in process (2).

The purpose of this first section is to introduce the concepts of yield learning, the defect reduction cycle, and yield management tools and systems as they relate to rapid yield learning and the association of defects (referred to as *sourcing*) to tools and processes. Overall, it is the goal of this chapter to present and tie together the different components of integrated yield management (IYM), beginning with the very basic measurement and collection of process data at the source in Sec. II, Data Sources. Section III, Analysis and Information, describes the extraction of additional process information (i.e., what might be called meta-data) from the source data for the purpose of reducing the data to smaller, information-bearing quantities. These analysis techniques and strategies represent relatively new research and development that address the issue of increasing data volumes in the manufacturing process. Finally, Sec. IV, Integrated Yield Management, describes the integration of the various sources of data and information for the purpose of yield learning and prediction.

## A.  Yield Learning

Yield management is applied across different phases of the yield learning cycle. These phases are represented in the top portion of Figure 1, beginning with exploratory research and development (R&D) and process development, followed by a yield learning phase during the yield ramp, and finally yield monitoring of the mature manufacturing process.

   The nature and quantity of data available to the manufacturer vary greatly, depending on the development stage of the process. In the first stage of exploratory research, relatively few measurements are made due to the very low volume required to support feasibility studies and experiments. As manufacturability matures from the process development stage to the yield learning stage, automated data collection and test routines are designed and put into place to maximize yield learning while maintaining or increasing wafer throughput (3). At these stages of manufacturing the number of device measurements reaches its maximum, possibly several thousand per chip (3), and encompasses both random and systematic defect sources.

   For the purposes of this discussion, *random defects* are defined as particles that are deposited on the wafer during manufacturing that come from contamination in process gases, tool chambers, wafer-handling equipment, and airborne particulates in the fabrication environment. Random particles are characterized statistically in terms of expected defect densities, and are the limiting source in the theoretical yield that can be achieved for an integrated circuit device. Systematic defects are associated with discrete events in the manufacturing process, such as scratches from wafer-handling equipment, contamination deposited in a nonrandom pattern during a deposition process, microscratches resulting from planarization processes, and excessive pattern etch near the edge of a wafer. Figure 2

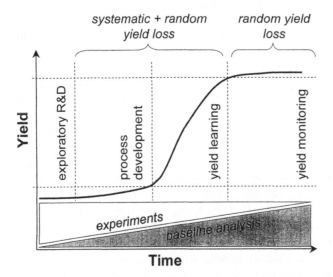

**Figure 1**  Yield curve representing the different phases of semiconductor manufacturing (top), and the tradeoff between experimental process design and baseline analysis as the process matures (bottom).

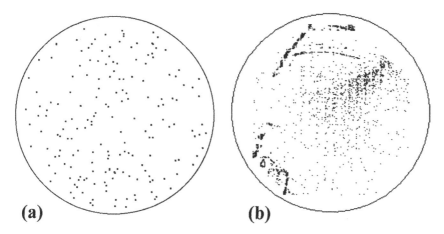

**Figure 2** Example of a semiconductor wafermap showing (a) a random defect distribution and (b) several systematic distributions, such as scratches and a deposition pattern.

shows examples of (a) a random particle distribution versus (b) a systematic distribution. During yield learning, random yield loss and systematic yield loss both occur to various extents, with systematic yield loss dominant early on and random defect yield loss dominant later. As manufacturing approaches the yield monitoring phase of mature production, systematic yield loss becomes more rare and random defects become the dominant and limiting source of yield loss in the process.

The amount of experimental design versus baseline analysis varies across the yield learning cycle as well. This is represented in the bottom portion of Fig. 1.

*Experimentation* refers to the process design sequence and the design of appropriate tool parameters (i.e., recipes) required to achieve a desired product specification, e.g., linewidth, film thickness, dopant concentration. Experiments are performed by varying many operational parameters to determine an optimal recipe for a process or tool.

*Baseline analysis* refers to the establishment of an average expectation for a process or tool. The baseline operating parameters will produce an average wafer of a given yield. As yield learning is achieved, the baseline yield will be upgraded to accommodate lessons learned through process and equipment recipe modifications. As the process matures for a given product, process and tool experiments are replaced by baseline analysis until a stable and mature yield is achieved.

## B. The Defect Reduction Cycle

It has been estimated that up to 80% of yield loss in the mature production of high-volume integrated circuits can be attributed to visually detectable, random, process-induced defects (PIDs) such as particulates in process equipment (4,5). Yield learning in the semiconductor manufacturing environment can therefore be closely associated with the process of defect reduction. Figure 3 shows the process by which yield learning is approached by many semiconductor manufacturers today (6).

At the top of the cycle is the yield model that is used to predict the effects of process-induced defectivity on the function and yield of devices. The model is used to predict process yield and to allocate defect budgets to semiconductor process equipment (7). Defect detection encompasses a group of critical inspection methods for evaluating and

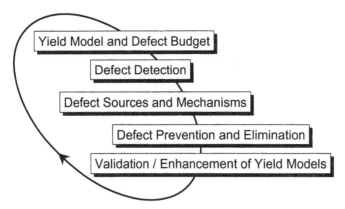

**Figure 3**  Typical defect reduction cycle strategy for controlling process-induced defectivity levels in processes and equipment.

estimating the efficacy of manufacturing on devices that cannot be tested for electrical function at early production stages. Inspection can be broken into two major categories, in-line and off-line. In-line inspection takes place in the fab and typically depends on optical microscopy and laser-scattering systems to scan large areas of the wafer. The result of in-line inspection is a wafermap file containing information about the defect location and size along with process information such as layer, lot number, and slot position. The wafermap information is stored in the data management system (DMS) and contains an electronic roadmap of defect sites that are used to relocate defects for detailed analysis during off-line review. Off-line review is a materials characterization and failure analysis process and methodology that includes many inspection modalities, such as high-resolution color optical microscopy, confocal optical microscopy, scanning electron microscopy (SEM), atomic force microscopy (AFM), and focused ion beam (FIB) cross-sectional analysis. In-line review is typically nondestructive and relatively timely (i.e, keeps up with the manufacturing process through the application of computer vision), whereas off-line techniques are typically destructive (e.g., SEM or FIB) and are expensive, tedious, and time consuming.

The main purpose for collecting defect, parametric, and functional test data is to facilitate the sourcing and discovery of defect creation mechanisms, i.e., isolating the tools and processes that are damaging the wafer and investigating and correcting these errant conditions as rapidly as possible. Much of the day-to-day yield management activities are related to this process. Defect sourcing and mechanism identification represents a tactical approach to addressing yield loss issues. The learning that takes place in conjunction with this day-to-day process is used to develop a strategic approach to defect prevention and elimination, i.e., reducing the likelihood of yield loss from reoccurring in the future by the modification or redesign of processes and products. Finally, the reduction and elimination of the various sources of defects and parametric yield loss mechanisms is fed back into the yield model, effectively closing the defect reduction cycle.

## C.  Yield Management Tools and Systems

The variety, extent, and rate of change of both manufacturer-developed and commercially available yield management systems in the field today precludes an exhaustive description of these capabilities. The types of data that are measured and maintained in the yield management database are also varied, but they include a common subset that we will refer to throughout this discussion.

*Defect Metrology*—defect data collected from in-line inspection and off-line review microscopy and laser-scattering equipment. This data is typically generated across a whole wafer, and an electronic wafermap, i.e., digital record, is generated that maintains information on the location and size of detected defects. There may also be defect classification information in this record supplied through manual engineer classification or automatic defect classification systems during off-line review or in-line on-the-fly defect classification.

*Equipment metrology*—includes measurements that represent physical characteristics of the device or wafer, such as linewidth, location of intentionally created fiducial features, film thickness, and overlay metrology. Imagery can also be created by metrology inspection, as described later.

*Imagery*—images collected from off-line review tools corresponding to defects detected in-line that are also maintained in the yield management database. These images come from many different imaging modalities, such as optical microscopy, confocal microscopy, SEM, AFM, and FIB cross-sectional microscopy. Included in this category of data can be images that represent physical characteristics of the wafer, such as critical-dimension and overlay metrology. The latter two categories are related not to defect and pattern anomalies but rather to geometric characteristics of the patterns and layers.

*Parametric/binmap and sort*—a category of data commonly referred to as *electrical test data*. Electrical testing is performed to verify operational parameters such as input and output voltage, capacitance, frequency, and current specifications. The result of parametric testing is the measurement and recording of a real-valued number, whereas a bin or sort test results in the assignment of a pass/fail code for each parametric test designated as a bin code. The bin codes are organized into a whole-wafer record called a *binmap*, analogous to the wafermap described earlier. The binmap is used to characterize the manufacturing process in terms of functional statistics, but it is also used to determine which devices will be sorted for pass or fail, i.e., which devices yield and will be sold. For this reason, binmap data is also referred to as *sort data* and is a fundamental measurement of yield. It should be noted that die sort based on chip processing speed is critical, since current in-line critical-dimension and dopant control does not ensure that in-line binning is the same as final sort. Parametric testing in the form of electrical testing is also used to infer other nonelectrical parameters, such as linewidth and film thickness.

*Bitmap*—electrical testing of memory arrays to determine the location of failed memory bits resulting in a whole-wafer data record analogous to the wafermap described earlier.

*In situ sensors*—tool-based sensors that measure a given characteristic of a process, such as particle counts, moisture content, or endpoint detection in an etch process. In situ sensors can be advantageous in that they measure properties of the process, potentially before a drifting tool causes significant yield impact on the product. In situ sensor data is inherently different in its structure and form, since it does not spatially describe product quality like visual or electrical inspection. In situ data is time based and describes the state of the process over

a period of time. An in situ measurement may encompass a single-wafer process or a wafer-lot process.

*Tool condition/tool health*—a monitor signal generated by every process tool that is
used for local tool control, e.g., temperature, pressure, gas flow rate, or radio
frequency power level. This data is increasingly being maintained for use in
prognostics, diagnostics, preventive maintenance, and health assessment of the
process tools and equipment. As with in situ sensors, tool health data is time
based and describes the state of the process over a period of time.

*Work-in-process (WIP)*—corresponds to the wafer tracking system. This data
describes the processes, tooling, and recipes required to manufacture a desired
product. It is also used to track wafers and lots while in process. It represents
the initial planning for processing wafers in the plant, and it contains a wafer
history of which tools and processes the wafer was exposed to during production. This data is key to associating yield loss with specific processes and
equipment, i.e., tool isolation.

*Computer-aided design (CAD)*—data that contains the electronic layout for the
integrated circuit. Layout is a complicated process by which a composite picture, or record, of the circuit is generated, layer by layer, and supplied to the
manufacturer to create lithography masks and WIP information. It should be
noted that the design data represented by the electronic CAD drawing is not
typically integrated into the yield management environment but holds much
promise for providing feedback and design modification to optimize the layout
to mitigate the effects of random and systematic defectivity impact on the
device.

More detailed descriptions of the nature of these and related data sources are provided in Sec. II, Data Sources. A typical yield management database contains various
proportions of the data types just described, and this data is maintained within the database for various lengths of time. Table 1 shows the average distribution of these data types
across yield management systems in use today, along with the average time the data is
stored and the range of storage capacity that these systems maintain (8).

There are several reasons that some data is maintained for longer periods than other
data. Binmap, parametric, and process data is typically retained for 6 months to 2 years;
other types of data are usually kept for 2–6 months. Storage capacity is the overriding
reason for limiting data retention time. In addition to capacity, factors relating to the
lifetime of some aspects of manufacturing (e.g., cycle time, lot lifetime, product lifetime)

**Table 1**  Distribution of Data Types, Storage Time, and Storage Capacities Within Today's
Yield Management System

| Data type | Storage distribution | Duration of storage |
|---|---|---|
| Images | 29% | 2–6 months |
| Wafermap | 22% | 2–6 months |
| Binmap | 17% | 6 months to 2 years |
| In situ/tool condition | 15% | 6 months to 2 years |
| Parametric | 10% | 6 months to 2 years |
| Bitmap | 7% | 2–6 months |
| Range of database capacities: 10–300 Gbytes | | |

are important factors. As an example, parametric data can be useful in locating the cause of reliability issues, i.e, device failures after leaving the fab. A parametric data record or data system is also relatively compact (e.g., versus imagery) and easier to maintain for longer periods. Image data generally requires a large block of storage space per record and can quickly fill the data system to capacity. Therefore this data type is typically maintained for only a few months. Other factors related to data retention times include the availability of software tools for analyzing the volume of data, the cost of ownership of an increasingly integrated and complex yield management system, and the lack of standards for acquiring and storing information such that it can be efficiently retrieved and used at a later date.

Figures 4 and 5 represent a simplified description of two extremes in current yield management architectures and philosophies across a broad category of semiconductor manufacturers (8). In Fig. 4, each independent database is represented according to the data type maintained. The shaded regions represent areas where commercial yield management systems are finding the highest acceptance to date. Other, nonshaded regions represent technologies that tend to be designed and implemented in-house by the yield engineering team.

Figure 5 represents the highest level of database integration observed to date (8). This configuration is not as common due to the requirement that data from legacy database systems need to be replaced with newer technologies to achieve high levels of integration. To implement a configuration such as that shown in the figure requires that older databases and systems be ported to newer technology platforms. Therefore, the general trend in yield management system technology is to move toward distributed systems while attempting to integrate more of the data from these systems for engineering (i.e., investigative) analysis. Facilities to measure, store, and maintain in situ process data and tool-condition data are the least mature, while the ability to measure, store, and maintain wafer-based data (e.g., defect, parametric, binmap, and bitmap) are the most advanced. The primary issue with in

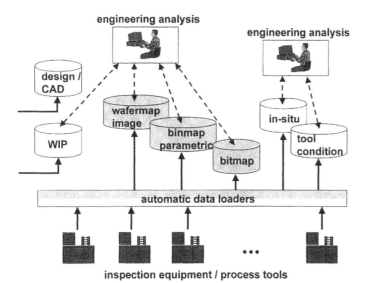

**Figure 4** Yield management systems in today's fabrication environment tend to consist of many separate systems developed over time to address differing needs. These are subsequently joined together in a virtual environment for data sharing and analysis. The grey shading represents areas where commercial systems are having the most impact.

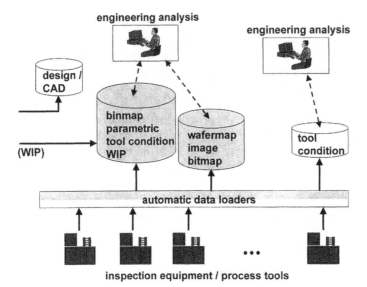

**Figure 5**  Some companies have achieved a high level of integration of the many separate systems shown in the previous figure. These companies have had to integrate their yield management processes and systems from the ground up. This configuration is not as common as the virtual environment.

situ and tool-health data is that it is time based, not wafer-based. Correlating time-based data with wafer-based data for yield analysis is difficult to implement.

Although yield management systems and capabilities are continuing to mature at a rapid pace, there are many areas of standards, infrastructure, and technology that are continuing to be addressed in an evolutionary sense. Figure 6 represents a roadmap of several of the most pressing issues that are being addressed today by yield engineers, information technology teams, and standards organizations regarding the evolution of semiconductor DMSs.

## II.  DATA SOURCES

This section will describe in more detail many of the data sources initially listed in Sec. I.C, Yield Management Tools and Systems, and will enable a discussion of the uses of this data for analysis in Sec. IV, Integrated Yield Management. The character of semiconductor manufacturing is noteworthy for the number and variety of data sources that can be collected and used for yield and product performance enhancement. Aside from WIP data and final test data, which are collected as a by-product of the fabrication process, many data sources are explicitly created at substantial expense as an investment in accelerating yield learning. The primary examples of additional data sources in this category are defect metrology, equipment metrology, laboratory defect analysis, and parametric electrical test.

### A.  Defect Metrology

Defect metrology data can be described as the identification and cataloging of physical anomalies, even on the wafer at intermediate operations during manufacturing. Individual

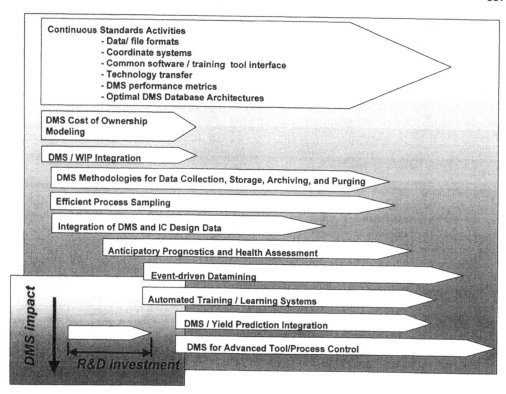

**Figure 6** Although DMSs are continuing to evolve in analysis capability, functionality, and capacity, there are still many issues that are continuing to be addressed by researchers, manufacturers, and suppliers. This figure presents the top issues in the field today.

detectable defects are not guaranteed to cause functional failures (e.g., an organic particle that is later removed during an etch operation), nor are all defects that cause failures guaranteed to be detected by defect metrology equipment during manufacturing (e.g., nonfunctional transistors caused by inadequate ion implantation). The key challenge in optimizing a defect metrology scheme is to maximize the detection of the defects that are likely to cause functional failures (commonly called *killer defects*) while minimizing the resources that detect nonkiller (or *nuisance*) defects. Due to this objective and the complexities of defect metrology equipment, defect metrology data collection has historically been divided into two phases: inspection and review. *Inspection* is the automated identification and collection of data such as defect size, imagery, and automatic categorization, while *defect review* is typically a time-intensive, manual process during which additional data and imagery are collected for targeted defects of interest identified during the inspection process. Although this data is critical for yield learning, it is expensive to collect. In practice, only a fraction of the total number of wafers in a fab are inspected, and of those inspected only a smaller fraction are reviewed.

## B. Laboratory Defect Analysis

Once wafer fabrication has been completed or once a specific device structure has been covered by another layer (e.g., transistor gates that are physically covered by the first metallization layer), further analysis of defects is possible only with significant difficulty. The main complication is the requirement to remove the covering material to expose the defects of interest. Once the desired material is removed through a variety of resource-intensive processes, the target area can be analyzed with laboratory equipment. This equipment is capable of taking extremely high-resolution imagery (e.g., on the order of 10–100 nm) in conjunction with material composition analysis through the use of techniques including focused ion beam etching, SEM, and energy-dispersive x-ray spectroscopy (EDX). These techniques allow the collection of even more detailed data than is available from other types of metrology for the purpose of identifying the root cause of a particular defect.

## C.  Process Metrology

Different types of data-collection tools are used to determine if the width, thickness, and physical placement of intentionally created features meet specification limits. The most common examples of such metrology are critical-dimension (CD) measurement of lines, trenches, and vias, thin-film metrology (i.e., measurement of the thickness of deposited, etched, or polished film layers), and registration (i.e., measurement of the relative alignment between two layers of structures, e.g., between a via at metal level 2 and the corresponding landing pad at metal level 1). Such metrology can be used to characterize critical nondefect-related contributors to yield loss (e.g., overly thick transistor gates that lead to unacceptable device performance).

## D.  Parametric Electrical Testing

The ability to measure the electrical behavior of the silicon is invaluable in yield analysis. The collection of such data relies on the intentional creation of electrical test structures. In the layout process, simple circuits are created that enable the measurement of parametric (i.e., noncategorical real-valued) quantities such as sheet resistance and transistor-off current. The collection of the data from these structures is performed by electrical test equipment that places electrical probes on special contact pads, creates specific test inputs on several probes, and reads the electrical conditions present at other probes. Probe outputs can then be input to analysis equations based on the test circuit design to determine the value of the desired parametric value. There can be several hundred different parametric electrical tests that are collected at several sites across a wafer, such as capacitance and current. These values can be used to identify wafer-to-wafer or across-wafer variability of critical electrical properties, as well as links between physical properties of the silicon and functional device behavior.

## E.  Sort Testing

Sort testing is the final assessment of whether a specific die performs as desired and should be targeted for assembly as a packaged part for sale. Sort test equipment makes electrical contact with the output pads of the die, applies specific input patterns to some pads, reads the outputs off other pads, and determines from these outputs whether the circuitry performs the desired functions at the desired clock speed. There are three types of sort data that can be obtained, based on three different methodologies: bitmap testing, functional testing, and structural testing. All sort testing applies input electrical patterns to a set of electrical contacts and reads outputs from a set of output contacts, but the test methodology dictates the character of data obtained.

*Bitmap testing* is possible only on memory circuitry, such as SRAM, DRAM, or memory cache. This testing performs read and write operations on each memory cell to determine precisely which memory cells or groups of cells are functional and which are flawed. Not only do these data provide detailed information regarding which devices are functional enough for sale, but the data can also pinpoint the locations of the electrical faults to enable laboratory failure analysis.

In contrast, *functional testing* subjects the die to a set of test input patterns that are designed to exercise the main functions of the circuitry. Based on the specific failures (i.e., actual output patterns that do not match target output patterns), each die is classified as being a member of a particular "sort bin" that defines a general category of functional behavior. For example, die categorized in bin 1 may be fully functional at the desired clock speed, bin 2 may be for fully functional die at a less profitable clock speed, bin 3 for die with an unacceptable number of cache failures, etc. Depending on how the bins are defined, categorization of die in certain bins may indicate the functional block of the die that is nonfunctional (e.g., arithmetic logic unit failure), but such electrical fault localization is typically not at sufficient resolution to pinpoint the exact location of the failure.

*Structural testing* is a methodology that is designed to enable reliable localization of electrical faults, even in random circuitry (i.e., it is not limited to memory circuitry) while requiring less expensive test equipment than is required for functional testing. Ideally, structural testing data will include not only sort bin data, indicating which functions are inoperable, but also a list of electrical nodes or specific circuit structures that are faulty. This fault localization can be used for laboratory failure analysis of random logic circuits. The ability to accurately localize fault locations is a critical capability on the *International Technology Roadmap for Semiconductors* (9), and structural testing is expected to play an increasingly significant role in meeting that challenge.

## F. Work-in-Process Data

Work-in-process data is a general term that describes the processing chronology or history of a wafer. This data consists of a list of all manufacturing operations to which a wafer was subjected, along with the specifics of the processing configuration at each operation. These specifics include the time at which the processing occurred, the relative positions of each wafer in the processing tool (e.g., slot position or processing order), and the exact tool settings or recipe used.

The source of this data is the factory automation system, whose primary function is to ensure that each wafer is processed exactly as specified. The unique process specification is combinatorially complex, given the hundreds of individual processing operation, the tens of processing tools that can be used to execute each operation, and the hundreds of configurations specified at each operation. Although the primary function of the automation system is to ensure correct processing, the storage of WIP data is required to identify a specific piece of process equipment as the root cause of yield loss.

## G. Industry Formats for Data Transmission and Storage

Although data-collection methodologies and basic informational content for the foregoing data types are largely common across the industry, data file formats are not. In practice, data file formats are specific to each tool that generates the data. As an example, there are completely different defect metrology data formats for each of the separate tool types sold by a single tool vendor. This multiplicity of data formats is one of the major factors complicating yield analysis. This issue was discussed in Sec. I.C, Yield Management Tools and Systems, and represented as a main issue in Figure 6 under "Continuous Standards Activities."

There are two mechanisms for the establishment of industry standards in the future. Standards can emerge either by de facto adoption by tool suppliers or by official establishment by a standards committee, such as Semi or I300I. However, neither mechanism for standards establishment is yet to have yielded significant results. The impact of this lack of standardization will be described in Sec. IV, Integrated Yield Management.

## III. ANALYSIS AND INFORMATION

Semiconductor yield analysis makes use of multiple sources of data collected from the manufacturing process, sources that are continuing to grow in volume due to increasing wafer size and denser circuitry. This section begins with a review of the fundamental techniques of statistical yield analysis. Yield is based on a measure of the fraction of shippable product versus total input. This is typically determined at functional test, when each die on the wafer is electrically determined to pass or fail a set of operating parameters. It is important to understand what is happening in the manufacturing process prior to final test; therefore, there are a number of techniques for estimating instantaneous device yield based on measurements of physical and parametric defects. Due to increased wafer dimensions and decreasing line-width, there are huge quantities of data being collected in the fab environment. To accommodate this situation, there are new levels of automation coming on-line that result in the reduction of data for informational purposes. Automatic defect classification (ADC), spatial signature analysis (SSA), and wafer tracking techniques represent a few of these techniques that are described next in relation to yield management, analysis, and prediction.

### A. Yield Prediction

*Yield* can be defined *as the fraction of total input transformed into shippable output.* Yield can be further subdivided into various categories, such as (10):

> *Line yield*—the fraction of wafers not discarded prior to reaching final electrical test
> *Die yield*–the fraction of die on yielding wafers that are not discarded before reaching final assembly and test
> *Final test yield*—the fraction of devices built with yielding die that are deemed acceptable for shipment

Yield modeling and analysis is designed as a means of proactive yield management versus the traditional sometimes "reactive" approach that relies heavily on managing yield crashes (i.e, "fire fighting"). A yield management philosophy that promotes the detection, prevention, reduction, control, and elimination of sources of defects contributes to fault reduction and yield improvement (11).

Semiconductor yield analysis encompasses developing an understanding of the manufacturing process through modeling and prediction of device yield based on the measurement of device function. Historically, the modeling of process yield has been based on the fundamental (and simple) assumptions of binomial or Poisson statistics (12,13), specifically the following.

The yield is binary; i.e., a device either functions or it does not function.

The number of occurrences of failures is small relative to the population of devices on the wafer.

Failure of a device on the wafer is uncorrelated to the failure of any other device on the wafer (or lot, etc.); i.e., device failures are uncorrelated and random.

Yield is a simple function of active device area, $A$, and the average wafer defect density, $D$, i.e., $Y = e^{-AD}$ i.e., the Poisson distribution.

These assumptions typically hold true, to a reasonable approximation, for mature processes, where the yield is high and is limited primarily by random events. But during the process development and yield learning stage, these models do not correlate well with observed yields. To account for these inaccuracies, there have been some attempts to incorporate defect clustering relationships and/or systematic defect processes into the analysis models. It is at this point that the measurement of the spatial distributions of defect/fault events begins to address correlated populations of defects as unique systematic and repeatable signature events.

Yield modeling has application to yield prediction, i.e., an estimate of position on the yield curve of Fig. 1, device architecture design, process design, and the specification of allowable defectivity on new process tools necessary to achieve desired future yield goals. In relation to process control, yield analysis has applicability to process characterization, e.g., in relation to improving the rate of yield learning. To achieve this last goal, it is important that yield modeling accommodate both systematic and random defects. Once on top of the yield curve during the yield monitoring phase, systematic mechanisms are a small portion of the overall defect source issue, but it should be noted that manufacturing can remain in the yield learning phase for several years (15).

To accommodate systematic defect and fault distributions, researchers have modeled concepts of defect "clustering" on wafermaps. The well-known negative binomial yield model (12) recognizes defect clustering by integrating the simple Poisson yield model over an effective defect density distribution function $f(D)$, i.e., $Y = \int e^{-AD} f(D) \, dD$. This result is the compound Poisson distribution model, well known as the negative binomial relationship, $Y = (1 + AD/\alpha)^{-\alpha}$, where $\alpha$ is defined as a clustering parameter that accounts for the variability in defect densities from wafer to wafer or lot to lot. Different values for $\alpha$ attempt to facilitate different models of clustering in the defect distributions measured across a wafer and result in the relationships commonly used for yield prediction that are shown in Table 2.

When clustering becomes very severe, i.e., the distributions of defects become more dense and systematic, variations of the models shown in Table 2 are derived by partitioning the wafer into discrete, independent zones, e.g., stepper fields, and/or quadrant or radial zones (12,14). While this can improve the performance of the model, partitioning methods are still susceptible to the limitations of the Poisson model; e.g., the model assumes a random, uncorrelated distribution of defects in each zone and a generally small population of defects. An approach of this nature has been put forth by SEMATECH for a 250-nm-yield model (15) and a 150-nm-yield model (16).

Simply detecting the onset of a systematic defect creation mechanism can be helpful to catching a process that is drifting or moving rapidly out of control. Kaempf (17) has shown how application of a simple binomial distribution model can be used to detect the onset of systematic distributions, e.g., a reticle-induced defect (repetitive) pattern or an

**Table 2** Yield Models Derived from the Poisson Probability Distribution

| Degree of clustering | Yield model |
|---|---|
| *Poisson*: no clustering, $\alpha > 7$ | $Y = e^{-AD}$ |
| *Murphy's*: minor degree of clustering, $\alpha = 4.5$ | $Y = \left[ (1 - e^{-AD}) AD \right]^2$ |
| *Negative binomial*: moderate clustering, $\alpha = 2$ | $Y = \left( 1 - \dfrac{AD}{\alpha} \right)^{-\alpha}$ |
| *Seed's*: large degree of clustering, $\alpha = 1$ | $Y = \dfrac{1}{1 + AD}$ |

Each model accommodates a varying degree of clustering.

edge ring pattern. A plot of yield probability as a function of device yield will deviate from a binomial-shaped distribution as systematic events take precedence over random ones. This technique, although simple to implement, requires a fairly large number of data points, e.g., wafers and/or lots, before a determination can be made; and the method cannot resolve one type of systematic event from another; i.e., it is primarily an alarm for process drift or excursion.

## B. Automatic Defect Classification

Automatic defect classification (ADC) has been developed to provide automation of the tedious manual inspection processes associated with defect detection and review. Although the human ability to recognize patterns in data exceeds the capabilities of computers in general, effectively designed ADC can provide a more reliable and consistent classification result than can human classification under well-defined conditions. These conditions are typified by highly manual and repetitive tasks that are fatiguing and prone to human error. Figure 7 shows representative examples of the variety of defect imagery that arises in semiconductor manufacturing. These include examples of individual pattern and particle defects sensed using optical and electron microscopy.

Automatic defect classification was initially developed in the early 1990s to automate the manual classification of defects during off-line optical microscopy review (18,19,20). Since that time, ADC technologies have been extended to include optical in-line defect analysis (21) and SEM off-line review. For in-line ADC, a defect may be classified "on-the-fly," i.e., during the initial wafer scan of the inspection tool, or during a revisit of the defect after the initial wafer scan. During in-line detection the defect is segmented from the image using a die-to-die comparison or a "golden template" method, as shown in Figure 8 (22,5). This figure shows an approach to defect detection based on a serpentine scan of the wafer using a die-to-die comparison, first showing A compared to B, B compared to C, etc., and ultimately building a map of the entire wafer. During off-line review the defect is redetected using the specified electronic wafermap coordinates and die-to-die or golden template methods. The classification decision derived from the ADC process is maintained in the electronic wafermap for the wafer under test and will be used to assist in the rapid sourcing of yield-impacting events and for predicting device yield through correlation with binmap and bitmap data if available.

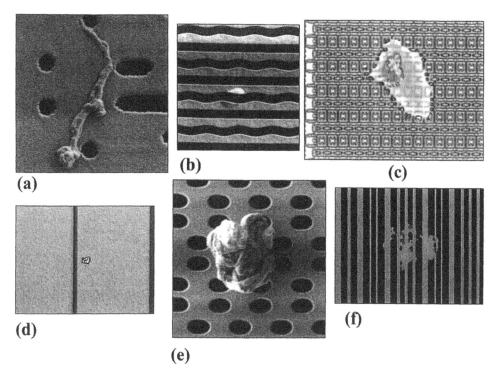

**Figure 7** Examples of typical defects collected in the manufacturing environment by inspection equipment: (a) extra material; (b) embedded particle; (c) missing pattern; (d) poly-flake; (e) surface particle; (f) missing pattern.

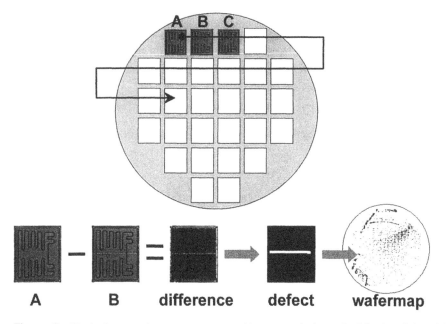

**Figure 8** Typical serpentine scan pattern and image analysis method for localizing defects in wafer imagery. The result is an electronic wafermap, which logs defect locations and structural characteristics.

Figure 9 shows an example of a frequency distribution of defects that can occur in a manufacturing process. This particular data set came from a deep trench process and shows the distribution of 18,840 classified defects across 314 wafers (23). In the figure, the defect classes are labeled as RE, SF, LI, HO, LS, etc., and the height of each bin is the frequency of occurrence of each class. It is apparent from the data that 97% of the defects that occurred in the deep trench process are of class LS (67%) and ST (30%). If the cause of the defined defect categories is sufficiently characterized and maintained a-priori in the yield management system, frequency distributions such as this are useful in directing the engineer to the highest-priority issues. In this case, the highest-priority issue may not be the one most frequently occurring. In fact, it should be noted that not all detected defects cause electrical failures. The ratio of defects that cause failures to the total defects detected is termed the *kill ratio* (5). A kill ratio can be determined for each class of defect as well, thereby, giving a relative measure of importance to each frequency bin shown in the figure. For example, if the kill ratio for category LS was 0.25 and the kill ratio for category ST was 0.9, then a prediction can be made that $67\% \times 0.25 = 17\%$ of the total defect population are of the category killer-LS and $30\% \times 0.9 = 27\%$ of the total population are killer-ST. Therefore, if the class-dependent kill ratio is known, the ST category would be the more immediate yield-detracting mechanism to address. This type of statistical data also provides a methodology to estimate device yield prior to electrical test.

## C.  Spatial Signature Analysis

It has been widely recognized that although knowledge of process yield is of primary concern to semiconductor manufacturers, spatial information is necessary to distinguish between systematic and random distributions of defects. Recall that Fig. 2 shows a whole-wafer view of a random distribution of defects in (a) and a systematic pattern of defects exhibiting both clustered events (e.g., scratches) and distributed systematic events (e.g., a chemical vapor deposition contamination problem) in (b). Knowing that a distribution has a spatial effect provides further information about the manufacturing process, even if the distribution has little effect on device yield. Therefore, focusing only on process yield and ignoring the systematic and spatial organizations of defects and faults represents a lost opportunity for learning more about the process (24).

When a high level of clustering is observed on wafermaps, simple linear regression models can lead to inaccurate analysis of the yield data, such as negative yield predictions. Ramirez et al. discuss the use of logistic regression analysis with some modifications to

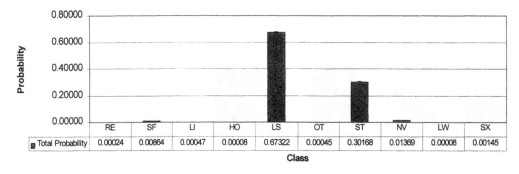

**Figure 9** Example of the frequency distribution of defined defect classes (RE, SF, LI, etc.) across several manufactured lots for a particular process layer.

account for this negative yield prediction effect, known as *overdispersion* (25). This technique attempts to accommodate clustering indirectly in the model.

A more direct approach to handling defect clustering is demonstrated through the work of Taam and Hamada. While defect clustering yield analysis has historically been treated by application of the compound Poisson distribution model discussed in the previous section (which imbeds the defect clustering in the yield model), Taam and Hamada use a measure of clustering based on the idea of nearest neighbors (24). This method makes a direct measure of the spatial relationships of good and bad die on a wafer. The method uses join-count statistics and the log-odds ratio as a measure of spatial dependence. While the technique does identify the occurrence of spatial events, wafermaps and join-count maps (which reveal the joined clusters of die) are required to be analyzed manually to learn more about the root cause.

Collica et al. (26,27) incorporate the methods of Ramirez et al. (25) and Taam and Hamada (24) using join-count statistics and spatial log-odds ratios to organize and describe spatial distributions of good and bad die through the use of CUMSUM charts. The two main objectives of the approach are to recognize and understand common causes of variation, which are inherent in the process and identify special causes of variation that can be removed from the process.

An interesting extension of this approach has been applied to the analysis of spatial bit-fail patterns in SRAM devices, where Kohonen self-organizing maps and perceptron back-propagation neural networks have been used to classify the spatial patterns. These patterns occur, e.g., in vertical bit-line faults, horizontal single-word faults, diagonal doublets, and clustered bit faults (28,29).

The analysis of spatial patterns of defects across whole wafers is described earlier as a means to facilitate yield prediction in the presence of systematic effects. Tobin et al. have developed an automated whole-wafer wafer analysis technique called SSA to address the need to intelligently group, or cluster, wafermap defects together into spatial signatures that can be uniquely assigned to specific manufacturing processes and tools (30–33). This method results in the rapid resolution of systematic problems by assigning a label to a unique distribution, i.e., signature, of defects that encapsulate historical experience with processes and equipment. Standard practice has been to apply proximity clustering that results in a single events being represented as many unrelated clusters. Spatial signature analysis performs data reduction by clustering defects together into extended spatial groups and assigning a classification label to the group that reflects a possible manufacturing source. Figures 10 and 11 show examples of clustered and distributed defect distributions, respectively, that are isolated by the SSA technique. The SSA technology has also been extended to analyze electrical test binmap data (i.e., functional test and sort) to recognize process-dependent patterns (34) in this data record.

The SSA and ADC technologies are also being combined to facilitate intelligent wafermap defect subsampling for efficient off-line review and improved ADC classifier performance (23,35,36). The integration of SSA with ADC technology can result in an approach that improves yield through manufacturing process characterization. It is anticipated that SSA can improve the throughput of an ADC system by reducing the number of defects that must be automatically classified. For example, the large number of defects that comprise a mechanical scratch signature that is completely characterized by SSA will not need to be further analyzed by an ADC system. Even if a detected signature cannot be completely characterized, intelligent signature-level defect sampling techniques can dramatically reduce the number of defects that need to be sent to an ADC system for subsequent manual or automated analysis (e.g., defect sourcing and tool isolation).

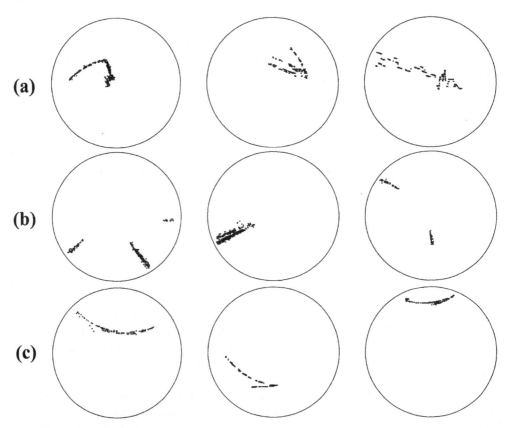

**Figure 10**   Examples of spatial clusters, or signatures, of defects across a series of wafers: (a) scratches; (b) streaks from rotating equipment such as spin coaters; (c) double-slot patterns caused by scraping one wafer against another during transport. These patterns were isolated using SSA technology.


The accuracy of an ADC system can potentially be improved by using the output of the SSA wafermap analysis to perform focused ADC. Focused ADC is a strategy by which the SSA results are used to reduce the number of possible classes that a subsequent ADC system would have to consider for a given signature. And SSA signature classification can be used to eliminate many categories of potential defects if the category of signature can be shown a priori to consist of a limited number of defect types. This prefiltering of classes reduces the possible alternatives for the ADC system and, hence, improves the chance that the ADC system will select the correct classification. It is anticipated that this will result in improved overall ADC performance and throughput.

Another yield management area where SSA can provide value is in statistical process control (SPC). Today, wafer-based SPC depends highly on the tracking of particle and cluster statistics; primarily to monitor the contribution of random defects. Recall that random defects define the theoretical limit to yield, and controlling this population is a key factor in achieving optimal fabrication performance. A cluster is defined as a group of wafer defects that reside within a specified proximity of each other. Current strategies typically involve removing cluster data from the population and tracking the remaining particle data under the assumption that these are random, uncorrelated defects. Field

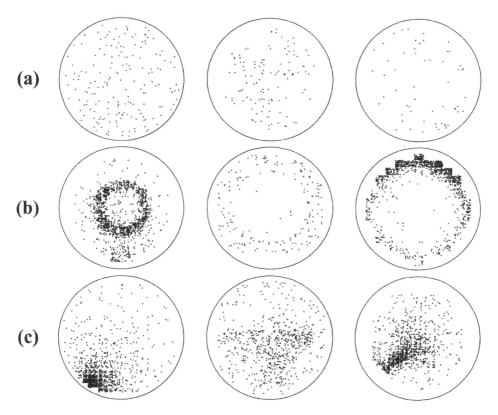

**Figure 11** Examples of random and systematic distributions of defects across a series of wafers: (a) random distributions of various densities; (b) rotationally symmetric patterns caused by processes such as etch; (c) asymmetric distributions caused from processes such as deposition. These patterns were isolated using SSA technology.

testing of the advanced clustering capabilities of SSA has revealed that this basic approach can be modified dramatically to reveal much information regarding systematic defect populations during the yield ramp.

For example, the last wafermap shown in row (a) of Fig. 10 contains a long, many-segmented scratch that commonly used proximity clustering algorithms would categorize as multiple clusters. The ability of SSA to isolate and analyze this event as one single scratch removes ambiguity from the clustering result (i.e., the event is accurately represented by a single group of defects, not many independent clusters). It allows the user to assign process-specific information via the automatic classification procedure to facilitate SPC tracking of these types of events to monitor total counts, frequency of occurrence, etc. Care must also be taken in analyzing random events on a wafer. Row (a) of Fig. 11 shows random populations of defects that are uncorrelated, while rows (b) and (c) show distributed (i.e., disconnected) populations that are systematic, i.e., nonrandom, and can be related to a specific manufacturing process. If the pattern is determined to be systematic, it is virtually impossible to separate random defects from the distributed, systematic event. The current practice of filtering clusters based on proximity alone would result in the counting of these systematic distributions as random defects. Unless a yield engineer happens to view these maps, the count errors could go undetected indefinitely, resulting in the spurious rise and fall of random particle population estimates. Using an approach such as SSA results in the separation of wafer events into random and systematic events (i.e., both clustered and distributed) that provides a higher level of information about the manufacturing process. Using this informational content to distinguish and then monitor, random defects from systematic distributions from scratches, streaks, and other clusters provides the yield engineer a much clearer window into the manufacturing process.

## D.  Wafer Tracking

A contemporary semiconductor process may consist of more than 500 intricate process steps (5,9). A process drift in any of these discrete steps can result in the generation of pattern or particle anomalies that affect other, downstream processes and ultimately reduce yield. Mechanisms for rapidly detecting and isolating the offending process step and specific tools are required to perform rapid tool isolation. One such technique that is becoming commonplace in the fab is wafer tracking. *Wafer tracking* involves monitoring the location of each wafer in the process by reading the laser-etched serial number from the flat or notch of the wafer that is provided by the silicon manufacturer. Tracking requires that an optical character recognition system and wafer sorter be located at each critical step in the process. The serial number is then mapped to a precise equipment location that is subsequently maintained by the DMS (37,38). This allows the wafer to be followed down to the specific slot number or position in the carrier or process tool. Using the silicon manufacturer's serial number also allows the device manufacturer to correlate process patterns with the supplier's silicon wafer parameters.

Yield and process engineers can refer to the wafer-tracking information in the DMS to resolve yield loss issues within the manufacturing line. For example, if an engineer suspects a faulty furnace operation, a report can be generated from the DMS detailing the deviating parameter (e.g., a parametric test result or yield fraction) for wafers versus their location in the furnace tube. Wafer-level data also provides evidence of difficult process problems when a hypothesis of root cause is not initially apparent. In the case of the tube furnace, the engineer may generate a plot that shows the particular step or steps where the impacted wafers were processed together. This discernment can be made, because at each wafer-reading station the wafer positions are randomized by the automatic handler prior to the subsequent processing step. The randomization takes place at every process step and facilitates the isolation of particular tool and positional dependencies in the data. This is typically viewed in a parameter-versus-position plot that will be ordered or random, depending on the tool where the process impacted the lot. For example, a two-dimensional plot with high yield on one end and low yield on the other would implicate a specific tool versus a plot revealing a random yield mix that shows no correlation to that tool.

Historically, wafer tracking has relied on the comparison of whole-wafer parameters such as yield with positional information to determine correlations. A recent development in wafer tracking incorporates spatial signature analysis to track the emergence of particular signature categories and to correlate those events back to specific processes and tools (39). Recall that SSA maps optical defect clusters and electrical test failure wafermap patterns to predefined patterns in a training library. Wafer tracking with SSA captures a wafer's position/sequence within various equipment throughout the fab, and correlates observational and yield results to positional signatures. By frequently randomizing wafer order during lot verification and processing, positional information provides a unique signature of every process step. Integrating SSA with wafer tracking helps to resolve the root causes of multiple yield loss mechanisms associated with defect and sort test wafer-map and positional patterns. This is accomplished by isolating individual defect clusters (i.e., signatures) and identifying which process step most strongly correlates with yield loss. It is anticipated that this capability will facilitate rapid yield learning, particularly during the introduction of new processes.

## IV. INTEGRATED YIELD MANAGEMENT

As integrated circuit fabrication processes continue to increase in complexity, it has been determined that data collection, retention, and retrieval rates increase geometrically. At future technology nodes, the time necessary to source manufacturing problems must at least remain constant, i.e., approximately 50% of the cycle time on average during yield learning. In the face of this increased complexity, strategies and software methods for integrated yield management (IYM) have been identified as critical for maintaining productivity. And IYM must comprehend integrated circuit design, visible defect, parametric, and electrical test data to recognize process trends and excursions to facilitate the rapid identification of yield-detracting mechanisms. Once identified, the IYM system must source the product issue back to a point of occurrence. The point of occurrence is defined to be a process tool, design, test, or process integration issue that resulted in the defect, parametric problem, or electrical fault. Thus, IYM will require a merging of the various data sources that are maintained throughout the fabrication environment. This confluence of data will be accomplished by both the physical and virtual merging of data from currently independent databases. The availability of multiple data sources and the evolution of automated analysis can provide a mechanism to convert basic defect, parametric, and electrical test data into useful process information.

    With the continued increase in complexity of the fabrication process, the ability to detect and react to yield-impacting trends and excursions in a timely fashion will require a larger dependence on passive data. This will be especially true during yield learning, where maximum productivity and profit benefits will be achieved. *Passive* data is defined as defect, parametric, and electrical test data collected in-line from the product through appropriate sampling strategies. The additional time required to perform experiments, e.g., short-loop testing, will not be readily available at future nodes. The time necessary to trend potential problems and/or identify process excursions will require the development of sampling techniques that maximize the signal-to-noise ratio inherent in the measured data. The goal of IYM is to identify process issues in as few samples as possible. Analysis techniques that place product data in the context of the manufacturing process provide a stronger "signal" and are less likely to be impacted by measurement noise, since they comprehend various levels of process history and human experience, i.e., lessons learned (9).

### A. Rapid Yield Learning Through Resource Integration

One of the few commonalities among virtually all semiconductor yield analyses is the requirement to integrate multiple data sources. One of the simplest cases to illustrate this point is the analysis required to identify a single piece of equipment (e.g., a diffusion furnace) that has deposited an unusually large number of killer defects on the wafers processed by that tool in a specific time frame. To identify the root cause, one would use sort data to first identify a subset of wafers output from a fab that had poor die yield.

Next, one would compare WIP data for the low yielding wafers with similar data for high-yielding wafers and identify the diffusion furnace as being correlated with yield. Third, one would analyze the defect metrology data collected for the low-yielding lots and attempt to identify the specific failure mode of the furnace by the spatial signature of any defect clusters. If defect images were available, they could be used to confirm the root cause by matching the defect images with known failure modes of the diffusion furnace. If defect metrology imagery had not been collected, then it might be necessary to send some of the low-yielding wafers to the laboratory for detailed defect analysis. In this simple example from the semiconductor industry, no fewer than four data sources (die yield, WIP, defect metrology, and defect analysis) must be integrated to identify the root cause.

A different example of data integration is the hypothetical analysis of an experiment designed to identify the best lithography tool settings to optimize yield by minimizing the number of bad memory cells in the cache. In this case, an experiment is specified so that each wafer is subjected to a different set of lithography settings at a specific operation; these configurations are stored as WIP data. Once the wafers have been fabricated, sort bitmap data can be extracted to measure the number of failed memory cells on the die of each wafer. At this point, one could combine only two sets of data and naively assume that the best set of processing conditions is the one whose wafers show the best yield (i.e., fewest cache failures), and only two data sources have been combined. However, in practice one must perform additional analysis to ensure that the differences in processing are the root cause of the differences in yield. This analysis may include parametric electrical test results or process metrology data to validate that the lithography configurations have had the anticipated effect on device structures, such as transistor gate width, that adequately explains the reduction in cache memory cell failure. In this example as well, four data sources (WIP, sort bitmap, parametric electrical test, and process metrology) must be analyzed together for yield learning from an experiment.

## B.  Virtual Database

Collecting data from multiple sources as just described should be a simple task of executing a database query and retrieving all of the desired data for analysis. Unfortunately, the state of the industry is characterized by either fragmented, inconsistent, or nonexistent data storage for many of the data sources that may be required (8). This makes some data collection difficult, requiring multiple data queries followed by complex operations to merge the data. Consider our second data integration example, concerning analysis of a lithography experiment. When cache memory cell failures are compared with defect data, it may be desirable to compare the physical locations of the cache failures with those of the defects detected during processing. This comparison can be made much more difficult if, for example, the spatial coordinate systems for bitmap and defect data are significantly different. In such cases, the conversion of the data to a common coordinate system may represent significant additional effort required to execute the desired analysis.

For analysis purposes, the ideal database would be a single database that stores all fab data sources (i.e., defect metrology, equipment metrology, WIP, sort, etc.) for an infinite amount of time and with a common data referencing structure. In more practical terms, this is as of yet unachievable due to finite computing and storage capacity, the difficulties of incorporating legacy systems into a single (i.e., virtual) integrated environment, and the lack of the standard data-referencing structures required to facilitate the storage, transmission, and analysis of yield information.

## C.  Data Mining and Knowledge Discovery

Beyond database infrastructure and merging issues come new methods that attribute informational content to data, e.g., the assignment of defect class labels through ADC, or unique signature labels in the population of defects distributed across the wafer using SSA. These methods put the defect occurrence into a context that can later be associated with a particular process, a material characteristic, or even a corrective action. For example, a defect coordinate in a wafermap file contains very little information, but a tungsten particle (ADC) within a deposition signature (SSA) is placed in the context of a specific manufacturing process and contamination source. Later reporting of this information can lead to rapid yield learning, process isolation, and correction.

Effective data mining represents the next frontier in the evolution of the yield management system. *Data mining* refers to techniques used to discover correlation between various types of input data. For example, a table containing parametric data and functional yield for a given lot (or lots) would be submitted to a data-mining process. The result would be data correlation indicating which parametric issues are impacting yield.

Knowledge discovery refers to a higher level of query automation that can incorporate informational content and data-mining techniques to direct the yield engineer towards a problem solution with minimal human-level interaction. For example, an autosourcing SPC approach may evolve that tracks defects, signatures, or parametric issues and initiates a data-mining query once a set of control limits has been exceeded. The result of the data-mining process would be the correlation of all associated SPC parameters, resulting in a report of a sequence of recommended actions necessary to correct the errant tool or process. These recommendations for corrective actions would be based on historical experience with the process, i.e., through an encapsulation and retention of human expertise.

The highest level of automation is associated with system-level control. System-level control represents a much more complex and potentially powerful control scenario than single or cluster-tool control. System-level control encompasses many tools and processes and can be open-loop (human in the loop) or closed-loop, depending on the reliability and potential impact of the decision-making process. System-level control is currently far down the road in terms of current technical capabilities but represents the future direction of automated yield management. It will be both deterministic and heuristic in its implementation and will be highly dependent on historical human-level responses to wafermap data.

## V.  CONCLUSION

We have presented an overview of the motivation, goals, data, techniques, and challenges of semiconductor yield enhancement. The financial benefits of yield enhancement activities make it of critical importance to companies in the industry. However, yield improvement challenges will continue to increase (9) due to data and analysis complexity, requiring major improvements in yield analysis capability. The most basic challenge of yield analysis is the ability to effectively collect, store, and make use of extremely large amounts of disparate data. Obstacles currently facing the industry include the existence of varied data file formats, legacy systems and incompatible databases, insufficient data-storage capacity or processing power, and an explosion of raw data volumes at a rate exceeding the ability of process engineers to analyze that data. Strategic capability targets include continued development of algorithms to extract information from raw data (e.g., spatial signature analysis), standardization of data formats and data-storage technology to facilitate the integration of multiple data sources, automated tools such as data mining to identify information hidden in the data that is not detectable by engineers due to lack of adequate tools or resources, and fully automated closed-loop control to automatically eliminate many yield issues before they happen.

## REFERENCES

1. Semiconductor Industry Association. Semiconductor Industry Association World Semiconductor Forecast 1999–2002. The association, San Jose, CA, June 1999.
2. C Weber, V Sankaran, G Scher, KW Tobin. Quantifying the value of ownership of yield analysis technologies. 9th Annual SEMI/IEEE Advanced Semiconductor Manufacturing Conference, Boston, September 23–25, 1999.
3. G Freeman, Kierstead, W Schweiger. Electrical parameter data analysis and object-oriented techniques in semiconductor process development. IEEE BCTM, 0-7803-3616-3, 1996, p 81.
4. T Hattori. Detection and identification of particles on silicon surfaces. In: KL Mittal, ed. Particles on Surfaces, Detection, Adhesion, and Removal. New York: Marcel Dekker, p 201.
5. V Sankaran, CM Weber, KW Tobin, Jr., F Lakhani. Inspection in Semiconductor Manufacturing. Webster's Encyclopedia of Electrical and Electronic Engineering. Vol. 10. New York: Wiley, pp 242–262, 1999.
6. D Jensen, C Gross, D Mehta. New industry document explores defect reduction technology challenges. Micro 16(1):35–44, 1998.
7. DL Dance, D Jensen, R Collica. Developing yield modeling and defect budgeting for 0.25 mm and beyond. Micro 16(3):51–61, 1998.
8. KW Tobin, TP Karnowski, F Lakhani. A survey of semiconductor data management systems and technology. Technology Transfer 99073795A-ENG, SEMATECH, Austin, TX, August 1999.
9. Semiconductor Industry Association. International Technology Roadmap for Semiconductors: The Association, 1999.
10. SP Cunningham, JS Costas. Semiconductor yield improvement results and best practices. IEEE Transactions Semiconductor Manufacturing 8(2):103, 1995.
11. C Weber, B Moslehi, Manjari Dutta. An integrated framework for yield management and defect/fault reduction. IEEE Transactions Semiconductor Manufacturing 8(2):110, 1995.
12. CH Stapper. Fact and fiction in yield modeling. Microelectronics J. 20(1–2):129, 1989.
13. AV Ferris-Prabhu. On the assumptions contained in semiconductor yield models. IEEE Transactions Computer-Aided Design II(8):966, 1992.
14. AY Wong. A statistical Parametric and Probe Yield Analysis Methodology. IEEE, 1063-6722/96, p 131.
15. D Dance, F Lakhani. SEMATECH 0.25-micron yield model final report with validation tool targets. Technology Transfer No. 97033263A-TR, SEMATECH, Austin, TX, May 1997.
16. D Dance, C Long. SEMATECH 150-nm random defect limited yield (RDLY) model final report with validated tool targets. Technology Transfer No. 9908308A-ENG, SEMATECH, Austin, TX, August 1999.
17. U Kaempf. The binomial test: a simple tool to identify process problems. IEEE Transactions Semiconductor Manufacturing 8(2):160, 1995.
18. MM Slama, MH Bennettt, PW Fletcher. Advanced in-line process control of defects. Integrated Circuit Metrology, Inspection, Process Control VI. SPIE 1673:496, June 1992.
19. SS Gleason, MA Hunt, H Sari-Sarraf. Semiconductor yield improvement through automatic defect classification. ORNL92-0140, Oak Ridge National Laboratory, Oak Ridge, TN, September 1995.

20. B Trafas, MH Bennett, M Godwin. Meeting advanced pattern inspection system requirements for 0.25-μm technology and beyond. Microelectronic Manufacturing Yield Reliability and Failure Analysis. SPIE 2635:50, September 1995.

21. R Sherman, E. Tirosh, Z Smilansky. Automatic defect classification system for semiconductor wafers. Machine Vision Applications in Industry Inspection. SPIE 1907:72, May 1993.

22. PB Chou, AR Rao, MC Sturzenbecker, FY Wu, VH Becher. Automatic defect classification for semiconductor manufacturing. Machine Vision Applications 9(4):201, 1997.

23. KW Tobin, TP Karnowski, TF Gee, F Lakhani. A study on the integration of spatial signature analysis and automatic defect classification technologies. Technology Transfer No. 99023674A-ENG, SEMATECH, Austin, TX, March 1999.

24. W Taam, M Hamada. Detecting spatial effects from factorial experiments an application from integrated-circuit manufacturing. Technometrics 35(2):149, 1993.

25. JG Ramirez, B Cantell. An analysis of a semiconductor experiment using yield and spatial information. Quality Reliability Engineering Int. 13:35, 1997.

26. JG Ramirez, RS Collica, BS Cantell. Statistical analysis of particle defect data experiments using poisson and logistic regression. International Workshop on Defect and Fault Tolerance in VLSI Systems, IEEE, 1063–6722, 1994, p 230.

27. RS Collica, JG Ramirez. Process monitoring in integrated circuit fabrication using both yield and spatial statistics. Quality Reliability Engineering Int. 12:195, 1996.

28. R Collica. A logistic regression yield model for SRAM bit fail patterns. International Workshop on Defect and Fault Tolerance in VLSI Systems. IEEE, 0-8186-3502-9, 1993, p. 127.

29. RS Collica, JP Card, W Martin. SRAM bitmap shape recognition and sorting using neural networks. IEEE Transactions Semiconductor Manufacturing 8(3):326, 1995.

30. KW Tobin, SS Gleason, F Lakhani, MH Bennett. Automated analysis for rapid defect sourcing and yield learning. Future Fab Int. 4: , 1997.

31. KW Tobin, SS Gleason, TP Karnowski, SL Cohen. Feature analysis and classification of manufacturing signatures on semiconductor wafers. SPIE 9th Annual Symposium on Electronic Imaging: Science and Technology, San Jose, CA, February 9–14, 1997.

32. KW Tobin, SS Gleason, TP Karnowski, SL Cohen, F Lakhani. Automatic classification of spatial signatures on semiconductor wafermaps. SPIE 22nd Annual International Symposium on Microlithography, Santa Clara, CA, March 9–14, 1997.

33. KW Tobin, SS Gleason, TP Karnowski, MH Bennett. An image paradigm for semiconductor defect data reduction. SPIE 1996 International Symposium on Microlithography, Santa Clara, CA, March 10–15, 1996.

34. SS Gleason, KW Tobin, TP Karnowski. Rapid yield learning through optical defect and electrical test analysis. SPIE Metrology, Inspection, and Process Control for Microlithography II, Santa Clara, CA, February 1998.

35. SS Gleason, KW Tobin, TP Karnowski. An integrated spatial signature analysis and automatic defect classification system. 191st Meeting of the Electrochemical Society, May 1997.

36. KW Tobin, TP Karnowski, SS Gleason, D Jensen, F Lakhani. Integrated yield management. 196th Meeting of the Electrochemical Society, Honolulu, HI, Oct. 17–22, 1999.

37. GM Scher. Wafer tracking comes of age. Semiconductor Int. May 1991, p 127.

38. G Scher, DH Eaton, BR Fernelius, J Sorensen, JW Akers. In-line statistical process control and feedback for VLSI integrated circuit manufacturing. IEEE Transactions Components, Hybrids, Manufacturing Technol. 13(3): , 1990.

39. G Scher. Feasibility of integrating capabilities of spatial signature analysis (SSA) and water sleuth: interim report. Technology Transfer No. 9083814A-ENG, SEMATECH, Austin, TX, August 31, 1999.

# 24

# Statistical Metrology, with Applications to Interconnect and Yield Modeling

**Duane S. Boning**
*Massachusetts Institute of Technology, Cambridge, Massachusetts*

## I. INTRODUCTION

Statistical metrology is the body of methods for understanding variation in microfabricated structures, devices, and circuits. The majority of this book focuses on the fundamental science, technology, and applications of metrology or measurement. Given the ability to make a "unit" measurement, a remaining challenge is to determine exactly what should be measured and how those measurements should be analyzed. The underlying principle of *statistical metrology* is that, in addition to the measurement and study of nominal processes, devices, or circuits, it is also critical to measure and study the "statistics," or variation, inherent in advanced semiconductor technology.

In this chapter, we begin with an overview of the background of statistical metrology. The sources or categories of variations are reviewed along with the key elements or stages in using statistical metrology. In Sec. II we focus on the application to statistical process and device characterization and modeling. Here, key issues include the characterization of process variation, the generation of statistical device models, and the problem of spatial matching and intradie variation of devices. In Sec. III we examine the application to statistical interconnect modeling, where variation issues are increasingly important as clock and signal skew begin to limit the performance in high-speed-clock and long-signal paths. This mapping of measured or characterized variation sources into the resulting circuit impact is a key element of statistical metrology. In Sec. IV, we consider the relationship between statistical metrology methods and yield modeling. Finally, in Sec. V we note the future trends and directions for statistical metrology of semiconductor technology.

### A. Background and Evolution of Statistical Metrology

Statistical modeling and optimization have long been a concern in manufacturing. Formal methods for experimental design and optimization, for example, have been developed and presented by Box et al. (8) and Taguchi (30), and application to semiconductor manufacturing has been presented by Phadke (24). More recently, these methods are seeing development in the form of "statistical metrology." Bartelink (1) introduces statistical metrology, emphasizing the importance of statistical metrology as a "bridge" between manufacture and simulation. Reviews of the defining elements of statistical metrology are presented in Refs. 2, 3, and 4. These elements include an emphasis on the characterization of variation, not only temporal (e.g., lot-to-lot or wafer-to-wafer drift) but also spatial variation (e.g., within-wafer and particularly within-chip or within-die). A second important defining element and key goal of statistical metrology is to identify the systematic elements of variation that otherwise must be dealt with as a large "random" component through worst-case or other design methodologies. The intent is to isolate the systematic, repeatable, or deterministic contributions to the variation from sets of deeply confounded measurements. Such variation identification and decomposition, followed by variation modeling and impact assessment, is critical in order to focus technology development on variation-reduction efforts, as well as to help in defining device or circuit design rules in order to minimize or compensate for the variation.

## B. Variation Sources

*Variation* can be defined as any deviation from designed or intended manufacturing targets. In semiconductor manufacturing, deviations can be broadly categorized as resulting in either defect or parametric variation. Defects, particularly those caused by the presence of particles, may be responsible for functional yield loss by creating undesired shorts, opens, or other failures. In contrast, *parametric* variation arises from "continuous" deviation from intended values of device or circuit performance goals. Parametric variation can also result in yield loss (e.g., circuits that fail due to timing variations or that do not perform at required speeds) or may result in substandard product performance or reliability.

While defect-limited yield (and the rate of yield improvement) has long received attention in semiconductor manufacturing, parametric variation is an increasing concern in integrated circuit fabrication. Stringent control of both device and interconnect structures, such as polysilicon critical dimension (channel length) or copper metal line thicknesses is critical not only for adequate yield but also to meet increasingly aggressive performance and reliability requirements. Understanding and assessing such variation, however, is difficult: variation may depend on the process, equipment, and specifics of the layout patterns all confounded together. Here we summarize the nature and scope of parameter variation under study.

Variation in some physical or electrical parameter may manifest itself in several ways. One key characteristic of variation is its scope in time and in space, as shown in Figure 1, where we see that the variation appears at a number of different scales. The separation of variation by unique signatures at different scales is a key feature enabling one to analyze such variation. Process control has often been concerned with the variation that occurs from lot-to-lot or wafer-to-wafer. That is, some measure of a parameter for the lot may vary from one lot to the next as the equipment, incoming wafer batch, or consumable material drifts or undergoes disturbances. In addition to temporal variation, different spatial variation occurs at different scales. In batch processes, for example, the spatial variation from one wafer to the next (e.g., along a polysilicon deposition tube) may be a concern. In equipment design and process optimization, spatial uniformity across the wafer is a typical goal and specification. For example, in most deposition or etch processes, uniformity on the order of 5% across the wafer can be achieved; if one examines the value for some structural parameter taken at the same point on every die on the wafer, a fairly tight distribution results. At a smaller scale, on the other hand, additional variation issues may arise. In particular, the variation within an individual die on the wafer is emerging as a major concern, in large part because of the potential yield and circuit

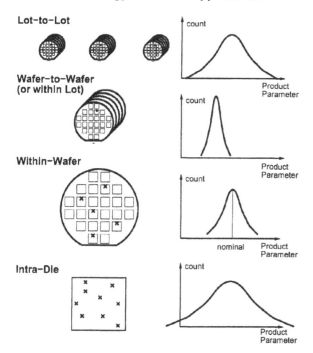

**Figure 1**  Spatial and temporal variation scales.

performance degradation. An important observation is that knowing something about one scale of variation says little about the variation at the other scales. This is because different physical causes are at work at each scale; e.g., wafer-level uniformity in plasma etch is driven by macroscopic tool design issues, while die-level pattern dependencies (which may in fact be larger than wafer variation) in the same etch process may arise through details of microscopic etch loading.

The second key characteristic of variation is systematic versus random constituents in the parameter distribution. An important goal is to isolate those systematic, repeatable, or deterministic contributions to the variation from a set of deeply confounded measurements. A set of dielectric thickness measurements from many locations on one wafer following planarization by chemical-mechanical polishing is collected; without detailed understanding of the individual contributions, the resulting distribution, shown in Figure 2, might be considered to be "random" and large (26). Better understanding of the specific contributions to the distribution enables one to focus variation-reduction efforts more appropriately or to design the device or circuit to compensate for the expected variation. Using decomposition approaches, for example, one might find that a repeatable "die pattern," or signature, within the die is responsible for the oxide thickness variation observed in Fig. 2, focusing improvement efforts on the pattern dependency rather than on wafer-scale uniformity.

## C.  Elements of Statistical Metrology

Given the foregoing family of temporal and spatial variation, the key statistical metrology goal—understanding variation—is accomplished through several elements. These elements are briefly introduced here; examples of these techniques in different application areas then

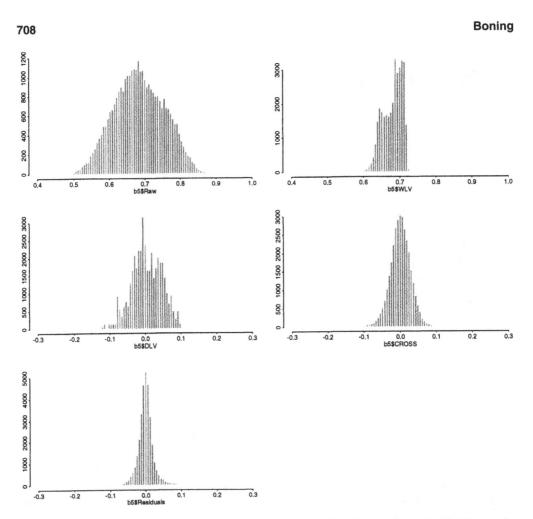

**Figure 2** Distributions of interlevel dielectric thickness after chemical-mechanical polishing. At the upper left is the distribution of all wafer measurements. After spatial decomposition, we see that only a portion of the variation is due to wafer scale nonuniformity (upper right), while the majority of variation is due to within-die pattern dependencies (middle left) and interaction of that dependency with the wafer nonuniformity (middle right). The remaining "random" or unexplained variation (at lower left) is relatively small. (From Ref. 26.)

follow in later sections. The key phases and techniques in statistical metrology are, broadly speaking, (a) variation exploration, (b) variation decomposition, (c) variation modeling, and (d) impact assessment. Statistical metrology can be seen to draw heavily on statistical methods but also to go beyond statistical modeling in seeking an integrated methodology for understanding variation and its impact in semiconductor manufacturing. In particular, statistical metrology has found a natural focus on spatial variation, and layout-pattern-dependent variation in particular, in integrated circuits.

## 1. Variation Exploration

The key purpose of variation exploration is to understand what the potential sources of variation are. An experimental approach, coupled with fundamental metrology, is typically required. Variation exploration and identification involves (a) the development of test structures and measurement strategy, (b) the definition of short flow process experiments (when possible), and (c) the application of experimental design methods to search for the key factors responsible for the observed variation. In keeping with a common focus of statistical metrology on layout pattern dependencies, the application of experimental design methods often takes an unusual twist. Rather than focus simply on "process factors" (e.g., equipment or process variables), the experimental designs often involve layout or spatial factors. For example, the geometric orientation of devices or other structures on the chip, the density of features, and a host of related geometric and spatial issues are often combined into test patterns for experimental process fabrication. In a similar fashion, the test structure and measurement strategies used in statistical metrology are often somewhat unusual or specialized: the requirement is often to gather a large number of measurements so that the variation itself can be studied. Many experimental design methods in use in the semiconductor industry seek to understand first- or second-order functional dependencies for the purpose of process/device design or optimization. For example, one often seeks to generate a low-order polynomial model relating the response to the underlying factors, in order to select an appropriate process operating point. In the case of statistical metrology, the problem is not simply to model the nominal response of the process; rather, the *variation* in the response must be modeled. This typically requires an order of magnitude more experimental data, and the development of test structure and measurement strategies reflects this requirement.

## 2. Variation Decomposition

The analysis of variation in semiconductor manufacturing is complicated by the often deeply nested structure of that variation and the variety of length scales involved in spatial variation. Temporal variation often arises on a particular tool, but may have both lot and wafer variation sources: the postpolish average thickness of an oxide layer across a wafer, for example, may trend downward from one wafer to the next (as the polishing pad wears), but may also exhibit different averages from one lot to the next due to differences in the incoming wafer oxide thickness for one lot versus another.

In the case of spatial variation, the dependence at different length scales can be deeply intertwined, and statistical metrology methods have evolved for the decomposition of spatial variation sources. In particular, variation across the wafer, from die to die, within the die, and from feature to feature may all depend on different physical sources and need to be separated from each other. The spatial characteristics of the variation can thus provide substantial insight into the physical causes of the variation. Variation decomposition methods have been developed that successively serve to extract key components from measurement data (26). First, wafer-level trends are found, which are themselves often of key concern for process control. The trend is removed from the data and then methods are employed to extract the "die-level" variation (that is, the component of variation that is a clear signature of the die layout) using either 2-D Fourier techniques or modified analysis of variance approaches (33,26). This is especially important in order to enable further detailed analysis of the causal feature- or layout-dependent effects on a clean set of data. Third, examination of wafer–die interaction is needed. For example, if the pattern dependencies and wafer-level variation are highly coupled, then process control decisions (usually made based on wafer-level objectives) must also be aware of this die-level impact (5).

### 3. Variation Modeling

Given basic metrology capability in combination with experimental designs guided by variation exploration and decomposition requirements, a large volume of statistical data can be generated. Models of both the random and systematic sources of variation can then be formulated. A great deal of experience exists in generation of random-variation models; indeed, the traditional approach is simply to "lump" all variation into a single "random" component and estimate statistical moments (e.g., mean, variance) from the data. The most common model is a normal distribution, where the variation in some parameter $P$ of interest is summarized using the sample variance $\sigma_p^2$. As a result of variance decomposition, either decoupled or nested models of variance can be generated, e.g.,

$$\sigma_p^2 = \sum_i \sigma_i^2 \tag{1}$$

where $\sigma_i^2$ is due to variation source $i$ in the case of independent variation sources. In the case of nested variance structures, much more careful analysis and model assumptions are needed (12). For example, the value of a parameter $y$ for a particular structure $k$ within a given die $j$ on a particular wafer $i$ might need to be modeled as (6):

$$y = \mu + W_i + D_{j(i)} + S_{k(ji)} \tag{2}$$

where wafer variation might be distributed as $W_i \sim N(0, \sigma_w^2)$, the die variation $D$ depends on which die is examined within wafer $i$, and the structure variation $S$ depends on which structure $k$ within that die is considered.

In addition to "random" variation sources, systematic variation effects are often the object of statistical metrology study. In these cases, empirical models of the effect as a function of layout, spatial, or process parameters are often developed. As physical understanding of the effect improves, these models become more detailed and deterministic in nature and ultimately can be included in the process/device/circuit design flow.

### 4. Variation Impact Assessment

Given models of systematic and random variation, the final key element of statistical metrology is to understand the impact of this variation on product parameters of concern. These product parameters might be performance related (e.g., clock or circuit speed) or yield related. Such impact assessment is often carried out using statistical analysis methods (e.g., Monte Carlo sampling) in conjunction with process/device/circuit simulation tools. Ultimately, such impact assessment techniques should integrate into statistically based design and optimization methods so as to maximize performance, yield, or other objectives.

## II. STATISTICAL METROLOGY OF PROCESS AND DEVICE VARIATION

Integrated circuits have always had some degree of sensitivity to manufacturing process variations. The traditional approach to addressing such variation in device and circuit design has been worst-case consideration of the bounds of variation: If circuits consisting not only of the nominal (target) device but also of extremal devices continue to function and meet specifications, then the design is considered robust. As the push to higher performance continues, however, such worst-case design can be overly pessimistic and result in designs with inferior performance. Coupled with the need for faster yield ramps, this push for statistically realistic design has helped spawn statistical metrology methods for more accurate and detailed understanding of process variation and its effects. In this section, we overview the key elements of statistical methods as applied to process/device modeling. Additional details on statistical process/device modeling and optimization can be found in the recent reviews by Boning and Nassif (7), Nassif (20), and Duvall (13).

## A. Exploration and Decomposition of Process/Device Variation

Statistical metrology of process/device variation begins with study of the physical sources of variation. These include all device parameters of interest, such as threshold voltage $V_T$, gate oxide thickness $T_{ox}$, and sheet resistivity $R_S$. In general, these parameters may vary from one device to the next in a systematic and/or random fashion across the entire wafer. From the perspective of circuit design, however, the sources of variation (deviation from nominal value $P_0$) in some parameter $P$ of interest can be broken into three key families: interdie variation $\tilde{P}_{interdie}$ (also referred to as wafer-level or die-to-die variations), intradie variation $\tilde{P}_{intradie}$ (also referred to as within-die variations), and random residuals $\tilde{P} \sim N(0, \sigma_\varepsilon^2)$:

$$P = P_0 + \tilde{P}_{interdie} + \tilde{P}_{intradie} + \tilde{P}_\varepsilon \tag{3}$$

An example of the decomposition of a dielectric thickness parameter into interdie and intradie variation sources is shown in Fig. 2. In studies of device variation, one of the most important parameters considered is MOSFET channel-length (polysilicon critical-dimension) variation. In Ref. 28, for example, the effects of optical proximity are compared to the wafer-scale variation in order to understand the impact on an SRAM cell, and a similar decomposition technique is used. The construction of models (either lumped "random" or detailed systematic spatial models) depends strongly on experimental design methods. For example, the understanding of poly CD variation has been explored using test patterns replicated across the die; layout factors might include isolated versus nested lines, line orientations, density of nearby structures, line width, and other factors (14). Based on electrical, optical, or other measurements of the resulting geometric or device response, decomposition of the sources of variation can then be undertaken. A sophisticated decomposition method is used by Yu et al. to study linewidth variation involving spatial frequency analysis to separate out reticle and lens-distortion effects from random leveling and local layout-pattern effects (33,34).

## B. Variation Modeling

For the purposes of device and circuit modeling, "compact models" of device behavior are typically required that express device outputs (e.g., drain current) as a function of the operating environment (e.g., applied voltages), the device geometry, and device model parameters (e.g., $T_{ox}$, $V_T$). Such models, embodied in various Spice or Spice-like MOSFET models, define the nominal characterization of a device and are extracted from experimental data.

Statistical variants of these models can be generated in two ways. In the first case, large numbers of devices are fabricated and direct measurement of the variations in device parameters are captured and statistical moments directly estimated (10). In other cases, modeling of the underlying geometric variation is pursued (e.g., linewidths, film thicknesses) or models generated of fundamental process parameters (e.g., measurement or assumption of variation in diffusion temperature, ion implant energies and doses). When more basic models are used, the underlying variation can be propagated through to produce device distributions using process and device simulation tools (25).

In the case of either direct device parameters (e.g., $V_T$) or process/geometric variation (e.g., $L_{eff}$), the variation may have both deterministic and random components. For example, the effect of a gentle wafer-scale process nonuniformity projected onto any one die might be well expressed as a linear function of position $(x, y)$ in the die:

$$\tilde{P}_{intradie}(x, y) = W(w, x, y) = w_0 + w_x \cdot x + w_y \cdot y \tag{4}$$

In this case, however, the particular orientation of "slanted plane" within the die will itself vary from one die to the next, depending on the location of the die on the wafer. From the perspective of the circuit, then, one might consider the foregoing variables $w_x$, $w_y$, and $w_0$ as random variables (21).

The systematic effect of layout patterns on a parameter has become especially important and hence the focus of much statistical metrology effort. In these cases, the parameter $P$ may be modeled as

$$P = P_0 + \tilde{P}_{interdie} + F(x, y, \theta) + \tilde{P}_\varepsilon \tag{5}$$

with $F(x, y, \theta)$ expressing a potentially mixed deterministic and random function of position on the die, where $\theta$ represents a vector of random parameters of the function. This position dependence can be modeled in several ways.

In analog circuit design, the spatial matching of device parameters is quite important. Models of parameter variation as a function of device size and separation distance have been found to be quite effective for some variation sources. For example, if a random process is at work that is dependent on the area $A$ of the device, then larger device sizes can reduce the effect of that variation. If $\Delta P = P_1 - P_2$ is the mismatch between two different devices with areas $A_1$ and $A_2$, the variation in mismatch can be modeled by

$$\Delta P \sim N(0, \sigma_A^2/A_1 + \sigma_A^2/A_2) \tag{6}$$

where we see that larger devices average or reduce the uncorrelated spatial variance $\sigma_A^2$. In a similar fashion, some parameters may have a length dependence rather than an area dependence, so mismatch between devices (or interconnect lines) of length $L_1$ and $L_2$ could be modeled as

$$\Delta P \sim N(0, \sigma_L^2/L_1 + \sigma_L^2/L_2) \tag{7}$$

In other cases, the degree of mismatch between device parameters has been found to increase as a function of the separation distance $D$ between those devices (hence leading to the practice of placing devices that must be well matched very close to each other). The typical model used is to assume variance proportional to distance squared, or

$$\Delta P \sim N(0, K \cdot D^2) \tag{8}$$

where $K$ is an extracted or estimated parameter.

## C.   Process/Device Variation Impact and Statistical Circuit Design

The last key element of statistical metrology is using measurements and models of process/device variation in order to understand the importance and impact of that variation on circuit or product performance. In the case of analog circuits, the foregoing mismatch models are typically used in conjunction with analytic models or Spice simulations to study the effect on circuit parameters, e.g., frequency response or gain in an operational amplifier (19).

In the case of digital circuits, statistical analysis has traditionally focused on die-to-die variation. That is to say, most analysis has proceeded assuming that all of the devices in a circuit see the same disturbance or deviation from nominal, due to die-to-die variation, or

$$P = P_0 + \tilde{P}_{\text{interdie}} \tag{9}$$

where $\tilde{P}_{\text{interdie}} \sim N(0, \sigma_j^2)$ is assumed equal for all devices on die $j$.

With a new emphasis on statistical metrology of within-die (or intradie) variation, deviations from device to device spatially within the die have been considered. The first stage of such impact analysis has been to use uncorrelated-device models; i.e., an assumption that each device is drawn from a random distribution is used to study the sensitivity of a canonical digital circuit to different sources of variation. For example, Zarkesh-Ha et al. (35) develops a model of clock skew in a balanced H distribution network, and then uses first-order sensitivity to variation source $x$ to compare the impact of a set of process variations on that skew:

$$T_{CSK}(x) = \Delta T_{\text{Delay}} \approx \left| \frac{\partial T_{\text{Delay}}}{\partial x} \right| \Delta x \tag{10}$$

An analytically defined expression for clock path delay, $T_{\text{Delay}}$, can be differentiated to approximate the sensitivity of skew and multiplied by some percentage measure of variation (e.g., in-line capacitance $C_L$) to determine the total clock skew $T_{CSK}$ due to that variation:

$$T_{CSK}(C_L) = 0.7 R_{tr} C_L \frac{\Delta C_L}{C_L} \tag{11}$$

With digital circuits pushing performance limits, increasingly tight timing constraints are forcing the examination of more detailed, spatially dependent within-die-variation models and use of those models to study the impact on the circuit (31,29). Nassif demonstrates a worst-case analysis of clock skew in a balanced H clock tree distribution network under different types of device and interconnect variation models (21). For example, he shows that a worst-case configuration for random MOSFET channel-length variation [using $\Delta L \sim N(0, (0.035\,\mu\text{m})^2)$] can be found using a statistical procedure, as shown in Figure 3. Other analyses of circuit sensitivity to spatial and pattern variation are beginning to emerge (15,17).

## III.  STATISTICAL METROLOGY OF INTERCONNECT VARIATION

In previous reviews of statistical metrology (3,4), the variation arising during the planarization of interlevel dielectric layers has been discussed as a case study of statistical metrology (9). In this section, we consider a different case arising in advanced interconnect: the variation in the thickness of metal lines arising from chemical-mechanical polishing (CMP) of copper Damascene wiring. The key elements of statistical metrology (as discussed in Sec. I) applied to this problem include (a) copper CMP variation exploration and decomposition, (b) variation modeling of dishing and erosion, and (c) circuit analysis to understand the impact of copper interconnect variation.

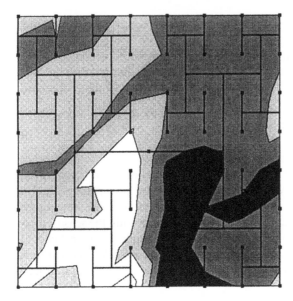

**Figure 3** Channel lengths for worst-case skew in a clock distribution network. The shading indicates the channel length in fanout buffer devices within the clock net. Large skew occurs at the bottom center of the chip, where spatially near devices suffer large skews due to device variations along left-side (white) and right-side (black) paths in the balanced H distribution network where transistors switch at different speeds due to channel-length variation. (From Ref. 21.)

## A.   Copper Chemical-Mechanical Polishing Variation Exploration and Decomposition

A key concern in the use of CMP for the formation of copper Damascene lines is the undesirable thinning of copper lines due to dishing (loss of thickness within a copper feature) and erosion (loss of both oxide in an array of lines and spaces), as illustrated in Figure 4. Based on early experimentation (22), the degree of dishing and erosion is known to have a strong pattern effect, depending on the density of features as well as linewidth and line spacing. In applying a statistical metrology approach, the first element is the design of test structures and test masks that can explore these pattern-dependent variation sources in detail (23). The basic test structure for single-level copper dishing and erosion evaluation consists of an array of lines and spaces, as shown in Figure 5. An experimental design in which the density, linewidth, and line space are varied systematically across the chip is pictured in Figure 6. Based on electrical and physical measurements of the dishing and erosion profiles (such as that pictured in Figure 7), large volumes of data for a given CMP process can be experimentally gathered.

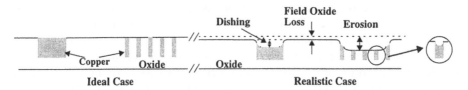

**Figure 4** Pattern-dependent problems of dishing and erosion in copper CMP. (From Ref. 23.)

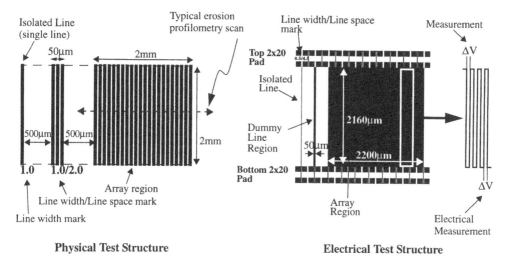

**Figure 5** Line and array test structures for study of copper dishing and erosion. (From Ref. 23.)

## B. Copper Chemical-Mechanical Polishing Variation Models

Two phases in modeling of the systematic pattern-dependent variation of copper lines can be pursued. In the first phase, empirical plots and general trends of dependencies in dishing and erosion can be generated. For example, the plots in Figure 8 show that dishing increases as a function of linewidth, saturating to some maximum dishing for very large lines; in contrast, erosion increases for very thin line spaces (at which point the oxide between copper lines is unable to support the pressure of the polishing pad and process). Such trends can be extremely useful in defining design rules or other compensation approaches.

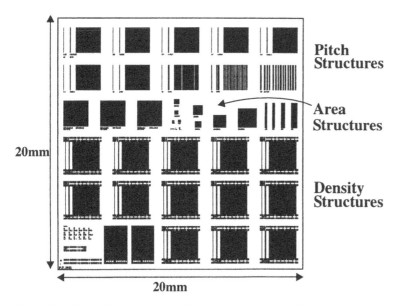

**Figure 6** Test chip consisting of array test structures for study of copper dishing and erosion. (From Ref. 23.)

**Surface Profiles (in Å)**

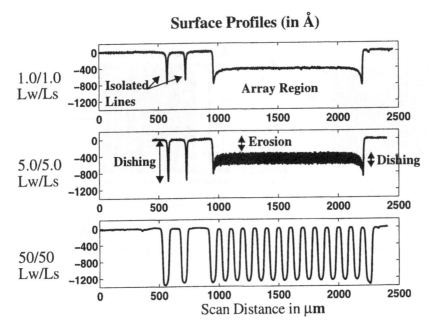

**Figure 7**  Surface profiles showing dishing and erosion trends in line/array structures, as a function of linewidth (Lw) and line space (Ls). (From Ref. 23.)

The second phase of variation modeling is to understand the functional relationship between the copper line thickness and layout parameters. In such physical or semi-empirical models, key model parameters are extracted based on experimental results, and the model can then be applied to different conditions (e.g., to predict the line loss for arrays in a new layout). Such copper CMP models are just emerging (32), and more work remains to be done.

## C.  Copper Interconnect Variation Analysis

Given empirical or physical models of copper-thickness variation, it then becomes possible to explore the potential impact of such variation on circuit yield or performance. In

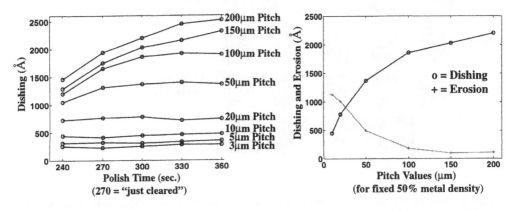

**Figure 8**  Copper dishing and erosion dependencies on polish time and pitch. (From Ref. 23.)

the case of loss of copper line thickness, one might worry that the increased line resistance could have a significant effect on circuit speed. If the line-thickness loss is different for different regions of the chip (e.g., for different branches of a balanced clock), then one might be concerned about variation-induced clock skew. The impact of this variation source can also be compared against expected variation in other technologies (e.g., aluminum interconnect) or other variation sources (e.g., transistor channel-length variation). An analysis of clock skew by Mehrotra et al. (18) compares the inherent clock skew (due to transistor loading differences) against the skew with interconnect or device variation sources. As summarized in Table 1, in this case it is found that copper lines in clock distribution trees can be relatively well controlled such that device variation is a much bigger concern (18). Examination of a data bus, on the other hand, shows that pattern-dependent copper line-thickness variation may be a significant concern for signal-delay variation as a function of bit position in an array. Using erosion profile data shown earlier, Mehrotra et al. find that bit position can contribute up to 35% delay variation, as shown in Figure 9. Such analyses may lead to improved design-rule or dummy-fill-patterning practices similar to those that have emerged for aluminum interconnect (27). The analysis of both interconnect and device variation will increasingly be needed in order to decrease clock and signal skews and achieve future performance improvements.

## IV. STATISTICAL METROLOGY FOR YIELD

An excellent example illustrating the development and application of statistical metrology methods to yield characterization and modeling has been demonstrated by Ciplikas et al. (11). A comprehensive methodology is shown for predictive yield modeling in which key elements are (a) metrology and measurements using specifically designed "characterization vehicles" in conjunction with in-line defect detection and imaging metrology; (b) identification of systematic and random yield-loss mechanisms; (c) statistical modeling of defect density and size distributions as well as killer-defect dependencies; and (d) coupling to layout and design-rule practices to calculate the of expected amount of product yield loss due to each type of potential killer defect. In this section we recap the key elements of statistical metrology developed and used by Ciplikas et al. in addressing the yield modeling problem (11).

**Table 1** Maximum Clock Skew Arising from Variation Analysis of a Balanced H Clock Tree in Both Aluminum and Copper Interconnect Technologies

| Interconnect | Variation Source | Max skew (ps) |
|---|---|---|
| Cu | None (nominal metal thickness) | 34 |
| Cu | Metal thickness (dishing + erosion) | 40 |
| Al | None (nominal oxide thickness) | 59 |
| Al | Pattern-dependent oxide thickness | 62 |
| Cu | Poly CD (0–5% by quad.) | 83 |

*Source*: Ref. 18.

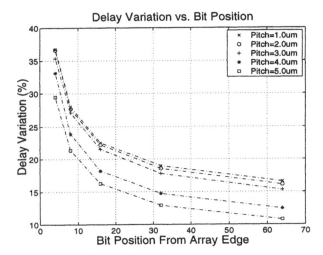

**Figure 9** Simulated percentage delay variation for 5-mm interconnect (bus lines) with copper CMP. (From Ref. 18.)

## A. Yield Variation Exploration and Decomposition

A key issue in yield evaluation is to disaggregate random defect-limited yield and systematic yield loss mechanisms. The fundamental metrology tools (e.g. bright- and dark-field optical imaging and laser scanning apparatus) are extremely useful in the detection and categorization of discrete defect events. Defect decomposition or classification is used. The goal is to build models for each individual defect mechanism so that aggregate yield loss can later be estimated. Here, manual and automatic defect review and classification are important, and the results are fed forward to defect size modeling of the inspected defects.

To complement defect inspection metrology tools, the development of specialized full- and short-flow electrical test masks ("characterization vehicles," or CVs) are found to be crucial to separate out and characterize the sources of systematic (nonparticle) and random defects. While scribe line or small portions of integration test vehicles are often used, dedicated short-flow layouts provide the full opportunity for experimental design and exploration of yield loss mechanisms (11). An example of such a pattern is shown in Figure 10. For example, via chain, short, and open test structures at, below, and above the minimum design rule can be designed in large enough areas to accurately model yield loss.

## B. Yield Modeling

Yield modeling has a long history; from the viewpoint of statistical metrology, the innovation in Ref. 11 is to use couple strongly to experimental and metrology methods to extract defect density and size distribution parameters. For example, defect size distribution (DSD) models, as in Eq. (12), can be extracted from both inspection and characterization vehicle data:

$$\text{DSD}(x) = D_0 f(x) \frac{k}{x_p} \tag{12}$$

where $x$ is the defect size, $D_0$ is the measured defect density, $k/x^p$ is the size distribution function, and $f(x)$ is a scaling function used to extrapolate observed defect densities down

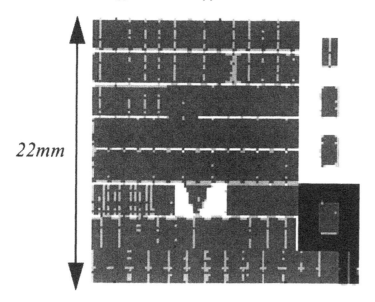

**Figure 10** Via characterization vehicle (CV) floorplan, used to extract via layer defect and failure mechanism statistical model parameters. Multiple regions on the chip are designed using different layout practices (e.g., via chains with nominal, slightly enlarged, slightly shrunk vias; different via densities) to study their impact on defect and failure statistics. (From Ref. 11.)

to the smaller-size defects that are not directly observable. An additional innovation is to create "nest" test structures (16) to aid in the extraction of $D_0$ and $p$ parameters for poly and metal defect distributions.

In addition to traditional defect distributions, the electrical characterization vehicles also help assess contact and via-hole failure rates. These types of failures are often process related as opposed to particle induced, and characterization of these loss mechanisms is critical but extremely difficult using optical methods. Given large test structures, each loss mechanism can be modeled with a straightforward Poisson model, where the (e.g., contact hole) failure rate parameter $\lambda$ can be extracted from a via chain with $N$ vias in series:

$$Y = e^{-\lambda_{via} N} \tag{13}$$

Many other yield modeling and extraction methods have been developed. For example, memory bitmap analysis can be very helpful in identifying process modules that may be responsible for particular yield losses.

## C. Impact Assessment—Yield Loss Prediction

The connection and application of statistical models with the circuit-level impact is a hallmark of statistical metrology. In the yield prediction case, the process defectivity rates and models extracted in the earlier steps can be applied to a given layout and product to predict yield. In this case, different circuit blocks can be analyzed for the sensitivity to different loss mechanisms and to explore the component of total yield loss to be expected for that module. A block failure rate $\lambda_a$ can be associated with a particular defect type $a$, and the fraction of chips that survive the defect (the defect-limited yield) calculated as

$Y_a = e^{-\lambda_a}$. The failure rate can be calculated using traditional critical area (CA) analysis for a given block $b$ and layer $l$ with minimum feature size $x_0$; e.g.,

$$\lambda_{b,l} = \int_{x_0}^{\infty} [\text{CA}_{b,l}(x)][\text{DSD}_l(x)] \, dx \tag{14}$$

Finally, failure rates for process- rather than defect oriented-parameters can be estimated, for example, as $Y_{b,l_{\text{via}}} = e^{-\lambda_{l_{\text{via}}} N_{b_{\text{via}}}}$ given the number of vias on the given layer $l$ in block $b$.

All together, the impact on a given product can thus be decomposed and predicted, forming a yield impact matrix as in Table 2. Taken together, statistical metrology methods of experimental design, test structures, and characterization vehicles, variation decomposition methods, yield modeling, and yield impact analysis can combine to effectively focus yield-ramp and -optimization efforts (11).

## V. CONCLUSIONS AND FUTURE DIRECTIONS

Statistical and spatial variation, particularly that with systematic elements, will become an increasingly important concern in future scaled integrated circuit technologies. Statistical metrology methods are needed to gather data and analyze spatial variation, both to decompose that variation and to model functional dependencies in the variation. Methods are also needed to understand and minimize the impact of such variation. In this chapter, we have described the development and application of a statistical metrology methodology to understanding variation in three key areas: process/device variation, interconnect variation, and yield modeling.

Much work is still needed on the individual elements and methods that statistical metrology draws upon. Test structure and spatial experimental design methods that can clearly identify sources of variation remain a challenge. Statistical modeling, particularly spatial and pattern-dependent modeling will continue to become more important as semiconductor technology is pushed to its limits. Finally, efficient statistical circuit optimization methods are needed to better understand and address the impact of variation on circuit performance as the size and complexity of circuit design continues to grow.

**Table 2**  Defect Limited Yields for Random Defect Mechanisms in Process Modules

|        | Poly  | Contact | M1    | Via1  | M2    | Via2  | M3    | Total |
|--------|-------|---------|-------|-------|-------|-------|-------|-------|
| Block1 | 98.0% | 99.0%   | 98.0% | 99.0% | 99.0% | 97.0% | 99.9% | 90.3% |
| Block2 | 96.0% | 97.0%   | 97.0% | 97.0% | 98.0% | 95.0% | 98.9% | 80.7% |
| Block3 | 97.0% | 98.0%   | 97.0% | 98.0% | 98.0% | 96.0% | 98.9% | 84.1% |
| Block4 | 95.0% | 97.0%   | 95.0% | 96.0% | 96.0% | 94.1% | 96.9^ | 73.4% |
| Block5 | 99.0% | 99.0%   | 99.0% | 99.9% | 99.9% | 98.0% | 99.9% | 94.8% |
| Total  | 85.8% | 90.4%   | 86.7% | 90.1% | 91.1% | 81.6% | 94.7% | 42.7% |

The matrix shows the impact of each process loss on different blocks and aggregate process module yield loss, enabling focus on sensitive blocks as well as problematic process modules.
*Source*: Ref. 11.

## REFERENCES

1. D Bartelink. Statistical metrology: At the root of manufacturing control. J. Vac. Sci. Tech. B 12:2785–2794, 1994.
2. D Boning, J Chung. Statistical Metrology: Understanding spatial variation in semiconductor manufacturing. Manufacturing yield, reliability, and failure analysis session, SPIE 1996 Symposium on Microelectronic Manufacturing, Austin, TX, 1996.
3. D Boning, J Chung. Statistical metrology: tools for understanding variation. Future Fab Int. Dec. 1996.
4. D Boning, J Chung. Statistical metrology—measurement and modeling of variation for advanced process development and design rule generation. 1998 Int. Conference on Characterization and Metrology for ULSI Technology, Gaithersburg, MD, 1998.
5. D Boning, J Chung, D Ouma, R Divecha. Spatial variation in semiconductor processes: modeling for control. Electrochem. Society Meeting, Montreal, CA, 1997.
6. D Boning, J Stefani, SW Butler. Statistical methods for semiconductor manufacturing. In: JG Webster, ed. Encyclopedia of Electrical and Electronics Engineering 20:463–479 New York: Wiley, 1999.
7. D Boning, S Nassif. Models of process variations in device and interconnect. Design of high performance microprocessor circuits. In: A Chandrakasan, W Bowhill. F Fox, eds. IEEE Press, 2000.
8. GEP Box, WG Hunter, JS Hunter. Statistics for Experimenters—An Introduction to Design, Data Analysis and Model Building. New York: Wiley, 1978.
9. E Chang, B Stine, T Maung, R Divecha, D Boning, J Chung, K Chang, G Ray, D Bradbury, S Oh, D Bartelink. Using a statistical metrology framework to identify random and systematic sources of intra-die ILD thickness variation for CMP processes. 1995 International Electron Devices Meeting, Washington DC, 1995, pp 499–502.
10. JC Chen, C Hu, C-P Wan, P Bendix, A Kapoor. E-T based statistical modeling and compact statistical circuit simulation methodologies. 1996 International Electron Devices Meeting, San Francisco, 1996, pp 635–638.
11. DJ Ciplikas, X Li, AJ Strojwas. Predictive Yield Modeling of VLSIC's. Fifth International Workshop on Statistical Metrology, Honolulu, HI, 2000.
12. D Drain. Statistical Methods for Industrial Process Control. New York: Chapman & Hall, 1997.
13. S Duvall. Statistical circuit modeling and optimization. Fifth International Workshop on Statistical Metrology, Honolulu, HI, 2000.
14. DD Fitzgerald. Analysis of Polysilicon Critical Dimension Variation for Submicron CMOS Processes. Master's Thesis, MIT, Cambridge, MA, 1994.
15. M Hatzilambrou, A Neureuther, C Spanos. Ring oscillator sensitivity to spatial process variation. First International Workshop on Statistical Metrology, Honolulu, HI, 1996.
16. C Hess, D Stashower, BE Stine, G Verma, LH Weiland. Fast extraction of killer density and size distribution using a single layer short flow NEST structure. Proc. of the 2000 ICMTS, Monterey, CA, 2000, pp 57–62.
17. W Maly, M Bollu, E Wohlrab, W Weber, J Vazquez. A study of intra-chip transistor correlations. First International Workshop on Statistical Metrology, Honolulu, HI, 1996.
18. V Mehrotra, SL Sam, D Boning, A Chandrakasan, R Valishayee, S Nassif. A methodology for modeling the effects of systematic within-die interconnect and device variation on circuit performance. Design Automation Conference, 2000.
19. C Michael, M Ismail. Statistical Modeling for Computer-Aided Design of MOS VLSI Circuits. Boston: Kluwer, 1992.
20. S Nassif. Modeling and forecasting of manufacturing variation. Fifth International Workshop on Statistical Metrology, Honolulu, HI, 2000.
21. S Nassif. Within-chip variability analysis. 1998 International Electron Devices Meeting, 1998.

22.  T Park, T Tugbawa, J Yoon, D Boning, R Muralidhar, S Hymes, S Alamgir, Y Gotkis, R Walesa, L Shumway, G Wu, F Zhang, R Kistler, J Hawkins. Pattern and process dependencies in copper damascene chemical mechanical polishing processes. VLSI Multilevel Interconnect Conference, Santa Clara, CA, 1998.
23.  T Park, T Tugbawa, D Boning. An overview of methods for characterization of pattern dependencies in copper CMP. Chemical Mechanical Polish for ULSI Multilevel Interconnection Conference (CMP-MIC 2000), Santa Clara, 2000, pp 196–205.
24.  MS Phadke. Quality Engineering Using Robust Design. Englewood Cliffs, NJ: Prentice Hall, 1989.
25.  S Sharifzadeh, JR Koehler, AB Owen, JD Shott. Using simulators to model transmitted variability in IC manufacturing. IEEE Trans. on Semi. Manuf. 2:82–93, 1989.
26.  B Stine, D Boning, J Chung. Analysis and decomposition of spatial variation in integrated circuit processes and devices. IEEE Trans. on Semi. Manuf. 10:24–41, 1997.
27.  BE Stine, DS Boning, JE Chung, L Camilletti, F Kruppa, ER. Equi, W Loh, S Prasad, M Muthukrishnan, D Towery, M Berman, A Kapoor. The physical and electrical effects of metal fill patterning practices for oxide chemical mechanical polishing processes. IEEE Trans. on Electron Devices, 45:665–679, 1998.
28.  BE Stine, DS Boning, JE Chung, D Ciplickas, JK Kibarian. Simulating the impact of poly-CD wafer-level and die-level variation on circuit performance. Second International Workshop on Statistical Metrology, Kyoto, Japan, 1997.
29.  BE Stine, V Mehrotra, DS Boning, JE Chung, DJ Ciplickas. A methodology for assessing the impact of spatial/pattern dependent interconnect parameter variation on circuit performance. 1997 International Electron Devices Meeting, Washington, DC, 1997, pp 133–136.
30.  G Taguchi. Introduction to Quality Engineering. Asian Productivity Organization. Dearborn, MI: American Supplier Institute, 1986.
31.  M Terrovitis, C Spanos. Process variability and device mismatch. First International Workshop on Statistical Metrology, Honolulu, HI, 1996.
32.  T Tugbawa, T Park, D Boning, T Pan, P Li, S Hymes, T Brown, L Camilletti. A mathematical model of pattern dependencies in Cu CMP processes. CMP Symposium, Electrochemical Society Meeting, Honolulu, HA, 1999.
33.  C Yu, T Maung, C Spanos, D Boning, J Chung, H-Y Liu, K-J Chang, D Bartelink. Use of short-loop electrical measurements for yield improvement. IEEE Trans. on Semi. Manuf. 8:150–159, 1995.
34.  C Yu, C Spanos, H Liu, D Bartelink. Lithography error sources quantified by statistical metrology. Solid State Tech. 38:93–102, 1995.
35.  P Zarkesh-Ha, T Mule, JD Meindl. Characterization and modeling of clock skew with process variations. Proc. of the IEEE Custom Integrated Circuits Conf., 1999, pp 441–444.

# 25

# Physics of Optical Metrology of Silicon-Based Semiconductor Devices

**Gerald E. Jellison, Jr.**
*Oak Ridge National Laboratory, Oak Ridge, Tennessee*

## I. INTRODUCTION

Optical metrology has played a fundamental role in the development of silicon technology and is routinely used today in many semiconductor fabrication facilities. Perhaps one of the most useful techniques is ellipsometry, which is used primarily to determine the thicknesses of films on silicon. However, many other optical techniques have also had an impact on the semiconductor industry. For example, infrared transmission has been used for decades to determine the carbon and oxygen content of silicon crystals, and the infrared transmission properties are also sensitive to the carrier concentration. Many electrical characterization techniques, such as minority carrier lifetime and quantum efficiency measurements, use light to create electron-hole pairs in the material. Silicon is an ideal material for these techniques, since the average depth of the created electron-hole pairs depends significantly on the wavelength of light. Light with a wavelength shorter than $\sim 375$ nm will create electron-hole pairs very near the surface, while photons with a wavelength between 375 and 1150 nm will create electron-hole pairs with depths ranging from a few tenths of a micron to millimeters.

In order to interpret any optical measurement involving silicon, it is essential to know the values of the optical functions and how these properties can change with a variety of external and internal factors. The optical functions of any material are wavelength dependent and can be described by either the complex refractive index or the complex dielectric function. Because crystalline silicon is a cubic material, its optical functions are normally isotropic; that is, they do not depend upon the direction of light propagation in the material. However, the optical functions of silicon in the visible and near-ultraviolet will exhibit certain features (called *critical points*) in their spectra due to the crystallinity of the material. The temperature of crystalline silicon is an extremely important external parameter, changing the position and the shape of the critical points and the position of the bandgap and altering the optical functions in other parts of the spectrum. The carrier concentration also affects the optical functions of silicon in the near-infrared due to free carrier absorption. Strained silicon is no longer isotropic and will have optical functions that depend upon the direction of light propagation in the material. The degree of crystallinity can also alter the optical functions of silicon: Both polycrystalline and amorphous silicon have quite different optical functions from those of crystalline silicon. All of these factors can be important in any optical diagnostic technique and may have to be taken into account.

Many other materials are also important to silicon technology. Silicon dioxide ($SiO_2$) has been used as a gate dielectric, as an insulator, and as an encapsulant. Bare silicon exposed to air will grow a very stable oxide that is $\sim 1$ nm thick in about a day; the thickness of the oxide increases to $\sim 2$ nm over several days, but is then very stable. The optical functions of thin-film $SiO_2$ depend marginally on temperature and strain, but also can depend upon film thickness and deposition conditions. Generally, it is not valid to assume that the optical functions of thin-film $SiO_2$ are the same as bulk fused silica, although often the differences are small. With some film deposition techniques, a suboxide can be formed ($SiO_x$, where $x$ is less than 2), which can have considerably different optical properties than fused silica. Silicon nitride ($Si_xN_y$:H) is another important material in semiconductor manufacture. The nominal stoichometric ratio for silicon nitride is $x/y = 3/4$, but a whole range of materials can be fabricated with quite different silicon-to-nitrogen ratios, depending on the method and the details of film growth. The deposition conditions can also affect the amount of hydrogen in the films. Changes in the silicon-to-nitrogen ratio have profound effects on the optical functions of thin-film $Si_xN_y$:H, while the hydrogen content appears to matter most in the infrared.

Much of the physics concerning the optical functions of silicon and related materials has been known since the late 1950s. Although modern experimental techniques often result in more accurate measurements, the basic understanding of the optical physics has not changed significantly over the last 20–30 years. In this chapter, the emphasis will be on a review of the understanding of the physics of optical measurement and on the optical properties of silicon and related materials. There are many good references that describe, in detail, many of the subjects covered in this chapter (1–11). The interested reader is referred to these works for many of the details.

In this chapter, the fundamental aspects of optical measurements are reviewed in the interest of providing the background required for understanding the detailed applications described in other chapters in this volume. As can be imagined, several different optical techniques are used in commercial metrology instrumentation. Even for a standard technique like ellipsometry, several different arrangements of the optical components can be used, and the resulting polarization-sensitive information can be quite different. Therefore, several of the basic instrumental configurations will be discussed to compare and contrast the information content of each configuration.

One of the most important quantities obtained from optical measurements in the semiconductor industry is the thickness of a thin film. However, many optical measurements, such as ellipsometry and reflectivity, do not measure film thickness directly, and the near-surface region must be properly modeled to obtain this information. As film thicknesses become thinner, the modeling becomes more important, and the errors associated with improper modeling become larger. For example, it is well known that the refractive index of a thin film of silicon oxynitride will depend on the ratio of silicon, oxygen, and nitrogen atoms. However, the refractive index of the film may well also depend upon the concentration of hydrogen atoms and on the details of the deposition conditions. Since the thickness of a silicon oxynitride film as measured from an ellipsometry experiment depends strongly on the value used for the refractive index of the film, it is critical to use its correct value. Unfortunately, the refractive index may not be very well known, dramatically affecting the value and error of the film thickness that is obtained from the measurement. One possible solution to this dilemma is to perform an ellipsometry experiment at many different wavelengths of light (spectroscopic ellipsometry) and then to parameterize the film optical functions using standard optical dispersion models. Some of the common models used to parameterize optical functions of materials will be discussed in this chapter. If spectroscopic ellipsometry measurements are used in conjunction with an appropriate model, it is possible to determine parameters other than film thickness, which can then be related to film quality.

The chapter will be organized as follows. Section II will outline the physics of the optical measurements common to many instruments used in the semiconductor industry. This section will discuss the optical functions of materials, the physics of light reflection, light transmission, and light polarization. In addition, many common optical configurations used in semiconductor instrumentation will be compared and contrasted. The optical properties of silicon will be discussed in Sec. III, as well as the perturbation in the optical functions due to temperature and doping. In addition, we will briefly discuss the changes in the optical functions that occur when silicon is amorphous, polycrystalline, or liquid. Section IV will discuss other materials used in the semiconductor industry. As already mentioned, the optical functions of many of these materials cannot be known a priori, and so they must be modeled. Many of the common models used will be discussed in Sec. IV.A. Representative values of the optical functions of many of the materials will be presented in Sec. IV.B as taken from the literature.

## II. PHYSICS OF OPTICAL REFLECTION AND TRANSMISSION

### A. Definitions of Optical Functions

Before we can discuss the optical physics of silicon and related materials, it is important to define the parameters that will be used to describe the physics. The reader is referred to a standard text (such as Refs. 1–11) for more details.

The optical response of an isotropic material (such as unstrained silicon) is expressed by its complex refractive index

$$\tilde{n}(E) = n(E) + ik(E) \tag{1}$$

where $n$ is the refractive index, $k$ is the extinction coefficient, and $i$ is $\sqrt{-1}$ (note that some authors use the definition $\tilde{n}(E) = n(E) - ik(E)$ for the complex refractive index). The refractive index is a measure of the speed of light in the medium, while the extinction coefficient is a measure of the attenuation of a light beam as it passes through this same medium. Perfectly transparent materials will have $k = 0$. The expression of the complex refractive index is shown in Eq. (1) explicitly in terms of the photon energy $E$, which is inversely proportional to the wavelength of light $\lambda$. If the wavelength is expressed in nanometers and the energy in terms of electron volts, then

$$E\,(\text{ev}) = \frac{1239.8}{\lambda\,(\text{nm})} \tag{2}$$

If a light beam goes from a medium of lower refractive index into one of higher refractive index (such as from air into water), then the light beam will be refracted toward the normal of the surface. The angles of the light beams in the two media are related by Snell's law:

$$\tilde{n}_0 \sin \phi_0 = \tilde{n}_1 \sin \phi_1 \tag{3}$$

where $\phi_0$ is the angle of incidence in air and $\phi_1$ is the angle of light propagation in the material. If both materials are isotropic, then the incident beam, the reflected beam, and the transmitted beam will all lie in the same plane, called the *plane of incidence*.

Many geometric optics expressions [such as Snell's law, Eq. (3)] are most naturally expressed in terms of the refractive index. However, many physical phenomena are often best expressed in terms of the dielectric function, which is given by

$$\varepsilon = \varepsilon_1 + i\varepsilon_2 = \tilde{n}^2 = (n^2 - k^2) + 2ink \tag{4}$$

(If the $n - ik$ convention is used, then $\varepsilon_2 \to -\varepsilon_2$.) The dielectric function is the constant of proportionality between the electric field vector and the displacement vector used in Maxwell's equations. Therefore, $\varepsilon$ is closely related to the fundamental physics of light interaction with materials. In this chapter, we will express all optical functions in terms of the complex refractive index, with the understanding that Eq. (4) can be used to convert the complex refractive index to complex dielectric functions.

All optical functions, whether expressed in terms of the complex refractive index $\tilde{n}$ or in terms of the complex dielectric function $\varepsilon$, must obey the Kramers–Kronig relations. This is a basic physical limitation on the optical functions based on the law of causality. For nonmagnetic systems, the two parts of the complex refractive index are related by the Kramers–Kronig relations (see Ref. 8, p. 240):

$$n(E) = 1 + \frac{2}{\pi} P \int_0^\infty \frac{\xi k(\xi)}{\xi^2 - E^2} \, d\xi \tag{5a}$$

$$k(E) = -\frac{2E}{\pi} P \int_0^\infty \frac{n(\xi) - 1}{\xi^2 - E^2} \, d\xi \tag{5b}$$

Obviously, the Kramers–Kronig relations impose a very important constraint on the complex refractive index. If either $n(E)$ or $k(E)$ is known over the entire range of $E$, then it is possible to calculate the other. Similar expressions can be constructed for the complex dielectric function $\varepsilon$ as well as the reflectivity.

If the light beam goes from a medium of higher refractive index to one of lower refractive index, then the exiting light beam will be bent at a larger angle than the entrant light beam. In fact, if the angle of incidence is greater than a certain value, the light will be totally internally reflected, and no light will propagate in the lower-index region. The angle of incidence at which this occurs is given by

$$\phi_{\text{tr}} = \sin^{-1}\left(\frac{n_1}{n_0}\right) \tag{6}$$

If the extinction coefficient of a material is greater than zero at some wavelength, then the light will be partially absorbed by the material. A common way of describing this effect is in terms of the optical absorption coefficient $\alpha$, which is given by

$$\alpha(\lambda) = \frac{4\pi k(\lambda)}{\lambda} \tag{7}$$

If a light beam is initially at intensity $I_0$ and penetrates a length $d$ of a material, then the intensity of the light beam at $d$ is given by

$$I(d, \lambda) = I_0 \exp(-\alpha(\lambda)d) \tag{8}$$

The extinction coefficient $k$ is a dimensionless quantity, and $\alpha$ is in units of inverse length, with $\text{cm}^{-1}$ being the most common unit.

## B. Reflection Coefficients

In general, light beams go in straight lines unless they encounter regions of changing refractive index. If a light beam does encounter a region where refractive index changes, some of the light will penetrate the interfacial region and some of the light will be reflected (see Figure 1). Moreover, the transmitted and reflected light beams will also be shifted in phase with respect to the incident light beam. The amount of light reflected and transmitted, as well as the associated phases, will depend upon the polarization state of the incident light beam.

For a single interface, the expression for the complex reflection ratio can easily be derived from Maxwell's equations. These expressions, called the *Fresnel reflection coefficients*, are given by:

$$r_p = \frac{E_p^{\text{out}}}{E_p^{\text{in}}} = \frac{\tilde{n}_1 \cos(\phi_0) - \tilde{n}_0 \cos(\phi_1)}{\tilde{n}_1 \cos(\phi_0) + \tilde{n}_0 \cos(\phi_1)} \tag{9}$$

$$r_s = \frac{E_s^{\text{out}}}{E_s^{\text{in}}} = \frac{\tilde{n}_0 \cos(\phi_0) - \tilde{n}_1 \cos(\phi_1)}{\tilde{n}_0 \cos(\phi_0) + \tilde{n}_1 \cos(\phi_1)} \tag{10}$$

In Eqs. (9) and (10), the subscripts $s$ and $p$ refer to the polarization states perpendicular and parallel to the plane of incidence, respectively. The electric fields are denoted by $E$ (not to be confused with the photon energy $E$, which is not subscripted) and are complex to represent the phase of the input and output light. If both media are not absorbing ($k = 0$), then the quantities $r_p$ and $r_s$ are real.

Very often, one is interested in the normal-incidence reflectivity $R$ of a material, which is the fraction of incident light that is reflected from a sample surface when the incident light is perpendicular to the plane of the surface. In this case, the reflection coefficients given in Eqs. (9) and (10) are particularly simple, since $\cos(\phi_0) = \cos(\phi_1) = 1$. Aside from a sign change, $r_p = r_s$. The intensity of the reflected beam $I_R$ is then given by

$$I_R = I_0 R = \frac{I_0(r_p r_p^* + r_s r_s^*)}{2} \tag{11}$$

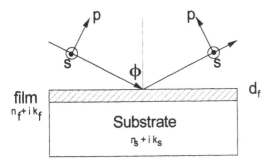

**Figure 1** Schematic diagram showing optical reflection experiment. The plane of incidence is defined as the plane containing both the incoming light beam and the normal to the sample surface. The p-polarized direction is parallel of the plane of incidence, while the s-polarized direction is perpendicular to the plane of incidence.

where the asterisk represents complex conjugation. When medium 0 is air ($n_0 = 1$, $k_0 = 0$) and medium 1 is transparent ($k_1 = 0$), then the expression for the sample reflectivity $R$ is particularly simple and is given by

$$R = \left(\frac{n_1 - 1}{n_1 + 1}\right)^2 \tag{12}$$

If the light beam is incident upon a more complicated structure that contains one or more thin films, then interference occurs. This will alter both the intensity and the phase of the reflected and transmitted beams. If there is only one film, such as is shown in Fig. 1, then the expressions of the s- and p-components of the reflection coefficient are given by:

$$r_p = \frac{r_{01,p} + r_{12,p}e^{-i\beta}}{1 + r_{01,p}r_{12,p}e^{-i\beta}} \tag{13a}$$

$$r_s = \frac{r_{01,s} + r_{12,s}e^{-i\beta}}{1 + r_{01,s}r_{12,s}e^{-1\beta}} \tag{13b}$$

In Eqs. (13), the phase factor is

$$\beta = \frac{4\pi d_f \tilde{n}_f \cos(\phi_f)}{\lambda} \tag{13c}$$

where the thickness and refractive index of the film are $d_f$ and $n_f$, respectively. The angle of light propagation in the film, $\phi_f$, is calculated using Snell's law [Eq. (3)]. The factors $r_{01,p}$, $r_{12,p}$, $r_{01,s}$, and $r_{12,s}$ are the complex reflection coefficients for the first and second interfaces, where (9) is used to calculate for p-polarized light and Eq. (10) is used to calculate for s-polarized light. Due to the complex phase factor $i\beta$, the value of $r_p$ and $r_s$ can be complex, even if $r_{01,p}$, $r_{12,p}$, $r_{01,s}$, and $r_{12,s}$ are real. Clearly, $r_p$ and $r_s$ depend very strongly on the thickness of the film $d_f$. This gives rise to the differences in the apparent color of samples with different thicknesses of films, since blue light (with a shorter wavelength) will have a larger value of $\beta$ than will red light.

For multiple layers, the calculation of the complex reflection coefficients is more complicated, but it can easily be performed using a matrix method. The most common matrix method is that due to Abelés (12), which uses $2 \times 2$ complex matrices, one for the p-polarization and the other for the s-polarization. In this method, the $j$th layer is represented by two transfer matrices:

$$\mathbf{P}_{j,p} = \begin{pmatrix} \cos(\beta_j) & -i\dfrac{\cos(\phi_j)}{\tilde{n}_j}\sin(\beta_j) \\[2mm] i\dfrac{\tilde{n}_j}{\cos(\phi_j)}\sin(\beta_j) & \cos(\beta_j) \end{pmatrix} \tag{14a}$$

$$\mathbf{P}_{j,s} = \begin{pmatrix} \cos(\beta_j) & i\dfrac{\sin(\beta_j)}{\tilde{n}_j\cos(\phi_j)} \\[2mm] i\tilde{n}_j\cos(\phi_j)\sin(\beta_j) & \cos(\beta_j) \end{pmatrix} \tag{14b}$$

where $\beta_j$ is the phase factor for the $j$th layer given in Eq. (13c) and $\phi_j$ is the complex angle in the $j$th layer as given by Snell's law [Eq. (3)]. If a film is extremely thin (where $\beta_j \ll 1$), then its Abelés matrices simplify to

$$\mathbf{P}_{j,p} = \begin{pmatrix} 1 & -i\dfrac{\cos(\phi_j)}{\tilde{n}_j}\beta_j \\ i\dfrac{\tilde{n}_j}{\cos(\phi_j)}\beta_j & 1 \end{pmatrix} \tag{14c}$$

$$\mathbf{P}_{j,s} = \begin{pmatrix} 1 & i\dfrac{\beta_j}{\tilde{n}_j\cos(\phi_j)} \\ i\tilde{n}_j\cos(\phi_j)\beta_j & 1 \end{pmatrix} \tag{14d}$$

The characteristic matrix for the layer stack consisting of $N$ films is determined by matrix multiplication:

$$\mathbf{M}_p = \chi_{0,p}\left(\prod_{j=1}^{N}\mathbf{P}_{j,p}\right)\chi_{\text{sub},p} \tag{15a}$$

$$\mathbf{M}_s = \chi_{0,s}\left(\prod_{j=1}^{N}\mathbf{P}_{j,s}\right)\chi_{\text{sub},s} \tag{15b}$$

The $\chi_0$ and $\chi_{\text{sub}}$ matrices are the characteristic matrices for the ambient and the substrate, respectively, and are given by

$$\chi_{0,p} = \frac{1}{2}\begin{pmatrix} 1 & \dfrac{\cos(\phi)}{\tilde{n}_0} \\ -1 & \dfrac{\cos(\phi)}{\tilde{n}_0} \end{pmatrix} \tag{16a}$$

$$\chi_{0,s} = \frac{1}{2}\begin{pmatrix} 1 & \dfrac{1}{\tilde{n}_0\cos(\phi)} \\ 1 & \dfrac{-1}{\tilde{n}_0\cos(\phi)} \end{pmatrix} \tag{16b}$$

$$\chi_{\text{sub},p} = \begin{pmatrix} \dfrac{\cos(\phi_{\text{sub}})}{\tilde{n}_{\text{sub}}} & 0 \\ 1 & 0 \end{pmatrix} \tag{16c}$$

$$\chi_{\text{sub},s} = \begin{pmatrix} \dfrac{1}{\tilde{n}_{\text{sub}}\cos(\phi_{\text{sub}})} & 0 \\ 1 & 0 \end{pmatrix} \tag{16d}$$

The complex reflection coefficients are then calculated from the elements of the characteristic matrices using the following relations:

$$r_p = \frac{M_{21,p}}{M_{11,p}} \tag{17a}$$

$$r_s = \frac{M_{21,s}}{M_{11,s}} \tag{17b}$$

If the film stack consists of two very thin films [see Eqs. (14c) and (14d)] then their representative Abelés matrices commute to first order in the parameter $\beta_j$. This means that it is not possible to use optical techniques to distinguish between the film combination $AB$ and the film combination $BA$, in the limit where $\beta_j$ is very small. In order to distinguish between the film combination $BA$ and that of the combination $AB$, the $\beta_j$ parameter must be large enough so that the first-order expansion [Eqs. (14c) and (14d)] is not valid and the Abelés matrices do not commute. One way to increase the $\beta_j$ factor is to use ultraviolet light (where the wavelength $\lambda$ is shorter), but this can introduce problems associated with light absorption from the thin films.

## C. Polarized Light and Its Mathematical Representation

Light beams can be described in a variety of ways. Most people are familiar with the intensity or power of a light beam and the wavelength or photon energy of a monochromatic beam. Another very important characteristic of a light beam is its polarization. In the classical picture of a propagating light beam as described by electromagnetic theory, the beam consists of oscillating electric and magnetic fields, where the direction of the magnetic field is perpendicular to the direction of the electric field. This beam is linearly polarized light if the electric field oscillates in only one direction. If the electric field oscillates in a random direction, the light beam is said to be unpolarized.

One of the most useful optical metrology tools for the semiconductor industry is ellipsometry, which uses polarized light to measure quantities such as film thickness and optical functions of materials. These measurement techniques will be described in detail in another chapter, but here we will briefly outline the physics of polarization measurements (see Refs. 1–5 for a more detailed account). To do this, we first need to describe how polarized light is created and how to handle it mathematically.

Linearly polarized light is created by a polarizer. There are many different ways to polarize light, depending upon many factors, including the wavelength of light and the intensity of the light beam to be polarized. However, any polarizer will transform any form of light (such as unpolarized light, partially polarized light, or elliptically polarized light) to linearly polarized light. No polarizer is perfect, but crystal polarizers (such as those made from calcite, quartz, or magnesium fluoride) in the visible and near-ultraviolet can produce linearly polarized light with as little as $10^{-6}$ leakage. In the infrared, the polarization process is considerably more difficult, and much higher values of leakage must be tolerated. Here, we will assume that the polarization process is perfect, with zero leakage. The angle of linear polarization is important, and it can be changed by altering the azimuthal angle of the polarizer with respect to the principal frame of reference, such as the plane of incidence in a reflection experiment.

A second important optical element in polarization-sensitive experiments is a compensator or retarder. In the near-infrared to the near-ultraviolet region of the spectrum, compensators can easily be made. Many commercial compensators are made from birefringent crystals (such as crystalline quartz or mica). There are two parameters that are important in describing a compensator: the degree of retardation $\delta$, and the angle of the fast axis of the compensator with respect to the principal frame of reference $\theta_c$. A compensator used in conjunction with a linear polarizer produces elliptically polarized light; if $\theta_c = \pm 45°$ and $\delta = \pi/2$, then circularly polarized light is produced. For the crystalline compensators just mentioned, the degree of retardation is a strong function of the wavelength of light and will have a value of $\delta = \pi/2$ at only one wavelength. Other types of compensators (such as Fresnel rhombs, where the retardation occurs from multiple internal reflections) will have a value of $\delta$ that is not as strong a function of wavelength.

One of the most convenient and accurate ways to describe the polarization effects of optical elements in ellipsometers is by the Mueller matrix–Stokes vector representation. The polarization characteristics of a light beam are described by its polarization state, which is best quantified by its Stokes vector. The Stokes vector is a four-element vector of real quantities that can describe any polarization state, even if the light beam is partially depolarized. The matrices that are used to describe the transformations of the polarization state due to particular optical elements are Mueller matrices, which are $4 \times 4$ real matrices. Only a brief summary of the Mueller–Stokes formalism can be given here; the interested reader is referred to Refs. 2 and 4.

The Stokes vector is defined by

$$
\mathbf{S} = \begin{pmatrix} I_o \\ Q \\ U \\ V \end{pmatrix} = \begin{pmatrix} I_o \\ I_0 - I_{90} \\ I_{45} - I_{-45} \\ I_{rc} - I_{lc} \end{pmatrix}
\tag{18}
$$

where $I_o$ is the intensity of the light beam, and $I_0$, $I_{45}$, $I_{90}$, and $I_{-45}$ are the light intensities for linearly polarized light at $0°$, $45°$, $90°$, and $-45°$, respectively, with respect to the plane of incidence. $I_{rc}$ and $I_{lc}$ are the intensities of right- and left-circularly polarized light, respectively. All elements of the Stokes vector are intensities and therefore are real. The total light intensity is

$$
I_o \geq \left( Q^2 + U^2 + V^2 \right)^{1/2}
\tag{19a}
$$

where the equality holds only if the light beam is totally polarized. If the light beam were totally unpolarized, then its Stokes vector would be

$$
\mathbf{S} = I_o \begin{pmatrix} 1 \\ 0 \\ 0 \\ 0 \end{pmatrix}
\tag{19b}
$$

When a light beam interacts with an optical device such as a polarizer or compensator or is reflected off a sample, its polarization state changes. The $4 \times 4$ matrix that describes this change in polarization state is called a *Mueller matrix*. The Mueller matrices for polarizers and compensators are given by:

$$
\mathbf{M_P} = \frac{1}{2} \begin{pmatrix} 1 & 1 & 0 & 0 \\ 1 & 1 & 0 & 0 \\ 0 & 0 & 0 & 0 \\ 0 & 0 & 0 & 0 \end{pmatrix} \quad \text{(Polarizer)}
\tag{20a}
$$

$$
\mathbf{M_C} = \begin{pmatrix} 1 & 0 & 0 & 0 \\ 0 & 1 & 0 & 0 \\ 0 & 0 & \cos\delta & \sin\delta \\ 0 & 0 & -\sin\delta & \cos\delta \end{pmatrix} \quad \text{(Compensator)}
\tag{20b}
$$

The Mueller matrices $\mathbf{M_P}$ and $\mathbf{M_C}$ represent the polarization state changes caused by a polarizer and a compensator, respectively, where the optical element is aligned in the principal frame of reference (such as the plane of incidence). The degree of retardation of the compensator is given by the angle $\delta$.

Rotations of optical elements are *unitary transformations*. Therefore, if the optical element is rotated about the light beam, then one must apply the rotation matrix $\mathbf{R}(-\theta)$ before the Mueller matrix for the optical element, followed by the rotation matrix $\mathbf{R}(\theta)$, where the rotation matrix is given by

$$\mathbf{R}(\theta) = \begin{pmatrix} 1 & 0 & 0 & 0 \\ 0 & \cos(2\theta) & \sin(2\theta) & 0 \\ 0 & -\sin(2\theta) & \cos(2\theta) & 0 \\ 0 & 0 & 0 & 1 \end{pmatrix} \quad \text{(Rotation)}$$

For example, if a polarizer is rotated $\theta$ degrees, the resultant Mueller matrix in the principal frame of reference is just the matrix multiplication $\mathbf{R}(-\theta)\mathbf{M_P}\mathbf{R}(\theta)$. If only linear polarizers are used in an optical instrument, then it is not possible to create light with a circularly polarized component (that is, the 4th element of the Stokes vector is always 0). A polarizer-compensator combination, however, can create light with a circularly polarized component where the 4th element of the Stokes vector is nonzero.

The Mueller matrix for light reflected off an isotropic mirror or sample is given by $\mathbf{M_R}$:

$$\mathbf{M_R} = R \begin{pmatrix} 1 & -N & 0 & 0 \\ -N & 1 & 0 & 0 \\ 0 & 0 & C & S \\ 0 & 0 & -S & C \end{pmatrix} \quad \text{(Reflection sample)} \tag{20d}$$

The elements of this matrix are the associated ellipsometry parameters, and they are given by

$$N = \cos(2\psi) \tag{21a}$$

$$S = \sin(2\psi)\sin\Delta \tag{21b}$$

$$C = \sin(2\psi)\cos\Delta \tag{21c}$$

The angles $\psi$ and $\Delta$ are the traditional ellipsometry parameters, which are defined as

$$\rho = \frac{r_p}{r_s} = \tan\psi \exp(i\Delta) = \frac{C + iS}{1 + N} \tag{21d}$$

The parameters $N$, $S$, and $C$ are not independent, since $N^2 + S^2 + C^2 = 1$.

The Mueller matrices are extremely useful, in that they can be used to express the final intensity of most polarization-sensitive optical experiments. There are other ways of performing these calculations (such as the Jones matrix formulation, described Refs. 2 and 4), but the Mueller–Stokes approach uses real matrices and can deal with depolarization. As a result, the Mueller–Stokes approach is the more powerful technique.

As an example, let us consider a polarized reflectivity measurement, described in more detail in the next section. In this measurement, the input light is linearly polarized, reflected off a sample (which changes the polarization state of the light), and then analyzed (using another linear polarizer) before being detected. If it can generally be assumed that the input light is unpolarized, then the Stokes vector for the input beam is given by Eq. (19b). If it can be assumed that the detector is also insensitive to polarization state, the intensity of the polarized reflectivity measurement is given by

$$I = I_0 \begin{pmatrix} 1 & 0 & 0 & 0 \end{pmatrix} \mathbf{R}(-\theta_1)\mathbf{M_P}\mathbf{R}(\theta_1)\mathbf{M_R}\mathbf{R}(-\theta_0)\mathbf{M_P}\mathbf{R}(\theta_0)\begin{pmatrix} 1 & 0 & 0 & 0 \end{pmatrix}^T \tag{22a}$$

$$I = I_o \frac{R}{4} \begin{pmatrix} 1 & \cos(2_1) & \sin(2\theta_1) & 0 \end{pmatrix} \begin{pmatrix} 1 & -N & 0 & 0 \\ -N & 1 & 0 & 0 \\ 0 & 0 & C & S \\ 0 & 0 & -S & C \end{pmatrix} \begin{pmatrix} 1 \\ \cos(2\theta_0) \\ \sin(2\theta_0) \\ 0 \end{pmatrix} \tag{22b}$$

The intensity for many other polarization-sensitive measurements can be obtained in a similar manner. Although the final value of the intensity may be very complicated algebraically, the Mueller–Stokes formalism allows us to express the intensity quite succinctly.

## D.  Experiments to Measure Reflection and Transmission

There are many ways to measure the reflection and transmission properties, some of which are displayed schematically in Figure 2. In some of these experiments, one is interested only in the intensity change of the incident light beam; these measurements are usually

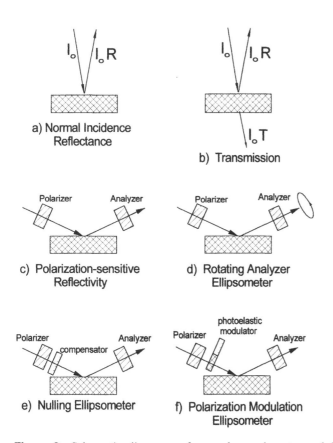

**Figure 2**  Schematic diagrams of several experiments used in semiconductor diagnostics. The reflection and transmission experiments (a and b) use unpolarized light, while the remaining experiments utilize polarized light.

carried out at normal incidence. However, one can usually obtain more information about the sample if polarization-sensitive measurements are made. These types of measurements, normally called *ellipsometry measurements*, are usually carried out at a large angle of incidence and are more complicated to perform, but they yield more information than simple reflection or transmission experiments. The reader is referred to the review article by Hauge (5) that describes in more detail all ellipsometry configurations discussed here.

## 1. Reflectance

Reflectance measurements have a very long history in semiconductor physics, and they have been very useful in identifying critical points in the optical spectra of many semiconductors. Since this measurement technique does not require polarization optics, reflectance can be carried out over a very wide range in wavelength. Figure 2a shows a schematic of a simple optical reflectance measurement carried out at near-normal incidence, where the quantity determined is the fraction of light reflected from the sample surface $R$.

At near-normal incidence and for isotropic samples, there is very little difference between s- and p-polarized light, so Eq.(11) can be used to analyze the results. If there is no film on the surface and the material is transparent ($k = 0$), Eq. (12) can be used to determine the refractive index from the measured value of $R$, although this is not a very accurate way to do it. If there is a thin film on the surface, then the Airy formula [Eqs. (13)] must be used to interpret the results, where $R = rr^*$. If the film is very thin, then the phase factor $\beta$ is very small, and the effect of the film on the value of $R$ is small. It is very difficult to measure the characteristics of very thin films ($\sim 5$–10 nm) using a simple reflectance measurement, but it can be a very useful tool to study thicker films. Increased sensitivity can be obtained by performing the measurement in the UV, since the smaller wavelength increases the factor $\beta$.

If the reflectance measurement can be carried out over a wide range of wavelengths, then Kramers–Kronig analysis can be used to determine both the refractive index $n$ and the extinction coefficient $k$. The Kramers–Kronig integration is performed from 0 energy to infinite energy, so reasonable extrapolations must be made from the measured spectra to 0 and infinity.

## 2. Transmission

Optical transmission measurements have also been used extensively over the last several decades in semiconductor physics studies. These measurements are very useful in measuring very small values of the optical absorption coefficient of materials, and therefore they have been applied mostly to regions of the spectrum near and below the optical bandedge. This technique is shown schematically in Fig. 2b. If the incident beam has intensity $I_0$, then the intensity of the transmitted beam is given by

$$I_T = I_0 \frac{(1 - R_1)(1 - R_2)e^{-\alpha d}}{1 - R_1 R_2 e^{-2\alpha d}} \qquad (23)$$

The quantities $R_1$ and $R_2$ are the reflectivities of the front and back surfaces, respectively (including the effects of any thin films on the front and back surfaces). The absorption coefficient $\alpha$ is usually the quantity to be determined, where the sample thickness is given by $d$. Equation (23) assumes that any films on the front and back surfaces do not absorb light; if they do, then a more complicated formalism must be employed to interpret the experimental results (see Ref. 1).

As shown in Fig. 2b, the transmission measurement is used to determine the absorptive properties of the bulk material, where light reflection from the back surface is assumed to be incoherent. If one is interested in the absorption coefficient of a thin film, then interference effects must be taken into account in determining the transmission coefficients. See a standard text (such as Ref. 1) for the details.

### 3. Polarized Reflectance

Reflectance and transmission measurements are normally performed at near-normal incidence, so there is very little dependence of the results on the polarization state of the incoming and outgoing light if the sample is isotropic. Consequently, only a single value is determined at each wavelength. More information can be obtained if the incoming and outgoing light beams are polarized. Figure 2c shows one of the simplest examples of an experiment that will give polarization information. The incident light is first polarized and then interacts with the sample at a large angle of incidence. The reflected beam is then repolarized (using the polarizer commonly called the analyzer) before the intensity is detected. At a large angle of incidence, there is a significant difference in the reflectance for s- and p-polarized light [see Eqs. (19) and (20)], so there are two pieces of information that can be obtained about the sample at each wavelength: $R_p = r_p r_p^*$ and $R_s = r_s r_s^*$ These quantities are measured by aligning both polarizers either in the s or the p orientation (see Fig. 1). If the sample is isotropic, then the crossed polarizer reflection coefficients (where one polarizer is aligned in the s direction and the other in the p direction) are zero.

This experiment can also be used as an example for the calculation using Mueller matrices, described earlier. For the p measurement, each polarizer is aligned parallel to the plane of incidence, and so $\theta_0 = \theta_1 = 0$. Performing the Mueller matrix multiplication given in Eqs. (22) results in

$$I_p = \frac{I_0 R(1 - N)}{4} = I_0 R \sin^2 \frac{\psi}{2} = I_0 R_p \tag{24a}$$

For the s measurement, each polarizer is aligned perpendicular to the plane of incidence, and so $\theta_0 = \theta_1 = 90°$, resulting in

$$I_s = \frac{I_0 R(1 + N)}{4} = I_0 R \cos^2 \frac{\psi}{2} = I_0 R_s \tag{24b}$$

Therefore, polarized reflectivity measurements can measure the average reflectance $R$ and the ellipsometric angle $\psi$. The ellipsometric angle $\Delta$ cannot be measured using this technique.

### 4. Rotating Polarizer/Analyzer Ellipsometry

A common type of spectroscopic ellipsometer is the rotating analyzer (or rotating polarizer) ellipsometer, first constructed by Aspnes and Studna (13). This instrument is readily available from several commercial vendors and is shown schematically in Fig. 2d. In this experiment, the second polarizer, called the analyzer, is physically rotated so that the azimuthal angle of the analyzer is given by $\theta_1 = \omega t$. (For the rotating polarizer ellipsometer, the polarizer is rotated, and $\theta_0 = \omega t$.) Equation (22b) can be used to calculate the final intensity, which is given by

$$I(t) = I_{dc}[1 + a_1 \cos(2\omega t) + a_2 \sin(2\omega t)] \tag{25a}$$

where

$$I_{dc} = 1 - N \cos(2\theta_p) \tag{25b}$$

$$a_1 = \frac{\cos(2\theta_p) - N}{1 - N \cos(2\theta_p)} \tag{25c}$$

$$a_2 = \frac{C \sin(2\theta_p)}{1 - N \cos(2\theta_p)} \tag{25d}$$

There are three possible measured parameters for this type of instrument: $I_{dc}$, $a_1$, and $a_2$. Usually, the signal is normalized to the $I_{dc}$ term and it is not directly measured. The two Fourier components $a_1$ and $a_2$ are functions of the fixed polarizer angle and the sample parameters $N$ and $C$. This instrument does not measure the parameter $S$ and so is very insensitive whenever $\Delta$ is close to $0°$ or $180°$. This is a direct consequence of only linear polarizers being used in this optical experiment; the measurement of $S$ requires the use of a compensating optical element. This will be described later in Sec. II.D.6.

## 5. Nulling Ellipsometer

The nulling ellipsometer has been in use for nearly a century and is shown schematically in Fig. 2e. This is a very common instrument in semiconductor fabrication facilities, and it has been described in detail in Refs. 2 and 3. It usually consists of a polarizer–compensator pair before the sample and a polarizer after the sample, called the analyzer. Using the Mueller–Stokes formalism, we can easily construct the matrix multiplication chain required to calculate the intensity incident upon the detector:

$$I = I_0\begin{pmatrix} 1 & 1 & 0 & 0 \end{pmatrix} R(\theta_1) M_R R(-\theta_c) M_C R(\theta_c - \theta_0) \begin{pmatrix} 1 & 1 & 0 & 0 \end{pmatrix}^T \tag{26a}$$

Measurements are made with a nulling ellipsometer by rotating the azimuthal angles of the polarizer ($\theta_0$) and the analyzer ($\theta_1$) to minimize the intensity of light incident upon the detector. Nulling ellipsometry measurements are made by fixing the azimuthal angle of the compensator (such as at $\theta_c = 45°$) and the degree of retardation of the compensator $\delta = \pi/2$. Under these assumptions, the intensity at the detector is given by

$$I = \left(\frac{I_0 R}{4}\right)\{1 - N \cos(2\theta_1) + \sin(2\theta_1)[C \sin(2\theta_0) + S \cos(2\theta_0)]\} \tag{26b}$$

This expression for the intensity is 0 if

$$\psi = \theta_1 \quad \text{and} \quad \Delta = 270° - 2\theta_0 \tag{26c}$$

or

$$\psi = 180° - \theta_1 \quad \text{and} \quad \Delta = 90° - 2\theta_0 \tag{26d}$$

Since a compensating element is used, it is possible to obtain an accurate value of $\Delta$ regardless of its value.

As mentioned earlier, most compensators used in nulling ellipsometry experiments are designed for a specific wavelength (such as 633 nm) where its retardation is precisely $\pi/2$. While this type of ellipsometer can be used at wavelengths other than its principal design wavelength, the mathematical interpretation is much more complicated, so nulling ellipsometers are generally used only at the design wavelength of the compensator. These instruments have been available commercially for several decades.

## 6. Rotating Analyzer Ellipsometer with a Compensator

One of the major limitations of the standard rotating analyzer ellipsometer is that it is insensitive to the $S$ parameter, and therefore cannot measure $\Delta$ accurately when $\Delta$ is near $0°$ or $180°$. One solution to this is to include a compensating element in the light path (14); this is implemented in many commercial ellipsometers. A rotating analyzer ellipsometer with a compensating element after the polarizer is shown schematically in Fig. 2e, modified only in the respect that the analyzer is rotated. Usually a quasi-achromatic compensator is used, where $\delta \sim \pi/2$. (As mentioned previously, there are no perfect achromatic compensators, so $\delta$ always is a function of wavelength. However, $\delta(\lambda)$ can be measured and the results corrected accordingly.) The intensity at the detector is then given by Eq. (26a), where the angle $\theta_1 = \omega t$. This can be expanded, resulting in a dc term and two Fourier coefficients [(see Eq. (25)]. If it is assumed that the polarizer is set at $45°$ with respect to the fast axis of the compensator, then the coefficients are:

$$1_{dc} = 1 - N \sin(2\theta_c) \cos(\delta) \tag{27a}$$

$$a_1 = \frac{-\sin(2\theta_c) \cos(\delta) - N}{1 - N \sin(2\theta_c) \cos(\delta)} \tag{27b}$$

$$a_2 = \frac{C \cos(2\theta_c) \cos(\delta) - S \sin(\delta)}{1 - N \sin(2\theta_c) \cos(\delta)} \tag{27c}$$

If the phase retardation of the compensator $\delta = \pi/2$, then $a_1 = -N$ and $a_2 = -S$. Of course, if $\delta = 0$, the results are the same as for the rotating analyzer ellipsometer, described earlier. This points to the fact that the inclusion of a precise $\delta = \pi/2$ compensator will allow the ellipsometer to measure $S$, but $C$ is not measurable. For other values of the retardation, a linear combination of $S$ and $C$ is measured.

## 7. Rotating Compensator Ellipsometer

Some modern ellipsometers use the scheme of nulling ellipsometers (shown schematically in Fig. 2e), but where the two polarizers are fixed and the compensator is rotated (15). The single-wavelength rotating compensator ellipsometer has been readily available for many years; just recently, a spectroscopic version has also become available. The Mueller–Stokes representation of the intensity given in Eq. (26a) above is still valid, but $\theta_c = \omega t$. The light intensity incident upon the detector is now a function of time, much like the rotating analyzer ellipsometer described previously. Because the compensator is rotating between the sample and the polarizer, there are two frequency components: $2\omega t$ and $4\omega t$. The intensity of the light beam incident upon the detector is given by:

$$I = I_{dc} + a_{2S} \sin(2\omega t) + a_{2C} \cos(2\omega t) + a_{4S} \sin(4\omega t) + a_{4C} \cos(4\omega t) \tag{28a}$$

The values of the five coefficients shown in Eq. (28a) depend upon the retardation of the compensator $\delta$ (which will also be a function of wavelength), as well as the azimuthal angles $\theta_p$ and $\theta_a$. For this to be a complete ellipsometer (that is, where $N$, $S$, and $C$ are all measured), $\theta_a$ must not be close to $0°$ or $90°$ and $\delta$ must be significantly different from $0°$ or $180°$ (see Ref. 16 for a discussion of the errors involved when $\delta$ is other than $90°$). If it is assumed that $\theta_a = 45°$, then the five coefficients are given by:

$$I_{dc} = 1 + \tfrac{1}{2}(1 + \cos \delta)\left[C \sin(2\theta_p) - N \cos(2\theta_p)\right] \tag{28b}$$

$$a_{2S} = S \sin \delta \cos(2\theta_p) \tag{28c}$$

$$a_{2C} = S \sin \delta \sin(2\theta_p) \tag{28d}$$

$$a_{4S} = \tfrac{1}{2}(1 - \cos \delta)\left[C \cos(2\theta_p) - N \sin(2\theta_p)\right] \tag{28e}$$

$$a_{4C} = -\tfrac{1}{2}(1 - \cos \delta)\left[C \sin(2\theta_p) + N \cos(2\theta_p)\right] \tag{28f}$$

Clearly, this ellipsometer is capable of measuring $N$, $S$, and $C$ for all values of $\theta_p$

## 8. Polarization Modulation Ellipsometer

Polarization modulation ellipsometers are usually defined as ellipsometers that use a modulation technique other than the physical rotation of one or more optical elements. Commonly, a photoelastic modulator (PEM) is used to generate a time-dependent polarization state, although it is also possible to use electro-optic modulators. The spectroscopic ellipsometer of Jasperson and Schnatterly (17) was of this type. Ellipsometers utilizing PEMs are also available commercially.

Although there are several different designs of PEMs, they all share a common description. The PEM is just a time-dependent compensator, where the retardation $\delta(t) = A \sin(\omega t)$, the modulation amplitude $A$ being proportional to the driving force of the modulator. In its operation, a nominally isotropic optical element is set into physical oscillation by some external driving force. If the optical element of the PEM is momentarily in compression, then the refractive index along the compressive direction is higher than the refractive index would be in the unstrained optical element, while the refractive index perpendicular to the compressive direction is lower. Similarly, if the optical element of the PEM is momentarily in expansion, then the refractive index along the expansive direction is lower, but the refractive index perpendicular to the expansive direction is higher than in the unstrained optical element. The Mueller matrix for a PEM is just the Mueller matrix for a compensator [(Eq. (20b)] where $\delta(t) = A \sin(\omega t)$.

The basis functions for rotating element ellipsometers are of the form $\sin(n\omega t)$ and $\cos(n\omega t)$, where $n$ is an integer, and therefore standard Fourier analysis can be used. The basis functions for polarization modulation ellipsometers are of the form $X = \sin [(A \sin(\omega t)]$ and $Y = \cos [A \sin(\omega t)]$. These basis functions can be expressed in terms of an infinite series of sines and cosines, using integer Bessel functions as coefficients:

$$X = \sin[A \sin(\omega t)] = 2 \sum_{j=1} J_{2j-1}(A) \sin[(2j - 1)\omega t] \tag{29a}$$

$$Y = \cos[A \sin(\omega t)] = J_0(A) + 2 \sum_{j=1} J_2(a) \cos(2j\omega t) \tag{29b}$$

As can be seen from Eq. (29), the $Y$ basis function has no dc term if $J_0(A) = 0$, which happens if $A = 2.4048$ radians. Often the modulation amplitude is set to this value to simplify the analysis. In normal operation, the first polarizer is set to $\pm 45°$ with respect to the PEM and the analyzer is set to $\pm 45°$ with respect to the plane of incidence (we will assume $+45°$ for both angles). Using the Mueller–Stokes formalism described earlier, the intensity incident upon the detector can be given as

$$I(t) = I_0\left(\frac{R}{4}\right)\{1 - SX - [\cos(2\theta_c)C + \sin(2\theta_c)N]Y\} \tag{30a}$$

From this expression, it can be seen that two sample parameters can be measured at any one time. The $\sin(\omega t)$ Fourier coefficient is always proportional to the $S$ parameter, and the $\cos(2\omega t)$ is proportional to either $N$ or $C$, depending on the azimuthal angle of the PEM ($\theta_c$).

It is possible to make a single PEM ellipsometer complete (that is, to measure $N$, $S$, and $C$). One way to do this is via the 2-channel spectroscopic polarization modulation ellipsometer (18) (2-C SPME), where the single analyzer polarizer is replaced with a Wollaston prism. The Wollaston prism deviates the incident light beam into two mutually orthogonal linearly polarized light beams, both of which are detected with the 2-C SPME. If the azimuthal angle of the PEM is set to $\theta_c = \pm 22.5°$ and if the Wollaston prism set at $\theta_a = \pm 45°$, then it is possible to measure $N$, $S$, and $C$.

## 9. Other Forms of Ellipsometers

There are obviously many other forms of ellipsometers, some of which will use two or more rotating or oscillating elements. Although the analysis of the time-dependent intensity waveform for ellipsometers using two components is far more complicated than for a single rotating or oscillating element, much more information can be obtained. If the sample is not isotropic, then many of the off-block diagonal elements of the sample Mueller matrix [Eq. (20d)] are nonzero. This is a direct result of the cross-polarization terms, which are nonzero for anisotropic surfaces (that is, pure s-polarized light, reflecting off an anisotropic surface, may have a p-polarized component and vice versa). Several of these ellipsometers have been reported in the literature, (19–22) but are not yet available commercially.

In this section, we have dealt only with variations in the polarization optics that differentiate measurement schemes. Obviously many other factors are also very important. For example, spectroscopic ellipsometry requires a broadband light source and a means to select out wavelengths for measurement. In some cases, the light source is made quasi-monochromatic before it is polarized. In other cases, the light beam is made quasi-mono-chromatic after the analyzer, using either a spectrograph arrangement, where the light beam is incident upon an array detector, or a monochromator arrangement, where a small range of wavelengths is incident upon a detector. Many of these arrangements are available commercially.

## III. OPTICAL PROPERTIES OF SILICON

As seen in Sec. II, the results of reflection and transmission experiments depend on the optical functions of the materials involved, since the complex refractive index is necessary to calculate the complex reflection coefficients [see Eqs. (9) and (10)]. For the semiconductor industry, this means that understanding the optical functions of crystalline silicon is extremely important. Of course, this is complicated by the fact that these optical functions can depend on a variety of factors, such as the wavelength of light, the temperature, the stress, and the carrier concentration.

Figure 3 shows a plot of the optical functions [$n$, $k$, and $\alpha$, taken from many sources (23–33)] of undoped, unstrained, crystalline silicon at room temperature from the mid-infrared (IR) to the deep ultraviolet (UV). The plot in Fig. 3 is shown in terms of the photon energy, expressed in electron volts (eV). At the top of the figure, several corresponding wavelengths are shown. In this figure, as well as others in this chapter, the refractive index $n$ and the extinction coefficient $k$ are plotted linearly, but the absorption

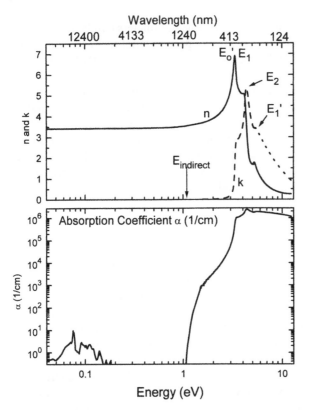

**Figure 3** Refractive index *n*, extinction coefficient *k*, and absorption coefficient α of silicon from 0.01 to 20 eV at room temperature. (Data from Refs. 23–33.)

coefficient α is plotted logarithmically. Thus, features in the α spectrum may not be observed directly in this plot of the *k* spectrum (such as near 0.1 eV and from 1 to 3 eV in Fig. 3), even though *k* is finite and α and *k* are related through Eq. (7).

Referring to Fig. 3, the refractive index *n* in the IR is relatively constant, but the absorption coefficient α shows structure near 0.1 eV that is due to two-phonon lattice absorption. (Note: The absorption in this wavelength region is very weak for covalently bonded semiconductors such as silicon and germanium since the optical phonons are not infrared active. The optical phonons of III-V semiconductors, such as GaAs and InP, are infrared active, resulting in much higher absorption coefficients in this region.) The band gap of silicon is ∼ 1.1 eV at room temperature, as evidenced by the very low value of the absorption coefficient below this photon energy. Between ∼ 1.1 and 3 eV (which corresponds to the near-IR and the visible portions of the spectrum: 1127–413 nm), optical absorption in silicon is determined by indirect transitions, which results in relatively small values of the absorption coefficient. Above 3.4 eV (< 364 nm or the UV part of the spectrum), the dominant optical transitions are direct, resulting in much higher values of absorption coefficient and structure in the refractive index and extinction coefficient. At much higher energies, the extinction coefficient asymptotically approaches 0 while the refractive index asymptotically approaches 1 (x-rays and gamma rays tend to go through materials, requiring an absorption coefficient that goes to 0 as the photon energy gets large). In the next subsections, individual parts of the spectrum will be discussed, as well as the effects of temperature, stress, and doping on the optical functions.

## A.  Band Structure

Many of the features seen in the visible and UV portions of the optical spectrum of silicon can be understood from the band structure of silicon, shown in Figure 4 (taken from Ref. 34). The band structure is a theoretical construct, and it relates the energy of electrons with the wave vector **k** (see Refs. 6–8 for a detailed discussion). Band structure calculations are quantum mechanical in nature and have been performed for decades. All band structure calculations use the one-electron model (meaning that they average the electron–electron interactions), but they rely on the translational, rotational, and reflection symmetries that all crystals possess. Consequently, band structure calculations cannot be carried out for amorphous materials. In order to simplify the calculation, many band structure calculations will ignore core electrons or approximate them by the introduction of pseudo-potentials.

Generally, the electronic wavefunctions in a crystalline material will have the form

$$\Phi_{\mathbf{k}}(\mathbf{x}) = \exp{(i\mathbf{kx})}u_{\mathbf{k}}(\mathbf{x}) \tag{31}$$

where $u_k(\mathbf{x})$ is a periodic function with the same periodicity as the crystal lattice. The function $\Phi_{\mathbf{k}}(\mathbf{x})$ is known as the Bloch function of the electron in the crystal. For a lattice translation **R** that takes the lattice from one equivalent position to another, $u_{\mathbf{k}}(\mathbf{x} + n\mathbf{R}) = u_{\mathbf{k}}(\mathbf{x})$ where $n$ is an integer. In one dimension, an equivalent Bloch function is obtained if $k_x$ is replaced with $k_x + (2n_x\pi/R_x)$.. Band structure calculations are usually carried out in **k**-space or reciprocal space, where the coordinates are $\mathbf{k} = (k_x, k_y, k_z)$. The symmetry of reciprocal space is directly related to the symmetry of real space [where the coordinates are $\mathbf{x} = (x, y, z)$], so group

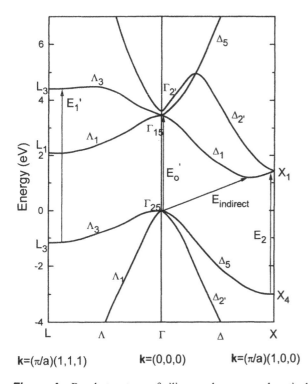

**k**=(π/a)(1,1,1)          **k**=(0,0,0)          **k**=(π/a)(1,0,0)

**Figure 4**  Band structure of silicon, where several optical transitions are shown. (Data from Ref. 34.)

theoretical techniques can be used to simplify band structure calculations. The region of **k**-space where $n_x$, $n_y$, and $n_z$ are all zero is referred to as the *first Brillouin zone*, and the resulting band structure is then given in the reduced zone scheme.

Several points in the first Brillouin zone are points of high symmetry and are assigned special notation for ease of identification. There are several different schemes, but we will describe the most common. Since silicon is a cubic crystal, we can restrict the discussion to cubic symmetry, where the parameter $a$ is used to denote the size of the unit cube (in units of length). The center of the Brillouin zone is referred to as the $\Gamma$ point and corresponds to $\mathbf{k} = (0, 0, 0)$. The $X$ point is one edge of the Brillouin zone, and it corresponds to $\mathbf{k}(\pi/a)(1, 0, 0)$, and intermediate points between the $\Gamma$ point and the $X$ point are $\Delta$ points, where $\mathbf{k} = (\pi/a)(b_x, 0, 0)$, $b_x < a$. Similarly, the $L$ point corresponds to $\mathbf{k} = (\pi/a)(1, 1, 1)$, and intermediate points between the $\Gamma$ point and the $L$ point are $\Lambda$ points, where $\mathbf{k} = (\pi/a)(b, b, b)$, $b < a$.

The band structure of a semiconductor consists of a valence band (usually denoted by energies less than 0) and a conduction band. The energy difference between the highest point in the valence band and the lowest point in the conduction band is called the *band gap*. The electron states in the valence band are normally occupied, while the electron states in the conduction band are normally unoccupied. If the **k**-vector of the highest point in the valence band corresponds to the **k**-vector of the lowest point in the conduction band, the semiconductor is called a *direct gap semiconductor* (such as GaAs). If this is not the case, it is called an *indirect gap semiconductor* (such as Si or Ge).

The crystal momentum of an electron in the lattice is given by $\mathbf{p} = h\mathbf{k}/2\pi$, where $h$ is Planck's constant. Since photons in the visible part of the spectrum have nearly no momentum, direct optical transitions can be represented on the band structure diagram as vertical transitions, as shown in Fig. 4. Direct optical absorption of a photon of energy $E$ can occur between a state in the valence band and a state in the conduction band if the vertical separation of these two states is just the energy $E$. The probability that this absorption will occur depends upon many factors, including the symmetry of the wavefunctions of the states in the valence band and the conduction band, and on the joint density of states. In particular, high probabilities of absorption exist where there are large regions in **k**-space where the energy separation $E$ is nearly constant. These regions of **k**-space result in large joint density of states and are referred to as critical points in the Brillouin zone.

Features in the visible and near-ultraviolet optical spectra of materials are often a result of critical points in the Brillouin zone. In silicon, one such critical point exists near the $\Gamma$ point, which is called the $E_o'$ point; since this is the smallest energy for a direct transition, the energy of this transition ($\sim 3.4$ eV) is called the direct bandgap. Another feature seen in the optical spectrum can be seen at $\sim 4.25$ eV ($\sim 292$ nm), called the $E_2$ critical point. Although this feature is not as well identified as the $E_o'$ point, it is believed to be due to transitions along the $\Lambda$ direction. The $E_1'$ critical point is $\sim 5.5$ eV and arises from transitions near the $L$ point in the Brillouin zone.

For photon energies less than the direct bandgap, there are no direct transitions from the top of the valence band to the bottom of the conduction band. Yet it is known that silicon absorbs light in the visible part of the spectrum (otherwise, we could see through Si wafers!). In the near-IR and visible parts of the spectrum ($\sim 1.1$–$3.4$ eV or 1127–364 nm), optical absorption in silicon occurs through indirect transitions, where a phonon, or quantum of lattice vibration, is either absorbed or created to conserve momentum. These indirect optical transitions are denoted by a diagonal arrow in the band structure diagram, as is shown in Fig. 4.

At this point, various parts of the optical spectrum of silicon will be discussed, with reference, when appropriate, to the band structure of Fig. 4. There are several external factors that will also affect the band structure (and therefore the optical spectra) of a crystalline material, such as temperature, doping, and lattice strain. These factors will also be discussed later.

## B. Ultraviolet and Visible Optical Functions: Critical Points

The optical functions of silicon in the visible and near-UV part of the spectrum are shown in Figure 5 as a function of temperature (data taken from Refs. 28, 30, and 35; see also Refs. 36 and 37). Also shown are the critical-point identifications mentioned in the previous subsection (see Ref. 37). Most authors discuss these features in terms of dielectric functions $\varepsilon$ since it is related directly to the physics of optical response.

The highest-energy feature shown is the $E_2$ critical point, which is a peak in the extinction coefficient $k$ spectrum and a high-energy shoulder in the refractive index $n$ spectrum. As the temperature increases, this feature broadens and moves to lower photon energy.

The next major feature is the $E_o'$ critical point, which is a low-energy shoulder in the $k$ spectrum and a peak in the $n$ spectrum. This critical point also decreases in energy and broadens as the temperature is increased. Associated with the $E_o'$ critical point is an

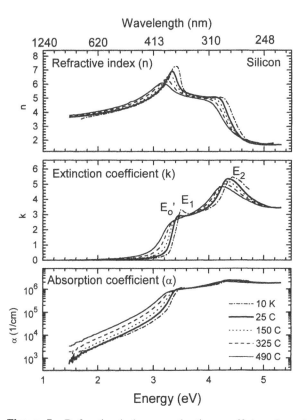

**Figure 5** Refractive index $n$, extinction coefficient $k$, and the absorption coefficient $\alpha$ of silicon from 1.5 to 5.3 eV at several different temperatures. (Data from Refs. 28, 30, and 35.)

excitonic feature, which is a peak in the $k$ spectrum superimposed on the $E_o'$ shoulder. Excitons are excited electron-hole pairs that are bound together by a very small energy. Because excitons represent an interaction between electrons (or holes, which are just the absence of an electron), they cannot be represented using the noninteracting electron model used to calculate band structure. At very low temperatures, the exciton peak in the $k$ spectrum is very strong, but it is still noticeable at higher temperatures. Above $\sim 300°C$, all evidence of the excitonic peak disappears.

It can be noted that the optical absorption coefficient, also plotted semilogrithmically in Fig. 5, varies between $1 \times 10^6$ and $2 \times 10^6$/cm for photon energies above the direct bandgap. This corresponds to optical penetration depths of 50–100 Å, which does not depend significantly on the temperature of the silicon. For most optical experiments, this has the very practical consequence that light above $\sim 3.4$ eV will be absorbed in the near-surface region of a silicon device, regardless of the temperature of the silicon.

Below the direct bandedge, all optical absorption takes place by phonon-assisted transitions. Because of the requirement for the presence of a phonon for absorption to take place, this type of optical absorption is generally less efficient than direct optical absorption, resulting in much smaller values of the absorption coefficient. Moreover, since the population of phonons is extremely temperature dependent, the absorption coefficient for indirect transitions is also very sensitive to temperature.

Figure 5 also shows the values of the absorption coefficient at several temperatures above and below room temperature. Moreover, the refractive index is also somewhat dependent on temperature. A precise theoretical understanding of this data is not yet available. However, the $n$ and $k$ data from 1.5 to 3.2 eV and from 25 to 490°C has been fit to an empirical formulation (35):

$$n(E, T) = n_o(E) + a(E)T \tag{32a}$$

where

$$n_o(E) = \sqrt{4.565 + \frac{97.3}{E_g^2 - E^2}} \tag{32b}$$

$$a(E) = 10^{-4}\left(-1.864 + \frac{53.94}{E_g^2 - E^2}\right) \tag{32c}$$

A similar expression can be obtained for $k$:

$$k(E, T) = k_o(E)\exp\left(\frac{T}{T_o(E)}\right) \tag{33a}$$

where

$$k_o(E) = -0.0805 + \exp\left(-3.1893 + \frac{7.946}{E_g^2 - E^2}\right) \tag{33b}$$

$$T_o(E) = 369.9 - \exp\left(-12.92 + 5.509E\right) \tag{33c}$$

The expressions given in Eqs. (32) and (33) use $E_g = 3.648$ eV. As with any empirical formulation, these expressions must not be used outside their range of validity. See Ref. 35 for a complete discussion.

At room temperature, a more accurate parameterization of the optical functions of silicon have been obtained by Geist (32). Whereas the parameterization presented earlier used only one data set, the Geist formulation uses much of the data available in the literature and includes a critique of this data.

In addition to temperature, both doping level and strain can alter the optical functions of silicon. Figure 6 shows a plot of the optical functions of silicon for several different doping densities, where the samples were made by ion implantation followed by pulsed laser annealing (39). This fabrication technique has been used because higher doping densities can be obtained than with traditional diffusion techniques. If very high doping densities of arsenic are used (generally greater than 1%), then the $E_o'$ and the $E_2$ features are broadened and are moved to lower energies; furthermore, the excitonic structure in $k$ is eliminated, but the indirect optical absorption is dramatically increased. Similar results are observed if the dopant is phosphorus or antimony (all n-type dopants), but a considerably reduced effect is observed for boron-doped silicon (p-type). Doping the silicon with up to $\sim 1\%$ Ge (forming a silicon–germanium alloy) does not have an appreciable effect on the optical functions of silicon.

If the silicon lattice is strained, the optical functions will also be altered (40). Under strain, the crystal structure of silicon is no longer cubic, and the optical functions must now be expressed in terms of the anisotropic optical functions $\tilde{n}_o$ and $\tilde{n}_e$, where the subscript $o$ indicates the ordinary direction (perpendicular to the strain) and $e$ indicates the extraordinary direction (parallel to the strain). The strain-optic coefficients are relatively

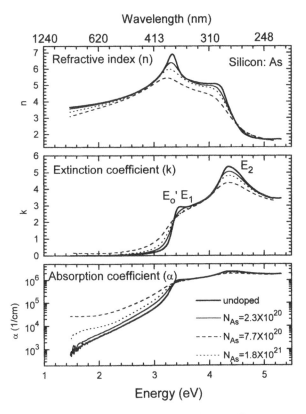

**Figure 6**  Refractive index $n$, extinction coefficient $k$, and absorption coefficient $\alpha$ of silicon from 1.5 to 5.3 eV at several different doping densities where arsenic is the dopant atom. (Data from Ref. 39.)

small, except for photon energies near the direct bandedge (the $E_0'$ and the $E_1$ features), where they can be quite large. Generally, one needs pressures of $\sim 1$ Gpa in order to affect the optical functions appreciably (see Ref. 40 for a complete discussion).

## C. Near the Bandedge: Phonon-Assisted Absorption

The optical bandedge of silicon is near 1.16 eV at 0 K, and decreases to $\sim 1.1$ eV at room temperature (see Figure 7); at higher temperatures, the decrease is even more dramatic. Since silicon is an indirect bandgap material, the values of the absorption coefficient $\alpha$ are very small near the bandedge, so it is relatively easy to make very accurate measurements of $\alpha$. The early measurements of McFarlane et al. (41) were performed at room temperature and below. Later, measurements (42,43) were made at elevated temperatures.

   The work of McFarlane et al. (41) at low temperatures showed that the primary optical absorption mechanism in silicon near the bandedge involves phonon-assisted transitions from the top of the valence band at the $\Gamma$ point to the minimum of the conduction band along the $\Delta$ branch. There are several phonons involved, but the two primary ones are the transverse and longitudinal acoustical phonons with **k**-vectors in the (100) direction. The expression for the optical absorption coefficient is given by (41,43).

$$\alpha(E, T) = \sum_{i=1}^{2} \sum_{l=1}^{2} (-1)^l \frac{\alpha_i [E - E_g(t) + (-1)^l k\theta_i]}{\exp\left[(-1)^l \frac{\theta_i}{T}\right] - 1} \tag{34a}$$

where the contribution from the transverse acoustical phonon ($i = 1$) is given by

$$\alpha_1(E) = 0.504\sqrt{E} + 392(E - 0.0055)^2, \qquad \theta_1 = 212\,\text{K} \tag{34b}$$

The contribution from the interaction with the longitudinal acoustical phonon ($i = 2$) is given by

$$\alpha_2(E) = 18.08\sqrt{E} + 5760(E - 0.0055)^2, \qquad \theta_2 = 670\,\text{K} \tag{34c}$$

The temperatures $\theta_1$ and $\theta_2$ are expressed in terms of degrees Kelvin. The optical bandgap $E_g$ is a function of temperature and is given by the Varshni (44) expression as modified by Thurmond (45):

$$E_g(T) = 1.155 - \frac{4.73 \times 10^{-4} T^2}{635 + T} \tag{35}$$

where the temperature is expressed in degrees Kelvin. The temperature dependence of the bandgap is plotted in Fig. 7, and the values of the absorption coefficient are plotted against energy for several temperatures in Figure 8.

   These expressions are valid only within a few tenths of an electron volt of the indirect bandedge. For absorption coefficients greater than $\sim 500\,\text{cm}^{-1}$, additional phonons contribute to the optical absorption, and the parabolic band approximation is no longer valid. It is for these reasons that only empirical relations, such as discussed earlier and that given by Geist (32), can be used to accurately describe the optical absorption coefficient of silicon in the region from $\sim 1.5$ eV to $\sim 3.4$ eV.

## D. Infrared Properties

As can be seen from Fig. 7, the indirect bandgap of silicon varies from $\sim 1.1$ eV ($\sim 1.1$ microns) at room temperature to $\sim 0.6$ eV at 1200°C. For photon energies less than this,

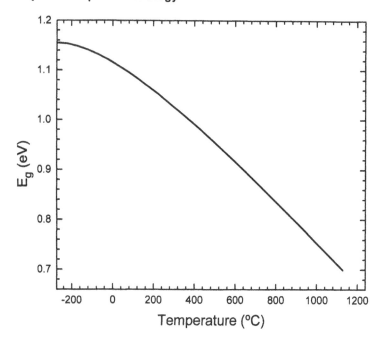

**Figure 7** Indirect bandgap of silicon plotted against the lattice temperature.

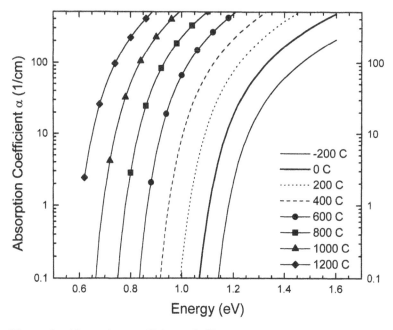

**Figure 8** Absorption coefficient of silicon versus photon energy near the bandedge for several different temperatures.

there can be no creation of electron-hole pairs and therefore no optical absorption from interband transitions. However, there are two other mechanisms that can absorb light in silicon: phonon-assisted absorption, and free carrier absorption.

Phonons, or quanta of lattice vibration, can interact with light and thereby absorb the light. For many crystals (such as III-V semiconductors), single phonons can interact with photons, creating a significant mechanism of optical absorption. However, the optical phonons in diamond, silicon, and germanium are not infrared active, so this mechanism for light absorption is not allowed. Infrared photons can be absorbed in these materials, but the mechanism requires two phonons to conserve energy and wavevector. Consequently, the absorption coefficient associated with this mechanism in silicon is very small. The peaks observed in the absorption coefficient near 0.1 eV (see Fig. 1 and Figure 9) are a result of this interaction.

A more important infrared absorption mechanism in silicon results from the interaction of the light with free carriers. The free carriers can be either electrons or holes, and they can be created either by doping or by the thermal generation of electron-hole pairs.

If there are free carriers in a semiconductor, then very small-energy photons can interact with an electron in an occupied state and raise it up in energy to an unoccupied state. This interaction can take place either in the valence band for holes or in the conduction band for electrons. The theoretical understanding of this mechanism originates

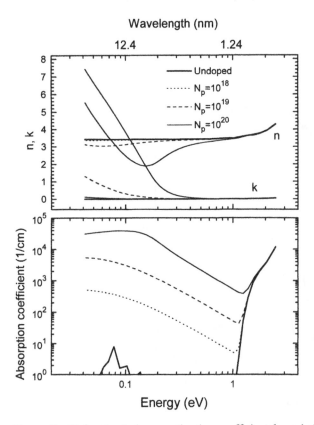

**Figure 9**   Refractive index $n$, extinction coefficient $k$, and absorption coefficient $\alpha$ of silicon in the infrared portion of the spectrum plotted for several hole concentrations.

with Drude (see Refs. 1 and 8 for a more complete discussion), who showed that the complex refractive index could be expressed as

$$\tilde{n}(E)^2 = [n(E) + ik(E)]^2 = \varepsilon(E) = \varepsilon_1(E) + i\varepsilon_2(E) = \varepsilon_\infty \left[ 1 - \frac{E_p^2}{E^2 + i\hbar E/\tau} \right] \quad (36)$$

In Eq. (36), the term $\varepsilon_\infty$ is the dielectric function of the crystal without free carrier effects, the quantity $\tau$ is the lifetime of the carrier in the excited state, and $E_p$ is the plasma energy, given by

$$E_p^2 = \frac{4\pi N_c e^2}{m^* \varepsilon_\infty} \quad (37)$$

The quantity $N_c$ is the free carrier concentration, $e$ is the charge of the electron, $h$ is Planck's constant divided by $2\pi$, and $m^*$ is the effective mass of the carrier.

In silicon, the effective mass $m^*$ and the relaxation time $\tau$ are different for electrons and holes, and may change if the carrier concentration $N_c$ gets very high (39). Table 1 lists approximate values for these parameters. Clearly, the most important factor in Eqs. (36) and (37) is the carrier concentration $N_c$, which can vary several orders of magnitude in crystalline silicon. Figure 9 shows a plot of the values of the refractive index, extinction coefficient, and absorption coefficient for silicon in the infrared for several concentrations of holes, using the values of the relative effective mass and relaxation time given in Table 1. As can be seen, the absorption coefficient increases with decreasing photon energy, and can be significantly larger than the optical absorption resulting from two-phonon mechanisms. In fact, the two-phonon peaks cannot be seen in silicon that is doped more than $\sim 10^{16}/cm^3$. Similar curves can be generated for electrons if the different values of the effective mass and the relaxation times are used.

The most common way to generate free carriers in silicon is to dope the material n- or p-type. Depending on the dopant atom, it is possible to get carrier densities in excess of $10^{21}$ carriers/cm$^3$. However, it is also possible to generate free carriers thermally. In this case, the dielectric function of the material is also given by Eq. (36), but both electrons and holes contribute to the free carrier absorption.

## E. Polycrystalline Silicon

Thin-film silicon can exist in several different forms. If much care is taken in the preparation, it is possible to create very nearly single-crystal silicon even in thin-film form. More commonly, thin-film silicon is either polycrystalline or amorphous. However, a wide range

**Table 1** Values of Relative Effective Masses of Electrons and Holes in Silicon with Respect to the Mass of the Electron, and Relaxation Time of Free Carriers

|  | Effective mass $(m^*/m_o)$ | Relaxation time $(\tau, s)$ |
|---|---|---|
| Conduction band (electrons) | 0.26[a] | $1.8 \times 10^{-15}$[b] |
| Valence band (holes) | 0.39[a] | $7 \times 10^{-15}$[c] |

[a] Ref. 46.
[b] Ref. 39.
[c] Ref. 47.

of material can be produced, depending on deposition conditions. Polycrystalline silicon can exist as large-grain material, where the grain size is approximately the thickness of the film, or as small-grain material, where the lateral extent of the grains is 10–20 nm. In other film growth regimes, a wide range of amorphous silicon can be made.

These variations in film morphology are reflected also in the optical properties of the films. Figure 10 shows a plot of the optical functions for various forms of thin-film silicon determined from spectroscopic ellipsometry measurements (48). As can be seen, the optical functions of large-grain polycrystalline silicon resemble the optical functions of single-crystal silicon. But there are significant differences: The value of $\varepsilon_2$ near the $E_2$ peak is significantly reduced, and the $E_1$ feature near the direct bandedge is gone. Below the direct bandedge, the absorption coefficient is increased. These effects are even more pronounced in fine-grain polycrystalline silicon. The $E_2$ peak of $\varepsilon_2$ is further reduced and moves to lower photon energy, the shoulder in $\varepsilon_2$ near the $E_o'$ feature is broadened, and the absorption coefficient below the direct bandedge is increased. The optical functions of amorphous silicon films, on the other hand, are completely different from the optical functions of single-crystal or polycrystalline silicon.

It is extremely important, when performing optical metrology experiments on thin-film silicon, to be aware of these possible changes in the optical functions. It is not sufficient to know the atomic constituents of a film; one must also know the morphology of the film.

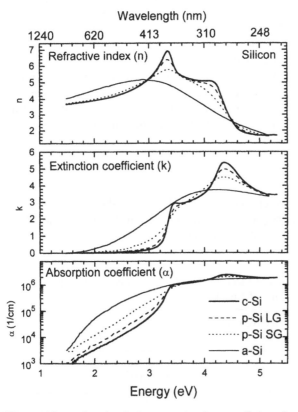

**Figure 10**  Refractive index $n$, extinction coefficient $k$, and absorption coefficient $\alpha$ for several forms of thin-film silicon. (Data from Ref. 48.)

## F. Liquid Silicon

Silicon undergoes a solid- to liquid-phase transition at 1410°C. Whereas crystalline silicon is a cubic semiconductor, where each atom is fourfold coordinated, liquid silicon is metallic, with sixfold coordination. These large structural changes upon melting indicate that the optical functions of liquid silicon are considerably different from the optical functions of the solid phase.

Aside from the very high temperatures involved, liquid silicon is very reactive in air. Any surface layer that forms on the surface during an optical reflectivity measurement can affect the resulting values of the optical functions obtained from the measurement. One way to reduce the surface overlayer, and therefore improve the accuracy of the measurements, is to keep the silicon molten for a very short period of time. This has been done by laser-irradiating the silicon surface with a pulsed excimer laser (49,50) or from a YAG laser (51), melting the surface for ~ 100 ns. During this time, a time-resolved ellipsometry measurement can be made using a cw probe laser as a light source, from which the optical functions of liquid silicon can be obtained. The results of these measurements are shown in Figure 11.

## IV. OPTICAL PROPERTIES OF RELATED MATERIALS

Many of the constituent materials in silicon microelectronics are amorphous or polycrystalline. One of the most common is $SiO_2$, but SiN, TiN, and amorphous Si are also used.

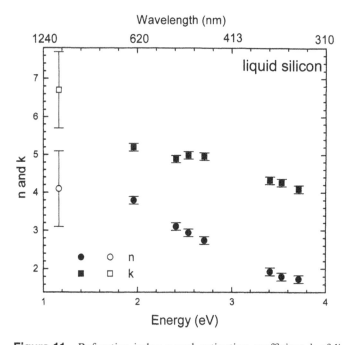

**Figure 11** Refractive index $n$ and extinction coefficient $k$ of liquid silicon obtained using time-resolved ellipsometry. The silicon was liquified using laser irradiation from an excimer laser and was molten for only ~ 100 ns. (Solid data points from Ref. 49 and 50; open data points taken from Ref. 51.)

As has been discussed in the previous section and shown in Fig. 10, the optical functions of amorphous and polycrystalline silicon are considerably different from those of crystalline silicon. This statement can be generalized to many other materials. However, many diagnostic experiments, such as ellipsometric techniques, rely on accurate values of these optical functions, which may not be known a priori. Often, the best solution to this dilemma is to model the optical functions of a material using a realistic and accurate parameterization of these quantities.

## A. Models of Optical Functions

Like crystalline materials, amorphous materials can have a bandedge, above which photons are absorbed by interband transitions. Well below the bandedge, the absorption coefficients can be very small. For example, absorption in optical fibers is often measured in units of decibels per kilometer,, indicating that the extinction coefficient $k < 10^{-10}$. Often, it is a good approximation to assume that there is no optical absorption in the wavelength region between the infrared vibrational absorption bands and the bandedge of the material. In this section, several models of optical functions will be discussed. It is particularly important to understand that these models often have a range of wavelengths for which the models are valid. Application of a model outside this range of validity will very often result in wrong values of the dielectric function.

### 1. Lorentz Oscillator Dispersion Models

One of the oldest and most useful models for the dielectric functions of materials is the Lorentz oscillator model (see Refs. 1 and 8). This model assumes that any light interaction with a solid will generate a time-dependent internal polarization of the solid by oscillating the positions of the individual electrons with respect to the stable nuclei of the solid. The equation describing this interaction is just the classical expression for a damped harmonic oscillator, where the driving force is proportional to the local electric field of the light beam. The dielectric function for this model is given by

$$\varepsilon(\omega) = \tilde{n}(\omega)^2 = 1 + \frac{4\pi e^2}{m} \sum_j \frac{N_j}{(\omega_{oj}^2 - \omega^2) - i\Gamma_j\omega} \tag{38a}$$

where $\omega_{oj}$ is the natural frequency of the $j$th oscillator, $\omega$ is the driving field frequency, and $\Gamma_j$ is the resonance width. The frequency $\omega = c/2\pi\lambda$, where $\lambda$ is the wavelength, and $c$ is the speed of light. The sum is over $j$ resonances, where $N_j$ is the number of electrons per unit volume associated with the $j$th resonance. The electronic charge is given by $e$ and the electronic mass by $m$. The same functional form can be deduced through quantum mechanical considerations, but the parameters involved have different meanings. In particular, the resonant frequency $\omega_j$ of the $j$th oscillator becomes the energy difference $E_{oj} = \hbar\omega_{oj}/2\pi$ between the initial and final quantum oscillator states. In a single Lorentz oscillator model of a typical semiconductor or insulator, this is just the energy gap $E_g$. The quantum mechanical calculation replaces $N_j$ with $Np_j$, where $p_j$ represents the transition probability between the initial and final states, while the width of the resonance $\Gamma_j$ becomes the broadening of the initial and final states due to finite lifetimes. In order to facilitate comparison of this model with the discussion of the dielectric function of silicon, Eq. (38a) can be written directly in terms of photon energy:

$$\varepsilon(E) = \tilde{n}(E)^2 = 1 + \sum_j \frac{B_j}{E_{oj}^2 - E^2 - i\Gamma_j E} \tag{38b}$$

where the prefactors have been incorporated into the parameter $B_j$. Note this expression can also be expressed in terms of the wavelength of light:

$$\varepsilon(\lambda) = \tilde{n}(\lambda)^2 = 1 + \sum_j \frac{A_j \lambda^2}{\lambda^2 - \lambda_{o,j}^2 - i\zeta_j \lambda} \tag{38b}$$

where the resonance wavelength $\lambda_{oj} = c/(2\pi\omega_{0j})$, $\zeta_j = 2\pi\lambda_{oj}^2 \Gamma_j / c$, and $A_j = 4\pi^2 \lambda_{oj}^2 N_j / c$.

The Lorentz oscillator model is valid only for photon energies well below the band-gap of the material. If the photon energy is near the bandgap, then it is possible to generate electron-hole pairs, which will create unbound electrons. Since the Lorentz model assumes that the electrons are bound, the model is not valid in this region. However, the Lorentz model can still be valid in regions of the spectrum where there is significant absorption of light by the material. One of the main applications of the full Lorentz model is the interaction of infrared light with lattice vibrations in both crystalline and amorphous materials, where the Lorentz model is very accurate (see Ref. 8, p. 288).

One of the principal applications of the Lorentz oscillator model has been has been to parameterize the refractive index of transparent materials, such as optical glasses. If the material is an insulator, then it generally is a good approximation to ignore the absorptive term and set $\Gamma_j = 0$. This approximation, known as the *Sellmeier approximation*, is given by

$$\varepsilon(\lambda) = n(\lambda)^2 = 1 + \sum_j \frac{A_j \lambda^2}{\lambda^2 - \lambda_{o,j}^2} \tag{39}$$

The Sellmeier approximation is valid only for materials and regions of the spectrum where there is very little absorption. Therefore it is usually restricted to the spectral region below the bandgap of the material and above the lattice phonon absorption band (generally in the IR region of the spectrum). Because the Sellmeier approximation arbitrarily sets the imaginary part of the dielectric function to 0, it cannot be Kramers–Kronig consistent; that is, the Sellmeier approximation is not consistent with Eqs. (5a) and (5b).

Another parameterization based loosely on the Sellmeier approximation is the Cauchy expansion, again where it is assumed that $k = 0$:

$$n = B_0 + \sum_j \frac{B_j}{\lambda^{2j}} \tag{40}$$

As with the Sellmeier approximation, the Cauchy expansion is not Kramers–Kronig consistent. However, both the Sellmeier and Cauchy approximations have been very successfully used for many decades to parameterize the refractive indices of glasses.

The Drude expression, which is used primarily to express the optical functions of metals and the free carrier effects in semiconductors (see Sec. III.D), is given by setting $E_{oj} = 0$ in Eq. (38):

$$\varepsilon(E) = 1 - \sum_j \frac{B_j}{E} \left( \frac{1}{E + i\Gamma_j} \right) \tag{41}$$

This is exactly the same expression given in Eqs. (32) for a single oscillator, where $\varepsilon_\infty = 1$, $B_1 = E_p^2$, and $\Gamma_1 = h/2\pi\tau$, where , $h$ is Planck's constant.

All the dielectric functions shown in Eqs. (38) to (41) incorporate several terms in the summation. If one is interested in very accurate values of the refractive index over a wide range of wavelengths, then two or more terms must be included. However, if the accuracy required is not that large and the wavelength region is restricted, a single term may be all that is required. For example, Malitson (52) used a three-term expansion to fit the refractive index of bulk $SiO_2$ to the 6th place from $\sim 210$ to $\sim 2500$ nm. If one is interested only in three-place accuracy in the near-visible part of the spectrum ($\sim 250$–850 nm), then a single term is all that is required to fit $n$ of bulk $SiO_2$.

Over the last several decades, workers have used several other models to fit the refractive indices of transparent materials. The models are generally based on the models already presented, but it must be emphasized that these models are empirical. The workers are interested in fitting observed refractive indices with an expression that has a few fittable parameters. A common mistake is to extrapolate these models outside the valid wavelength region. Even the complete Lorentz oscillator model as given in Eqs. (38), which are Kramers–Kronig consistent, is a very poor approximation for solids near the fundamental bandedge, as discussed in Sec. III.

## 2. Tauc–Lorentz Model for Amorphous Materials

If the photon energy is close to or above the bandgap of an amorphous material, neither the Lorentz oscillator model nor any of its derivatives work very well. It is well known that optical transitions in amorphous materials lack **k**-vector conservation (8,53), so one expects that the optical functions of amorphous materials will not have very sharp features that are characteristic of the optical function spectra of crystalline materials. Moreover, the optical functions of amorphous materials can vary considerably with growth conditions. In a very real sense, all amorphous materials, and particularly thin-film amorphous materials, are different.

Near the bandedge of an amorphous material, Tauc and coworkers found that the imaginary part of the dielectric function is given by (53)

$$\varepsilon_2(E) = A_T \Theta(E - E_g) \frac{(E - E_g)^2}{E^2} \tag{42}$$

where $\Theta(E - E_g)$ is the Heaviside function [$\Theta(E) = 1$ for $E \geq 0$ and $\Theta(E) = 0$ for $E < 0$] and $E_g$ is the bandgap of the amorphous material. Equation (42) has been used extensively to model the bandedge region of amorphous semiconductors, but it was not used much beyond this region. In particular, Eq. (42) gives no information concerning $\varepsilon_1$. To interpret optical measurements such as ellipsometry experiments, it is quite useful to have an expression for $\varepsilon(E)$ that corresponds to Eq. (42) near the bandedge, but it also extends beyond the immediate vicinity of $E_g$. Furthermore, it is important that the expression be Kramers–Kronig consistent.

One such parameterization that meets these criteria is the Tauc–Lorentz expression (54). This model combines the Tauc expression [Eq. (42)] near the bandedge and the Lorentz expression [Eq. (38b)] for the imaginary part of the complex dielectric function. If only a single transition is considered, then

$$\varepsilon_2(E) = 2n(E)k(E) = \frac{A(E - E_g)^2}{(E^2 - E_o^2)^2 + \Gamma^2} \frac{\Theta(E - E_g)}{E} \tag{43a}$$

The real part of the dielectric function is obtained by Kramers–Kronig integration

$$\varepsilon_1(E) = \varepsilon_1(\infty) + \frac{2}{\pi} P \int_{R_g}^{\infty} \frac{\xi \varepsilon_2(\xi)}{\xi^2 - E^2} \, d\xi \qquad (43b)$$

which can be integrated exactly. There are five parameters that are used in this model: the bandgap $E_g$, the energy of the Lorentz peak $E_o$, the broadening parameter $\Gamma$, the value of the real part of the dielectric function $\varepsilon_1(\infty)$, and the magnitude $A$.

This expression has been examined for many different thin-film systems, including amorphous silicon, amorphous silicon nitride, and amorphous carbon. As will be shown shortly. The optical properties of all of these materials vary considerably with deposition conditions, but the Tauc–Lorentz formulation works well in most cases.

The Tauc–Lorentz parameterization describes only interband transitions in an amorphous semiconductor. Since additional effects (such as free carrier absorption or lattice absorption) that might contribute to absorption below the bandedge are not included in the model, $\varepsilon_2(E) = 0$ for $E < E_g$. Furthermore, it can be seen that $\varepsilon_2(E) \rightarrow 0$ as $1/E^3$ as $E \rightarrow \infty$. This corresponds to the general observation that $\gamma$-rays and x-rays are not absorbed very readily in any material. Any realistic parameterization of optical functions of amorphous semiconductors must satisfy these two limitations. (NB: An earlier parameterization proposed by Forouhi and Bloomer (55) does not meet these criteria and is therefore unphysical.)

## 3. Effective Medium Approximations (EMAs)

If a material consists of a mixture of two or more materials, the composite dielectric function can sometimes be approximated using an effective medium approximation. The general expression for the composite dielectric function $\varepsilon$ for many EMAs is given by

$$\frac{\varepsilon - \varepsilon_h}{\varepsilon + 2\varepsilon_h} = \sum_j f_j \frac{\varepsilon_j - \varepsilon_h}{\varepsilon_j + 2\varepsilon_h}, \qquad \sum_j f_j = 1 \qquad (44)$$

The dielectric function of the host material is expressed by $\varepsilon_h$, $\varepsilon_j$ is the dielectric function for the $j$th constituent, and $f_j$ is the fraction of the $j$th constituent. The sum of the constituent fractions is constrained to be equal to 1. There are two primary EMAs used in the interpretation of ellipsometry data: the Maxwell–Garnett (56) theory, where the major constituent is used as the host material ($\varepsilon_h = \varepsilon_1$), and the Bruggeman (57) theory, which is a self-consistent approximation ($\varepsilon_h = \varepsilon$). The Maxwell–Garnett theory is most useful to describe materials where one constituent predominates, while the Bruggeman EMA is most useful describing materials where two or more materials have significant constituent fractions.

All EMAs are only as good as the dielectric functions that are used for the constituents. For example, many people have used the Maxwell–Garnett theory to describe nanocrystals imbedded in a matrix, since the host material (the matrix) is usually the predominant material. This is valid only if it is understood that the matrix material may very well be somewhat different optically from the same matrix material without the inclusions. Furthermore, the optical functions of the inclusion material are probably significantly different from the similar bulk material. (For example, it can be expected that nanocrystalline silicon has quite different optical properties than does single-crystal materials.) The Bruggeman effective medium theory is quite useful in describing surface roughness on sample surfaces, as long as the thickness of the surface roughness region is not very large.

One of the explicit assumptions in any EMA is that any feature size in the composite material must be much smaller than the wavelength of light. If any feature size is on the order of the wavelength of light, then it is improper to use the traditional effective medium approximation.

## B.  Some Materials Used in Semiconductor Processing

Figure 12 shows the refractive index $n$, the extinction coefficient $k$, and the absorption coefficient $\alpha$ for several materials used in semiconductor processing (taken from Refs. 9, 52, and 58–62). All of the materials shown are amorphous, with the exception of metallic TiN, which is polycrystalline. It must be emphasized that these are just representative data sets; measurements on different samples will most likely result in quite different values.

There are several general features in the refractive index spectra that apply to these materials, as well as any other materials:

1.   The values of both $n$ and $k$ of metallic materials (such as TiN shown in Fig. 12) will increase as the photon energy gets smaller. This is a direct consequence of free carrier absorption that was discussed in Sec. III.D. Qualitatively, the Drude expression [(Eq. (37)] can be used to describe $n$ and $k$ very well in the infrared, but the optical interaction with metals is more complicated in the visible, since band-structure effects can result in critical points in the optical spectra.

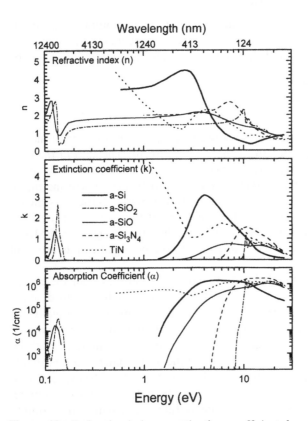

**Figure 12**   Refractive index $n$, extinction coefficient $k$, and absorption coefficient $\alpha$ for several other materials used in semiconductor manufacture.

2.   Amorphous semiconductors and insulators will have a bandedge (see earlier). This is characterized by a sharp increase in $k(E)$ and $\alpha(E)$, with photon energy at the bandedge $E_g$ and a large feature in $n(E)$. Amorphous materials will also have one or more broad peaks in $k(E)$. Since amorphous materials do not have an identifiable band structure (such as discussed in Sec. III.A.), features in their optical spectra cannot be associated with critical points in the Brillouin zone. However, features that do occur can be associated with peaks in the joint density of states. As with any material, any sharp transition in either $n(E)$ or $k(E)$ will result in a feature in the other spectrum; this is a consequence of the Kramers–Kronig constraints discussed in Sec. II.A.

3.   Multiple-element amorphous materials will have a feature in the infrared part of the spectrum (near 0.1 eV, or 10,000 nm) due to lattice absorption. Single-element amorphous materials (such as amorphous silicon) have very weak features in this region of the spectrum, since the interaction is second order, much the same as for crystalline silicon, as discussed in Sec. III. However, hydrogen can be a major constituent of amorphous silicon films, and the hydrogen-silicon vibration can result in significant lattice absorption in this region of the spectrum. The resonant absorption seen in Fig. 12 for the silicon nitride film is due primarily to a silicon-nitrogen stretch mode, while the similar feature in $SiO_2$ or in the $SiO_x$ film is due primarily to a silicon-oxygen stretch mode. The amorphous nature of the material broadens this peak, and its precise position in energy will depend upon the material and the way in which it was made.

The plot of the optical functions of a-$SiO_2$ has been taken from the chapter by Philipp (58) in *Handbook of Optical Constants I*. These measurements were made on fused silica, which will have optical functions that do not vary significantly with growth conditions. (NB: Water content can change the optical absorption in the near-IR, but the increase in absorption coefficient is generally less than the lower bound in Fig. 12.) Although this data is very accurate for bulk fused silica, it is only an approximation for the thin-film $SiO_2$ that is used so extensively in the semiconductor industry. It has been observed by some workers (63,64) that the refractive index of thin-film $SiO_2$ increases with decreasing film thickness (possibly caused by the additional strain associated with very thin cover layers), although others (65) have disagreed with this result.

If thin-film silicon oxide is not stoichiometric (where there is insufficient oxygen in the film), a silicon suboxide is obtained. A representative plot of the optical functions of a silicon suboxide is shown in Fig. 12, taken from Philipp (60). As can be seen, the suboxide has considerably higher refractive index in the infrared as well as a significantly lower bandgap. This points to the sensitivity of the optical functions on the atomic composition.

Another important thin-film amorphous material used in semiconductor device fabrication is silicon nitride (a-$Si_xN_y$:H). Silicon nitride can be deposited in many different ways, which can result in quite different values of the silicon-to-nitrogen ratio $(x/y)$. Stoichometric silicon nitride has a silicon-to-nitrogen ratio of 3 to 4, but both silicon-rich and nitrogen-rich films can be formed. Since most of the silicon nitride used in the semiconductor industry is made from gases that contain hydrogen (such as silane and ammonia), a significant amount of hydrogen can be incorporated into the structure of the thin film. The amount and bonding configurations of the hydrogen in a-$Si_xN_y$:H can also be expected to have an effect on the optical functions of the silicon nitride film. The data for silicon nitride shown in Fig. 12 was taken from Philipp (59) for nominally stoichiometric silicon nitride, where its refractive index at 633 nm is $\sim 2.02$ and its band gap is $\sim 4$ eV. If silicon-rich silicon nitride is made, higher refractive indices and lower bandgaps can be obtained (see Figure 13, which combines data from Refs. 57 and 64). Claassen (67) has measured the refractive index of various silicon nitride films with different silicon-to-

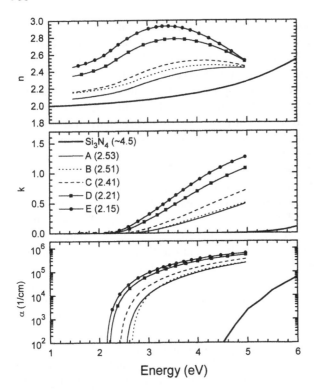

**Figure 13** Refractive index $n$, extinction coefficient $k$, and absorption coefficient $\alpha$ for several forms of silicon nitride, formed using different deposition conditions. The number in the parentheses is the bandgap energy in electron volts. (Data from Refs. 59 and 66.)

nitrogen ratios and found that $n(633\,nm) \sim 1.38 + 0.70\,(Si/N)$. Clearly, the optical functions of this material vary considerably with deposition conditions; it is not even approximately correct to assign a set of optical functions to silicon nitride until more is known about its stoichiometry and the way in which it has been made.

Another amorphous material used extensively in semiconductor manufacture is amorphous silicon. This material can be made very cheaply and can cover very large areas, so it has found applications in photovoltaics and displays. Although amorphous silicon contains principally only one atom (silicon), its optical properties also are a function of deposition conditions (one such data set is shown in Fig. 12 and has been taken from Ref. 61). Clearly, the amount of hydrogen incorporated into the amorphous silicon structure can be an important consideration in the determination of the optical functions, but many other deposition factors can also affect the optical functions of the final material.

Thin-film titanium nitride is another material that is used extensively in semiconductor manufacture. TiN is metallic and polycrystalline, and a representative data set of its optical functions is shown in Fig. 12 and taken from Ref. 62, where the inflection in $n$ and $k$ near 3–4 eV is due to a critical-point absorption in the material and is therefore due to the crystalline nature of the material.

It can be generally stated that the optical functions of amorphous and polycrystalline thin-film materials will depend significantly on deposition conditions. Therefore, it is inappropriate to assign optical functions to a material without knowledge of its deposition conditions. This is at once a problem and an opportunity. It is a problem because one cannot blindly use a value of the refractive index in an experiment (such as an ellipsometry experiment) and expect to interpret the results with a high level of accuracy. On the other hand, this variation in the optical functions is an opportunity, in that the optical functions are sensitive indications of the average atomistic arrangement in the film and can therefore be used for quality control, once the dependence of the optical functions on various growth parameters is understood.

## ACKNOWLEDGMENTS

The author would like to thank F. A. Modine and R. F. Wood for reading and commenting on this manuscript. This research was sponsored in part by the Division of Materials Sciences, U.S. Department of Energy, under contract No. DE-AC05-96OR22464 with Lockheed Martin Energy Research, Inc.

## REFERENCES

1.  M Born, E Wolf. Principals of Optics. 6th ed. New York: Pergamon, 1975.
2.  RMA Azzam, NM Bashara. Ellipsometry and Polarized Light. New York: North Holland, 1977.
3.  HG Tompkins. A User's Guide to Ellipsometry. New York: Academic Press, 1993.
4.  DS Kliger, JW Lewis, CE Randall. Polarized Light in Optics and Spectroscopy. New York: Academic Press, 1990.
5.  PS Hauge. Surf Sci 96:108–140, 1980.
6.  JI Pankove. Optical Processes in Semiconductors. New York: Dover, 1971.
7.  C Kittel. Introduction to Solid State Physics. 7th ed. New York: Wiley, 1995.
8.  P Yu, M Cardona. Fundamentals of Semiconductors. New York: Springer-Verlag, 1996.
9.  ED Palik, ed. Handbook of Optical Constants I. New York: Academic Press, 1985.
10. ED Palik, ed. Handbook of Optical Constants II. New York: Academic Press, 1991.
11. ED Palik, ed. Handbook of Optical Constants III. New York: Academic Press, 1998.
12. F Abelès. Ann de Physique 5:596, 1950.
13. DE Aspnes, AA Studna. Appl Opt 14:220–228, 1975.
14. P Chindaudom, K Vedam. Appl Opt 32:6391–6397 1993.
15. PS Hauge, FH Dill. Opt Commun.14: 431, 1975.
16. J Opsal, J Fanton, J Chen, J Leng, L Wei, C Uhrich, M Senko, C Zaiser, DE Aspnes. Thin Solid Films 313–314:58–61, 1998.
17. SN Jasperson, SE Schnatterly. Rev Sci Instrum. 40:761–767, 1969; errata Rev Sci Instrum 41:152, 1970.
18. GE Jellison Jr, FA Modine. Appl Opt 29:959–974, 1990.
19. DH Goldstein. Appl Opt 31:6676–6683, 1992.
20. RC Thompson, JR Bottinger, ES Fry. Appl Opt 19:1323–1332, 1978.
21. GE Jellison Jr, FA Modine Appl Opt 36:8184–8189, 1997; Appl Opt 36:8190–8198, 1997.
22. J Lee, PI Rovira, I An, RW Collins. Rev Sci.Instrum 69:1800–1810, 1998.
23. WC Dash, R. Newman. Phys Rev 99:1151–1155, 1955.
24. CD Salzberg, JJ Villa. J Opt Soc Am 47:244–246, 1957.
25. W Primak. Appl Opt 10:759–763, 1971.
26. HR Philipp. J Appl Phys 43:2835–2839, 1972.
27. HH. Li. J. Phys. Chem Ref Data 9:561, 1980.
28. GE Jellison Jr., FA. Modine. J Appl Phys 53:3745–3753, 1982.

29. DE Aspnes, AA Studna. Phys Rev B 27:985–1009, 1983.
30. GE Jellison Jr. Optical Materials 1:41–47, 1992.
31. DF Edwards. In: ED Palik, ed. Handbook of Optical Constants I. New York: Academic Press, 1985, pp 547–569.
32. J Geist. In: ED Palik, ed. Handbook of Optical Constants III. New York: Academic Press, 1998, pp 519–529.
33. DF Edwards. In: ED Palik, ed. Handbook of Optical Constants III. New York: Academic Press, 1998, pp 531–536.
34. ML Cohen, J Chelikowsky. Electronic Structure and Optical Properties of Semiconductors. 2nd ed. Berlin: Springer, 1989, p 81.
35. GE Jellison Jr., FA Modine. J Appl Phys 76:3758–3761, 1994.
36. GE Jellison Jr., FA Modine. Appl Phys Lett 41:180–182, 1982.
37. P Lautenschlager, M Garriga, L Vina, M Cardona. Phys Rev: B36:4821–4830, 1987.
38. GE Jellison Jr, HH Burke. J Appl Phys 60:841–843, 1986.
39. GE Jellison Jr., SP Withrow, JW McCamy, JD Budai, D Lubben, MJ Godbole, Phys Rev B:52:14607–14614, 1995.
40. P Etchegoin, J.Kircher, M Cardona. Phys Rev B47:10292–10303, 1993.
41. GG McFarlane, TP McLean, JE Quarrington, V Roberts. Phys Rev 111:1245, 1958.
42. HA Weakliem, D Redfield. J Appl Phys 50:1491–1493, 1979.
43. GE Jellison Jr., DH Lowndes. Appl Phys Lett 41:594–596, 1982.
44. YP Varshni. Physica 34:149–154, 1967.
45. CD Thurmond. J Electrochem Soc 122:1133, 1973.
45. WG Spitzer, HY Fan. Phys Rev 106:882–890, 1957.
47. H Engstrom. J Appl Phys 51:5245, 1980.
48. GE Jellison Jr., MF Chisholm, SM Gorbatkin, Appl Phys Lett 62:3348–3350, 1993.
49. GE Jellison Jr., DH Lowndes. Appl Phys Lett 47:718–720, 1985.
50. GE Jellison Jr., DH Lowndes. Appl Phys Lett 51:352–354, 1987.
51. KD Li, PM Fauchet. Solid State Commun 61:207, 1986.
52. IH Malitson, J Opt Soc Am 55:1205-1209, 1965.
53. J Tauc, R Grigorovici, A Vancu. Phys Stat Solid 15:627, 1966.
54. GE Jellison Jr., FA Modine. Appl Phys Lett 69:371–373, 1996; Appl Phys Lett 69:2137 1996.
55. AR Forouhi, I Bloomer. Phys Rev B 34:7018–7026, 1986.
56. JC Maxwell Garnett. Philos Trans R Soc London 203:385, 1904; Philos Trans R Soc London 205:237, 1906.
57. DAG Bruggeman. Ann Phys (Leipzig) 24:635, 1935.
58. HR Philipp. In: ED Palik, ed. Handbook of Optical Constants I. New York: Academic Press, 1985, pp 749–763.
59. HR Philipp. In: ED Palik, ed. Handbook of Optical Constants I. New York: Academic Press, 1985, pp 771–774.
60. HR Philipp. In: ED Palik, ed. Handbook of Optical Constants I. New York: Acadmic Press, 1985, pp 765–769.
61. H Piller. In: ED Palik, ed. Handbook of Optical Constants I. New York: Academic Press, 1985, pp 571–586.
62. J Pfluger, J. Fink. In: ED Palik, ed. Handbook of Optical Constants II. New York: Academic Press, 1991, p 307.
63. GE Jellison Jr. J Appl Phys 69:7627–7634, 1991.
64. E Taft, L Cordes. J Electrochem Soc 126:131, 1979.
65. CM Herzinger, B Johs, WA McGahan, W Paulson. Thin Solid Films 313–314:281–285, 1998.
66. GE Jellison Jr., FA Modine, P Doshi, A Rohatgi. Thin Solid Films 313-314:193–197, 1998.
67. WAP Claassen, WGJN Valkenburg, FHPM Habraken, Y Tamminga. J Electrochem Soc 130:2419, 1983.

# 26

# Ultraviolet, Vacuum Ultraviolet, and Extreme Ultraviolet Spectroscopic Reflectometry and Ellipsometry

**J. L. Stehle, P. Boher, C. Defranoux, P. Evrard, and J. P. Piel**
*SOPRA, Bois-Colombes, France*

## I. INTRODUCTION

This chapter describes various aspects of the application of spectroscopic ellipsometry in the UV range. After a short introduction on the electromagnetic spectrum and the definition of UV, VUV, PUV, and EUV ranges, general features of the optical indices versus wavelength are presented. Specific instrumental problems related to the wavelength range are discussed, in particular the problem of sources, polarizers, and detectors. The optical mounting selected to work in the PUV range is described in detail with the problems of stray light rejection and carbon contamination. Various measurement results at 157 nm concerning oxynitrides, antireflective coatings, masks, phase shifters, and photoresists are also presented. Finally, the specific problem of EUV lithography and the associated metrology is introduced.

## A. Definitions

One of the last regions of the electromagnetic spectrum to be routinely used in the IC industry is that between ultraviolet and x-ray radiation, generally shown as a dark region in charts of the spectrum. From the experimental point of view, this region is one of the most difficult, for various reasons. Firstly, the air is absorbent, and, consequently, experimental conditions need vacuum or purged glove boxes. Secondly, the optics needed in the range are generally not available. In fact, exploitation of this region of the spectrum is relatively recent, and the names and spectral limits are not yet uniformly accepted.

Figure 1 shows that portion of the electromagnetic spectrum extending from the infrared to the x-ray region, with wavelengths across the top and photon energies along the bottom. Major spectral regions shown are the infrared (IR), which can be associated with molecular resonances and heat (1 ev = 1.24 µm); the visible region from red to violet, which is associated with color and vision; the ultraviolet (UV), which is associated with sunburn and ionizing radiation; the vacuum ultraviolet (VUV) region, which starts where the air becomes absorbent; the extreme ultraviolet (EUV) region, which extends from 250

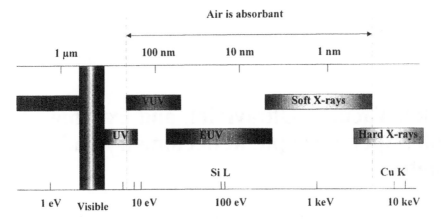

**Figure 1** Electromagnetic spectrum with the different regions of interest. The Cu K-alpha line at 0.154 nm and the Si L-alpha line at 12.5 nm are indicated.

eV to about 30 eV; the soft x-ray region, which extends from about 250 eV (below carbon K-alpha edge) to some keV, and the hard x-ray region, where the air again becomes transparent.

## B.  Spectroscopic Ellipsometry Technique

Spectroscopic ellipsometry is now well established as a key metrology method for characterization of thin films and multilayers. In ellipsometry, one measures the change of polarization state after nonnormal reflection from the sample. The polarization state is defined by two parameters so that each ellipsometric measurement provides two independent parameters. The term *ellipsometry* derives from the fact that linearly polarized light after oblique reflection becomes elliptically polarized due to the different reflection coefficients for s- and p-polarized light. By measuring the polarization ellipse, one can determine the complex dielectric function of the sample. The fact that two independent parameters are obtained at each wavelength is an important advantage compared with photometric measurements like reflectometry. The measurement is also self-calibrated, with no need for baseline measurement, and the accuracy is generally better than for reflectometry.

Spectroscopic ellipsometry traditionally works in the visible and UV region (1). Some extension is possible into the IR region using a Fourier transform spectrometer (2). Special attention to the optical components is needed to reach the limit of the air absorption around 190 nm (3). Below this value, two possibilities can be used to work in the VUV region. Using a purged glove box to reduce the oxygen and water content to the parts per million range, it is possible to reach 135 nm (4). Below this, vacuum systems are needed, as, for example, the system mounted on the BESSY synchrotron (5). The EUV region requires a special experimental setup (see Section IV, New Applications).

## C.  Optical Indices

Optical indices of all the materials give a picture of the energetic structure of the material under study. From the hard x-ray region to the infrared, the photons interact successively with the electrons in the core levels of the atoms, with the electrons involved in the band structure of the material, with the vibration modes of the molecules or phonons in a solid, and with the weak bonds of the material. Some optical indices measured for different reference materials are shown in Figures 2, 3, and 4. These indices are taken from Ref. 6 and have been measured with various methods, depending on the wavelength range. They are associated by type of materials, such as semiconductors (crystalline silicon, crystalline germanium, and gallium arsenide in Fig. 2), insulators (silicon carbide, silicon nitride, aluminum oxide, and silicon dioxide in Fig. 3), and metals (copper, molybdenum, and tungsten in Fig. 4).

So in the hard x-ray region, the soft x-ray region, and the EUV region the optical properties of all the materials depend only on the atoms of the material. The refractive index of the materials can be written as (7)

$$n = 1 - \delta + i\beta = 1 - \frac{\lambda 2 r_e}{2\pi} \sum_a (Z_a + f'_a + if''_a)N_a$$

Here, $r_e$ is the classical electron radius and is a measure of the scattering strength of a free electron, and $\lambda$ is the wavenumber of the x-rays, and $\lambda$ is the x-ray wavelength. The atomic number of the atomic species a is denoted $Z_a$, and the correction for anomalous dispersion is $f'_a$ and $f''_a$. $\delta$ and $\beta$ are always in the range $10^{-5}$–$10^{-6}$ in the hard x-ray region, and all the materials are transparent to x-rays in this range. This can be seen for all the materials reported in Figs. 2, 3, and 4. Against wavelength, the absorption increases as a function of $\lambda^2$, with some discontinuities due to the absorption edges of the atoms (Si L-alpha edge at 12.4 nm, for example, as reported in Fig. 2 for silicon and in Fig. 3 for $SiO_2$).

In the VUV and UV range the insulators become rapidly transparent (cf. Fig. 3). The interesting part is located in the VUV (the case of $Si_3N_4$ and $Al_2O_3$) or the UV (the case of the SiC, which can be taken as a semiconductor). The decrease of the extinction coefficient is followed by the appearance of optical transitions, well defined in particular on the refractive index. For semiconductors like silicon, germanium, and GaAs (cf. Fig. 2), the interesting part corresponding to the conduction band is located in the visible range, and this is why standard ellipsometry has long been used to characterize such materials. For both the insulators and the semiconductors, there is a range in the infrared where the material is completely transparent, which is useful to measure the layer thickness of thick films. This does not apply to the case of metals, as shown in Fig. 4. In this case, the material stays completely absorbent in the entire wavelength range from VUV to IR. This is called the Drude tail for conductive materials.

## II. OPTICAL MOUNTING FOR PUV

### A. Air Transmission

Owing to air's strong molecular absorption band, transmission in air below 190 nm is very poor. It is therefore necessary to install measurement tools such as ellipsometers in a carefully controlled purged environment where the oxygen and water vapor levels must be maintained at the parts per million range.

### B. Adsorbed Carbon Layer

Contaminants adsorbed on substrate or film surfaces can strongly affect measurement at 157 nm. It is estimated that 1 nm of organic contaminants such as alkanes reduces transmission by 1%. Similarly a 1-nm-thick layer of water reduces transmission by 2%.

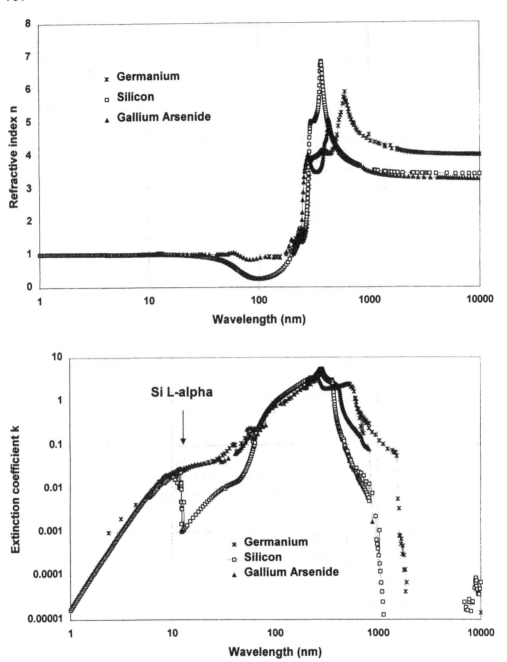

**Figure 2** Optical indices of some semiconductors in the range from soft x-rays to infrared. (From Ref. 6.)

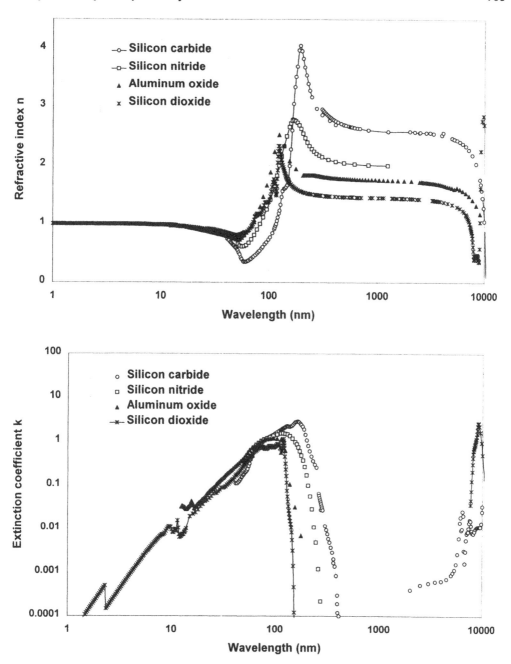

**Figure 3** Optical indices of some insulators in the range from soft x-rays to infrared. (From Ref. 6.)

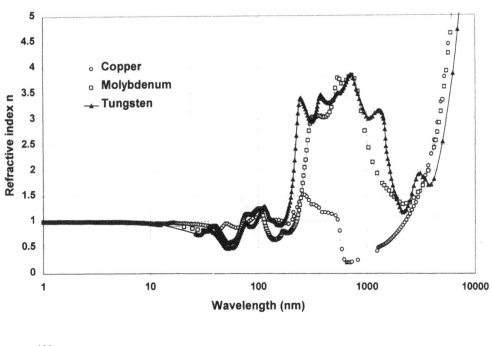

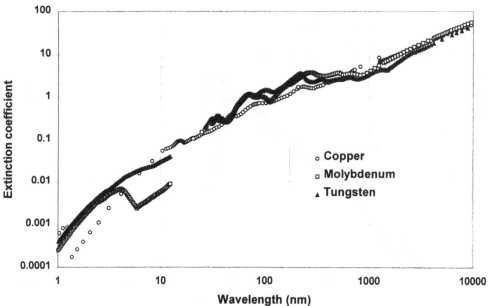

**Figure 4** Optical indices of some metals in the range from soft x-rays to infrared. (From Ref. 6.)

Contamination can occur during brief exposure to the ambient or as a result of outgassing from chamber walls and samples as photoresists in the measurement. The surface contaminants play an increasing role in transmission measurements as the wavelength is reduced further, below 157 nm. To minimize contamination it is possible to clean the surface prior to the measurements using a 157-nm excimer laser beam.

Measurement of ambient is also very important, and it is reported (9) that after 30 minutes in vacuum, the transmission of $CaF_2$ samples drops significantly across the measured spectrum from 90.2% to 87% at 157 nm. Using laser-cleaning in $N_2$ atmosphere, it is possible to restore the transmission to its original values. Moreover purging the chamber with nitrogen rather than maintaining it under vacuum reduces the level of surface contamination. It is clear that this effect strongly depends on the pumping arrangement used.

## C. Stray Light Rejection

All light that reaches the spectrometer exit slit from anywhere other than the disperser (prism or grating) by any means other than diffraction is called *stray light*. All components in an optical system contribute to the stray light, such as apertures, partially reflecting surfaces, and even the grating itself. In this last case, it could come from imperfections in the shape and spacing of the grooves and micro-roughness on the surface of the grating (diffuse stray light). We can also observe what are called *ghosts*, which are caused by periodic errors in the groove spacing. Reflection from instrument chamber walls and mounting hardware also contributes to the redirection of unwanted energy toward the image plane. Light incident on the detector may be reflected backward to the grating and rediffracted. Since the angle of incidence may now be different, light rediffracted along a given direction will generally be of a different wavelength from that of the light originally diffracted along the same direction. Stray light can be reduced greatly by the use of a double spectrometer, in which light diffracted from one grating is incident on a second disperser.

The signal-to-noise ratio (SNR) could be defined in this case as the ratio of diffracted useful energy to unwanted light energy. Stray light usually plays the limiting role in the achievable SNR for a grating system.

## D. Calibrations

We distinguish four kinds of calibrations for the spectroscopic ellipsometer (SE). The first one consists of the spectrometer calibration using well-known spectral lines (from a Hg lamp). The second calibration is used to evaluate the nonlinearity of the photomultiplier tube. The third one consists of the calibration of the residual polarization induced by the photomultiplier window. The last one allows evaluation of the offset of the polarizer and analyzer main axes versus the plane of incidence, which is the reference plane of the instrument. This procedure is applied at installation and when the user changes the lamp, so the stability of the calibration can be checked every quarter, as for visible ellipsometers. Before a daily use, the user can verify the level of the signal in the critical range to be sure that nothing special occurred as gas absorption or no counting occurred because of a lamp or detector failure.

## 1.   Wavelength Calibration

First of all, it is necessary to determine the mirror position of the prism (4% of reflection on the surface of the prism). To perform this calibration, it is necessary to scan the prism around this mirror position for a fixed position of the grating. The maximum of signal corresponds to the mirror position (Figure 5). This position is then used for the calibration of the grating, independent of the prism. The grating is then calibrated using a Hg lamp with the prism fixed in mirror position. The different emission lines and the zero order of the grating are selected. For each theoretical position, a scan of the grating is made. The difference between the theoretical position and the experimental one is fitted with a polynomial of degree 2 and the parameters saved (Figure 6). Finally, it is necessary to perform the calibration of the prism. For this purpose, different wavelengths are selected by the grating, and the prism is scanned around each of these wavelengths (Figure 7).

## 2.   Calibration of the Nonlinearity of the Photomultiplier Tube

The aim of this calibration is to evaluate the nonlinearity and the residual polarization of the detector. In the fixed polarizer and rotating analyzer mounting used, we need to take into account their effect in terms of residual polarization.

The software measures the residual and the phase of the signal versus the intensity in a straight line at different wavelengths and analyzer positions. The intensity of the signal measured versus the polarizer position is given by the following equation when there is no nonlinearity effect:

$$I_{ideal} = K(1 + \alpha \cos 2A + \beta \sin 2A)$$

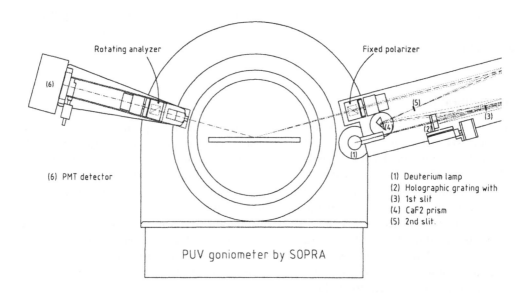

**Figure 5**   General optical layout of PUV-SE to be installed in purged box.

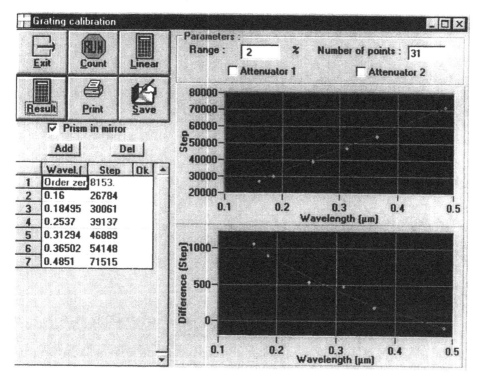

**Figure 6**  Grating calibration window with calibration already made.

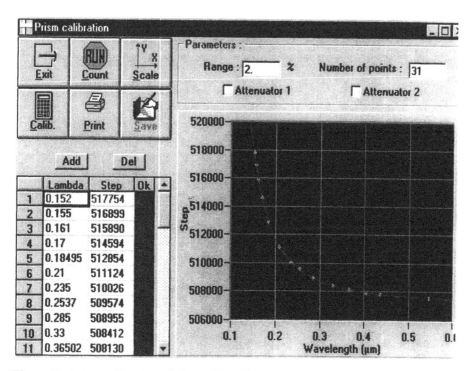

**Figure 7**  Prism calibration window with calibration already made.

When there is a nonlinear effect of the detector, the measured intensity can be assumed to be given by

$$I = I_{ideal} + pI_{ideal}^2 = K'(1 + \alpha' \cos 2A + \beta' \sin 2A)$$

The Hadamard detection measures the $\alpha'$ and $\beta'$, and the corrected parameters $\alpha$ and $\beta$ can be recalculated using the following equations (Figure 8):

$$\alpha = \frac{2(f+1)\alpha'}{(2f+1) + \left[(2f+1)^2 - 2f(f+1)(\alpha'^2 + \beta'^2)\right]^{1/2}}$$

$$\beta = \beta' \frac{\alpha}{\alpha'}$$

The $f$ parameter is characteristic of the nonlinearity. It is given by:

$$f = \frac{2\eta - 2}{4 - 3\eta}$$

where $\eta = \sqrt{\alpha'^2 + \beta'^2}$

### 3. Calibration of the Residual Polarization of the Detector

If the photomultiplier $MgF_2$ window induces anisotropic light with a small linearly polarized contribution of amplitude $e$ with an angle $\phi$ with regard to the reference plane, the electric field is given by

$$E = E_p(1 + e_x \cos \phi + e_y \sin \phi)$$

Then the intensity measured by the Hadamard detection is given by

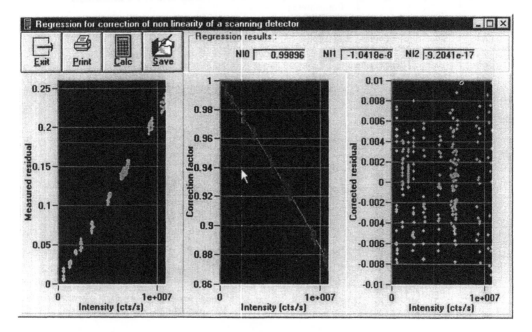

**Figure 8**  Nonlinearity correction of the photomultiplier tube.

$$I = K(1 + e_x \cos A + e_y \sin A)^2 (1 + \alpha \cos 2A + \beta \sin 2A)$$

The measured Fourier coefficients can be expressed by:

$$\alpha' = \frac{2(\alpha + p_x)}{2 + p_x \alpha + p_y \beta} \qquad \text{and} \qquad \beta' = \frac{2(\beta + p_y)}{2 + p_x \alpha + p_y \beta}$$

where

$$p_x = \frac{e^2}{2 + e^2} \cos 2\phi \qquad \text{and} \qquad p_y = \frac{e^2}{2 + e^2} \sin 2\phi$$

The $p_x$ and $p_y$ values can be calibrated versus the wavelength using the variation of the phase with the analyzer position. It is given by

$$\Delta \theta = \arctan\left(-\frac{\beta'}{\alpha'}\right) - 2A = \arctan\left(-\frac{\beta + p_y}{\alpha + p_x}\right) - 2A$$

In practice, we make a regression on the experimental points at each wavelength using a Levenberg–Marquard procedure (Figure 9).

## 4. Polarizer ($P_0$) and Analyzer ($A_0$) Offset Calibration

In the measurement procedure, precise calibration of the position of the plane of incidence is a direct part of the measurement. Two methods can be used to calibrate the position of

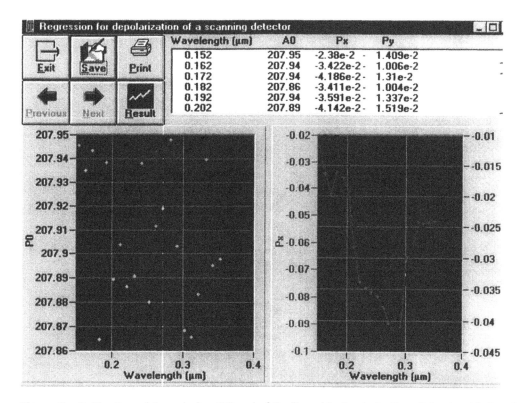

**Figure 9**   Calibration of the polarizer PO and of the Px residual polarization of the signal induced by the detector.

the plane of incidence: the residual method (3), described next, and the regression method (Figure 10).

The idea is to make a scan around the expected position of the zero of the analyzer and to determine precisely its zero position and the corresponding phase using the measured parameters. In this respect, we use the residual value $R$ obtained by

$$R = 1 - \alpha^2 - \beta^2$$

where $\alpha$ and $\beta$ are the Fourier coefficients of the measured signal. In the ideal case, the residual value is given by

$$R = 1 - \frac{(\tan^2 \psi - \tan^2 P) + (2\cos \Delta \tan P)}{(\tan^2 \psi + \tan^2 P)^2}$$

where $P$ is the polarizer position. The minima of the residual values are then obtained when $P = k\pi/2$ with $k$ an integer value.

## E.  Instrumental Function

In order to check the accuracy of the SE instrument, it is very useful to perform $\tan \psi$ and $\cos \Delta$ measurements without sample. To do this, the system is set in a straight line such that the polarizer arm is at $90°$ from the normal to the sample (same for the angle of the analyzer). In this configuration, the theoretical values for $\tan \psi$ and $\cos \Delta$ must be 1. The measured values give useful information about the goodness of the system itself (Figure 11).

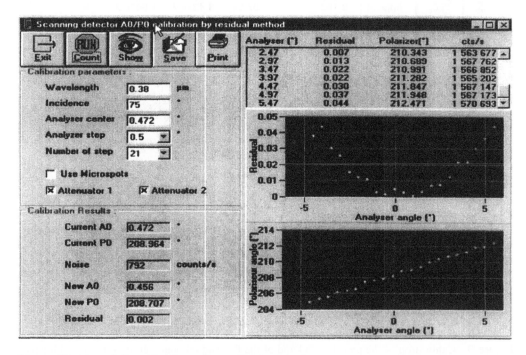

**Figure 10**   $A_0$ and $P_0$ calibration using residual method.

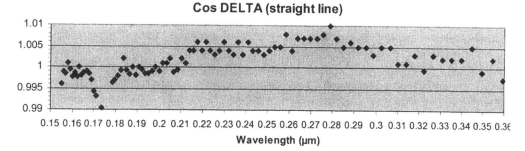

**Figure 11** Measured Cos delta without sample in straight line.

For the evaluation of wavelength accuracy and resolution, it is necessary to use another approach, such as a calibrated thick sample or an interference filter or an atomic spectral lamp.

## F. Limitations

The transmission of the flux is limited by the presence of $H_2O$ and $O_2$ gases in the glove box. Even for a concentration of few ppm of these gases, the transmission decreases significantly below 175 nm. Contamination of the mirror surfaces and the polarizer surfaces can easily reduce the light flux arriving at the detector.

The sensitivity of the detector itself is optimized for the PUV range but consequently is limited to a shorter spectral range, 125–750 nm.

The spectrometer resolution is another limitation. Due to the fact that we need to optimize the light flux arriving on the sample, it is necessary to compromise between the flux and the spectrometer so its resolution is around 1.0 nm.

## III. APPLICATIONS

Applications of UV, DUV, and PUV ellipsometry are numerous. Extensions to the lower wavelength have been driven by photolithography, but many other applications have also benefited. The determination of optical characteristics such as refractive index of materials in the UV has permitted the definition and calculation of properties of the material using SE. The UV-SE applications will be presented with the following plan:

    A. Photolithography
        1. Photoresist, antireflective coating, swing ratio simulation
            At 365 nm
            At 248 nm and 193 nm
        2. Mask and phase-shift masks
        3. Measurements for 157-nm lithography
    B. Thin oxide, thin ONO, and nitrided oxides
    C. Optical gaps
    D. Roughness and surface effects

## A. Photolithography

Ultraviolet spectroscopic ellipsometry for photolithography materials is certainly one of the most complete and useful applications of ellipsometry. Not only can we measure thickness or refractive indices of materials like photoresist (PR) and antireflective coating (ARC), we can also predict by simulation the interaction and behavior of the light with the resist during the photolithography process. Calculating the best thickness for each layer of a structure, like ARC/PR/ARC, at any stepper wavelength will become easy with the accurate knowledge of their refractive indices at the stepper wavelength.

### 1. Photoresist (PR), Antireflective Coating (ARC), Swing Ratio Simulation

Spectroscopic ellipsometry is the best optical technique to determine the thickness and refractive index of materials like PR and ARC at the same time and using the same instrument. These materials are transparent in the visible range and become absorbent in the UV part. Calculating the refractive index in the UV range requires a high-precision instrument and a well-defined thickness (the transparent part of the materials will help to determine N and T, where the $k$ value is 0).

At 248 and 193 nm and 157 nm, the sensitivity of DUV PR requires measurement in a photo environment and the use of a premonochromator to avoid bleaching the resist. Figure 12 shows the refractive indices and the extinction coefficients of two types of DUV 248-nm resists.

Measuring at 193 nm has been a technical challenge. In 1992, a good signal-to-noise ratio was obtained from measurement of photoresist. Since 1992, the accuracy of the determination of the optical properties at these wavelengths has reached the quality required by the PR manufacturers and users (11). Figure 13 presents a measurement

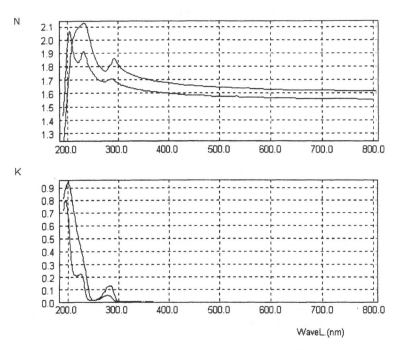

**Figure 12**   Refractive indices and extinction coefficients of two types of DUV 248-nm resists.

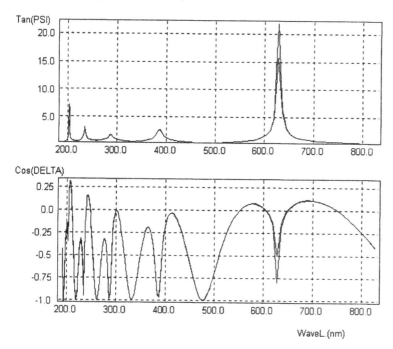

**Figure 13**   Measured and fitted curves of DUV 193-nm photoresist.

and a fit on a DUV 193-nm photoresist sample. The measurement and fitted curves are in good agreement.

Figure 14 shows the optical properties determined using this last measurement. Accurate determination of $n$ and $k$ can be reached down to 190 nm.

All these measurements and determinations on PR are done without any effect on the PR by using a special ellipsometer setup. The light source and the monochromator are before the sample. By using another setup, photoresist behavior versus exposure dose can be investigated (19).

In today's photolithography the extension to the deep UV (KrF excimer laser line at 248 nm and ArF excimer laser at 193 nm) leads to very difficult problems. The film's interference effect can lead to periodic behavior of the resist sensitivity *swing effect*, and the stability is linked to the minimum of reflectance, which depends on the optical indices. On nonplanar substrates, the effect of reflection on the vertical sidewalls of the steps can cause an increasing coupling of the energy into the resist volume and a reduction in linewidth (16,17). The best solution that has been chosen so far is the use of top and/or bottom antireflective coatings. One needs a nondestructive technique to characterize not only the thickness of the layers but also their optical indices. Their characterization must be done at the working wavelength. And SE has been chosen as the best method (12,13).

$SiO_xN_y$:H films make appropriate ARCs. Such films are obtained by PECVD using $SiH_4$ and $N_2O$ gases, whose flow ratio can be adjusted for the deposition process, changing the optical indices at the same time (5). Figure 15 shows the refractive indices of different $SiO_xN_y$:H films.

Organic ARCs are also used in the lithography process. Figure 16 presents organic top ARC and bottom ARC and photoresist refractive index.

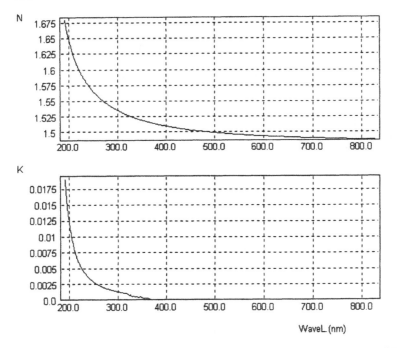

**Figure 14** Refractive index and extinction coefficient of DUV 193-nm PR.

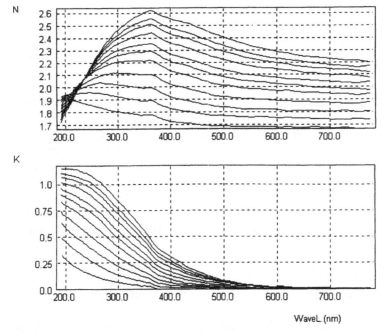

**Figure 15** ARC: $SiO_xN_y$:H. Refractive indices and extinction coefficients.

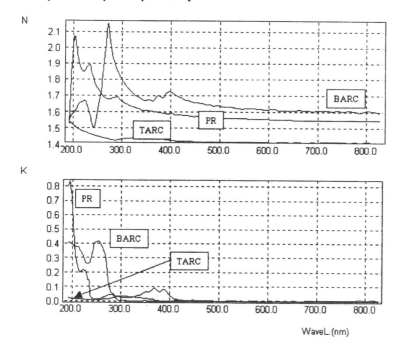

**Figure 16**  $N$ and $k$ of top ARC, PR, and bottom ARC for DUV 248 nm.

Using the optical indices and appropriate lithography simulation software, such as WINELLI (from SOPRA S.A.) or PROLITH (from FINLE Technologies), we can simulate the reflectivity of the material for any undefined parameters (12,13).

Figure 17 shows the effect of ARC on the reflectivity versus PR thickness.

## 2. Mask and Phase-Shift Mask

The decrease of the linewidth is also linked to the progress in photomask development. The need for mask characterization is important, especially in the new design of phase-shift masks and in all multilayer mask techniques. Materials like TiN, Chromium, Mo, and SiON are some examples of such layers (Figure 18).

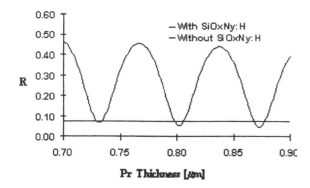

**Figure 17**  Reflectivity calculation of $PR/SiO_xN_y:H$ films versus PR thickness.

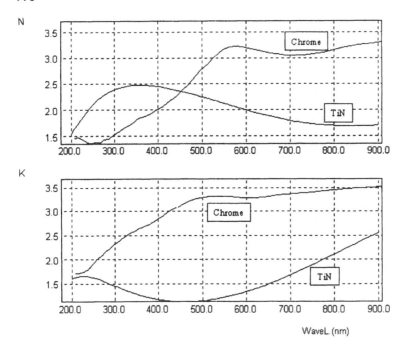

**Figure 18** *N* and *k* spectra for Cr and TiN.

## 3. Applications for 157-nm Lithography

Recently, the increasing interest in 157-nm laser source in projection lithography as successor to 193-nm-based systems has pushed SE manufacturers to find technical solutions to be able to measure the optical properties at this wavelength. One of the first applications for these ellipsometers is to characterize the refractive indices and extinction coefficients of the optics of the future steppers. Primarily lens material for projection systems will be metal fluorides, such as $CaF_2$, $MgF_2$, and LiF, and these require intensive characterization as well as their coatings.

The optical indices of photoresist and antireflective coatings also need to be known at 157 nm with the same accuracy as for 248 or 193 nm. The layer thickness is generally smaller than for the previous lithographic generations. In this case, ellipsometry is also one of the best techniques (Figure 19).

## B. Thin ONO and Nitrided Oxides

In the DRAM industry, upgrading storage capacity used to be one of the major goals of R&D. For its achievement, materials with high dielectric constant (high $\zeta$) became the state of the art. The related technology now focuses on the use of thin ONO (oxide/nitride/oxide) stacks, which necessarily tend to be thinner and thinner to enhance capacitive performances.

Until now, most optical techniques could only measure the thickness of the stack as a whole or determine the different thicknesses by fixing one of the oxides in the fit. Since the thickness of each layer is difficult to measure independently, the ONO structures show a tremendous need for measurement using spectroscopic ellipsometry, especially in the deep UV range (Figure 20). Indeed, in this part of the spectrum, the extinction coefficient *k* for

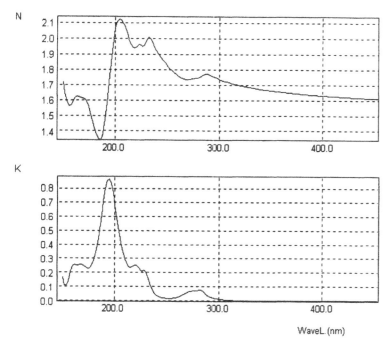

**Figure 19** *N* and *k* of photoresist between 150 and 450 nm.

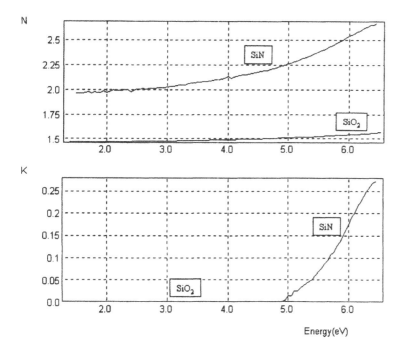

**Figure 20** *N* and *k* of SiN and $SiO_2$ versus energy (eV = 1.24/μm).

the nitride increases. Thus, the three thicknesses of the ONO stack can be decorrelated. The refractive index of the nitride has been determined independently on a single-layer sample.

Figure 21 shows the difference between three ONO samples. The main differences are in $\cos \Delta$. On $\tan \psi$ the differences occur in the UV range. Figure 22 presents the measurement and the regression curves on the 40/50/40 sample.

As the semiconductor industry continues to follow Moore's law, gate dielectric thickness has become one of the most rapidly scaled dimensions. The SIA roadmap states that a sub-0.25-μm process requires equivalent oxide values of less than 50 Å. Some leading-edge manufacturers are already pushing 0.18-μm process technologies with gate oxide thickness in the 30–40-Å range and for 0.13 μm near 20Å.

Nitride/oxide gate dielectrics have already been widely used. In this structure the reliability of a high electric field depends not only on the layer thickness but also on the nitridation level of the $SiO_xN_y$ layer. It is important to characterize precisely the layer thickness and composition of this structure in order to control the process quality (20).

## C.  Optical Gaps

By measuring the extinction coefficient of silicon nitride down to the UV range, we are able to calculate the optical gap of the material. This property is used in the flat panel display industry to control the deposition of the SiN around 5.2 eV (Figure 20).

## D.  Roughness Effects

Because silicon and polysilicon are absorbent in the UV range, the only interface of the Si seen in this range by the reflected beam is the top one. On a blank wafer or on a polysilicon

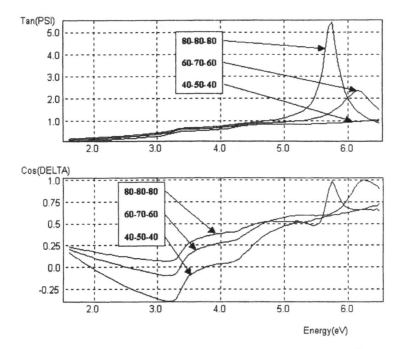

**Figure 21**  Measurement of three ONO samples. Thicknesses are in angstroms.

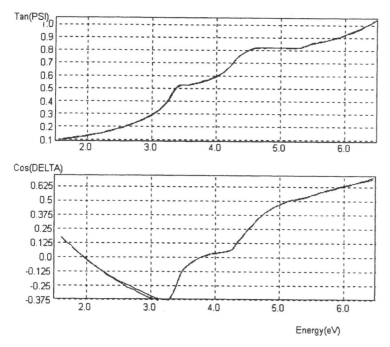

**Figure 22** Fit on a 40-50-40 Å ONO samples.

without any top layers, the quality of surface, the roughness, and /or the crystallinity of the material are easy to investigate. The following Table compares two similar structures:

| SiO$_2$ 2 nm |
| --- |
| Polysilicon 150 nm |
| SiO$_2$ 100 nm |
| Si Substrate |

| SiO$_2$ 2 nm |
| --- |
| Roughness 2 nm |
| Polysilicon 150 nm |
| SiO$_2$ 100 nm |
| Si Substrate |

We add 2 nm roughness between polysilicon and native oxide. The roughness is composed of a mixture of polysilicon and SiO$_2$ 50%–50% (Figure 23).

## IV. NEW APPLICATIONS

### A. Extreme UV (EUV) for Lithography

The next-generation lithography may be using x-rays, electron beams, or ion beam projection and probably will be based on EUV photons with wavelength of 13.4 nm (another wavelength at 11.34 nm using a MoBe mirror is less probable). If EUV becomes the next-generation lithography, this will require intensive metrology, including defect inspection, reflectometry, and spectroscopic ellipsometry. The stepper at 13.4 nm will require a source

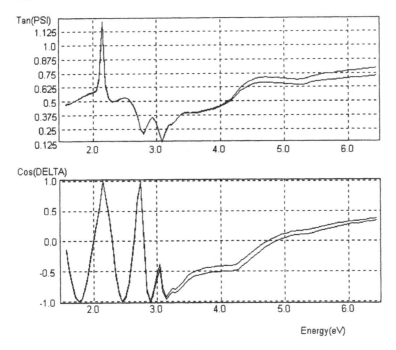

**Figure 23**  Roughness effect on ellipsometric spectra of polysilicon/SiO$_2$/Si.

of photons specially developed to generate the spectral wavelength and the spectral purity that will match with the mirror's bandwidth.

There are several potential sources for EUV lithography. Because the design of the stepper is based on a step-and-scan mounting, the source is essential in terms of throughput, and it must run at very high frequency, such as 6 KHz, because of the scanning of the slit image on the field. The source may be a plasma of Xe or water created by focusing a powerful laser beam at a high repetition rate. A capillary electrical discharge in O$_2$ or Xe may be an alternate, cheaper source than the synchrotron radiation used today. Other sources are under development as plasma pinch and plasma focus.

The source for metrology is also an issue at this wavelength, but the requirements are different from those for lithography. It requires a broad emission to cover a minimum spectral range around the maximum reflectivity of the layers, because the matching of reflectance is very important in terms of the total transmittance of the system composed of at least nine reflecting optics. Another characteristic is the stability. The frequency of the repetition rate can be reduced to 10 Hz if a normalization is used (23,24). The debris generated by the plasma creates a real production problem . It can decrease the quality of the first collection mirrors. But for metrology, the number of shots for a laser plasma source is reduced and the lifetime of the optics is not such an issue.

### 1.   Importance of the Reflectance of the Multilayers

In this spectral range, around 13.4 nm, the materials are not transparent, but it is possible to use catioptric optics, polished mirrors coated with multilayers. The refractive index is very close to 1, and the absorption coefficient must not be large because the wavelength is very short. There are a couple of materials that enable us to achieve these conditions, with a large enough difference in absorption, Mo and Si. When their respective thicknesses are 2.8 and 4.2 nm, then by coating 40 bilayers or more, reflectivity near normal incidence has been measured at over 68% for one reflectance mirror.

The projection optics (four or six mirrors) added to the collimating optics (four mirrors) and the reflecting mask must have the highest reflectivity possible and must match themselves very accurately to avoid losing costly light and to maximize the throughput of the stepper. The large-aperture optics are illuminated with different angles of incidence from the center to the edge, requiring the multilayers to gradually adapt their thickness with the angle of incidence; the tolerance of thickness is less than 1 Å.

The instrument must be enclosed in a high-vacuum environment to avoid absorption from the residual gas. To separate the source vessel from the fragile projection optics, one very thin membrane is used, losing 50% of the light. Another membrane of SiN or Si may be used to protect the costly optics from the degassing of the resists of the wafer, again losing 50% of light. The consortium called EUV-LLC has designed a prototype of the first stepper, called ETS, for Engineering Test System, to be used as an alpha tool for testing the concepts and the specifications and to prove the manufacturability of the EUV stepper (Figure 24).

The new metrology developed for lithography will also be used for surface and layer characterization in semiconductors, as well as to detect defects on the masks. Due to the wavelengths used and the sensitivity to the surface, very thin films will be analyzed. The range from 10 to 15 nm needs vacuum and covers the absorption peak of the silicon at 12.4 nm, corresponding to the L line of the x-ray spectrum of Si, so these wavelengths are very sensitive to layers including Si, such as poly-Si, silicon oxide, and silicon nitride, as well as silicon oxinitride and new high-$k$ materials for next-generation gate oxides.

The wavelength is 30 times shorter than the visible ones, so the sensitivity to the thickness is then enhanced 30 times, except that the refractive index is close to 1 for all materials. But the absorption can be used to characterize the materials.

The roughness will be very large because its amplitude becomes comparative to the wavelength. The scattering effect of the light is proportional to lambda E-4; the scattering will be enhanced by E6 compared to visible, so very smooth surfaces must be used and the scattering induced by very small amplitudes, even the atomic steps, will play an important role in the complex reflectance of light at these wavelengths, which can be considered as soft x-rays. A model must be used that will not be a layer mixture for the roughness but may be closer to the Debye–Wahler used in hard x-rays.

The detectors currently used are MCP (microplates) and photodiodes and channeltrons or photomultiplier tubes or proportional counters. The efficiency of these detectors is usually high, because one such photon has an energy of 100 eV and the noise is low, particularly when the signal is pulsed. So the signal-to-noise ratio can be very good. The fluctuation is due mainly to quantum noise associated with the low number of photons (1 sigma = (number of photons)E-1/2. To reach a fluctuation less than 1%, one needs at least 10 000 photons.

The detection can be digital, by counting the photons one by one, if the source is continuous, or at very high repetition rate if one collects only one photon maximum at each pulse, such as, for instance, synchrotron radiation or high-voltage discharges on a cathode of Si. In the case of pulsed sources, a reference is required to normalize the amplitude fluctuation on a pulse-to-pulse basis. In this case, a 1% fluctuation may be achieved in a single large pulse. The reference detector must be representative of the signal for the numerical optical aperture, the spectral bandwidth, and the image of the source.

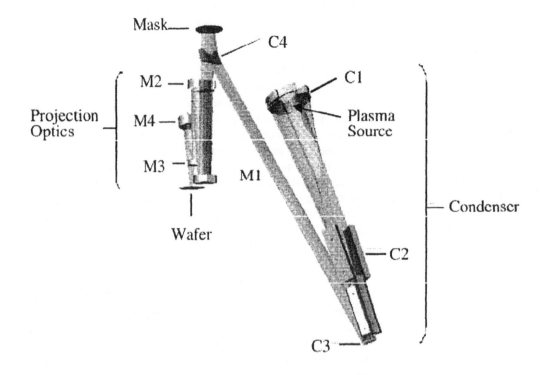

**Figure 24**  Optical mounting of the stepper in EUV. (From Charles Gwyn, EUV LLC.)

The resists at this wavelength are the same as for either e-beam writing or DUV. The sensitivity is below 10 mJ/cm$^2$, and isolated lines of 35 nm have already been demonstrated by using an interferometer.

## B.  Combination of GXR and DUV SE

It is advantageous to combine a very-short-wavelength source, such as Cu K-alpha 1.54 Å, with the UV-visible-wavelength spectroscopic ellipsometer (Figure 25). The models are the same optically, except for the wavelength, the effects are complementary, and no vacuum is required.

The refractive indices for hard x-rays are approximately 1 within $10^{-5}$, and the absorption is very low, about $10^{-4}$, so all materials are transparent at this energy, including metals, dielectrics, semiconductors, and organic layers. The optical beams are not refracted as in the visible because the $N$ values are always the same and equal to 1; the Fresnel (Snell) law applies in a straight line.

The thickness of the layers is measured by the angle position of the interference's after-reflectance at grazing incidence, between 0.2 and 4° typically, relative to the sample surface. This measurement is very accurate for the optical thickness, and it is independent of the material because $N = 1$, so the GXR interference values, linked to angle positions, give the thickness without ambiguity. The accuracy is given by the angular position of the

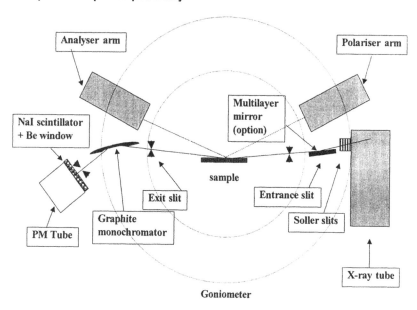

**Figure 25**   Schematic view of the combined GXR/SE system.

interference, which is very high. When the thickness is known with precision, after grazing x-rays reflectance, it can be used in the data regression in UV-visible spectroscopic ellipsometry to extract the refractive indices $N$ and $k$ with the same relative precision as the thickness measured. This is particularly useful for thin layers, when the thickness and refractive indices are strongly corelated. In the case of multilayers, an addition of reflective interference will give beats with amplitudes proportional to the respective density or refractive indices of each layer.

The roughness is measured with precision and without ambiguity by the rapid decrease of reflectance and can be modeled using the Debye–Wahler factor. In the case of a layer it is even possible to discriminate the top layer roughness from that of the interface.

The advantage of combining the two techniques in the same equipment is that one can measure the same area of the same sample on the same day by using only one sample holder, one goniometer, one electronic rack, and one computer with the same software for modeling. This is not only a saving of time and money but it also gives a better and more accurate result with fewer uncertainties. Nontransparent layers in the UV-visible can be measured with grazing x-rays reflectance, as long the thickness is less than 200 nm. For thicker, flat, and smooth layers, the interference pattern is so tight that the contrast disappears, and the measure is lost. The critical angle of incidence, at which the light penetrates into the top layer, is related directly to the density of the material; for example, for $SiO_2$ it is $0.26°$, because of its density of 2.8 g/cc

The limitations linked to the GXR are the need for a large and very flat surface of at least 1 cm$^2$ with a very smooth surface. If the roughness value is larger than 5 nm, the reduction in the reflectance is so large that no layer interference can be measured and the critical angle itself is also affected.

The measurement is traditionally made by a sequential scanning of angles of incidence and of reflectance. But recently a multichannel detector has been used to detect the reflectance coefficients in parallel. This has been made possible by the use of multilayer graded concave mirrors. In this case it is very difficult to measure all the dynamic range required—7 decades—and the angular resolution is limited to resolving very thin films and very thick films as well as multilayers. This system is well suited to measure the thickness of barrier layers such as TiN and TaN, semimetals such as TiSi, and seeds for Cu where the thickness varies from 10 nm to 100 nm and the densities are very different from the Si or $SiO_2$. Monitor wafers are still needed, due to the spot larger than 1 mm. It is not easy to focus the beam, and if the source size was $40\,\mu$, then the grazing image should cover a length of $2400\,\mu$ for an angle of incidence of $1°$.

The same x-ray source, goniometer and detector can be used to measure x-ray diffraction. Eventually, with a solid-state SiLi detector, the layers and surface x-ray fluorescence spectrum can be measured to analyze the materials that fluoresce under Cu K-alpha line excitation. Another source would give better sensitivity and a larger spectrum. The atmosphere can reduce the range for light elements.

## V. CONCLUSION

Ultraviolet, VUV, and EUV are spectral ranges where interesting features in optical indices are present. These ranges are very sensitive to very thin films for absorption and roughness, due to the shorter wavelengths compared to visible and infrared. Many difficulties must be overcome, such as the need for vacuum, the degassing of materials, and the availability of polarizers and sources. Because it is new, there is a need for characterization of the new materials now available to build a good database. This range will be investigated further in the future for its special attractive qualities for analysis, particularly for the photolithography application, which always uses shorter wavelengths.

## REFERENCES

1. See, for example, the proceedings of the Second International Conference on Spectroscopic Ellipsometry, Charleston, SC, 12–15 May 1997, Thin Solid Films, Vols. 313–314 (1998).
2. P Boher, JP Piel, P Evrard, JL Stehle. A new combined instrument with UV-Visible and far infrared spectroscopic ellipsometry: application to semiconductor characterization. EMRS Conference, paper PVII2/P6 (1999).
3. P Boher, JP Piel, C Defranoux, JL Stehle, L Hennet. SPIE 2729: , 1996.
4. P Boher, JP Piel, P Evrard, C Defranoux, M Espinsa, JL Stehle. A new purged UV spectroscopic ellipsometer to characterize thin films and multilayers at 157 nm. SPIE's 25th Annual International Symposium, 27 February–3 March 2000.
5. J Barth, RL Johnson, M Cardona. Spectroscopic ellipsometry in the 6–35eV region. In: E D Palik, ed. Handbook of Optical Constants of Solids II. San Diego, CA: Academic Press, 1991, ch. 10, p 123.
6. ED Palik, ed. Electronic Handbook of Optical Constants of Solids, ScVision. San Diego, CA: Academic Press, 1999.
7. BL Henke. Low-energy x-ray interactions: photoionization, scattering, specular and Bragg reflection. Proceedings of the American Physical Society Topical Conference on Low-Energy X-Ray Diagnostics, Monterey, CA, June 8–10, 1981.
8. ED Palik, ed. Handbook of Optical Constants of Solids II. San Diego, CA: Academic Press, 1991.
9. Bloomstein et al. Proceedings of SPIE. 3676:342, March 1999.

10. DE Aspnes, AA Studna. Rev. Sci. Instruments. 49:291, 1978.
11. P Boher, C Defranoux, S Bourtault, JL Stehle. High accuracy characterization of ARC and PR by SE: a new tool for 300 mm wafer technology. 24th Annual SPIE Meeting on Microlithography,
12. P Boher, C Defranoux, JP Piel, JL Stehle, L Hennet. Characterization of resists and ARC by SE in the UV and deep UV. SPIE 2729: ,1996.
13. P Boher, C Defranoux, JP Piel, JL Stehle, L Hennet. Precise ARC optical indices in the deep UV range by variable angle SE. SPIE's 1997 Symposium on Microlithography,
14. C Defranoux, JP Piel, JL Stehle. Deep UV measurements of SiON ARC by SE. Thin Solid Films 313–314:742–744, 1998.
15. T Ogawa, H Nakano, T Gocho, T Tsumori. SiOxNy: high performance ARC for the current and future optical lithography. SPIE 2197:722–732,
16. RR Dammel. Antireflective coating: sultans of the swing curve. Semiconductor Fabtech. 2nd ed., 1995, pp 215–221.
17. JE Lamb III. "Organic ARC, ARC Application within the Microelectronics industry. Semi conductor Fabtech. 2nd ed. 1995, pp 223–227.
18. P Boher, JP Piel, P Evrard, JL Stehle. A new Purged UV spectroscopic ellipsometer to char-acterize 157-nm nanolithographic materials. MRS 1999 Fall Meeting,
19. P Boher, C Defranoux, JP Piel, JL Stehle. A new method of determination of the photoresist dill parameters using SE. SPIE 3678:126, 1999.
20. P Boher, JP Piel, JL Stehle. Precise characterization of ultrathin nitride/oxide gate dielectrics by grazing x-ray reflectance and spectroscopic ellipsometry. MRS 1999 Fall Meeting,
21. D Attwood. Soft X-Rays and Extreme Ultraviolet Radiation: Principles and Applications. New York: Cambridge University Press.
22. C Gwyn et al. EUV LLC. Livermore, CA, November, 1998. PO Box 969, MS-9911, Livermore, CA 94551-0069.
23. M Yamamoto, K Mayama, H Kimura, Y Goto, and M Yanagihara. Thin film ellipsometry at a photon energy of 97 eV. J Electron Spectroscopy Related Phenomena 80:465–468, 1996.
24. JH Underwood, EM Gullikson. Beam line for measurement and characterization of multi-layers optics for EUV lithography. Proc. SPIE 3331:52, 1998.

# 27

# Analysis of Thin-Layer Structures by X-Ray Reflectometry

**Richard D. Deslattes**
*National Institute of Standards and Technology, Gaithersburg, Maryland*

**Richard J. Matyi**
*University of Wisconsin, Madison, Wisconsin*

## I. INTRODUCTION

The well-established structural methods of x-ray specular and diffuse scattering are less widely used in semiconductor metrology than their capababilities would suggest. We describe some technical enhancements that make these highly useful tools even more accessible and productive. These enhancements include improvements in beam-forming optics combined with channel-crystal analysis of reflected/scattered x-rays and high-rate detectors to allow more efficient and informative x-ray measurements covering a wide range of thin-layer structures. We also introduce some methods of wavelet analysis that appear to offer useful qualitative structural insights as well as providing effective denoising of observed data.

## II. BASICS OF X-RAY STRUCTURAL PROBES

Subnanometer wavelengths and weak (i.e., nonperturbing) interactions make x-ray probes a nearly ideal way of delineating the geometry of the layer and multilayer structures that underlie much of modern semiconductor manufacturing. The intensity of x-rays reflected by such thin films or layer stacks varies periodically in response to changes of either the angle of incidence or the x-ray wavelength. These periodic variations arise from the interference of radiation reflected by each of the interlayer interfaces, in a manner analogous to the formation of "Newton's rings" in visible light.

It is generally most convenient to use the fixed wavelengths of characteristic x-ray lines (e.g., Cu Kα radiation) and measure the reflectivity as a function of incidence angle. The small wavelengths of the x-ray probe (typically of the order of 0.05–0.80 nm) allow even very thin layers ($\sim$ 2–3 nm) to be conveniently investigated. Because x-rays interact principally with electrons in the inner shells of atoms, the index of refraction of the films is easy to calculate and can be shown to be (1–3)

$$n = 1 - \frac{\lambda^2 r_0 \rho_e}{2\pi} - i\frac{\lambda \mu_x}{4\pi} = 1 - \delta - i\beta \tag{1}$$

where $\lambda$ is the x-ray wavelength, $r_0 = e^2/mc^2$ is the classical electron radius ($2.818 \times 10^{-15}$ m), $\rho_e$ is the electron density ($Z \times \rho$, electrons m$^{-3}$), and $\mu_x$ is the linear x-ray absorption coefficient.* In most materials and for conventional laboratory x-ray wavelengths, the value of $\beta$ is one to two orders of magnitude smaller than $\delta$; hence the imaginary term can usually be neglected. The real $\delta$ term differs only slightly from unity (typically a few parts in $10^6$), so the index of refraction for x-rays in solids is very close to the $n = 1$ value of the vacuum (or normal atmosphere). The small refractive index decrement confines the region of appreciable x-ray reflectivity to grazing angles (typically in the range of $0°$–$3°$). Owing to the algebraic sign of the decrement, this high-reflectivity region occurs on the external side of the interface.

In analogy with conventional light optics, the fact that the index of refraction differs from unity implies that there will be a critical angle for total specular reflection of the incident x-ray beam. Snell's law shows that in the absence of absorption ($\beta \to 0$), the critical angle is given by $\theta_{crit} = n \approx 1 - \delta$. In the case of silicon irradiated with Cu K$\alpha$ radiation, the critical angle for total external reflection is about $0.21°$. Under zero absorption conditions the specular reflectivity will be perfect for all angles less than the critical angle. If, on the other hand, absorption cannot be neglected, then the specular reflectivity will monotonically decrease with increasing angle as the critical angle is approached.

Above the critical angle the incident x-ray beam with wavelength $\lambda$, impinging on a surface at an angle $\theta$, will generate both reflected and refracted waves at the surface; the refracted wave will be likewise split at the next interface, etc. This generates an angle-dependent interference between the various reflected waves that is manifested as a periodic variation in the total reflected intensity. In the case of a single layer of thickness $d$ on a substrate, the spacing of the fringes (known as "Kiessig fringes" [4]) far from the critical angle is related to the thickness by

$$\Delta\theta = \frac{\lambda}{2d} \tag{2}$$

Because the wavelengths, $\lambda$, of the characteristic x-ray lines have been accurately established (see later), the film thickness can be determined to a high accuracy. Note that a determination of film thickness by measurement of the angular separation of the fringes can be made without precise knowledge of the film composition.

---

* Modern x-ray reflectivity studies began with the work of Parratt in 1954 (1). With simple x-ray sources and primitive optics, his group measured metal oxide films ($\sim 20$ nm thickness) and developed a theory of reflectivity that remained essentially unchanged until the 1980s. In the early 1970s, Segmüller (2) used x-ray reflectivity to determine film thicknesses of silicon on sapphire, but the method did not find application in the semiconductor industry. Not only was the method little known, but it was likely unneeded owing to the film thickness ranges of interest at the time. Also there was a very limited and stable materials base with aluminum conductors and silicon oxide dielectrics that could be characterized by other means, such as ellipsometry. Nowadays, however, semiconductor manufacturing emphasizes film thickness in the nanometer range and a much wider range of materials. When these trends are coupled with an increasingly rapid development schedule and the continued drive toward smaller feature size, there is an evident need for some of the unique capabilities of x-ray reflectivity.

In the case of multiple thin layers, the angular dependence of the reflected x-rays becomes more complicated. The reflectivity profile must then be calculated by determining the conditions for continuity of the electric and magnetic fields at each interface in the scattering medium. In the case of x-rays, this was first done by Parratt (1), who defined the reflectivity ratio $R_{n,n+1} = a_n^2(E_n^R/E_n)$ in terms of the electric vector of the incident $(E_n)$ and reflected $(E_n^R)$ beams in the $n$th layer of an $N$-layer medium, and $a_n$ is the amplitude factor for the $n$th layer, given by

$$a_n = \exp\left(\frac{-ik_1 f_n d_n}{2}\right) \tag{3}$$

where $d_n$ is the layer thickness, $k_1 = 2\pi/\lambda$, and $f_n = [\theta^2 - 2(\delta + i\beta]^{1/2}$. This leads to a recursion relation

$$R_{n-1,n} = a_{n-1}^4 \left[\frac{R_{n+1,n} + F_{n-1,n}}{R_{n+1,n}F_{n-1,n} + 1}\right] \tag{4}$$

where $R_{n,n+1}$ is as defined earlier and

$$F_{n-1,n} = \frac{f_{n-1} - f_n}{f_{n-1} + f_n} \tag{5}$$

Starting from the bottom layer (the substrate), where $R_{n,n+1}$ is zero (since the thickness of the substrate is infinite), the recursive relation can be evaluated at each interface moving up to the top surface. The final ratio of reflected to incident intensity is given by

$$\frac{I_R}{I_0} = \left|\frac{E_1^R}{E_1}\right|^2 \tag{6}$$

The foregoing expressions hold strictly only in the case of ideally smooth interfaces. Real interfaces, however, are not ideally sharp (e.g., roughness and interdiffusion); moreover, they may possess a grading in chemical composition. If an interface has a finite width due to roughness or grading, the overall reflectivity envelope will decrease with increasing angle more rapidly than expected. The effect of grading can be approximated by modeling the interface reflectivity as a stack of layers of uniform composition, but with the composition varying from layer to layer. In contrast, the effect of interfacial roughness can be most easily considered by the inclusion of a Debye–Waller factor, where the perfect interface Fresnel reflectivity $R_F$ is damped by a Gaussian height distribution according to (5,6):

$$R_F^{\text{rough}} = R_F(Q_Z)\exp(-Q_Z^2\sigma^2) \tag{7}$$

where $\sigma$ is the root mean square (rms) roughness of the interface and $Q_Z$ is the difference between the scattered and incident wavevectors.

The scattered x-ray intensity will also be distributed into nonspecular directions when the interface is rough; however, this does not happen if the interface is merely graded. The calculation of the nonspecular diffuse scatter due to roughness is much more difficult than the specular case; it is usually performed using the distorted-wave Born approximation (DWBA), where the roughness is treated as a perturbation to the electric field within an ideal multilayer. Although computationally intensive, the DWBA approach allows the inclusion of a correlation function $C_{l,l'}(R) = \langle\delta z_l(0)\delta z_{l'}(R)\rangle$ that contains all the information about the roughness at individual interfaces and the way that the roughness replicates from one interface to another. One of the most common forms of the correlation function is that of a self-affine, fractal interface with a correlation length $\xi$ and a fractal dimension $D = 3\text{-}H$. In this case the correlation function can be written as:

$$C(R) = \sigma^2 \exp\left[-\left(\frac{R}{\xi}\right)^{2H}\right] \tag{8}$$

where $\sigma$ again is the rms roughness. Increasing $\sigma$ results in an increase in the off-specular diffuse scatter, while increasing $\xi$ or decreasing $H$ concentrates the diffuse intensity about the specular region.

Specular scattering at small angles (often called grazing incidence x-ray reflectivity, or GIXR) has a number of features that favor its use in systems of interest to semiconductor manufacturing. For example, GIXR is a nondestructive probe that provides thickness, density, and interface information from even fairly complex thin-film stacks. And GIXR is easily adaptable to different thin-film systems, particularly for film stacks in the range from micrometer to nanometer thicknesses. Finally, GIXR is a simple technique to use, and its data is easy to analyze.

Thicknesses emerging from GIXR are accurate without reference to artifact standards or reference materials, because the wavelength used (typically Cu K$\alpha$ radiation) is highly reproducible (at the level of $10^{-6}$) (3). This wavelength has been linked to the base unit of length in the International System of Units (SI) by an unbroken chain with sufficient accuracy to fully exploit this reproducibility. The weak x-ray interaction allows simple representations of x-ray indices of refraction with the needed accuracy, thereby eliminating a large source of uncertainties associated with the use of optical probes. To a good first approximation, x-ray reflectivity profiles measure geometrical path lengths that are closely equal to film stack thicknesses. In contrast, dimensions obtained in the near-visible are normally altered by sample-dependent optical constants to the extent of 20%–50%.

The issues surrounding the term *accuracy* deserve further comment. One refrain that finds echoes in many forms of manufacturing (including the semiconductor sector) is that accuracy is a secondary concern—all that really matters, it is said, is that the product be reproducible and that the required functionality (electronic performance in the case of semiconductor manufacturing) be delivered. In contrast, "accuracy" entails coupling of the measurement in question to an invariant primary standard, or at least to a highly reproducible secondary standard that is, itself, well connected to a primary invariant. Certain conveniently available characteristic x-ray emission lines embody all of the needed features. These lines are highly reproducible (within 1 part in $10^6$), and they have been linked to the base unit of length in the International System of Units (SI) by an unbroken measurement chain that does not degrade the indicated reproducibility. Even a brief experience in the area of thin-film metrology is enough to convince a practitioner that the only stable measurement systems are those that are accurate.*

## III.  ELEMENTARY X-RAY REFLECTOMETRY

Figure 1 shows the geometry of a simple reflectometer. Competent reflectivity measurements require an incident x-ray beam having both a narrow spatial extent (in the plane of the figure)

---

* This point is often reiterated by T. Quinn, Director of the International Bureau of Weights and Measures (BIPM).

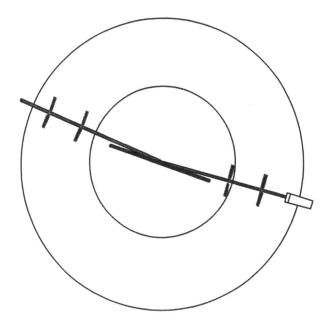

**Figure 1** Diagram of an elementary x-ray reflectometer.

and a small angular divergence, typically of the order of 0.01° or less. This input beam is scanned in an angle with respect to the sample surface over a range of grazing angles, from zero through a few degrees. Reflected radiation is detected by the counter shown through a pair of slits, with a small spatial extent (in the plane of the figure) positioned so as to realize the $\theta$–$2\theta$ geometry corresponding to specular reflection. The reflectivity, $R$, is the ratio of reflected to incident intensities. For our purposes, a reflectivity profile is obtained at fixed incident x-ray wavelength as a function of the (grazing) angle of incidence.

## A. Some Experimental Details of Higher-Performance Reflectometers

Although there are several commercial systems available for x-ray reflectometry, instruments such as those at NIST can be either assembled from components of diffractometers or built for this application. The flexibility offered by this approach has allowed consideration of a variety of x-ray optical arrangements. The main design challenge arises from the large dynamic range of reflectivity encountered in the analysis of thin films. This reflectivity range leads to a correspondingly large range of x-ray counting rates. Since most of the important structural information is found in the high-angle (and corresponding low-intensity) region of the reflectivity profile, the x-ray source and beam-forming optics need to aim for high incident flux on the sample. At the same time, the rapid falloff of reflectivity with angle ($\sim \theta^{-4}$) means that the beam-forming optics as well as the analysis optics need to have collimated profiles in which the wings of the window function are highly suppressed, i.e., fall off much faster than $\theta^{-4}$.

The collimation requirements can be effectively met by carefully designed slit arrangements if the angular resolution requirements are modest. In contrast, to achieve the angular resolution needed to map relatively thick layered materials ($\sim 1$ micrometer), multiple crystal reflections are needed. Such multiple-reflection x-ray optics are conveniently realized by using channel-cut crystals, such as is illustrated in Figure 2a where we use "four-bounce" geometry for both the preparation and analysis optics. One simple modification of the symmetric channel in Fig. 2a is realized by introducing an asymmetric initial reflection. It can be shown from the dynamical (perfect crystal) theory of x-ray diffraction that such an asymmetric reflection increases the angular range accepted by the collimator. This results in a considerable gain in the counting rate for the configuration shown in Fig. 2b. One less desirable aspect, however, is that it also broadens the beam spatially, thereby limiting the system's utility at very small incidence angles. Still further intensity gain (with still greater spatial broadening) is realized by introducing a graded-spacing parabolic mirror, as illustrated in Fig. 2c.

High x-ray flux levels can be obtained either by combining the optical design exercises described earlier with a stationary anode x-ray tube or a more powerful (and costly) rotating anode source. In either case, data rates in the region of high reflectivity rapidly exceed the counting-rate capability of the usual NaI scintillation detectors and the cap-

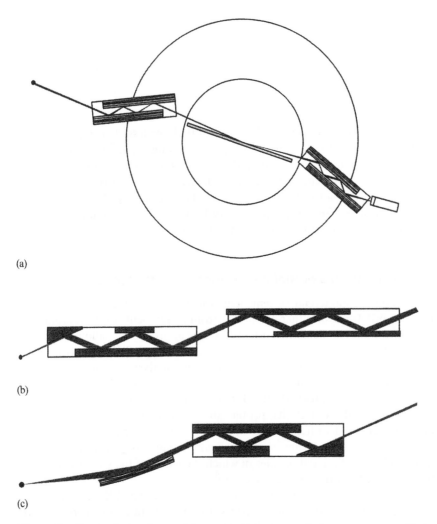

(a)

(b)

(c)

**Figure 2** Comparison of several beam-conditioning optical arrangements offering improvements over simple slit collimators.

abilities of even the newer high-speed counters that use YAP crystals. For the present, we have chosen to reduce the x-ray tube power and operate within the rate capacity of the available detector using a single dead-time correction function, as indicated in Figure 3. A simple upgrade providing automatic insertion of x-ray attenuators will allow full utilization of the available dynamic range of $10^8$.

The characteristics of the reflectometer are also of great importance for high resolution and high accuracy. Many instruments (including early devices at NIST, whose results are described later) are worm-gear driven. The local angle scale is uniform near the 0.001° level, but it is not well controlled for large-angle excursions. Other, more modern systems have encoders directly on the detector and sample axes, as does the next generation of NIST instrumentation. There has been little attention paid to the potential need for in situ calibration of the angle scale, a problem we intend to address. For most purposes this is not a problem, on account of the rather limited precision needed. On the other hand, the issue of traceability does need to be addressed by either calibration or the introduction of an externally calibrated standard.

Among the many practical issues, the problem of sample fixation is particularly troublesome. Good practice in other x-ray measurements often involves having three-point contact support for the wafer. This is generally unacceptable for semiconductor wafers because of contamination and inevitable marking of the processed surface. It has been more acceptable to use a vacuum chuck that contacts the wafer only on its "backside." Such a fixation scheme may tend to flatten a significantly bowed wafer or distort one that is initially flat. The tradeoff needs careful consideration.

## B. Some Examples Provided by Simulations

We begin with some very simple examples of the kinds of reflectivity profiles produced by single-component smooth surfaces and primitive film structures assembled on such ideal

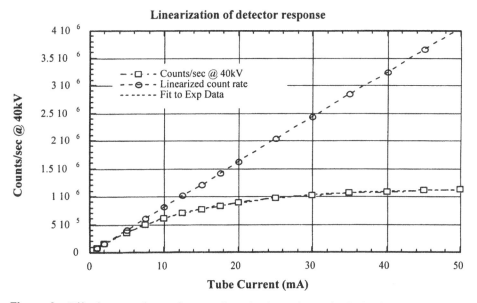

**Figure 3**  Effectiveness of counting-rate linearization using a single dead-time model.

surfaces. For readers interested in more quantitative detail there is a recent book (7) as well as several reviews (8–10). Our purpose here is to convey a sense of qualitative expectations while recognizing that these will need refinement as we proceed toward quantitative studies of real systems.

Figure 4 illustrates simulated reflectivity profiles for several ideal elementary solids. The linear representation in Fig. 4a emphasizes the range of "total reflection" as it increases

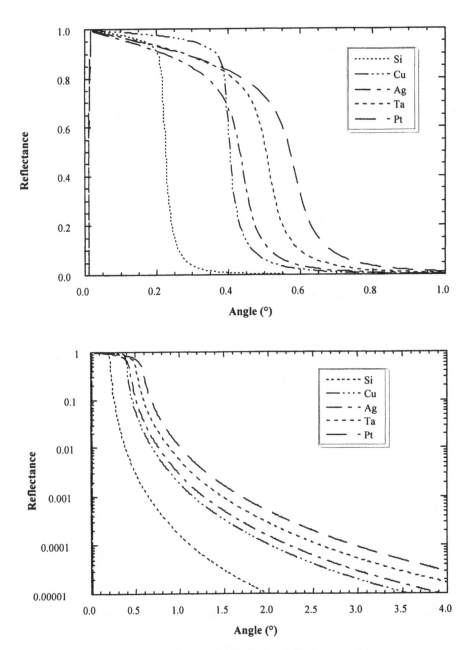

**Figure 4** Elementary reflectivity profiles for several elemental solids shown on linear (top) and logarithmic (bottom) scales.

with increasing atomic number for normal bulk densities. These profiles are conveniently parameterized in terms of the critical angle associated with the "corner" of the reflectivity profile (in the case of nonabsorbing films). Note that the effect of finite absorption is manifested by a decrease in the reflected intensity, even at angles smaller than the critical angle.

In thin films, the atomic or molecular microstructure often leads to density values differing from the bulk density value, a situation in which changes in the critical angle allow use of reflectivity curves to extract film density (for known chemical composition). Figure 4b shows the same data on a log scale to emphasize the rapid ($\theta^{-4}$) decrease of reflectivity with angle and the greater range of reflection for heavier elements.

The following figures illustrate the effects of added complexity and deviations from ideality. Figure 5 shows the effect of adding a single 100-Å layer of TiN with a 30-Å cap layer of amorphous $SiO_2$. Figure 6 shows what happens when a diffuse boundary is introduced in the structure of Fig. 5 and the additional effect of 15 Å of added roughness. It is important to note that specular reflectivity cannot distinguish between the loss of contrast shown in Fig. 6 arising from a diffuse interface and a similar loss of contrast arising from a rough interface. These alternatives can, however, be separated by nonspecular scattering data, as will be seen later.

## IV. X-RAY REFLECTIVITY CHARACTERIZATION OF THIN FILMS FOR DEVICE APPLICATIONS

A number of candidate films for Cu diffusion barriers, gate dielectrics, memory elements, and interconnects are being investigated at the present time. Because of the simplicity of x-ray interactions, x-ray reflectivity methods can be applied to these materials with considerable ease. In systems encountered up to the present there is no need to develop a

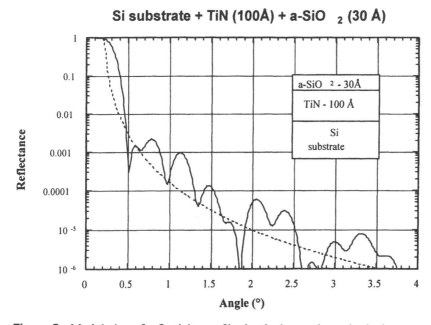

**Figure 5** Modulation of reflectivity profiles by the layers shown in the inset.

## Effect of 15Å roughness

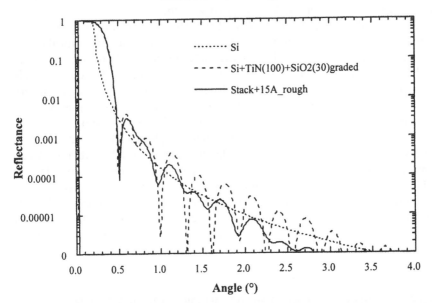

**Figure 6** Effect of grading the interface between the top layer pair and adding roughness as indicated.

material-specific approach to characterizing any of these systems. We have had opportunities to analyze a number of these thin-film structures[*] and their changes in response to processing conditions. This section gives results from several of these studies and examines process-dependent structural changes. We have not as yet had the chance to be involved in connecting such data with device performance, but we look forward to doing so when we can introduce the x-ray measurements at a sufficiently early stage of process development.

### A.    Process Improvement Seen in TaN Films

One early group of $TaN_x$ samples with varying amounts of nitrogen had sheet resistivities that were too high for their intended application. Samples of GIXR data from this group of 16 $TiN_x$ on Si wafers are shown in Fig. 7. Analysis of these profiles showed that with increasing nitrogen concentration, densities decreased from a starting value near that of cubic Ta ($16.6 \text{ g/cm}^3$) to a much smaller value near $10.0 \text{ gm/cm}^3$. Such behavior was not consistent with the expectations for the process being used. Subsequent powder diffraction studies showed the dominance of a tetragonal phase with a high degree of preferred orientation, suggesting the presence of columnar growth. Reflectivity modeling also suggested the presence of a thin high-density layer beneath the $TaN_x$ film. Subsequent diffraction study showed the presence of a 1-nm-thick layer of (cubic) Ta so highly oriented as to suggest epitaxial material.

A year later we received a new group of $TaN_x$ wafers that had been obtained using improved growth procedures. Examples from this improved group are shown in Figure 8. The analysis and modeling lead in this case to density values near the $15.5\text{-g/cm}^3$

---

[*] Unless otherwise noted, all samples were provided by International SEMATECH.

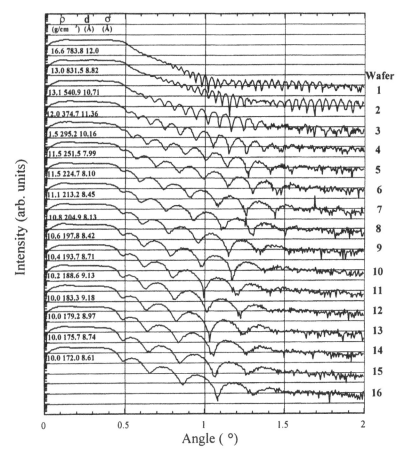

**Figure 7** Reflectivity profiles for an early group of $TaN_x/Si$ Cu diffusion barrier films with excessive sheet resistivity.

density of TaN, and to improved limits on the roughness parameter. The modeled structure is a stack consisting of a top layer of $Ta_2O_5$, atop the TaN layer with a $SiO_2$ interface layer between the $TaN_x$ and the silicon substrate. Because of the large thickness of the silicon oxide layer and the low contrast between it and the underlying bulk silicon, its parameters are not determined in the modeling, nor is there a need for such modeling in this application.

## B. Measurement Challenges from Ultrathin-Gate Dielectrics

While the use of silicon oxide gate dielectrics remains dominant, much effort is currently being directed toward alternatives. Among these, thin silicon oxynitride is compatible with the existing manufacturing environment. In addition, there is some gain in dielectric constant and good resistance to breakdown. Several processes have been developed that result in samples with $SiO_xN_y$ layers in the 2–4 nm range. The combination of light atoms and small thicknesses leads to low-contrast profiles in the region where the reflectivity is very small. To partly address this problem, the profiles shown in Figure 9 were obtained under conditions in which the counting rate in the region of total reflection was entirely beyond the capabilities of our fast detector. Accordingly the curves in Fig. 9 begin well

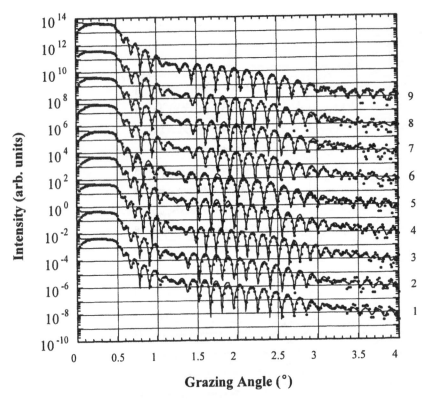

**Figure 8** Reflectivity profiles for a later group of $TaN_x$ copper diffusion barrier films $(Ta_2O_5/TaN_x/SiO_2/Si)$ with improved process control.

past the critical angle and do not allow the global fitting toward which our analysis is usually oriented.

In spite of the technical limitations just noted, the results have several interesting and useful aspects. Although the observed structure has limited contrast and is seen only in regions where counting rates are low, the fitted structural parameters are obtained with satisfactory precision. Graphical indications of model-parameter sensitivity are shown in Figure 10. The separate panels indicate the effect of displacing the thickness parameter from the value obtained by our analysis by $\pm 0.3$ nm, the effect of changing the diffusion/roughness parameter by $\pm 0.2$ nm, and the effect of changing the average film density by $\pm 10\%$

### C.  High-*K* Dielectric Films, Barium Strontium Titanate (BST), and Tantalum Pentoxide ($Ta_2O_5$)

As part of a round-robin characterization,[*] 68 BST wafers were analyzed (11). In some cases the BST had been grown on a Pt-coated substrate; in others, growth had taken place on a thick silicon oxide layer. As is evident from the sample of raw data shown in the upper part of Figure 11, the sample quality was highly variable. While the precise pre-

---

[*] The BST samples were supplied by T. Remmel of Motorola.

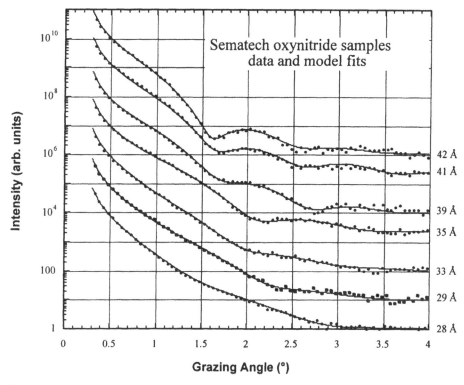

**Figure 9** X-ray reflectivity profiles for a group of oxynitride films. Curves are labeled according to approximate $SiO_xN_y$ layer thicknesses.

parative procedures used were not provided, it appears that some change in growth conditions, or a spontaneous morphological change, produced a structural boundary during the early stage of film formation, leading to the strong beating period evident in some samples. In addition, a number of films had been grown with a Pt precoat; these showed more extended single period-oscillation, indicating more sharp boundaries and freedom from polymorphism. In spite of this wide diversity, relatively good modeling has been possible, as is shown in the lower panel of Fig. 11.

Tantalum pentoxide is another prominent candidate for high-$K$ dielectric applications. Unlike the case of silicon oxide and oxynitride, layers of $Ta_2O_5$ contrast sharply with the silicon or a silicon oxide substrate. The strong oscillations shown in Figure 12 reflect this fact. The model underlying the fitted curves has an oxynitride layer between the silicon substrate and the $Ta_2O_5$ main layers. Aside from layer thicknesses, the main additional parameters are the densities assigned to the oxynitride and $Ta_2O_5$ layers. The densities obtained for the oxynitride layers are similar to those seen in previous oxynitride studies. The tantalum oxide densities vary over a considerable range, with two of the samples giving values of 6.31 $g/cm^3$ and 5.08 $g/cm^3$, respectively. Aside from these, the remaining samples have densities approaching those quoted for bulk material, namely, 8.73 $g/cm^3$. On the other hand, a theoretical density of 8.316 $g/cm^3$ is expected for the hexagonal form ($\delta$ phase) produced in CVD process, while the alternative orthorhombic form ($\beta$ phase), produced when Ta is subjected to oxygen radical oxidation during growth, has a closely equal theoretical density. Crystallographic databases give structural information on a number other oxides and an additional tetragonal form of $Ta_2O_5$.

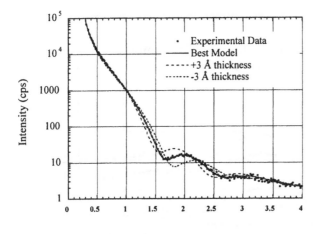

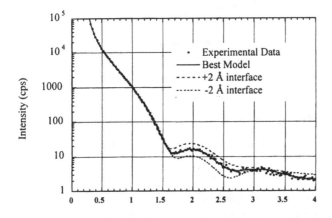

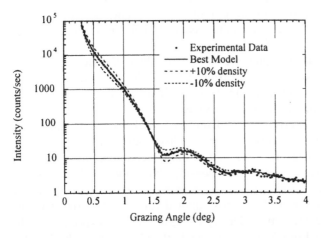

**Figure 10** X-ray reflectivity analyses of Sematech Si oxynitrides; thickness values are given in angstroms, densities in grams per cubic centimeter. The plots indicate effects of varying thickness, interface (roughness/interdiffusion), and density.

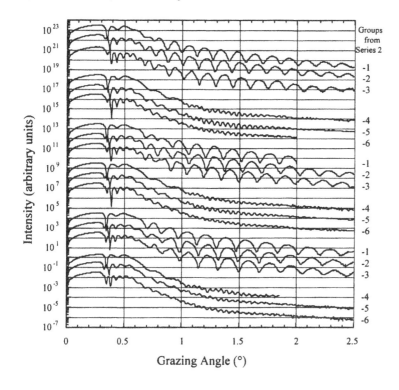

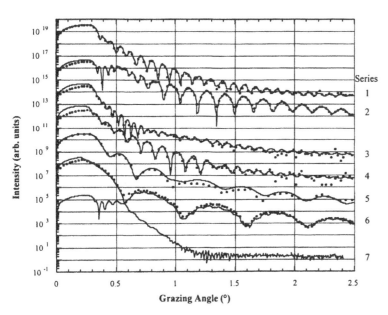

**Figure 11** X-ray reflectivity profiles of Sematech BST high-$K$ films: series with Pt precoat (top); modeled data from 7 series (bottom), most on $SiO_2$ directly.

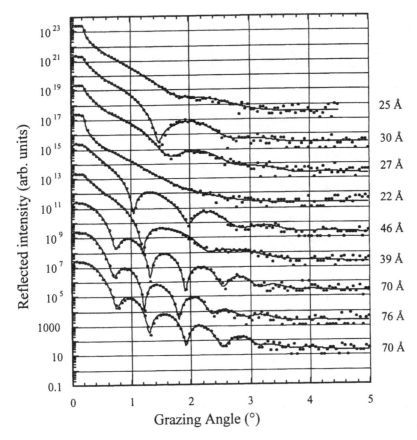

**Figure 12**  Experimental and calculated x-ray reflectivity profiles from Sematech $Ta_2O_5$ layers.

## D.  Extracting Structural Data from Specular Reflectivity Profiles

The process of going from the measured data to the extracted parameters, as illustrated in Figs. 7 through 12, requires some discussion to which we now turn. In general, there is prior knowledge of the material characteristics of the stack components. This prior knowledge is used to define an initial model stack made up of assumed component materials that are initially assigned densities corresponding to bulk values. It is convenient to apply a fast Fourier transform (FFT) to see what the predominant angular periods are. The resulting periods are not particularly accurate and do not correspond to component layer thicknesses. Instead, the observed periods correspond to the cumulative paths from the top surface to successive interfaces. From these pathlengths, it is easy to see that the component layer thicknesses emerge as pairwise differences between the observed angular frequencies, beginning with that corresponding to the largest path. Hence the reflectivity cycles are not uniformly spaced in angle but are periodic in the scattering vector $q_z = \lambda^{-1} \sin(\theta)$, except for significant corrections in the region of the critical angle.

Typically the determination of the structure of an object from its x-ray scattering signature involves (a) construction of a model that is believed to describe the structure, (b) calculation of the scattering profile anticipated from that structure, (c) the comparison of the calculated scattering profile with the experimental data, and (d) the modification of the original model. This process is repeated until the difference between the calculated and experimental data is judged to be sufficiently small by the practitioner; at this point the model is accepted to be a reasonable representation of the true material structure.

The problems associated with this process are obvious. The interpretation of x-ray reflectivity data is, in essence, a nonlinear curve-fitting problem, albeit one involving complicated functional forms and special constraints. This process is notoriously inefficient, since the experimenter often resorts to increasingly complex model structures (interfacial layers, composition or strain gradients, etc.) in order to rationalize the observed scattering profile. An additional difficulty is that the comparison is usually performed by visual inspection, with the input to the model adjusted according to the experience of the individual performing the analysis and the "quality" of the fit between model and experiment based on a subjective assessment. While often accurate, the visual method is dependent on the presentation of the data (for example, the use of a linear or logarithmic scale) and the expertise of the individual in deciding the relative importance of the various features in the pattern.

Although these processes may be "automated," that fact does not eliminate the need for good judgment, a basic understanding of materials, and an experienced practitioner. One goal of some of our future work, as described later, is to extract more guidance from the dataset itself by applying methods of wavelet analysis. For the present we have available several modeling packages, including a locally built toolbox, the IMD software,[*] and one commercial package.[†] In addition, each of the equipment manufacturers includes modeling and analysis software in the systems they supply. Naturally, the fact that these are "closed codes" means that at some stage validation procedures will need to be developed in order to meet the requirements of ISO-900X and Guide 25.

For any multivariate modeling procedure, the assignment of meaningful uncertainties to the model parameter values is a formidable task. Effective least squares programs do return estimated standard deviations for the derived parameters, but these are insufficient, for at least two reasons. First, in most cases the derived parameters do not have symmetric confidence levels, since most thickness values are more tightly bounded from below than from above. Second, these computationally generated statistical measures do not include estimates of systematic uncertainties.

Recently, the use of "evolutionary algorithms" has shown great promise for the fitting of x-ray reflectivity data (12). This approach achieves the evolution of parameter vectors by a repeated process of mutation, reproduction, and selection. With a surprising degree of computational efficiency, small random changes (mutation) in the population of parameter vectors generate a diversity in the population; selection guarantees that the "best-fitting" parameter vectors will propagate into future generations. Evolutionary algorithms appear to be one of the most efficient methods for the automated fitting of reflectivity profiles, and it is likely that their use in this regard will only increase in time.

---

[*] Developed by David Windt (Lucent).

[†] Bede Scientific Incorporated. Commercial items are identified in this report in order to specify the experimental procedure adequately. Such identification is not intended to imply recommendation or endorsement by the National Institute of Standards and Technology, nor is it intended to imply that the materials or equipment identified are necessarily the best available for the purpose.

## V. NON-SPECULAR SCATTERING, AN ESSENTIAL VALUE-ADDED MEASUREMENT

All that has been considered up to this point is scattering under the condition of equal incident and scattering angles, i.e. specular reflection. When the sample surface, internal interfaces, and/or the substrate are rough, appreciable scattering takes place at angles other than those corresponding to specular reflection. Conversely, in the absence of roughness, there is very little nonspecular scattering, although some will take place due to thermal diffuse scattering and to Compton scattering. More generally, the procedure is to make such scans for a large number of scattering angles, recording many curves such as those shown in Figure 13. Such an ensemble of nonspecular profiles can be disposed with respect to the scattering angle, as suggested by the simulation of a $WN_2/Si$ single-layer structure shown in Fig. 13. A surface can be associated with the profile envelope, or the data can be flattened into a plot of equal-intensity contours. These representations are generally referred to as *reciprocal space maps* (or *mappings*). Although it is time consuming to obtain such maps and analyze their content, there are cases in which this effort is justified by the valuable information they can be shown to contain. For a single rough interface, the spectral power density of the height–height correlation function can be derived over a region of spatial wavelengths that overlaps that of scanning probe microscopy on the high-spatial-frequency end and overlaps that of optical figure mapping in the low-spatial-frequency region. If more interfaces are involved, it is possible to distinguish correlated-roughness distributions from uncorrelated-roughness distributions.

On a more immediate level, simple measurements of the integrated diffuse scattering gives an approximate measure of the root mean square roughness. As mentioned in an

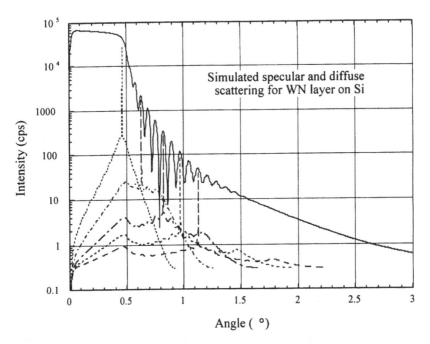

**Figure 13** Simulated specular (solid) and diffuse (dashed) scattering curves from a single-layer structure of $WN_2/Si$ with $d = 50$ nm, with interface parameters $\sigma(WN_2) = 1.5$ nm, $\sigma(Si) = 0.3$ nm.

earlier section, this roughness produces an effect on the specular reflectivity profile that is indistinguishable from the effect produced by a diffuse interface with approximately the same-scale parameter. The most important use of the diffuse scattering information is in its comparison with the roughness or diffusion parameter that emerges from analysis of the specular scattering data. If the diffuse scattering is small compared to what is needed to account for the loss of contrast seen in the specular scans, then there would appear to be diffuse interfaces. In practice, we find that the combination of roughness/diffusion parameter from specular reflectivity with the root mean square roughness from diffuse scattering suggests a reasonable picture of the film stacks so far investigated.

## VI. NEW WAYS OF LOOKING AT WHAT WE SEE—THE WAVELET APPROACH

Scattering phenomena are conventionally described in terms of a Fourier transform of the electron density (in the case of diffraction) or the index of refraction profile (in the case of reflectometry). In a typical Fourier synthesis, a function $f(x)$ that is periodic with period $2\pi$ can be expressed as the sum

$$f(x) = a_0 + \sum_{k=1}^{\infty} (a_k \cos kx + b_k \sin kx) \tag{9}$$

where the coefficients are given by (13)

$$a_0 = \frac{1}{2\pi} \int_0^{2\pi} f(x)\, dx; \quad a_k = \frac{1}{2\pi} \int_0^{2\pi} f(x) \cos(kx) dx; \quad b_k = \frac{1}{2\pi} \int_0^{2\pi} f(x) \sin(kx) dx$$

While Fourier transforms are widely used, the sine and cosine functions are "nonlocal"—that is, they extend to infinity. As a result, Fourier analysis often does a poor job in approximating data that is in the form of sharp spikes.

Wavelet transforms differ from the more familiar Fourier transforms principally because of their "compact support," i.e., the wavelet basis does not extend beyond the data array, while the Fourier basis lacks such localization (14). Other than that, they share the completeness and orthogonality properties of the Fourier transform. The closest similarity is to the moving-window Fourier transform, which has already been applied to reflectivity analysis in at least one instance (15).

Wavelet analysis also provides systematic denoising of low statistics data, unambiguous diagnostics of layer ordering, and useful visualization enhancement. Denoising, as shown in the example of Figure 14, represents an important application of wavelets, where the time-frequency representation becomes an angle-frequency mapping. The input data is decomposed according to the angular frequency partitions indicated by the ordinate tracks, beginning with the highest sampling rate at the bottom and proceeding to successively coarser partitions along the vertical axis. Evidently, the highest angular frequency bands are dominated by shot noise and contain no useful structural information since they have higher frequencies than that corresponding to the total sample thickness. The denoising operation consists in removing these frequency ranges before applying the inverse wavelet transform. Note that this is a decidedly nonlinear operation, very different from a smoothing operation.

The way in which layer ordering information emerges with wavelet analyses (in contrast with Fourier transforms) is evident by considering a simple example. In the

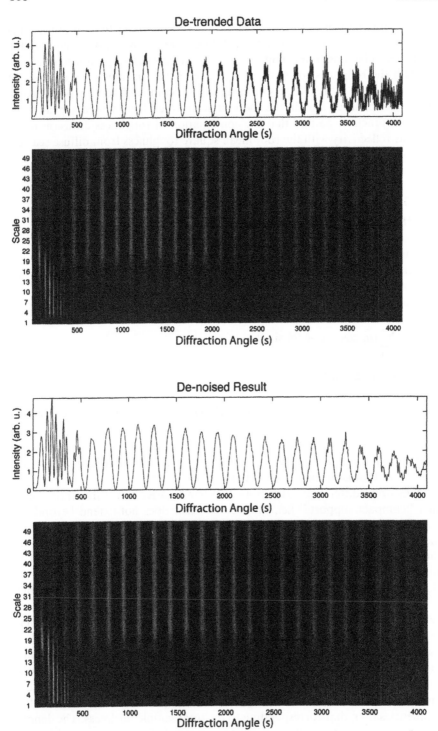

**Figure 14** Example of wavelet transformation and denoising of reflectivity data, as described in the text.

Fourier transform process one obtains very good values for the frequency parameters but no sense of where these frequencies lie along the angle scale. In the wavelet analysis, one exploits the fact that the x-ray penetration increases with increasing angle, meaning that the effects of layers near the surface are manifest at smaller angles than larger ones. This effect is particularly prominent for layers with diffuse boundaries such as were chosen for this didactic example.

## VII. CONCLUSIONS

Grazing incidence x-ray reflectivity is a useful tool for the structural characterization of thin film structures. It is nondestructive and relatively straightforward to implement, and the theoretical basis for x-ray scattering is very well founded. A wide variety of technologically important thin-film structures have been quantitatively analyzed by GIXR; the list is expected to grow with the implementation of new, complex thin-film materials systems. Advances in data-analysis methods (particularly through advanced modeling and fitting procedures and in the implementation of wavelet methods) ensure that GIXR will only grow in importance as an analytical tool.

## REFERENCES

1.  LG Parratt. Physical Rev. 95:359–369, 1954.
2.  A Segmüller, P Krishna, L Esaki. J. Appl. Cryst. 10:1–6, 1977.
3.  AJC Wilson, E Prince, eds. International Tables for Crystallography. Vol. C. Dordrecht, Netherlands: Klewer, 1999.
4.  H Kiessig. Ann. Phys. Leipzig. 10:769–778, 1931.
5.  DE Savage, J Kleiner, N Schimke, YH Phang, et al. J. Appl. Phys. 69:1411–1424, 1991.
6.  SK Sinha, EB Sirota, S Garoff, HB Stanley. Phys. Rev. B 38:2297–2311, 1988.
7.  J Daillant, K Quinn, C Gourier, F Rieutord, J. Chem. Soc. Faraday Trans. 92:505–513, 1996.
8.  DKG d Boer, AJG. Leenaers, WW vd Hoogenhof. X-Ray Spectrometry 24:91–102, 1995.
9.  M Tolan, W Press. Z. Kristallogr. 213:319–336, 1998.
10. H Zabel Appl. Phys. A 58:159–168, 1994.
11. T Remmel, M Schulbert, K Singh, S Fujimura, et al. Barium strontium titanate thin film analysis. Presented at Denver x-ray Conference, Advances in x-ray Analysis, Denver, CO, 1999.
12. M Wormington, C Panaccione, KM Matney, DK Bowen. Phil. Trans. R. Soc. Lond. A 357:2827–2848, 1999.
13. E Aboufadel, S Schlicker. Discovering Wavelets. New York: Wiley, 1999.
14. A Graps. IEEE Comput. Sci. Eng. 2:50–61, 1995.
15. R Smigiel, A Knoll, N Broll, A. Cornet. Materials Science Forum 278–281:170–176, 1998.

# 28
# Ion Beam Methods

**Wolf-Hartmut Schulte, Brett Busch, Eric Garfunkel, and Torgny Gustafsson**
*Rutgers University, Piscataway, New Jersey*

## I. INTRODUCTION

The continued scaling of microelectronic devices now requires gate dielectrics with an effective oxide thickness of 1–3 nm (1–6). This establishes new and very strict material requirements demanding advanced analysis techniques with high depth resolution. Atomic transport processes leading to the growth of such ultrathin dielectric films and the resulting growth kinetics are modified when compared to thicker films. The physical proximity of surface and interface changes the defect structures that determine atomic transport and electrical performance. New materials, such as silicon oxynitrides, are replacing silicon oxide as the gate dielectric in some applications, with new issues arising about the formation mechanisms of such films. Furthermore, deuterium ($^2$H) transport through oxide, nitride, and oxynitride films, and its incorporation near the interface, recently became a subject of intensive investigation, due to a new giant-isotope effect on transistor lifetime. Alternatives to replace silicon oxide by high-dielectric-constant materials, such as $HfO_2$, $ZrO_2$, $La_2O_3$, and $Al_2O_3$ or their silicates, are currently being investigated (7–11). Ion beams have been widely used to determine the thickness and composition of these thin films and to increase our knowledge of the physical and chemical processes that occur during their growth, in particular in the gate-thickness range where the classical Deal–Grove (12) model of oxidation breaks down.

The ion beam techniques described here can be separated into two different general approaches: ion scattering techniques, such as the well-known Rutherford backscattering spectrometry (RBS), and nuclear reaction analysis (NRA). Both approaches yield critical and complementary information that help researchers develop an atomic-scale picture of the structure and growth dynamics. For example, the total amount and the depth distribution of the atomic species contained in thin and ultrathin silicon oxide, nitride, and oxynitride films on silicon can be determined with high precision. When combined with isotope tracing techniques, both approaches also permit one to examine atomic transport in the films.

The strength of the ion beam techniques discussed here is that they are *quantitative*. Under appropriate circumstances, the measured signals can be converted directly to highly precise atomic concentrations of the species under study. Further, when performed with a suitable narrow nuclear resonance (in the case of NRA) or with a well-designed high-resolution spectrometer (in the case of RBS), the depth dependence of the concentration of the species can be determined with subnanometer resolution. The sensitivity of the techniques varies, but submonolayer quantities of the most important elements are usually easily achieved. However, because energetic ions are used, a dedicated accelerator is needed, and the techniques are not suitable for real-time monitoring of processing.

## II.  ION SCATTERING TECHNIQUES

In this review we concentrate on experimental techniques to examine thin films involving ion beams in the energy range above about 100 keV. We note in passing that below approximately 10 keV, low-energy ion scattering (LEIS) (13,14) can be applied to determine the surface composition of thin films. Little progress, however, has been made in applying this technique to quantitatively study thin-film growth, in part due to the nature of the scattering cross section and charge exchange effects in this energy range. Another more widely used technique for thin-film analysis is secondary ion mass spectroscopy (SIMS). This low-energy ion beam technique has superior detection limits for many isotopes compared to RBS and NRA. It provides very high sensitivity (in some cases on the order of 0.001 atomic %) and can be performed rapidly (15,16). However, for many of the applications discussed later, conventional SIMS techniques do not offer the depth resolution needed, due to matrix effects and ion beam mixing. There are several excellent reviews of SIMS methods (17,18)

Rutherford backscattering spectrometry is a well-established method to analyze the composition and elemental depth distributions of thin films (19). It allows an absolute quantification of the amount of elements contained in thin films. The sensitivity for the detection of heavy elements is in the range of $10^{13}$ atoms/cm$^2$ but is significantly higher for light elements. Depth profiles of elements can be obtained with a resolution of a few nanometers (depending on energy, detector, geometry, composition, etc.). The standard arrangement for RBS uses collimated 1–3-MeV He ions at a beam current in the 10 nA range to irradiate a sample (Figure 1). Due to Coulomb repulsion with the nuclei in the sample, the incident ions are elastically scattered. The energy of the backscattered projectiles (typically at scattering angles $\theta \sim 160°$) is measured by means of a silicon surface barrier detector. Information on the sample composition and its depth dependence can be obtained from the energy spectrum. In a variant of the technique, light elements (e.g., hydrogen) can be investigated by studying the recoil nuclei, which are detected at forward angles (elastic recoil detection—ERD).

## A.  Kinematics

An ion (mass $M_1$, energy $E_0$) scattered from a surface loses energy by elastic collision with a surface atom (Fig. 1). This energy $E_1$ of the scattered projectile is defined by the kinematic factor $K_M$, which is a function of sample atom mass $M_2$ and scattering angle $\theta$. This factor can easily be calculated using elementary classical momentum and energy conservation in a two-body collision (19):

$$K_M = \frac{E_1}{E_0} = \left[ \frac{\sqrt{M_2^2 - M_1^2 \sin^2 \theta} + M_1 \cos \theta}{M_2 + M_1} \right]^2 \tag{1}$$

Because $K_M$ is a unique function of target mass, this permits the conversion of the backscattering energy into a mass scale. The energy of the scattered particle increases with

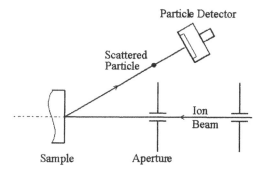

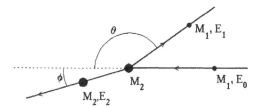

**Figure 1** Schematic of a standard RBS spectrometer and the Rutherford scattering process.

the mass of the sample atom. This dependence of the energy $E_1$ from the mass $M_2$ allows the identification of the sample atom ($M_2$). In samples containing heavy elements, with only slightly different $K_M$ values, signals from different elements may be difficult to separate, while the situation for lower-mass elements is easier. The mass resolution is maximum at $\theta = 180°$. This equation also states that scattering to backward angles $\theta > 90°$ is only possible for $M_2 > M_1$.

The energy $E_2$ transferred to the sample atom, which is scattered at an angle $\phi$, is given by (20):

$$\frac{E_2}{E_0} = 1 - K_M = \frac{4M_1M_2}{(M_1 + M_2)^2}\cos^2\phi \tag{2}$$

The cross section for a scattering process is described by the Rutherford formula. This formula allows a quantitative analysis of element distributions. The scattering cross section in the center of mass (CM) system in cgs units is given by:

$$\left(\frac{d\sigma}{d\Omega}\right)_{CM} = \left(\frac{Z_1Z_2e^2}{4E_{CM}}\right)^2 \frac{1}{\sin^4(\theta_{CM}/2)} \tag{3}$$

Here, $E_{CM}$ and $\theta_{CM}$ are, respectively, the projectile energy and the scattering angle in the center of mass system. For the laboratory system one finds (20):

$$\left(\frac{d\sigma}{d\Omega}\right)_{Lab} = \left(\frac{Z_1Z_2e^2}{4E_0}\right)^2 \frac{4}{\sin^4\theta} \frac{(\sqrt{1 - (M_1/M_2)^2\sin^2\theta} + \cos\theta)^2}{\sqrt{1 - (M_1/M_2)^2\sin^2\theta}} \tag{4}$$

Only projectiles scattered into a solid angle defined by the size of the particle detector and its distance from the sample are counted. The cross section averaged over the solid angle $\Omega$ is given by (20):

$$\sigma(\theta) = \frac{1}{\Omega} \int_{\Omega} \left(\frac{d\sigma}{d\Omega}\right) d\Omega \tag{5}$$

Thus the counting rate per incident projectile Y for a thin sample layer (thickness $t_d$) is given by:

$$Y = (n_p t_d)\varepsilon\sigma(\theta)\Omega \tag{6}$$

with $n_p$ being the density of the sample atoms and $\varepsilon$ the detection efficiency of the particle detector. For the commonly used silicon surface barrier detectors, $\varepsilon \approx 1$.

The incident (and scattered) ions lose energy when traveling through the solid, via collisions with the electrons in the target. Due to these electronic energy losses, scattering from deeper layers occurs at lower energies than the incident energy ($E_z < E_0$). This results in a larger cross section for subsurface layers. Since the energy dependence of the cross section is precisely known, this effect can easily be taken into account during analysis.

Due to the $Z^2$ dependence of the cross section, RBS is very sensitive for the detection of heavy elements, while detecting light elements, especially in a heavy-atom matrices, is more difficult. Small fractions of one atomic layer of heavy atoms, such as gold or tantalum, result in a reasonably strong RBS yield. Light elements, such as boron, give a counting rate more than two orders of magnitude lower, and their detection often suffers from the underlying signal resulting from scattering on heavy constituents located in deeper layers of the sample. For example, the stoichiometry of thin $SiO_2$ films can precisely be determined, whereas the spectrum from $Ta_2O_5$ films may be completely dominated by the intense Ta signal, not allowing a sensitive detection of oxygen. However, measurements in channeling and blocking geometry (see later) can significantly improve this situation.

The Rutherford cross section [Eq. (3)] is valid with high precision for a wide range of projectile energies and masses. However, in some cases corrections that involve either electron screening and/or nuclear reactions have to be considered:

> Due to tightly bound inner shell electrons, the Coulomb field of the sample nucleus is screened in the collision process, such that projectiles "see" an effectively reduced Coulomb potential. In the case of low projectile energies and high $Z_1$ or $Z_2$, this leads to a reduced cross section compared to that given by Eq. (3). The "electron screening effect" is important for RBS measurements (21–23). The experimental results concerning electron screening can often be described by the theory developed by Lenz and Jensen (24,25). For very low energies, further deviations occur and empirical correction terms (23,26,27) are used. With a good precision for light projectiles at energies as low 10 keV, the ratio of the screened Rutherford cross section and [Eq. (3)] is given by (27):

$$\frac{\left(\dfrac{d\sigma}{d\Omega}\right)_{\text{Scr}}}{\left(\dfrac{d\sigma}{d\Omega}\right)_{\text{CM}}} = \frac{1}{1 + V_1^{LJ}/E_{CM}} \tag{7}$$

with

$$V_1^{LJ} = 48.73 Z_1 Z_2 (Z_1^{2/3} + Z_2^{2/3})^{1/2} \text{eV} \tag{8}$$

As an example, one finds for the case of scattering 100-keV protons from Si that electron screening reduces the cross section by somewhat less than 5%. For precise measurements this effect clearly has to be taken into account, especially at subMeV projectile energies and for heavy sample nuclei.

Nuclear resonances may also be present for low-$Z_1$ projectiles at high projectile energies. During the scattering process, in this case, the projectile and target form a compound nucleus in an excited state. The compound nucleus can decay into different reaction products, including decay back to the projectile and sample nucleus. This process may result in drastic changes of the cross section and a strongly non-Rutherford angular distribution of the products. Since the cross sections of hydrogen and helium projectiles have been measured over a wide range of energies, this effect can be accounted for in RBS spectra simulation. The resonant increase of the cross section is frequently used to identify and quantify light elements in heavy-element matrices. Resonant scattering has been used, for example, by Foster et al. (28) to identify nitrogen in metals. To enhance the sensitivity of oxygen in RBS measurements for analysis of thin $SiO_2$ films, $^{16}O(\alpha,\alpha)^{16}O$ resonant scattering using 3.045-MeV $^4$He incident ions (29) is frequently used (30,31).

## B. Energy Resolution and Sensitivity

Besides determining absolute quantities of elements contained in thin films, RBS can be used to determine depth distributions of atomic species. The critical parameter in such applications is the energy resolution of the specific experimental setup, because it is directly related to the attainable depth resolution.

When passing through matter, projectiles lose energy due to interactions with the sample electrons such that at a given depth $z_0$, $E(z_0) < E$. Analogous energy losses occur on the projectile's way out of the sample. Thus, ions backscattered deeper in a film are detected with an energy $E - \Delta E$. By means of the differential energy loss ($dE/dz$) (also called the *stopping power*), $z_0$ can be determined. This is the crucial step for using the RBS technique for depth profiling applications.

The energy loss is due predominantly to the collision of projectiles with sample electrons. The energy loss due to collisions with sample nuclei can be neglected for light projectiles at energies above a few hundred kiloelectron-volts. At these energies, the "nuclear" energy loss contributes less than 1% to the total energy loss. However, it becomes significant or even dominant for low projectile energies in the kiloelectron-volt range and/or heavy projectiles. In contrast to collisions with nuclei, the collisions involving sample electrons do not contribute to a significant change in the ion path, due to the large mass difference between the ion and the electron.

Due to the large number of ion–electron interactions, the energy loss is assumed to be a quasi-continuous process and is described by an averaged differential energy loss ($dE/dz$) that itself is a function of energy, projectile, and sample. The statistical distribution of the energy losses is taken into account by the "energy straggling." A minimum energy is necessary to excite electrons in different electron shells; thus, deep inner-shell electrons do not contribute to the stopping power at low projectile energies. One finds that ($dE/dz$) increases with the projectile energy until reaching a maximum, and then slowly decreases at higher energies due to lower excitation probabilities. For He ions penetrating a $SiO_2$ film, the maximum of $dE/dz$ is at about $E = 600$ keV ($\approx 35$ eV/Å). In the energy range 400–1200 keV, the stopping-power value changes by less than 10%. For H ions, the maximum stopping power in $SiO_2$ is at $E = 100$ keV ($\approx 13$ eV/Å), changing by less than

10% in the energy interval 50 keV $< E <$ 150 keV. The depth resolution $\Delta d$ is (in a rough estimate) given by:

$$\Delta d = \frac{\delta E}{2}\left(\frac{dE}{dz}\right)^{-1} \tag{9}$$

with the energy resolution $\delta E$. Working with projectile energies near this maximum helps in obtaining optimal depth resolution.

A number of authors developed theoretical models to describe the energy loss; however, a rigorous description is not possible (32). Reviews may be found in Refs. 33 and 34. For converting measured energy losses into a depth scale, tabulated values or simulation codes for $dE/dz$ are usually used (35,36). It should be noted that although the stopping of megaelectron-volt He ions in Si and $SiO_2$ has been used for several decades to obtain RBS depth profiles, the precision of the stopping power (as well as straggling) values is not as good as desirable (37). The stopping-power cross section of a compound is generally obtained by adding the stopping-power values of the various constituents, taking corrections due to the influence of the chemical state into account (38–40). Recent work on the stopping of He ions in $SiO_2$ has been reported by Bauer et al. (41).

From the energy loss, the depth in the sample where the scattering process occurred can be calculated (19):

$$\Delta E = \int_z \frac{dE}{dz}(E)\, dz \tag{10}$$

Assuming the differential energy loss to be constant in the incoming and outgoing channels ("surface energy approximation"):

$$\Delta E = \left[\frac{K_M}{\cos\alpha_1}\frac{dE}{dz}\bigg|_{in} + \frac{1}{\cos\alpha_2}\frac{dE}{dz}\bigg|_{out}\right]z_0 = [S]z_0 \tag{11}$$

or:

$$z_0 = \frac{\Delta E}{[S]} \tag{12}$$

$\alpha_1$ and $\alpha_2$ take into account the rotation of the sample with respect to the beam axis and the sample-detector axis. In general $dE/dz$ for the incoming and for the outgoing channel are approximated by using averaged energies:

$$\overline{E_{in}} = E_0 - \frac{1}{4}\Delta E \qquad \text{and} \qquad \overline{E_{out}} = E_1 + \frac{1}{4}\Delta E \tag{13}$$

This is a good approximation if the energy losses are comparable for both channels.

Rutherford backscattering experiments are often performed in a channeling geometry, not only to study the structure and the lattice parameters of crystalline samples, but also to reduce the background counting rate from crystalline substrates. In the channeling geometry, the ion beam is aligned along a crystallographic direction of the target substrate. This helps to significantly reduce the background scattering from the substrate and to increase sensitivity to the overlayer. Aligning sample and detector position in such a way that the scattered projectile also travels along a crystallographic axis (so-called channeling/blocking or double alignment geometry) leads to a further reduction in the background.

A projectile traveling along a crystal axis suffers a smaller energy loss per unit distance relative to transport in a random direction (42). Furthermore, small-angle scattering processes enhance channeling of the ions through the crystal. This effect can be used to characterize crystal structures and defects (20). When depth-profiling samples, channeling may result in additional complications to the analysis. Experiments are sometimes performed in both random and channeling scattering geometries to obtain maximum information.

The statistical distribution of energy losses (straggling) limits the energy resolution and may for near-surface layers be even larger than the energy loss $E$. A well-established approximation for this energy straggling is based on Bohr's theory (43), which assumes a Gaussian distribution described by the straggling parameter (20):

$$\Omega_B^2 = 4\pi Z_1^2 e^4 n_e z \tag{14}$$

with $z$ being the path length in the sample and $n_e$ the electron density. The FWHM of this distribution is given by

$$\delta E_s = 2.355\Omega_B \sim \sqrt{z} \tag{15}$$

In this model the energy straggling is independent of the projectile energy and scales as $\sim z^{1/2}$. The calculated values are too large for low projectile energies, very thin films, and heavy sample atoms (44). A more detailed investigation shows deviations from the Bohr straggling. Several authors (e.g., Landau, Symon, Vavilov) developed models to describe realistic energy straggling distributions (44,45). All these models are characterized by an asymmetric shape of the energy loss distribution, resulting in different values for the most probable energy loss and the average energy loss. For example, Vavilov described the energy distribution by means of a Poisson distribution. In 1953 Lindhard and Scharff developed a model to describe the deviations from Bohr straggling (46). Simulations based on this model have often been used to interpret RBS data (especially for the energy straggling of protons) (47,48). Although there have been some experimental and theoretical studies on heavy-ion straggling, it still can not be reliably described (48).

Detailed studies of energy loss and straggling of H and He ions in carbon and silicon have recently been performed by Konac et al. (49). They found significant deviations from earlier experimental data (50), semiempirical results (48), and calculations (51) for projectile energies in the submegaelectron-volt range. Lennard et al. (37) showed that the stopping power data reported by Konac et al. gave good agreement in simulating RBS spectra from $SiO_2$ and $Si_3N_4$ samples obtained at He energies of 2 MeV, but $\sim$ 8% disagreement was found for 1-MeV He ions.

Although a complete theory that includes a precise description of energy losses and charge exchange processes is not available, computer codes have been developed to simulate RBS spectra (34). While the most often-used code is the RUMP program developed by Doolittle (52,53) other codes, such as SENRAS (54), RBX (55), GISA (56), GRACES (57), and DEPTH (58), should be mentioned. Climent-Font et al. (59,60) have investigated the agreement between experiment and simulation, particularly for RUMP. They found that the $^4$He stopping powers must be scaled in order to achieve agreement between simulation and experiment for several samples. A recent paper by Bianconi et al. (61) to define an RBS standard based on the plateau height of the RBS silicon signal in the near-surface region represents an important attempt to improve the present situation for Si.

The precision in which depth profiles of elements in thin films can be obtained in RBS is dependent primarily on the detector or spectrometer used to identify the scattered particles. Both depth resolution and mass separation scale with energy resolution. A silicon surface barrier detector (generally used in standard RBS instrumentation) with a typical energy resolution of 12–20 keV limits the depth resolution at perpendicular incidence ($\alpha_{1,2} = 0$) to 10–30 nm (19). For heavy projectiles the resolution decreases mainly due to the energy straggling in the contact layer of the detector. Although the stopping power of heavy projectiles in the sample can be significantly larger than that of light projectiles, the depth resolution of RBS cannot be improved when using surface barrier detectors. In the case of ERD measurements, the surface barrier detector has to be protected from the intense flux of the primary beam by means of a thin film in the beam path, unless it is used in combination with an electrostatic or magnetic spectrometer (62). This thin foil contributes to additional energy loss and straggling and a reduced depth resolution.

Independent of the type of detector used, the depth resolution of RBS can be improved by rotating the sample perpendicular axis to an angle $\alpha$ with respect to the incident ($\alpha_1$) and/or exit ($\alpha_2$) beam axis. This permits a longer path length for the ions in the sample. Compared to perpendicular incident/exiting beam, the path length is increased by $f_\alpha^{-1}/\cos\alpha_{1,2}$. By rotating the sample out of the scattering plane, both the path in and the path out of the sample can be increased simultaneously in an optimized way. However, the energy straggling increases with increasing $\alpha_{1,2}$, such that the depth resolution for deeper layers cannot be improved drastically. Even assuming an absolutely flat sample surface, the increasing number of small-angle scattering events (multiple scattering) limits the maximum $\alpha_{1,2}$ values to about 80° ($f_\alpha \approx 6$). Using a surface barrier detector, Barradas et al. (63) obtained a depth resolution $\Delta z_0 \approx 1$ nm at $\alpha_{1,2} = 84°$ ($f_a \approx 10$) for near-surface layers.

Besides the scattering geometry, various factors may limit the achievable depth resolution in RBS measurements:

*Energy spread of the ion beam*: The energy spread $\delta E_b$ of a 1-MeV He ion beam is typically in the range of 1 keV. It increases with the projectile energy. The resulting energy distribution can generally be described by a Gaussian shape and has to be added to the energy straggling: $\delta E_t^2 = \delta E_b^2 + \delta E_S^2$.

*Kinematic broadening*: Due to the finite diameter of the ion beam and the solid angle of the detector, scattered projectiles are detected at different angles. Due to the angular dependence of $K_M$ this results in a kinematic broadening of the spectrum (64).

*Different ion trajectories*: In grazing incidence geometry the divergence of the ion beam trajectories can result in different energy losses in the sample.

*Multiple scattering*: Especially for deeper layers or at grazing incidence, multiple scattering processes may occur. In these cases the direct relation between the detected energy and the scattering angle is not valid (65).

*Charge dependence of the energy loss*: The energy loss depends on the charge state of the projectile. The charge state of the projectile is changed when penetrating through the sample and approaches an average value depending on the projectile energy. For heavy projectiles this effect leads to energy losses that are difficult to calculate. Special care has to be taken in ERD experiments using heavy ions (66,67).

*Doppler broadening*: The thermal vibrations of the sample atoms result in an additional energy broadening of the scattering spectra. For light ions this effect contributes only a small amount to the energy spread (68). However, when detecting light elements using heavy projectiles this effect may become significant.

*Discrete energy losses*: For near-surface layers and ultrathin films with thickness < 1 nm investigated with high-resolution ion beams, the assumption of continuous slowing of the projectile is not valid. Instead, a concept taking into account details of single energy-loss processes has to be used that allows the description of the Lewis effect (68–71).

*Sample damage*: Especially when identifying trace elements in samples or when using magnetic or electrostatic spectrometers, the high ion dose necessary to obtain spectra can, if left unattended, introduce significant radiation damage in the sample. This results in modification of the depth distribution of the elements (ion beam mixing) and sputtering of the sample (53,72,73). Local heating of the sample can lead to additional diffusion of elements. Finally, the accumulation of ions in the sample may also lead to blistering effects (used beneficially in the smart-cut process). Deleterious effects from radiation damage can be overcome by averaging data from many different beam spots on the sample.

## C. Applications of Standard Rutherford Backscattering Spectrometry Techniques

For analyzing thin dielectric films, RBS techniques using silicon surface barrier detectors to detect scattered He ions are extremely useful to determine the total amounts of elements contained in the films. This information can then be used to determine the chemical composition and (when knowing the density) the thickness of the corresponding films. Thus it is not surprising that "standard" RBS has become a major analytical tool in many industrial and university laboratories. The knowledge of the chemical composition and the sample thickness is used predominantly to support other analytical approaches. However, in some specific cases RBS analysis by itself has been used to provide important information on the interface structure of thin films.

The interface and the near interface region of $Si/SiO_2$ and related films have received intensive scrutiny for several decades. In addition to suboxides, which may be considered essentially a compositional transition, structural transition layers are formed on both sides of the oxide/Si interface in order to relax stresses. As an example for such investigations involving the RBS technique, the work of Jackman et al. (74) will be discussed. In this work, He ions in the energy range 0.8–1.0 MeV were used. Figure 2 illustrates the results obtained with high-energy ion scattering in a channeling geometry, which allowed the determination of the Si and O areal densities in the oxide ($N_{Si}$ and $N_O$). An extrapolation to $N_O = 0$ in a $N_{Si}$ vs. $N_O$ plot (slope = 2 corresponds to an ideal $SiO_2$ stoichiometry) indicates that in the near-interface region there is a zone with excess of Si, corresponding either to strain in the crystalline substrate or excess Si in the $SiO_2$ transition region. The thickness of the $SiO_2$ transition region was determined (74–76) to be approximately 0.2–0.4 nm for conventional, heated furnace oxidation, depending on the substrate orientation.

The ERD technique has frequently been used to profile depth distributions of hydrogen isotopes in thin films (77,78). Habraken et al. used this technique to study the out-diffusion of hydrogen from silicon oxynitride films (79). Different mechanisms describing the reactivity and diffusion were discussed. However, their depth resolution was limited, due to use of a surface barrier detector. Thus these studies could not be extended to investigations of interfacial hydrogen in ultrathin films.

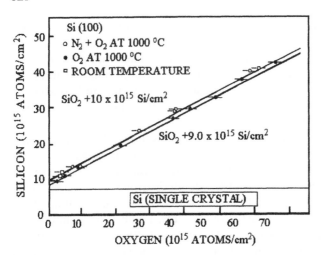

**Figure 2** Measured areal densities of Si and O for oxides grown on Si(100) under different conditions. The "visible" Si from the substrate is represented by the darkened area. The straight line fit to the data assumes stoichiometric $SiO_2$ and indicates slightly different values in excess Si for different growth parameters. (Adapted from Ref. 74.)

## D. Use of Magnetic Spectrometers

In RBS and ERD, different techniques have been developed to detect scattered particles with improved energy resolution when compared to surface barrier detectors. Magnetic spectrometers, electrostatic spectrometers, and time-of-flight techniques have all been applied to improve the depth resolution.

In nuclear physics the momentum of reaction products is frequently measured using high-resolution magnetic spectrometers. These spectrometers are designed mostly for ions in the energy range above 500 keV with both high momentum and energy resolution. In most cases double-focusing 90° sector magnets are used. Due to the homogeneous magnetic field, the ions describe orbits inside the magnet (deflection radius $r$). By means of entrance and exit slits a radius can be defined such that only particles in a certain energy range can reach the detector. Due to the entrance and exit fringe fields of the sector magnet, ions with different divergences are focused onto the exit slit. The energy resolution $\Delta E$ is determined mainly by the slit widths $\Delta r$: $\Delta E/E = 2\Delta r/r$.

However, the use of these spectrometers in materials analysis is not widespread. This is due partly to their size (deflection radius of 0.5–1 m) and large weight (high magnetic field = several tons). Furthermore, they are much more expensive than surface barrier detectors. Due to the narrow entrance and exit slits necessary to obtain a high resolution, the sensitivity of magnetic spectrometers is relatively small. To obtain a spectrum of scattered particles over a wide range of energies, the magnetic field has to be varied over a wide range. This results in a high ion dose to record a spectrum and in radiation damage problems. Arnoldbik et al. (80) achieved an energy resolution of 1.4 keV for scattering of He at 2.135 MeV. Magnetic spectrometers have been built to increase the solid angle and to detect a wide range of energies simultaneously. In this case, the exit slit is replaced by a position-sensitive detector. By optimizing the focusing properties of the different ion optical lenses and correction elements, relatively good energy resolution can be achieved for different ion trajectories.

Dollinger et al. (66,67) used the Q3D magnetic spectrograph at the University of München for materials analysis, in particular for the characterization of ultrathin oxynitrides (81). This spectrometer consists of three sector magnets and a multipole deflector placed between the first and the second sector magnets. Due to the large solid angle of 5–12 msr, it allows one to detect a wide range of energies. An energy resolution of $7 \times 10^{-4}$ has been obtained. This spectrometer is used for ERD studies, where, in contrast to the standard RBS setup, no foils in the beam path are necessary to protect the detector from the intense flux of the primary beam. Scattering a 60-MeV $^{58}$Ni beam at grazing incidence on a graphite crystal a depth resolution of about 3 Å has been obtained. Depth profiles of nitrogen and oxygen were obtained using an incident 80-MeV $^{197}$Au beam at a scattering angle of 15° (81). The 5+ (6+) charge state of the nitrogen (oxygen) recoil ion was analyzed with a large solid angle of 7.5 msr, resulting in an estimated sensitivity of about $10^{13}$ atoms/cm$^2$. Besides total amounts of oxygen ($1.0 \pm 0.1 \times 10^{16}$ atoms/cm$^2$) and nitrogen ($8.8 \pm 1.0 \times 10^{14}$ atoms/cm$^2$), surface contaminants such as hydrogen ($2.5 \pm 0.8 \times 10^{15}$ atoms/cm$^2$) and carbon ($2.5 \pm 0.8 \times 10^{15}$ atoms/cm$^2$) were identified in a thin oxynitride film. High-resolution ERD depth profiles of oxygen and nitrogen are shown in Figure 3. The results indicate that the nitrogen is situated between the oxide and the silicon substrate. The maximum of the layer coincides with the tail of the oxygen profile, and ends at a similar depth as the oxygen profile. Both the slope of the oxygen profile at the surface and that at the interface are significantly wider than the experimental resolution. The interfacial width may also be influenced by thickness variations of the layer. However, the quantitative analysis of the ERD spectra is complicated due to the different charge states of the recoiled ions (67). Another problem is the dependence of the differential energy loss on the charge state as discussed by Sigmund (34), accompanied by imprecisely known stopping-power values (82). The radiation damage during analysis—for example, hydrogen, nitrogen, and oxygen loss from silicon oxynitrides—also has to be carefully considered (72,73).

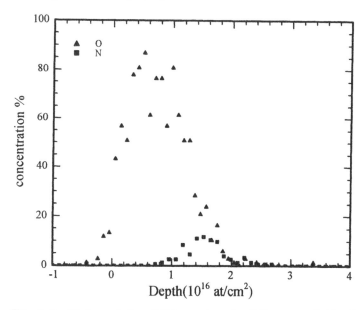

**Figure 3** High-resolution ERD spectra of a 2.5-nm oxynitride. Only the oxygen and nitrogen profiles are shown, where $1.3 \times 10^{16}$ atoms/cm$^2$ corresponds to a thickness of about 2 nm. (From Ref. 81.)

A similar spectrometer for He ions at lower projectile energies has been described by Lanford et al. (83). It consists of two sector magnets and is optimized for scattering studies using ions of a few megaelectron-volts (max. 5 MeV for He$^{2+}$ ions). An energy resolution of $\delta E < 1$ keV is postulated.

Kimura et al. (84,85) developed a high-resolution RBS system using 300-keV incident He ions and a 90° magnetic sector magnet. They used this spectrometer with He ions scattered at a PbSe(111) surface in grazing incidence geometry to demonstrate a depth resolution of about one monolayer (85). This system also has been used for high-resolution depth profiling in thin oxynitride films (81). In order to reduce the signal from the silicon substrate, the experiments were performed in a channeling geometry. The detection of nitrogen is hampered by the small scattering cross section and the large Si background signal. The observed nitrogen profile was rather weak and noisy, but it was peaked at the interface between the SiO$_2$ and the silicon substrate, in agreement with the other techniques applied. Again the observed interface width ($\approx 1.5$ nm) was larger than the depth resolution of the magnetic-sector RBS experiment, which was estimated to be 0.8 nm at the interface using the formula of Lindhard and Scharff (46, 86).

## E.  Use of Electrostatic Spectrometers

Only a few reports have been published on the application of electrostatic spectrometers to improve the depth resolution of conventional RBS using projectiles in the megaelectron-volt range. Carstanjen (87) used a cylindrical-sector field analyzer (deflection radius 0.7 m, angle 100°) in combination with an electrostatic lens system to increase the solid angle and to correct ion optical errors. This system is designed for ions at energies of up to 2 MeV. For He ions at 1 MeV a resolution of 1.44 keV was obtained, corresponding to monolayer depth resolution when performing the experiments in tilted-sample geometry. A similar depth resolution has been obtained when detecting recoiled deuterons and protons with the same spectrometer. In an investigation of charge exchange phenomena in ion scattering, charge-state-dependent energy spectra of ions were studied using Ar scattering on Au samples. It was found that the spectra using backscattered $1+$ charged ions showed a flat distribution near the surface region, while spectra of higher charge states ($> 3$) exhibit pronounced peaks. This behavior has to be understood when evaluating monolayer resolution RBS and ERD spectra.

A very different situation occurs at lower ion beam energies (below 500 keV). Here, intense efforts have been made to develop high-resolution elastic scattering methods using an ultrahigh vacuum electrostatic spectrometer (88,89). This technique has become a valuable and more widely used tool for materials analysis and will now be discussed in more detail.

Since the stopping power of light-impinging particles (i.e., H and He ions) is relatively small in the megaelectron-volt range, energy shifts of the backscattered peaks due to energy loss of the scattered ions are small, and the depth resolution is therefore poor. This situation, however, can be improved significantly when working at lower energies (typically 100–200 keV, H or He ions). The abbreviation MEIS (medium-energy ion scattering) is used to describe RBS measurements in this energy range (89). A collimated, monoenergetic ion beam is accelerated toward a sample under ultrahigh vacuum, and the backscattered fraction is energy analyzed and detected. In this energy range the energy loss is near its maximum for protons or $\alpha$ particles in solids, improving depth resolution. For example, the stopping power ($dE/dz$) for $\sim$ 100-keV protons is 13 eV/Å for Si, 12 eV/Å for

$SiO_2$, 18 eV/Å for $Ta_2O_5$, and 20 eV/Å for $Si_3N_4$. Even more important, the lower ion energies mean that the backscattered ion energies can be measured with an electrostatic energy analyzer ($\delta E/E \sim 0.1\%$, i.e., $\delta E \sim 100$ eV for 100-keV ions), greatly increasing the energy resolution (and therefore the depth resolution) relative to silicon surface barrier detectors. Neglecting the effects of energy straggling for a moment, atoms located at a depth $d$ below the surface of a $SiO_2$ film may then be profiled with a depth resolution $\Delta d$:

$$\Delta d = \frac{\delta E}{f_d \dfrac{dE}{dx}} \tag{16}$$

When assuming the typical travel length $l$ ($f_d d$) of the projectiles in the sample to be $l = 3d$ (due to angular factors), and taking into account the value of the stopping power, a depth resolution of about 3 Å can be obtained in the near-surface region. As an example, it has been demonstrated that 0.5-nm accuracy in nitrogen depth profiles can be achieved near the surface or for ultrathin films (90).

The instrumentation used in MEIS is sketched in Figure 4. It consists of an ion beam source and an accelerator, a transport line with beam defining, focusing, and steering components (Fig. 4a), a goniometer for sample orientation, an electrostatic toroidal energy analyzer, and a position-sensitive detector (Fig. 4b). Data-acquisition and -processing units complete the system. The detector energy resolution is such that isotopes of light nuclei can be separated. As in standard RBS, detailed information about atomic depth distributions can be gained through spectral simulations, taking into account energy loss and straggling. And MEIS allows one to simultaneously monitor absolute concentrations and depth profiles of N, O, Si and other elements in dielectric films (91). The areas under the MEIS peaks are proportional to the total amounts of the elements in the probed region of the sample, and the shape of each peak contains information about the depth distribution of the corresponding element. When investigating thin amorphous structures on crystalline substrates such as Si, MEIS experiments are usually performed in a channeling and blocking (double-alignment) geometry. The reason for this is that the channeling effect hides the Si atoms in the substrate from the ion beam, so backscattering from these is drastically reduced. The (weak) signal from the lighter atoms then becomes easy to study over the (much-reduced) background. The small solid angle for the detection of the backscattered nuclei requires long data-acquisition times. Thus, high ion doses are required to obtain satisfactory statistics. In practice, because the beam size is small, the position of the beam spot hitting the sample is changed frequently during the measurement in order to minimize radiation damage. Since the detection limit is proportional to the square of the atomic number, it is very high for heavy elements and decreases with atomic number. As a benchmark for a light element, the detection limit of nitrogen in $SiO_2$ is about $2–3 \times 10^{14}$ atoms/$cm^2$, depending on film thickness and the width of the nitrogen distribution in the film.

Such MEIS systems have been set up at different laboratories to study surface and thin-film structures and compositions. In contrast to most other research groups that use commercially obtained MEIS systems (92–94), Yamashita et al. developed an electrostatic analyzer that allowed a good depth resolution in tilted-sample geometry. This spectrometer has been used to investigate copper overlayers on Si(111) (95). The spectrograph built by Kido et al. employs a very efficient detection system (96).

Very light elements such as hydrogen and boron can also be detected using MEIS instrumentation in the elastic recoil configuration, as has been demonstrated (93). In order to suppress the intense signal resulting from scattered projectiles, Copel and Tromp generated a pulsed ion beam by periodic deflection. Taking advantage of the difference in

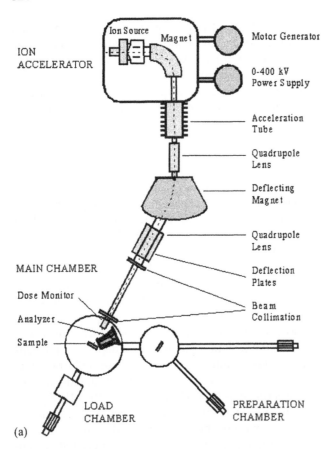

**Figure 4** Block diagram of (a) MEIS system and (b) high-resolution electrostatic detector.

flight times of scattered projectiles and recoiled ions, the data acquisition was triggered only for detection of recoil ions. A depth resolution of $\sim 10$ Å for hydrogen and boron was obtained. The total hydrogen coverage of HF Si surfaces could be measured using an incident 220-keV Li$^+$ beam. As an example for the depth-profiling capabilities, Figure 5 shows hydrogen signals obtained from an amorphous Si overlayer on HF-etched Si(001). Since the experiment was performed in channeling geometry, the thickness of the Si overlayer could be monitored by the signal from the Si surface peak (Figure 5b). The hydrogen signal becomes broader with the thickness of the Si overlayer (Figure 5a).

The MEIS technique has been used extensively to study the growth mechanism of thin $SiO_2$ films on Si (91,92,97) using an incident proton beam at energies of about 100 keV. Figure 6 shows MEIS oxygen spectra obtained in a channeling/blocking geometry for a thin oxide on Si(100) taken at two different scattering angles. The simulation of the data with the O profile, shown by the solid lines, fits both experimental curves, demonstrating the reliability of the modeling (98). In MEIS investigations special benefit has been taken from the capability to distinguish between proton scattering from $^{16}O$ and from $^{18}O$ atoms in thin-film structures, thus allowing the use of isotope-labeling techniques (Figure 7) (98). Besides clarifying the picture of $SiO_2$ growth in the ultrathin film regime (where the classical Deal–Grove model of oxidation breaks down), the oxygen surface exchange and interface structure to the Si substrate have been investigated in detail. These studies are complemented by the investigations involving nuclear reactions discussed later.

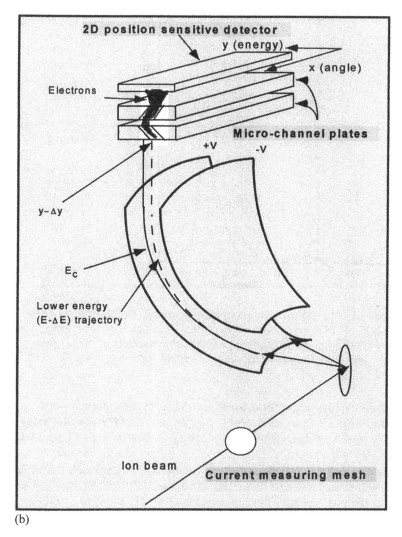

(b)

**Figure 4**   (*continued*)

Figure 8 shows oxygen MEIS spectra before (solid symbols) and after (open symbols) reoxidation of initial 45-Å $Si^{16}O_2/Si$ film in $^{18}O_2$. $^{18}O$ incorporation at/near the interface and at the outer oxide surface is observed. The loss of $^{16}O$ at the surface is seen. The $^{16}O$ peaks becomes broader on the lower-proton-energy side after the reoxidation, indicative of the isotopic mixing at the interface (99).

The thickness of the transition region $SiO_2/Si$ was also determined by MEIS (92,94,97,99). We note that the region between the oxide and the silicon substrate was found not to be sharp. Although the thickness of the (compositional) transition region estimated from this MEIS experiment appears to be about 1 nm, the real compositional transition region width cannot yet be accurately determined (to the angstrom level, as one would like). Roughness, strain, and density fluctuations (in both Si substrate and $SiO_2$ overlayer) and uncertainties in straggling for ultrathin films complicate this determination (94).

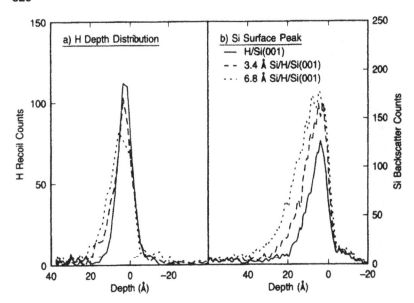

**Figure 5** Hydrogen signals (a) obtained from an amorphous Si overlayer on HF-etched Si(001), demonstrating the capability of MEIS to detect light nuclei while maintaining good depth resolution. Since the experiment was performed in the channeling geometry, the thickness of the Si overlayer could be monitored by the signal from the Si surface peak (b). (From Ref. 93.)

At the Korean Research Institute of Standards and Science, the compositional and structural changes in the transition layer of the $SiO_2$–Si(100) interface for thermal oxides was studied (94,100). The results of their channeling/blocking studies indicated crystalline Si lattices in this layer.

The growth mechanism and composition of silicon oxides and oxynitrides, including the various atomic reaction processes, have been studied using MEIS (90,91,101). As an example, Figure 9 shows the nitrogen and oxygen sections of a MEIS spectrum taken in

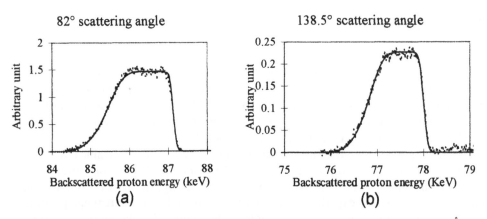

**Figure 6** MEIS oxygen spectra obtained in the channeling/blocking geometry for a 45-Å oxide on Si(100) taken at two different scattering angles, demonstrating the increased depth resolution for increased travel length. The simulation of the data (solid lines) fits well the two experimental curves, which renders the modeling reliable. (From Ref. 98.)

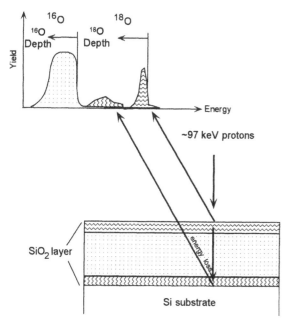

**Figure 7** Schematic MEIS spectrum for protons backscattering from oxygen in a thin $SiO_2$ film (top). The darkened areas in the spectrum correspond to $^{18}O$-enriched layers at the sample surface and at the interface to the Si substrate (as shown at the bottom). Due to energy-straggling effects, the $^{18}O$ peak from the interface layer is broadened. (From Ref. 98.)

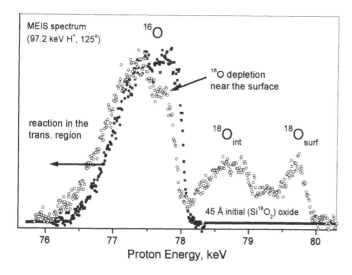

**Figure 8** Oxygen MEIS spectra before (solid symbols) and after (open symbols) reoxidation of initial 45-Å $Si^{16}O_2$/Si film in $^{18}O_2$. $^{18}O$ incorporation at/near the interface and at the outer oxide surface is observed. The loss of $^{16}O$ at the surface is seen. The $^{16}O$ peaks becomes broader on the lower-proton-energy side after the reoxidation, indicative of isotopic mixing at the interface. (From Ref. 99.)

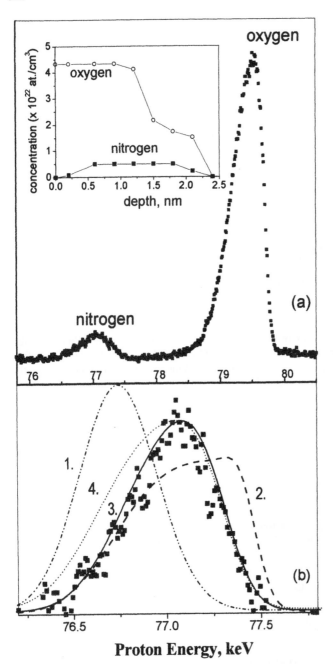

**Figure 9** The upper panel shows a MEIS spectrum for a thin (∼ 2 nm) oxynitride grown on Si(100) in a furnace at 950°C for 60 min in NO. The scattering angle is 125°. The insert shows the corresponding nitrogen and oxygen profiles (as a function of distance from the oxide surface). The lower panel shows a close-up of the spectrum in the nitrogen region, along with the results of simulations for different nitrogen distributions: (1) all nitrogen, $8.8 \times 10^{14}$ atoms/cm², is located at the SiO₂/Si interface; (2) all nitrogen is uniformly distributed in the film; (3) nitrogen profile shown in the inset (best fit); and (4) nitrogen profiles same as in the inset extended 0.3 nm into the substrate. (From Ref. 90.)

double alignment for an ultrathin (2.2-nm) silicon oxynitride film grown on Si(001) (90). To each element, there corresponds a peak, the lighter nitrogen ($8.8 \pm 0.7 \times 10^{14}$ atoms/cm$^2$) appearing at lower backscattering energies.

## F. Time-of-Flight Techniques

Time-of-flight (TOF) techniques are widely applied in nuclear physics and chemistry (102,103). They are based on a determination of the velocity of atoms, ions, and molecules. In materials analysis this method has only recently been applied to RBS as a higher-resolution alternative to surface barrier detectors. Projectiles backscattered off the sample at a certain scattering angle are detected using an arrangement of channel-plate detectors. Two detectors, placed at the beginning and end of a defined length, are used to produce timing signals such that the velocity of projectiles can be determined from their time of flight (Figure 10). The "start" detector consists of a thin foil from which an impinging ion produces secondary electrons, which are detected by a channel-plate detector. The ion penetrates the foil and travels to the "stop" detector. The time delay between the two signals is measured. The energy resolution, and accordingly the attainable depth resolution, is significantly limited by energy straggling and thickness inhomogeneity in the "start" foil (104); time resolution of the electronics and particle detection are additional parameters deteriorating the resolution.

Four time-of-flight RBS systems that have recently been used to investigate the composition of thin films will briefly be discussed next.

Mendenhall and Weller (105–107) used a time-of-flight spectrometer in combination with H$^+$, He$^+$, Li$^+$, and C$^{2+}$ ions ($E < 500$ keV) for RBS and ERD investigations. They obtained a time resolution of about 1 ns, corresponding to a depth resolution of about 2 nm. Weller et al. (105) demonstrated the potential of using this spectrometer to investigate the structure and composition of gate oxides, taking advantage of the large stopping power for light medium energy

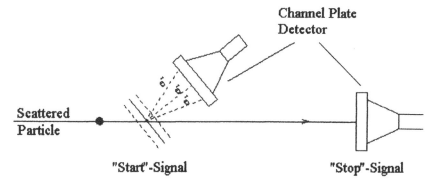

**Figure 10** Schematic drawing of a time-of-flight RBS setup. Two channel-plate detectors, placed at the beginning and end of a defined length, are used to produce timing signals such that the velocity of projectiles can be determined. The "start" detector consists of a thin foil from which a transmitting ion produces secondary electrons; the electrons are counted by means of a channel-plate detector. The ion penetrates the foil and travels to the "stop" detector. The time interval between "start" and "stop" signals is measured.

projectiles. They used 270-keV He ions (beam current 30–40 nA) to investigate $Si_3N_4$ and $SiO_2$ multilayer films in the 10-nm thickness range.

TOF-RBS was applied to quantify oxygen and nitrogen in ultrathin oxynitrides and to determine their depth profiles. Experiments were performed at the ETH Zürich with a 1-MeV $^4$He beam, with the sample normal tilted by 82° with respect to the incoming beam and at a backscattering angle of 175°. The data revealed an $SiO_2/Si$ interface more than 1 nm thick. Poor statistics prevented a reliable assignment of the location of N atoms (81,108).

Knapp et al. (109) used a heavy-ion backscattering spectrometer at medium energy projectile energies, taking advantage of the large cross sections in heavy-ion backscattering for trace element analysis. Their time-of-flight spectrometer was designed to increase the counting rate. The associated energy resolution, however, is not as good as in other RBS techniques.

Interference between the foil and the ions and the related unwanted energy loss and straggling can be avoided by exploiting the time structure of a pulsed ion beam to extract the "start" signal. Due to the low duty cycle inherent in the use of pulsed beams, the data-accumulation times increase significantly (110–112). Presently this technique is not used in routine materials analysis.

## III. NUCLEAR REACTION ANALYSIS

Nuclear reaction techniques have been used for many years in materials science and have proven to be reliable and accurate for determining the total amount of elements/isotopes in thin films. Special techniques have been developed to extend the use of nuclear reactions for depth profiling with a resolution reaching the sub-nanometer range for certain isotopes. For studies of thin dielectric films on silicon, two different methods can be distinguished:

Charged-particle-induced nuclear reactions with energy-independent cross sections to measure total amounts of isotopes in thin films (Sec. III.A)

Resonance enhancement in proton-induced nuclear reactions to measure isotope depth profiles (nuclear resonance profiling, NRP; Sec. III.B).

### A. Determination of Total Amounts of Isotopes in Thin Films

To determine the total amount of one isotope in a thin film, one prefers a nuclear reaction with a cross section that is large (giving a high counting rate) and independent of the projectile energy over the thickness of the film. Although many projectiles and ion energies can in principle be exploited, the most frequently used reactions are proton and deuteron beam induced at energies ranging from 0.5 to 2 MeV. This is due mainly to the availability of the projectile at many accelerator facilities. It is of course particularly attractive that these projectiles allow the detection of two or more species at the same time. The cross sections of these reactions have been measured by different groups under well-defined experimental conditions with good precision and have been standardized for specific projectile energies and experimental geometries. The number of atoms of the specific isotope in the film can easily and precisely be obtained, either by comparing the reaction yield to a reference sample or by a calculation using the cross section, the number of incident projectiles, angular acceptance of the detection system, and its efficiency. This direct proportionality between the measured yield and the number of isotopes in the sample makes nuclear reaction analysis (NRA) unique in its potential to precisely quantify light isotopes in thin films. In general the precision of NRA is not hampered by matrix effects or the chemical environment. In contrast to RBS, light elements are identified in a heavy substrate without introducing severe background problems. The isotope-specific sensitivity can be as low as $10^{12}$ atoms/cm$^2$, with a precision in the 1% range. The repulsive Coulomb interaction between the projectile and the sample nucleus limits NRA for this energy range to light sample nuclei (for example, nitrogen or oxygen). Excellent review articles describing the microanalysis of light elements $1 \leq Z \leq 10$ can be found in Ref. 113. Due to the large cross section of these reactions, sputter damage during analysis can generally be neglected. However, in the case of hydrogen isotopes (e.g., for diffusion studies), radiation-induced loss of sample atoms from the surface has to be considered (see later). Table 1 gives an overview of some frequently used nuclear reactions. Some NRA studies to determine total amounts of carbon, nitrogen, oxygen, and hydrogen isotopes are discussed in the following paragraphs.

Hydrogen is present in most thin- and ultrathin-film structures. Significant concentrations are found even in thermal oxides grown on silicon in dry, nominally pure oxygen. The relatively high background level of hydrogen-containing molecules, such as $H_2$, $H_2O$, $Si_2H_6$, $NH_3$, and HF, in Si semiconductor processing ambients (5) and the high mobility of hydrogen introduce a significant amount of hydrogen into devices. Investigations of the electrically active dangling bonds at the oxide/Si interface have shown that hydrogen seems to bind to dangling bonds at and near the oxide–Si interface, passivating them (75,114). This phenomenon motivated an additional processing step, annealing in forming gas (a mixture of $H_2$ and $N_2$) to passivate electrical interface defects. Under MOS device operation conditions, some electrons may obtain sufficient energy (hot electrons) to break the Si–H bonds, releasing hydrogen and degrading the device electrical characteristics. These processes (among many others) show the importance of quantifying the hydrogen content in thin-film structures and at their interfaces. Since it is desirable to differentiate between the hydrogen in the film and the hydrogen in the outer surface layer (arising from sample exposure to ambient), information on the total amount of hydrogen is not particularly helpful. Only in combination with depth information does the hydrogen signal becomes meaningful. A very sensitive nuclear reaction for detecting hydrogen is the nuclear resonance reaction $^{15}N + p$ (Sec.III.B).

**Table 1**  Nuclear Reactions and Resonances Used for Characterizing Thin Films

| Isotope | Nuclear reaction | Projectile energy (keV) | Sensitivity (atoms/cm$^2$) | Remark |
|---------|------------------|--------------------------|-----------------------------|--------|
| H | $H(^{15}N, \alpha\gamma)^{12}C$ | 6400 | $\sim 10^{13}$ | Resonance |
| D | $D(^3He,p)^4He$ | 700–900 | $\sim 10^{12}$ | |
| $^{12}C$ | $^{12}C\,(d,p)^{13}C$ | 800–1000 | $\sim 10^{13}$ | |
| $^{14}N$ | $^{14}N(d,\alpha)^{12}C$ | 800–1100 | $\sim 5 \times 10^{13}$ | |
| $^{15}N$ | $^{15}N(p,\alpha\gamma)^{12}C$ | 900–1000 | $\sim 10^{12}$ | |
| $^{16}O$ | $^{16}O(d,p)^{17}O$ | 800–900 | $\sim 10^{14}$ | |
| $^{18}O$ | $^{18}O(p,\alpha)^{15}N$ | 700–800 | $\sim 10^{12}$ | |
| $^{18}O$ | $^{18}O(p,\alpha)^{15}N$ | 151 | $\sim 10^{13}$ | Resonance |
| $^{29}Si$ | $^{29}Si(p,\gamma)^{30}P$ | 324 | $\sim 5 \times 10^{13}$ | Resonance |
| | | 417 | $\sim 10^{13}$ | Resonance |

Due to its chemical similarity to hydrogen and its low natural abundance (hence low background levels in processing environments), it is very convenient to use deuterium as a "marker" to elucidate basic chemical and physical properties of hydrogen in thin films. Recently it has been demonstrated that replacing hydrogen with deuterium during forming-gas anneal at 450°C reduces hot-electron degradation effects in MOS transistors by factors of 10–50 (115). Following this observation, deuterium has been examined in ultrathin $SiO_2$ films using NRA techniques (116). Because the hydrogen (deuterium) solubility in crystalline silicon is extremely low (117), all measured deuterium is in the dielectric film or at its interfaces. This assumption has been experimentally verified by depth profiling using HF etching. To determine the etch rate, the total amount of oxygen in the film was measured after each etching step with the $^{16}O(d,p)^{17}O$ reaction at 730 keV. For measuring the total deuterium amount in the samples, the reaction $d(^3He,p)^4He$ (at $E = 700$ keV) was used. Figure 11 shows the experimental setup and a corresponding proton spectrum (116). A mylar foil (13 μm) inserted in front of the surface barrier detector stops the backscattered $^3He$ ions and the $\sim$ 2-MeV α-particles produced by the $d(^3He,p)^4He$ reaction: only the 13-MeV protons can reach the detector.

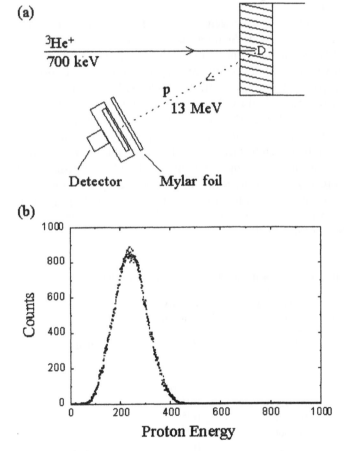

**Figure 11** Experimental setup for determining total amounts of deuterium in thin films using the $(^3He,p)$ nuclear reaction (a). A proton spectrum obtained at a $^3He$ projectile energy of 700 keV is shown in (b).

The use of the ($^3$He,p) nuclear reaction at $E = 700$ keV brought three important consequences. First, since there is no other nuclear reaction induced in the samples, the selectivity to deuterium of the analysis is ensured. Second, the background of the proton spectra is essentially zero. Last, since the cross section of the d($^3$He,p)$^4$He nuclear reaction presents a broad peak at $E \approx 700$ keV (FWHM > 200 keV), and only thin deuterium containing SiO$_2$ films were examined, the cross section may be considered constant within the thickness of the film. All these facts lead to a sensitivity of $10^{12}$ atoms/cm$^2$ and an accuracy in the determination of the total amounts of D in the films of better than 5% (by calibrating against a deuterated silicon nitride reference sample) (116). The SiO$_2$ film samples loaded with deuterium presented a high areal density ($6 \times 10^{14}$ atoms/cm$^2$), almost three orders of magnitude above the solubility limit of in bulk silica, consistent with previous measurements using thicker films. Figure 12 shows the total amount of deuterium left in a 5.5-nm-thick deuterated SiO$_2$ film after vacuum-annealing at different temperatures. One can see that the amount of deuterium drops below the sensitivity limit ($10^{12}$ atoms/cm$^2$) after annealing at 600°C for 20 min (116,118).

The nuclear reaction $^{12}$C(d,p)$^{13}$C at a projectile energy of about 1 MeV is routinely used for the detection of carbon (119), with a detection limit of about $10^{13}$ atoms/cm$^2$. Carbon is frequently found in thin-film structures or on sample surfaces as a contaminant, and care should be taken to distinguish between these two cases. In some cases, simple chemical cleaning or flash of the sample to 200–300°C is sufficient to remove surface hydrocarbon impurities.

$^{14}$N isotopes can be detected with a detection limit of about $10^{14}$ atoms/cm$^2$ using the $^{14}$N(d,$\alpha$)$^{12}$C reaction at projectile energies around 1 MeV (120–122). Alternatively, the reaction $^{14}$N(d,p)$^{15}$N may be used (122,123). A sensitive reaction for detecting $^{15}$N isotopes is the nuclear resonance in the $^{15}$N(p,$\alpha$)$^{12}$C reaction. This reaction is most frequently used for nitrogen analysis because of its large cross section (122,124). The resonance at a projectile energy of 429 keV provides information not only on the total amount but also on the depth distribution of nitrogen atoms. Due to its very large cross section, this resonance can in some cases even be applied to measure the nitrogen distribution in

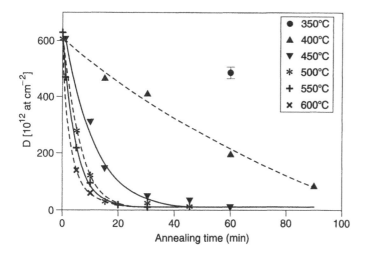

**Figure 12**  Total amount of deuterium left in a 5.5-nm-thick deuterated SiO$_2$ film after vacuum-anneal at different temperatures. The lines through the data points are only to guide the eye. (From Ref. 116.)

thin films that are not enriched in $^{15}$N (see later). Alternatively, the $^{15}$N(p,$\alpha\gamma$)$^{12}$C reaction can be used at a projectile energy of about 750 keV, where a flat plateau is exhibited in the cross-sectional curve, to measure total amounts of nitrogen.

The standard reaction for measuring the total amount of $^{16}$O in thin films is the $^{16}$O(d,p)$^{17}$O reaction at a projectile energy below 1 MeV (29). Figure 13 shows the energy spectrum of the detected protons obtained using deuterons as the incident ion impinging on a silicon oxide thin film, under scattering and sample conditions shown in the inset. Since the same projectile is used, carbon may be detected simultaneously. Figure 14 shows the O content in silicon oxide films, measured by NRA, as a function of growth time in ultradry $O_2$ with different levels of water vapor content on Si(001), Si(110), and Si(111) (125). The oxide film thickness can be approximately estimated from the measured oxygen areal density, using the equivalent thickness based on the bulk density of SiO$_2$ (2.21 g/cm$^3$): $1 \times 10^{15}$ atoms/cm$^2$ = 0.226 nm of SiO$_2$. For the driest gas (LIII), the initial stages of growth follow a slower regime, whereas for the gas with higher content of water vapor (LI) — still not higher than a few ppm — a much faster growth regime is established at the initial stages. The results shown in Fig. 14 demonstrate the important role played by water vapor in growth kinetics (125). Gosset et al. (126) used the reactions $^{16}$O(d,p)$^{17}$O at 810 keV, $^{18}$O(d,$\alpha$)$^{15}$N at 730 keV, and $^{15}$N(p,$\alpha\gamma$)$^{12}$C at 1 MeV and NRP techniques to characterize the composition and to study the depth profiles in isotopically ($^{15}$N, $^{18}$O) enriched ultrathin silicon oxynitrides.

In the case of determining the areal density of $^{18}$O in silicon oxide films, the nuclear reaction $^{18}$O(p,$\alpha$)$^{15}$N induced by protons at about 760-keV is adequate. At $E \approx 760$ keV, the cross section shows a flat plateau, indicating a useful energy region for nuclear reaction

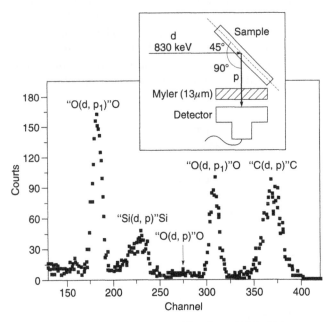

**Figure 13** Proton spectrum resulting from 830-keV deuterons incident on a silicon oxide film. The areal density of $^{16}$O in the sample can be determined from the area under the peaks corresponding to the nuclear reaction $^{16}$O(d,p)$^{17}$O. Carbon constituents or contaminants can be quantified simultaneously by analyzing the counting rate from the $^{12}$C(d,p)$^{13}$C reaction. A sketch of the experimental arrangement designed to maximize the proton yield of these reactions is shown in the inset. (From Ref. 5.)

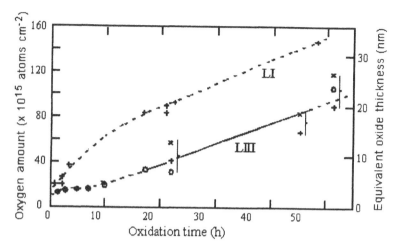

**Figure 14** Oxygen content in silicon oxide films (grown at 13.8 mbar, 930°C in dry oxygen) measured by nuclear reaction analysis (NRA) as a function of growth time. LI denotes a higher water pressure level compared to LIII. Samples with three different substrate orientations (+ (100), x (110), o (111)) were investigated. The dashed and solid lines are for guidance only. (From Ref. 125.)

analysis. A spectrum of detected α-particles obtained by 760-keV incident protons on a silicon oxide film rich in $^{18}O$ is shown in Figure 15, spectrum A. The same proton beam can induce the $^{15}N(p,\alpha\gamma)^{12}C$ reaction with the $^{15}N$ atoms also present in the sample. In Figure 15, spectrum B corresponds to the α particles obtained by 760-keV protons incident on a silicon nitride film rich in $^{15}N$ ($^{15}N$ standard). This spectrum appears at higher energies, well resolved from the α particles originating from the $^{18}O(p,\alpha)^{15}N$ reaction. Spectrum C demonstrates the presence of both $^{18}O$ and $^{15}N$ species in a thin oxynitride film. Stedile et al. investigated the thickness of the $SiO_2/Si$ interface and the composition of thin silicon films depending on the preoxidation cleaning procedure for rapid thermally grown oxides (RTOs) using RBS and nuclear reaction techniques (127).

The total amount of silicon isotopes can be detected using $(d,\gamma)$ reactions at projectile energies around 1 MeV (128). The resulting γ-ray spectra also contain information on the oxygen and carbon content. Due to the substrate silicon, such measurements become meaningful in combination with Si-isotope–enriched layers.

Nuclear reaction analysis has superior sensitivity over scattering techniques in detecting light trace elements. Examples are the detection of fluorine and boron. Due to the low-scattering cross sections and the overlapping yield due to scattering from silicon atoms, small amounts of either element are not easily detected in RBS. For example, NRA can be used to examine fluorine contamination from sample cleaning procedures using the $^{19}F(p,\alpha\gamma)^{16}O$ reaction. Boron diffusion in a film can be studied using the $^{11}B(p,\alpha)^{8}B$ nuclear reaction.

## B. Narrow Nuclear Resonance Profiling Technique for High-Depth-Resolution Determination of Composition

Isolated nuclear resonances are characterized by a large cross section in a narrow energy window around a well-defined projectile energy (resonance energy $E_R$). The cross section $\sigma(E_p)$ of the nuclear reaction around this resonance energy is described by a Breit–Wigner

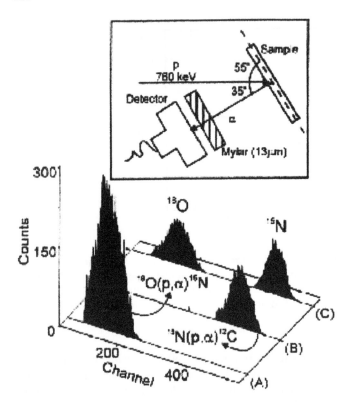

**Figure 15** α Particle spectra obtained with the nuclear reactions $^{18}O(p,\alpha)^{15}N$ and $^{15}N(p,\alpha\gamma)^{12}C$ at 760 keV from (A) a $Si^{18}O_2$ standard, (B) a $Si_3^{15}N_4$ standard, and (C) a typical silicon oxynitride sample. The geometry employed in this analysis is represented in the inset. (From Ref. 5.)

shape (FWHM = Γ), and may be many orders of magnitude larger than the nonresonant cross section (129):

$$\sigma(E_p) \propto \frac{1}{(E_p - E_R)^2 + (\Gamma/2)^2} \tag{17}$$

High-resolution nuclear resonance profiling (NRP) is based on the fact that certain nuclear reaction resonances can have an intrinsic width that can be as small as a few electron-volts. The principle is illustrated in Figure 16. If a sample is bombarded with monoenergetic projectiles with $E_p < E_R$, no reaction products are detected. At $E_p = E_R$ the reaction product yield is proportional to the probed nuclide concentration in a thin layer near the surface of the sample. If the beam energy is raised, the yield from the surface vanishes, and the resonance energy is reached only after the beam has lost part of its energy by inelastic interactions with the sample. Then the reaction yield is proportional to the nuclide concentration at that depth. The higher the energy of the beam above the reso-nance energy, the deeper is the probed region located in the sample. In essence, the high depth resolution is a consequence of the narrow resonance acting as a high-resolution energy filter on a beam that is continuously losing energy as it travels through a sample. Interpretation of the data under the stochastic theory of energy loss in matter can lead to a depth resolution of about 1 nm, with sensitivities of about $10^{13}$ atoms/cm$^2$.

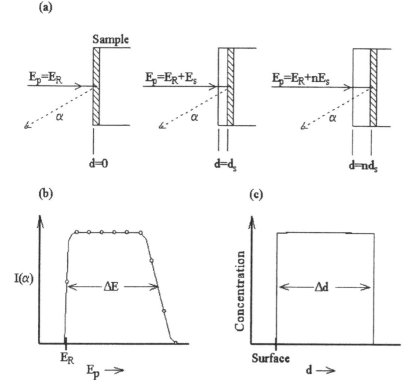

**Figure 16** Principle of nuclear resonance depth profiling discussed for the example of an α particle emitting narrow resonance. At different incident projectile energies $E_p \geq E_R$ the resonance reaction excites nuclei at a different depth in the sample (a). The yield curve, i.e., the number of detected α particles as a function of the projectile energy (b), can be converted to the concentration of the specific isotope as a function of depth using the stopping power and straggling data (c).

To determine depth profiles of isotopes in thin films, techniques involving such narrow resonances at low projectile energies (i.e., below 500 keV) have been developed during the past few years. This energy region is preferable to the more fully explored megaelectron-volt range, due to the low proton stopping power in the that range. In contrast to RBS, the depth resolution of NRP experiments is determined not by the energy resolution of the detection system but ultimately by the intrinsic resonance width $\Gamma$. Some resonances in proton-induced nuclear reactions at projectile energies in the range 150–500 keV have a resonance width $\Gamma \approx 1$ eV. Taking into account a stopping power of typically > 10 eV/Å for $SiO_2$, and neglecting the energy straggling for the moment, a nominal depth resolution of less than 1 Å could be expected. Although the detector used for detecting the reaction products does not influence the resolution, other parameters play an important role: the ion beam energy width (for the ion-implantation type of accelerators typically in the range of 50 eV and larger) and the thermal movement of the sample atoms (leading to an energy spread of about 50 eV). Thus the depth resolution of such experiments is limited to several angstroms for near-surface layers. As in the case of RBS, the depth resolution also deteriorates with the analyzing depth due to straggling. Analogous to ion scattering experiments, the depth resolution in NRP can be improved by tilting the sample with respect to the ion beam direction.

A key aspect of NRP is that the reaction product yield versus beam energy curve (called the excitation curve) is not the probed nuclide concentration profile itself but is related to it by the integral transform (5)

$$I(E_p) = \int_0^\infty C(x)q_0(x; E_p)\, dx \tag{18}$$

in which $I(E_p)$ is the number of detected reaction products as a function of beam energy (excitation curve); $C(x)$ is the nuclide concentration at depth $x$; and $q_0(x; E_p)\, dx$ is the energy loss spectrum at depth $x$ (the probability for an incoming particle to produce a detected event in the vicinity $dx$ of $x$ for unit concentration). The width of the energy loss spectrum defines the depth resolution and depends on the resonance shape and width, the ion beam energy spread, the incoming ion energy loss, and straggling processes. In the case of light projectiles, straggling dominates the resolution. The stochastic theory of energy loss as implemented in the SPACES program has been used to calculate $q_0(x; E_p)dx$ and accurately interpret experimental data (69). Nuclide concentration profiles are assigned on an iterative basis, by successive calculation of a theoretical excitation curve for a guessed profile followed by comparison with experimental data. The process is repeated with different profiles until satisfactory agreement is achieved.

As in the case of nonresonant nuclear reactions, one advantage in applying nuclear reactions is their isotope sensitivity. The isotope of interest in the thin film can be detected essentially independent of all other isotopes. Problems due to the high scattering cross sections of heavy atoms are not present. This isotope sensitivity, however, can also be contemplated as a disadvantage, because one nuclear resonance provides only depth profiles of one specific isotope. When combining NRP and isotope labeling, valuable information on atomic transport processes can be obtained (see Figure 17).

Due to the nuclear-level schemes and the Coulomb barrier, narrow resonances at low projectile energies with sufficient strength enabling high-resolution NRP are present in only a limited number of collision systems. For materials and surface science applications, proton beams at energies ranging from 150 to about 500 keV are used to excite narrow (1–150 eV wide) resonances on $^{15}N$, $^{18}O$, $^{21}Ne$, $^{23}Na$, $^{24,25,26}Mg$, $^{27}Al$, $^{28,29,30}Si$. Proton-induced resonances in this energy range are also available to detect $^{11}B$ and $^{19}F$ isotopes. However, these resonances are more than 2 keV wide and do not allow high-resolution depth profiling.

The cross section of nuclear reactions at the relevant projectiles energies is much (in many cases, orders of magnitude) smaller than the elastic scattering (Rutherford) cross section, which can lead to severe radiation damage problems in NRP. Radiation damage leads to losses of light elements from the sample due to sputtering, blistering of the sample, and enhanced diffusion processes. Thus the ion dose necessary to perform an analysis has to be kept as small as possible, demanding a high detection efficiency in collecting the reaction products. But even with improved detection systems (such as large solid-angle particle detectors or a $4\pi\gamma$-ray spectrometer (131)), the sensitivity of high-resolution nuclear resonance profiling remains lower than that of many scattering techniques. An exception is the strong resonance in the reaction $^{15}N + p$, which allows very sensitive detection of $^{15}N$ isotopes (or using the reaction in inverse kinematics, depth profiling of hydrogen by $^{15}N$ in thin films). It should be noted that due to the high stability of the $^{16}O$ nucleus, no resonance for high-resolution NRP exists. Thus high-resolution NRP of oxygen has to involve the rare isotope $^{18}O$.

In the following paragraphs a few examples of NRP profiling of nitrogen, oxygen, silicon, and hydrogen in thin dielectric films are discussed.

(a) Thermal oxidation in $^{16}O_2$ :

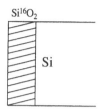

(b) Subsequent oxidation in $^{18}O_2$ :

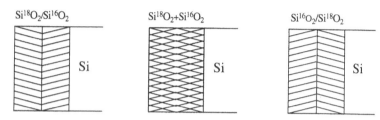

(c) Corresponding $^{18}O$-Profiles :

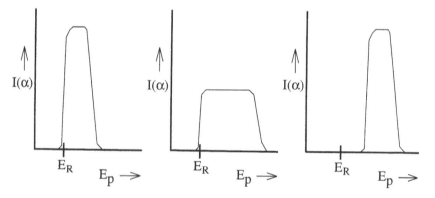

**Figure 17** The principle of isotope tracing is discussed for the thermal oxidation of silicon. A first oxidation in $^{16}O_2$ results in a thin oxide layer (a). Subsequent oxidation in $^{18}O_2$ may result in different $^{18}O$ depth distributions in the oxide film. A few possible distributions are sketched in (b). These depth distributions result in different yield curves of a $^{18}O(p,\alpha)^{15}N$ resonance reaction (c).

The reaction $^{15}N + p$ exhibits a strong resonance at $E_p$ of 429 keV. This resonance has a width $\Gamma \approx 120$ eV. The easily detected $\approx 4.4$-MeV $\gamma$-rays emitted from this reaction and the low (five orders of magnitude smaller) off-resonant cross section permit one even to measure nitrogen depth profiles in films that are not isotopically enriched. Due to the large cross section of this resonance reaction, it is one of the few cases where the sensitivity of low-energy nuclear resonance depth profiling is superior to the scattering techniques. The large cross section combined with the isotope selectivity allows one to detect $^{15}N$ amounts of less than $10^{13}$ atoms/cm$^2$. A depth resolution of about 1 nm can be achieved when investigating near-surface layers. This resonance has frequently been applied to characterize nitrogen in thin silicon nitrides (132) and silicon oxynitride films, for example Refs. 126,133 and 134. As an example to demonstrate the good sensitivity and depth resolution, Figure 18a shows the yield of this resonance obtained from an ultrathin oxynitride film processed using $^{15}$N-enriched NO gas (101). The inset shows the corresponding depth profile of $^{15}$N isotope calculated using the SPACES code. One can conclude that the nitrogen in this oxynitride film is dominantly located near the interface to the Si substrate.

The reaction of protons with $^{18}$O exhibits an $\alpha$ particle emitting resonance at $E_p = 151$ keV (135,136). The width $\Gamma \approx 120$ eV makes this resonance favorable to study

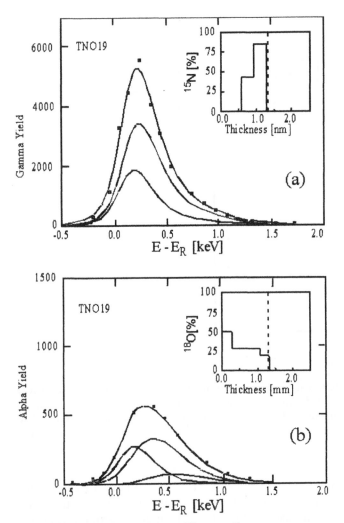

**Figure 18** Yield of the 429-keV $^{15}$N(p,$\alpha\gamma$)$^{12}$C resonance obtained from an ultrathin oxynitride film processed using $^{15}$N-enriched NO gas (a). The inset shows the corresponding depth profile of $^{15}$N isotope calculated using the SPACES code. The nitrogen in this oxynitride film was predominantly located near the interface to the Si substrate. The depth resolution that can be obtained when using the 151-keV resonance in the $^{18}$O(p,$\alpha$)$^{15}$N reaction is demonstrated in (b): the yield of this resonance obtained from an ultra-thin oxynitride film processed using $^{18}$O enriched NO gas is shown. The inset shows the corresponding depth profile of $^{18}$O isotope. The oxygen concentration decreases with depth, inverse to the depth profile of nitrogen, which is dominantly located near the Si–substrate interface (a). (From Ref. 142.)

the depth distribution of oxygen in isotope tracing experiments. The megaelectron-volt-α particles can be detected with large solid-angle silicon surface barrier detectors with almost no background counting rate. It has been applied by several research groups to study the atomic transport processes taking place in thin $SiO_2$ films, for example, Refs. 137–140, and silicon oxynitrides, for example, Refs. 126, 133, and 134. All these experiments were performed using isotope-labeling methods. The depth resolution is about 1 nm for near-surface layers. This is by far superior to the resolution achieved when applying resonances at higher energies. For example, earlier isotope tracing studies used the resonance in $^{18}O$ + p at $E_R = 629$ keV ($\Gamma \approx 2100$ eV), resulting in a depth resolution of $\sim 10$ nm (141). However, the strength of the 629-keV resonance is much larger compared to the 151-keV resonance, thus problems due to radiation damage are significantly smaller. The depth resolution that can be obtained when using the 151-keV resonance is demonstrated in Fig. 18b. It shows the yield of this resonance obtained from an ultrathin oxynitride film processed using $^{18}O$-enriched NO gas (142). The inset shows the corresponding depth profile of $^{18}O$ isotope calculated using the SPACES code. It is shown that the oxygen concentration decreases with depth. This is complementary to the depth profile of nitrogen in the same sample, where the nitrogen is predominantly located near the interface to the Si substrate (Fig. 18a).

Transport during oxidation to form thin silicon dioxide films has mostly been studied indirectly by tracing the oxygen atoms (143). Low-energy resonances in the reaction $^{29}Si$ + p (at $E_R = 324$ keV and at $E_R = 416$ keV) also allow one to directly trace the movement of silicon isotopes with a depth resolution of about 1 nm. Silicon-NRP combined with isotopic labeling has recently been used to study the mobility of silicon atoms during thermal nitridation (144) and oxidation of silicon (130,145,146). Figure 19 shows a $^{29}Si$ excitation curve of the 416-keV $^{29}Si(p,\gamma)^{30}P$ resonance obtained from an epitaxial $^{29}Si$-enriched layer deposited on Si substrate after oxidation in oxygen gas. The simulation of the $^{29}Si$ depth distribution shows that the $^{29}Si$ stays at the surface after oxidation (i.e., is

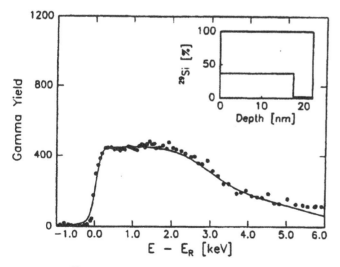

**Figure 19** $^{29}Si$ excitation curve of the 429-keV resonance in $^{29}Si(p,\gamma)^{30}P$ obtained from an epitaxial $^{29}Si$-enriched layer deposited on a Si-substrate oxidized in dry oxygen gas. The simulation of the $^{29}Si$ depth distribution (inset) shows that the $^{29}Si$ stays at the surface after oxidation, i.e., is not buried under the newly growing oxide. (From Ref. 130.)

not buried under the newly growing oxide) (130). These measurements gave the first direct evidence that Si atoms do not diffuse through SiO$_2$ during thermal growth of ultrathin oxides.

The very strong resonance in the $^{15}$N + p reaction has a sharp width and has been applied in the inverse kinematic configuration to study hydrogen in thin films (147). In inverse kinematics a $^{15}$N projectile of 6.4 MeV is needed to excite the resonance reaction. The dynamics of H bonding and dissociation in silicon oxide films on Si has been intensively researched for over 20 years. Although transport properties of hydrogen play an important role in film growth dynamics, it has been far less studied than oxygen, because of the experimental difficulties in hydrogen detection and profiling. Recently, studies of the depth distribution of H in thin oxide films on Si have been performed using this resonance (148,149). Krauser et al. found the concentration of hydrogen at the interface of thicker (95 nm) silicon oxide films to be as high as $10^{15}$ atoms/cm$^2$, well above the solubility limit (149). Hydrogen depth profiles in silicon oxide films on Si are shown in Figure 20. Apart from a surface peak due to surface H adsorption in air, in all cases the H peak concentrations are found within the oxide, approximately 2–3 nm away from the oxide/Si interface. It should be noted that in experiments involving heavy ion beams, significant losses of hydrogen from the surface and accumulation of hydrogen in deeper layers may occur (148). Due to its strength and narrow width, the $^1$H($^{15}$N,$\alpha\gamma$)$^{12}$C resonance at 6.4 MeV can also be used to characterize the hydrogen bonds on the surface. Such measurements use the broadening of the yield curve of the resonance due to the thermal movement of the hydrogen atoms to define the bonding character (Doppler spectroscopy) (150). Corresponding measurements have been performed by Zinke-Allmang et al. (151), Iwata et al. (152), and Jans et al. (153). This method has recently been extended to study hydrogen on Si(111) surfaces (154). Special care has to be taken in these extremely sensitive experiments to obtain reliable results (see, for example, Ref. 148).

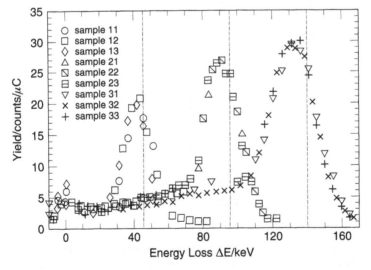

**Figure 20** Excitation curves of the $^1$H($^{15}$N,$\alpha\gamma$)$^{12}$C reaction around the resonance at 6.40 MeV in a SiO$_2$ film on Si that was loaded in H$_2$ atmosphere. (From Ref. 149.)

## IV.  SUMMARY

The ion scattering and nuclear reaction techniques discussed in this chapter have proven to be powerful tools to quantitatively determine the total amount of isotopes contained in thin films. In some cases, depth profiles of atomic species may be obtained with near monolayer resolution. This section compares the different techniques with respect to their ability to analyze the major constituents of ultrathin silicon oxides and oxynitrides (in the 1–5-nm thickness range). This discussion provides only a guideline to evaluate the potential of the different techniques. The parameters given in Table 2 depend strongly on the characteristics of the individual experimental setup. Recent developments—for exam-

**Table 2**  Detection and Depth Profiling of Elements in Ultrathin Silicon Oxide and Oxynitride Films Using Different Ion Beam Techniques.

| Element/Isotope | Technique | Sensitivity (atoms/cm$^2$) | Depth resolution (Å) |
|---|---|---|---|
| Hydrogen/H | ERD (MEIS) | $2 \times 10^{13}$ | 5–20 |
| | ERD $^a$ | $\sim 10^{13}$ | $\sim 10$ |
| | NRA (NRP) | $\sim 10^{13}$ | $\sim 30$ |
| Hydrogen/D | ERD (MEIS) | $2 \times 10^{13}$ | 5–20 |
| | ERD $^a$ | $\sim 10^{13}$ | $\sim 10$ |
| | NRA | $< 10^{12}$ | —$^b$ |
| Nitrogen/$^{14}$N | RBS | $\sim 10^{14}$ | — |
| | High-res. RBS $^c$ | $> 10^{14}$ | $\sim 5$ |
| | MEIS | $5 \times 10^{13}$ | 3–8 |
| | ERD $^a$ | $\sim 10^{14}$ | 2–5 |
| | NRA | $\sim 5 \times 10^{13}$ | —$^b$ |
| Nitrogen/$^{15}$N | High-res. RBS | $> 10^{14}$ | $\sim 5$ |
| | MEIS | $5 \times 10^{13}$ | 3–8 |
| | ERD $^a$ | $\sim 10^{14}$ | 2–5 |
| | NRA (NRP) | $\sim 10^{13}$ | $\sim 10$ |
| Oxygen/$^{16}$O | RBS | $\sim 10^{14}$ | — |
| | High-res. RBS | $> 10^{14}$ | $\sim 5$ |
| | MEIS | $2 \times 10^{13}$ | 3–8 |
| | ERD $^a$ | $\sim 10^{14}$ | 2–5 |
| | NRA | $\sim 10^{13}$ | —$^b$ |
| Oxygen/$^{18}$O | RBS | $\sim 10^{14}$ | — |
| | High-res. RBS | $> 10^{14}$ | $\sim 5$ |
| | MEIS | $2 \times 10^{13}$ | 3–10 |
| | ERD$^a$ | $\sim 10^{14}$ | 2–5 |
| | NRA | $< 10^{12}$ | —$^b$ |
| | NRA (NRP) | $\sim 10^{13}$ | 3–10 |
| Silicon/$^{29,30}$Si | NRA (NRP) | $\sim 10^{13}$ | 5–10 |

$^a$ high-energy, heavy-ion ERD.
$^b$ depth profiling in combination with chemical etch-back techniques.
$^c$ magnetic spectrometer and time-of-flight RBS.
The numbers given for sensitivity and depth resolution depend strongly on individual experimental conditions, so they should be considered as only an estimate.

ple, those related to ERD-based techniques—may result in improved sensitivity and depth resolution for future applications.

It should be noted that ion scattering techniques may simultaneously provide both total amounts and depth information on different elements contained in the thin film. The NRA and NRP techniques are selective for a certain isotope. This is an advantage, compared to ion scattering, with respect to the background counting rate, especially for lighter elements. And NRA can provide accurate depth information (e.g., for $^{16}O$ or $^2H$) when combined with step-by-step chemical etch-back techniques.

The amount of oxygen in ultrathin silicon oxides and oxynitrides can be determined precisely (to within a few percent) using many of the different techniques presented earlier. In standard RBS, the limited energy resolution broadens the corresponding oxygen peak, leading to a reduced sensitivity.

The NRA technique is very precise in determining oxygen amounts. The yield obtained from reference samples has been compared at many different laboratories. Reliable oxygen amounts with an error less than ±3% are routinely determined. Combined with chemical etch-back, NRA also provides high-resolution depth profiles.

For depth profiling oxygen (of natural isotopic abundance), MEIS is well established. The depth resolution can be as high as 3 Å at the surface and degrades somewhat with depth.

Magnetic spectrometers and time-of-flight RBS are currently being investigated to push the depth resolution of standard RBS to the subnanometer range. The recently applied ERD technique using heavy ions at high energies offers an excellent depth resolution (of a few angstroms), but special care concerning the charge state dependence of spectra and radiation damage must be taken.

In investigations involving oxygen-labeling techniques (i.e., $^{18}O$ isotope–enriched samples), the peaks from $^{16}O$ and $^{18}O$ are separated in MEIS spectra for oxide thickness below ~ 3 nm, and a depth resolution of 3–8 Å is obtained. The NRP technique (using the 151-keV p + $^{18}O$ resonance) provides comparable precision in determining total amounts and depth profiles.

High background counting rates and the limited resolution commonly deteriorate detection and depth profiling capabilities for nitrogen when using standard RBS. Amounts in the range of a few times $10^{14}$ atoms/cm$^2$ present in ultrathin oxynitrides are near the detection limit.

The sensitivity of MEIS for nitrogen is somewhat reduced compared to oxygen detection. However, it provides nitrogen depth profiles with subnanometer resolution at a sensitivity as good as ~ $5 \times 10^{13}$ atoms/cm$^2$ in certain cases. Again, ERD techniques provide very good depth resolution of a few angstroms and a sensitivity in the range of $10^{14}$ atoms/cm$^2$.

As in the oxygen case, nitrogen amounts ($^{14}N$ and $^{15}N$) can be determined precisely using NRA. Depth profiles of $^{15}N$ isotopes can be measured with good depth resolution and very high sensitivity using NRP (using the 429-keV p + $^{15}N$ resonance).

The structure of the interface layer between the Si substrate and a thin silicon oxide or oxynitride can be studied using ion scattering in a channeling geometry. These measurements provide quantitative information on displaced layers and/or nonstoichiometric interface layers using standard RBS or MEIS (although distinguishing between the two is difficult).

Isotope tracing of silicon can be performed using NRP by taking advantage of the isotope selectivity of nuclear techniques. Due to the small differences in the kinematic factor between the Si isotopes, they are generally not resolved in ion scattering spectra.

Detection and profiling of hydrogen can be performed with very good sensitivity using NRP ($^{15}N$ + p resonance in inverse kinematics). The depth resolution obtained is about 30 Å. Improved depth resolution for hydrogen isotopes can be obtained using ERD techniques, both at megaelectron-volt projectile energies as well as in the medium-energy range. Deuterium can be detected with very good sensitivity using NRA (better than $10^{12}$ atoms/$cm^2$), but no depth resolution is provided (except by etch-back methods).

## ACKNOWLEDGMENTS

The authors acknowledge our interactions with G. Amsel, I. J. R. Baumvol, M. Copel, L. C. Feldman, J.-J. Ganem, M. L. Green, E. P. Gusev, W. N. Lennard, H. C. Lu, S. Rigo, F. Rochet, F. C. Stedile, and others, whose work and ideas are incorporated in the text and figures. We also acknowledge the financial support of the NSF and SRC.

## REFERENCES

1. D.A. Buchanan. IBM J. Res. Develop. 43:245, 1999.
2. G. Timp et al. In: IEEE International Electron Devices Meeting, Washington, D.C., 1997.
3. G.D. Wilk, Y. Wei, H. Edwards, and R.M. Wallace, Appl. Phys. Lett. 70:2288, 1997.
4. R.D. Isaac. IBM J. Res. Develop. 44:369, 2000.
5. I.J.R. Baumvol. Surface Science Reports 36:5, 1999.
6. L.C. Feldman, E.P. Gusev, E. Garfunkel. In: E. Garfunkel, E.P. Gusev, A. Vul', eds. Fundamental Aspects of Ultrathin Dielectrics on Si-based Devices. Dordrecht, Netherlands: Kluwer Academic, 1998, p 1.
7. G. Wilk, R.M. Wallace, J.M. Anthony, J. Appl. Phys. 87:484, 2000.
8. B.H. Lee, L. Kang, R. Nieh, W.J. Qi, J.C. Lee. Appl. Phys. Lett. 76:1926, 2000.
9. M.E. Hunter, M.J. Reed, N.A. El-Masry, J.C. Roberts, S.M. Bedair. Appl. Phys. Lett. 76:1935, 2000.
10. M. Copel, M. Gribelyuk, E.P. Gusev. Appl. Phys. Lett. 76:436, 2000.
11. E.P.Gusev, M. Copel, E. Cartier, I.J.R. Baumvol, C. Krug, M. Gribelyuk. Appl. Phys. Lett. 76:176, 2000.
12. B.E. Deal, A.S. Grove. J. Appl. Phys. 36:3770, 1965.
13. G.J.A. Hellings, H. Ottevanger, S.W. Boelens, C.L.C.M. Knibbeler, H.H. Brongersma. Surf. Sci. 162:913, 1985.
14. A.W.D. van der Gon, M.A. Reijme, R.F. Rumphorst, A.J.H. Maas, H.H. Brongersma. Rev. Sci. Instr. 70:3910, 1999.
15. M.R. Frost, C.W. Magee. Appl. Surf. Sci. 104/105:379, 1996.
16. C.J. Han, C.R. Helms. J. Electrochem. Chem. 135:1824, 1988.
17. A. Benninghoven, F. G. Rudenauer, H. W. Werner. Secondary Ion Mass Spectrometry: Basic Concepts, Instrumental Aspects, Applications and Trends. New York: Wiley, 1987.
18. K. Wittmack. Surf. Sci. 112:168, 1981.
19. W.K. Chu, J.W. Mayer, M.-A. Nicolet. Backscattering Spectrometry. London: Academic Press, 1978.
20. L.C. Feldman, J.W. Mayer. Fundamentals of Surface and Thin Film Analysis. Amsterdam: North-Holland, 1986.
21. J.L'Ecuyer, J.A. Davies, N. Matsunami. Nucl. Inst. Meth. 160:337, 1979.
22. J.L'Ecuyer, J.A. Davies, N. Matsunami. Rad. Eff. 47:229, 1980.
23. H.H. Anderson, F. Besenbacher, P. Loftager, W. Möller. Phys. Rev. A21:1891, 1980.
24. W. Lenz. Z. Phys. 77:713, 1932.
25. H. Jensen. Z. Phys. 77:722, 1932.

26. E. Huttel, W. Arnold, J. Baumgart, G. Clausnitzer. Nucl. Instr. Meth. B12:193, 1985.
27. S.R. Lee, R.R. Hart. Nucl. Instr. Meth. B79:463, 1993.
28. L.A. Foster, J.R. Tesmer, T.R. Jervis, M. Nastasi. Nucl. Instr. Meth. B79:454, 1993.
29. D.D. Cohen, E.K. Rose. Nucl. Instr. Meth. B66:158, 1992.
30. M. Watamori, K. Oura, T. Hirao, K. Sasabe. Nucl. Instr. Meth. B118:228, 1996.
31. W. de Coster, B. Brijs, R. Moons, W. Vandervorst. Nucl. Instr. Meth. B79:483, 1993.
32. M.A. Kumakhov, F.F. Komarov. Energy Loss and Ion Ranges in Solids. Gordon & Breach, 1991.
33. H. Bichsel. Rev. Mod. Phys. 60:663, 1988.
34. P. Sigmund. In: A. Gras-Marti, ed. Interaction of Charged Particles with Solids and Surfaces. New York: Plenum, 1991.
35. J.F. Ziegler, J. Biersack, U. Littmack. The Stopping Power and Range of Ions in Solids. New York: Pergamon, 1985.
36. J.F. Ziegler. TRIM97. Yorktown Heights, NY: IBM, 1997.
37. W.N. Lennard, G.R. Massoumi, T.W. Simpson, I.V. Mitchell. Nucl. Instr. Meth. B152:370, 1999.
38. W.H. Bragg, R. Kleemann. Phil. Mag. 10:318, 1905.
39. D.I. Thwaites. Nucl. Instr. Meth. B12:1985, 1985.
40. D.I. Thwaites. Nucl. Inst. Meth. B69:53, 1992.
41. P. Bauer, R. Golser, D. Semrad, P. Maier-Komor, F. Aumayr, A. Arnau. Nucl. Instr. Meth. B136–138:103, 1998.
42. J. Bird, J.S. William. Ion Beams for Materials Analysis. Academic Press, 1989.
43. N. Bohr. Mat. Fys. Medd. Dan. Vid. Selsk. 18:No. 8, 1948.
44. W.K. Chu. Phys. Rev. A13:2057, 1976.
45. E. Bonderup, P. Hvelplung. Phys. Rev. A4:562, 1971.
46. J. Lindhard, M. Scharff. Mat. Fys. Medd. Dan. Vid. Selsk. 27:No.15, 1953.
47. F. Besenbacher, J.U. Anderson, E. Bonderup. Nucl. Instr. Meth. 168:1, 1980.
48. Q. Yang, D.J. O'Connor, Z. Wang. Nucl. Instr. Meth. B61:149, 1991.
49. G. Konac, S. Kalbitzer, C. Klatt, D. Niemann, R. Stoll. Nucl. Instr. Meth. B136–138:159, 1998.
50. J. Schuchinsky, C. Peterson. Rad. Eff. 81:221, 1984.
51. T. Kaneko, Y. Yamamura. Phys. Rev. A33:1653, 1986.
52. L.R. Doolittle. Nucl. Instr. Meth B9:344, 1985.
53. L.R. Doolittle. Nucl. Inst. Meth. B15:227, 1986.
54. G Vizkelethy. Nucl. Instr. Meth. B45:1, 1990.
55. E. Kotai. Nucl. Instr. Meth. B85:588, 1994.
56. J. Saarilahti, E. Rauhala. Nucl. Instr. Meth. B64:734, 1992.
57. Q.Yang. Nucl. Instr. Meth. B90:602, 1994.
58. E. Szilágyi, F. Pászti, G. Amsel. Nucl. Instr. Meth. B100:103, 1995.
59. A. Climent-Font, U. Wätjen, H. Bax. Nucl. Instr. Meth. B71:81, 1992.
60. A. Climent-Font, M.T. Fernández-Jiménez, U. Wätjen, J. Pierrìere. Nucl. Instr. Meth. A353:575, 1994.
61. M. Bianconi, F. Abel, J.C. Banks, et al. Nucl. Instr. Meth. B161–163:293, 2000.
62. J. Genzer, J.B. Rothman, R.J. Composto.Nucl. Instr. Meth. B86:345, 1994.
63. N.P. Barradas, J.C. Soares, M.F. da Silva, E. Pászti, E. Szilágyi. Nucl. Instr. Meth. B94:266, 1994.
64. D. Dieumegard, D. Dubreuil, G. Amsel. Nucl. Injstr. Meth. 166:431, 1979.
65. L.S. Wielunski, E. Szilágyi, G.L. Harding. Nucl. Instr. Meth. B136–138:713, 1998.
66. G. Dollinger. Nucl. Instr. Meth. B79:513, 1993.
67. G. Dollinger, M. Boulouednine, A. Bergmaier, T. Faestermann. Nucl. Instr. Meth. B136–138:574, 1998.
68. A. Bergmaier, G. Dollinger, C.M. Frey. Nucl. Instr. Meth. B136–138:638, 1998.
69. I. Vickridge, G. Amsel. Nucl. Instr. Meth. B45:6, 1990.

70. W.H. Schulte, H. Ebbing, H.W. Becker, M. Berheide, M. Buschmann, C. Rolfs, G.E. Mitchell, J.S. Schweitzer. J. Phys. B. At. Mol. Opt. Phys. 27:5271, 1994.

71. I. Vickridge, G. Amsel. Nucl. Instr. Meth. B64:687, 1992.

72. R. Behrisch, V.M. Prozesky, H. Huber, W. Assmann. Nucl. Instr. Meth. B118:262, 1996.

73. S.R. Walker, J.A. Davies, J.S. Foster, S.G. Wallace, A.C. Kockelkoren. Nucl. Instr. Meth. B136–138:707, 1998.

74. T.E. Jackman, J.R. McDonald, L.C. Feldman, P.J. Silverman, I. Stensgaard. Surf. Sci. 100:35, 1980.

75. L.C. Feldman, P.J. Silverman, J.S. Williams, T.E. Jackman, I. Stensgaard. Phys. Rev. Lett. 41:1396, 1978.

76. R.L. Kaufman, L.C. Feldman, P.J. Silverman, R.A. Zuhr. Appl. Phys. Lett. 32:93, 1978.

77. W.A. Lanford. Nucl. Instr. Meth. B66:65, 1992.

78. J.F. Browning, R.A. Langley, B.L. Doyle, J.C. Banks, W.R. Wampler. Nucl. Instr. Meth. B161–163:211, 2000.

79. F.H.P.M. Habraken, E.H.C. Ullersma, W.M. Arnoldbik, A.E.T. Kuiper. In: E. Garfunkel, E.P. Gusev, A. Vul', eds. Fundamental Aspects of Ultrathin Dielectrics on Si-based Devices. Dordrecht, Netherlands: Kluwer Academic, 1998, p 411.

80. W.M. Arnoldbik, W. Wolfswinkel, D.K. Inia, V.C.G. Verleun, S. Lobner, J.A. Reinder, F. Labohm, D.O. Boersma. Nucl. Instr. Meth. B118:566, 1996.

81. B. Brijs, J. Deleu, T. Conrad, et al. Nucl. Instr. Meth. B161–163:429, 2000.

82. G. Dollinger, M. Boulouednine, A. Bergmaier, T. Faestermann, C.M. Frey. Nucl. Instr. Meth. B118:291, 1996.

83. W.A. Lanford, S. Bedell, S. Amadon, A. Haberl, W. Skala, B. Hjorvarsson. Nucl. Instr. Meth. B161–163:202, 2000.

84. K. Kimura, K. Ohshima, M.H. Mannami. Appl. Phys. Lett. 64:2233, 1994.

85. K. Kimura, M. Mannami. Nucl. Instr. Meth. B113:270, 1996.

86. J. Lindhard, M. Scharff. Phys. Rev. 124:128, 1961.

87. H.D. Carstanjen. Nucl. Instr. Meth. B136–138:1183, 1998.

88. R.G. Smeenk, R.M. Tromp, H.H. Kersten, A.J.H. Boerboom, F.W. Saris. Nucl. Instr. Meth. 195:581, 1982.

89. J.F. van der Veen. Surf. Sci. Rep. 5:199, 1985.

90. H.C. Lu, E.P. Gusev, T. Gustafsson, E. Garfunkel, M.L. Green, D. Brasen, L.C. Feldman. Appl. Phys. Lett. 69:2713, 1996.

91. E. Garfunkel, E.P. Gusev, H.C. Lu, T. Gustafsson, M. Green. In: E. Garfunkel, E.P. Gusev, and A. Val', eds. Fundamental Aspects of Ultrathin Dielectrics on Si-based Devices. Dordrecht, Netherlands: Kluwer Academic, 1998, p 39.

92. E.P. Gusev, H.C. Lu, T. Gustafsson, E. Garfunkel. Phys. Rev. B52:1759, 1995.

93. M. Copel, R.M. Tromp. Rev. Sci. Instr. 64:3147, 1993.

94. Y.P. Kim, S.K. Choi, H.K. Kim, D.W. Moon. Appl. Phys. Lett. 71:3504, 1997.

95. K. Yamashita, T. Yasue, T. Koshikawa, A. Ikeda, Y. Kido. Nucl. Instr. Meth. B136–138:1086, 1998.

96. Y. Kido, H. Namba, T. Nishimura, A. Ikeda, Y. Yan, A. Yagashita. Nucl. Instr. Meth. B136–138:798, 1998.

97. E.P. Gusev, H.C. Lu, T. Gustafsson, E. Garfunkel, M.L. Green, D. Brasen. J. Appl. Phys. 82:896, 1996.

98. H.C. Lu. PhD dissertation, Rutgers University, Piscataway, NJ, 1997.

99. E.P. Gusev, H.C. Lu, T. Gustafsson, E. Garfunkel. Appl. Surf. Sci. 104/105:329, 1996.

100. A. Kurakowa, K. Nakamura, S. Ichimura, D.W. Moon. Appl. Phys. Lett. 76:493, 2000.

101. E.P. Gusev, H.C. Lu, E. Garfunkel, T. Gustafsson, M.L. Green. IBM J. Res. Develop. 43, 1999.

102. G. Gabor, W. Schimmerling, D. Greiner, F. Bieser, P. Lindstrom. Nucl. Instr. Meth. 130:65, 1975.

103. R.E. Heil, J. Drexler, K. Huber, U. Kneissl, G. Mang, H. Ries, H. Ströher, T. Weber, W. Wilke. Nucl. Instr. Meth. A239:545, 1985.
104. K. McDonald, R.A. Weller, V.K. Liechtenstein. Nucl. Instr. Meth. B152:171, 1999.
105. R.A. Weller, K.McDonald, D. Pedersen, J.A. Keenan. Nucl. Instr. Meth. B118:556, 1996.
106. M.H. Mendenhall, R.A. Weller. Nucl. Instr. Meth. B47:193, 1990.
107. M.H. Mendenhall, R.A. Weller. Nucl. Instr. Meth. B59/60:120, 1991.
108. M. Döbeli, R.M. Ender, V. Liechtenstein, D. Vetterli. Nucl. Instr. Meth. B142:417, 1998.
109. J.A. Knapp, D.K. Brice, J.C. Banks. Nucl. Instr. Meth. B108:324, 1996.
110. T. Kobayashi, G. Dorenbos, S. Shimoda, M. Iwaki, M. Aono. Nucl. Instr. Meth. B118:584, 1996.
111. N. Piel, H.W. Becker, J. Meijer, W.H. Schulte, C. Rolfs. Nucl. Instr. Meth. B138:1235, 1998.
112. N. Piel, H.W. Becker, J. Meijer, W.H. Schulte, C. Rolfs. Nucl. Instr. Meth. A437:521, 1999.
113. G. Demortier. In: Third International Conference on Chemical Analysis, Namur, Belgium, 1991, Nucl. Instr. Meth. B66, 1992.
114. K. Vanheusden, R.A.B. Devine. Appl. Phys. Lett. 76:3109, 2000.
115. J.W. Lyding, K. Hess, I.C. Kizilyalli. Appl. Phys. Lett. 68:2526, 1996.
116. I.J.R. Baumvol, F.C. Stedile, C. Radke, F.L. Freire Jr., E. Gusev, M.L. Green, D. Brasen. Nucl. Instr. Meth. B136–138:204, 1998.
117. H. Park, C.R. Helms. J. Electrochem. Soc. 139:2042, 1992.
118. I.J.R. Baumvol, E.P. Gusev, F.C. Stedile, F.L. Freire Jr., M.L. Green, D. Brasen. Appl. Phys. Lett. 72:450, 1998.
119. R.D. Vis. Nucl. Instr. Meth. B66:139, 1992.
120. G. Amsel. Nucl. Instr. Meth. 92:481, 1971.
121. K. Bethge. Nucl. Instr. Meth. B10/11:633, 1985.
122. K. Bethge. Nucl. Instr. Meth. B66:146, 1992.
123. J.A. Davies. Nucl. Instr. Meth. 218:141, 1983.
124. K.M. Horn and W.A. Lanford. Nucl. Instr. Meth. B34:1, 1998.
125. F. Rochet, S. Rigo, M. Froment, C. D'Anterroches, C. Maillot, H. Roulet, G. Dufour. Adv. Phys. 35:237, 1986.
126. L.G. Gosset, J.-J.Ganem, I. Trimaille, S. Rigo, F. Rochet, G. Dufour, F. Jolly, F.C. Stedile, I.J.R. Baumvol. Nucl. Instr. Meth. B136–138:521, 1998.
127. F.C. Stedile, I.J.R. Baumvol, I.F. Oppenheim, I. Trimaille, J.-J. Ganem, S. Rigo. Nucl. Instr. Meth. B118:493, 1996.
128. H.V. Bebber, L. Borucki, K. Farzin, A.Z. Kiss, W.H. Schulte. Nucl. Instr. Meth. B138:1998.
129. R.D. Evans. The Atomic Nucleus. New York: McGraw-Hill, 1955.
130. I.J.R. Baumvol, C. Krug, F.C. Stedile, F. Gorris, W.H. Schulte. Phys. Rev. B60:1492, 1999.
131. M. Mehrhoff et al. Nucl. Instr. Meth. B133:671, 1997.
132. A. Markwitz, H. Baumann, E.F. Krimmel, M. Rose, K. Bethge, P. Misaelides, S. Logothetides. Vaccum 44:367, 1992.
133. J.-J. Ganem, S. Rigo, I. Trimaille, I.J.R. Baumvol, F.C. Stedile. Appl. Phys. Lett. 68:2366, 1996.
134. I.J.R. Baumvol, F.C. Stedile, J.-J. Ganem, I. Trimaille, S. Rigo. Appl. Phys. Lett. 70:2007, 1997.
135. G. Battistig, G. Amsel, E. d'Artemare, I. Vickridge. Nucl. Instr. Meth. B61:369, 1991.
136. G. Battistig, G. Amsel, E. d'Artemare, I. Vickridge. Nucl. Instr. Meth. B66:1, 1992.
137. J.-J. Ganem, G. Battistig, S. Rigo, I. Trimaille. Appl. Surf. Sci. 65/66:647, 1993.
138. G. Battistig, G. Amsel, I. Trimaille, J.-J. Ganem, S. Rigo, F.C. Stedile, I.J.R. Baumvol, W.H. Schulte, H.W. Becker. Nucl. Instr. Meth. B85:326, 1994.
139. T. Åkermark, L.G. Gosset, J.-J. Ganem, I. Trimaille, I. Vickridge, S. Rigo. J. Appl. Phys. 86:1153, 1999.
140. F.C. Stedile, I.J.R. Baumvol, J.-J. Ganem, S. Rigo, I. Trimaille, G. Battistig, W.H. Schulte, H.W. Becker. Nucl. Instr. Meth. B85:248, 1994.
141. F. Rochet, B. Agius, S. Rigo. J. Electrochem. Soc. 131:914, 1984.

142. I. Trimaille, J.J. Ganem, L.G. Gosset, S. Rigo, I.J.R. Baumvol, F.C. Stedile, F. Rochet, G. Dufour, F. Jolly. In: E. Garfunkel, E. Gusev, A. Vul', eds. Fundamental Aspects of Ultrathin Dielectrics on Si-based Devices. Dordrecht, Netherlands: Kluwer Academic, 1998, p 165.
143. M.P. Murrell, C.J. Sofield, S. Sudgen. Phil, Mag. B63:1277, 1991.
144. I.J.R. Baumvol et al. Nucl. Instr. Meth. B118:499, 1996.
145. F. Gorris, C. Krug, S. Kubsky, I.J.R. Baumvol, W.H. Schulte, C. Rolfs. Phys. Stat. Sol. A173:167, 1999.
146. I.C. Vickridge, O. Kaitasov, R.J. Chater, J.A. Kilner, Nucl. Instr. Meth. B161–163:441, 2000.
147. A.E.T. Kuiper, M.F.C. Williamsen, J.M.L. Mulder, J.B.O. Elfebrink, F.H.P.M. Habraken, W.F.v.d. Weg. J. Vac. Sci. Technol. B7:465, 1989.
148. K.H. Ecker, J. Krauser, A. Weidinger, H.P. Weise, K. Maser. Nucl. Instr. Meth. B161–163:682, 2000.
149. J. Krauser, A. Weidinger, D. Bräuning. In: H.Z. Massoud, E.H. Poindexter, C.R. Helms, eds. The Physics and Chemistry of $SiO_2$ and the $Si$–$SiO_2$ Interface-3. Pennington, NJ: Electrochemical Society, 1996, p 184.
150. L. Borucki, H.W. Becker, F. Gorris, S. Kubsky, W.H. Schulte, C. Rolfs. Eur. Phys. J. A5:327, 1999.
151. M. Zinke-Allmang, S. Kalbitzer, M. Weiser. Z. Phys. A:183, 1986.
152. Y. Iwata, F. Fujimoto, E. Vilalta, A. Ootuka, K. Komaki, H. Yamashita, Y. Murata. Nucl. Instr. Meth. B33:574, 1988.
153. S. Jans, S. Kalbitzer, P. Oberschachtsiek, J.P.F. Sellschop. Nucl. Instr. Meth. B85:321, 1994.
154. B. Hartmann, S. Kalbitzer, M. Behar. Nucl. Instr. Meth. B103:494, 1995.

# 29

# Electron Microscopy–Based Measurement of Feature Thickness and Calibration of Reference Materials

**Alain C. Diebold**
*International SEMATECH, Austin, Texas*

## I. INTRODUCTION

Scanning electron microscopy (SEM), transmission electron microscopy (TEM), and scanning transmission electron microscopy (STEM) have all been used to calibrate or check other thickness measurements. As the thickness of many films approaches a few atomic planes, application of these methods requires careful attention to the details of both the physics of electron microscopy and the physics of the method under comparison. For example, the use of high-resolution TEM for calibrating optical film thickness is a questionable practice, considering that atomic-level interfacial properties are not being averaged over the larger area measured by ellipsometry. In addition, the TEM image is a two-dimensional projection of three-dimensions worth of data, and contrast in the image is a complex function of beam–sample interactions where sample thickness and focal conditions can cause contrast reversal. In this chapter, the use of electron beam methods for calibration is reviewed.

Since its inception, electron microscopy has provided pictures that have often been considered representative of true feature dimensions. Proper care needs be taken to account for the physical limitations of these methods in terms of accuracy and precision when these images are used to calibrate film thickness or linewidth. Every measurement method has limitations; through understanding these, metrology can be improved. In this short chapter, the advantages and disadvantages of electron beam methods as applied to film thickness measurement are described. Issues associated with linewidth by SEM are covered in both Chapter 14, on critical dimension (CD)-SEM, and Chapter 15, on CD–atomic force microscopy (CD-AFM). Some of the same phenomena observed during CD-SEM measurement result in artifacts in SEM-based thin-film thickness images. Therefore, those chapters provide important reference for the discussion in this chapter.

Integrated circuits have many features that could be described as short segments of films that are not parallel to the surface of the silicon wafer. Barrier layers on sidewalls are one example of this. Dual-beam FIBs (systems with both an ion beam and an electron beam) have been used for process monitoring of contact/via and trench capacitors.

This chapter is divided into three sections. First, SEM, TEM, and STEM film thickness measurements are briefly described. Then, SEM-based film thickness is described along with the necessary SEM calibrations. Next, TEM- and STEM-based measurement of on-chip interconnect film thickness is discussed. Finally, TEM and STEM measurement of gate dielectric film thickness is presented. This chapter does not review SEM, TEM, and STEM in detail. The reader is referred to several outstanding books reviewing the technology and application of these methods (1–3).

## II.  SEM, TEM, AND STEM FILM THICKNESS MEASUREMENT

In SEM, a focused electron beam is scanned over the sample while the detected secondary electron intensity at each point in the scan is used to form the image. Other signals, such as the detected backscattered electron intensity, can be used to form images. Magnification is determined by selecting the area over which the beam is rastered, and resolution is determined by the beam diameter as well as the stability of the scan coils and sample stage.

There are several types of imaging modes that can be used during TEM analysis. In TEM, a fine beam is transmitted through a very thin sample and then focused onto the image plane. The sample is typically less than 100 nm thick so that the electrons can pass through the sample. Image contrast is determined by the amount of diffraction or scattering, and there are several imaging modes. All of the TEM imaging modes can be useful for determining physical dimensions of features or layers in a reference sample. A bright-field (BF) image is obtained when the electrons that pass directly through the sample are used to form the image. The contrast in a BF image comes from diffraction and/or scattering from either local density or mass (Z) differences (4). A dark-field (DF) image is formed by selecting electrons that are diffracted (4). Several DF images can be formed by selecting one or more diffracted beams using specific diffraction conditions (e.g., 001 or 011), and phase-contrast images are formed when two or more of the diffracted beams interfere to form an image (4,5). The BF and DF TEM imaging modes are shown in Figure 1. Phase-contrast images are formed when two or more of the diffracted beams interfere to form an image. In high-resolution TEM characterization of transistor cross sections, the "on-axis lattice fringe image" from the silicon substrate can be used as a high-resolution calibration of the device dimensions. One is observing not atomic positions, but the interference of several beams, one of which is the directly transmitted and the other(s) is(are) beam(s) diffracted from the lattice planes of interest. In Figure 2, we show how this works for a two-beam phase-contrast image (5). The term *on-axis lattice-fringe image* refers to a phase-contrast image formed when the sample is tilted so that the lattice planes used to form the diffraction beams are parallel to the electron optics axis. A special case occurs when the sample is aligned so that the electron beam is along a low-index-zone axis (such as 001 or 002; 110 or 220; or 111) direction in the crystal (5). If many beams are used, an atomic-like image forms with interference maximums whose spacing corresponds to a lattice spacing that can be smaller than the BF TEM resolution. This process is shown in Figure 3. It is the many-beam, phase-contrast image that is most commonly used to calibrate dimensions and check dielectric thickness in a transistor cross section. Since interconnect structures are displaced far enough from the silicon substrate, it is more difficult to use this calibration method to determine dimensions for barrier layers. Elemental and some chemical analysis can be done in a small spot in the TEM image using x-ray analysis by energy-dispersive spectroscopy (EDS) or by electron energy-loss spectroscopy.

## Bright Field TEM Image          Dark Field TEM Image

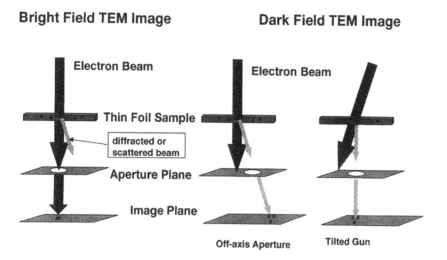

**Figure 1** Amplitude bright field and dark field imaging in TEM and STEM. In a bright-field image, the contrast arises from differences in local electron scattering (i.e., diffraction and/or mass density differences). Electrons diffract from the lattice planes in crystalline sections of the sample. In TEM, diffracted electrons focus at the plane of the aperture, and a subset of the diffracted beams can be selected for image formation by varying the size and location of the aperture. A dark-field image is obtained from contrast differences that occur when one or more diffracted beams are selected for transmission to the image plane. The TEM and sample can be aligned so that the diffracted beams are along the optical axis, as shown in the far right. This is the most common dark-field imaging method. But the BF and DF TEM modes shown are based on intensity changes in the image, which are due to the amplitude of the electron wave that reaches the detector.

Although scanning TEM (STEM) is done in a TEM equipped with scanning capability, the physics of image formation is different from the high-resolution, phase-contrast images discussed earlier. In STEM, a finely focused electron beam is scanned across the imaged area of a very thin sample. Again, the sample must be thin enough so that the electrons can pass through it. Both BF and DF images can be formed in STEM. One advantage of STEM is that materials characterization maps or linescans by EDS and EELS can be done at high resolution (6). When quantitative or semiquantitative analysis is required, TEM/STEM systems equipped with thermal field emission sources have the required brightness and current stability. The latest-generation STEM systems are equipped with *high-angle* annular dark-field (HA-ADF) detectors that provide very high-resolution images (6, 7). Although the contrast in these images is a result of atomic number differerences, image interpretation requires consideration of subtle physics (6,7). The objective is to separate the elastically scattered electrons from the diffracted beams that interfere with the incoherently scattered electrons and make image interpretation difficult. Because the Bragg condition is not met for high angles, the HA-ADF was introduced as a means of atomic-level imaging free from the interference of diffraction beams. The HA-AFD STEM image is a map of the atomic scattering factor, which is a strong function of the atomic number (6). In Figure 4, HA-ADF STEM operation is shown. As others have stated, HA-ADF imaging *may* become the microscopy mode of choice for high-resolution imaging (6). Careful studies of silicon dioxide–based transistor-gate dielectric thickness and interfacial layer thickness have used HA-ADF (7).

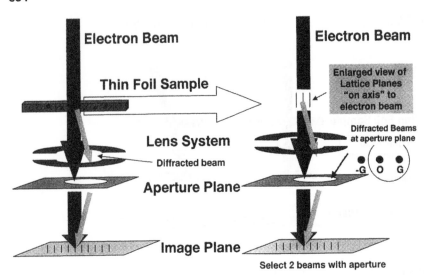

**Figure 2**  Phase-contrast imaging for crystalline samples—two-beam method. On-axis, phase-contrast image is formed when the lattice planes of interest are parallel to the (electron) optical axis of the TEM. The O beam is transmitted directly through the sample. When the $(h,k,l)$ lattice planes and the electron beam are "on-axis," i.e., parallel to the optical axis, the diffraction spots from electrons that meet the Bragg reflection criteria are focused onto the aperture plane. Although diffraction spots from other lattice planes also arrive at the aperture plane, the aperture can be reduced and moved to select the G, O, and −G reflections. The G refers to the diffraction spot from the $(h, k, l)$ lattice plane, and −G refers to diffraction in the opposite direction from the same $(\bar{h}, \bar{k}, \bar{l})$ lattice planes. Some of the diffracted beams are selected by aperture and then refocused at the image plane. The interference of the O and G beams results in the lattice-fringe image observed at the image plane. Although the diffraction images at the aperture is in reciprocal space, the interference pattern at the image plane is in real space. The fringe spacing is $1/(h, k, l)$, i.e., the lattice plane spacing in real space. The contrast at the image plane from the noncrystalline part of the sample forms a bright-field image. This is how the gate dielectric and amorphous silicon of a transistor or capacitor stack are imaged while the lattice fringes of the crystalline silicon are used to calibrate the dimensions.

Many TEM systems that have filed emission sources are being equipped with biprism lenses for electron holography. Recently, there have been great advances in using electron holography for imaging two-dimensional dopant profiles in transistor cross sections (8).

## III.  SEM-BASED FILM THICKNESS MEASUREMENT AND CALIBRATION

SEM has provided useful measurements of film thickness for many years. The thick films found in older generations of integrated circuits, before the 1-μm IC node (generation), were relatively easy to measure once sample preparation was developed. SEM is still capable of measuring the thicker films found in research and development of even the most advanced IC. Some examples are the metal and dielectric used for interconnect that is either an aluminum alloy (with trace Cu and Si) with silicon dioxide or copper with one of several possible dielectric materials. The barrier layer used in interconnect is usually difficult to accurately measure using SEM.

The issues facing SEM-based measurement start with calibration of the SEM itself. Magnification in both the $x$ and $y$ directions must be checked using a NIST-traceable

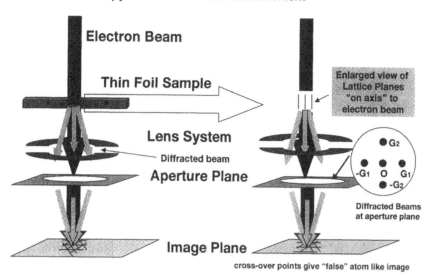

**Figure 3** Low-index phase-contrast images for crystalline samples—many beam method: An on-axis, phase-contrast image is formed when a low-index direction of the crystal lattice is parallel to the (electron) optical axis of the TEM. Again, the O beam is transmitted directly through the sample. The interference of the O and $G_1$ beams results in one set of lattice fringes, and the interference of the O and $G_2$ beams results in another set of lattice fringes. The intersection of several sets of lattice fringes results in a pointlike interference pattern that appears to be a picture of atoms when observed at the image plane.

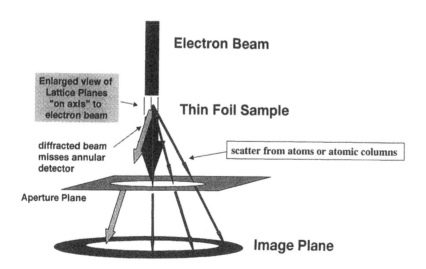

**Figure 4** High-angle—annular dark field STEM high-resolution imaging: The HA-ADF detector is shown. In order to avoid interference from diffraction peaks, the HA-ADF detector is placed outside of other detectors. Annular rings maximize detection of the anisotropic signal that results from atomic scattering of the electron beam.

magnification standard, as shown in Figure 5. Frequent checks of system calibration is a necessity when a system is used for metrology purposes. As SEM systems become equipped with Fourier transform–based image processing and special calibration samples, the resolution and astigmatism of the systems will improve and become much more stable. Chapter 14, by Postek and Vladar, provides further details. Another approach to SEM calibration is the use of a reference material with layers have thicknesses close to those of interest that is glued to the appropriate place on the sample (9). This can be done using the standard reference materials for oxide layers available from the National Institute of Standards and Technology.

Another source of error comes from edge effects such as those observed in linewidth measurements. The edge of a metal or poly(crystalline) silicon line will have a greatly increased secondary electron intensity due to the larger exposed surface area. This effect is discussed in Chapters 14 and 15. Occasionally, a cross section must be prepared by means of polishing and etching methods. Since different materials chemically etch at different rates, it is possible for a film to protrude above another in a cross section. There will be enhanced secondary electron emission from these edges, and the true edge of the film will be difficult to determine from the SEM image. Software that corrects for these effects has been developed (Chapter 14). The measurement of thin-film ($< 500$-nm) thickness by SEM is susceptible to error due to edge effects. Again, TEM and STEM are more appropriate for the accurate evaluation of thin-film thickness.

Lower-magnification SEM images have the advantage that they show a relatively large part of the sample of interest. A 1-micron-thick film can be observed over a width of more than 100 microns with a low-resolution view of film thickness. As one increases magnification, the width of the film that is observed is decreased, and film nonuniformities are easily missed. Another factor is the repeatability of the thickness measurement. When one considers multiple measurement of a single image, the thickness values will vary with each analysis of the image by different individuals. The location of the film edge will be different each time someone draws a line at the interface. A well-designed image analysis algorithm would use signal intensity information that is not discernable to the human eye, and the measurement variation on a single image would be reduced to those resulting from noise. Therefore, there is some uncertainty in the thickness from the nonautomated process of using an image to obtain a value.

## IV.  TEM AND STEM MAGNIFICATION/DIMENSIONAL CALIBRATION

Calibration of TEM and STEM thickness measurement can be done using a special reference material or through the spacings of lattice fringes in phase-contrast images. One exciting new reference material is an epitaxial multilayer sample of $Si_{0.81}Ge_{0.19}$, shown in Figure 6 (10). The advantage of this reference material is that both layer spacings and lattice fringes can be used for calibration. Transistor cross sections have the advantage of having the silicon lattice available for internal calibration.

## VI.  SEM, TEM, AND STEM MEASUREMENT OF INTERCONNECT FILM THICKNESS

The materials used to fabricate on-chip interconnect structures continue to evolve. Traditionally, interconnect lines were made from silicon dioxide insulators and aluminum

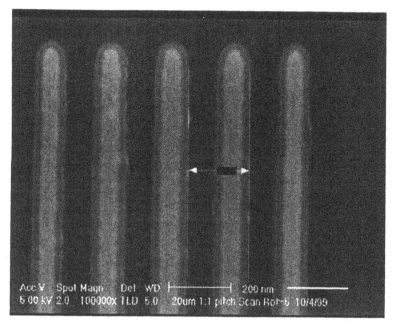

(a)

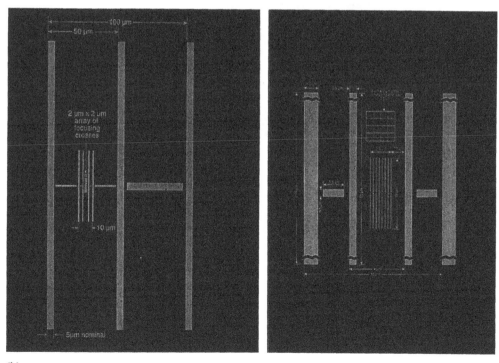

(b)

**Figure 5** SEM magnification calibration standard: (a) NIST 8090 SEM standard shown at a magnification of 100,000×. The 200-nm pitch is used to calibrate SEM. (Figure courtesy David Griffith.) (b) View of NIST 8090 SEM standard at lower magnification showing other features. (Figure courtesy Mike Postek and rights retained by NIST.)

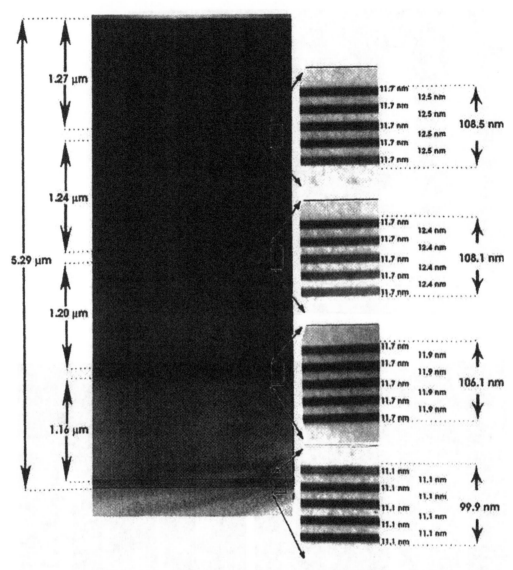

**Figure 6** Mag-I-Cal™ TEM/STEM references material. (Figure courtesy of Structure Probe, Inc.)

metal lines with titanium nitride barrier layers. Patterning of the metal lines was done using etch processing, and then the silicon dioxide insulator was deposited in the open areas. Planarization was done using doped silicon dioxide that flowed after heating. More recently, chemical-mechanical polishing of the oxide layer has been used to achieve planarization. Today, copper metal lines with suitable barrier layers such as tantalum nitride are deposited into open trench and contact structures. This is known as the damascene process. The insulator is evolving from fluorinated silicon dioxide to materials with lower dielectric constants. Eventually, porosity will be used to further lower the dielectric constant. *The critical point to this discussion is that each new materials set requires further development of the process used for preparation of cross sections.* Metal lines adhere differently to each insulator material. Each new insulator will fracture and polish differently. Copper is much more difficult to cleave and polish. Focused ion beam (FIB) systems are frequently used for SEM and TEM sample preparation. Although FIB is indispensable, it does introduce its own artifacts (11). Sample preparation artifacts must be reduced (11).

Examples of feature measurement will illustrate the capabilities of each microscopy. Film thickness determination using the SEM is shown in Figure 7. Copper Damascene structures are shown in the SEM cross section shown in Figure 8. A TEM image of a uniform barrier layer thickness along sidewall of a filled contact is shown in Figure 9.

## VI. TEM AND STEM MEASUREMENT OF TRANSISTOR FEATURE DIMENSIONS AND RELATED FILM STACKS

Transistor features such as gate dielectric thickness continue to shrink to dimensions that require the high resolution of TEM and STEM. Gate dielectric film thickness is below 3 nm for transistors having a gate length of 180 nm. It is often useful to employ the lattice fringes of phase-contrast TEM images or the lattice images of HA-ADF STEM to calibrate dimensions of the TEM image. The lattice spacings of silicon are well known and are visible when the silicon substrate is aligned along a low-index-zone axis, as discussed earlier. An example of this is shown in Figure 10. It is most useful to have samples with the gate electrode layer on top of the dielectric film when using this method. The deposition of a top layer during sample preparation can cause artifacts that alter thickness. Great care must be used when applying this method to calibration of optical measurements. The sampling area is small compared to optical methods, so differences in interfacial roughness

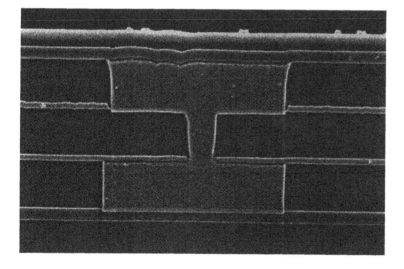

**Figure 7** Film thickness evaluation using SEM cross section: A dual Damascene structure is shown at 40,000 times magnification using a 2-keV probe beam. The via connecting the two metal lines is deposited at the same time as the deposition of the upper metal line.

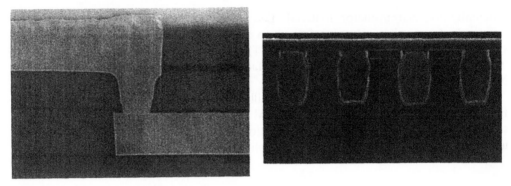

**Figure 8** SEM cross-sectional image of a low-k copper Damascene structure: Two different Damascene structures are shown.

are not accounted for. In addition, the act of determining the location of the interface is not reproducible unless it is done using a digital image and appropriate software. At the time of writing, this software was not available. Other transistor features are shown in Figure 11.

Other measurements are possible. Electron holographic images can be used to observe two-dimensional dopant profiles, thus determining electrical junction depth and the shape of the low-dose drain (also called *drain extension*). An example of this is shown in Figure 12. Another potential use of TEM is the determination of the location and thickness of a silicon onynitride layer in a silicon dioxide gate dielectric. Venables and Maher have described this work (12). It should be noted that this method is considered

0.20 μm

**Figure 9** TEM image showing barrier layer on the sidewall of a via.

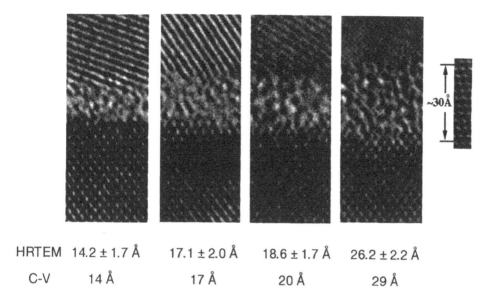

| HRTEM | 14.2 ± 1.7 Å | 17.1 ± 2.0 Å | 18.6 ± 1.7 Å | 26.2 ± 2.2 Å |
| C-V | 14 Å | 17 Å | 20 Å | 29 Å |

**Figure 10** Use of lattice fringe images to calibrate gate dielectric thickness: The silicon (111) lattice fringes are spaced at 0.314 nm and provide a direct calibration of film thickness. It is important to note that the error is approximate 0.2 nm. (Figure courtesy Dennis Maher.)

**Figure 11** TEM image of a transistor cross section.

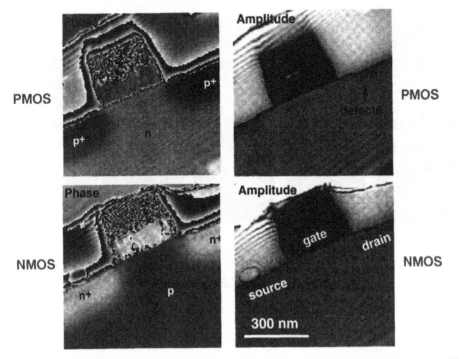

**Figure 12** Electron hologram showing a two-dimensional dopant profile taken in a TEM: Phase- and amplitude-contrast imagesof 300-nm P-MOS and C-MOS transistors. The dopant profiles can be observed in the phase-contrast image. (From Ref. 13. Figure courtesy Abbas Ourmazd and W. D. Rau, © IEEE 1998, and reproduced with permission.)

difficult, and EELS imaging of nitrogen in an HA-ADF STEM will provide the same information with more certainty. Using a single beam in the bright-field mode and a specimen thickness of 50 nm, the higher density of silicon atoms in silicon nitride than in silicon oxide will scatter more, and a layer of silicon nitride above or below silicon dioxide can be observed. The poly silicon gate electrode must be present for this method to work. The oxynitride layer in a oxynitride/oxide stack is visible in Figure 13.

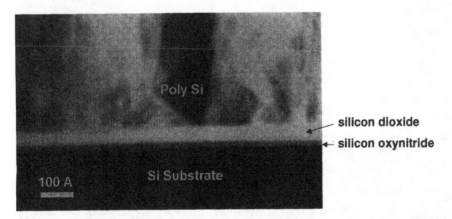

**Figure 13** Bright-field TEM image showing the contrast difference between silicon nitride and silicon dioxide.

## ACKNOWLEDGMENTS

I gratefully acknowledge the complete review of this chapter by Tom Shaffner. A number of other people are also gratefully acknowledged for helpful discussions, incuding Dennis Maher, Fred Shaapur, Bryan Tracy, and David Venables. I also thank David Mueller, whose participation in Ref. 4 was very useful during my writing of this manuscript. I gratefully acknowledge the review and suggestions of the TEM discussion by Fred Shaapur and Brendan Foran.

## REFERENCES

1. DB Williams, CB Carter. Transmission Electron Microscopy. New York: Plenum, 1996.
2. JI Goldsten, DE Neuburyr, P Echlin, JC Joy, AD Romig Jr, CE Lyman, C Fiori, E Lifshin. Scanning Electron Microscopy and X-Ray Microanalysis. 2nd ed. New York: Plenum, 1992.
3. S Amelinckx, D van Dyck, J van Landuyt, G van Tendeloo, eds. Handbook of Microscopy—Applications in Materials Science, Solid State Physics and Chemistry. Methods IV. Weinheim, CH Vergasgesellschaft mbH, Germany: 1997.
4. DB Williams, CB Carter. Transmission Electron Microscopy. New York: Plenum, 1996, pp 351–353.
5. DB Williams, CB Carter. Transmission Electron Microscopy. New York: Plenum, 1996, pp 441–444.
6. DB Williams, CB Carter. Transmission Electron Microscopy. New York: Plenum, 1996, pp 358–361.
7. AC Diebold, DA Venables, Y Chabal, D Mueller, M Weldon, E Garfunkel. Mat Sci Semicond Processing 2:103–147, 1999.
8. E Völkl, LF Allard. Introduction to Electron Holography. Dordrecht, Netherlands: Kluwer Academic, 1999.
9. B Tracy. Private communication.
10. Mag-I-Cal Reference Sample, Structure Probe, Inc., 569 East Gay Street, West Chester, PA.
11. LA Gianuzzi, JL Drown, SR Brown, RB Irwin, FA Stevie. Specimen Preparation for Transmission Electron Microscopy of Materials IV. In: RM Anderson, SD Walck, eds. Materials Research Society Proceedings 480:19–28, 1997.
12. D Venables and D Maher. Private Communication.
13. WD Rau, FH Baumann, HH Vuong, B Heinemann, W Hoppner, CS Rafferty, H Rucker, P Schwander, A Ourmazd. IEDM Tech Dig 713–716, 1998.

# 30

# Status of Lithography Metrology as of the End of 2000

**Alain C. Diebold**
*International SEMATECH, Austin, Texas*

Several advances in critical dimension metrology technology have occurred during 2000. When this volume was initiated, scatterometry-based control of critical dimensions was just beginning to be recognized as a production capable method. By the end of 2000, scatterometry is reported to be used in pilot line and moving into production (1). An IBM group published its third paper on optical critical dimension measurement and lithography process control (2–5). Critical dimension-scanning electron microscopy (CD-SEM) is facing several issues such as loss of depth of focus (6) and damage to the photoresist used for 193 nm lithography (7–9). CD-SEM improvements that evolved during 2000 include not only precision improvements but also software and hardware to enable 3-D information determination such as sidewall angle. These issues are briefly reviewed in this chapter.

## I. MULTIWAVELENGTH SCATTEROMETRY OR SPECULAR SPECTROSCOPIC SCATTEROMETRY

Single-wavelength, multiangle scatterometry is reviewed by Raymond in this volume (Chapter 18). Another approach is multiwavelength, single angle scatterometry which is also known as Specular Spectroscopic Scatterometry (10, 11). A similar approach using polarized reflection as a function of wavelength has also been described (12, 13). Scatterometry determines the average CD and lineshape over the illuminated area. Scatterometry is moving from a research and development activity into production CD control. The multiwavelength method can be done using commercially available software and in-line spectroscopic ellipsometers. As with the multiangle approach, a library of simulated data is compared to the experimental data and the best fit determines the line shape and CD. The library must have a simulation set for CD and lineshape variations with the resolution (e.g., 0.1-nm CD changes) required for the process control application. Nonplanar sidewalls can be measured with this method, which is very useful for the undercut gate structures. It can be applied to measuring patterned photoresist for focus–exposure control as well as to etched poly silicon lines. Application to Damascene-patterned structures is under development.

In multiwavelength scatterometry, the ellipsometric parameters $\Psi$ and $\Delta$ are calculated assuming a lineshape and pitch (10). The dielectric function of each layer in the materials stack is determined for blanket films. The light that is transmitted through the materials stack strongly affects $\Psi$ and $\Delta$. A library must be compiled for each materials stack. Recent publications suggest that the rigorous coupled wave analysis method is used to generate libraries for specular spectroscopic scatterometry (10, 11). The precision of multiwavelength scatterometry for 100-nm lines has been reported to be 1 nm or better. The next step is to extend this technology to contact arrays.

## II.  OPTICAL CD

Ausschnitt has published three papers that show how an overlay metrology tool can be used to control the focus and exposure, and thus CD variation, during lithography processing (2–5). The method is based on a fact that is well known in the lithography community. The change in length of a line is more sensitive measure of linewidth changes than the linewidth of an individual line. In the final implementation of this method, two arrays of lines printed at the smallest critical dimension for the IC and two arrays of spaces are printed. This is shown in Figure 1. The use of both arrays and spaces allows decoupling of focus (defocus) and exposure (dose). Instead of measuring line lengths, the spacing between the arrays of lines (L) and the spacing between arrays of spaces (S) is measured for each set of focus–exposure pair. L–S provides dose control and L + S provides focus control (3). Once the focus and exposure ranges are quantified, they can be monitored by measuring a test area in the scribe lines of a product wafer. The relationship between the L and S with the focus and exposure is determined and stored for each lithography process (3). CD is measured for the individual lines and spaces using CD-SEM when the process is developed, and a parameterized equation relating the observed L and S provides CD information.

Thus, a single measurement of the distance between two arrays of lines (L) and two arrays of spaces (S) with an overlay measurement system can be used to ensure that focus and exposure are within allowed process range. According to Ausschnitt, 200-nm CD processes require control of dose to 10 and focus to 0.5 microns ranges. This method has the precision and sensitivity to measure 1 dose changes and 50-nm focus changes (2).

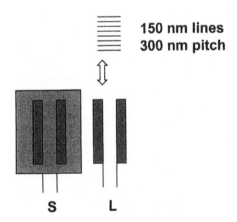

150 nm lines
300 nm pitch

S      L

**Figure 1**  Optical CD test structure showing arrays of lines on the right and spaces on the left. The distance between the arrays of lines (L) and spaces (S) is used to control CD. The 150 nm lines shown here represent the smallest CD on the IC. (From Refs. 2–5.)

## III. CD-SEM MEASUREMENT OF 193-NM PHOTORESIST

Measurement of the critical dimensions in the photoresist used for 193-nm lithography is very difficult due to electron beam induced shrinkage (6–8). Although electron beam induced shrinkage of photoresist is not a new phenomena, the magnitude of the problem is much greater for 193-nm photoresists. The solution to this issue involves reducing the total dose of electrons used during imaging. This approach can result in reduced signal to noise in the image, and can decrease precision. The metrology community is beginning to determine measurement precision under reduced current conditions, and one might find such information in the proceedings of the SPIE conference on Metrology, Inspection, and Process Control for Microlithography held in 2001.

## IV. SUMMARY

It is expected that both CD-SEM and scatterometry will be used for production control of critical dimensions in 2001 and for several years thereafter.

## REFERENCES

1. AC Diebold. Metrology strategy for next generation semiconductor manufacturing. Proceedings of the 2000 International Symposium on Semiconductor Manufacturing, pp.
2. CP Ausschnitt, W Chu, L Hadel, H Ho, P Talvi, Process window metrology. In: Metrology, Inspection, and Process Control for Microlithography XIV, Sullivan, Neil T ed., SPIE Vol. 3998, 2000, pp 158–166.
3. CP Ausschnitt, Distinguishing dose from defocus for in-line lithography control, In Metrology, Inspection, and Process Control for Microlithography XIII, SPIE Vol. 3677, 1999, pp 140–147.
4. CP Ausschnitt, ME Lagus. Seeing the forest from the trees: a new approach to CD control. In: Metrology, Inspection, and Process Control for Microlithography. SPIE Vol. 3332, 1998, pp 212–220.
5. CP Ausschnitt. Rapid optimization of the lithographic process window. In: Optical / Laser Microlithography II. SPIE Vol. 1088, 1989, pp 115–123.
6. M Sato, F Mizuno. Depth of field at high magnification of scanning electron microscope. Proceedings of 2000 Electron, Ion, and Proton Beams and Nanotechnology Conference, Palm Springs, 2000.
7. T Sarubbi, M Neisser, T Kocab, B Beauchemin, S Wong, W Ng. Mechanism studies of scanning electron microscope measurement effects on 193 nm photoresists and the development of improved line width measurement methods. Interface 2000, Nov 5–7, 2000, proceedings published by Arch Chemicals, Inc.
8. L Pain, N Monti, N Martin, V Tirard, A Gandolfi, M Bollin, M Vasconi. Study of 193 nm resist behavior under SEM inspection: how to reduce line-width shrinking effect? Interface 2000, Nov 5–7, 2000, proceedings published by Arch Chemicals, Inc.
9. B Su and A Romano. Study on 193 nm photo resist shrinkage after electron beam exposure. Interface 2000, Nov 5–7, 2000, proceedings published by Arch Chemicals, Inc.
10. X Niu, N Jakatdar, J Bao, C Spanos, and S Yedur. Specular spectroscopic scatterometry in DUV lithography. Proceedings of Metrology, Inspection, and Process Control for Microlithography XIII. SPIE Vol. 3677, 1999, pp 159–168.

11.  N Jakatdar, X Niu, J Bao, C Spanos, S Yedur, A Deleporte. Phase profilometry for the 193 nm lithography gate stack. Proceedings of Metrology, Inspection, and Process Control for Microlithography XIV. SPIE Vol. 3998, pp 116–124.

12.  J Algair, D Benoit, R Hershey, LC Litt, I Abdulhalim, B Braymer, M Faeyrman, JC Robinson, U Whitney, Y Xu, P Zalicki, J Seligson, Manufacturing considerations for implementation of scatterometry for process monitoring. Proceedings of Metrology, Inspection, and Process Control for Microlithography XIV. SPIE Vol. 3998, pp. 125–134.

13.  W Kong, H Huang, FL Terry, Jr. A hybrid analysis of ellipsometry data from patterned structures. In: Seiler, AC Diebold, TJ Shaffner, R McDonald, WM Bullis, PJ Smith, E Secula, eds., Characterization and Metrology for ULSI Technology. AIP Conference Proceedings, 2001, in press.

# Index

Printed and bound by CPI Group (UK) Ltd, Croydon, CR0 4YY

23/10/2024

01778254-0016